Progress in Mathematics
Volume 150

Series Editors
Hyman Bass
Joseph Oesterlé
Alan Weinstein

Masahiro Shiota

Geometry of Subanalytic and Semialgebraic Sets

Birkhäuser
Boston • Basel • Berlin

Masahiro Shiota
Department of Mathematics
Nagoya University
Furocho, Chikusa
Nagoya 464, Japan

Library of Congress Cataloging-In-Publication Data

Shiota, Masahiro, 1947-
    Geometry of subanalytic and semialgebraic sets / Masahiro Shiota.
       p.   cm. -- (Progress in mathematics ; v. 150)
    Includes bibliographical references (p. -) and index.
    ISBN 0-8176-4000-2 (Boston : alk. paper). -- ISBN 3-7643-4000-2
    (Basel : alk. paper)
      1. Semialgebraic sets. 2. Semianalytic sets. I. Title.
    II. Series: Progress in mathematics (Boston, Mass.) ; vol. 150.
    QA564.S46   1997   97-9061
    516.3--dc21   CIP

AMS Classification:
03c, 14p, 32c, 32s, 58a

Printed on acid-free paper
© 1997 Birkhäuser Boston

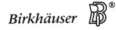

ISBN 0-8176-4000-2
ISBN 3-7643-4000-2

Reformatted from author's disk in $\mathcal{A}_{\mathcal{M}}\mathcal{S}$-TeX by Texniques, Inc., Boston, MA
Printed and bound by Quinn-Woodbine, Woodbine, NJ
Printed in the U.S.A.

9 8 7 6 5 4 3 2 1

CONTENTS

Introduction

Real analytic sets in Euclidean space (i.e., sets defined locally at each point of Euclidean space by the vanishing of an analytic function) were first investigated in the 1950's by H. Cartan [Car], H. Whitney [W$_{1-3}$], F. Bruhat [W-B] and others. Their approach was to derive information about real analytic sets from properties of their complexifications. After some basic geometrical and topological facts were established, however, the study of real analytic sets stagnated. This contrasted the rapid development of complex analytic geometry which followed the groundbreaking work of the early 1950's. Certain pathologies in the real case contributed to this failure to progress. For example, the closure of—or the connected components of—a constructible set (i.e., a locally finite union of differences of real analytic sets) need not be constructible (e.g., $\mathbf{R} - \{0\}$ and $\{(x, y, z) \in \mathbf{R}^3 : x^2 = zy^2,\ x^2 + y^2 \neq 0\}$, respectively). Responding to this in the 1960's, R. Thom [Th$_1$], S. Lojasiewicz [L$_{1,2}$] and others undertook the study of a larger class of sets, the *semianalytic sets*, which are the sets defined locally at each point of Euclidean space by a finite number of analytic function equalities and inequalities. They established that semianalytic sets admit Whitney stratifications and triangulations, and using these tools they clarified the local topological structure of these sets. For example, they showed that the closure and the connected components of a semianalytic set are semianalytic.

The principal tools in the study of Whitney stratifications are the first and second isotopy lemmas of Thom. Semianalytic versions of these lemmas have never been proved. As a consequence, Whitney stratifications did not play a significant role in semianalytic geometry after [Th$_1$]. Indeed, there were few cases in which real Whitney stratifications were used outside of research on C^∞ maps (e.g., Mather's work on stability of C^∞ maps).

Semianalytic sets suffer some of the same pathologies exhibited by real analytic sets. In particular, the image of a semianalytic set under a proper analytic map is not necessarily semianalytic. One may seek to characterize those good analytic maps that have semianalytic images by trying to find analytic sets which define the semianalytic sets and then complexifying the analytic sets and the map. Unfortunately, this process is non-canonical (and laborious). This makes it extremely difficult to study parameterized semianalytic sets—and hence semianalytic maps. (As an example of how little is known, it has not been established whether semianalytic functions are semianalytically triangulable.)

A.M. Gabrielov [Ga] and H. Hironaka [H$_1$] resolved the trouble by introducing the category of subanalytic sets and maps. The family of all *subanalytic sets* is defined to be the smallest family $\mathfrak{S}$ of subsets of all Euclidean spaces which satisfies the following axioms:

(i)$_{\mathfrak{S}}$ Every analytic set in any Euclidean space is an element of $\mathfrak{S}$.

(ii)$_{\mathfrak{S}}$ If $X_1 \subset \mathbf{R}^n$ and $X_2 \subset \mathbf{R}^n$ are elements of $\mathfrak{S}$, then $X_1 \cap X_2$ and $X_1 - X_2$ are elements of $\mathfrak{S}$.

(iii)$_{\mathfrak{S}}$ If $p: \mathbf{R}^m \to \mathbf{R}^n$ is any linear map and $X \subset \mathbf{R}^m$ is an element of $\mathfrak{S}$ such that the restriction of p to the closure of X is proper, then $p(X)$ is an element of $\mathfrak{S}$.

A *subanalytic map* between subanalytic sets is a continuous map with subanalytic graph.

Much of the theory of subanalytic sets is facilitated by the following result, which is a consequence of Hironaka's desingularization theorem (see [H$_1$]): A set $X \subset \mathbf{R}^n$ is subanalytic if and only if X is a finite union of sets of the form $\operatorname{Im} f_1 - \operatorname{Im} f_2$, where f_1 and f_2 are proper analytic maps whose domains are analytic manifolds.

Let $\mathfrak{X}$ be a family of subsets of Euclidean spaces, which satisfies the axioms:

Axiom (i) Every algebraic set in any Euclidean space is an element of $\mathfrak{X}$.

Axiom (ii) If $X_1 \subset \mathbf{R}^n$ and $X_2 \subset \mathbf{R}^n$ are elements of $\mathfrak{X}$, then $X_1 \cap X_2$, $X_1 - X_2$ and $X_1 \times X_2$ are elements of $\mathfrak{X}$.

Axiom (iii) If $X \subset \mathbf{R}^n$ is an element of $\mathfrak{X}$ and $p : \mathbf{R}^n \to \mathbf{R}^m$ is a linear map such that the restriction of p to $\overline{X}$ is proper, then $p(X)$ is an element of $\mathfrak{X}$.

Axiom (iv) If $X \subset \mathbf{R}$ and $X \in \mathfrak{X}$, then each point of X has a neighborhood in X which is a finite union of points and intervals.

An $\mathfrak{X}$-*set* is an element of $\mathfrak{X}$, and an $\mathfrak{X}$-*map* is a continuous map between $\mathfrak{X}$-sets with $\mathfrak{X}$-graph. We investigate $\mathfrak{X}$-sets and $\mathfrak{X}$-maps. The family $\mathfrak{S}$ satisfies these axioms and hence is an example of $\mathfrak{X}$. The smallest example of $\mathfrak{X}$ is the family of semialgebraic subsets of Euclidean spaces. Other non-trivial examples of $\mathfrak{X}$ are: (1) the 0-minimal Tarski system generated by summation, multiplication and the exponential function, and (2) the 0-minimal Tarski system generated by summation, multiplication and a finite number of Pfaffian functions. (See the below for the definition of an 0-minimal Tarski system. The existence of (1) and (2) was shown by A. Wilkie [Wi$_{1,2}$].)

Many mathematicians have considered axiomatically defined families of subsets of Euclidean spaces. L. van den Dries, J.F. Knight, A. Pillay and C. Steinhorn, [Dr$_{1,2}$], [P-S] and [K-P-S], introduced and considered the concept of an 0-minimal Tarski system, which is, by definition, a sequence $\{S_n\}_{n \in \mathbf{N}}$ such that for each $n \in \mathbf{N}$,

(i) S_n is a boolean algebra of subsets of $\mathbf{R}^n$,
(ii) if $X \in S_n$, then $\mathbf{R} \times X$ and $X \times \mathbf{R}$ are elements of S_{n+1},
(iii) $\{(x_1, \dots, x_n) \colon x_1 = x_n\} \in S_n$,
(iv) if $X \in S_{n+1}$, then the image of X under the projection of $\mathbf{R}^{n+1}$ onto the first n coordinates is an element of S_n,
(v)   any point of $\mathbf{R}$ is an element of S_1,
(vi) $\{(x,y) \in \mathbf{R}^2 \colon x < y\} \in S_2$, and
(vii) an element of S_1 is a finite union of intervals and points.

These authors and others proved some elementary properties of 0-minimal Tarski systems. For example, each element has a finite number of connected components and these components are also elements of the system, and if the system contains all semialgebraic sets in Euclidean spaces then each compact element admits a triangulation. Any 0-minimal Tarski system containing all semialgebraic sets is an $\mathfrak{X}$, but an $\mathfrak{X}$ is not necessarily an 0-minimal Tarski system, for example, $\mathfrak{S}$ is not. This is the reason why we treat systems $\mathfrak{X}$ rather than 0-minimal Tarski systems.

There are two properties of subanalytic geometry which $\mathfrak{X}$ (including 0-minimal Tarski systems) geometry loses. One is the Lojasiewicz inequality (see [H$_1$]), which states that for compact subanalytic sets X and Y there exist positive numbers c and d such that

$$\mathrm{dis}(x, X \cap Y) \le c(\mathrm{dis}(x,Y))^d \quad \text{for} \quad x \in X.$$

This inequality does not necessarily hold for $\mathfrak{X}$. For example, if $\mathfrak{X}$ is the above 0-minimal Tarski system (1) of Wilkie, the sets

$$X = [0,1] \times 0 \quad \text{and} \quad Y = \{0\} \cup \{(x,y) \in \mathbf{R}^2 \colon y = \exp(-1/x),\ 0 < x \le 1\}$$

are $\mathfrak{X}$-sets that do not satisfy the inequality. This failure is unlucky for us.

The other lost property is infinite differentiability and analyticity as we now explain. By M. Tamm [Ta], if X is a bounded subanalytic set in $\mathbf{R}^n$, then $\Sigma_\infty X$ and $\Sigma_\omega X$—which denote, respectively, the C^∞ and C^ω singular point sets of X—are identical and subanalytic. Moreover, they equal $\Sigma_r X$— the C^r singular point set—for some integer r. But this is not the case for $\mathfrak{X}$. Let $\mathfrak{X}$ be the same as in Example 1. The above Y does not satisfy the equality $\Sigma_\infty Y = \Sigma_\omega Y$. Let Z denote the graph of the function:

$$f(x_1, x_2) = \begin{cases} x_1^{x_2} & \text{for} \quad x_1 > 0,\ x_2 > 0 \\ 0 & \text{for} \quad x_1 \le 0,\ x_2 > 0. \end{cases}$$

This is an $\mathfrak{X}$-set because

$$Z = \{(x_1, x_2, 0) \in \mathbf{R}^3 \colon x_1 \le 0,\ x_2 > 0\}$$
$$\cup \{(x_1, x_2, x_3) \in \mathbf{R}^3 \colon \log x_3 = x_2 \log x_1,\ x_1, x_2, x_3 > 0\}.$$

Then, the singularities of Z fail to be stationary in the above sense because (see [D-M$_1$])
$$\Sigma_r Z = 0 \times]0, r], \quad r = 1, 2, \dots .$$

(It is not known whether $\Sigma_\infty X$ and $\Sigma_\omega X$ are $\mathfrak{X}$-sets for any $\mathfrak{X}$ and any $\mathfrak{X}$-set X). This failure is acceptable for us because we work in the context of C^r manifolds and maps, r a non-negative integer. However, since many known proofs of theorems on subanalytic sets and maps use analyticity, we need to develop new proofs.

Chapter I is preliminary for the following chapters. Readers are required to know elementary PL and differential topology (e.g., Grassmannian manifolds, the characteristic map of a fibre bundle) and they need at least to have heard of the concepts of subanalytic and semialgebraic sets. Most of the results presented here are new.

In Chapter II, fundamental facts and theorems on $\mathfrak{X}$ are stated and proved. They are mainly logical consequences of the definition of $\mathfrak{X}$, that is to say, the proofs use model-theoretic methods. We have avoided the terminology of model theory in order to make our exposition accessible to as wide an audience as possible. Some of the results and proofs in the first three sections may seem somewhat technical and esoteric, but we have included them because they are necessary later.

§II.2 and §II.3 contain a proof of the unique triangulability of $\mathfrak{X}$-sets and $\mathfrak{X}$-functions, assuming uniqueness of triangulations of compact $\mathfrak{X}$-sets. In §II.4, we let $\mathfrak{X}$ be an 0-minimal Tarski system containing all algebraic sets. We show that the results from the preceding sections can, in this case, be strengthened to "globally finite" ones. In the case where $\mathfrak{X}$ is the family of semialgebraic sets, all the results hold for any real closed field.

In §II.5, we consider C^r $\mathfrak{X}$-manifolds and maps. In [S$_3$], we constructed a theory of Nash (i.e., semialgebraic C^ω) manifolds and maps. We shall generalize some of the results of that work, using different methods of proof—in particular, Hironaka's desingularization theorem and the Nash approximation theorem (which states that a C^r Nash map can be approximated by a Nash map). In the $\mathfrak{X}$-case, we use the so called "pasting method" and the $\mathfrak{X}$-version of Thom's transversality theorem. The latter is of fundamental importance in the following sections (and beyond) and is new even in the semialgebraic case. Unfortunately, we have no results about C^ω $\mathfrak{X}$-manifolds or the $\mathfrak{X}$-versions of the above approximation theorem or the desingularization theorem.

In §II.6, we prove the $\mathfrak{X}$-versions of Thom's first and second isotopy lemmas and related theorems. The semialgebraic version of the first isotopy lemma was proved by [C-S$_1$]. The original second isotopy lemma of Thom has long been one of the most important tools in singularity theory, but the only proofs known previously involved integration of vector fields. As explained in the preface, these methods are not applicable in the present context. Taken together, the results of this section provide an alternative to integration methods, using only constructions that can be carried out within the $\mathfrak{X}$-category. The $\mathfrak{X}$-version of the second isotopy lemma is new even in the semialgebraic case. The crux of the proofs is II.6.8, which enables us to avoid the method of integration. It is perhaps worth mentioning here that in practice, singularity theorists tend to think in terms of polynomial maps and map germs, although they state and prove theorems in the C^∞ category. For this reason, the semialgebraic version is more natural and more consistent with some of the underlying motivations of singularity theory than is the original.

$\mathfrak{X}$-singularity theory aims at completing classification of $\mathfrak{X}$-sets, $\mathfrak{X}$-functions, $\mathfrak{X}$-maps and their germs by $\mathfrak{X}$-equivalence relations. §II.7 shows some fundamental facts and indicates future research, though far from perfect. For reasons that we elaborate in this section, when dealing with polynomial maps and map germs, it is more natural to use semialgebraic equivalence relations than C^0 ones. Apart from the construction of examples, the results of this section are applications of the Triangulation Theorem and the $\mathfrak{X}$-version of the second isotopy lemma. Though not included, another application is a semialgebraic version of Mather's C^0 stability theorem.

In Chapter III, we take up the problem of deciding when the existence of a homeomorphism between compact polyhedra implies the existence of a PL homeomorphism (Hauptvermutung). Our task amounts to showing that if $\pi\colon X \to Y$ is an $\mathfrak{X}$-homeomorphism between compact polyhedra, then π is isotopic to a PL homeomorphism ([S-Y] in the subanalytic case). Actually, we shall assume conditions on π which are weaker than—but are implied by—being of class $\mathfrak{X}$. The first condition we impose upon π is that there are Whitney stratifications $\{X_i\}$ of X and $\{Y_i\}$ of Y such that $\pi(X_i) = Y_i$, $\pi|_{X_i}$ are diffeomorphisms and $\{\text{graph}\,\pi|_{X_i}\}$ is a Whitney stratification of graph π. We call such π an isomorphism. This does not alone seem to imply that X and Y are PL homeomorphic, as Remark I.1.12 indicates. Therefore, we impose a second condition, which relates to the directions of grad $\rho_i^X|_{X_j}$ and $\text{grad}(\rho_i^Y \circ \pi)|_{X_j}$ for all i, j such that $X_i \cap \overline{X_j} = \varnothing$. We call such π a strong isomorphism. (Here ρ_i^X and ρ_i^Y are the functions which measure distance from X_i and Y_i, respectively.) We conjecture that any strong isomorphism is isotopic to PL homeomorphism, though even this is not proved. Ultimately, we find it necessary to add certain additional conditions related to the use of the Alexander trick in order to arrive at a theorem we can prove.

Chapter IV aims to determine when a stratified map is triangulable. This is one step toward a proof of Thom's conjecture that a stratified map *sans éclatement* is triangulable. We give certain necessary and sufficient conditions for $\mathfrak{X}$-maps to be triangulable, and we show that any complex analytic function satisfies the conditions. (The local triangulability of a complex analytic function was already proved in $[S_4]$.) This chapter uses results from all the previous sections, with the exception of the last part of chapter II and the idea of sheaf theory.

The simplest family of subsets of Euclidean spaces that we can treat systematically is the family of *semilinear sets*, i.e., sets defined by a finite number of equalities and inequalities of linear functions. In Chapter V, we generalize the concept of $\mathfrak{X}$ so that it includes this family. Let $\mathfrak{Y}$ be a family of subsets of Euclidean spaces satisfying the following four axioms.

Axiom (i)′   All rational semilinear sets in Euclidean spaces are elements of $\mathfrak{Y}$.

Axiom (ii)   If $X_1 \subset \mathbf{R}^n$ and $X_2 \subset \mathbf{R}^n$ are elements of $\mathfrak{Y}$, then $X_1 \cap X_2$, $X_1 - X_2$ and $X_1 \times X_2$ are elements of $\mathfrak{Y}$.

Axiom (iii)   If $X \subset \mathbf{R}^n$ is an element of $\mathfrak{Y}$, and $p \colon \mathbf{R}^n \to \mathbf{R}^{n-1}$ is the projection which forgets the last factor such that $p|_{\overline{X}}$ is proper, then $p(X)$ is an element of $\mathfrak{Y}$.

Axiom (iv)   If $X \subset \mathbf{R}$ and $X \in \mathfrak{Y}$, then each point of X has a neighborhood in X which is a finite union of points and intervals.

We show that our theory covers such families. If every element of $\mathfrak{Y}$ is *locally semilinear*, i.e., semilinear locally at each point of ambient Euclidean spaces then problems in $\mathfrak{Y}$ can be reduced to corresponding problems on PL topology. If $\mathfrak{Y}$ does not satisfy the $\mathfrak{X}$-axioms and there is at least one element of $\mathfrak{Y}$ which is not locally semilinear, then there exists a family $\mathfrak{X}$ and a homeomorphism $\pi \colon I \to J$ between intervals of $\mathbf{R}$ such that a set $Y \subset I^n$ is an element of $\mathfrak{Y}$ if and only if $\pi \times \cdots \times \pi(Y) \subset J^n$ is an element of $\mathfrak{X}$ ([Pe-S], [Pi-S]). Hence problems on $\mathfrak{Y}$ may be translated to $\mathfrak{X}$.

It seems very possible that the results in this book hold over any ordered field.

Chapter I.   Preliminaries

§I.1. Whitney stratifications

In this section, we give definitions of a Whitney stratification, a tube system, a vector field on a stratification, isomorphisms between Whitney stratifications, etc., and show their properties needed later, particularly I.1.13, with complete proofs. Some of our definitions are a little different from the usual ones, e.g. [G-al]. We modify the definitions to suit them to our purpose. We treat special topics unknown even to singularity specialists. We use them in Chapter III. For this, we need the method of integration of vector fields, which may contradict our philosophy. This is because the theorems in Chapter III are stated in a more general situation than $\mathfrak{X}$. The $\mathfrak{X}$-versions of the results of this section and Chapter III, except I.1.6 and I.1.7, can be proved without the method of integration. Note that the $\mathfrak{X}$-versions work in the C^r category, r a positive integer (see Chapter II).

Let M be a C^∞ submanifold of $\mathbf{R}^n$. (A *submanifold* always means a regular submanifold.) A *tube* at M is a triple $T = (|T|, \pi, \rho)$, where $|T|$ is an open neighborhood of M in $\mathbf{R}^n$, $\pi \colon |T| \to M$ is a submersive C^∞ retraction, and ρ is a non-negative C^∞ function on $|T|$ such that $\rho^{-1}(0) = M$ and each point x of M is a unique and nondegenerate critical point of the restriction of ρ to $\pi^{-1}(x)$. An example of a tube T at M is given by $|T| =$ a tubular neighborhood of M in $\mathbf{R}^n$, $\pi =$ the orthogonal projection, and $\rho =$ the square of the function which measures distance from M.

Lemma I.1.1 (Uniqueness of tube). *Let $T_i = (|T_i|, \pi_i, \rho_i)$, $i = 1, 2$, be tubes at a C^∞ submanifold M of $\mathbf{R}^n$. Shrink $|T_i|$. There exists a C^∞ diffeomorphism τ from $|T_1|$ to $|T_2|$ such that*

$$\tau|_M = \mathrm{id}, \qquad \pi_1 = \pi_2 \circ \tau \quad and \quad \rho_1 = \rho_2 \circ \tau.$$

Consequently, each point a of M has a local coordinate system $(x_1, \ldots, x_n)$ of $\mathbf{R}^n$, regarded as a C^∞ manifold, such that near a

$$(x_1(a), \ldots, x_n(a)) = 0, \qquad M = \{x_1 = \cdots = x_m = 0\} \quad for\ some\ integer\ m,$$

$$\pi(x_1, \ldots, x_n) = (0, \ldots, 0, x_{m+1}, \ldots, x_n), \quad and \quad \rho(x_1, \ldots, x_n) = x_1^2 + \cdots + x_m^2.$$

It is clear by the following proof that the $\mathfrak{X}$-version holds.

Proof. First we reduce the problem to the case where $\pi_1 = \pi_2$ on $|T_1| \cap |T_2|$. For this it suffices to find a C^∞ diffeomorphism τ' from a neighborhood of

M in $|T_1|$ to a neighborhood of M in $|T_2|$ such that

$$\tau'|_M = \mathrm{id} \quad \text{and} \quad \pi_1 = \pi_2 \circ \tau'.$$

Let U be a small neighborhood of the set $\Delta = \{(x, y) \in |T_2| \times M : \pi_2(x) = y\}$ in $|T_2| \times M$, and let $p \colon U \to |T_2|$ denote the map such that for each $(x, y) \in U$, $p(x, y)$ is the image of x under the orthogonal projection onto $\pi_2^{-1}(y)$. Then p is a C^∞ submersion such that $p(x, y) = x$ for $(x, y) \in \Delta$. Set

$$\tau'(x) = p(x, \pi_1(x)) \quad \text{for} \quad x \in |T_1| \quad \text{near} \quad M,$$

and shrink $|T_1|$ and $|T_2|$. Then τ' is a C^∞ diffeomorphism from $|T_1|$ to $|T_2|$ with the required properties. Thus we assume $\pi_1 = \pi_2$ on $|T_1| \cap |T_2|$.

Case where $M = \{x_1 = \cdots = x_m = 0\}$. We can suppose π_1 and π_2 are the orthogonal projection and

$$\rho_1(x_1, \ldots, x_n) = x_1^2 + \cdots + x_m^2.$$

It is easy to find C^∞ functions $\alpha_{i,j}$ on $|T_2|$, $i, j = 1, \ldots, m$, such that

$$\rho_2(x_1, \ldots, x_n) = \sum_{i,j=1}^{m} \alpha_{i,j} x_i x_j.$$

Note that $\alpha_{r,r} > 0$ on M for every positive integer $r \le m$ because ρ_2 is non-negative and its restriction to $\pi_2^{-1}(x)$ for each $x \in M$ has a non-degenerate critical point x. By induction, it suffices to prove the following statement.

Assume

$$\rho_2 = \sum_{i=1}^{r-1} x_i^2 + \sum_{i,j=r}^{m} \alpha_{i,j} x_i x_j.$$

Shrink $|T_1|$ and $|T_2|$. Then there exists a C^∞ diffeomorphism $\tau_r \colon |T_1| \to |T_2|$ such that $\pi_1 = \pi_2 \circ \tau_r$ and $\rho_2 \circ \tau_r$ is of the form $= \sum_{i=1}^{r} x_i^2 + \sum_{i,j=r+1}^{m} \beta_{i,j} x_i x_j$ for C^∞ functions $\beta_{i,j}$.

This is clear. Indeed, define τ_r so that

$$\tau_r^{-1}(x_1, \ldots, x_n) = (x_1, \ldots, x_{r-1},$$

$$\alpha_{r,r}^{1/2} x_r + \sum_{i=r+1}^{m} (\alpha_{i,r} + \alpha_{r,i}) x_i / 2\alpha_{r,r}^{1/2}, x_{r+1}, \ldots, x_n).$$

Then τ_r fulfills the requirements.

Note 1.   In the above case, if $\rho_1 = \rho_2$ on $\pi_1^{-1}(U) \cap \pi_2^{-1}(U)$ for an open subset U of M, then τ can be chosen so that $\tau = \mathrm{id}$ on $\pi_1^{-1}(U)$. This is easy to see as follows. We can reduce the problem to the above case of π_1, π_2 and ρ_1. Then automatically

$$\alpha_{i,i} = 1 \quad \text{and} \quad \alpha_{i,j} = 0 \quad \text{for} \quad i \neq j \quad \text{on} \quad \pi_1^{-1}(U) \cap \pi_2^{-1}(U).$$

By the above method of construction of τ_r, there exists the required τ.

Note 2.   In Note 1, let W_1, W_2 be open subsets of M with $\overline{W_1} \cap \overline{W_2} = \varnothing$. There exists a C^∞ diffeomorphism $\tau' \colon |T_1| \to |T_2|$ such that

$$\pi_1 = \pi_2 \circ \tau' \quad \text{on} \quad |T_1|, \qquad \rho_1 = \rho_2 \circ \tau' \quad \text{on} \quad \pi_1^{-1}(W_1), \text{ and}$$
$$\tau = \mathrm{id} \quad \text{on} \quad \pi_1^{-1}(U \cup W_2).$$

This follows from the method of construction of τ_r and a C^∞ partition of unity subordinate to $\{M - \overline{W_1}, M - \overline{W_2}\}$.

General case of M.   Let $\{U_i\}_{i=1,2\ldots}$ be a locally finite covering of M by coordinate neighborhoods. For a positive integer s, set $V_s = \cup_{i=1}^s U_i$. By induction, assuming $\rho_1 = \rho_2$ on $\pi_1^{-1}(V_{s-1})$ $(= \pi_2^{-1}(V_{s-1}))$ and shrinking U_s, $\pi_1^{-1}(U_s)$ and $\pi_2^{-1}(U_s)$, we only need to find a C^∞ diffeomorphism $\tau_s \colon |T_1| \to |T_2|$ such that

$$\pi_1 = \pi_2 \circ \tau_s \quad \text{and} \quad \rho_1 = \rho_2 \circ \tau_s \quad \text{on} \quad \pi_1^{-1}(V_s), \text{ and}$$
$$\tau_s = \mathrm{id} \quad \text{on} \quad \pi_1^{-1}(V_{s-1}).$$

By the special case and Note 1, we obtain a C^∞ diffeomorphism τ_s' from a shrunk $\pi_1^{-1}(U_s)$ to a shrunk $\pi_2^{-1}(U_s)$ such that

$$\pi_1 = \pi_2 \circ \tau_s' \quad \text{and} \quad \rho_1 = \rho_2 \circ \tau_s' \quad \text{on} \quad \pi_1^{-1}(U_s), \text{ and}$$
$$\tau_s = \mathrm{id} \quad \text{on} \quad \pi_1^{-1}(U_s \cap V_{s-1}).$$

Moreover, by Note 2, we shrink U_s and extend τ_s' to $\tau_s \colon |T_1| \to |T_2|$.   □

Let $G_{n,m}$ denote the *Grassmannian* of m-dimensional subspaces of $\mathbf{R}^n$. We give an affine non-singular algebraic manifold structure to $G_{n,m}$ as follows. Let each $\lambda \in G_{n,m}$ correspond to the orthogonal projection of $\mathbf{R}^n$ onto λ. By this correspondence, we regard $G_{n,m}$ as a subset of L_n, the space of all linear transformations of $\mathbf{R}^n$. (We assume every transformation in L_n carries 0 to 0, but we use the term "linear" in the affine sense without this property unless otherwise specified.) Then $G_{n,m}$ is a non-singular algebraic submanifold of $L_n \cong \mathbf{R}^{n^2}$ (see [Pa]). The *universal vector bundle* over the Grassmannian is $(E_{n,m}, \pi_G, G_{n,m})$, where

$$E_{n,m} = \{(\lambda, x) \in G_{n,m} \times \mathbf{R}^n : x \in \lambda\}$$

and $\pi_G : E_{n,m} \to G_{n,m}$ is the projection. We see that $E_{n,m}$ is a non-singular algebraic submanifold of $L_n \times \mathbf{R}^n$ and the algebraic structure on $G_{n,1}$ coincides with the usual algebraic structure of the projective space $\mathbf{P}_n(\mathbf{R})$.

A *stratification* of a set $X \subset \mathbf{R}^n$ is a partition of X into C^∞ submanifolds X_i of $\mathbf{R}^n$ such that the family $\{X_i\}$ is locally finite at each point of X (not of $\mathbf{R}^n$). Moreover, if each stratum X_i is an analytic submanifold of $\mathbf{R}^n$, we call the stratification *analytic*. For a positive integer r, we define also a C^r *stratification*. A stratification $\{X_i\}$ of X is called a *Whitney stratification* if each pair of strata X_i and X_j, $i \neq j$, satisfy the following *Whitney condition*. If $\{a_k\}$ and $\{b_k\}$ are sequences in X_i and X_j, respectively, both converging to a point b of X_j, if the sequence of the tangent spaces $\{T_{a_k} X_i\}$ converges to a subspace $T \subset \mathbf{R}^n$ in $G_{n,m}$, where $m = \dim X_i$, and if the sequence $\{\overrightarrow{a_k b_k}\}$ of lines containing 0 and $a_k - b_k$ converges to a line $L \subset \mathbf{R}^n$ in $G_{n,1}$, then $L \subset T$. If this is the case for a given point $b \in X_j$ and for any $\{a_k\}$ and $\{b_k\}$, we say that X_i and X_j satisfy the *Whitney condition* at b. Here we note that $T_b X_j \subsetneq T$, hence $\dim X_j < \dim X_i$, and that the Whitney condition does not depend on choice of a coordinate system of $\mathbf{R}^n$. In other words, the condition is invariant under a diffeomorphism of $\mathbf{R}^n$. We say that a stratification $\{X_i\}$ satisfies the *frontier condition* or the *weak frontier condition* if the condition $(\overline{X_i} - X_i) \cap X_{i'} \neq \varnothing$ implies $\overline{X_i} \supset X_{i'}$ or if the family of the connected components of all X_i satisfies the frontier condition, respectively.

Remark I.1.2. Let $\{X_1, X_2\}$ be a Whitney stratification of $X \subset \mathbf{R}^n$ with $X_1 \cap \overline{X_2} \neq \varnothing$, and let $(|T_1|, \pi_1, \rho_1)$ be a tube at X_1. The map $X_2 \ni x \to (\pi_1(x), \rho_1(x)) \in X_1 \times \mathbf{R}$ is C^∞ regular on $U \cap X_2$ (i.e., the rank of its differential at each point of $U \cap X_2$ equals $\dim X_1 + 1$) for some neighborhood U of X_1 in X. This follows from I.1.1.

Let M_i and N_i, $i = 1, 2, \ldots$, be C^∞ submanifolds of $\mathbf{R}^n$. We say that M_i, $i = 1, 2, \ldots$, are *transversal to each other* if for each subset $\lambda \subset \{1, 2, \ldots\}$, and for each $i \notin \lambda$, the intersection $\cap_{j \in \lambda} M_j$ is a C^∞ submanifold which intersects transversally with M_i. We say that M_i, $i = 1, 2, \ldots$, are *transversal* to N_i, $i = 1, 2, \ldots$, if for each pair i and j, M_i and N_j are transversal to each other.

Let $\{X_i\}_{i=1,\ldots,k}$ be a Whitney stratification of a compact set $X \subset \mathbf{R}^n$ such that $\dim X_1 < \cdots < \dim X_k$, and let ρ_i, $i = 1, \ldots, k$, denote the squares of the functions which measure distance from X_i on $\mathbf{R}^n$. By the

Whitney condition we easily obtain a positive number δ_1 and positive functions $\delta_i \colon \mathbf{R}^{i-1} \to \mathbf{R}$, $i = 2, \ldots, k-1$, which satisfy the following conditions. Let $1 \leq i \leq l \leq k$ be integers, let $\varepsilon_1, \ldots, \varepsilon_{k-1}$ be positive numbers with $\varepsilon_1 \leq \delta_1$, $\varepsilon_2 \leq \delta_2(\varepsilon_1), \ldots$, and $\varepsilon_{k-1} \leq \delta_{k-1}(\varepsilon_1, \ldots, \varepsilon_{k-2})$, and let $E_j = \varepsilon_j$ or $=]\varepsilon_j, \infty[$ for each $j = 1, \ldots, i-1$. Then the set

$$E(i-1) = \rho_1^{-1}(E_1) \cap \cdots \cap \rho_{i-1}^{-1}(E_{i-1})$$

is a C^∞ submanifold of $\mathbf{R}^n$, $E(i-1)$, $\rho_i^{-1}(\varepsilon_i)$ and X_l are transversal to each other, and the restrictions of ρ_i to $\rho_i^{-1}(]0, \varepsilon_i]) \cap E(i-1)$ and $\rho_i^{-1}(]0, \varepsilon_i]) \cap E(i-1) \cap X_l$ are C^∞ regular. Note that the sets $X_i \cap \rho_1^{-1}(E_1') \cap \cdots \cap \rho_{i-1}^{-1}(E_{i-1}')$ are compact C^∞ manifolds possibly with boundary and corners, where

$$E_j' = \varepsilon_j \quad \text{or} \quad = [\varepsilon_j, \infty[, \quad j = 1, \ldots, i-1.$$

We call such $\delta = \{\delta_i\}_{i=1,\ldots,k-1}$ a *removal data* of $\{X_i\}_{i=1,\ldots,k}$, and we write $\varepsilon \leq \delta$ for such a sequence $\varepsilon = \{\varepsilon_i\}_{i=1,\ldots,k-1}$.

If X is not compact, a removal data does not always exist. We can generalize it as follows when X is closed in $\mathbf{R}^n$. By the Whitney condition we have a positive proper C^∞ function f on $\mathbf{R}^n$ such that I—which denotes the common C^∞ regular values of f, $f|_{X_1}, \ldots, f|_{X_k}$—is not bounded above. (If $\dim X_1 = 0$, we do not call a point of $f(X_1)$ a C^∞ regular value.) Let I' be a subset of I which is not bounded above and locally is a finite union of closed intervals at each point of $\mathbf{R}$, and let F denote the family of positive C^∞ functions on $\mathbf{R}$ which are locally constant on $\mathbf{R} - I'$. Then there exist as in the compact case an element $\delta_1 \in F$ and maps $\delta_i \colon F^{i-1} \to F$, $i = 2, \ldots, k-1$, with the following properties. Let $1 \leq i \leq l \leq k$ be integers, let $t \in I'$, let $\varepsilon_1, \ldots, \varepsilon_{k-1}$ be elements of F with $\varepsilon_1 \leq \delta_1$, $\varepsilon_2 \leq \delta_2(\varepsilon_1), \ldots, \varepsilon_{k-1} \leq \delta_{k-1}(\varepsilon_1, \ldots, \varepsilon_{k-2})$, and let $*_j \in \{=, >\}$, $j = 1, \ldots, i-1$. Set

$$E(i-1) = \{x \in \mathbf{R}^n \colon \rho_j(x) *_j \varepsilon_j \circ f(x), \ j = 1, \ldots, i-1\}.$$

Then $E(i-1)$ is a C^∞ submanifold of $\mathbf{R}^n$, $E(i-1)$, $f^{-1}(t)$, $\{\rho_i(x) = \varepsilon_i \circ f(x)\}$ and X_l are transversal to each other , and the restrictions of ρ_i to $\{0 < \rho_i(x) \leq \varepsilon_i \circ f(x)\} \cap E(i-1)$, $\{0 < \rho_i(x) \leq \varepsilon_i \circ f(x)\} \cap E(i-1) \cap X_l$, $\{f(x) = t, \ 0 < \rho_i \leq \varepsilon_i \circ f(x)\} \cap E(i-1)$ and $\{f(x) = t, \ 0 < \rho_i(x) \leq \varepsilon_i \circ f(x)\} \cap E(i-1) \cap X_l$ are C^∞ regular. In this case also, we call δ a *removal data* and write $\varepsilon \leq \delta$. We shall use a removal data for non-compact X only in the case where X and X_i are $\mathfrak{X}$-sets (see Chapter II). We can choose then $f(x) = |x|^2$ by II.1.8. Hence we assume this special f unless otherwise specified.

A *tube system* for a stratification $\{X_i\}$ consists of one tube $T_i = (|T_i|, \pi_i, \rho_i)$ at each X_i. We call a tube system $\{T_i\}$ *controlled* if for each pair i and j with $(\overline{X_j} - X_j) \cap X_i \neq \varnothing$, the following property holds true:

$$\left.\begin{array}{l} \pi_i \circ \pi_j(x) = \pi_i(x), \\[2mm] \rho_i \circ \pi_j(x) = \rho_i(x) \end{array}\right\} \quad \text{for} \quad x \in |T_i| \cap |T_j|. \qquad \text{ct}(T_i, T_j)$$

When only the first equality holds, we call $\{T_i\}$ *weakly controlled*. Here we assume $\pi_i \circ \pi_j$ and $\rho_i \circ \pi_j$ are well-defined on $|T_i| \cap |T_j|$, namely,

$$|T_i| \cap \pi_j^{-1}(|T_i|) = |T_i| \cap |T_j|.$$

Note that the condition $\text{ct}(T_i, T_j)$ implies the inequality $\dim X_i < \dim X_j$ and that for a (possibly non-controlled) tube system $\{T_i\}$ for a stratification $\{X_i\}$, we can shrink $\{|T_i|\}$ so that

$$|T_i| \cap \pi_j^{-1}(|T_i|) = |T_i| \cap |T_j|$$

for all i and j with $(\overline{X_j} - X_j) \cap X_i \neq \varnothing$.

In the above definition of a removal data, we can replace the squares of the functions which measure distance from X_i with ρ_i of a tube system $\{T_i = (|T_i|, \pi_i, \rho_i)\}$. (Here we extend ρ_i to $\mathbf{R}^n$ by setting $\rho_i = 1$ on $\mathbf{R}^n - |T_i|$.) Then we say a *removal data* of $\{X_i\}$ for $\{T_i\}$. But we sometimes omit $\{T_i\}$ because, for two tube systems $\{T_i\}$ and $\{T_i'\}$, there is a common removal data for $\{T_i\}$ and for $\{T_i'\}$.

Lemma I.1.3 (Existence of controlled tube system). *If $\{T_i = (|T_i|, \pi_i, \rho_i)\}_{i=1,2,\ldots}$, is a tube system for a Whitney stratification $\{X_i\}_{i=1,2,\ldots}$ with $\dim X_1 < \dim X_i$, $i \neq 1$, then there exists a controlled tube system $\{T_i' = (|T_i'|, \pi_i', \rho_i')\}$ for $\{X_i\}$ such that for each i,*

$$|T_i'| \subset |T_i|, \qquad \rho_i = \rho_i' \quad \text{on} \quad |T_i'| \quad \text{and}$$
$$\pi_1 = \pi_1' \quad \text{on} \quad |T_1'|.$$

See II.6.10 for the $\mathfrak{X}$-versions of I.1.3 and I.1.3$'$.

Proof. By replacing each X_{i+1}, $i > 0$, with the union of the strata of dimension $= i + \dim X_1$, we assume $\dim X_1 < \dim X_2 < \cdots$. We prove the lemma by double induction as follows. Assume that for some k, the given tube system already satisfies the above condition $\text{ct}(T_i, T_j)$ for all $i < j < k$. It suffices to shrink $|T_i|$, $i \leq k$, and modify π_k so that $\text{ct}(T_i, T_k)$ for all $i < k$ are satisfied. Moreover, consider the following downward induction. Assume we have shrunk $|T_i|$ and we have modified π_k so that the condition $\mathrm{ct}(T_i, T_k)$, $l < i < k$, is satisfied for some $l < k$. Keeping $\mathrm{ct}(T_i, T_k)$, $l < i < k$, let us obtain $\mathrm{ct}(T_l, T_k)$.

Define a C^∞ map $\sigma_l \colon |T_l| \to X_l \times \mathbf{R}$ by

$$\sigma_l(x) = (\pi_l(x), \rho_l(x)) \quad \text{for} \quad x \in |T_l|.$$

The condition $\mathrm{ct}(T_l, T_k)$ is restated as:

$$\sigma_l \circ \pi_k(x) = \sigma_l(x) \quad \text{for} \quad x \in |T_l| \cap |T_k|.$$

Shrink $|T_l|$ if necessary. By I.1.2 we can assume $\sigma_l|_{|T_l| \cap X_k}$ is C^∞ regular. Hence for each $(y, t) \in X_l \times \mathbf{R}$, the set

$$X_{k,y,t} = \sigma_l^{-1}(y, t) \cap |T_l| \cap X_k$$

is a C^∞ submanifold of the manifold $|T_l| \cap X_k$. Let $p_{y,t} \colon U_{y,t} \to X_{k,y,t}$ denote the projection of a tubular neighborhood of $X_{k,y,t}$ in $|T_l| \cap X_k$. Here we choose $U_{y,t}$, $(y, t) \in X_l \times \mathbf{R}$, so that the set $\cup_{(y,t) \in X_l \times \mathbf{R}} U_{y,t} \times (y, t)$ is an open subset of $(|T_l| \cap X_k) \times X_l \times \mathbf{R}$, which is possible because $\{X_{k,y,t}\}_{y,t}$ is a C^∞ foliation of $|T_l| \cap X_k$. Then, by shrinking $|T_l|$ and $|T_k|$, we obtain the property:

$$\pi_k(x) \in U_{\sigma_l(x)} \quad \text{for} \quad x \in |T_l| \cap |T_k|,$$

because for $x \in |T_l| \cap X_k$, $\pi_k(x) = x$ and hence $\pi_k(x) \in U_{\sigma_l(x)}$.

We want to replace the π_k with the C^∞ retraction $\pi_{kl} = p_{\sigma_l} \circ \pi_k \colon |T_l| \cap |T_k| \to |T_l| \cap X_k$. We may suppose that π_{kl} is submersive. Note for each $l < i < k$,

$$\pi_{kl} = \pi_k \quad \text{on} \quad |T_l| \cap |T_i| \cap |T_k|,$$

because for $x \in |T_l| \cap |T_i| \cap |T_k|$,

$$
\begin{aligned}
\pi_{kl}(x) = p_{\sigma_l(x)} \circ \pi_k(x) = p_{\sigma_l \circ \pi_i(x)} \circ \pi_k(x) \quad && \text{by} \quad \mathrm{ct}(T_l, T_i) \\
= p_{\sigma_l \circ \pi_i \circ \pi_k(x)} \circ \pi_k(x) \quad && \text{by} \quad \mathrm{ct}(T_i, T_k) \\
= p_{\sigma_l \circ \pi_k(x)} \circ \pi_k(x) \quad && \text{by} \quad \mathrm{ct}(T_l, T_i) \\
= \pi_k(x) \quad && \text{by definition of } p_{y,t}.
\end{aligned}
$$

(For the last two equalities we need to shrink $|T_l|$.) Hence we can take the modification of π_k on $|T_l| \cap |T_k|$ to be π_{kl}.

It remains to extend the π_{kl} to $|T_k|$. For that we use a C^∞ partition of unity. Let θ be a C^∞ function on $\mathbf{R}^n - (\overline{X_l} - X_l)$ such that $0 \le \theta \le 1$,

$\theta = 1$ on a small neighborhood of X_l in $|T_l|$ and $\theta = 0$ outside of another one. Replace the π_k with the map:

$$\pi_k(\theta \pi_{kl} + (1 - \theta)\pi_k),$$

and shrink $|T_k|$. The condition $\mathrm{ct}(T_l, T_k)$ is satisfied for this new π_k, which proves the lemma. $\square$

Remark I.1.4. Let $\{T_i = (|T_i|, \pi_i, \rho_i)\}_{i=1,2,\dots}$ be a controlled tube system for a Whitney stratification $\{X_i\}_{i=1,2,\dots}$. Assume that

$$X_1 = \{(x_1, \dots, x_n) \in \mathbf{R}^n : x_1 = \cdots = x_m = 0\}, \;\; 1 \le m \le n,$$

and $\pi_1 \colon |T_1| \to X_1$ is the orthogonal projection. The image of a point $x = (x_1, \dots, x_n) \in |T_i| \cap |T_1|$, $i > 1$, under π_i is of the form $(y_1, \dots y_m, x_{m+1}, \dots, x_n)$ because if we set $\pi_i(x) = (y_1, \dots, y_n)$, then by the condition $\mathrm{ct}(T_1, T_i)$,

$$(0, \dots, 0, y_{m+1}, \dots, y_n) = \pi_1(y_1, \dots, y_n)$$
$$= \pi_1 \circ \pi_i(x) = \pi_1(x) = (0, \dots, 0, x_{m+1}, \dots, x_n).$$

Let $f \colon X \to Y$ be a continuous map between subsets of $\mathbf{R}^n$. A (*Whitney*) *stratification* of f is a pair of (Whitney) stratifications $\{X_i\}$ of X and $\{Y_j\}$ of Y such that for each i, the image $f(X_i)$ is included in some Y_j and the map $f|_{X_i} \colon X_i \to Y_j$ is a C^∞ submersion ($\{$graph $f|_{X_i}\}$ is a Whitney stratification, respectively). We call $f \colon \{X_i\} \to \{Y_j\}$ a *stratified map*.

Note that if f is proper and if $\{X_i\}$ satisfies the weak frontier condition, the image under f of each connected component of each X_i coincides with some connected component of some Y_j for the following reason. By restricting $\{X_i\}$ and f to $f^{-1}(Y_j)$ for one j, we assume that $\{Y_j\}$ consists of one element Y and then Y is connected. Moreover, we suppose that Y is a simple curve because if for one X_i, $f(X_i)$ includes every simple C^∞ curve in Y, then $f(X_i) = Y$. We can assume also that the X_i are all connected. Since the image of a connected set under a continuous map is connected, $f(X_i)$ are connected subcurves of Y. Therefore, it suffices to prove that $f(X_i)$ are dense in Y. But this is clear because $f(\overline{X_i} \cap X)$ are closed in Y by properness and because $f(\overline{X_i} \cap X)$ are open in Y by the weak frontier condition.

A C^1 Whitney stratified map $f \colon \{X_i\} \to \{Y_j\}$ is called a stratified map *sans éclatement* if the following condition is satisfied. Let X_i and $X_{i'}$ be distinct strata with $\overline{X_i} \supset X_{i'}$. If $\{a_k\}$ is a sequence of points of X_i convergent to a point b of $X_{i'}$, and if the sequence of the tangent spaces $\{T_{a_k}(f|_{X_i})^{-1}f(a_k)\}$ converges to a space $T \subset \mathbf{R}^n$ in $G_{n,m}$, $m = \dim(f|_{X_i})^{-1}f(a_k)$, then $T_b(f|_{X_{i'}})^{-1}f(b) \subset T$.

A C^∞ *function* on a subset $X \subset \mathbf{R}^n$ is the restriction to X of a C^∞ function defined on a neighborhood of X in $\mathbf{R}^n$. A C^∞ *map* from a subset $X \subset \mathbf{R}^n$ to another $Y \subset \mathbf{R}^n$ is defined in the same way.

Let $f\colon X \to Y$ be a C^0 map between subsets of $\mathbf{R}^n$, let $(\{X_i\}, \{Y_k\})$ be its stratification, and let $\{T_i = (|T_i|, \pi_i, \rho_i)\}$, $\{T'_k = (|T'_k|, \pi'_k, \rho'_k)\}$ be tube systems for $\{X_i\}$, $\{Y_k\}$, respectively. We call $\{T_i\}$ *controlled over* $\{T'_k\}$ if $\{T_i\}$ satisfies the first equality of the condition $\mathrm{ct}(T_i, T_j)$, the latter equality for each pair X_i, X_j with $f(X_i) \cup f(X_j) \subset Y_k$ for some k, and the condition:

$$f(|T_i|) \subset |T'_k|, \quad f \circ \pi_i = \pi'_k \circ f \quad \text{on} \quad X \cap |T_i| \quad \text{for} \quad i, k \quad \text{with} \quad f(X_i) \subset Y_k.$$

We generalize I.1.3 as follows. As we can prove it in the same way, we omit the proof.

Lemma I.1.3′. *Let $f\colon X \to Y$ be a C^∞ map between subsets of $\mathbf{R}^n$, let f admit a Whitney stratification sans éclatement $(\{X_i\}, \{Y_j\})$, and let $\{T'_j\}$ be a weakly controlled tube system for $\{Y_j\}$. Then there exists a tube system $\{T_i\}$ for $\{X_i\}$ controlled over $\{T'_j\}$.*

Let $\{T_i\}$ be a controlled tube system for a stratification $\{X_i\}$ of a set $X \subset \mathbf{R}^n$. A *vector field* ξ on $\{X_i\}$ consists of one C^∞ vector field ξ_i on each X_i. We call ξ *semicontrolled* if for each pair X_i and X_j with $(\overline{X_j} - X_j) \cap X_i \neq \varnothing$, we have the condition $\mathrm{scv}(T_i, T_j)$, defined as

$$d(\rho_i|_{X_j})\xi_{jx} = 0 \qquad \text{for} \quad x \in X_j \cap U_i, \qquad \mathrm{scv}(T_i, T_j)$$

where $U_i \subset |T_i|$ is some neighborhood of X_i in $\mathbf{R}^n$. If, in addition, we have the condition $\mathrm{cv}(T_i, T_j)$ defined as

$$d(\pi_i|_{X_j})\xi_{jx} = \xi_{i\pi_i(x)} \quad \text{for} \quad x \in X_j \cap U_i, \qquad \mathrm{cv}(T_i, T_j)$$

then ξ is called *controlled*. If ξ is continuous as a map from X to $\mathbf{R}^n$, we call it *continuous*.

Lemma I.1.5 (Lift of vector field). *Let $\{T_i = (|T_i|, \pi_i, \rho_i)\}_{i=1,2,\ldots}$ be a controlled tube system for a Whitney stratification $\{X_i\}_{i=1,2,\ldots}$ with $\dim X_1 < \dim X_i$, $i \neq 1$. Let ξ_1 be a C^∞ vector field on X_1. There exists a continuous controlled vector field $\xi = \{\xi_i\}_{i=1,2,\ldots}$ on $\{X_i\}$.*

We call ξ a *lift* of ξ_1 to $\{X_i\}_{i=1,2,\ldots}$. We can prove the $\mathfrak{X}$-version but it is not very useful because we can not integrate the vector field. (We will use the $\mathfrak{X}$-version in the proof that III.2.4 implies III.1.1, but we can avoid it. See the note after I.1.9.)

Proof. We can assume the index i runs from 1 to $k > 1$ and $\dim X_1 < \cdots < \dim X_k$. First we reduce the problem to the case $k = 2$ by induction on k. Assume the lemma for $k - 1$ and for $k = 2$. There exist a continuous controlled vector field $\{\xi_i\}_{i=1,\ldots,k-1}$ on $\{X_i\}_{i=1,\ldots,k-1}$ and C^∞ vector fields ξ_{ik}, $i = 1,\ldots,k-1$, on X_k such that for each $1 \le i < k$, $\{\xi_i, \xi_{ik}\}$ is a continuous controlled vector field on $\{X_i, X_k\}$.

By using $\{\xi_{ik}\}$, we want to construct a C^∞ vector field ξ_k which, together with the $\{\xi_i\}_{i=1,\ldots,k-1}$, constitutes a continuous controlled vector field on $\{X_i\}_{i=1,\ldots,k}$. As $\{\xi_i\}_{i=1,\ldots,k-1}$ is continuous, we have small open neighborhoods U_i of X_i in $\mathbf{R}^n$ for all $1 \le i \le k$ such that for each $1 \le j < k$, $\{\xi_i\}_{i=1,\ldots,j} \cup \{\xi_{jk}|_{U_j \cap X_k}\}$ is a continuous vector field on $\{X_i\}_{i=1,\ldots,j} \cup \{U_j \cap X_k\}$. Moreover, if we shrink each U_j, this vector field is controlled, because for each $1 \le i < j$,

$$
\begin{aligned}
d(\rho_i|_{X_k})\xi_{jkx} &= d(\rho_i \circ \pi_j|_{X_k})\xi_{jkx} && \text{by} \quad \mathrm{ct}(T_i, T_j) \\
&= d(\rho_i|_{X_j}) \circ d(\pi_j|_{X_k})\xi_{jkx} = d(\rho_i|_{X_j})\xi_{j\pi_j(x)} && \text{by} \quad \mathrm{cv}(T_j, T_k) \\
&= 0 \quad \text{for} \quad x \in U_i' \cap U_j' \cap X_k && \text{by} \quad \mathrm{scv}(T_i, T_j), \\
d(\pi_i|_{X_k})\xi_{jkx} &= d(\pi_i \circ \pi_j|_{X_k})\xi_{jkx} && \text{by} \quad \mathrm{ct}(T_i, T_j) \\
&= d(\pi_i|_{X_j}) \circ d(\pi_j|_{X_k})\xi_{jkx} = d(\pi_i|_{X_j})\xi_{j\pi_j(x)} && \text{by} \quad \mathrm{cv}(T_j, T_k) \\
&= \xi_{i\pi_i \circ \pi_j(x)} && \text{by} \quad \mathrm{cv}(T_i, T_j) \\
&= \xi_{i\pi_i(x)} \quad \text{for} \quad x \in U_i' \cap U_j' \cap X_k && \text{by} \quad \mathrm{ct}(T_i, T_j),
\end{aligned}
$$

where $U_i' \subset |T_i|$ and $U_j' \subset U_j \cap |T_j|$ are some neighborhoods of X_i and X_j, respectively, in $\mathbf{R}^n$. Hence we assume that for each $j = 1,\ldots,k-1$, $\{\xi_i\}_{i=1,\ldots,j} \cup \{\xi_{jk}|_{U_j \cap X_k}\}$ is a continuous controlled vector field on $\{X_i\}_{i=1,\ldots,j} \cup \{U_j \cap X_k\}$.

Consider the open covering $\{U_j \cap X_k\}_{j=1,\ldots,k-1} \cup \{X_k\}$ of X_k. Let $\{V_i\}_{i=1,\ldots,k}$ be its refinement defined by

$$V_{k-1} = U_{k-1} \cap X_k,$$

$$V_{k-2} = U_{k-2} \cap X_k - (\text{a small closed neighborhood of } X_{k-1}),$$

$$\vdots$$

$$V_1 = U_1 \cap X_k - (\text{a small closed neighborhood of } \bigcup_{i=2}^{k-1} X_i), \quad \text{and}$$

$$V_k = X_k - (\text{a small closed neighborhood of } \bigcup_{i=1}^{k-1} X_i),$$

and let $\{\varphi_i\}_{i=1,\dots,k}$ be a C^∞ partition of unity on X_k subordinate to $\{V_i\}_{i=1,\dots,k}$. The vector field on X_k

$$\xi_k = \sum_{i=1}^{k-1} \varphi_i \xi_{ik}$$

is what we wanted. Thus we have reduced the problem to the case $k = 2$.

Assume $k = 2$. By virtue of a C^∞ partition of unity, the problem of construction of ξ_2 becomes local at each point of X_1 as shown above. (We call this method the *pasting method*.) Hence by I.1.1, we can assume

$$X_1 = \{x = (x_1, \dots, x_n) \in \mathbf{R}^n : x_1 = \cdots = x_m = 0\},$$
$$|T_1| = \mathbf{R}^n, \quad \pi_1(x_1, \dots, x_n) = (0, \dots, 0, x_{m+}, \dots, x_n), \quad \text{and}$$
$$\rho_1(x_1, \dots, x_n) = x_1^2 + \cdots + x_m^2.$$

By I.1.2 $\rho_1|_{X_2}$ is C^∞ regular on $U \cap X_2$ for some open neighborhood U of X_1 in $\mathbf{R}^n$. Hence for every $a \in U \cap X_2$, $(\rho_1|_{U \cap X_2})^{-1}(\rho_1(a))$ is a C^∞ submanifold of $\mathbf{R}^n$ of dimension $m' = \dim X_2 - 1$. Let L_a denote its tangent space at a regarded as a linear subspace of $\mathbf{R}^n$ of dimension m'. Clearly the map $U \cap X \ni x \to L_x \in G_{n,m'}$ is of class C^∞. Let us consider the following statement:

(1)   If $\{a_i\}$ is a sequence of points in X_2 converging to a point $a \in X_1$, and if the sequence $\{L_{a_i}\}$ converges to a subspace L in $G_{n,m'}$, then $L \supset T_a X_1 = X_1$.

The lemma in the case $k = 2$ follows from this statement for the following reason. Assume this and let $e_1, \dots, e_n$ be the unit vectors of $\mathbf{R}^n$ subordinate to the coordinate system of $\mathbf{R}^n$. Let us consider the cases $\xi_1 = e_{m+1}, \dots,$ or e_n. For each $x \in U \cap X_2$, let $v_{m+1x}, \dots, v_{nx}$ denote the orthogonal projection images of $e_{m+1}, \dots, e_n$, respectively, onto L_x. By (1) there are C^∞ functions $f_{i,j}$, $i = m+1, \dots, n$, $j = 1, \dots, n$, on $U \cap X_2$ such that for $x \in U \cap X_2$,

$$v_{ix} = e_i + \sum_{j=1}^n f_{i,j}(x) e_j, \quad i = m+1, \dots, n,$$

and $f_{i,j}(x) \to 0$ as x converges to a point of X_1. Let E, A and A' denote the unit $(n-m)$-matrix, the $(n-m, n-m)$- and $(n-m, m)$-matrices, respectively, whose (i, j)-components are $f_{m+i,m+j}$ and $f_{m+i,j}$. Then

$$\begin{pmatrix} v_{m+1} \\ \vdots \\ v_n \end{pmatrix} = (E + A) \begin{pmatrix} e_{m+1} \\ \vdots \\ e_n \end{pmatrix} + A' \begin{pmatrix} e_1 \\ \vdots \\ e_m \end{pmatrix}.$$

Let U' be an open neighborhood of X_1 in $\mathbf{R}^n$ smaller than U where $E + A$ is invertible, and set

$$\begin{pmatrix} v'_{m+1} \\ \vdots \\ v'_n \end{pmatrix} = (E + A)^{-1} \begin{pmatrix} v_{m+1} \\ \vdots \\ v_n \end{pmatrix} = \begin{pmatrix} e_{m+1} \\ \vdots \\ e_n \end{pmatrix} + (E + A)^{-1} A' \begin{pmatrix} e_1 \\ \vdots \\ e_m \end{pmatrix}$$

on $U' \cap X_2$. For each $i = m + 1, \ldots, n$, $\{e_i, v'_i\}$ is a continuous controlled vector field on $\{X_1, U' \cap X_2\}$. Its continuity follows from convergence of $A(x)$ and $A'(x)$ to 0 as x converges to a point of X_1 and its controlledness follows from the equality $\pi_1(x_1, \ldots, x_n) = (0, \ldots, 0, x_{m+1}, \ldots, x_n)$. We can extend the v'_i to X_2 by the pasting method.

For a general C^∞ vector field ξ_1 on X_1 of the form $\sum_{i=m+1}^n g_i e_i$, g_i being C^∞ functions on X_1, the vector field $\{\xi_1, \xi_2 = \sum_{i=m+1}^n (g_i \circ \pi_1) \cdot v'_i\}$ on $\{X_1, X_2\}$ is continuous and controlled.

It remains to prove (1). For this, it suffices to show:

(2)   For any tangent vector w of X_1 at the a, there exists a sequence $\{w_i \in L_{a_i}\}_{i=1,2,\ldots}$ converging to w.

We want to construct the w_i of (2). Let u_i, for each $i = 1, 2, \ldots$, denote a unit vector in the line $\overrightarrow{a_i \pi(a_i)}$. Without loss of generality, we can assume the sequences $\{u_i\}$ and $\{T_{a_i} X_2\}$ converge to a vector u and a subspace $K \subset \mathbf{R}^n$, respectively. From the Whitney condition it follows that $u \in K$, $X_1 \subset K$. Hence we have vectors $u'_i, w'_i \in T_{a_i} X_2$, $i = 1, 2, \ldots$, such that $u'_i \to u$, $w'_i \to w$ and hence $u_i - u'_i \to 0$ as $i \to \infty$. Since $T_{a_i} X_2$ is the direct sum of $\mathbf{R} u'_i$ and L_{a_i} for each $i = 1, 2, \ldots$, we have unique $w_i \in L_{a_i}$ and $\alpha_i \in \mathbf{R}$ such that

$$w'_i = w_i + \alpha_i u'_i, \quad i = 1, 2, \ldots .$$

Then $w_i \to w$ as $i \to \infty$ for the following reason. Let us decompose each of u'_i and w'_i into the $T_{a_i} \rho_1^{-1}(\rho_1(a_i))$ factor and its orthogonal factor. By the assumption on ρ_1, all the u_i are orthogonal to $T_{a_i} \rho_1^{-1}(\rho_1(a_i))$. Hence we have $\beta_i \in \mathbf{R}$ and $u''_i \in T_{a_i} \rho_1^{-1}(\rho_1(a_i))$, $i = 1, 2, \ldots$, such that

$$u'_i = u''_i + \beta_i u_i$$

are the orthogonal decompositions. Then it follows from the convergence $u_i - u'_i \to 0$ as $i \to \infty$ that $\beta_i \to 1$ and $u''_i \to 0$ as $i \to \infty$, and for each i we have the decomposition of w'_i into the $T_{a_i} \rho_1^{-1}(\rho_1(a_i))$ factor and its orthogonal factor:

$$w'_i = (w_i + \alpha_i u''_i) + \alpha_i \beta_i u_i.$$

Recall that X_1 is a linear subspace of $T_{a_i} \rho_1^{-1}(\rho_1(a_i))$ which, together with the convergence $w'_i \to w$ as $i \to \infty$, implies the convergence $w_i + \alpha_i u''_i \to w$

and $\alpha_i \beta_i u_i \to 0$ as $i \to \infty$. Hence $\alpha_i \to 0$ as $i \to \infty$ because u_i are unit and $\beta_i \to 1$. Therefore, $w_i = w'_i - \alpha_i u'_i \to w$ as $i \to \infty$ because $u'_i \to u$. □

Let $\xi = \{\xi_i\}$ be a vector field on a stratification $\{X_i\}$. For each i let $\theta_i \colon D_i \to X_i$, $D_i \subset X_i \times \mathbf{R}$, be the maximal C^∞ flow defined by ξ_i. Set $D = \cup D_i$ and define a map $\theta \colon D \to X$ by $\theta|_{D_i} = \theta_i$ for each i. We call θ the *flow* of ξ.

Lemma I.1.6 (Integrability of vector field). *Let $\{X_i\}$ be a stratification of a locally closed set $X \subset \mathbf{R}^n$, let $\{T_i\}$ be a controlled tube system for $\{X_i\}$, let $\xi = \{\xi_i\}$ be a semicontrolled vector field on $\{X_i\}$, and let C be a compact subset of one X_i. Fix one i. Let θ denote the flow of ξ, and let ε be a positive number such that for any j with $(\overline{X_j} - X_j) \cap X_i \neq \varnothing$, the condition $\mathrm{scv}(T_i, T_j)$ holds true on $X_j \cap \pi_i^{-1}(C) \cap \rho_i^{-1}([0, \varepsilon[)$. Then for any $x \in \pi_i^{-1}(C) \cap X$ with $\rho_i(x) < \varepsilon$, ρ_i is constant on the connected component of the set $\theta(D \cap (x \times \mathbf{R})) \cap \pi_i^{-1}(C)$ containing x.*

Moreover, if ξ is controlled, then D is open in $X \times \mathbf{R}$, θ is continuous, and there exist neighborhoods U of C in X and I of 0 in $\mathbf{R}$ such that

$$U \subset |T_i|, \qquad U \times I \subset D, \qquad \pi_i(U) \times I \subset D,$$

$$\pi_i \circ \theta(x, t) = \theta(\pi_i(x), t) \quad for \quad (x, t) \in U \times I,$$

$$\left.\begin{array}{l} \theta(x, t) \in |T_j|, \\ (\pi_j(x), t) \in D, \\ \pi_j \circ \theta(x, t) = \theta(\pi_j(x), t), \\ \rho_j \circ \theta(x, t) = \rho_j(x) \end{array}\right\} \quad \begin{array}{l} for \quad x \in (a \ neighborhood \ of \ X_j \cap U \ in \ U), \\ t \in I, \ j \neq i. \end{array}$$

Proof. The first half. Let $x \in X_j$ and let $\theta(x \times I)$ be the connected component of the set $\theta(D \cap (x \times \mathbf{R})) \cap \pi_i^{-1}(C)$ containing x. Then I is either 0 or an interval in $\mathbf{R}$ containing 0. Consider the function ρ_i^x on I defined by $\rho_i^x(t) = \rho_i \circ \theta(x, t)$. Of course,

$$\rho_i^x(0) = \rho_i(x) < \varepsilon,$$

ρ_i^x is of class C^∞, and from the condition $\mathrm{scv}(T_i, T_j)$ on $X_j \cap \pi_i^{-1}(C) \cap \rho_i^{-1}([0, \varepsilon[)$, it follows that ρ_i^x is constant on $\rho_i^{x-1}([0, \varepsilon[)$. Hence $\rho_i^x = \rho_i^x(0)$ on I, which proves the first half.

The latter half. Assume that ξ is controlled. For openness of D and continuity of θ, it suffices to prove them locally. Hence by I.1.1 we can assume X

is closed in $\mathbf{R}^n$,

$$X_i = \{x = (x_1, \dots, x_n) \in \mathbf{R}^n : x_1 = \cdots = x_m = 0\},$$
$$|T_i| = \mathbf{R}^n, \qquad \pi_i(x_1, \dots, x_n) = (0, \dots, 0, x_{m+1}, \dots, x_n),$$
$$\rho_i(x_1, \dots, x_n) = x_1^2 + \cdots + x_m^2,$$
$$X_i \cap (\overline{X_j} - X_j) \neq \varnothing, \quad j \neq i,$$

and the conditions $\mathrm{scv}(T_i, T_j)$ and $\mathrm{cv}(T_i, T_j)$ are satisfied on $X_j \cap U_i$, $j \neq i$, where $U_i = \rho_i^{-1}([0, 1[)$. It suffices to consider the problem around each point $(a, c) \in D \cap (X_i \times \mathbf{R})$.

Proof of openness of D. It suffices to prove that for each $(a, c) \in D \cap (X_i \times \mathbf{R})$, there exist a neighborhood C' of a in X and an open interval I in $\mathbf{R}$ containing c such that

$$C' \times I \subset D. \tag{1}$$

Consider this condition (1) on $X_i \times \mathbf{R}$. Since ξ_i is of class C^∞ on X_i, for some neighborhood C'' of a in X_i and for an open interval I in $\mathbf{R}$ containing c, we have the inclusion:

$$C'' \times I \subset D.$$

Set $C' = \pi_i^{-1}(C'') \cap U_i$. We want to prove the inclusion (1) for these C' and I by reduction to absurdity. Assume (1) does not hold true for these C' and I, namely,

$$a' \times I \not\subset D$$

for some $a' \in C'$. Let $c_l \to c$, $l = 1, 2, \dots$, be a convergent sequence in I such that $(a', c_l) \in D$, $l = 1, 2, \dots$, and $(a', c) \notin D$. We have $(\pi_i(a'), c) \in D$. Hence

$$\theta(\pi_i(a'), c_l) \to \theta(\pi_i(a'), c) \quad \text{as} \quad l \to \infty. \tag{2}$$

Let $a' \in X_j \neq X_i$. From the first half of the lemma it follows that

$$\rho_i = \rho_i(a') \quad \text{on} \quad \theta(D \cap (a' \times \mathbf{R})), \tag{3}$$

which implies

$$\theta(D \cap (a' \times \mathbf{R})) \subset U_i,$$

and hence by the condition $\mathrm{cv}(T_i, T_j)$, we have

$$\pi_i \circ \theta(a', c') = \theta(\pi_i(a'), c') \quad \text{for} \quad (a', c') \in D.$$

In particular,

$$\pi_i \circ \theta(a', c_l) = \theta(\pi_i(a'), c_l), \quad l = 1, 2, \dots.$$

Hence by (2),

$$\pi_i \circ \theta(a', c_l) \to \theta(\pi_i(a'), c) \quad \text{as} \quad l \to \infty.$$

Therefore, by (3) we can assume $\{\theta(a', c_l)\}_{l=1,2,\dots}$ converges to a point b because X is closed in $\mathbf{R}^n$. This b is contained in some $X_k \neq X_j$ because $(a', c) \notin D$. Once more, apply the first half of the lemma to the X_k and a compact neighborhood of b in X_k. We see that the sequence $\{\rho_k(\theta(a', c_l))\}_{l=1,2,\dots}$ is constant for large l, which is a contradiction. Thus (1) holds true.

Proof of continuity of θ. It suffices to prove that for a sequence $\{(a_l, t_l)\}_{l=1,2,\dots}$ in D converging to a point $(a, c) \in D \cap (X_i \times \mathbf{R})$, the sequence $\{\theta(a_l, t_l)\}_{l=1,2,\dots}$ converges to $\theta(a, c)$. We can assume all a_l are contained in U_i. The above arguments show

$$\left. \begin{aligned} \rho_i(\theta(a_l, t_l)) &= \rho_i(a_l) \to \rho_i(a) = 0, \\ \pi_i(\theta(a_l, t_l)) &= \theta(\pi_i(a_l), t_l) \to \theta(a, t) \end{aligned} \right\} \quad \text{as} \quad l \to \infty,$$

which proves the convergence.

The last statements in the lemma were already shown or are clear. $\qquad\square$

Remark I.1.7 (compare with I.1.4). Let $\{T_i = (|T_i|, \pi_i, \rho_i)\}_{i=1,2,\dots}$ be a controlled tube system for a Whitney stratification $\{X_i\}_{i=1,2,\dots}$ of a locally closed set $X \subset \mathbf{R}^n$. Let $\xi = \{\xi_i\}_{i=1,2,\dots}$ be a controlled vector field on $\{X_i\}$, let $\theta \colon D \to X$ be the flow of ξ, and let $C \subset X_1$ and $C' \subset \mathbf{R}$ be compact sets. Assume

$$X_1 = \{(x_1, \dots, x_n) \in \mathbf{R}^n \colon x_1 = \dots = x_m = 0\}, \quad 1 \le m < n,$$
$$\pi_1(x_1, \dots, x_n) = (0, \dots, 0, x_{m+1}, \dots, x_n) \quad \text{for} \quad (x_1, \dots, x_n) \in |T_1|, \quad \text{and}$$
$$\xi_1 = \frac{\partial}{\partial x_n}.$$

There exist neighborhoods U of C in X and I of C' in $\mathbf{R}$ such that

$$U \subset |T_1|, \qquad U \times I \subset D, \quad \text{and}$$
$$\theta((U \cap \{x_j = 0\}) \times I) \subset \{x_j = 0\}, \quad j = m+1, \dots, n-1.$$

This is immediate by the latter half of I.1.6 and the fact that ξ_1 is integrable.

Let M be a C^∞ submanifold of $\mathbf{R}^n$ (always with the Riemannian metric induced by $\mathbf{R}^n$ unless otherwise specified), and let f_1 and f_2 be C^∞ functions on M. Let $\xi(f_1, f_2)$ denote the C^0 vector field

$$|\operatorname{grad} f_1| \operatorname{grad} f_2 + |\operatorname{grad} f_2| \operatorname{grad} f_1$$

on M. We note that $\xi(f_1, f_2)$ is of class C^∞ at points where $\xi \neq 0$, and that f_1 and f_2 are monotone on every integral curve of $\xi(f_1, f_2)$. We call f_1 and f_2 *friendly* at a point $a \in M$ if $\xi(f_1, f_2) \neq 0$ at a (i.e., $\operatorname{grad} f_{1a} \neq 0$, $\operatorname{grad} f_{2a} \neq 0$ and they do not point in the opposite directions), and we call f_1 and f_2 *friendly* if so everywhere. Note that this definition does not depend on choice of a Riemannian metric on M. If M has corners, we define friendliness as follows. As the problem is local, we can assume $M = [0, \infty[^m \times \mathbf{R}^{m'}$ and $a = 0$. Then we call f_1 and f_2 are *friendly* at 0 if $f_1|_{0 \times \mathbf{R}^{m'}}$ and $f_2|_{0 \times \mathbf{R}^{m'}}$ are friendly at 0. The vector field $\xi(f_1, f_2)$ and its flow are the most important tools in the proof of triangulations of strong isomorphisms (III.1.1 and III.1.2) and its application to the $\mathfrak{X}$-Hauptvermutung (III.1.4). The following lemma and remark explain how to use these tools.

Lemma I.1.8. *Let $\{X_i\}_{i=1,\dots,k}$ be a Whitney stratification of a set $X \subset \mathbf{R}^n$, let $\{T_i\}_{i=1,\dots,k}$ be a controlled tube system for $\{X_i\}$, and let $U' \subset U$ be open neighborhoods of $\cup_{i=1}^l X_i$ in X with $\overline{U'} \subset U$, $l \leq k$. Let f_1 and f_2 be C^0 functions on X of class C^∞ on $\cup_{i=l+1}^k X_i$ such that*

$$f_1^{-1}(0) = f_2^{-1}(0) = \bigcup_{i=1}^{l} X_i,$$

$$f_1 > 0, \ f_2 > 0 \quad on \quad \bigcup_{i=l+1}^{k} X_i,$$

and $f_1|_{X_i \cap U}$ and $f_2|_{X_i \cap U}$ are friendly for any $i > l$. There exists a continuous controlled vector field $\xi = \{\xi_i\}_{i=l+1,\dots,k}$ on $\{X_i\}_{i=l+1,\dots,k}$ such that for each $i = l+1, \dots, k$, the functions $\xi_i(f_1|_{X_i})$ and $\xi_i(f_2|_{X_i})$ are positive on $X_i \cap U'$.

The proof works in the $\mathfrak{X}$-category, but the lemma is useless.

Proof. Let $\xi' = \{\xi'_i\}_{i=l+1,\dots,k}$ be a vector field on $\{X_i\}_{i=l+1,\dots,k}$ such that

$$\xi'_i = \xi(f_1|_{X_i}, f_2|_{X_i}) \quad \text{on} \quad X_i \cap \overline{U'}, \ i = l+1, \dots, k.$$

For each $i = l+1, \dots, k$, the functions $\xi'_i(f_1|_{X_i})$ and $\xi'_i(f_2|_{X_i})$ are positive on $X_i \cap \overline{U'}$, but ξ' is not necessarily continuous nor controlled. So we need to modify ξ'.

We proceed in the same way as in the proof of I.1.5. By I.1.5, for each $l < j \leq k$, there exists a continuous controlled vector field $\xi^j = \{\xi_i^j\}$ on $\{X_i \colon \dim X_i \geq \dim X_j,\ i > l\}$ such that $\xi_j^j = \xi_j'$. Since f_1 and f_2 have C^∞ extensions to some neighborhood of $\cup_{i=l+1}^k X_i$ in $\mathbf{R}^n$, and since ξ^j are continuous, there are open neighborhoods U_j of X_j in $\mathbf{R}^n$ for all $j = l+1, \ldots, k$ such that

$$\xi_i^j(f_1|_{X_i}) > 0, \ \xi_i^j(f_2|_{X_i}) > 0 \quad \text{on} \quad X_i \cap U_j \cap \overline{U'},$$
$$i, j = l+1, \ldots, k \quad \text{with} \quad \dim X_i \geq \dim X_j.$$

In the same way as in the proof of I.1.5, we can construct a suitable C^∞ partition of unity $\{\varphi_j\}_{j=l+1,\ldots,k}$ so that the vector field

$$\xi = \sum_{j=l+1}^k \varphi_j \xi^j = \left\{ \sum_{j=l+1}^k \varphi_j \xi_i^j \right\}_{i=l+1,\ldots,k}$$

fulfills the requirements, where we set $\xi_i^j = 0$ if $\dim X_i < \dim X_j$. $\qquad\square$

Remark I.1.9. In I.1.8, assume X is compact. For a small number $\varepsilon > 0$ there exists a homeomorphism τ of X such that

$$\tau(f_1^{-1}([0, \varepsilon])) = f_2^{-1}([0, \varepsilon]),$$
$$\tau = \mathrm{id} \quad \text{on} \quad \bigcup_{i=1}^l X_i,$$
$$\tau(X_i) = X_i, \quad i = 1, \ldots, k,$$

and the restriction of τ to each X_i is a C^∞ diffeomorphism.

The $\mathfrak{X}$-version of this remark can be proved by the elementary methods stated in the preface as in the proof of II.6.1' without the method of integration of vector fields. The key lemma of the proof of II.6.1' is II.6.8. By II.6.8, we can lift a flow on a stratum to a "controlled" flow on strata of larger dimension. Thus we do not need the method of integration. We omit the proof because it is long and similar to the proof of II.6.1'.

Proof. Let $\{T_i\}$, $\{\xi_i\}$ and U' be the same as in I.1.8. Let $0 < \varepsilon_- < \varepsilon < \varepsilon_+$ be numbers such that

$$f_1^{-1}([0, 2\varepsilon_-]) \subset f_2^{-1}([0, \varepsilon]) \subset f_1^{-1}([0, \varepsilon_+/2]) \subset f_1^{-1}([0, \varepsilon_+]) \subset U'.$$

Let $\theta \colon D \to X - f_1^{-1}(0)$, $D \subset (X - f_1^{-1}(0)) \times \mathbf{R}$, be the flow of ξ. For each $a \in U'$, the restrictions of f_1 to $\theta(D \cap (a \times \mathbf{R})) \cap f_1^{-1}(]0, \varepsilon_+])$ and of f_2 to $\theta(D \cap (a \times \mathbf{R})) \cap f_2^{-1}(]0, \varepsilon])$ are C^∞ regular. Hence by I.1.6 there exists a homeomorphism:

$$\sigma \colon f_1^{-1}(\varepsilon_-) \times [\varepsilon_-, \varepsilon_+] \to f_1^{-1}([\varepsilon_-, \varepsilon_+])$$

such that

$$f_1 \circ \sigma(x, t) = t \quad \text{for} \quad (x, t) \in f_1^{-1}(\varepsilon_-) \times [\varepsilon_-, \varepsilon_+],$$

$$\sigma(x \times [\varepsilon_-, \varepsilon_+]) = \theta(D \cap (x \times \mathbf{R})) \cap f_1^{-1}([\varepsilon_-, \varepsilon_+]) \quad \text{for} \quad x \in f_1^{-1}(\varepsilon_-),$$

$$\sigma((X_i \cap f_1^{-1}(\varepsilon_-)) \times [\varepsilon_-, \varepsilon_+]) = X_i \cap f_1^{-1}([\varepsilon_-, \varepsilon_+]), \quad i = l+1, \dots, k,$$

and the restriction of σ to each manifold with boundary $(X_i \cap f_1^{-1}(\varepsilon_-)) \times [\varepsilon_-, \varepsilon_+]$ is a C^∞ diffeomorphism.

By the friendliness assumption, for each $x \in f_1^{-1}(\varepsilon_-)$, the composite $f_2 \circ \sigma$ is C^∞ regular and increasing on $x \times [\varepsilon_-, \varepsilon_+]$, and $f_2 \circ \sigma(x, t) = \varepsilon$ for some unique $2\varepsilon_- < t < \varepsilon_+/2$. This t will be denoted $t(x)$. The function $x \to t(x)$ is continuous, and it is of class C^∞ on each $X_i \cap f_1^{-1}(\varepsilon_-)$. Then we easily construct a homeomorphism τ' of the set $f_1^{-1}(\varepsilon_-) \times [\varepsilon_-, \varepsilon_+]$ such that

$$\tau' = \text{id} \quad \text{on a neighborhood of} \quad f_1^{-1}(\varepsilon_-) \times \{\varepsilon_-, \varepsilon_+\},$$

$$\tau'(x, \varepsilon) = (x, t(x)) \quad \text{for} \quad x \in f_1^{-1}(\varepsilon_-),$$

$$\tau'(x \times [\varepsilon_-, \varepsilon_+]) = x \times [\varepsilon_-, \varepsilon_+] \quad \text{for} \quad x \in f_1^{-1}(\varepsilon_-),$$

and the restriction of τ' to each $(X_i \cap f_1^{-1}(\varepsilon_-)) \times [\varepsilon_-, \varepsilon_+]$ is a C^∞ diffeomorphism. We can extend the homeomorphism $\sigma \circ \tau' \circ \sigma^{-1}$ of the set $f_1^{-1}([\varepsilon_-, \varepsilon_+])$ to a homeomorphism of X by setting it equal to the identity on the complement $X - f_1^{-1}([\varepsilon_-, \varepsilon_+])$. The extension is what we wanted. $\square$

We shall need more than one couple of friendly functions (see I.1.13), but by I.1.10' we can reduce the problem to the case of one couple. First we explain I.1.10' in a simpler form as follows.

Lemma I.1.10. *Let $\{X_i\}_{i=1,2,\dots}$ be a Whitney stratification of a locally closed subset $X \subset \mathbf{R}^n$, and let $\{T_i\}_{i=1,2,\dots}$ be a controlled tube system for $\{X_i\}$ such that*

$$X_1 \subset \overline{X_i} - X_i, \quad i = 2, \dots, \quad and$$

$$X_1 = \{x_1 = \cdots = x_m = 0\}.$$

Let Λ denote the family of all subsets of the set $\{m+1,\dots,n\}$ (we assume $\varnothing \in \Lambda$), and let U be an open neighborhood of X_1 in $\mathbf{R}^n$ such that all the sets $X_i \cap \{x_j = 0, j \in \lambda\} \cap U$, $i = 2,\dots,\lambda \in \Lambda$, are C^∞ submanifolds of $\mathbf{R}^n$. Let f_1 and f_2 be C^0 functions on X of class C^∞ on $X - X_1$ such that

$$f_1^{-1}(0) = f_2^{-1}(0) = X_1;$$
$$f_1 > 0, \ f_2 > 0 \quad on \quad X - X_1;$$

and for each $i > 1$, and for each $\lambda \in \Lambda$, the restrictions of f_1 and f_2 to $X_i \cap \{x_j = 0, j \in \lambda\} \cap U$ are friendly. There exist an open neighborhood $U' \subset U$ of X_1 in $\mathbf{R}^n$ and a semicontrolled vector field $\xi = \{\xi_i\}_{i=2,3,\dots}$ on $\{X_i\}_{i=2,3,\dots}$ such that for each $i > 1$, the functions $\xi_i(f_1|_{X_i \cap U'})$ and $\xi_i(f_2|_{X_i \cap U'})$ are positive; and for each $m < j \le n$, the vector field $\xi_i|_{\{x_j=0\} \cap X_i \cap U'}$ is tangent to $\{x_j = 0\} \cap X_i \cap U'$.

In I.1.10 only the last condition on ξ_i is not clear, and we do not know existence of continuous ξ. But we can easily give an example of $\{X_i\}$ and $\{T_i\}$ where no controlled vector fields with the required properties exist. We shall not need either continuity or controlledness of this ξ later, because when we consider several couples of friendly functions we shall require a good C^∞ diffeomorphism of only one stratum X_i but not a homeomorphism of X (see I.1.13).

The $\mathfrak{X}$-versions of II.1.10 and II.1.10$'$ can be proved in the same way but are not useful.

To prove this lemma, we need the following lemma.

Lemma I.1.11 (Local C^0 triviality of Whitney stratifications). *For the same $\{X_i\}$, X and $\{T_i\}$ as in I.1.10, there exist small open neighborhoods U and V of 0 in $\mathbf{R}^n$ and in $\mathbf{R}^m \times 0$ ($\subset \mathbf{R}^m \times \mathbf{R}^{n-m} = \mathbf{R}^n$), respectively, small open intervals $I_{m+1},\dots,I_n \subset \mathbf{R}$ containing 0 and a homeomorphism $\alpha \colon V \times I_{m+1} \times \cdots \times I_n \to U$ of the form $\alpha(x, y_{m+1}, \dots, y_n) = (\alpha_1(x, y_{m+1}, \dots, y_n), y_{m+1}, \dots, y_n)$ such that*
(i) *for each i, $\alpha|_{X_{i,0} \times I_{m+1} \times \cdots \times I_n}$ is a C^∞ diffeomorphism onto $X_i \cap U$, where $X_{i,0} = X_i \cap V$;*
(ii) *for each i,*

$$\rho_i \circ \alpha(x, y_{m+1}, \dots, y_n) = \rho_i(x) \quad for \quad x \in (a \ neighborhood \ of \ X_{i,0} \ in \ V)$$
$$and \quad y_j \in I_j, \ j = m+1, \dots, n;$$

and
$$\alpha|_{V \times 0} = \mathrm{id}. \tag{iii}$$

The $\mathfrak{X}$-version of I.1.11 also holds true although we do not prove it. See the note after I.1.9.

Proof of I.1.10. If we prove the lemma locally at each point of X_1, the lemma in the global form follows from the pasting method in the proof of I.1.5. The local lemma at each non-zero point of X_1 is easier than that at 0. Hence we prove the lemma locally at $0 \in \mathbf{R}^n$.

For each $\lambda \in \Lambda$, by using I.1.11, we shall construct a semicontrolled vector field $\xi^\lambda = \{\xi_i^\lambda\}_{i=2,3,\dots}$ on $\{X_i \cap U'\}_{i=2,3,\dots}$ which satisfies the following conditions, where U' is a small open neighborhood of 0 in $\mathbf{R}^n$.

(1) For each i, and for each $j \in \lambda$, the vector field $\xi_i^\lambda|_{\{x_j=0\} \cap X_i \cap U'}$ is tangent to $\{x_j = 0\} \cap X_i \cap U'$.

(2) For each i the functions $\xi_i^\lambda(f_1|_{X_i \cap U_i^\lambda})$ and $\xi_i^\lambda(f_2|_{X_i \cap U_i^\lambda})$ are positive, where U_i^λ is a small open neighborhood of $\{x_j = 0, j \in \lambda\} \cap X_i \cap U'$ in U'.

If we have such ξ^λ, $\lambda \in \Lambda$, we can construct the ξ as follows, which is a precise version of the pasting method. We assume the above neighborhoods U' of 0 do not depend on the λ, by replacing them with their intersection. For each $i > 1$, set

$$U_i^{\lambda'} = U_i^\lambda - \bigcup_{j \notin \lambda} \{x_j = 0\} \quad \text{for} \quad \lambda \in \Lambda.$$

The family $\{U_i^{\lambda'}\}_{\lambda \in \Lambda}$ is an open covering of $X_i \cap U'$. Let $\{\varphi_i^\lambda\}_{\lambda \in \Lambda}$ be a C^∞ partition of unity subordinate to $\{U_i^{\lambda'}\}_{\lambda \in \Lambda}$, and set

$$\xi_i = \sum_{\lambda \in \Lambda} \varphi_i^\lambda \xi_i^\lambda.$$

The family $\xi = \{\xi_i\}_{i=2,3,\dots}$ is a semicontrolled vector field on $\{X_i \cap U'\}_{i=2,3,\dots}$. (We cannot require continuity nor controlledness of ξ from such a construction.) Moreover, for each $i > 1$, and for each $m < j \le n$, the vector field $\xi_i|_{\{x_j=0\} \cap X_i \cap U'}$ can be tangent to $\{x_j = 0\} \cap X_i \cap U'$ because if $j \in \lambda \in \Lambda$, then $\xi_i^\lambda|_{\{x_j=0\} \cap X_i \cap U'}$ is tangent to $\{x_j = 0\} \cap X_i \cap U'$ by the condition (1), and if $j \notin \lambda \in \Lambda$, then we can choose φ_i^λ so that $\varphi_i^\lambda = 0$ on $\{x_j = 0\} \cap X_i \cap U'$. It follows from the condition (2) that

$$\xi_i(f_1|_{X_i \cap U'}) > 0 \quad \text{and} \quad \xi_i(f_2|_{X_i \cap U'}) > 0, \quad i = 2, 3, \dots .$$

Thus ξ is a local solution at 0 of I.1.10.

For each $\lambda \in \Lambda$ we want to construct the ξ^λ with the properties (1) and (2). Without loss of generality, we can assume $\lambda = \{m' + 1, \dots, n\}$, $m \le m' \le n$. Set

$$R^\lambda = \{x_{m'+1} = \cdots = x_n = 0\}, \qquad X^\lambda = X \cap R^\lambda,$$

$$\{X_i^\lambda\}_{i=1,2,\dots} = \{X_i \cap R^\lambda\}_{i=1,2,\dots}, \quad \text{and} \quad f_l^\lambda = f_l|_{X^\lambda}, \quad l = 1, 2.$$

We shall define ξ^λ first on $X^\lambda - X_1^\lambda$ near 0 by using I.1.8 and then extend it to $(X - X_1) \cap U'$ by using I.1.11. For an open neighborhood V^λ of 0 in R^λ, $\{X_i^\lambda \cap V^\lambda\}_{i=1,2,...}$ is a Whitney stratification of $X^\lambda \cap V^\lambda$;

$$f_1^{\lambda-1}(0) = f_2^{\lambda-1}(0) = X_1^\lambda;$$
$$f_1^\lambda > 0, \ f_2^\lambda > 0 \quad \text{on} \quad X^\lambda - X_1^\lambda;$$

and for each $i > 1$, $f_1^\lambda|_{X_i^\lambda \cap V^\lambda}$ and $f_2^\lambda|_{X_i^\lambda \cap V^\lambda}$ are friendly. Moreover, there exists a controlled tube system $\{T_i^\lambda = (|T_i^\lambda|, \pi_i^\lambda, \rho_i^\lambda)\}_{i=1,2,...}$ for $\{X_i^\lambda \cap V^\lambda\}_{i=1,2,...}$ such that

$$|T_i^\lambda| \subset |T_i| \cap R^\lambda, \quad \text{and} \quad \rho_i^\lambda = \rho_i|_{|T_i^\lambda|}, \ i = 1, 2, \ldots.$$

Indeed, if we choose as $|T_i^\lambda|$ a sufficiently small open neighborhood of $X_i^\lambda \cap V^\lambda$ in $\mathbf{R}^n$ and as π_i^λ the orthogonal projection $|T_i^\lambda| \to X_i^\lambda \cap V^\lambda$, then $(|T_i^\lambda|, \pi_i^\lambda, \rho_i|_{|T_i^\lambda|})$ satisfies the condition of a tube. Hence by I.1.3, by shrinking $|T_i^\lambda|$ and modifying π_i^λ, we can assume $\{T_i^\lambda\}$ is controlled.

Apply I.1.8 to $\{X_i^\lambda \cap V^\lambda\}_{i=1,2,...}$, $\{T_i^\lambda\}_{i=1,2,...}$, $f_1^\lambda|_{X^\lambda \cap V^\lambda}$ and $f_2^\lambda|_{X^\lambda \cap V^\lambda}$, and let $\xi^\lambda = \{\xi_i^\lambda\}_{i=2,3,...}$ be a resulting continuous controlled vector field. (We need only its semicontrolledness.) Then
(3) for each $i > 1$ the functions $\xi_i^\lambda(f_1|_{X_i^\lambda \cap V^\lambda})$ and $\xi_i^\lambda(f_2|_{X_i^\lambda \cap V^\lambda})$ are positive, here we shrink V^λ if necessary;
and
(4) for each $i, j > 1$ with $(\overline{X_j^\lambda} - X_j^\lambda) \cap X_i^\lambda \cap V^\lambda \neq \varnothing$,

$$d(\rho_i|_{X_j^\lambda \cap V^\lambda})\xi_{jx}^\lambda = 0 \ \text{for} \ x \in X_j^\lambda \cap V^\lambda \cap \text{(a neighborhood of } X_i^\lambda \cap V^\lambda \text{ in } V^\lambda)$$

by semicontrolledness.

Thus we define ξ^λ on $\{X_i^\lambda \cap V^\lambda\}_{i=2,3,...}$. We need to extend it.

Next apply I.1.11 to $\{X_i\}$, X and $\{T_i\}$. By using the resulting homeomorphism, we easily obtain open neighborhoods U', V and W of 0 in $\mathbf{R}^n$, $\mathbf{R}^{m'} \times 0 \ (\subset \mathbf{R}^{m'} \times \mathbf{R}^{n-m'})$ and $0 \times \mathbf{R}^{n-m'} \ (\subset \mathbf{R}^m \times \mathbf{R}^{n-m'})$, respectively, and a homeomorphism $\alpha \colon V \times W \to U'$ which satisfies the following conditions:
(5) For each i, $\alpha|_{(X_i^\lambda \cap V) \times W}$ is a C^∞ diffeomorphism onto $X_i \cap U'$.
(6) For each i,

$$\rho_i \circ \alpha(x, y) = \rho_i(x) \ \text{for} \ x \in \text{(a neighborhood of } X_i^\lambda \cap V \text{ in } V) \ \text{and} \ y \in W.$$

(7) For each $m' < j \le n$,

$$U' \cap \{x_j = 0\} = \alpha(V \times (W \cap \{x_j = 0\})).$$
$$\alpha|_{V \times 0} = \text{id}. \tag{8}$$

We replace V^λ and V with their intersection, and we keep the notation V for it.

We extend ξ^λ to $\{X_i \cap U'\}_{i=2,3,\dots}$ as follows. First choose the vector field $(\xi^\lambda, 0) = \{(\xi_i^\lambda, 0)\}_{i=2,3,\dots}$ on $\{(X_i^\lambda \cap V) \times W\}_{i=2,3,\dots}$. Next transmit this vector field to a vector field on $\{X_i \cap U'\}_{i=2,3,\dots}$ by the α, which is possible because of condition (5). By (8) the new vector field is an extension of ξ^λ to $\{X_i \cap U'\}_{i=2,3,\dots}$. We keep the same notation ξ^λ for it. Then ξ^λ is what we wanted. Its semicontrolledness follows from (4) and (6). The property (1) is immediate by (7) because $(\xi_i^\lambda, 0)$ is tangent to $(X_i^\lambda \cap V) \times (W \cap \{x_j = 0\})$. Existence of U_i^λ with the property (2) follows from (3) and from the fact that the functions $\xi_i^\lambda(f_1|_{X_i \cap U'})$ and $\xi_i^\lambda(f_2|_{X_i \cap U'})$ are continuous. $\square$

We shall apply I.1.10 in the form of I.1.10$'$. The proof of I.1.10$'$ is immediate by I.1.10, the pasting method and by the note following the proof of I.1.11.

Lemma I.1.10$'$. *Let $\{X_i\}_{i=1,2,\dots}$ be a Whitney stratification of a locally closed set $X \subset \mathbf{R}^n$, let $\{T_i\}_{i=1,2,\dots}$ be a controlled tube system for $\{X_i\}_{i=1,2,\dots}$, let U be an open neighborhood of X_1 in $\mathbf{R}^n$, let $\{Y_j\}_{j=1,\dots,k}$ be a family of C^∞ submanifolds of U closed in U such that X_i, $i = 1, 2, \dots$, and Y_j, $j = 1, \dots, k$, are transversal to each other, let Z be the closure in X of the union of some connected components of $X - \cup_{j=1}^k Y_j$. Let f_1 and f_2 be C^0 functions on Z of class C^∞ on $(Z \cap U) - X_1$ such that*

$$f_1^{-1}(0) = f_2^{-1}(0) = Z \cap X_1,$$
$$f_1 > 0, \ f_2 > 0 \quad on \quad Z - X_1,$$

for each $i > 1$, and for each subset λ of $\{1, \dots, k\}$, $f_1|_{Y_{\lambda,i} \cap Z \cap U}$ and $f_2|_{Y_{\lambda,i} \cap Z \cap U}$ are friendly, where $Y_{\lambda,i} = \cap_{j \in \lambda} Y_j \cap X_i$. There exists an open neighborhood $U' \subset U$ of X_1 in $\mathbf{R}^n$ and a semicontrolled vector field $\xi = \{\xi_i\}_{i=2,3,\dots}$ on $\{X_i\}_{i=2,3,\dots}$ such that for each $i > 1$, and for each λ, the functions $\xi_i(f_1|_{X_i \cap Z \cap U'})$ and $\xi_i(f_2|_{X_i \cap Z \cap U'})$ are positive, and the vector field $\xi_i|_{Y_{\lambda,i} \cap Z \cap U'}$ is tangent to $Y_{\lambda,i}$.

Proof of I.1.11. In I.1.11 α does not depend on choice of $\{\pi_i\}$ of $\{T_i\}$, in other words, we can replace $\{\pi_i\}$ with suitable submersive C^∞ retractions. Hence by I.1.3 we assume π_1 is the orthogonal projection. Moreover, by replacing $\{X_i\}$ with $\{X_i\} \cup \{\mathbf{R}^n - \overline{X}\}$, we assume X is open in $\mathbf{R}^n$. We prove the lemma by induction on n. If $n = 1$ or $m = n$, the lemma is trivial. Hence assume $m < n$ and that the lemma holds for $n - 1$.

By I.1.5 we have a controlled vector field $\xi = \{\xi_i\}_{i=1,2,\dots}$ on $\{X_i\}_{i=1,2,\dots}$ such that $\xi_1 = \frac{\partial}{\partial x_n}$. Let $\theta \colon D \to X$ denote the flow of ξ. By I.1.6 and I.1.7

there exist an open neighborhood V' of 0 in $\mathbf{R}^{n-1} \times 0$ ($\subset \mathbf{R}^{n-1} \times \mathbf{R}$) and an open interval $I_n \subset \mathbf{R}$ containing 0 such that $\{X_i \cap V'\}_{i=1,2,\ldots}$ is a Whitney stratification,

$$V' \subset |T_1|, \qquad V' \times I_n \subset D, \qquad \theta(V' \times I_n) \subset |T_1|,$$
$$\pi_1 \circ \theta(x, y) = \theta(\pi_1(x), y) \quad \text{for} \quad (x, y) \in V' \times I_n, \text{ and} \qquad (1)$$
$$\rho_i \circ \theta(x, y) = \rho_i(x) \quad \text{for} \quad x \in \text{(a neighborhood } A \text{ of } X_i \cap V' \text{ in } V') \qquad (2)$$
$$\text{and} \quad y \in I_n.$$

Set

$$U' = \theta(V' \times I_n), \qquad \alpha' = \theta|_{V' \times I_n}.$$

Then α' is a homeomorphism of the form $\alpha'(x, y) = (\alpha'_1(x, y), y)$ for $(x, y) \in V' \times I_n$ by (1), and we also have

(3) for each i, $\alpha'|_{(X_i \cap V') \times I_n}$ is a C^∞ diffeomorphism onto $X_i \cap U'$; and clearly

$$\alpha'|_{V' \times 0} = \text{id}. \qquad (4)$$

Apply the induction hypothesis to $\{X_i \cap V'\}_{i=1,2,\ldots}$, $X \cap V'$ and some controlled tube system $\{T_i^0 = (|T_i^0|, \pi_i^0, \rho_i^0)\}_{i=1,2,\ldots}$ for $\{X_i \cap V'\}$ such that π_1^0 is the orthogonal projection,

$$|T_i^0| \subset |T_i| \quad \text{and} \quad \rho_i^0 = \rho_i|_{|T_i^0|}, \quad i = 1, 2, \ldots .$$

We have open neighborhoods U'' and V of 0 in $V' \subset \mathbf{R}^{n-1} \times 0$ and in $\mathbf{R}^m \times 0$, respectively, open intervals $I_{m+1}, \ldots, I_{n-1} \subset \mathbf{R}$ containing 0 and a homeomorphism $\alpha'' \colon V \times I_{m+1} \times \cdots \times I_{n-1} \to U''$ of the form $\alpha''(x, y_{m+1}, \ldots, y_{n-1}) = (\alpha''_1(x, y_{m+1}, \ldots, y_{n-1}), y_{m+1}, \ldots, y_{n-1})$ such that

(5) for each i, $\alpha''|_{X_{i,0} \times I_{m+1} \times \cdots \times I_{n-1}}$ is a C^∞ diffeomorphism onto $X_i \cap U''$, where $X_{i,0} = X_i \cap V$;

(6) for each i and for some neighborhood B of $X_{i,0}$ in V,

$$\rho_i \circ \alpha''(x, y_{m+1}, \ldots, y_{n-1}) = \rho_i(x)$$
$$\text{for} \quad x \in B \quad \text{and} \quad y_j \in I_j, \; j = m+1, \ldots, n-1;$$

and

$$\alpha''|_{V \times 0} = \text{id}. \qquad (7)$$

Set

$$\alpha(x, y_{m+1}, \ldots, y_n) = \alpha'(\alpha''(x, y_{m+1}, \ldots, y_{n-1}), y_n)$$
$$\text{for} \quad (x, y_{m+1}, \ldots, y_n) \in V \times I_{m+1} \times \cdots \times I_n,$$

$$U = \alpha(V \times I_{m+1} \times \cdots \times I_n).$$

The map $\alpha\colon V \times I_{m+1} \times \cdots \times I_n \to U$ is what we wanted. Indeed, it is a homeomorphism of the required form because so are α' and α''. Property (i) in the lemma follows from (3) and (5). If we shrink $I_{m+1},\ldots,I_{n-1}$ and B so that the image of $B \times I_{m-1} \times \cdots \times I_{n-1}$ under α'' is included in the A, then Property (ii) holds true by (2) and (6). Property (iii) is immediate by (4) and (7).   □

In the above proof we did not use the condition $X_1 \subset \overline{X_i} - X_i$, $i = 2,\ldots$. Hence I.1.11 shows the following fact. Any Whitney stratification of a locally closed set in $\mathbf{R}^n$ satisfies the weak frontier condition.

Let $\{X_i\}_{i=1,2,\ldots}$ and $\{Y_i\}_{i=1,2,\ldots}$ be Whitney stratifications of sets X and Y in $\mathbf{R}^n$, respectively. An *isomorphism* $f\colon \{X_i\} \to \{Y_i\}$ is a homeomorphism $f\colon X \to Y$ such that for each i, $f(X_i) = Y_i$, $f_i = f|_{X_i}$ is a diffeomorphism onto Y_i, and the family $\{\text{graph } f_i\}_{i=1,2,\ldots}$ is a Whitney stratification of graph $f \subset \mathbf{R}^{2n}$. We call $\{X_i\}$ and $\{Y_i\}$ *isomorphic*.

Remark I.1.12. The classification of Whitney stratifications by isomorphisms may seem natural. But it is too rough to investigate sets that admit Whitney stratifications as shown below. Let us consider a Whitney stratification where only one natural triangulation exists, because we may reduce the problem of classification in that case to a PL one. (See §I.3 for the terminology PL.) A trivial case is that X is smooth and its stratification consists of one stratum. Then we can take its natural triangulation to be its C^∞ triangulation (see §I.3) by uniqueness of C^∞ triangulations (I.3.13). In this case we have no problems.

A non-trivial case is a Whitney stratification $\{X_i\}_{i=1,2,\ldots}$ of a closed set $X \subset \mathbf{R}^n$ such that $\dim X_1 = 0$, each point x of X_1 has a neighborhood in X which is a cone with vertex x, and for each $i > 1$, $\overline{X_i} - X_i$ is included in X_1. (We call such $\{X_i\}$ a *Whitney stratification with solid triangulation*.) A natural triangulation of $\{X_i\}$ is defined as follows. We assume X is compact for simplicity. First choose a C^∞ triangulation of $X - $ (the open ε-neighborhood of X_1 in X) for a small number $\varepsilon > 0$. Next extend the triangulation to X by the cone structure of X near X_1. This triangulation is not an invariant of an isomorphism class as follows. Let M and N be compact C^∞ manifolds of dimension ≥ 5 which are h-cobordant but not homeomorphic (see [Mi₂]), let X and Y denote their suspensions, let X_1 and Y_1 denote the respective suspension point sets, and set $X_2 = X - X_1$, $Y_2 = Y - Y_1$. Then $\{X_1, X_2\}$ and $\{Y_1, Y_2\}$ are Whitney stratifications with solid triangulation which are weakly isomorphic in the following sense by the weak h-cobordism theorem, and hence isomorphic by the following fact (i), and whose natural triangulations are not PL homeomorphic.

We call $\{X_i\}$ and $\{Y_i\}$ *weakly isomorphic* if there exists a homeomorphism $f\colon X \to Y$ such that for each i, $f(X_i) = Y_i$, and $f_i = f|_{X_i}$ is a C^∞ diffeomorphism onto Y_i. We call $f\colon \{X_i\} \to \{Y_i\}$ a *weak isomorphism*. It seems very possible that if Whitney stratifications $\{X_i\}$ and $\{Y_i\}$ are weakly isomorphic, they are isomorphic. This holds true by (II.1.13) in the $\mathfrak{X}$-case. We can prove in general:

(i) If $\{X_i\}$ and $\{Y_i\}$ are weakly isomorphic Whitney stratifications with solid triangulation, they are isomorphic.

Proof of (i). Let $f\colon \{X_i\}_{i=1,2,\ldots} \to \{Y_i\}_{i=1,2,\ldots}$ be a weak isomorphism between Whitney stratifications with solid triangulation of sets X, $Y \subset \mathbf{R}^n$, respectively. Assume $0 \in X_1$, $0 \in Y_1$ and $f(0) = 0$, and consider the problem only on small neighborhoods of 0 in X and in Y. Let ξ be the vector field on X_2 defined by $\xi_x = x/|x|$, $x \in X_2$. It suffices to modify $f|_{X_2}$ so that $\{0, \mathrm{graph}f|_{X_2}\}$ satisfies the Whitney condition at 0, which follows if

$$|df\xi_x| \to 0, \quad \text{and} \quad |f(x)|/|x| \to 0 \quad \text{as} \quad x \in X_2 \to 0.$$

Let α be a positive C^0 function on $]0, \infty[$ such that

$$\alpha(|f(x)|)|f(x)||df\xi_x| \to 0, \quad \text{and} \quad \alpha(|f(x)|)|f(x)|/|x| \to 0 \quad \text{as} \quad x \in X_2 \to 0. \tag{1}$$

It is easy to find a C^∞ diffeomorphism β of $]0, \infty[$ such that

$$\beta = \mathrm{id} \quad \text{on} \quad [\varepsilon, \infty[\quad \text{for a small number} \quad \varepsilon > 0,$$
$$\beta(t)/t \le \alpha(t), \quad \text{and} \quad \beta'(t)/t \le \alpha(t) \quad \text{on} \quad]0, \varepsilon/2]. \tag{2}$$

Let $\gamma\colon \mathbf{R}^n - 0 \to \mathbf{R}^n - 0$ be the diffeomorphism defined by

$$\gamma(x) = \beta(|x|)x/|x| \quad \text{for} \quad x \in \mathbf{R}^n - 0, \tag{3}$$

which carries the δ-sphere for each number $\delta > 0$ to the $\beta(\delta)$-sphere. Then $f_1 = \gamma \circ f(x)$ fulfills the requirements. Indeed, as $x \in X_2 \to 0$,

$$|df_1\xi_x| \le |d\gamma(f(x))||df\xi_x| = (\max\{\beta(|f(x)|), \beta'(|f(x)|)\}) \times |df\xi_x|$$
$$\le \alpha(|f(x)|)|f(x)||df\xi_x| \to 0, \quad \text{and}$$

$$|f_1(x)|/|x| = \beta(|f(x)|)/|x| \le \alpha(|f(x)|)|f(x)|/|x| \to 0$$

by (1), (2) and (3). □

By I.1.12 we need to introduce a stronger class of isomorphism between Whitney stratifications. Consider when natural triangulations of two Whitney stratifications with solid triangulation $\{X_i\}_{i=1,2,\ldots}$ of a compact set

$X \subset \mathbf{R}^n$ and $\{Y_i\}_{i=1,2,\dots}$ of $Y \subset \mathbf{R}^n$ are PL homeomorphic. A sufficient condition is the following, as shown in the introduction. Let $f: \{X_i\} \to \{Y_i\}$ be an isomorphism, and let $\rho_1^X: X \to \mathbf{R}$ and $\rho_1^Y: Y \to \mathbf{R}$ be the squares of the functions which measure distance from X_1 and Y_1, respectively. Assume the condition:

(∗) the restrictions of ρ_1^X and $\rho_1^Y \circ f$ to each $U_1 \cap X_i$ are friendly for some neighborhood U_1 of X_1 in $\mathbf{R}^n$.

Their natural triangulations are PL homeomorphic. Indeed, from I.1.9, it follows that the C^∞ manifolds with boundary $X-$ (the open ε-neighborhood of X_1) and $Y-$ (the open ε-neighborhood of Y_1) are C^∞ diffeomorphic for a small number $\varepsilon > 0$. Hence by uniqueness of C^∞ triangulations (I.3.13), their C^∞ triangulations are PL homeomorphic, and so are their cone extensions.

The above sufficient condition (∗) leads us to introduce the following concept of a strong isomorphism. Let $f: \{X_i\}_{i=1,\dots,k} \to \{Y_i\}_{i=1,\dots,k}$ be an isomorphism between Whitney stratifications of closed sets X and Y in $\mathbf{R}^n$. For simplicity of notation, we assume that $\dim X_1 < \cdots < \dim X_k$, and X is compact. Set

$$Z = \operatorname{graph} f, \quad \text{and} \quad \{Z_i\} = \{\operatorname{graph} f|_{X_i}\},$$

and let $p_1: Z \to X$ and $p_2: Z \to Y$ denote the projections. For each i, let ρ_i^X, ρ_i^Y and ρ_i^Z denote the squares of the functions which measure distance from X_i, Y_i and Z_i, respectively. We call f a *strong isomorphism* if there exists a common removal data $\delta = \{\delta_i\}_{i=1,\dots,k-1}$ of $\{X_i\}$, $\{Y_i\}$ and $\{Z_i\}$ (not necessarily continuous) such that for each $1 \le i < j \le k$ and for any sequence of positive numbers $\varepsilon = \{\varepsilon_i\}_{i=1,\dots,k-1}$ with $\varepsilon \le \delta$, the restrictions of ρ_i^Y and $\rho_i^Z \circ p_2^{-1}$ to the manifold (possibly with boundary and corners):

$$(Y_j \cap \rho_i^{Y-1}([0,\varepsilon_i[)) - \bigcup_{l<i} \rho_l^{Y-1}([0,\varepsilon_l[)$$

are friendly, and so are the restrictions of ρ_i^X and $\rho_i^Z \circ p_1^{-1}$ to the manifold (possibly with boundary and corners):

$$(X_j \cap \rho_i^{X-1}([0,\varepsilon_i[)) - \bigcup_{l<i} \rho_l^{X-1}([0,\varepsilon_l[).$$

In such a case we say that $\{X_i\}$ and $\{Y_i\}$ are *strongly isomorphic*.

Note the following. Let the δ be lessened if necessary. We have a controlled tube system $\{T_l = (|T_l|, \pi_l, \rho_l)\}_{l=1,\dots,k}$ for $\{X_l\}_{l=1,\dots,k}$ with $\rho_l = \rho_l^X|_{|T_l|}$, $l = 1, \dots, k$, such that for each i, the restriction of π_i to the set $\rho_i^{X-1}([0,\varepsilon_i[) - \cup_{l<i}\rho_l^{X-1}([0,\varepsilon_l[)$ is a C^∞ fibre bundle with base space $X_i - \cup_{l<i}\rho_l^{X-1}([0,\varepsilon_l[)$.

In the above definition of a strong isomorphism, take ρ_i^X, ρ_i^Y and ρ_i^Z to be the functions appearing in given tube systems $\{T_i^X = (|T_i^X|, \pi_i^X, \rho_i^X)\}$, $\{T_i^Y = (|T_i^Y|, \pi_i^Y, \rho_i^Y)\}$ and $\{T_i^Z = (|T_i^Z|, \pi_i^Z, \rho_i^Z)\}$ for $\{X_i\}$, $\{Y_i\}$ and $\{Z_i\}$, respectively. We call f a *strong isomorphism* for $\{T_i^X\}$, $\{T_i^Y\}$ and $\{T_i^Z\}$, and we say that $\{X_i\}$ and $\{Y_i\}$ for $\{T_i^X\}$, $\{T_i^Y\}$ and $\{T_i^Z\}$ are *strongly isomorphic*.

The following proposition is the key of the proofs of III.1.1 and III.1.2, which show Hauptvermutung for a pair of strongly isomorphically stratified polyhedra under certain conditions. Remember the proof of the fact that if two isomorphic Whitney stratifications with solid triangulation satisfy the above friendliness condition $(*)$, their natural triangulations are PL homeomorphic. The proof suggests the reason why I.1.13 is important.

Proposition I.1.13. *Let* $f \colon \{X_i\}_{i=1,\ldots,k} \to \{Y_i\}_{i=1,\ldots,k}$ *be a strong isomorphism between Whitney stratifications of compact sets* $X \subset \mathbf{R}^n$ *and* $Y \subset \mathbf{R}^{n'}$, *and for each* i, *let* ρ_i^X *and* ρ_i^Y *be the squares of the functions which measure distance from* X_i *in* $\mathbf{R}^n$ *and from* Y_i *in* $\mathbf{R}^{n'}$, *respectively. Assume* $\dim X_1 < \cdots < \dim X_k$. *There exists a common removal data* $\delta = \{\delta_i\}_{i=1,\ldots,k-1}$ *of* $\{X_i\}$ *and of* $\{Y_i\}$ *such that for any sequence of positive numbers* $\varepsilon = \{\varepsilon_i\}_{i=1,\ldots,1-k} \leq \delta$, *the sets*

$$X_\varepsilon = X - \bigcup_{i=1}^{k-1} \rho_i^{X-1}([0, \varepsilon_i[),$$

$$Y_\varepsilon = Y - \bigcup_{i=1}^{k-1} \rho_i^{Y-1}([0, \varepsilon_i[)$$

are C^∞ *diffeomorphic* C^∞ *manifolds possibly with corners.*

The $\mathfrak{X}$-version of I.1.13 is proved by the $\mathfrak{X}$-versions of I.1.9 and I.1.11 as in the proof of II.6.1′. We omit the proof. Note that if f is an $\mathfrak{X}$-isomorphism, it is a strong isomorphism (II.1.19).

Proof. First note $X_\varepsilon \subset X_k$ and $Y_\varepsilon \subset Y_k$. We already showed that X_ε and Y_ε are C^∞ manifolds possibly with corners. So we prove that they are C^∞ diffeomorphic. For that, it suffices to find C^∞ diffeomorphisms $Z_\varepsilon \to X_\varepsilon$ and $Z_\varepsilon \to Y_\varepsilon$, where Z_ε is defined in the same way for the graph Z of f. Hence we can reduce the problem to the case where f is the restriction to X of a C^∞ submersion from $\mathbf{R}^n$ to $\mathbf{R}^{n'}$. (We assume this for simplicity of notation, but C^∞ differentiability of f is sufficient.) As Y_ε is C^∞ diffeomorphic to the manifold possibly with corners $X - \cup_{i=1}^{k-1}(\rho_i^Y \circ f)^{-1}([0, \varepsilon_i[)$, by setting $\rho_i' = \rho_i^Y \circ f$, we can restate the problem as follows.

Assertion. Let $\{X_i\}_{i=1,\ldots,k}$ be a Whitney stratification of a compact set $X \subset \mathbf{R}^n$. For each i, let ρ_i be the square of the function which measures distance from X_i in $\mathbf{R}^n$, and let ρ_i' be a non-negative continuous function defined on $\mathbf{R}^n$ with $\rho_i'^{-1}(0) \cap X = \overline{X_i}$, of class C^∞ on an open neighborhood U_i of X_i in $\mathbf{R}^n$ and C^∞ regular on $U_i - \rho_i'^{-1}(0)$. Assume there exists a removal data $\delta = \{\delta_i\}_{i=1,\dots,k-1}$ of $\{X_i\}$ such that for each $1 \le i < j \le k$ and for any sequence of positive numbers $\varepsilon = \{\varepsilon_l\}_{l=1,\dots,k-1} \le \delta$, all X_l and all $\rho_{l'}'^{-1}(\varepsilon_{l'}) \cap U_{l'}$ are transversal to each other,

$$\rho_i'^{-1}([0, \varepsilon_i]) - \bigcup_{l<i} \rho_l'^{-1}([0, \varepsilon_l/2[) \subset U_i,$$

and the restrictions of ρ_i and of ρ_i' to the manifold possibly with corners

$$X_j \cap \rho_i'^{-1}([0, \varepsilon_i[) - \bigcup_{l<i} \rho_l'^{-1}([0, \varepsilon_l[)$$

are friendly. If we lessen the δ, the C^∞ manifolds possibly with corners

$$X_\varepsilon = X - \bigcup_{i=1}^{k-1} \rho_i^{-1}([0, \varepsilon_i[), \quad \text{and} \quad X_\varepsilon' = X - \bigcup_{i=1}^{k-1} \rho_i'^{-1}([0, \varepsilon_i[)$$

are C^∞ diffeomorphic.

Proof of Assertion.   Set

$$X_\varepsilon^i = X - \bigcup_{j=1}^{i} \rho_j'^{-1}([0, \varepsilon_j[) - \bigcup_{j=i+1}^{k-1} \rho_j^{-1}([0, \varepsilon_j[), \quad i = 0, \dots, k-1.$$

Then $X_\varepsilon^0 = X_\varepsilon$ and $X_\varepsilon^{k-1} = X_\varepsilon'$. Hence it suffices to construct C^∞ diffeomorphisms:

$$\varphi_i \colon X_\varepsilon^{i-1} \longrightarrow X_\varepsilon^i, \quad i = 1, \dots, k-1.$$

Let l be a positive integer smaller than k. By assuming $\varphi_1, \dots, \varphi_{l-1}$ and by fixing $\delta_1, \dots, \delta_{l-1}$ and $\varepsilon_1, \dots, \varepsilon_{l-1}$, we will lessen $\delta_l, \dots, \delta_{k-1}$ and construct the φ_l, which is parameterized by $\varepsilon_l, \dots, \varepsilon_{k-1}$ (essentially by only ε_l).

   We want a vector field whose flow induces the φ_l. To obtain the vector field, we shall apply I.1.10'. We put notation in order for the application. Set

$$P = X - \bigcup_{i<l} \rho_i'^{-1}([0, \varepsilon_i[),$$

$$P_i = \rho_i'^{-1}(\varepsilon_i), \quad i = 1, \dots, l-1,$$

$$P_\lambda = \bigcap_{i \in \lambda} P_i \quad \text{for} \quad \lambda \subset \{1, \dots, l-1\},$$

$$U = \rho_l'^{-1}([0, \varepsilon_l[), \quad \text{and}$$

$$Q = \mathbf{R}^n - \bigcup_{i<l} \rho_i'^{-1}([0, \varepsilon_i/2[).$$

By the assumption in the assertion we see the following: all $Q \cap X_i$ and all $Q \cap P_i$ are transversal to each other because $Q \cap P_i \subset U_i$, all $Q \cap P_i$ are C^∞ submanifolds of Q and are closed in Q, and for each $i = l+1, \ldots, k$ and each subset $\lambda \subset \{1, \ldots, l-1\}$, $\rho_l|_{X_i \cap P_\lambda \cap U}$ and $\rho'_l|_{X_i \cap P_\lambda \cap U}$ are friendly C^∞ functions by the definition of friendly C^∞ functions, which is equivalent to the fact that $\rho_l|_{X_i \cap P \cap U}$ and $\rho'_l|_{X_i \cap P \cap U}$ are friendly C^∞ functions. Clearly P is the closure in X of the union of some connected components of the set $X - \cup_{i=1}^{l-1} P_i$. Lessen the δ_l so that there exists a controlled tube system $\{T_i = (|T_i|, \pi_i, \rho_i|_{|T_i|})\}_{i=l, \ldots, k}$ for $\{Q \cap X_i\}_{i=l, \ldots, k}$ such that $P \cap U \subset |T_l| \subset U_l$. We can apply I.1.10' to $\{Q \cap X_i\}_{i=l, \ldots, k}$, $\{T_i\}_{i=l, \ldots, k}$, $Q \cap U$, $\{Q \cap U \cap P_i\}_{i=1, \ldots, l-1}$, $Q \cap X$, $\rho_l|_P$ and $\rho'_l|_P$. Hence we have a semicontrolled vector field $\xi = \{\xi_i\}_{i=l+1, \ldots, k}$ on $\{Q \cap X_i\}_{i=l+1, \ldots, k}$ and an open neighborhood $U' \subset U$ of X_l in $\mathbf{R}^n$ such that for each $i > l$ and for each subset $\lambda \subset \{1, \ldots, l-1\}$, the functions $\xi_i(\rho_l|_{X_i \cap P \cap U'})$ and $\xi_i(\rho'_l|_{X_i \cap P \cap U'})$ are positive, and the vector field $\xi_i|_{Q \cap X_i \cap P_\lambda \cap U'}$ is tangent to $X_i \cap P_\lambda$.

Lessen δ_l again so that we assume $U' = U$, which is possible, because $X_l \cap P$ is compact and $\rho_l'^{-1}(0) \cap X = \overline{X_l}$. Lessen also $\delta_{l+1}, \ldots, \delta_{k-1}$ so that

$$(P \cap V_i) - W \subset |T_i|, \quad i = l+1, \ldots, k-1, \quad \text{and} \qquad (\text{i})$$

$$d(\rho_i|_{X_k})\xi_{kx} = 0 \quad \text{for} \quad x \in X_k \cap P \cap V_i - W, \quad (\text{scv}(T_i, T_k))$$
$$i = l+1, \ldots, k-1,$$

$$\text{where} \quad V_{l+1} = \rho_{l+1}^{-1}([0, \varepsilon_{l+1}]),$$

$$V_i = \rho_i^{-1}([0, \varepsilon_i]) - \rho_{l+1}^{-1}([0, \varepsilon_{l+1}[) - \cdots - \rho_{i-1}^{-1}([0, \varepsilon_{i-1}[), \quad i = l+2, \ldots, k-1,$$

$$\text{and} \quad W = \rho_l'^{-1}([0, \varepsilon_l'[).$$

Here ε_l' is a positive number such that $\varepsilon_l' < \varepsilon_l$ and

$$P \cap \rho_l'^{-1}([0, 2\varepsilon_l']) \subset P \cap \rho_l^{-1}([0, \varepsilon_l]). \qquad (*)$$

This is possible, because the sets

$$X_l \cap P, \qquad (X_{l+1} \cap P) - W \quad \text{and}$$

$$(X_i \cap P) - W - \rho_{l+1}^{-1}([0, \varepsilon_{l+1}[) - \cdots - \rho_{i-1}^{-1}([0, \varepsilon_{i-1}[),$$
$$i = l+2, \ldots, k-1,$$

are compact.

In the same way as Remark I.1.9 and its proof, we produce the φ_l by ξ as follows. The remark showed that a controlled vector field induces a homeomorphism which carries a level of a function to a level of another function. In the present case, the controlledness condition of vector field

fails. However, we shall obtain the following statement in the same way as
the proof of I.1.9.

Let the δ_l be lessened more so that for each $0 < \varepsilon_l \leq \delta_l$, there exists
$\varepsilon_l'' > \varepsilon_l$ such that

$$P \cap \rho_l^{-1}([0,\varepsilon_l]) \subset P \cap \rho_l'^{-1}([0,\varepsilon_l''/2]) \subset P \cap \rho_l'^{-1}([0,\varepsilon_l'']) \subset P \cap U. \qquad (**)$$

Here U is fixed in advance so that the above-mentioned properties of ξ_i hold
true. Then for such $\varepsilon_l' < \varepsilon_l < \varepsilon_l''$, the flow $\theta\colon D \to P - X_l$ of $\xi|_{P-X_l}$ induces
a bijective map (not necessarily a homeomorphism):

$$\tau\colon (P \cap \rho_l'^{-1}(\varepsilon_l')) \times [\varepsilon_l', \varepsilon_l''] \to P \cap \rho_l'^{-1}([\varepsilon_l', \varepsilon_l''])$$

such that

$$\rho_l' \circ \tau(x,t) = t \quad \text{for} \quad (x,t) \in (P \cap \rho_l'^{-1}(\varepsilon_l')) \times [\varepsilon_l', \varepsilon_l''],$$
$$\tau(x \times [\varepsilon_l', \varepsilon_l'']) = \theta(D \cap (x \times \mathbf{R})) \cap \rho_l'^{-1}([\varepsilon_l', \varepsilon_l'']) \quad \text{for} \quad x \in P \cap \rho_l'^{-1}(\varepsilon_l'),$$

and for each $i > l$, the restriction of τ to $(X_i \cap P \cap \rho_l'^{-1}(\varepsilon_l')) \times [\varepsilon_l', \varepsilon_l'']$ is a
C^∞ diffeomorphism onto $X_i \cap P \cap \rho_l'^{-1}([\varepsilon_l', \varepsilon_l''])$.

The reason why the above statement follows is now shown. It suffices to
prove that for each point $x \in P \cap \rho_l'^{-1}(\varepsilon_l')$, there exists a number $t > 0$ such
that

$$x \times [0,t] \subset D, \qquad \theta(x \times [0,t]) \subset P,$$

and $\rho_l' \circ \theta|_{x \times [0,t]}$ is a C^∞ diffeomorphism onto $[\varepsilon_l', \varepsilon_l'']$. The inclusion condition
$\theta(x \times [0,t]) \subset P$ follows from the other conditions and from the fact that
$\xi_i|_{X_i \cap P_\lambda \cap U}$ is tangent to $X_i \cap P_\lambda$. Hence, if there were not such a t for some
x, there would exist some t_0 which is either a positive number or ∞ such
that

$$x \times [0,t_0[\subset D, \qquad x \times t_0 \notin D \quad \text{unless} \quad t_0 = \infty,$$
$$\text{and} \quad \rho_l' \circ \theta(x \times [0,t_0[) = [\varepsilon_l', \varepsilon_l'''[$$

for some $\varepsilon_l''' \leq \varepsilon_l''$. Let $x \in X_i$. Then there would exist a sequence $\{t_j\}_{j=1,2,\ldots}$
in $[0,t_0[$ converging or diverging to t_0 such that the sequence $\{\theta(x,t_j)\}_{j=1,2,\ldots}$
converges to a point y of $X_{i'}$. Clearly $y \in P \cap U$ and hence $\xi_{i'y} \neq 0$, which
implies $i' \neq i$. This inequality contradicts, however, semicontrolledness of ξ
and the first half of I.1.6.

Moreover, the τ for small $\delta_{l+1}, \ldots, \delta_{k-1}$ has the properties:

$$\rho_i \circ \tau(x,t) = \rho_i(x) \text{ for } (x,t) \in (X_k \cap P \cap \rho_l'^{-1}(\varepsilon_l') \cap V_i) \times [\varepsilon_l', \varepsilon_l''], \text{ and}$$
$$\tag{i$'$}$$
$$i = l+1, \ldots, k-1,$$

and, consequently,

$$\tau((X_k \cap P \cap \rho_l'^{-1}(\varepsilon_l') \cap V_i) \times [\varepsilon_l', \varepsilon_l'']) = X_k \cap P \cap \rho_l'^{-1}([\varepsilon_l', \varepsilon_l'']) \cap V_i \qquad (i)''$$

for the following reason.

The property $(l+1)'$ follows from the property $(l+1)$, the condition $\mathrm{scv}(T_{l+1}, T_k)$ on $X_k \cap P \cap V_{l+1} - W$ and the first half of I.1.6, because the closure of $\pi_{l+1}(X_k \cap P \cap \rho_l'^{-1}([\varepsilon_l', \varepsilon_l'']) \cap V_{l+1})$ in X_{l+1} is compact for small δ_{l+1}. Argue $(l+2)'$ in the same way. We can assume that for each point $x \in X_k \cap P \cap \rho_l'^{-1}(\varepsilon_l') \cap V_{l+2}'$, the function $\rho_{l+2} \circ \tau$ is constant on each connected component of $(x \times [\varepsilon_l', \varepsilon_l'']) \cap \tau^{-1}(X_k \cap P \cap \rho_l'^{-1}([\varepsilon_l', \varepsilon_l'']) \cap V_{l+2}')$, where

$$V_{l+2}' = \rho_{l+2}^{-1}([0, 2\varepsilon_{l+2}]) - \rho_{l+1}^{-1}([0, \varepsilon_{l+1}/2[) \supset V_{l+2}.$$

Hence for $(l+2)'$, it suffices to prove that for each $x \in X_k \cap P \cap \rho_l'^{-1}(\varepsilon_l') \cap V_{l+2}$,

$$x \times [\varepsilon_l', \varepsilon_l''] \subset \tau^{-1}(X_k \cap P \cap \rho_l'^{-1}([\varepsilon_l', \varepsilon_l'']) \cap V_{l+2}).$$

Set

$$x \times I = (x \times [\varepsilon_l', \varepsilon_l'']) \cap \tau^{-1}(X_k \cap P \cap \rho_l'^{-1}([\varepsilon_l', \varepsilon_l'']) \cap V_{l+2}), \quad \text{and}$$

$$x \times I' = (x \times [\varepsilon_l', \varepsilon_l'']) \cap \tau^{-1}(X_k \cap P \cap \rho_l'^{-1}([\varepsilon_l', \varepsilon_l'']) \cap V_{l+2}').$$

Then I' is a neighborhood of I in $[\varepsilon_l', \varepsilon_l'']$, and I and I' are both closed in $[\varepsilon_l', \varepsilon_l'']$ because the map $\tau|_{(X_k \cap P \cap \rho_l'^{-1}(\varepsilon_l')) \times [\varepsilon_l', \varepsilon_l'']}$ is a C^∞ diffeomorphism to $X_k \cap P \cap \rho_l'^{-1}([\varepsilon_l', \varepsilon_l''])$. Hence if I were not equal to $[\varepsilon_l', \varepsilon_l'']$, there would exist numbers $\varepsilon_l' \leq \alpha < \beta \leq \varepsilon_l''$ with the properties:

$$\alpha \in I, \qquad]\alpha, \beta[\subset [\varepsilon_l', \varepsilon_l''] - I \quad \text{and} \quad [\alpha, \beta] \subset I'.$$

Here the inclusion $]\alpha, \beta[\subset [\varepsilon_l', \varepsilon_l''] - I$ means that

$$\rho_{l+1} \circ \tau < \varepsilon_{l+1} \quad \text{on} \quad x \times]\alpha, \beta[,$$

because by the inclusion $[\alpha, \beta] \subset I'$, the function $\rho_{l+2} \circ \tau$ is constant on $x \times [\alpha, \beta]$. On the other hand, the property $\alpha \in I$ means that

$$\tau(x, \alpha) \in V_{l+2} \quad \text{and hence} \quad \rho_{l+1} \circ \tau(x, \alpha) \geq \varepsilon_{l+1}.$$

These inequalities contradict the property $(l+1)''$. Therefore, $I = [\varepsilon_l', \varepsilon_l'']$, namely, $(l+2)'$ holds true. For $(i)'$, $i = l+3, \ldots, k-1$, we make the same arguments by induction on i. We omit the details.

Now the rest of the production of φ_l by τ proceeds in exactly the same way as the proof of I.1.9. Let us sketch the procedure. By τ we identify the set $X_k \cap P \cap \rho_l'^{-1}([\varepsilon_l', \varepsilon_l''])$ with the product $(X_k \cap P \cap \rho_l'^{-1}(\varepsilon_l')) \times [\varepsilon_l', \varepsilon_l'']$. By using (∗) and (∗∗), we construct a C^∞ diffeomorphism of $(X_k \cap P \cap \rho_l'^{-1}(\varepsilon_l')) \times [\varepsilon_l', \varepsilon_l'']$ which carries $(X_k \cap P \cap \rho_l'^{-1}(\varepsilon_l')) \times [\varepsilon_l', \varepsilon_l]$ to $\tau^{-1}(X_k \cap P \cap \rho_l'^{-1}([\varepsilon_l', \varepsilon_l'']) \cap \rho_l^{-1}([0, \varepsilon_l]))$, and we have a C^∞ diffeomorphism ψ_l of $X_k \cap P$ such that

$$\psi_l(X_k \cap P \cap \rho_l^{-1}([0, \varepsilon_l[)) = X_k \cap P \cap \rho_l'^{-1}([0, \varepsilon_l[).$$

Then by (i)″, ψ_l satisfies the equalities:

$$\psi_l(X_k \cap P \cap V_i) = X_k \cap P \cap V_i, \;\; i = l+1, \ldots, k-1.$$

By these we see that

$$\psi_l\left((X_k \cap P) - \rho_l^{-1}([0, \varepsilon_l[) - \bigcup_{i=l+1}^{k-1} \rho_i^{-1}([0, \varepsilon_i])\right)$$

$$= (X_k \cap P) - \rho_l'^{-1}([0, \varepsilon_l[) - \bigcup_{i=l+1}^{k-1} \rho_i^{-1}([0, \varepsilon_i])$$

because

$$\bigcup_{i=l+1}^{k-1} V_i = \bigcup_{i=l+1}^{k-1} \rho_i^{-1}([0, \varepsilon_i]).$$

Clearly the closures of $(X_k \cap P) - \rho_l^{-1}([0, \varepsilon_l[) - \cup_{i=l+1}^{k-1} \rho_i^{-1}([0, \varepsilon_i])$ and of $(X_k \cap P) - \rho_l'^{-1}([0, \varepsilon_l[) - \cup_{i=l+1}^{k-1} \rho_i^{-1}([0, \varepsilon_i])$ in X_k are X_ε^{l-1} and X_ε^l, respectively. Hence we have

$$\psi_l(X_\varepsilon^{l-1}) = X_\varepsilon^l.$$

Thus the diffeomorphism φ_l, defined to be $\psi_l|_{X_\varepsilon^{l-1}}$, is what we wanted, which completes the proof.   □

Remark I.1.14. The C^∞ diffeomorphism $f_\varepsilon : X_\varepsilon \to Y_\varepsilon$ constructed in the above proof carries each $X_\varepsilon \cap \rho_i^{X-1}(\varepsilon_i)$ to $Y_\varepsilon \cap \rho_i^{Y-1}(\varepsilon_i)$. It is C^∞ isotopic to $f|_{X_\varepsilon} : X_\varepsilon \to Y_k$, and hence for a connected component C of X_ε, the images $f(C)$ and $f_\varepsilon(C)$ are included in a common connected component of Y_k. Moreover, for a connected component C_i of $X_\varepsilon \cap \rho_i^{X-1}(\varepsilon_i)$, the images $f \circ \pi_i^X(C_i)$ and $\pi_i^Y \circ f_\varepsilon(C_i)$ are included in a common connected component of Y_i.

The above proof did not depend on the special choice of ρ_i^X and ρ_i^Y. Hence I.1.13 holds true even if $f: \{X_i\} \to \{Y_i\}$ is a strong isomorphism for given tube systems $\{T_i^X = (|T_i^X|, \pi_i^X, \rho_i^X)\}$, $\{T_i^Y = (|T_i^Y|, \pi_i^Y, \rho_i^Y)\}$ and $\{T_i^Z = (|T_i^Z|, \pi_i^Z, \rho_i^Z)\}$ for $\{X_i\}$, $\{Y_i\}$ and $\{Z_i\}$, respectively.

Let r be a positive integer. For the rest of this section, we consider C^r Whitney stratifications. We modify the definition of a tube. Let M be a C^r submanifold of $\mathbf{R}^n$ of codimension m, let $|T|$ be an open neighborhood of M in $\mathbf{R}^n$, let $\pi: |T| \to M$ be a C^r map, and let ρ be a C^r function on $|T|$. We call $T = (|T|, \pi, \rho)$ a C^r *tube* at M if there exists a C^r imbedding τ of $|T|$ into $\mathbf{R}^n$ such that $\tau(M)$ is a C^∞ submanifold of $\mathbf{R}^n$ and $\tau_* T = (\tau(|T|), \tau \circ \pi \circ \tau^{-1}, \rho \circ \tau^{-1})$ is a tube at $\tau(M)$.

Examples. If $r > 1$, set $T = (|T|, \pi, \rho) = $ (a small tubular neighborhood of M in $\mathbf{R}^n$, the orthogonal projection, the square of the function which measures distance from M). It is a C^{r-1} tube at M for the following reason. Let $\varphi: M \to \mathbf{R}^n$ be a strong approximation of the inclusion map in the Whitney C^r topology such that the image $M' = \varphi(M)$ is a C^∞ submanifold of $\mathbf{R}^n$. Let a tube $T' = (|T'|, \pi', \rho')$ at M' be defined in the same way as T. Let $\xi_G = (E_{n,m}, \pi_G, G_{n,m})$ denote the universal vector bundle over the Grassmannian $G_{n,m}$. Set

$$U = \{(\lambda_1, x_1, \lambda_2) \in E_{n,m} \times G_{n,m} \subset G_{n,m} \times \mathbf{R}^n \times G_{n,m} \subset \mathbf{R}^{n^2} \times \mathbf{R}^n \times \mathbf{R}^{n^2}$$
$$: |\lambda_1 - \lambda_2| < a\}$$

for a small positive number a. There is a C^∞ map $p: U \to E_{n,m}$ which isomorphically carries $\pi_G^{-1}(\lambda_1) \times \lambda_2$ to $\pi_G^{-1}(\lambda_2)$ as metric vector spaces for each $(\lambda_1, \lambda_2) \in G_{n,m} \times G_{n,m}$ with $|\lambda_1 - \lambda_2| < a$. Let $f: M \to G_{n,m}$ and $f': M' \to G_{n,m}$ denote the characteristic maps of the normal bundles of M and of M' in $\mathbf{R}^n$, respectively. Set

$$\tau(x) = \varphi \circ \pi(x) + p(f \circ \pi(x), x - \pi(x), f' \circ \varphi \circ \pi(x)) \quad \text{for} \quad x \in |T|.$$

For small $|T|$ and $|T'|$ and a strong approximation φ, τ is a C^{r-1} diffeomorphism from $|T|$ to $|T'|$ such that $\tau_* T = T'$.

Assume next $r \geq 1$. A C^r tube at M is given as follows. Set

$$\rho_G(\lambda, x) = |x|^2 \quad \text{for} \quad (\lambda, x) \in E_{n,m} \subset G_{n,m} \times \mathbf{R}^n.$$

Let $\psi: M \to G_{n,m}$ be a C^r approximation of f in the C^{r-1} Whitney topology. Let $(\tilde{E}, \tilde{\pi}, M)$ and $\tilde{\rho}$ denote the induced vector bundle $\psi^* \xi_G$ of ξ_G by ψ and the induced function $\psi^* \rho_G$ of ρ_G by ψ, respectively. Define a map $\theta: \tilde{E} \to \mathbf{R}^n$ by

$$\theta(x, \psi(x), y) = x + y \quad \text{for} \quad (x, \psi(x), y) \in \tilde{E} \subset M \times E_{n,m} \subset M \times G_{n,m} \times \mathbf{R}^n.$$

Then $\theta = $ id on M and we can choose the above approximation so strong that the restriction of θ to a small tubular neighborhood of M in $\tilde{E}$ is a C^r imbedding. Set

$$|T| = \theta \text{ (the neighborhood)}, \qquad \pi = \theta \circ \tilde{\pi} \circ \theta^{-1} \quad \text{and} \quad \rho = \tilde{\rho} \circ \theta^{-1}.$$

As in the above case $r > 1$, we can prove that $T = (|T|, \pi, \rho)$ is a C^r tube at M.

We extend naturally the previous other notation and terminology to the C^r case, for example, a controlled tube system and a strong isomorphism. The C^r versions of I.1.1, I.1.2, I.1.3 and I.1.3' are clear. Moreover, by the next lemma we can reduce problems on C^r Whitney stratifications to ones on C^∞ Whitney stratifications.

Lemma I.1.15. *Let r be a positive integer. Let $\{X_i\}$ be a C^r Whitney stratification in $\mathbf{R}^n$ and let $\{T_i = (|T_i|, \pi_i, \rho_i)\}$ be a controlled C^r tube system for $\{X_i\}$. Shrink $|T_i|$. There is a C^r diffeomorphism τ of $\mathbf{R}^n$ such that the stratification $\{\tau(X_i)\}$ and the controlled tube system $\{\tau_* T_i = (\tau(|T_i|), \tau \circ \pi_i \circ \tau^{-1}|_{\tau(|T_i|)}, \rho_i \circ \tau^{-1}|_{\tau(|T_i|)})\}$ are of class C^∞.*

The $\mathfrak{X}$-version can be clearly proved in the same way as the following proof.

Proof. The proof follows the same procedure of double induction as the proof of I.1.3. For convenience we assume $\dim X_1 < \dim X_2 < \cdots$, $X_0 = |T_0| = \varnothing$, $\pi_0 \colon |T_0| \to X_0$ and $\rho_0 = 1$ on $|T_0|$. Let k be a positive integer. Assume $\{T_i\}_{i<k}$ is of class C^∞. It suffices to find a C^r diffeomorphism τ_k of $\mathbf{R}^n$ such that $\tau_{k*} T_i = T_i$, $i < k$, and $\tau_{k*} T_k$ is of class C^∞. By I.1.1 and by the above Examples we assume there exist a C^r map $\psi \colon X_k \to G_{n,m}$ and a C^r imbedding $\Phi \colon |T_k| \to \tilde{E}_k$ such that

$$\Phi = \text{id} \quad \text{on} \quad X_k, \qquad \pi_k = \Phi^{-1} \circ \tilde{\pi}_k \circ \Phi, \quad \text{and} \quad \rho_k = \tilde{\rho}_k \circ \Phi \quad \text{on} \quad |T_k|,$$

where

$$m = \operatorname{codim} X_k, \qquad (\tilde{E}_k, \tilde{\pi}_k, X_k) = \psi^* \xi_G,$$
$$\xi_G = (E_{n,m}, \pi_G, G_{n,m}) \quad \text{and} \quad \tilde{\rho}_k = \psi^* \rho_G.$$

Set $\tilde{F}_k = \Phi(|T_k|)$, and define a C^∞ map $\sigma_i \colon |T_i| \to X_i \times \mathbf{R}$ for each $i < k$ by

$$\sigma_i(x) = (\pi_i(x), \rho_i(x)) \quad \text{for} \quad x \in |T_i|.$$

We will give a good C^∞ structure to $\tilde{F}_k$ and approximate the Φ by a C^∞ diffeomorphism Φ' so that the natural extension τ_k of $\Phi'^{-1} \circ \Phi$ to $\mathbf{R}^n$ is a required diffeomorphism. (Here we define the natural extension τ_k by setting $\tau_k = \mathrm{id}$ outside of $|T_k|$.)

What we do first is to give C^∞ structures to $\tilde{F}_k$ and X_k so that X_k is a C^∞ submanifold of $\tilde{F}_k$, and the maps $\tilde{\pi}_k|_{\tilde{F}_k}$ and $\sigma_i \circ \Phi^{-1}|_{\Phi(|T_i| \cap |T_k|)}$ for all $i < k$ are of class C^∞. (We always identify X_k with the image of the zero cross-section of $(\tilde{E}_k, \tilde{\pi}_k, X_k)$.) For that, it suffices to consider a C^∞ structure only on X_k for the following reason. Assume there exists a C^∞ structure on X_k such that for each $i < k$, the map $\sigma_i|_{|T_i| \cap X_k}$ is of class C^∞. Let ψ' be a C^∞ approximation of ψ in the C^0 Whitney topology, and set $(\tilde{E}'_k, \tilde{\pi}'_k, X_k) = \psi'^* \xi_G$. It is known that the vector bundles $(\tilde{E}_k, \tilde{\pi}_k, X_k)$ and $(\tilde{E}'_k, \tilde{\pi}'_k, X_k)$ are C^r isomorphic. Hence $\tilde{F}_k$ admits a C^∞ structure such that X_k is a C^∞ submanifold of $\tilde{F}_k$ and $\tilde{\pi}_k|_{\tilde{F}_k}$ is of class C^∞. For such a C^∞ structure on $\tilde{F}_k$ and for each $i < k$, the map $\sigma_i \circ \Phi^{-1}|_{\Phi(|T_i| \cap |T_k|)}$ is automatically of class C^∞ because

$$\sigma_i \circ \Phi^{-1} = \sigma_i \circ \pi_k \circ \Phi^{-1} = \sigma_i \circ \Phi^{-1} \circ \tilde{\pi}_k.$$

We will give a C^∞ structure to X_k. We proceed by downward induction as follows.

Let $l < k$ be a non-negative integer. Assume we have already constructed a C^∞ structure on $X_k \cap \cup_{l<i<k}|T_i|$ so that all the maps $\sigma_i|_{|T_i| \cap X_k}$, $l < i < k$, are of class C^∞. We want to extend the structure to $X_k \cap \cup_{l \le i < k}|T_i|$. Note that the map $\sigma_l|_{|T_l| \cap X_k \cap \cup_{l<i<k}|T_i|}$ is of class C^∞ because all the maps $\pi_i|_{|T_i| \cap X_k \cap |T_i|}$, $l < i < k$ and $\sigma_l|_{|T_l| \cap X_i}$ are of class C^∞ and because

$$\sigma_l = \sigma_l \circ \pi_i \quad \text{on} \quad |T_l| \cap X_k \cap |T_i|.$$

Set

$$X_{kl} = X_k \cap |T_l|, \quad \text{and} \quad X_{kl-1} = X_{kl} \cap \bigcup_{l<i<k} |T_i|.$$

Let X'_{kl} be a C^∞ manifold, and let $f \colon X'_{kl} \to X_{kl}$ be a C^r diffeomorphism such that $f|_{f^{-1}(X_{kl-1})}$ is of class C^∞. (Existence of f is shown as follows. First find a C^∞ manifold X'_{kl} and a C^r diffeomorphism $g \colon X'_{kl} \to X_{kl}$, and approximate $g|_{g^{-1}(X_{kl-1})}$ by a C^∞ diffeomorphism in the C^r Whitney topology. The natural extension of the approximation to X'_{kl} is the required f.) It suffices to approximate f by a C^r diffeomorphism f' so that $f' = f$ on $f^{-1}(X_{kl-1})$ and $\sigma_l \circ f'$ is of class C^∞. Set $h = \sigma_l \circ f$, which is of class C^∞ on $f^{-1}(X_{kl-1})$. Let h' be a strong C^∞ approximation of h in the C^r Whitney topology such that $h' = h$ on $f^{-1}(X_{kl-1})$. Here we shrink $|T_i|$, $l < i < k$, if necessary. Then we can construct the C^r diffeomorphism $f' \colon X'_{kl} \to X_{kl}$ so that $\sigma_l \circ f' = h'$ as shown below. Set

$$\{Y_t\}_{t \in \sigma_l(X_{kl})} = \{\sigma_l^{-1}(t) \cap X_{kl} \colon t \in \sigma_l(X_{kl})\}.$$

For each point $t \in \sigma_l(X_{kl})$, let $p_t \colon U_t \to Y_t$ be a C^r projection of a tubular neighborhood of Y_t in X_{kl} such that $\cup_{t \in \sigma_l(X_{kl})} t \times U_t$ is open in $\sigma_l(X_{kl}) \times X_{kl}$ and the map $\cup_{t \in \sigma_l(X_{kl})} t \times U_t \ni (t,x) \to p_t(x) \in X_{kl}$ is of class C^r. (Existence of such p_t is shown in the same way as the construction of a C^r tube in Examples.) Note that by the above first condition, we have a continuous function ε on X_{kl} such that if points x and y in X_{kl} satisfy the condition $\mathrm{dis}(x,y) \le \varepsilon(x)$ and if $x \in Y_t$, then $y \in U_t$. Define the f' by

$$f'(x) = p_{h'(x)} \circ f(x) \quad \text{for} \quad x \in X'_{kl}.$$

It satisfies the required conditions. Thus we give C^∞ structures to $\tilde{F}_k$ and X_k.

Next we need a C^∞ approximation Φ' of Φ in the C^r Whitney topology under the C^∞ structure on $|T_k|$ induced by $\mathbf{R}^n$ and the above given C^∞ structure on $\tilde{F}_k$, such that

$$\sigma_i = \sigma_i \circ \Phi'^{-1} \circ \Phi \quad \text{on} \quad |T_i| \cap |T_k|, \quad i < k.$$

If there exists such Φ', the diffeomorphism τ_k, defined to be $\Phi'^{-1} \circ \Phi$ on $|T_k|$ and the identity on $\mathbf{R}^n - |T_k|$, fulfills the requirements at the beginning of the proof. For construction of Φ', it suffices to approximate Φ^{-1} by a C^∞ diffeomorphism Ψ so that

$$\sigma_i \circ \Phi^{-1} = \sigma_i \circ \Psi \quad \text{on} \quad \Phi(|T_i| \cap |T_k|), \quad i < k.$$

Recall that we gave a C^∞ structure to the $\tilde{F}_k$ so that all $\sigma_i \circ \Phi^{-1}|_{\Phi(|T_i| \cap |T_k|)}$, $i < k$, are of class C^∞. Hence we find Ψ in the same way as the above construction of a C^∞ structure on X_k, which completes the proof. $\square$

Proposition I.1.13′. *Let r be a positive integer. Let $\{X_i\}_{i=1,\dots,k}$ and $\{Y_i\}_{i=1,\dots,k}$ be C^r Whitney stratifications of compact sets $X \subset \mathbf{R}^n$ and $Y \subset \mathbf{R}^{n'}$, respectively. Assume there exists a strong C^r isomorphism $f \colon \{X_i\} \to \{Y_i\}$ for C^r tube systems $\{T_i^X = (|T_i^X|, \pi_i^X, \rho_i^X)\}$, $\{T_i^Y = (|T_i^Y|, \pi_i^Y, \rho_i^Y)\}$ and $\{T_i^Z = (|T_i^Z|, \pi_i^Z, \rho_i^Z)\}$ for $\{X_i\}$, $\{Y_i\}$ and $\{Z_i = \mathrm{graph}\, f|_{X_i}\}$, respectively. There exists a common removal data $\delta = \{\delta_i\}_{i=1,\dots,k-1}$ of $\{X_i\}$ for $\{T_i^X\}$ and $\{Y_i\}$ for $\{T_i^Y\}$ such that, for any sequence of positive numbers $\varepsilon = \{\varepsilon_i\}_{i=1,\dots,k-1} \le \delta$, the sets*

$$X_\varepsilon = X - \bigcup_{i=1}^{k-1} \rho_i^{X-1}([0,\varepsilon_i[) \quad \text{and}$$

$$Y_\varepsilon = Y - \bigcup_{i=1}^{k-1} \rho_i^{Y-1}([0,\varepsilon_i[)$$

are C^r diffeomorphic C^r manifolds possibly with corners.

The $\mathfrak{X}$-version of I.1.13′ coincides with that of I.1.13.

Proof. It suffices to prove that X_ε and Y_ε are C^1 diffeomorphic. If we make the same arguments as in the proof of I.1.13, we prove that X_ε and Y_ε are C^{r-1} diffeomorphic. Thus the case $r > 1$ is clear. So assume $r = 1$. Assume also $\dim X_1 < \cdots < \dim X_k$. Set $Z = \operatorname{graph} f$. Let $p\colon Z \to X$ denote the projection, and define Z_ε in the same way as X_ε and Y_ε. It suffices to show that Z_ε and X_ε are C^1 diffeomorphic. By I.1.15 we can suppose that $\{X_i\}$, $\{\rho_i^X\}$, $\{Z_i\}$ and $\{\rho_i^Z\}$ are of class C^∞. Then p, however, remains of class C^1.

Recall the assertion in the proof of I.1.13. If it holds true when adapted to the present case, we see that Z_ε and X_ε are C^1 diffeomorphic. In the present case, all the ρ_i' in the assertion are of class C^1 and we require only a C^1 diffeomorphism between the assertion X_ε and X_ε'. The only difficulty in following the same procedure as in the proof of the assertion is that the flow θ of the semicontrolled vector field $\xi = \{\xi_i\}_{i=l+1,\ldots,k}$ on $\{\theta \cap X_i\}_{i=l+1,\ldots,k}$ does not necessarily exist, or is not necessarily of class C^1 even if exists, because ξ_i are of class C^0. We need to modify ξ so that θ exists and is of class C^1. By giving another C^∞ structure to each X_i, we will approximate ξ_i by a C^∞ vector field ξ_i' with respect to the new structure. If the approximation is strong enough in the C^0 Whitney topology, the condition that the functions $\xi_i(\rho_l|_{X_i \cap P \cap U'})$ and $\xi_i(\rho_l'|_{X_i \cap P \cap U'})$ are positive holds. Hence it suffices to be careful of the semicontrolledness of ξ and of the condition that for each subset $\lambda \subset \{1,\ldots,l-1\}$, the vector field $\xi_i|_{Q \cap X_i \cap P_\lambda \cap U'}$ is tangent to $X_i \cap P_\lambda$. The reason why we consider another C^∞ structure on each X_i is that the manifolds, possibly with corners $Q \cap X_i \cap P_\lambda$ for all λ, are of class C^1. We want C^∞ structures on them. If we give a C^∞ structure to X_i so that $Q \cap X_i \cap P_\lambda \cap U'$ for all λ are of class C^∞ and $\rho_j|_{Q \cap X_i \cap (\text{neighborhood of } X_j)}$ for all $l < j < i$ continue to be of class C^∞, then we can approximate ξ_i by C^∞ ξ_i' keeping the required properties as in the proof of I.1.5. Let us state what we need to prove.

For each $i > l$, let X_i' be a C^∞ manifold, and let $f\colon X_i' \to X_i$ be a C^1 diffeomorphism such that for all $\lambda \subset \{1,\ldots,l-1\}$, $f^{-1}(Q \cap X_i \cap P_\lambda \cap U')$ are C^∞ submanifolds of X_i'. We can approximate f by a C^1 diffeomorphism f' so that the maps $\sigma_j \circ f'$ for all $l < j < i$ are of class C^∞ on $f^{-1}(Q \cap X_i \cap |T_j|)$ and

$$f^{-1}(Q \cap X_i \cap P_\lambda \cap U') = f'^{-1}(Q \cap X_i \cap P_\lambda \cap U') \quad \text{for} \quad \lambda \subset \{1,\ldots,l-1\}, \quad (*)$$

where $\sigma_j = (\pi_j, \rho_j)$.

Here we require C^∞ differentiability not only of $\rho_j \circ f'$ but also of $\sigma_j \circ f'$ for convenience of its proof. We proceed by double induction as follows. Let $k < l$ be another positive integer. Assume that for any subset

$\lambda \subset \{1, \ldots, l-1\}$ with $\sharp\lambda \geq k+1$, the maps $\sigma_j \circ f$ for all $l < j < i$ are of class C^∞ on $f^{-1}(Q \cap X_i \cap |T_j| \cap P_\lambda \cap U')$. It suffices to find a C^1 approximation f' of f such that the maps $\sigma_j \circ f'$ for all $l < j < i$ are of class C^∞ on $f^{-1}(Q \cap X_i \cap P_\lambda \cap U')$ for any $\lambda \subset \{1, \ldots, l-1\}$ with $\sharp\lambda \geq k$, and the condition $(*)$ holds. We can prove this by induction as in the proof of I.1.15. We omit the details. $\square$

Lemma I.1.16. *Let r be either a positive integer or $+\infty$, let $\{X_i\}_{i=1,\ldots,k}$ be a C^r Whitney stratification of a compact set $X \subset \mathbf{R}^n$ such that $\dim X_1 < \cdots < \dim X_k$, let $\{T_i = (|T_i|, \pi_i, \rho_i)\}$ and $\{T_i' = (|T_i'|, \pi_i', \rho_i')\}$ be C^r tube systems for $\{X_i\}$. Let $\delta = \{\delta_i\}_{i=1,\ldots,k-1}$ be a common removal data of $\{X_i\}$ for $\{T_i\}$ and for $\{T_i'\}$, and let $\varepsilon = \{\varepsilon_i\}_{i=1,\ldots,k-1}$ and $\varepsilon' = \{\varepsilon_i'\}_{i=1,\ldots,k-1}$ be sequences of positive numbers $\leq \delta$. The C^r manifolds, possibly with corners*

$$X_\varepsilon = X - \bigcup_{i=1}^{k-1} \rho_i^{-1}([0, \varepsilon_i[) \quad and$$

$$X_{\varepsilon'}' = X' - \bigcup_{i=1}^{k-1} \rho_i'^{-1}([0, \varepsilon_i'[),$$

are C^r diffeomorphic. We can choose the diffeomorphism so as to carry each $X_\varepsilon \cap \rho_i^{-1}(\varepsilon_i)$ to $X_{\varepsilon'}' \cap \rho_i'^{-1}(\varepsilon_i')$.

We can prove the $\mathfrak{X}$-version in the same way.

Proof. We can choose arbitrarily δ, ε and ε' for the following reason. We easily prove that for each $i = 1, \ldots, k-1$, the C^r manifolds, possibly with corners:

$$X - \bigcup_{j=1}^{i-1} \rho_j^{-1}([0, \varepsilon_j[) - \bigcup_{j=i}^{k-1} \rho_j^{-1}([0, \varepsilon_j'[) \quad and$$

$$X - \bigcup_{j=1}^{i} \rho_j^{-1}([0, \varepsilon_j[) - \bigcup_{j=i+1}^{k-1} \rho_j^{-1}([0, \varepsilon_j'[),$$

are C^r diffeomorphic. It follows that X_ε is C^r diffeomorphic to the manifold, possibly with corners $X_{\varepsilon'} = X - \cup_{i=1}^{k-1} \rho_i^{-1}([0, \varepsilon_i'[)$, which admits any choice of δ, ε and ε'. We assume $\varepsilon = \varepsilon'$. By I.1.3 and I.1.15 we can suppose that $\{X_i\}$ and $\{T_i'\}$ are of class C^∞, and $\{T_i'\}$ is controlled. Then the lemma follows from the next statement as in the proof of I.1.13.

Let $1 \le i < j \le k$ be integers, and lessen ε_i. The restrictions of ρ_i and of ρ_i' to the manifold, possibly with corners:

$$X_{j,i,\varepsilon} = (X_j \cap \rho_i^{-1}([0,\varepsilon_i[)) - \bigcup_{l=1}^{i-1} \rho_l'^{-1}([0,\varepsilon_l[),$$

are friendly.

Let us prove this. We need to prove friendliness of the restrictions of ρ_i and of ρ_i' to the interior $(P_\lambda \cap X_{j,i,\varepsilon})^\circ$ for each subset $\lambda \subset \{1,\dots,i-1\}$, where $P_\lambda = \cap_{l \in \lambda} \rho_l'^{-1}(\varepsilon_l)$. (Recall that $P_\lambda \cap X_{j,i,\varepsilon}$ is a C^∞ manifold possibly with corners.) For that, it suffices to find a (not necessarily continuous) nonsingular vector field v on $X_{j,i,\varepsilon}$ such that v is tangent to $(P_\lambda \cap X_{j,i,\varepsilon})^\circ$ at each point of $(P_\lambda \cap X_{j,i,\varepsilon})^\circ$ and the functions $v(\rho_i|_{X_{j,i,\varepsilon}})$ and $v(\rho_i'|_{X_{j,i,\varepsilon}})$ are everywhere positive.

Define v on $X_{j,i,\varepsilon}^\circ$ so that for each $x \in X_{j,i,\varepsilon}^\circ$, the vector v_x is the image of the vector $x - y$ under the orthogonal projection onto $T_x X_j$, where $y = \pi_i'(x)$ and $T_x X_j$ denote the tangent space of X_j at x. Then v is non-singular for small ε_i by the Whitney condition.

Moreover, v fulfills the other requirement as shown below. Set $m = \dim X_i$. For each point $x \in X_{j,i,\varepsilon}^\circ$, let S_x denote the tangent space of $\rho_i^{-1}(\rho_i(x)) \cap X_j$ at x. We want to see that S_x and v_x are independent. By the definition of a C^r tube there exists a C^r imbedding τ of $|T_i|$ into $\mathbf{R}^n$ such that $\tau_* T_i$ is a tube. As the problem is local at each point of X_i, we can assume

$$\tau(|T_i|) = \mathbf{R}^n, \qquad \tau(X_i) = 0 \times \mathbf{R}^{n-m} \subset \mathbf{R}^n,$$
$$\rho_i \circ \tau^{-1}(x_1,\dots,x_n) = x_1^2 + \cdots + x_m^2, \quad \text{and}$$
$$\tau \circ \pi_i \circ \tau^{-1}(x_1,\dots,x_n) = (0,\dots,0,x_{m+1},\dots,x_n).$$

We have

$$d\tau(S_x) = T_{\tau(x)}(\rho_i \circ \tau^{-1})^{-1}(\rho_i \circ \tau^{-1}(\tau(x))) \cap \tau(X_j \cup |T_i|), \quad \text{and}$$
$$\operatorname{dis}\left(\mathbf{R}d\tau v_x, \ \overrightarrow{\tau \circ \pi_i(x)\, \tau(x)}\right) \to 0 \quad \text{as} \quad x \to \text{a point of } X_i.$$

Hence it suffices to prove that $\tau(S_x)$ and $\overrightarrow{\tau \circ \pi_i(x)\, \tau(x)}$ are orthogonal. But this is clear because $\tau(S_x)$ is included in $T_{\tau(x)}(\rho_i \circ \tau^{-1})^{-1}(\rho_i \circ \tau^{-1}(\tau(x)))$, which is orthogonal to $\overrightarrow{\tau \circ \pi_i(x)\, \tau(x)}$.

The above arguments show, furthermore, that v_x points in the direction where ρ_i is positive. Hence the function $v(\rho_i|_{X_{j,i,\varepsilon}})$ is positive at x. In the

same way we prove that $v(\rho_i'|_{X_{j,i,\varepsilon}})$ is everywhere positive. We define v on
each $(X_{j,i,\varepsilon} \cap P_\lambda)^\circ$ in the same way.   □

§I.2. Subanalytic sets and semialgebraic sets

In this section, we define a subanalytic set and a subanalytic map and
then we show their properties. By the existence of subanalytic Whitney
stratifications of subanalytic sets (I.2.2), we can reduce problems on subanalytic sets and maps to problems on Whitney stratifications. By (I.2.7),
we can apply the lemmas and the propositions in §I.1 to subanalytic maps,
which cause uniqueness of subanalytic triangulations of subanalytic sets. We
also explain the semialgebraic case. Only in the case without good references
are proofs given.

We call a subset X of $\mathbf{R}^n$ *subanalytic* if each point of $\mathbf{R}^n$ has a neighborhood U such that $X \cap U$ is a finite union of sets of the form $\operatorname{Im} f_1 - \operatorname{Im} f_2$,
where f_1 and f_2 are proper analytic maps from analytic manifolds to $\mathbf{R}^n$.
Important examples of subanalytic sets are a semianalytic set in $\mathbf{R}^n$ (i.e., a
subset of $\mathbf{R}^n$ defined locally at each point of $\mathbf{R}^n$ by finitely many inequalities
of analytic functions) and a polyhedron imbedded and closed in $\mathbf{R}^n$. Note
that an analytic set contained $\mathbf{R}^n$ is not necessarily subanalytic in $\mathbf{R}^n$ because an analytic set is the zero set of an analytic function locally at each
point of the set. A *subanalytic stratification* $\{X_i\}$ of a subanalytic set in
$\mathbf{R}^n$ is an analytic stratification locally finite at each point of $\mathbf{R}^n$ (not only
of $\cup X_i$) such that each stratum is subanalytic in $\mathbf{R}^n$. Let f be a continuous map from a subanalytic set $X \subset \mathbf{R}^m$ to another $Y \subset \mathbf{R}^n$. We call f
subanalytic if the graph of f is subanalytic in $\mathbf{R}^m \times \mathbf{R}^n$. An example of a
subanalytic map is a PL map between polyhedra imbedded and closed in
Euclidean spaces.

By the above example, a subanalytic structure on a polyhedron X does
not depend on the choice of its closed imbedding in a Euclidean space. However, there may be plural subanalytic structures on X if the imbedding is not
closed. Indeed, by the arguments in I.1.12, there are two compact polyhedra
X and Y and points $x_i \in X$ and $y_i \in Y$, $i = 1, 2$, such that $X - x_1 - x_2$ and
$Y - y_1 - y_2$ are PL homeomorphic, and X and Y are homeomorphic but not
PL homeomorphic. Then $X - x_1 - x_2$ and $Y - y_1 - y_2$ are not subanalytically homeomorphic, because if they were, X and Y would be subanalytically
homeomorphic by (I.2.1.1) and hence PL homeomorphic by the subanalytic
Hauptvermutung (III.1.4). So, when we consider a subanalytic structure on
X, we imbed closedly X in a Euclidean space unless otherwise specified.

A *subanalytic homotopy* $f_t \colon X \to Y$, $t \in [0,1]$, is a homotopy such
that the map $X \times [0,1] \ni (x,t) \to f_t(x) \in Y$ is subanalytic. If $f_0 = f$,

we call the homotopy a *subanalytic homotopy* of f. We define naturally a *subanalytic isotopy*, a *subanalytic function*, a *subanalytic weak isomorphism* between *subanalytic Whitney stratifications*, a *subanalytic isomorphism* and a *subanalytic strong isomorphism*. Here we require diffeomorphisms between strata to be analytic and subanalytic. For $r = 1, 2, \ldots$, we define also the C^r version of the above concept, e.g., a *subanalytic C^r stratification*.

For a subanalytic set $X \subset \mathbf{R}^n$ and a point $a \in \mathbf{R}^n$, set

$$a * X = \{ ta + (1-t)x : t \in [0,1], \ x \in X \}$$

if $a \notin X$ and if every point of this set except a is uniquely described by t and x. We call this set the *cone* with vertex a and base X. Let $\{A_\alpha\}$ and $\{B_\beta\}$ be families of subsets of $\mathbf{R}^n$. We call $\{A_\alpha\}$ *compatible* with $\{B_\beta\}$ if for any α and β,

$$A_\alpha \subset B_\beta \quad \text{or} \quad A_\alpha \cap B_\beta = \varnothing.$$

For a simplicial (or cell) complex K in $\mathbf{R}^n$, we call K *compatible* with $\{B_\beta\}$ if the open simplex (or cell) family of K is also compatible.

I.2.1 (Properties of subanalytic sets and maps (see [Hi$_1$], [Ta], [V]). Let $X, Y \subset \mathbf{R}^n$ be subanalytic sets and let $f \colon X \to \mathbf{R}^n$ be a subanalytic map.

(I.2.1.1) The sets $X \times Y \subset \mathbf{R}^n \times \mathbf{R}^n$, $\overline{X}$, $X \cap Y$ and $X - Y$ are subanalytic, where $\overline{X}$ denotes the topological closure of X in $\mathbf{R}^n$.

(I.2.1.2) $\dim(\overline{X} - X) < \dim X$ if $X \neq \varnothing$ (define $\dim \varnothing = -1$).

(See I.2.2 for the dimension of a subanalytic set.)

(I.2.1.3) The family of the connected components of X is locally finite at each point of $\mathbf{R}^n$, and each connected component is subanalytic.

(I.2.1.4) If $f^{-1}(B)$ is bounded for any bounded set B in $\mathbf{R}^n$, $f(X)$ is subanalytic. If $f(X \cap B)$ is bounded for the same B, $f^{-1}(Y)$ is subanalytic.

(I.2.1.5) Let $r = 0, \ldots$, or ω. Set $m = \dim X$. Let $\Sigma_r X$ denote the C^r *singular point* set of X (we call a point x of X C^r *singular* if x has no neighborhood U in $\mathbf{R}^n$ such that $X \cap U$ is a C^r submanifold of $\mathbf{R}^n$ of dimension $= m$). The set is subanalytic and of dimension $< m$.

In the case $r = 0$, there are 4 kinds of singular point of X. Let x be a point of X. First the germ X_x of X at x is not homeomorphic to the germ $\mathbf{R}_0^m$. Second, the pair $(\mathbf{R}_x^n, X_x)$ is not homeomorphic to the pair $(\mathbf{R}_0^n, \mathbf{R}_0^m \times 0)$. Third, X_x is not subanalytically homeomorphic to $\mathbf{R}_0^m$. Fourthly, $(\mathbf{R}_x^n, X_x)$ is not subanalytically homeomorphic to $(\mathbf{R}_0^n, \mathbf{R}_0^m \times 0)$. For any kind of singular point set, the above result holds true.

If X is bounded, then for some integer r, $\Sigma_r X = \Sigma_{r+1} X = \cdots = \Sigma_\omega X$. In any case, $\Sigma_\infty X = \Sigma_\omega X$.

Proof. I find nowhere such an explicit expression in the case $r = 0$. So we prove the case. Let $\Sigma_0 X$ be any kind. By applying II.1.1″ to $\mathbf{R}^n$ and X, we can assume that there exists a simplicial complex K in $\mathbf{R}^n$ such that X is the union of some open simplices of K and the underlying polyhedron of K is closed in $\mathbf{R}^n$. Then if a point of an open simplex σ of K is contained in $\Sigma_0 X$, so are all the other points of σ. Therefore, $\Sigma_0 X$ is also the union of some open simplices of K and hence subanalytic. The inequality $\dim \Sigma_0 X < m$ is clear because an open simplex of K which is included in X and is of dimension m is included in $X - \Sigma_0 X$.   □

(I.2.1.6) Assume X and Y are C^1 submanifolds, $X \cap Y = \varnothing$ and $\overline{X} \cap Y \neq \varnothing$. Let Y' denote the set of the points of Y where the pair of X and Y does not satisfy the Whitney condition. Then Y' is subanalytic and of dimension $< \dim Y$.

Proof. The inequality $\dim Y' < \dim Y$ was shown in [H$_1$]. So we will prove subanalyticity of Y'. For each $(x, y) \in X \times Y$, let $f(x, y)$ denote the image of the vector $x - y$ under the orthogonal projection onto the tangent space $T_x X$ in $\mathbf{R}^n$. By (I.2.1.8), (I.2.1.10) and by the definition of $G_{n,m}$, where $m = \dim X$, the map $f \colon X \times Y \to \mathbf{R}^n$ is subanalytic. By the same reason, the function $g \colon X \times Y \to \mathbf{R}$, given by

$$g(x, y) = |\langle f(x, y), x - y \rangle| / |x - y|^2 \quad \text{for} \quad (x, y) \in X \times Y,$$

is subanalytic, where $\langle , \rangle$ denotes the inner product. By the definition of the Whitney condition, Y' equals the image under the projection $Y \times Y \times \mathbf{R} \to Y$ onto the first factor of the set:

$$\overline{\operatorname{graph} g} \cap (\Delta \times (\mathbf{R} - 1)),$$

where Δ denotes the diagonal of Y. Hence since $|g| \leq 1$, (I.2.1.1) and (I.2.1.4) imply subanalyticity of Y'.   □

(I.2.1.7) For each point $x \in \overline{X} - X$, there exists a subanalytic map $\varphi \colon [0, 1] \to \overline{X}$ such that $\varphi(]0, 1]) \subset X$, $\varphi(0) = x$, and $\varphi|_{]0,1]}$ is an analytic imbedding.

(I.2.1.8) Assume X is a C^1 submanifold of dimension m. The union $TX = \bigcup_{x \in X} x \times T_x X \subset \mathbf{R}^n \times \mathbf{R}^n$ of the tangent spaces of X and the set $\{(x, T_x X) \in X \times G_{n,m}\}$ are subanalytic.

(I.2.1.9) Let f_1 and f_2 be subanalytic functions on X such that $f_1(X \cap B)$ and $f_2(X \cap B)$ are bounded for any bounded set $B \subset \mathbf{R}^n$. The product $f_1 f_2$ is subanalytic by (I.2.1.4), because graph $f_1 f_2$ is the image of the set:

$$(\text{graph } f_1 \times \text{graph } f_2 \times \mathbf{R})$$
$$\cap \{(x_1, y_1, x_2, y_2, z) \in \mathbf{R}^n \times \mathbf{R} \times \mathbf{R}^n \times \mathbf{R} \times \mathbf{R} \colon x_1 = x_2, \ z = y_1 y_2\}$$

under the projection of $\mathbf{R}^n \times \mathbf{R} \times \mathbf{R}^n \times \mathbf{R} \times \mathbf{R}$ onto the first and last factors.

Let g_1 and g_2 be subanalytic functions on X. The sum $g_1 + g_2$ is subanalytic either if at least one of g_1 and g_2 satisfies the above boundedness condition or if g_1 and g_2 are non-negative.

(I.2.1.10) Assume $f(X) \subset Y$. Let $g \colon Y \to \mathbf{R}^m$ be a subanalytic map. If $f(X \cap B)$ is bounded for any bounded set $B \subset \mathbf{R}^n$, or if $g^{-1}(C)$ is bounded for any bounded set $C \subset \mathbf{R}^m$, then the composite $g \circ f$ is subanalytic by (I.2.1.4). This is because graph $g \circ f$ is the image of the set:

$$(\text{graph } f \times \text{graph } g) \cap \{(x, y_1, y_2, z) \in X \times Y \times Y \times \mathbf{R}^m \colon y_1 = y_2\}$$

under the projection $X \times Y \times Y \times \mathbf{R}^m \to X \times \mathbf{R}^m$.

(I.2.1.11) Assume $X \neq \varnothing$. The function g which measures distance from X in $\mathbf{R}^n$ and hence its square are subanalytic by (I.2.1.4). Indeed, we have

$$\text{graph } g = \{(y, t) \colon t \geq g(y)\} \cap \overline{\{(y, t) \colon t \geq (y)\}^c},$$

and the set $\{(y, t) \in \mathbf{R}^n \times \mathbf{R} \colon t \geq g(y)\}$ is the image of the closure of the set:

$$\{(x, y, t) \in X \times \mathbf{R}^n \times \mathbf{R} \colon t \geq |x - y|\}$$

under the projection $\overline{X} \times \mathbf{R}^n \times \mathbf{R} \to \mathbf{R}^n \times \mathbf{R}$.

(I.2.1.12) If X is closed in $\mathbf{R}^n$, it is the image of an analytic manifold M under a proper analytic map $M \to \mathbf{R}^n$.

(I.2.1.13) Let g be a subanalytic C^1 function on an open subanalytic subset U of $\mathbf{R}^n$ such that for any bounded set $B \subset \mathbf{R}^n$, $g(B \cap U)$ is bounded. The partial derivatives $\frac{\partial g}{\partial x_1}, \dots, \frac{\partial g}{\partial x_n}$ are subanalytic. Without the boundedness condition, the map $(g, \frac{\partial g}{\partial x_1}, \dots, \frac{\partial g}{\partial x_n})$ is subanalytic.

Proof. Consider only $\frac{\partial g}{\partial x_1}$. Let L denote the plane in $\mathbf{R}^{n+1}$ spanned by $(1, 0, \dots, 0)$ and $(0, \dots, 0, 1)$. For each $\lambda \in G_{n,n+1}$ with $\lambda \cap L \neq \{0\}$, let $p(\lambda)$ denote the number such that $(1, 0, \dots, p(\lambda)) \in \lambda \cap L$. Then p is a semialgebraic function on an open semialgebraic subset of $G_{n,n+1}$ (see the definition of a semialgebraic set). The graph of $\frac{\partial g}{\partial x_1}$ is the image under the projection of $U \times \mathbf{R} \times G_{n,n+1} \times \mathbf{R}$ onto the first and last factors of the set:

$$\{(x, g(x),\ T_{x,g(x)}\text{graph}\,g,\ p(T_{x,g(x)}\text{graph}\,g)) : x \in U\}.$$

Hence $\frac{\partial g}{\partial x_1}$ is subanalytic if g satisfies the bounded condition. $\qquad\qquad\square$

This proof shows also that given a subanalytic C^1 map g from a C^1 subanalytic manifold X to another Y, the differential $dg : TX \to TY$ is subanalytic.

(I.2.1.14) Let $g = (g_1, \ldots, g_m)$ be a map from X to $\mathbf{R}^m$. If g_i are all subanalytic, so is g because graph g is the image under the projection $X \times \mathbf{R} \times \cdots \times X \times \mathbf{R} \to X \times \mathbf{R} \times \cdots \times \mathbf{R}$, which forgets all X's except the first one, of the subanalytic set $\{(x_1, y_1, \ldots, x_n, y_n) \in \text{graph}\, g_1 \times \cdots \times \text{graph}\, g_n : x_1 = \cdots = x_n\}$. Conversely, if g is subanalytic and if for any bounded set B of $\mathbf{R}^n$, $g(X \cap B)$ is bounded, then all g_i are subanalytic by (I.2.1.10).

Let us give a partial order to a family F of stratifications in $\mathbf{R}^n$. For each $\{X_i\}, \{Y_j\} \in F$, write $\{X_i\} < \{Y_j\}$ if for some positive integer k, $\{X_i : \dim X_i \geq k\}$ coincides with $\{Y_j : \dim Y_j \geq k\}$ and if $\{X_i : \dim X_i = k-1\}$ is different from and compatible with $\{Y_j : \dim Y_j = k-1\}$. We call the maximal element of F *canonical* if exists.

Lemma I.2.2 (Existence of canonical subanalytic Whitney stratifications). *Given a family of subanalytic sets $\{A_\nu\}$ in $\mathbf{R}^n$, locally finite at each point of $\mathbf{R}^n$, any subanalytic set $X \subset \mathbf{R}^n$ admits the canonical subanalytic (C^r) Whitney stratification compatible with $\{A_\nu\}$, where r is a positive integer.*

Proof. We will prove the following statement by downward induction on a non-negative integer k, where the lemma is the case $k = 0$. Assume $X \neq \varnothing$

(k) Let F_k denote the family of subanalytic (C^r) Whitney stratifications $\{X_i\}$ of subsets of X compatible with $\{A_\nu\}$, such that each X_i is of dimension $\geq k$ and $X - \cup_i X_i$ is closed in X and of dimension $< k$. Then F_k is not empty and contains the maximal element if $k \leq \dim X$.

Proof of (k). This is trivial for $k > \dim X$. So assume that the statement $(k+1)$ holds and $k \leq \dim X$. Let $\{X_i\}$ be the maximal element of F_{k+1} if $k < \dim X$, and let $\{X_i\} = \varnothing$ if $k = \dim X$. Set $Y_k = X - \cup_i X_i$. It suffices to find a subanalytic (C^r) Whitney stratification $\{Z_j\}$ of a subset of Y_k compatible with $\{A_\nu\}$ such that each Z_j is of dimension $= k$, $Y_k - \cup_j Z_j$ is closed in Y_k and of dimension $< k$, the family $\{X_i\} \cup \{Z_j\}$ is a (C^r) Whitney stratification, and if $\{Z_{j'}'\}$ is another (C^r) Whitney stratification with the same properties as $\{Z_j\}$, then $\{Z'_{j'}\} \leq \{Z_j\}$, i.e., $\{Z'_{j'}\}$ is compatible with $\{Z_j\}$.

Let N denote the set of indices of $\{A_\nu\}$. It is easy to prove that the family

$$\left\{ \bigcap_{\nu \in N'} A_\nu \cap \bigcap_{\nu \in N-N'} A_\nu^c \cap Y_k \colon N' \subset N \right\}$$

is a subanalytic partition of Y_k locally finite at each point of $\mathbf{R}^n$ and that we can replace $\{A_\nu\}$ with this family. So we assume $\{A_\nu\}$ is a subanalytic partition of Y_k locally finite at each point of $\mathbf{R}^n$. By (I.2.1.5) and (I.2.1.6) each A_ν contains the minimal subanalytic subset B_ν closed in A_ν of dimension $< k$, such that $A_\nu - B_\nu$ is a possibly empty analytic (C^r) submanifold of $\mathbf{R}^n$ of dimension $= k$ and the family $\{X_i\} \cup \{A_\nu - B_\nu\}$ satisfies the Whitney condition. Take the Y_{k-1} to be the union of all B_ν and $Y_k \cap (\overline{A_\nu} - A_\nu)$, and take the $\{Z_j\}$ to be $\{A_\nu - Y_{k-1}\}_{\nu \in N}$. We easily prove that Y_{k-1} and $\{Z_j\}$ satisfy the above conditions. $\qquad \square$

We can generalize I.2.2 to a subanalytic homeomorphism case as follows.

Lemma I.2.3. *Let X, $Y \subset \mathbf{R}^n$ be subanalytic sets, let $\{A_\nu\}$ and $\{A_{\nu'}\}$ be families of subanalytic sets in $\mathbf{R}^n$ locally finite at each point of $\mathbf{R}^n$. Let $f \colon X \to Y$ be a subanalytic homeomorphism such that $f(X \cap B)$ and $f^{-1}(B)$ are bounded for any bounded set B in $\mathbf{R}^n$. There exists subanalytic (C^r) Whitney stratifications $\{X_i\}$ of X and $\{Y_i\}$ of Y compatible with $\{A_\nu\}$ and $\{A_{\nu'}\}$, respectively, such that $f \colon \{X_i\} \to \{Y_i\}$ is the canonical subanalytic (C^r) isomorphism, where r is a positive integer. Here the canonical isomorphism means that if $f \colon \{X'_{i'}\} \to \{Y'_{i'}\}$ is another isomorphism with the same properties, then $\{X'_{i'}\} \leq \{X_i\}$ and $\{Y'_{i'}\} \leq \{Y_i\}$.*

Proof. Let $\{A'_\mu\}$ denote the family of subanalytic subsets of $\mathbf{R}^n \times \mathbf{R}^n$ of the form $A_\nu \times A_{\nu'}$. Set $X' = \operatorname{graph} f \subset \mathbf{R}^n \times \mathbf{R}^n$, and let p_1 and p_2 denote the projections $\mathbf{R}^n \times \mathbf{R}^n \to \mathbf{R}^n$ onto the first and second factors, respectively. It suffices to find the canonical subanalytic (C^r) Whitney stratification $\{X'_i\}$ of X' compatible with $\{A'_\mu\}$, such that (i) for each i, the restrictions $p_1|_{X'_i}$ and $p_2|_{X'_i}$ are analytic (C^r) diffeomorphisms onto their images, and (ii) $\{p_1(X'_i)\}$ and $\{p_2(X'_i)\}$ are subanalytic (C^r) Whitney stratifications. Here we note that a subset C of X' is subanalytic if and only if both $p_l(C)$, $l = 1, 2$, are subanalytic by (I.2.1.4) and by the boundedness condition of f.

We construct a downward induction program whose last step coincides with the above requirement as we did in the proof of I.2.2. Let k be a non-negative integer. Assume $X' \neq \varnothing$. It suffices to prove the following statement.

$(k)'$ Let F'_k denote the family of subanalytic (C^r) Whitney stratifications $\{X'_i\}$ of subsets of X' compatible with $\{A'_\mu\}$, such that each X'_i is of dimension $\geq k$, $X' - \cup_i X'_i$, is closed in X' and of dimension $< k$, and the above conditions (i) and (ii) are satisfied. Then F'_k is not empty and contains the maximal element if $k \leq \dim X'$.

Proof of $(k)'$.   We prove $(k)'$ in the same way as the statement (k) in the proof of I.2.2. Assume that $k \leq \dim X$, there exists the maximal element $\{X'_i\}$ of F'_{k+1} if $k < \dim X$, and that $\{A'_\mu\}$ is a subanalytic partition of $X' - \cup_i X'_i$ locally finite at each point of $\mathbf{R}^n \times \mathbf{R}^n$. Let B'_μ be the minimal subanalytic subset of A'_μ defined in the same way as B_ν in the proof of I.2.2. We require B'_μ to satisfy, moreover, the conditions that for each $j = 1, 2$, the restriction $p_j|_{A'_\mu - B'_\mu}$ is an analytic (C^r) diffeomorphism onto the image and the family $\{p_j(X'_i)\} \cup \{p_j(A'_\mu - B'_\mu)\}$ is a (C^r) Whitney stratification. But by I.2.5 we solve the problem on the first condition, and we can check the second condition as we checked the Whitney condition on the family $\{X_i\} \cup \{A_\nu - B_\nu\}$ in the proof of I.2.2. The family $\{X'_i\} \cup \{A'_\mu - B'_\mu - \cup_{\mu'}(\overline{A'_{\mu'}} - A'_{\mu'})\}$ is the maximal element of F'_k. $\qquad\square$

Lemma I.2.4 (Cone extension of a subanalytic homeomorphism). *Let X and Y be compact subanalytic sets in $\mathbf{R}^n$, and let a and b be points in $\mathbf{R}^n$ such that the cones $a * X$ and $b * Y$ in $\mathbf{R}^n$ are well-defined.*
(i) *Then $a * X$ and $b * Y$ are subanalytic.*

  Moreover, let $f\colon X \to Y$ and $F\colon a*X \to b*Y$ be subanalytic homeomorphisms such that $F(X) = Y$, and let $g_t\colon X \to Y$, $t \in [0,1]$, be a subanalytic isotopy of $F|_X$.

(ii) *The cone extension homeomorphism $\varphi\colon a * X \to b * Y$ of f (i.e., $\varphi(sa + (1-s)x) = sb + (1-s)f(x)$ for $x \in X$ and $s \in [0,1]$) is subanalytic.*

(iii) *There exists a subanalytic isotopy F_t, $t \in [0,1]$, of F such that for each $t \in [0,1]$, $F_t = F$ on X, and F_1 is the cone extension of $F|_X$.*

(iv) *There exists also a subanalytic isotopy $G_t\colon a*X \to b*Y$, $t \in [0,1]$, of F such that for each $t \in [0,1]$, $G_t = g_t$ on X, and G_1 is the cone extension of g_1.*

Proof. (i): Let P and $Q\colon \mathbf{R}^n \times \mathbf{R} \to \mathbf{R}^n$ be the polynomial maps defined by

$$\left. \begin{array}{l} P(x, s) = sa + (1 - s)x \\[2mm] Q(x, s) = sb + (1 - s)x \end{array} \right\} \quad \text{for} \quad (x, s) \in \mathbf{R}^n \times \mathbf{R}.$$

The cones $a * X$ and $b * Y$ are the images of $X \times [0, 1]$ and of $Y \times [0, 1]$ under P and Q, respectively. Hence by (I.2.1.4), $a * X$ and $b * Y$ are subanalytic.

(ii): Let Φ denote the subanalytic homeomorphism $f \times \mathrm{id}\colon X \times [0,1] \to Y \times [0,1]$. Set

$$A_1 = \mathrm{graph}\,\Phi \times \mathbf{R}^n \times \mathbf{R}^n \subset \mathbf{R}^n \times \mathbf{R} \times \mathbf{R}^n \times \mathbf{R} \times \mathbf{R}^n \times \mathbf{R}^n,$$
$$A_2 = \{(x,s,y,t,P(x,s),z) \in X \times [0,1] \times \mathbf{R}^n \times \mathbf{R} \times \mathbf{R}^n \times \mathbf{R}^n\}, \quad \text{and}$$
$$A_3 = \{(x,s,y,t,z,Q(y,t)) \in \mathbf{R}^n \times \mathbf{R} \times Y \times [0,1] \times \mathbf{R}^n \times \mathbf{R}^n\}.$$

The graph of the φ is the image of $A_1 \cap A_2 \cap A_3$ under the projection: $\mathbf{R}^n \times \mathbf{R} \times \mathbf{R}^n \times \mathbf{R} \times \mathbf{R}^n \times \mathbf{R}^n \to \mathbf{R}^n \times \mathbf{R}^n$ onto the last two factors. Hence by (I.2.1.4) φ is subanalytic.

(iv): Set

$$X_1 = ((a * X) \times 0) \cup (X \times [0,1]) \subset \mathbf{R}^n \times \mathbf{R},$$
$$Y_1 = ((b * Y) \times 0) \cup (Y \times [0,1]) \subset \mathbf{R}^n \times \mathbf{R},$$
$$a_1 = (a,1), \quad \text{and} \quad b_1 = (b,1) \in \mathbf{R}^n \times \mathbf{R}.$$

Then $a_1 * X_1$ and $b_1 * Y_1$ are subanalytic cones by (i), and we have

$$a_1 * X_1 = (a * X) \times [0,1], \quad \text{and} \quad b_1 * Y_1 = (b * Y) \times [0,1].$$

Define a subanalytic homeomorphism G from X_1 to Y_1 by

$$G(x,t) = \begin{cases} (F(x),0) & \text{for} \quad (x,t) \in (a * X) \times 0 \subset \mathbf{R}^n \times \mathbf{R} \\ (g_t(x),t) & \text{for} \quad (x,t) \in X \times [0,1] \subset \mathbf{R}^n \times \mathbf{R}. \end{cases}$$

Let $H\colon a_1 * X_1 \to b_1 * Y_1$ denote the cone extension of G. By (ii) H is subanalytic, and for each $(x,t) \in (a * X) \times [0,1]$, $H(x,t)$ takes the form $(H'(x,t),t) \in (b*Y) \times [0,1]$. Hence $G_t(x) = H'(x,t)$ fulfills the requirements.

(iii): In the above proof of (iv), note that the G_1 is the cone extension of g_1. Hence it suffices to apply (iv) to F and $g_t = F|_X$, $t \in [0,1]$. $\qquad\square$

We remark that if X and Y in I.2.4 are polyhedra and if f is PL, then the cone extension $a * X \to b * Y$ of f is PL. In the case where X and Y are polyhedra, we call the method of construction of F_t and G_t in the proof of I.2.4 the *Alexander trick*. We shall use the Alexander trick many times.

Lemma I.2.5 (Singular point set of a subanalytic map). *Let X and Y be subanalytic C^1 submanifolds of $\mathbf{R}^n$, let k be a non-negative integer, and let $f\colon X \to Y$ be a subanalytic C^1 map such that $f(X \cap B)$ is bounded for any bounded set $B \subset \mathbf{R}^n$. The set of the points of X where the differential df has rank k is subanalytic.*

Proof. Set $Z = \mathrm{graph}\,f$. Let q_1 and $q_2\colon \mathbf{R}^n \times \mathbf{R}^n \to \mathbf{R}^n$ denote the projections onto the first and second factors, respectively. Then Z is a subanalytic

C^1 submanifold of $\mathbf{R}^n \times \mathbf{R}^n$ and $q_1|_{\overline{Z}}$ is proper. Let $e_1, \ldots, e_{2n}$ denote the unit vectors of $\mathbf{R}^n \times \mathbf{R}^n$ subordinate to its coordinate system, and for each $z \in Z$, let p_z denote the orthogonal projection $\mathbf{R}^n \times \mathbf{R}^n \rightarrow T_z Z$. For any element e of $\mathbf{R}^n \times \mathbf{R}^n$, the map $Z \ni z \rightarrow p_z(e) \in \mathbf{R}^n \times \mathbf{R}^n$ is subanalytic by (I.2.1.8) and by the definition of the algebraic structure on $G_{n,m}$. The differential df has rank k at $x \in X$ if and only if the vectors $q_2 \circ p_{x,f(x)}(e_1), \ldots, q_2 \circ p_{x,f(x)}(e_n)$ span a k-dimensional linear space in $\mathbf{R}^n$. Hence the set of the points $(x, f(x))$ of Z such that df has rank k at x is subanalytic by (I.2.1.9) and by elementary calculations of determinants, because $q_2 \circ p_{x,f(x)}(e_1), \ldots, q_2 \circ p_{x,f(x)}(e_n)$ are subanalytic maps on Z by (I.2.1.4) and (I.2.1.10). Therefore, by properness of $q_1|_{\overline{Z}}$, the set of the points of X where df has rank k is subanalytic. $\square$

Let $f \colon X \rightarrow Y$ be a subanalytic map between subanalytic subsets of $\mathbf{R}^n$. We define naturally a *subanalytic (Whitney) stratification* of f, a *subanalytic C^r (Whitney) stratification* of f and those with the (weak) frontier condition, $r = 1, 2, \ldots$. Let F be a family of stratifications of f. We define the *canonical* element of F to be the maximal one, as in the case of stratification of sets, by giving a partial order to F as follows. Let $(\{X_i\}, \{Y_j\})$ and $(\{X_{i'}'\}, \{Y_{j'}'\})$ be elements of F. Then $(\{X_i\}, \{Y_j\}) < (\{X_{i'}'\}, \{Y_{j'}'\})$ if for some positive integer k,

$$(\{X_i \colon \dim f(X_i) \geq k\}, \{Y_j \colon \dim Y_j \geq k\})$$
$$= (\{X_{i'}' \colon \dim f(X_{i'}') \geq k\}, \{Y_{j'}' \colon \dim Y_{j'}' \geq k\})$$

and if one of the following two conditions is satisfied:

$$\{Y_j \colon \dim Y_j = k - 1\} < \{Y_{j'}' \colon \dim Y_{j'}' = k - 1\}, \tag{i}$$
$$\{Y_j \colon \dim Y_j = k - 1\} = \{Y_{j'}' \colon \dim Y_{j'}' = k - 1\} \qquad \text{and}$$
$$\{X_i \colon \dim f(X_i) = k - 1\} < \{X_{i'}' \colon \dim f(X_{i'}') = k - 1\}. \tag{ii}$$

Lemma I.2.6 (Canonical stratification of a subanalytic map). *Let X and Y be subanalytic sets in $\mathbf{R}^n$, let $\{A_\nu\}$ and $\{A_{\nu'}\}$ be families of subanalytic sets in $\mathbf{R}^n$ locally finite at each point of $\mathbf{R}^n$. Let $f \colon X \rightarrow Y$ be a subanalytic map such that $f^{-1}(B)$ and $f(X \cap B)$ are bounded for any bounded set $B \subset \mathbf{R}^n$. Then f admits the canonical subanalytic (C^r) (Whitney) stratification (with the weak frontier condition) $(\{X_i\}, \{Y_j\})$ such that $\{X_i\}$ and $\{Y_j\}$ are compatible with $\{A_\nu\}$ and $\{A_{\nu'}\}$, respectively, $r = 1, 2, \ldots$.*

Proof. This proof is similar to the proof of I.2.2. We shall construct only a subanalytic Whitney stratification of f. Existence of other stratifications and of the canonical ones is immediate by the following construction. Let k be a non-negative integer. Assume $X \neq \varnothing$ and $Y \neq \varnothing$. Consider the next statement.

$(k)''$ Let F_k'' be the family of subanalytic Whitney stratifications $(\{X_i\}, \{Y_j\})$ of $f|_{\cup_i X_i} : \cup_i X_i \to \cup_j Y_j$ such that $\cup_i X_i$ and $\cup_j Y_j$ are subsets of X and Y, respectively, $\cup_i X_i = f^{-1}(\cup_j Y_j)$, $\{X_i\}$ and $\{Y_j\}$ are compatible with $\{A_\nu\}$ and $\{A_{\nu'}\}$, respectively, $Y - \cup_j Y_j$ is closed in Y and of dimension $< k$, and each Y_j is of dimension $\geq k$. Then F_k'' is not empty if $k \leq \dim Y$.

Proof of $(k)''$. Let $k \leq \dim Y$, and assume there exists an element $(\{X_i\}, \{Y_j\})$ of F_{k+1}'' if $k < \dim Y$. We want to find an element of F_k''. First we reduce the problem to the case where $Y - \cup_j Y_j$ is smooth. Apply the arguments in the proof of I.2.2 to $Y - \cup_j Y_j$ and $\{A_{\nu'}\}$. There exists a subanalytic stratification $\{Z_l\}$ of a subset of $Y - \cup_j Y_j$ compatible with $\{A_{\nu'}\}$ such that each Z_l is of dimension $= k$, the set $Y - \cup_j Y_j - \cup_l Z_l$ is closed in Y and of dimension $< k$, and the family $\{Y_j\}_j \cup \{Z_l\}_l$ satisfies the Whitney condition. We may restrict f to $\cup_i X_i \cup f^{-1}(Z_l)$ for each l. Indeed, if for each l, there exists an element $(\{X_i\}_i \cup \{X_{i'}^l\}_{i'}, \{Y_j\}_j \cup \{Z_l - Z_l'\})$ of F_k'' for the map $f|_{\cup_i X_i \cup f^{-1}(Z_l)} : \cup_i X_i \cup f^{-1}(Z_l) \to \cup_j Y_j \cup Z_l$, then $(\{X_i\}_i \cup \{X_{i'}^l\}_{i',l}, \{Y_j\}_j \cup \{Z_l - Z_l'\}_l)$ is an element of F_k'' for f, because the families $\{X_i\}_i \cup \{X_{i'}^l\}_{i',l}$ and $\{\text{graph } f|_{X_i}\}_i \cup \{\text{graph } f|_{X_{i'}^l}\}_{i',l}$ satisfy the Whitney condition. Set $Z = Y - \cup_j Y_j$. So we assume Z is a k-dimensional analytic submanifold of $\mathbf{R}^n$.

Next we stratify $f^{-1}(Z)$. By I.2.2 we can assume $\{A_\nu\}$ is a subanalytic stratification of $f^{-1}(Z)$. By I.2.5 each A_ν includes a closed subanalytic subset B_ν such that $\underline{f|_{A_\nu - B_\nu} \text{ is submersive into } Z}$ and $f(B_\nu)$ is of dimension $< k$. Set $Z' = Z \cap \cup_\nu f(B_\nu)$. Then Z' is closed in Z and of dimension $< k$, and hence we can neglect Z'. This implies that all A_ν are smooth and $f|_{A_\nu}$ are submersive into Z. We suppose, moreover, that $f^{-1}(Z) \neq \varnothing$ because if $f^{-1}(Z) = \varnothing$, then $(\{X_i\}_i, \{Y_j\}_j \cup \{Z\})$ is an element of F_k''.

It remains to modify $\{A_\nu\}_\nu$ so that $\{X_i\}_i \cup \{A_\nu\}_\nu$ and $\{\text{graph } f|_{X_i}\}_i \cup \{\text{graph } f|_{A_\nu}\}_\nu$ satisfy the Whitney condition. We carry this out by downward induction as follows. Let α be a non-negative integer.

$(\alpha)'''$ Let G_α be the family of pairs of subanalytic subsets Z^α of Z of dimension $< k$ and subanalytic Whitney stratifications $\{W_l\}$ of subsets of $f^{-1}(Z)$ compatible with $\{A_\nu\}$ such that each W_l is of dimension $\geq \alpha$, $f^{-1}(Z - Z^\alpha) - \cup_l W_l$ is of dimension $< \alpha$, each $f|_{W_l}$ is an analytic submersion into Z, and $\{X_i\} \cup \{W_l\}$ and $\{\text{graph } f|_{X_i}\} \cup \{\text{graph } f|_{W_l}\}$ satisfy the Whitney condition. Then G_α is not empty if $\alpha \leq \dim f^{-1}(Z)$.

If $(0)'''$ holds true, let $(Z^0, \{W_l\})$ be an element of G_0. Then $(\{X_i\}_i \cup \{W_l - f^{-1}(\overline{Z^0})\}_l, \{Y_j\}_j \cup \{Z - \overline{Z^0}\})$ is an element of F_k''.

Proof of $(\alpha)'''$. Let $\alpha \le \dim f^{-1}(Z)$ and assume there exists an element $(Z^{\alpha+1}, \{W_l\})$ of $G_{\alpha+1}$ if $\alpha < \dim f^{-1}(Z)$. For each ν, apply I.2.2 and I.2.5 to $A'_\nu = A_\nu - f^{-1}(Z^{\alpha+1}) - \cup_l W_l$. There exist subanalytic subsets B'_ν and B''_ν of A'_ν such that $A'_\nu - B'_\nu - B''_\nu$ is an analytic submanifold of $\mathbf{R}^n$ of dimension $= \alpha$, B'_ν is of dimension $< \alpha$, $f(B''_\nu)$ is of dimension $< k$, $\{X_i\}_i \cup \{W_l\}_l \cup \{A'_\nu - B'_\nu - B''_\nu\}$ and $\{\text{graph } f|_{X_i}\}_i \cup \{\text{graph } f|_{W_l}\}_l \cap \{\text{graph } f|_{A'_\nu - B'_\nu - B''_\nu}\}$ satisfy the Whitney condition, and $f|_{A'_\nu - B'_\nu - B''_\nu}$ is submersive into Z. Set $Z^\alpha = Z^{\alpha+1} \cup (Z \cap \cup_\nu \overline{f(B''_\nu)})$. Then $(Z^\alpha, \{W_l\}_l \cup \{A'_\nu - B'_\nu - B''_\nu\}_\nu)$ is an element of G_α. $\square$

Lemma I.2.7 (Friendliness of subanalytic functions). *Let* $\{X_1, X_2\}$ *be a subanalytic* C^1 *stratification in* $\mathbf{R}^n$, *and let* f_1, f_2 *be subanalytic functions on* X. *Assume that* $f_1^{-1}(0) = f_2^{-1}(0) = X_1$, *and the restrictions to* X_2 *of* f_1 *and of* f_2 *are positive* C^1 *functions. Then* $f_1|_{X_2 \cap U}$ *and* $f_2|_{X_2 \cap U}$ *are friendly for some open subanalytic neighborhood* U *of* X_1 *in* $\mathbf{R}^n$.

Proof. By considering the graph of (f_1, f_2), we can assume f_1 and f_2 are the restrictions to X of polynomial functions F_1 and F_2 on $\mathbf{R}^n$, respectively.

Let Y denote the set of the points of X_2 where f_1 and f_2 are not friendly. If we prove that Y is subanalytic, the lemma follows for the following reason. Assume there does not exist the U in the lemma, namely, $\overline{Y} \cap X_1 \ne \varnothing$. By (I.2.1.7) we have a subanalytic map $\varphi \colon [0, 1] \to \mathbf{R}^n$ such that

$$\varphi(]0, 1]) \subset Y, \qquad \varphi(0) \in X_1,$$

and $\varphi|_{]0,1]}$ is analytic. Consider the composites $f_1 \circ \varphi$ and $f_2 \circ \varphi$. By the definition of friendly functions, $f_1 \circ \varphi$ and $f_2 \circ \varphi$ are not friendly at every point of $]0, 1]$. Hence if $f_1 \circ \varphi$ is monotone increasing, $f_2 \circ \varphi$ is monotone decreasing. On the other hand, by I.2.5, the differentials $d(f_1 \circ \varphi)$ and $d(f_2 \circ \varphi)$ have rank 1 on the interval $]0, \varepsilon]$ for some $\varepsilon > 0$, because a 0-dimensional subanalytic set is locally finite. Hence both $f_1 \circ \varphi$ and $f_2 \circ \varphi$ are monotone increasing on $]0, \varepsilon]$, because f_1 and f_2 are positive on X_2 and because $f_1 = f_2 = 0$ on X_1. This is a contradiction.

We need to prove that the Y is subanalytic. For this we shall prove that the sets

$$Y_1 = \{(x, \text{grad } f_{1x}) \in X_2 \times \mathbf{R}^n\} \quad \text{and}$$
$$Y_2 = \{(x, \text{grad } f_{2x}) \in X_2 \times \mathbf{R}^n\}$$

are subanalytic. Indeed, if Y_1 and Y_2 are subanalytic, so is Y, because Y is the image of the subanalytic set:

$$Y_3 = (Y_1 \times \mathbf{R}^n \times \mathbf{R}^n \times \mathbf{R}^2) \cap (\mathbf{R}^n \times \mathbf{R}^n \times Y_2 \times \mathbf{R}^2) \cap Y_4$$

under the projection $p\colon \mathbf{R}^n \times \mathbf{R}^n \times \mathbf{R}^n \times \mathbf{R}^n \times \mathbf{R}^2 \to \mathbf{R}^n$ onto the first factor, and because $p|_{\overline{Y_3}}$ is proper, where

$$Y_4 = \{(x, u, y, v, a, b) \in \mathbf{R}^n \times \mathbf{R}^n \times \mathbf{R}^n \times \mathbf{R}^n \times \mathbf{R} \times \mathbf{R}$$
$$: x = y,\ a \geq 0,\ b \geq 0,\ a + b = 1,\ au + bv = 0\}.$$

Here properness of $p|_{\overline{Y_3}}$ follows from the assumption that f_1 and f_2 are the restrictions of polynomial functions on $\mathbf{R}^n$.

Thus it suffices to prove the following. Let F be a polynomial function on $\mathbf{R}^n$ and let f denote its restriction to X_2. The set

$$Z = \{(x, \operatorname{grad} f_x) \in X_2 \times \mathbf{R}^n\}$$

is subanalytic.

Set

$$Z_1 = \{(x, \operatorname{grad} F_x) \in \mathbf{R}^n \times \mathbf{R}^n\},$$
$$Z_2 = \{(x, T_x X_2) \in X_2 \times G_{n,m}\},\ m = \dim X_2,\ \text{and}$$
$$Z_3 = \{(x, L, y, u, v) \in X_2 \times G_{n,m} \times \mathbf{R}^n \times \mathbf{R}^n \times \mathbf{R}^n : x = y,\ Lu = v\}.$$

Here we regard an element L of $G_{n,m}$ as a linear transformation of $\mathbf{R}^n$. Then Z_1 is an algebraic set, and Z_2 and Z_3 are subanalytic by (I.2.1.8) and by the definition of the algebraic structure on $G_{n,m}$. Z is the image of the set:

$$(X_2 \times G_{n,m} \times Z_1 \times \mathbf{R}^n) \cap (Z_2 \times \mathbf{R}^n \times \mathbf{R}^n \times \mathbf{R}^n) \cap Z_3$$

under the projection $X_2 \times G_{n,m} \times \mathbf{R}^n \times \mathbf{R}^n \times \mathbf{R}^n \to X_2 \times \mathbf{R}^n$ onto the first and last factors. Hence Z is subanalytic by (I.2.1.4).   □

The following is immediate by (I.2.1.10), (I.2.1.11) and I.2.7.

Corollary I.2.8 (Strongness of a subanalytic isomorphism). *Let X, $Y \subset \mathbf{R}^n$ be compact subanalytic sets, let $\{X_i\}$ and $\{Y_i\}$ be subanalytic (C^r) Whitney stratifications of X and Y, respectively, and let $f\colon \{X_i\} \to \{Y_i\}$ be a subanalytic (C^r) isomorphism, where $r = 1, 2, \ldots$. Then f is a subanalytic strong (C^r) isomorphism.*

A *semialgebraic set* in $\mathbf{R}^n$ is a set of the form:

$$\bigcup_{\text{finite}} \{x \in \mathbf{R}^n : f_1(x) = 0,\ f_2(x) > 0, \ldots, f_m(x) > 0\}$$

for polynomial functions $f_1, \ldots, f_m$ on $\mathbf{R}^n$. An example of semialgebraic set is a compact polyhedron imbedded in a Euclidean space. A *semialgebraic stratification* of a semialgebraic set is a finite analytic stratification such that each stratum is semialgebraic. We define a *semialgebraic map*, a *semialgebraic homotopy*, a *semialgebraic isotopy*, a *semialgebraic function*, a *semialgebraic weak isomorphism*, a *semialgebraic isomorphism*, a *semialgebraic strong isomorphism*, a *semialgebraic Whitney stratification*, a *canonical semialgebraic stratification*, and C^r ones, $r = 1, 2, \ldots$, in the same way as in the subanalytic case. The following facts, lemmas and corollary follow in the same way. We omit the proofs.

I.2.9 (Properties of semialgebraic sets (see [L₂]). Let $X, Y \subset \mathbf{R}^n$ be semialgebraic sets, and let $f \colon X \to \mathbf{R}^n$ be a semialgebraic map.

(I.2.9.1) The sets $X \times Y \subset \mathbf{R}^n \times \mathbf{R}^n$, $\overline{X}$, $X \cap Y$ and $X - Y$ are semialgebraic.

(I.2.9.2) $\dim(\overline{X} - X) < \dim X$ if $X \neq \varnothing$ (define $\dim \varnothing = -1$).

(I.2.9.3) The family of the connected components of X is finite, and each connected component is semialgebraic.

(I.2.9.4) The sets $f(X)$ and $f^{-1}(Y)$ are semialgebraic.

(I.2.9.5) Let $r = 0, 1, \ldots$, or ω. Let $\Sigma_r X$ denote the C^r *singular point* set of X, namely, the set consisting of the points where the germ of X is either of dimension $< \dim X$ or not C^r smooth. The set is semialgebraic and of dimension $< \dim X$, and for some $r < \infty$, $\Sigma_r X = \Sigma_{r+1} X = \cdots = \Sigma_\omega X$. In the case $r = 0$, we can adopt the other definitions of singular point as in (I.2.1.5).

(I.2.9.6) Assume X and Y are C^1 submanifolds of $\mathbf{R}^n$, $X \cap Y = \varnothing$ and $\overline{X} \cap Y \neq \varnothing$. Let Y' denote the set of the points of Y where the pair of X and Y does not satisfy the Whitney condition. Then Y' is semialgebraic and of dimension $< \dim Y$.

(I.2.9.7) For each $x \in \overline{X} - X$ there exists a semialgebraic map $\varphi \colon [0, 1] \to \overline{X}$ such that $\varphi(]0, 1]) \subset X$, $\varphi(0) = x$ and $\varphi|_{]0,1]}$ is analytic.

(I.2.9.8) Assume X is a C^1 submanifold of $\mathbf{R}^n$ of dimension m. The union $TX = \cup_{x \in X} x \times T_x X \subset \mathbf{R}^n \times \mathbf{R}^n$ of the tangent spaces, and the set $\{(x, T_x X) \in X \times G_{n,m}\}$ are semialgebraic.

(I.2.9.9) Let f_1 and f_2 be semialgebraic functions on X. The product $f_1 f_2$ and the sum $f_1 + f_2$ are semialgebraic.

(I.2.9.10) Assume $f(X) \subset Y$. Let $g \colon Y \to \mathbf{R}^m$ be a semialgebraic map. The composite $g \circ f$ is semialgebraic.

(**I.2.9.11**) Assume $X \neq \varnothing$. The function which measures distance from X in $\mathbf{R}^n$ and its square are semialgebraic.

(**I.2.9.12**) If X is closed in $\mathbf{R}^n$, X is the image of an affine non-singular algebraic variety M under a proper polynomial map $M \to \mathbf{R}^n$. (We can prove this as $[\mathrm{H}_1]$ proved the subanalytic case.)

(**I.2.9.13**) Given a semialgebraic C^1 function g on $\mathbf{R}^n$, the partial derivatives $\frac{\partial g}{\partial x_1}, \ldots, \frac{\partial g}{\partial x_n}$ are semialgebraic.

(**I.2.9.14**) A map $g = (g_1, \ldots, g_m)$ from X to $\mathbf{R}^m$ is semialgebraic if and only if so is every g_i.

Lemma I.2.10 (Existence of canonical semialgebraic Whitney stratifications). *Given a finite family of semialgebraic sets $\{A_\nu\}$ in $\mathbf{R}^n$, any semialgebraic set in $\mathbf{R}^n$ admits the canonical semialgebraic (C^r) Whitney stratification compatible with $\{A_\nu\}$, $r = 1, 2, \ldots$.*

Lemma I.2.11. *Let $X \subset \mathbf{R}^n$ and $Y \subset \mathbf{R}^n$ be semialgebraic sets, let $\{A_\nu\}$ and $\{A_{\nu'}\}$ be finite families of semialgebraic sets in $\mathbf{R}^n$, and let $f \colon X \to Y$ be a semialgebraic homeomorphism. There exists the canonical semialgebraic (C^r) isomorphism $f \colon \{X_i\} \to \{Y_i\}$ such that $\{X_i\}$ and $\{Y_i\}$ are compatible with $\{A_\nu\}$ and $\{A_{\nu'}\}$, respectively, $r = 1, 2, \ldots$.*

Lemma I.2.12 (Cone extension of a semialgebraic homeomorphism). *Let X and Y be compact semialgebraic sets in $\mathbf{R}^n$, and let a and b be points in $\mathbf{R}^n$ such that the cones $a * X$ and $b * Y$ in $\mathbf{R}^n$ are well-defined.*
(i) *Then $a * X$ and $b * Y$ are semialgebraic.*

   Moreover, let $f \colon X \to Y$ and $F \colon a * X \to b * Y$ be a semialgebraic homeomorphism such that $F(X) = Y$, and let $g_t \colon X \to Y$, $t \in [0, 1]$, be a semialgebraic isotopy of $F|_X$.

(ii) *The cone extension homeomorphism $\varphi \colon a * X \to b * Y$ of f is semialgebraic.*

(iii) *There exists a semialgebraic isotopy F_t, $t \in [0, 1]$, of F such that for each $t \in [0, 1]$, $F_t = F$ on X, and F_1 is the cone extension of $F|_X$.*

(iv) *There exists also a semialgebraic isotopy $G_t \colon a * X \to b * Y$, $t \in [0, 1]$, of F such that for each $t \in [0, 1]$, $G_t = g_t$ on X, and G_1 is the cone extension of g_1.*

Lemma I.2.13 (Singular point set of a semialgebraic map). *Let X and Y be semialgebraic C^1 submanifold of $\mathbf{R}^n$, let k be a non-negative integer and let $f \colon X \to Y$ be a semialgebraic C^1 map. The set of the points of X where the differential df has rank k is semialgebraic.*

Lemma I.2.14 (Canonical stratification of a semialgebraic map). *Let X and Y be semialgebraic sets in $\mathbf{R}^n$, let $\{A_\nu\}$ and $\{A_{\nu'}\}$ be finite families of semialgebraic sets in $\mathbf{R}^n$, and let $f\colon X \to Y$ be a semialgebraic map. Then f admits the canonical semialgebraic (C^r) (Whitney) stratification (with the weak frontier condition) $(\{X_i\}, \{Y_j\})$ such that $\{X_i\}$ and $\{Y_j\}$ are compatible with $\{A_\nu\}$ and $\{A_{\nu'}\}$, respectively, $r = 1, 2, \ldots$.*

Lemma I.2.15 (Friendliness of semialgebraic functions). *Let $\{X_1, X_2\}$ be a semialgebraic C^1 stratification of a set $X \subset \mathbf{R}^n$, and let f_1 and f_2 be semialgebraic functions on X. Assume that $f_1^{-1}(0) = f_2^{-1}(0) = X_1$, and the restrictions to X_2 of f_1 and of f_2 are positive C^1 functions. Then $f_1|_{X_2 \cap U}$ and $f_2|_{X_2 \cap U}$ are friendly for some open semialgebraic neighborhood U of X_1 in $\mathbf{R}^n$.*

Corollary I.2.16 (Strongness of a semialgebraic isomorphism). *Let $\{X_i\}$ and $\{Y_i\}$ be semialgebraic (C^r) Whitney stratifications of compact semialgebraic sets, and let $f\colon \{X_i\} \to \{Y_i\}$ be a semialgebraic (C^r) isomorphism, where $r = 1, 2, \ldots$. Then f is a semialgebraic strong (C^r) isomorphism.*

Remark I.2.17. The above facts I.2.9, $\ldots$, I.2.16 hold true on an arbitrary real closed field. Here we replace the term "semialgebraic analytic" and "connected" in the statements by "semialgebraic C^∞", and "semialgebraically connected", respectively. See [B-C-R] for the definition of a real closed field and properties of semialgebraic sets and maps over it.

§I.3. PL topology and C^∞ triangulations

We introduce some terminology and techniques of PL topology used in this book, and develop a globally finite aspect of PL topology. We also give the definitions of subanalytic, semialgebraic and C^∞ triangulations. We show that a set which is locally C^∞ equivalent to a polyhedron is uniquely and globally C^∞ triangulable. This is a generation, answering our purpose, of the Cairns-Whitehead theorem, which states that a C^∞ manifold is uniquely C^∞ triangulable.

All the terminology and the results can be extended to the case of a real closed field except for subanalytic and C^∞ triangulations. In this extension, the collocation "a compact polyhedron" means the underlying polyhedron of a finite simplicial complex.

In this section, K and L always denote simplicial complexes in some Euclidean space. "Linear" means linear in the affine sense (e.g., a linear subspace of $\mathbf{R}^n$ is not necessarily meant to contain the origin). We denote by $|K|$ the underlying polyhedron of K. A *simplicial map* $f\colon K \to L$ is a continuous map $f: |K| \to |L|$ which carries linearly each simplex of K onto some simplex of L. If f is, in addition, a homeomorphism, we call f a *simplicial isomorphism*. For a simplex σ of K, let $\mathrm{st}(\sigma, K)$ and $\mathrm{lk}(\sigma, K)$ denote the star and the link of σ in K, respectively. We define also $\mathrm{st}(x, K)$ and $\mathrm{lk}(x, K)$ for $x \in |K|$. We denote by K^k the *k-skeleton* of K for a non-negative integer k. For a simplex σ, σ° and $\partial\sigma$ denote the interior and the boundary of σ, respectively. If $K \subset L$, the *simplicial neighborhood* $N(K, L)$ of K in L is the smallest subcomplex of K whose underlying polyhedron is a neighborhood of $|K|$ in $|L|$. We set

$$\partial N(K, L) = \{\sigma \in N(K, L): \sigma \cap |L| = \varnothing\}, \text{ and}$$
$$N^\circ(K, L) = N(K, L) - \partial N(K, L).$$

If $K \subset L$ with $|K| = X$, we write $K = L|_X$. We call a subcomplex K of L *full* if any simplex of L with vertices all in K belongs to K. For points $a_0, \dots, a_m$ of $\mathbf{R}^n$ which are independent (i.e., are not contained in any linear space of dimension $< m$), $\Delta a_0 \cdots a_m$ or $\Delta\{a_0, \dots, a_m\}$ denotes the m-simplex spanned by $a_0, \dots, a_m$, and $\nabla a_0 \cdots a_m$ or $\nabla\{a_0, \dots, a_m\}$ denotes the linear space spanned by $\Delta a_0 \cdots a_m$. For example, the line $\overrightarrow{a_0 a_1}$ in §I.1 coincides with $\nabla 0(a_0 - a_1)$. For each simplex σ of K, let a point v_σ in σ° be fixed. A *derived subdivision* of K consists of all the simplex $\Delta v_{\sigma_1} \cdots v_{\sigma_k}$ for $\sigma_1 \subset \cdots \subset \sigma_k \in K$. A typical example of a derived complex is the barycentric subdivision. We usually denote a derived subdivision by K'. Clearly K' is not unique. It is, however, simplicially unique in the following sense. Let K_1 and K_2 be derived subdivisions of K, and let $v_{i\sigma}$, $i = 1, 2$, denote the new vertices of K_i in σ° for each $\sigma \in K$. The correspondence $: v_{1\sigma} \to v_{2\sigma}$, $\sigma \in K$, can be extended to a simplicial isomorphism $K_1 \to K_2$. Note that if $K \subset L$ and if L' is a derived subdivision of L, then $L'|_{|K|}$ is a full subcomplex of L'. This is one of the reasons why we take a derived subdivision.

A *polyhedron* is a set in a Euclidean space which is a finite union of simplices locally at each point of the set (not necessarily of the Euclidean space). Let X, Y be polyhedra in $\mathbf{R}^n$. Their *join* $X * Y$ is the set:

$$\{tx + (1 - t)y: x \in X, \ y \in Y, \ t \in [0, 1]\},$$

provided $X \cap Y = \varnothing$, and every point of this set except points of X and Y is uniquely described by $x \in X$, $y \in Y$ and $t \in [0, 1]$. We say that K is a *simplicial decomposition* of X if $|K| = X$. A *PL map* $f: X \to Y$ is a continuous map such that for some simplicial decomposition K of X and for each $\sigma \in K$, $f|_\sigma$ is linear. We note that a PL map $f: X \to Y$ does not necessarily induce a simplicial map $f: K \to L$ for some simplicial decompositions K of X and L of Y. However, if f is proper in addition, there exist such K and L. A *PL homotopy* $f_t \colon X \to Y$, $t \in [0,1]$, is a homotopy such that the map $X \times [0,1] \ni (x,t) \to f_t(x) \in Y$ is PL. A *PL k-ball* or a *PL k-sphere* is a polyhedron PL homeomorphic to a k-simplex or to the boundary of a $(k+1)$-simplex, respectively. Let $X \supset Y$ be a polyhedron and a closed subpolyhedron. We call a polyhedral neighborhood U of Y in X a *regular neighborhood* if there exist a simplicial complex K, a full subcomplex L of K and a PL homeomorphism $\pi \colon |K| \to X$ such that

$$\pi(|L|) = Y, \quad \text{and} \quad U = \pi(|N(L', K')|)$$

for some derived subdivisions $K' \supset L'$ of $K \supset L$. The U admits a natural PL retraction to Y as follows, which is one reason why we introduce the concept of a regular neighborhood. For the above K and L, set

$$h(\text{the vertex of } K' \text{ in } \sigma^\circ) = \text{the vertex of } L' \text{ in } (\sigma \cap |L|)^\circ$$
$$\text{for } \sigma \in K \text{ with } \sigma \cap |L| \neq \varnothing,$$

and define a simplicial map from $N(L', K')$ to L' by h. Then the simplicial map induces a PL retraction of U to Y. Clearly the retraction is a *PL (strong) deformation retraction*, i.e., there is a PL homotopy of the identity map of U to the retraction (the homotopy is fixed on Y).

As a corollary, it follows that for polyhedra X, Y, a C^0 map f from X to Y can be approximated by a PL map in the compact-open C^0 topology as follows. Here the approximation map is fixed on a subset of X where f is PL. Assume $Y \subset \mathbf{R}^n$. If Y is open in $\mathbf{R}^n$, this is clear. For general Y, let U be a regular neighborhood of Y in $\mathbf{R}^n - (\overline{Y} - Y)$ and let $p \colon U \to Y$ be a PL retraction. Regard f as a map from X to $\operatorname{Int} U$, and approximate it by a PL map f'. Then $p \circ f' \colon X \to Y$ is a PL approximation of f.

We see also that for polyhedra $Y \subset X$, we can approximate a (*strong*) *deformation retraction* of X to Y (i.e., of class C^0) by a PL one. A *contractible* set is a set where there is a strong deformation retraction to some point. Note that the point to which the set is contractible is arbitrary, and if there is a deformation retraction of a set to a point, then it is contractible. For sets $Y \subset X$, if there is a (strong) deformation retraction of X to Y, we call Y a (*strong*) *deformation retract* of X. Note that if $Y \subset X$ are polyhedra, Y is closed in X and the inclusion map $Y \to X$ is a *weak homotopy equivalence* (i.e., the map induces isomorphisms of the homotopy groups), then Y is a strong deformation retract of X.

Let $A \subset B$ be a topological set and its subset. To avoid confusion, we denote by $\operatorname{Int} A$ and $\operatorname{bdry} A$ the interior and the boundary of A in B, respectively. For a manifold with boundary M (e.g. a PL ball), M° and ∂M denote the interior and the boundary, respectively.

For a subanalytic set $X \subset \mathbf{R}^n$, a *subanalytic triangulation* of X is a pair (K, π), where $|K|$ is closed in $\mathbf{R}^m$ and hence subanalytic, and π is a subanalytic homeomorphism from $|K|$ to X such that for each $\sigma \in K$, $\pi|_{\sigma^\circ}$ is a C^ω diffeomorphism onto the image. Note that if X admits a subanalytic triangulation, X is locally closed in $\mathbf{R}^m$. A *subanalytic triangulation* of a subanalytic function f on X is a subanalytic triangulation (K, π) of X such that $f \circ \pi$ is linear on each simplex of K. If X is a compact semialgebraic set, we define *semialgebraic triangulations* of X and of a semialgebraic function on X in the same way. For a triangulation of a semialgebraic set closed in $\mathbf{R}^n$, we introduce the following semialgebraic cell triangulation. We can define a (non-cell) semialgebraic triangulation of a non-compact semialgebraic set in a weak sense (see II.4.1). But when we treat triangulations of semialgebraic functions on non-compact semialgebraic sets, this concept is too narrow because a semialgebraically triangulable function in the weak sense is always bounded (see II.4.5). We need the concept of a semialgebraic cell triangulation.

A *cell* in $\mathbf{R}^n$ is a set σ of the form:

$$\{x \in \mathbf{R}^n : f_1(x) \geq 0, \dots, f_k(x) \geq 0, \ f_{k+1}(x) = \cdots = f_l(x) = 0\},$$

where f_i are linear functions. Here we do not assume compactness. If we replace some of the above inequalities $f_1 \geq 0, \dots, f_k \geq 0$ with the corresponding equalities $f_1 = 0, \dots, f_k = 0$, then we obtain a *face* of σ. A *vertex* of a cell is its face of dimension 0. Note that σ is a convex PL manifold possibly with boundary, the boundary $\partial\sigma$ of σ is the union of all the proper faces of σ, and the interior σ° is $\sigma - \partial\sigma$. We call a set of the form σ° for some cell σ an *open cell*. A *rational cell* is a cell defined by a linear function with rational coefficients and rational constant terms. A *(rational) semilinear* set in $\mathbf{R}^n$ is a finite union of open (resp., rational) cells in $\mathbf{R}^n$. For a cell σ of the above form we have

$$\{x \in \mathbf{R}^n : f_1(x) > 0, \dots, f_k(x) > 0, \ f_{k+1}(x) = \cdots = f_l(x) = 0\}$$
$$= \sigma^\circ \quad \text{or} \quad = \varnothing,$$

and we can choose $f_1, \dots, f_l$ so that the first case holds.

The following statement and the concept of a derived subdivision are important to prove the forthcoming lemmas and theorems, but they look unnatural.

Statement. Assume that a cell σ of the above form is non-compact and is included in the quadrant $\{x_1 \geq 0, \dots, x_n \geq 0\}$. Let $a_1, \dots, a_m$ denote its vertices, and let σ' be the cell spanned by $a_1, \dots, a_m$. There exists a unique compact cell σ'' in $\{x_1 + \cdots + x_n = 1\}$ such that

$$\sigma = \sigma' + \mathbf{R}_+ \sigma'' = \{a + tb : a \in \sigma', \ t \geq 0, \ b \in \sigma''\}, \tag{$*$}$$

where $\mathbf{R}_+$ denotes the non-negative reals. We write then

$$\sigma = \sigma(a_1, \dots, a_m; \sigma'') = \sigma(\sigma', \sigma'').$$

When σ is compact we also write $\sigma = \sigma(\sigma, \varnothing)$. We define $\sigma(\varnothing, \sigma'') = \varnothing$. Note that if $\sigma = \mathbf{R}_+ \sigma''$ for a compact cell σ'' in $\{x_1 + \cdots + x_n = 1\}$, then $\sigma = \sigma(0, \sigma'')$, $\sigma' = \{0\}$ and for any compact cells σ' and σ'' in $\mathbf{R}^n$, $\sigma' + \mathbf{R}_+ \sigma''$ is a cell (see the note after I.3.4).

Proof of Statement. The set

$$\sigma'' = \bigcap_{i=1}^{\infty} \overline{\{x/(x_1 + \cdots + x_n) : x = (x_1, \dots, x_n) \in \sigma, \ x_1 + \cdots + x_n \geq i\}}$$

satisfies the equality $(*)$. First we show $\sigma \subset \sigma' + \mathbf{R}_+ \sigma''$. Let $x_0 \in \sigma$. For a large number m, the intersection $\sigma \cap \{x_1 + \cdots + x_n \leq m\}$ is a compact cell and is spanned by σ' and $\sigma \cap \{x_1 + \cdots + x_n = m\}$, because a compact cell is spanned by its vertices and the vertices of $\sigma \cap \{x_1 + \cdots + x_n \leq m\}$ are included in $\sigma' \cup \{x_1 + \cdots + x_n = m\}$. Hence there exist $a \in \sigma'$, $b \in \sigma \cap \{x_1 \cdots + x_n = m\}$ and $t \in [0, 1]$ such that

$$x_0 = a + t(b - a) = a + tp(b - a)(b - a)/p(b - a),$$

where p denotes the function $\mathbf{R}^n \ni x = (x_1, \dots, x_n) \to x_1 + \cdots + x_n \in \mathbf{R}$. Therefore, we have sequences $a_1, a_2, \dots$ in σ', $b_1, b_2, \dots$ in $\{x_1 + \cdots + x_n = 1\}$ and $t_1, t_2, \dots$ in $\mathbf{R}_+$ such that for each i, $x_0 = a_i + t_i b_i$, and for some number $m_i > i$, $m_i b_i + a_i$ is an element of $\sigma \cap \{x_1 + \cdots + x_n = m_i + p(a_i)\}$. Here we can choose $a_1, a_2, \dots$, $b_1, b_2, \dots$ and $t_1, t_2, \dots$ to converge to a, b and t, respectively, because σ' and $\{x_1 + \cdots + x_n = 1, \ x_1 \geq 0, \dots, x_n \geq 0\}$ are compact, and if $t_i \to \infty$, then $p(t_i b_i) \to \infty$. Then

$$a \in \sigma', \quad b \in \sigma'', \quad t \in \mathbf{R}_+ \quad \text{and} \quad x_0 = a + tb.$$

Hence $\sigma \subset \sigma' + \mathbf{R}_+ \sigma''$.

For the converse inclusion $\sigma \supset \sigma' + \mathbf{R}_+ \sigma''$, it suffices to show

$$f_1 \geq f_1(0), \dots, f_k \geq f_k(0), \ f_{k+1} = f_{k+1}(0), \dots, f_l = f_l(0) \quad \text{on} \quad \sigma''.$$

By easy calculations we see, moreover, that

$$\sigma'' = \{x_1 + \cdots + x_n = 1, \ f_1(x) \geq f_1(0), \dots, f_k(x) \geq f_k(0),$$
$$f_{k+1}(x) = f_{k+1}(0), \dots, f_l(x) = f_l(0)\}.$$

Hence (∗) holds true.

A cell σ'' in $\{x_1 + \cdots x_n = 1\}$ which satisfies (∗) is unique. Indeed, if $\sigma_1 \neq \sigma_2$ are cells in $\{x_1 + \cdots + x_n = 1\}$, the inequality

$$\sigma' + \mathbf{R}_+\sigma_1 \neq \sigma' + \mathbf{R}_+\sigma_2$$

easily follows. Moreover, if a cell σ'' in $\{x_1 + \cdots + x_n = 1\}$ satisfies (∗), it should equal the set:

$$\{x \in \mathbf{R}^n \colon x_1 + \cdots + x_n = 1,\ \sigma' + \mathbf{R}_+x \subset \sigma\}. \qquad \square$$

The expression of a point of $\sigma(\sigma', \sigma'') - \sigma'$ by $a + tb$, $a \in \sigma'$, $t > 0$ and $b \in \sigma''$, is not unique. For example,

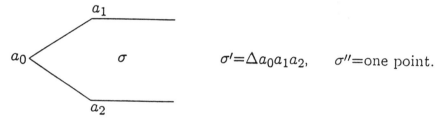

$$\sigma' = \Delta a_0 a_1 a_2, \qquad \sigma'' = \text{one point.}$$

The expression is unique if and only if $\dim \sigma' + \dim \sigma'' = \dim \sigma - 1$.

A *cell complex* C is a finite family of cells in some $\mathbf{R}^n$ satisfying the following conditions: (i) If $\sigma \in C$, all the faces of σ belong to C. (ii) If $\sigma \in C$ and $\sigma' \in C$, then $\sigma \cap \sigma'$ is empty or a face of both σ and σ'. A *rational cell complex* is a cell complex consisting of rational cells. For a cell complex C we define naturally the *underlying polyhedron* $|C|$, the *k-skeleton* C^k, a *cell map*, a *cell isomorphism*, a *cell subdivision* of C and a *cell decomposition*. A *semialgebraic cell triangulation* of a semialgebraic set X means a pair (C, π), where C is a cell complex and π is a semialgebraic homeomorphism from $|C|$ to X such that for each $\sigma \in C$, $\pi|_{\sigma^\circ}$ is a C^ω diffeomorphism onto the image. We define naturally a *semialgebraic cell triangulation* of a semialgebraic function.

Let C be a cell complex in $\mathbf{R}^n$ which is compatible with the hyperplanes $\{x_1 = 0\}, \ldots, \{x_n = 0\}$. We shall need a *derived subdivision* of C. Let us define a derived subdivision of C^k inductively on a non-negative integer k. A derived subdivision of C^0 is always C^0 itself. Assume there exists a derived subdivision $C^{k'}$ of C^k with the following properties.

(∗∗) Let $\sigma \in C^{k'}$ be in the quadrant $\{x_i \geq 0, i = 1, \ldots, n\}$, and set $\sigma = \sigma(\sigma', \sigma'')$. Then σ' is a simplex, and each point of $\sigma - \sigma'$ is uniquely described as

$$a + tb \quad \text{for} \quad a \in \sigma',\ t \geq 0 \ \text{and} \ b \in \sigma''.$$

If $\sigma \in C^k$ is contained in another quadrant, similar properties hold.

It follows from the unique expression property that if $0 \in \sigma'$, the linear space spanned by σ' and the linear space spanned by 0 and σ'' intersect only at the origin. Note that by this property, the set of faces of σ consists of all the cells of the form $\sigma(\text{a face of } \sigma', \text{a face of } \sigma'')$.

Let $\sigma \in C^{k+1} - C^k$, and let $\Sigma, \tilde{\Sigma}, \tilde{\Sigma}'$ denote the cell complexes generated by σ, $\Sigma - \sigma$, and the derived subdivision of $\tilde{\Sigma}$ induced by $C^{k'}$, respectively. We want to define a cell subdivision Σ' of Σ (which we call a *derived subdivision*) such that $\Sigma' = \tilde{\Sigma}'$ on $\partial\sigma$, and the above properties in $(**)$ are satisfied for Σ'. If the σ is compact, we define Σ' in the same way as in the simplicial complex case. So assume σ is non-compact. Since C is compatible with $\{x_1 = 0\}, \ldots, \{x_n = 0\}$, we suppose, in addition, that σ is included in the quadrant $\{x_i \geq 0, i = 1, \ldots, n\}$. Then we have $\sigma = \sigma(\sigma', \sigma'')$, where σ' is the cell spanned by the vertices of σ, and σ'' is a compact cell in $\{x_1 + \cdots + x_n = 1\}$. Let s be a point of σ°. Then

$$\sigma = \overline{s * \partial\sigma} \cup \sigma(s, \sigma'').$$

Hence it is natural to set

$$\Sigma' = \{\sigma(s, \sigma_1), \ \sigma_2, \ \overline{s * \sigma_2} : \sigma_1 \text{ face of } \sigma'', \ \sigma_2 \in \tilde{\Sigma}'\}.$$

We need to prove that the family Σ' is a cell complex with the properties in $(**)$. Let σ_1 be a face of σ'', and let σ_2, $\sigma_3 \in \tilde{\Sigma}'$.

Proof of $(**)$ and that every element of Σ' is a cell. We have $\sigma_2 = \sigma(\sigma_2', \sigma_2'')$ for some simplex σ_2' and some compact cell σ_2'' in $\{x_1 + \cdots + x_n = 1\}$. Since the point s is outside of the linear space spanned by σ_2, we have

$$\overline{s * \sigma_2} = \sigma(s * \sigma_2', \sigma_2''),$$

which is a cell and possesses the properties in $(**)$. Clearly every element of Σ' of other type is a cell and has $(**)$.   □

Proof that a face of a cell of Σ' belongs to Σ'. It suffices to consider only a face of the closure $\overline{s * \sigma_2}$. The face is of the form $\sigma(\text{a face of } s * \sigma_2', \text{a face of } \sigma_2'')$, and a face of $s * \sigma_2'$ is a face of σ_2' or of the form $s * (\text{a face of } \sigma_2')$. Hence the face belongs to Σ'.   □

Proof that the intersection of two cells of Σ' is a cell of Σ'. Let $\sigma_3 = \sigma(\sigma_3', \sigma_3'')$. It suffices to prove that the intersections:

$$\sigma(s, \sigma_1) \cap \overline{s * \sigma_2} \quad \text{and} \quad \overline{s * \sigma_2} \cap \overline{s * \sigma_3}$$

are elements of Σ' in the case where σ_2 and σ_3 are non-compact. Clearly

$$\overline{s * \sigma_i} - s * \sigma_i = \sigma(s, \sigma_i''), \quad i = 2, 3.$$

Since $\sigma(s, \sigma_1) \cap s * \sigma_2 = \varnothing$ (which follows from the property $\sigma(s, \sigma'') \cap s * \partial\sigma = \varnothing$), we have

$$\sigma(s, \sigma_1) \cap \overline{s * \sigma_2} = \sigma(s, \sigma_1) \cap (\overline{s * \sigma_2} - s * \sigma_2)$$
$$= \sigma(s, \sigma_1) \cap \sigma(s, \sigma_2'') = \sigma(s, \sigma_1 \cap \sigma_2'') \in \Sigma'.$$

Consider the intersection $\overline{s * \sigma_2} \cap \overline{s * \sigma_3}$. Divide it into the four sets: $s * \sigma_2 \cap s * \sigma_3$, $s * \sigma_2 \cap (\overline{s + \sigma_3} - s * \sigma_3)$, $(\overline{s * \sigma_2} - s * \sigma_2) \cap s * \sigma_3$ and $(\overline{s * \sigma_2} - s * \sigma_2) \cap (\overline{s * \sigma_3} - s * \sigma_3)$. We see that

$$s * \sigma_2 \cap s * \sigma_3 = s * (\sigma_2 \cap \sigma_3),$$
$$s * \sigma_2 \cap (\overline{s * \sigma_3} - s * \sigma_3) = s * \sigma_2 \cap \sigma(s, \sigma_3'') = \varnothing,$$
$$(\overline{s * \sigma_2} - s * \sigma_2) \cap s * \sigma_3 = \varnothing, \quad \text{and}$$
$$(\overline{s * \sigma_2} - s * \sigma_2) \cap (\overline{s * \sigma_3} - s * \sigma_3) = \sigma(s, \sigma_2'') \cap \sigma(s, \sigma_3'') = \sigma(s, \sigma_2'' \cap \sigma_3'').$$

There are four cases: (1) $\sigma_2 \cap \sigma_3 = \sigma_2'' \cap \sigma_3'' = \varnothing$; (2) $\sigma_2 \cap \sigma_3 = \varnothing$, $\sigma_2'' \cap \sigma_3'' \neq \varnothing$; (3) $\sigma_2 \cap \sigma_3 \neq \varnothing$, $\sigma_2'' \cap \sigma_3'' = \varnothing$; or (4) $\sigma_2 \cap \sigma_3 \neq \varnothing$, $\sigma_2'' \cap \sigma_3'' \neq \varnothing$. Case (1): $\overline{s * \sigma_2} \cap \overline{s * \sigma_3} = \varnothing \in \Sigma'$. Case (2): $\overline{s * \sigma_2} \cap \overline{s * \sigma_3} = \sigma(s, \sigma_2'' \cap \sigma_3'') \in \Sigma'$. Case (3): $\overline{s * \sigma_2} \cap \overline{s * \sigma_3} = s * (\sigma_2 \cap \sigma_3) = \overline{s * (\sigma_2 \cap \sigma_3)} \in \Sigma'$. Here the last equality follows from the fact that $\overline{s * \sigma_2} \cap \overline{s * \sigma_3}$ is closed in $\mathbf{R}^n$. Case (4): It suffices to prove that $\overline{s * (\sigma_2 \cap \sigma_3)}$ includes $\sigma(s, \sigma_2'' \cap \sigma_3'')$, that is, for every $t \geq 0$ and for every $b \in \sigma_2'' \cap \sigma_3''$, the point $s + tb$ is adherent to $s * (\sigma_2 \cap \sigma_3)$. Let $a \in \sigma_2 \cap \sigma_3$. For any number $u > 0$,

$$a + (t/u)b \in \sigma_2 \cap \sigma_3$$

by the definition of $\sigma = \sigma(\sigma', \sigma'')$. Hence, for every $0 < u < 1$,

$$(1 - u)s + u(a + (t/u)b) = s + tb + u(a - s) \in s * (\sigma_2 \cap \sigma_3),$$

which proves the adherence of $s + tb$. $\qquad\qquad\square$

A useful property of a derived subdivision is the unique expression property in (∗∗). We shall need the following stronger property in §V.1.

(∗∗∗) Let $\sigma = \sigma(\sigma', \sigma'')$, let $v_1, \dots, v_l$ denote the vertices of σ', and let $v_{l+1}, \dots, v_m$ denote the ones of σ''. The expression of a point of $\sigma - \sigma'$ by

$$\sum_{i=1}^{l} s_i v_i + t \sum_{i=l+1}^{m} s_i v_i, \quad s_i \leq 0, \ t > 0 \quad \text{with} \quad \sum_{i=1}^{l} s_i = 1 \text{ and } \sum_{i=l+1}^{m} s_i = 1$$

is unique.

Clearly (***) holds if and only if $\dim \sigma' + \dim \sigma'' = \dim \sigma - 1$, and σ' and σ'' are simplices . Hence, by the above proof, we can subdivide a derived subdivision so that (***) holds. But, in §V.1, we shall need to subdivide without introducing new vertices. We show that it is possible.

Let C be a cell complex in $\mathbf{R}^n$ compatible with $\{x_1 = 0\}, \ldots, \{x_n = 0\}$, and let C_1 denote the family of the cells σ'' and their faces for $\sigma = \sigma(\sigma', \sigma'') \in C$. Order the vertices of C and C_1 and suppose that we have inductively defined a cell subdivision $C^{k-1\prime}$ of C^{k-1} and a simplicial subdivision $C_1^{k-1\prime}$ of C_1^{k-1}, k a positive integer. Let $\sigma_1 \in C_1^k$ be a k-cell, and let v_1 be the first vertex of σ_1. Let F_1 be the union of faces of σ_1 which do not contain v_1. Then $F_1 \subset |C^{k-1}|$ and $\sigma_1 = v_1 * F_1$. Hence we can define $C_1^{k\prime}|_{\sigma_1}$ to be

$$\{v_1,\ v_1 * \sigma_1,\ \sigma_1 : \sigma_1 \in C_1^{k-1\prime}|_{F_1}\}.$$

By the ordering, $C_1^{k\prime}|_{\sigma_1}$ is an extension of $C_1^{k-1\prime}|_{\partial \sigma_1}$, and hence $C_1^{k\prime}$ is well-defined. We define C' inductively in the same way. Let $\sigma = \sigma(\sigma', \sigma'')$ be a k-cell of C and let v be the first vertex of σ. Let F be the union of faces of σ which do not contain v. Define $C^{k\prime}|_{\sigma}$ to be

$$\{v,\ \overline{v * \sigma_1},\ \sigma_1,\ \sigma(v, \sigma_2) : \sigma_1 \in C^{k-1\prime}|_F,\ \sigma_2 \in C_1'|_{\sigma''}\}.$$

In the same way as in the derived subdivision case, we see that $C^{k\prime}$ is a cell complex. Thus we obtain the required subdivision C'.

To avoid confusion, we call a usual cell complex with compact cells a *usual cell complex*. For a usual cell complex we do not assume finiteness, and we define some terminology in the same way as simplicial complexes and cell complexes. We remark that a derived subdivision of a usual cell complex is a simplicial complex.

Our cells and cell complexes have properties similar to usual ones and some advantageous properties. For example, σ° and $\partial \sigma$ are the interior and the boundary, respectively, of σ in L as a topological subset, where σ is a cell and L is the linear space spanned by σ; if $\sigma \subset \sigma'$ for another cell σ', then $\sigma^\circ \subset \sigma'^\circ$ or $\sigma^\circ \subset \partial \sigma'$; the inverse image of σ under a linear map p between Euclidean spaces is a cell, and the family of the faces of $p^{-1}(\sigma)$ consists of $p^{-1}(\sigma')$ for all the faces σ' of σ. The following lemmas and corollary I.3.1, ..., I.3.7 explain other properties which are not trivial.

Lemma I.3.1. *Let σ be a cell in $\mathbf{R}^n$, and let $p \colon \mathbf{R}^n \to \mathbf{R}^m$ be a linear map. The images $p(\sigma)$ and $p(\sigma^\circ)$ are a cell and its interior, respectively.*

Proof. We can assume that $\sigma \neq \varnothing$, $\dim \sigma = n = m + 1 > 0$, and p is the projection which forgets the first factor of $\mathbf{R}^n$. Then $\dim p(\sigma) = m$,

$$\sigma = \{f_1(x) \geq 0, \ldots, f_k(x) \geq 0\}, \quad \text{and}$$
$$\sigma^\circ = \{f_1(x) > 0, \ldots, f_k(x) > 0\},$$

for some linear functions $f_1, \ldots, f_k$. Set $x' = (x_2, \ldots, x_n)$. By changing the order of $f_1, \ldots, f_k$, we can assume, moreover, that the x_1-coefficients of f_i do not vanish for only $i = 1, \ldots, k'$ and there exist linear functions $g_1(x'), \ldots, g_{k'}(x')$ such that

$$\{f_i(x) \geq 0\} = \begin{cases} \{x_1 \geq g_i(x')\}, & 1 \leq i \leq k'' \\ \{x_1 \leq g_i(x')\}, & k'' + 1 \leq i \leq k'. \end{cases}$$

Then

$$p(\sigma) = \{g_i(x') \leq g_{i'}(x'), \ 1 \leq i \leq k'', k''$$
$$+ 1 \leq i' \leq k', \ f_{k'+1}(x') \geq 0, \ldots, f_k(x') \geq 0\} \quad \text{and}$$
$$p(\sigma^\circ) = \{g_i(x') < g_{i'}(x'), \ 1 \leq i \leq k'', k''$$
$$+ 1 \leq i' \leq k', \ f_{k'+1}(x') > 0, \ldots, f_k(x') > 0\}.$$

Hence the lemma follows because the last set is not empty. □

Lemma I.3.2. *Given cells $\sigma_1, \ldots, \sigma_k$ in $\mathbf{R}^n$, $\mathbf{R}^n$ admits a cell decomposition C such that each σ_i or σ_i° is the union of some cells or open cells, respectively, of C. If $\sigma_1, \ldots, \sigma_k$ are rational, we can choose rational C.*

Proof. For the former statement, it suffices to prove that each σ_i is the union of some cells of some C, because for each $\sigma \in C$, if $\sigma \subset \sigma_i$, then $\sigma^\circ \subset \sigma_i^\circ$ or $\sigma^\circ \subset \partial \sigma_i$. Let the cells $\sigma_1, \ldots, \sigma_k$ be defined by linear functions $f_1, \ldots, f_l$ on $\mathbf{R}^n$. Let α be a partition of the numbers $\{1, \ldots, l\}$ into three parts $\alpha_1, \alpha_2, \alpha_3$, and let A denote the family of such α's. For each $\alpha = \{\alpha_1, \alpha_2, \alpha_3\} \in A$, set

$$\sigma_\alpha = \{x \in \mathbf{R}^n \colon f_i(x) = 0 \text{ for } i \in \alpha_1, \quad f_i(x) \geq 0 \text{ for } i \in \alpha_2,$$
$$\text{and} \quad f_i(x) \leq 0 \text{ for } i \in \alpha_3\}.$$

The family C of all the σ_α, $\alpha \in A$, satisfies the conditions of cell complexes, and by the above proof each σ_i is the union of some cells of C. □

It clearly follows:

Corollary I.3.3. *Each connected component of a semilinear set $X \subset \mathbf{R}^n$ is semilinear. If X is rational, so is each connected component of X.*

Lemma I.3.4 (Cell decomposition of a semialgebraic polyhedron). *A semialgebraic closed polyhedron in $\mathbf{R}^n$ is a semilinear set.*

Proof. Let X be a semialgebraic closed polyhedron in $\mathbf{R}^n$. We prove the lemma by induction on $k = \dim X$. If $k = 0$, the lemma is trivial. So assume the lemma for any semialgebraic closed polyhedron of dimension $< k$. Let $\Sigma_\omega X$ denote the C^ω singular point set of X. This set is closed, semialgebraic and of dimension $< k$ by (I.2.9.5), and it is a polyhedron for the following reason.

Let K be a simplicial decomposition of X. It suffices to prove that if $\sigma^\circ \cap (X - \Sigma_\omega X) \neq \varnothing$, $\sigma \in K$, then $\sigma^\circ \subset X - \Sigma_\omega X$. Let $x \in \sigma^\circ \cap (X - \Sigma_\omega X)$. As X is non-singular at x, we have an open neighborhood U of x in X which is an open subset of a k-dimensional linear space Π in $\mathbf{R}^n$, because a connected polyhedral C^ω submanifold of $\mathbf{R}^n$ is an open subset of some linear space in $\mathbf{R}^n$. For proof of the inclusion $\sigma^\circ \subset X - \Sigma_\omega X$, it suffices to show that U can be chosen so that $\sigma^\circ \subset U$. We shall choose U to be $|\operatorname{st}(x, K)| - |\operatorname{lk}(x, K)|$. Note that $|\operatorname{st}(x, K)| = x * |\operatorname{lk}(x, K)|$. First we shrink U so small that $U \subset |\operatorname{st}(x, K)|$. It follows that

$$\Pi = \{tx + (1 - t)z : t \leq 1, \ z \in |\operatorname{lk}(x, K)|\}$$

because U is open in $|\operatorname{st}(x, K)|$ and because Π is the union of ∇xy, $y \in \partial U$. Therefore,

$$\Pi = \{tx + (1 - t)z : t \in \mathbf{R}, \ z \in |\operatorname{lk}(x, K)|\}$$

because Π is linear. Hence $|\operatorname{st}(x, K)| - |\operatorname{lk}(x, K)|$ is an open subset of Π. On the other hand, this set is open in X. Therefore, we can replace U by $|\operatorname{st}(x, K)| - |\operatorname{lk}(x, K)|$, which contains σ°. This shows that $\Sigma_\omega X$ is a polyhedron.

By the induction hypothesis $\Sigma_\omega X$ is semilinear in $\mathbf{R}^n$. By (I.2.9.3) $X - \Sigma_\omega X$ has finitely many connected components $W_1, \ldots, W_l$, which are polyhedral C^ω submanifolds of $\mathbf{R}^n$. As noted above, each W_i is an open subset of some linear space Π_i in $\mathbf{R}^n$. Clearly W_i is a connected component of $\Pi_i - \Sigma_\omega X$. Hence by I.3.3, each $\overline{W_i}$ and hence X are semilinear in $\mathbf{R}^n$. $\square$

Note. A convex semialgebraic polyhedron closed in $\mathbf{R}^n$ is a cell.

Proof. Let X be the polyhedron. Assume $X \neq \mathbf{R}^n$ and that $\mathbf{R}^n$ is the minimal linear space containing X. Then it is easy to show that $\dim X = n$. By the above proof it is semilinear. Let $\sigma_1, \ldots, \sigma_k$ be the cells of dimension $n - 1$, and for each i, let Π_i denote the linear space spanned by σ_i. For each i, σ_i° has a neighborhood U in $\mathbf{R}^n$ such that $X \cap U$ lies on one side of Π_i, i.e., we have a linear function f_i on $\mathbf{R}^n$ such that $f_i = 0$ on Π_i and $f_i \geq 0$ on $X \cap U$. Then by the convexity of X we see, moreover, that $f_i \geq 0$ globally on X. We want to prove $X = \{f_i \geq 0, \ i = 1, \ldots, k\}$. The inclusion $X \subset \{f_i \geq 0\}$ is clear, and it suffices to show $\{f_i > 0\} \subset X$. Assume there exists $x \in \{f_i > 0\} - X$. By convexity of X, for each $y \in X - \Sigma_\omega X$, there exists a point $\alpha(y) \in \Delta xy$ such that

$$\Delta x \alpha(y) - \alpha(y) \subset \mathbf{R}^n - X, \quad \text{and} \quad \Delta \alpha(y) y \subset X.$$

The union $\cup_{y \in X - \Sigma_\omega X} \alpha(y)$ is a subpolyhedron of $\Sigma_\omega X$ of dimension $n - 1$. Hence there exists $y \in X - \Sigma_\omega X$ such that $\alpha(y) \in \sigma_i^\circ$ for some i. Since $f_i \geq 0$ on $\Delta \alpha(y)y$ and $f_i > 0$ at y, we have $f_i < 0$ at x, which is a contradiction. $\square$

Lemma I.3.5 (Non-closed case of Lemma I.3.4). *A semialgebraic subset of $\mathbf{R}^n$ which is a finite union of open simplices locally at each point of $\mathbf{R}^n$ is semilinear.*

Let $Y \subset \mathbf{R}^n$ and $Z \subset \mathbf{R}^n$ be disjoint semialgebraic sets such that Y is closed in $\mathbf{R}^n$ and Z is a closed polyhedron in $\mathbf{R}^n - Y$. Then Z is a finite union of connected components of sets of the form $\sigma - Y$, where σ is a cell in $\mathbf{R}^n$.

Proof. Let X be a semialgebraic set in $\mathbf{R}^n$ which is a finite union of open simplices locally at each point of $\mathbf{R}^n$. Consider the family of sets:

$$X_0 = X, \quad \text{and} \quad X_i = \overline{X_{i-1}} - X_{i-1}, \quad i = 1, 2, \dots .$$

The sets X_i and $\overline{X_i}$, $i = 1, 2, \dots$, also are semialgebraic, they are finite unions of open simplices and simplices , respectively, locally at each point of $\mathbf{R}^n$, and the family of these sets is finite by (I.2.9.2). Moreover, we have

$$X = \overline{X_0} - X_1 = \overline{X_0} - (\overline{X_1} - X_2) = (\overline{X_0} - \overline{X_1}) \cup X_2 =$$
$$\cdots = (\overline{X_0} - \overline{X_1}) \cup (\overline{X_2} - \overline{X_3}) \cup \cdots .$$

Hence the first half of the lemma follows from I.3.2 and I.3.4 applied to the $\overline{X_0}, \overline{X_1}, \dots$.

We prove the latter half in the same way as the proof of I.3.4, as follows. We see that $\Sigma_\omega Z$, the C^ω singular point set of the Z, is disjoint to Y and it is a semialgebraic closed polyhedron in $\mathbf{R}^n - Y$. Hence by induction on the dimension of Z we can assume that $\Sigma_\omega Z$ is a finite union of connected components of sets of the form $\sigma - Y$, where σ is a cell in $\mathbf{R}^n$. Let $\{\sigma_\alpha\}_{\alpha \in A}$ denote the cells which appear here. We see also that each connected component W_i of $Z - \Sigma_\omega Z$ is an open subset of a linear space $\Pi_i \subset \mathbf{R}^n$. Then W_i is a connected component of $\Pi_i - Y - \Sigma_\omega Z$.

It suffices to prove that each W_i is a finite union of connected components of sets of the form $\sigma^\circ - Y$, where σ is a cell in $\mathbf{R}^n$. Apply I.3.2 to the family $\{\Pi_i \cap \sigma_\alpha\}_{\alpha \in A}$. We have a cell decomposition C_i of Π_i such that each $\Pi_i \cap \sigma_\alpha$ is the union of some cells of C_i. This implies that $\Pi_i \cap \Sigma_\omega Z$ is the union of some connected components of $\sigma - Y$, $\sigma \in C_i$. Therefore, W_i is the union of some connected components of $\sigma^\circ - Y$, $\sigma \in C_i$. $\square$

Lemma I.3.6 (Cell decomposition of a semialgebraic PL map). *Let $X \subset \mathbf{R}^m$ and $Y \subset \mathbf{R}^n$ be semialgebraic closed polyhedra, and let $f \colon X \to Y$ be a semialgebraic PL map. There exist cell decompositions C of X and D of Y such that f is a cell map from C to D.*

Proof. The graph of f is also a semialgebraic closed polyhedron in $\mathbf{R}^m \times \mathbf{R}^n$. Hence by I.3.2 and I.3.4 we have a cell decomposition E of the graph. Let p and q denote the projections $\mathbf{R}^m \times \mathbf{R}^n \to \mathbf{R}^m$ and $\mathbf{R}^m \times \mathbf{R}^n \to \mathbf{R}^n$, respectively. For each $\sigma \in E$, $p(\sigma)$ and $q(\sigma)$ are cells in $\mathbf{R}^m$ and $\mathbf{R}^n$, respectively, by I.3.1. $f(p(\sigma)) = q(\sigma)$, $f|_{p(\sigma)}$ is linear, and the image $\{p(\sigma) \colon \sigma \in E\}$ is a cell decomposition of X because for each $\sigma \in E$, $p|_\sigma \colon \sigma \to p(\sigma)$ is a linear homeomorphism. Apply, once more, I.3.2 and I.3.4 to the family $\{q(\sigma) \colon \sigma \in E\}$, and let D be a cell decomposition of Y such that each $q(\sigma)$ is the union of some cells of D. Set

$$C = \{f^{-1}(\sigma) \cap p(\sigma') \colon \sigma \in D,\ \sigma' \in E\}.$$

Then C is a cell decomposition of X which together with D fulfills the requirements of the lemma. $\square$

Lemma I.3.7 (Non-closed case of I.3.6). *Let $X \subset \mathbf{R}^m$ be a semilinear set, and let $f \colon X \to \mathbf{R}^n$ be a semialgebraic PL map. There exists a cell decomposition C of $\mathbf{R}^m$ such that X is the union of some open cells of C, and for each $\sigma \in C$, $f|_{\sigma \cap X}$ is linear.*

Let $Y \subset \mathbf{R}^m$ and $Z \subset \mathbf{R}^m$ be disjoint semialgebraic sets such that Y is closed in $\mathbf{R}^m$ and Z is a closed polyhedron in $\mathbf{R}^m - Y$. Let $g \colon Z \to \mathbf{R}^n$ be a semialgebraic PL map. There exists a cell decomposition D of $\mathbf{R}^m$ such that Z is the union of some connected components of $\sigma - Y$, $\sigma \in D$, and for each $\sigma \in D$, $f|_{\sigma \cap Z}$ is linear.

Proof. For the first half of the lemma, it suffices to prove that the graph F of f is a finite union of open simplices locally at each point of $\mathbf{R}^m \times \mathbf{R}^n$. Indeed, if F is so, by I.3.5, F is semilinear, and hence by I.3.2, $\mathbf{R}^m \times \mathbf{R}^n$ admits a cell decomposition C_1 such that F is the union of some open cells of C_1. Then a cell decomposition C of $\mathbf{R}^m$ which is compatible with $\{p(\sigma) \colon \sigma \in C_1\}$, fulfills the requirements, where p is the projection $\mathbf{R}^m \times \mathbf{R}^n \to \mathbf{R}^m$.

We will prove the above property of F. Let $(x_0, y_0) \in \mathbf{R}^m \times \mathbf{R}^n$. Since the problem is local at (x_0, y_0), we can assume X is a finite union of open simplices in $\mathbf{R}^m$. Moreover, since it suffices to consider each of these open simplices and the restriction of f to it, we assume $X = \sigma^\circ$ for a simplex σ in $\mathbf{R}^n$. Now $\partial \sigma \times \mathbf{R}^n$ and F satisfy the conditions in the latter half of I.3.5, namely, $\partial \sigma \times \mathbf{R}^n$ and F are disjoint, $\partial \sigma \times \mathbf{R}^n$ is a closed semialgebraic set in $\mathbf{R}^m \times \mathbf{R}^n$, and F is a closed polyhedron in $\mathbf{R}^m \times \mathbf{R}^n - \partial\sigma \times \mathbf{R}^n$. Hence by I.3.5 F is a finite union of open simplices locally at (x_0, y_0).

The latter half is immediate by the above arguments. $\qquad\square$

We state four facts of PL topology which are well-known to PL topologists. However, since we have no good reference, we give their proofs.

Lemma I.3.8. *Let A, $B \subset \mathbf{R}^n$ be compact polyhedra such that A and $A \cup B$ are PL n-balls, $A \cap B$ is a PL $(n-1)$-ball and contained in ∂A, and $(A \cap B)^\circ$ is contained in $(A \cup B)^\circ$. Then B is a PL n-ball.*

Proof. Recall two theorems of PL topology:

(i) Uniqueness of regular neighborhood, (3.8 in [R-S]). Let $X \supset Y$ be a polyhedron and a closed subpolyhedron, and let U_1 and U_2 be regular neighborhoods of Y in X. There exists a PL homeomorphism $h\colon X \to X$ which carries U_1 onto U_2 and is the identity on Y.

(ii) Simplicial neighborhood theorem, (3.11 in [R-S]). Let X be a compact polyhedron in a PL manifold M, and let U be a compact polyhedral neighborhood of X in M. Assume that U is a PL manifold with boundary and that there are a simplicial complex K, a full subcomplex L of K and a PL homeomorphism $\pi\colon |K| \to U$ such that $\pi(|L|) = X$, $N(L, K) = K$ and $\pi(|\partial N(L, K)|) = \partial U$. Then U is a regular neighborhood of Y in X.

These theorems lead to the following facts (iii), (iv) and (v). For each $i = 1, 2$, let $\sigma_i^n \supset \sigma_i^{n-1}$ denote an n-simplex and an $(n-1)$-face.

(iii) Let B_1 and B_2 be PL n-balls in $\mathbf{R}^n$ such that $B_1 \cap B_2$ is a PL $(n-1)$-ball. Then $B_1 \cup B_2$ is a PL n-ball.

Proof of (iii). We have the inclusion $\partial B_1 \supset B_1 \cap B_2$ because $\dim B_1^\circ \cap B_2 = n$ if $B_1^\circ \cap B_2 \neq \varnothing$. Apply the facts (i) and (ii) to ∂B_1, $B_1 \cap B_2$ and a point $x \in (B_1 \cap B_2)^\circ$. We see that the pair $(\partial B_1, B_1 \cap B_2)$ is PL homeomorphic to the pair $(\partial\sigma_1^n, \sigma_1^{n-1})$. Moreover, by cone extension, we can extend this PL homeomorphism to a PL homeomorphism $\varphi\colon (B_1, B_1 \cap B_2) \to (\sigma_1^n, \sigma_1^{n-1})$. In the same way we prove that the pairs $(B_2, B_1 \cap B_2)$ and $(\sigma_2^n, \sigma_2^{n-1})$ are PL homeomorphic. This implies that we can regard B_2 as a cone with base $B_1 \cap B_2$. Choose σ_1^n and σ_2^n in $\mathbf{R}^n$ so that $\sigma_1^n \cup \sigma_2^n$ is a PL n-ball and

$$\sigma_1^n \cap \sigma_2^n = \sigma_1^{n-1} = \sigma_2^{n-1}.$$

By the Alexander trick we can extend the PL homeomorphism $\varphi|_{B_1 \cap B_2}\colon B_1 \cap B_2 \to \sigma_1^{n-1} = \sigma_2^{n-1}$ to a PL homeomorphism: $B_2 \to \sigma_2^n$. Hence we have a PL homeomorphism: $B_1 \cup B_2 \to \sigma_1^n \cup \sigma_2^n$, namely, $B_1 \cup B_2$ is a PL n-ball.

(iv) Let $B^n \subset S^n$ be a PL n-ball in a PL n-sphere. The pair (S^n, B^n) is PL homeomorphic to $(\partial\sigma^{n+1}, \sigma^n)$, where σ^n is an n-face of an $(n+1)$-simplex σ^{n+1}. We can prove this in the same way as the above fact (iii).

(v) ($3.13_\mathbf{n}$ in [R-S]). In (iv), $\overline{S^n - B^n}$ is a PL n-ball. This is immediate by (iv).

Now we begin to prove the lemma. Let A and B be in a PL n-sphere S^n, which is possible because any compact polyhedron in $\mathbf{R}^n$ can be PL imbedded in S^n. Set $C = \overline{S^n - A - B}$. By the fact (v), C and $\overline{\partial A - (A \cap B)}$ are PL n- and $(n - 1)$-balls, respectively, and by (iv),

$$\partial C = \partial(A \cup B), \quad \text{and} \quad \partial(\overline{\partial A - (A \cap B)}) = \partial(A \cap B).$$

By (v) for proof of the lemma, it suffices to prove that
(vi) $A \cup C$ is a PL n-ball and
(vii) $B = \overline{S^n - A - C}$.
Moreover, by the fact (iii) and by the fact that C is a PL n-ball, the condition (vi) is equivalent to the one that
(viii) $A \cap C$ is a PL $(n - 1)$-ball.
If (vi) holds, then by (iv), the condition (vii) is equivalent to the condition:

$$(A \cup C) \cap B = \partial(A \cup C), \quad \text{and} \quad A \cup C \cup B = S^n. \tag{ix}$$

Proof of (viii). Since
$$(A \cup B)^\circ \cap C = \varnothing,$$
the inclusion $(A \cap B)^\circ \subset (A \cup B)^\circ$ implies that

$$(A \cap B)^\circ \cap (A \cap C) = \varnothing. \tag{x}$$

By the definition of C we have $A \cup B \cup C = S^n$ and hence $B \cup C \supset \overline{S^n - A}$. Hence

$$(A \cap B) \cup (A \cap C) = A \cap (B \cup C) \supset A \cap (\overline{S^n - A}) = \partial A.$$

By assumption, $A \cap B \subset \partial A$, and by the definition of C, $A \cap C \subset \partial A$. Therefore,
$$(A \cap B) \cup (A \cap C) = \partial A. \tag{xi}$$

From (iv), (x) and (xi), it follows that

$$A \cap C = \overline{\partial A - (A \cap B)}.$$

So the property (viii) follows from the fact (v) because $A \cap B$ is a PL $(n-1)$-ball in ∂A.

Proof of (ix). The latter equality is clear. From the facts (iv), (vi) and $B \cup (A \cup C) = S^n$ the inclusion $B \supset \partial(A \cup C)$ follows. Hence we have the inclusion:
$$(A \cup C) \cap B \supset \partial(A \cup C).$$

Therefore, it remains to prove that

$$(A \cup C) \cap B \subset \partial(A \cup C). \tag{xii}$$

For this, it suffices to prove that B is everywhere locally of dimension n. Indeed, if B is so and if (xii) were not true, $(A \cup C)^\circ \cap B$ is not empty and of dimension $= n$ because $(A \cup C)^\circ$ is an open subset of S^n. On the other hand,

$$\dim(A \cup C) \cap B = \max\{\dim(A \cap B), \dim(C \cap B)\} = n - 1,$$

because

$$C \cap B \subset (A \cup B) \cap C = \partial C.$$

That is a contradiction.

Let $x \in B$. We will prove that B is of dimension n locally at x. If $x \in B - A$, we have a small neighborhood U of x in the n-manifold with boundary $A \cup B$ such that $U \cap A = \varnothing$, namely, $U \subset B$. Hence the local dimension equals the dimension n of U. Assume $x \in A \cap B$. We can suppose x is contained in $(A \cap B)^\circ$. This is because $A \cap B$ is a PL manifold with boundary and hence x is adherent to $(A \cap B)^\circ$. By assumption $(A \cap B)^\circ$ is included in $(A \cup B)^\circ$, and the sets $A \cup B$ and A are a PL n-manifold with boundary and a PL n-submanifold with boundary. Hence if a small polyhedral neighborhood of x in $(A \cup B)^\circ$ is a cone with vertex x and base D, D is a PL $(n-1)$-sphere and $D \cap A$ is a PL $(n-1)$-ball. Consequently, $D \neq D \cap A$. This implies that $x \in \overline{B - A}$ and reduces the problem to the case $x \in B - A$. $\qquad\qquad\square$

Lemma I.3.9 (PL triviality). *Let K be a simplicial complex. Set $X = |K|$. Let $f \colon X \to \mathbf{R}$ be a PL function such that all the restrictions $f|_\sigma$, $\sigma \in K$, are linear. Assume $[0,1] \cap f(K^0) = \varnothing$. There exists a PL homeomorphism $\pi \colon f^{-1}(0) \times [0,1] \to f^{-1}([0,1])$ such that*

$$f \circ \pi(x,t) = t, \qquad \pi(\cdot, 0) = \mathrm{id} \quad and$$

$$\pi((\sigma \cap f^{-1}(0)) \times [0,1]) = \sigma \cap f^{-1}([0,1]) \quad for \quad \sigma \in K.$$

Proof. Set

$$L_1 = \{\sigma \cap f^{-1}(0) \colon \sigma \in K\} \times \{0,\ 1,\ [0,1]\}, \quad and$$
$$L_2 = \{\sigma \cap f^{-1}(0),\ \sigma \cap f^{-1}(1),\ \sigma \cap f^{-1}([0,1]) \colon \sigma \in K\}.$$

They are usual cell complexes, and the map $\varphi\colon L_1 \to L_2$, defined by

$$\varphi(\sigma \cap f^{-1}(0), \alpha) = \sigma \cap f^{-1}(\alpha) \quad \text{for} \quad \sigma \in K, \qquad \alpha = 0, \ 1 \text{ or } [0,1]$$

is bijective by the assumptions in the lemma. Define derived subdivisions L_1' and L_2' of L_1 and L_2, respectively, as follows. For each $\sigma \in L_i$, $i = 1, 2$, choose the new vertex v_σ of L_i' in σ° so that (i) $p \circ v_\sigma = 1/2$ if $\sigma \in L_1$ with $p(\sigma) = [0,1]$, (ii) $f \circ v_\sigma = 1/2$ if $\sigma \in L_2$ with $f(\sigma) = [0,1]$, and (iii) $q \circ v_\sigma = v_{\varphi(\sigma)}$ if $\sigma \in L_1$ with $p(\sigma) = 0$, where p and q are the projections $f^{-1}(0) \times [0,1] \to [0,1]$ and $f^{-1}(0) \times [0,1] \to f^{-1}(0)$, respectively. The one-to-one correspondence $L_1'^0 \ni v_\sigma \to v_{\varphi(\sigma)} \in L_2'^0$ between the 0-skeletons can be extended to a simplicial isomorphism $\pi\colon L_1' \to L_2'$. Clearly $f \circ \pi(x,t) = t$ and $\pi(\cdot, 0) = \mathrm{id}$ because

$$p \circ v_\sigma = f \circ v_{\varphi(\sigma)} \quad \text{for} \quad \sigma \in L_1, \quad \text{and}$$

$$v_\sigma = (v_{\varphi(\sigma)}, 0) \quad \text{for} \quad \sigma \in L_1 \quad \text{with} \quad p(\sigma) = 0.$$

The last equality in the lemma also is clear. $\square$

Let $X \supset Y$ be a polyhedron and a subpolyhedron, and let f_1 and f_2 be PL functions on X. We say that f_1 and f_2 are *locally PL equivalent* at Y if there exists a PL homeomorphism π of X such that $f_1 \circ \pi = f_2$ on a neighborhood of Y in X and $\pi = \mathrm{id}$ on Y. We call f_1 and f_2 *R-L PL equivalent* if there is PL homeomorphisms π_1 of X and π_2 of $\mathbf{R}$ such that $f_1 \circ \pi_1 = \pi_2 \circ f_2$.

Lemma I.3.10 (PL equivalence). *Let K be a simplicial complex. Set $X = |K|$. Let f_1 and f_2 be PL functions on X such that*

$$f_1^{-1}(0) = f_2^{-1}(0) = Y,$$

$$\{f_1 > 0\} = \{f_2 > 0\}, \quad \text{and} \quad \{f_1 < 0\} = \{f_2 < 0\}.$$

Then f_1 and f_2 are locally PL equivalent at Y. Moreover, the PL homeomorphism of equivalence π can be chosen so that

$$\pi = \mathrm{id} \quad \text{on} \quad Y \cup \overline{U^c}, \quad \text{and}$$

$$\pi(\sigma) = \sigma \quad \text{for} \quad \sigma \in K,$$

where U is a given neighborhood of Y in X.

Proof. Let K_1 and K_2 be simplicial decompositions of X such that all the restrictions $f_1|_{\sigma_1}$ and $f_2|_{\sigma_2}$, $\sigma_i \in K_i$, $i = 1, 2$, are linear. Replace K with a simplicial subdivision of the usual cell complex:

$$\{\sigma \cap \sigma_1 \cap \sigma_2 \cap \sigma_3 \colon \sigma \in K, \ \sigma_i \in K_i, \ i = 1, 2,$$

$$\sigma_3 \in \{f_1^{-1}(]-\infty, 0[), \ f_1^{-1}(0), \ f_1^{-1}(]0, \infty[)\}\}.$$

Then we can assume that all $f_i|_\sigma$, $i = 1, 2$, $\sigma \in K$, are linear and there is a subcomplex L of K such that $|L| = Y$. Moreover, after replacing K and U with some subdivision and a smaller neighborhood, respectively, we assume $U = |N(L, K)|$.

Case where X is compact. Let ε be a small positive number such that f_1 or f_2 take no values in $[-\varepsilon, \varepsilon] - 0$ on the vertices K^0. Note that a simplex σ of K is included in $Y \cup \overline{U^c}$ if and only if ε or $-\varepsilon$ is not contained in $f_i(\sigma)$ for each $i = 1, 2$. Choose derived subdivisions K_1 and K_2 of K as follows. For each $\sigma \in K$, let $v_{i\sigma}$ denote the new vertices of K_i, $i = 1, 2$, respectively, in the interior σ°. Choose $v_{i\sigma}$ so that $v_{1\sigma} = v_{2\sigma}$ if $\sigma \subset Y \cup \overline{U^c}$ and that $f_i(v_{i\sigma}) = \pm\varepsilon$ otherwise. The simplicial isomorphism $\pi\colon K_2 \to K_1$, defined by $\pi(v_{2\sigma}) = v_{1\sigma}$ for $\sigma \in K$, fulfills the requirements.

In the above arguments, if $f_1 = f_2$ on a simplex σ of K, let us choose the $v_{1\sigma}$, $v_{2\sigma}$ so that $v_{1\sigma} = v_{2\sigma}$. Then $\pi = \mathrm{id}$ on such σ.

General case. We want to construct the π on $|K^k|$ by induction on a nonnegative integer k. If $k = 0$, the construction is trivial. Hence we assume there exists a PL homeomorphism π_{k-1} of $|K^{k-1}|$ such that

$$f_1 \circ \pi_{k-1} = f_2 \quad \text{on a neighborhood of} \ \ Y \cap |K^{k-1}| \ \ \text{in} \ \ |K^{k-1}|,$$

$$\pi_{k-1} = \mathrm{id} \quad \text{on} \quad (Y \cup \overline{U^c}) \cap |K^{k-1}|, \ \ \text{and}$$

$$\pi_{k-1}(\sigma) = \sigma \quad \text{for} \quad \sigma \in K^{k-1}.$$

Since each simplex of K is the cone whose base is its boundary and whose vertex is a point in its interior, we can extend the π_{k-1} inductively by cone extension to a PL homeomorphism $\tilde\pi_{k-1}$ of X so that $\tilde\pi_{k-1}(\sigma) = \sigma$ for $\sigma \in K$, and $\tilde\pi_{k-1} = \mathrm{id}$ on $Y \cup \overline{U^c}$. Replace f_1 with $f_1 \circ \tilde\pi_{k-1}$. We can assume $f_1 = f_2$ on $|K^{k-1}|$. Here we have to note that f_1 is not necessarily linear on a simplex of $K - K^{k-1}$.

Let $\sigma \in K^k - K^{k-1}$. For proof of the lemma, it suffices to construct a PL homeomorphism π_σ of σ so that

$$f_1 \circ \pi_\sigma = f_2 \quad \text{on a neighborhood of} \ \ Y \cap \sigma \ \ \text{in} \ \ \sigma, \ \ \text{and}$$

$$\pi_\sigma = \mathrm{id} \quad \text{on} \quad (Y \cup \overline{U^c} \cup \partial\sigma) \cap \sigma.$$

Thus we have reduced the problem to the case of compact X. The new condition on π_σ is only that $\pi_\sigma = \mathrm{id}$ on $\partial\sigma$. But this condition is satisfied by the note at the end of the above proof in the special case because $f_1 = f_2$ on $\partial\sigma$. $\qquad\Box$

Lemma I.3.11 (Cardinal of PL equivalence classes). *Let X be a non-empty compact polyhedron of dimension > 0. There are only a countable number of R-L PL equivalence classes of PL functions on X.*

Proof. Clear by the following two facts.

For a positive integer k the potency of the set of simplicial isomorphism classes of all simplicial complexes in $\mathbf{R}^n$ consisting of at most k simplices is finite.

Let K be a finite simplicial complex in $\mathbf{R}^n$, and let f_1 and f_2 be PL functions on $|K|$ such that all the restrictions $f_i|_\sigma$, $i = 1, 2$, $\sigma \in K$, are linear. Assume that for every pair of vertices v_1 and v_2 of K, $f_1(v_1) < f_1(v_2)$ if and only if $f_2(v_1) < f_2(v_2)$. There exists a PL homeomorphism π of $\mathbf{R}$ such that $f_1 = \pi \circ f_2$. $\square$

We shall frequently use the following lemma, which was already proved in a construction of a cell subdivision.

Lemma I.3.12 (Simplicial subdivision). *A usual cell complex can be subdivided to a simplicial complex without introducing new vertices.*

We devote the rest of this section to C^∞ triangulations and to proving I.3.13, I.3.20 and I.3.21. A C^∞ *map* $f\colon K \to \mathbf{R}^n$ is a continuous map $f\colon |K| \to \mathbf{R}^n$ such that all the restrictions $f|_\sigma$, $\sigma \in K$, are of class C^∞. Let $b \in |K|$. We define $df_b\colon |\mathrm{st}(b, K)| \to \mathbf{R}^n$ by

$$df_b(x) = d(f|_\sigma)_b(x - b) \quad \text{for} \quad \sigma \in \mathrm{st}(b, K), \quad x \in \sigma.$$

We call f a C^∞ *imbedding* if f and df_b for all $b \in |K|$ are homeomorphisms onto the images. Let $X \subset \mathbf{R}^n$. A C^∞ *triangulation* of X is a pair of K and a C^∞ imbedding $f\colon K \to \mathbf{R}^n$ such that $f(|K|) = X$. We call X C^∞ *triangulable* if X admits a C^∞ triangulation. We define naturally also a C^∞ *cell triangulation*.

Examples of non C^∞ triangulable curves are the following. There we can construct C^∞ maps $K \to \mathbf{R}^2$ whose images equal the curves and which cannot be C^∞ imbeddings.

(1) One branch of a cusp, e.g.,

$$\{(x_1, x_2) \in \mathbf{R}^2 \colon x_1^2 = x_2^3, \ x_1 \geq 0\}.$$

(2) The union of two curves intersecting with the same tangent space, e.g.,

$$\{(x_1, x_2) \in \mathbf{R}^2 \colon x_2 = 0 \text{ or } x_2 = x_1^2\}.$$

Cairns and Whitehead proved that any C^∞ submanifold of $\mathbf{R}^n$ is C^∞ triangulable and the C^∞ triangulation is unique up to PL homeomorphism in the following sense. If (K_1, f_1) and (K_2, f_2) are C^∞ triangulations of the submanifold, $|K_1|$ and $|K_2|$ are PL homeomorphic. We need to generalize this result to the following proposition. A subset X of $\mathbf{R}^n$ is called *locally C^∞ equivalent* to a polyhedron at a point $x \in X$ if there is a C^∞ diffeomorphism germ φ of $\mathbf{R}^n$ at x such that $\varphi(X)$ is a polyhedron germ. A C^∞ submanifold M of $\mathbf{R}^n$ is clearly locally C^∞ equivalent to a polyhedron at each point of M.

Proposition I.3.13 (C^∞ triangulation). *If a set $X \subset \mathbf{R}^n$ is locally C^∞ equivalent to a polyhedron at each point of X, X is C^∞ triangulable uniquely up to PL homeomorphism.*

We will proceed with proof in the same way as [Mu] proves the above theorem of Cairns-Whitehead. Preparatory to it, we recall some term and results in [Mu]. Let $f\colon K \to \mathbf{R}^n$ be a C^∞ map and let δ be a positive continuous function on $|K|$. A *δ-approximation* of f is a C^∞ map $g\colon K' \to \mathbf{R}^n$ such that K' is a subdivision of K,

$$|f(x) - g(x)| \le \delta(x) \quad \text{for} \quad x \in |K|,$$

and

$$|df_b(x) - dg_b(x)| \le \delta(b)|x - b| \quad \text{for} \quad b \in |K|, \quad x \in |\mathrm{st}(b, K')|.$$

Let K' be a subdivision of K. For a C^∞ map $f\colon K \to \mathbf{R}^n$, the *secant map* on K' induced by f is the C^∞ map $K' \to \mathbf{R}^n$ which equals f on K'^0 and such that all the restrictions $f|\sigma$, $\sigma \in K'$, are linear.

Lemma I.3.14 (8.8 in [Mu]). *If $f\colon K \to \mathbf{R}^n$ is a C^∞ imbedding, then there exists a positive continuous function δ on $|K|$ such that any δ-approximation of f is a C^∞ imbedding.*

Lemma I.3.15 (9.6 in [Mu]). *Let $f\colon K \to \mathbf{R}^n$ be a C^∞ map and let δ be a positive number. Assume K is finite. There is an arbitrarily fine subdivision K' of K such that the secant map on K' induced by f is a δ-approximation of f.*

Here K' is called *fine* if $\max_{\sigma \in K'} \max_{x,y \in \sigma} \mathrm{dis}\,(x, y)$ is small. C^∞ triangulability in I.3.13 will follow from the next lemma.

Lemma I.3.16 (cf. 9.7 in [Mu]). *Let $f\colon K \to \mathbf{R}^n$ be a C^∞ imbedding, let K_1 be a finite subcomplex of K, and let δ be a positive number. Set*

$$K_2 = N(K_1, K), \quad \text{and} \quad \partial K_2 = \partial N(K_1, K).$$

Assume $f(|K_2| - |\partial K_2|)$ is a polyhedron. There is a C^∞ imbedding $g\colon K' \to \mathbf{R}^n$ which is a δ-approximation of f and such that

(i) *g equals the secant map induced by f on K_1';*

(ii) *g equals f outside of $|K_2|$;*

(iii) *K' equals K outside of K_2; and*

(iv) *$g(|K_2|)$ coincides with $f(|K_2|)$.*

Here K_1' is the subdivision of K_1 induced by K', and the condition (iii) means that every simplex of $K - K_2$ appears in K'.

If $f(|K_2| - |\partial K_2|)$ is not a polyhedron, then there exists g with the conditions (i), (ii) *and* (iii).

The theorem 9.7 in [Mu] is the latter half.

Proof. We have subdivisions $L_1 \subset L_2 \subset L$ of $K_1 \subset K_2 \subset K$ such that L equals K outside of $N^\circ(K_1, K)$, L_1 is a full subcomplex of L, and $|N(N(L_1, L), L)| = |K_2|$ as shown below. Set $k = \dim K$. For this, let us construct subdivisions $L(i)$ of the skeletons K^i, $i = 0, \ldots, k$, inductively on i so that $L = L(k)$. Set $L(0) = K^0$. Assume we have constructed $L(i - 1)$. For each $\sigma \in K^i - K^{i-1}$, it suffices to define $L(i)|_\sigma$. If $\sigma \notin N^\circ(K_1, K)$, set $L(i)|_\sigma = K|_\sigma$. If $\sigma \in N^\circ(K_1, K)$, choose a point v in σ° and set

$$L(i)|_\sigma = L(i - 1)|_{\partial\sigma} \cup \{v * \sigma' : \sigma' \in L(i-1)|_{\partial\sigma}\} \cup \{v\}.$$

Then $L = L(k)$ and $L_i = L|_{|K_i|}$, $i = 1, 2$, are what we want.

For the sake of arguments in the proof of I.3.17, we choose the v to be the barycenter of σ.

Set $L_3 = N(L_1, L)$. Then $L_2 = N(L_3, L)$. Let φ be the simplicial map $L \to \{0, 1, [0, 1]\}$ defined by $\varphi = 1$ on L_1^0 and $\varphi = 0$ on $L^0 - L_1^0$. By the above properties of L, $\varphi^{-1}(1)$ equals $|K_1|$, and $\varphi^{-1}(0)$ equals $\overline{|K| - |L_3|}$. Let δ' be a small positive number. By I.3.15 we have a subdivision L_3' of L_3 such that the secant map h on L_3' induced by $f|_{L_3}$ is a δ'-approximation of $f|_{L_3}$. Let L_2' denote the following standard extension of L_3' to L_2:

$$L_2' = L_3' \cup \partial N(L_3, L_2) \cup \{\sigma_1 * \sigma_2 : \sigma_1 \in L_3', \ \sigma_2 \in \partial N(L_3, L_2)$$

$$\text{for some } \sigma_3 \in L_2 \text{ with } \sigma_1 * \sigma_2 \subset \sigma_3\}.$$

Set

$$K' = L_2' \cup (K - K_2), \quad \text{and}$$

$$g = \begin{cases} \varphi h + (1 - \varphi)f & \text{on } |L_3| \\ f & \text{on } |K| - |L_3|. \end{cases}$$

By I.3.14, for sufficiently small δ', g is both a δ-approximation of f and a C^∞ imbedding of K' into $\mathbf{R}^n$ such that the conditions (i), (ii) and (iii) are satisfied. Hence the latter half holds true.

We want to prove the first half. It remains to prove that the condition (iv) holds. We proceed by induction on $m = \dim |K_2|$. If $m = 0$, (iv) is trivial by the definition of g. Suppose (iv) for the case of lower dimension. Set

$$X = f(|K_2| - |\partial K_2|).$$

Let X_1 denote the points x of X such that the germ of X at x is a linear m-space germ, i.e., $X - X_1$ is the C^∞ singular point set of X. Each connected component of X_1 is an open set of a linear m-space, and $X - X_1$ and $X \cap (\overline{X_1} - X_1)$ are polyhedra of dimension $< m$ (see the proof of I.3.4). It suffices to prove the following equalities:

$$g(f^{-1}(\overline{X_1})) = \overline{X_1}, \quad \text{and} \tag{1}$$
$$g(f^{-1}(\overline{X - X_1})) = \overline{X - X_1}. \tag{2}$$

Note $\overline{|K_2|} = |K_2|$ and $\overline{X} = f(|K_2|)$ because K_1 and hence K_2 are finite.

Proof of (2). We can apply the induction hypothesis to the restriction of f to the subcomplex of K:

$$\{\sigma \in K \colon f(\sigma) \subset \overline{X - X_1}\},$$

and then we obtain the equality (2) if the set $f^{-1}(\overline{X - X_1})$ is the union of some $\sigma \in K$. Hence it suffices to prove that:
(3) If $f(\sigma^\circ) \cap X_1 \neq \varnothing$ for $\sigma \in K$, then $f(\sigma^\circ) \subset X_1$.

Proof of (3). Let $\sigma \in K$ with $f(\sigma^\circ) \cap X_1 \neq \varnothing$. Clearly $\sigma \in K_2 - \partial K_2$. Note that all the domains $|\mathrm{st}(b, K)|$ of df_b, $b \in \sigma^\circ$, coincide with each other and are included in $|K_2|$. Denote by D the common domain, and set $E = (\mathrm{lk}(b, K))^\circ$. Then E is a finite point set, for each $b \in \sigma^\circ$, the image $df_b(D)$ is a polyhedron defined by 0 and the finite set $df_b(E)$, and the map $b \to df_b(E)$ is a C^∞ n'-valued map from σ° to $\mathbf{R}^n$ (i.e., of class C^∞ as a map from σ°

to $\overbrace{\mathbf{R}^n \times \cdots \times \mathbf{R}^n}^{n'}$), where $n' = \sharp E$. Hence $df_b(D)$ "moves" continuously as b moves in σ°. This together with the following fact (4) means that the set $f^{-1}(X_1) \cap \sigma^\circ$ is closed in σ°. On the other hand, $f^{-1}(X_1) \cap \sigma^\circ$ is open in σ° because f is continuous and X_1 is open in X. These prove $f(\sigma^\circ) \subset X_1$.

(4) For each $b \in \sigma^\circ$, $f(b) \in X_1$ if and only if the germ of $df_b(D)$ at 0 is a linear space germ of dimension $= m$.

This is easy to prove. Indeed, df_b is a PL homeomorphism from D to the image, and X is a polyhedron. Hence the germ of $df_b(D) + f(b)$ at $f(b)$ is included in the germ of X at $f(b)$. Moreover, the germs coincide with each other if at least one of them is a linear space germ, because f and df_b are homeomorphisms onto their images. From this (4) follows.

Proof of (1). Let $\{Y_i\}_{i=1,\ldots,k}$ be the family of the connected components of X_1, which is finite by (3) because K_2 is finite. Set

$$A_i = \{\sigma \in K_2 : f(\sigma) \subset \overline{Y_i}\}, \quad i = 1,\ldots,k, \quad \text{and}$$
$$B_i = \{\sigma \in A_i : f(\sigma) \subset \overline{Y_i} - Y_i\}, \quad i = 1,\ldots,k.$$

By (3), for each i, f maps bijectively $(|A_i|, |A_i| - |B_i|, |B_i|)$ to $(\overline{Y_i}, Y_i, \overline{Y_i} - Y_i)$, and the equality (1) is equivalent to the equalities:

$$(1)_i \qquad\qquad\qquad g(|A_i|) = \overline{Y_i}, \quad i = 1,\ldots,k.$$

So we prove $(1)_i$. As already noted, each Y_i is an open subset of some linear space Π_i. Moreover, we see that Y_i is a connected component of $\Pi_i - f(|B_i|)$. By the definition of g, $g(|A_i|)$ is included in Π_i, and $g|_{A_i'}$ is both a δ-approximation of $f|_{A_i}$ and a C^∞ imbedding, where A_i' denotes the subdivision of A_i induced by K'. Hence $(1)_i$ follows from the equality:

$$(2)_i \qquad\qquad\qquad g(|B_i|) = \overline{Y_i} - Y_i,$$

which we can prove by the induction hypothesis in the same way as (2). $\square$

We shall prove uniqueness in I.3.13 using the next lemma.

Lemma I.3.17. *Let f, K, K_1, K_2 and δ be the same as in I.3.16, and construct g as in its proof. Let $\hat{f}\colon \hat{K} \to \mathbf{R}^n$, $\hat{K}_1$, $\hat{K}_2$, $\hat{\delta}$ and $\hat{g}$ be given similarly. Let $J \subset K$ and $\hat{J} \subset \hat{K}$ be subcomplexes. Assume that*

$$f(|K|) = \hat{f}(|\hat{K}|), \qquad f(|J|) = \hat{f}(|\hat{J}|),$$
$$f(|J \cap K_i|) = \hat{f}(|\hat{J} \cap \hat{K}_i|), \quad i = 1,2,$$

and $\hat{f}^{-1} \circ f|_J\colon J \to \hat{J}$ is a simplicial isomorphism. We can construct g and $\hat{g}$ so that

$$\hat{g}^{-1} \circ g = \hat{f}^{-1} \circ f \quad \text{on} \quad |J|.$$

Proof. Recall the proof of I.3.16. We defined $L_1 \subset L_3 \subset L_2 \subset L$, $\varphi\colon L \to \{0, 1, [0, 1]\}$, $h\colon L_3' \to \mathbf{R}^n$ and $g\colon K' \to \mathbf{R}^n$. Let L_J and K_J' denote the subdivisions of J induced by L and K', respectively. We define $\hat{L}_1 \subset \hat{L}_3 \subset$

$\hat{L}_2 \subset \hat{L}$, $\hat{\varphi}$, $\hat{h}$, $\hat{g}$, $\hat{L}'_3$, $\hat{K}'$, $\hat{L}_J$ and $\hat{K}'_J$ in the same way for $\hat{f} \colon \hat{K} \to \mathbf{R}^n$, $\hat{K}_1$ and $\hat{J}$. We have the equality:

$$f(|L_J \cap L_3|) = \hat{f}(|\hat{L}_J \cap \hat{L}_3|). \qquad (*)$$

Indeed, clearly

$$\mathrm{st}(b, L) \cap L_J = \varnothing \quad \text{for} \quad b \in K_1^0 - J^0,$$
$$\mathrm{st}(\hat{b}, \hat{L}) \cap \hat{L}_J = \varnothing \quad \text{for} \quad \hat{b} \in \hat{K}_1^0 - \hat{J}^0,$$
$$L_3 = N(L_1, L) = \bigcup_{b \in K_1^0} \mathrm{st}(b, L),$$
$$\hat{L}_3 = N(\hat{L}_1, \hat{L}) = \bigcup_{\hat{b} \in \hat{K}_1^0} \mathrm{st}(\hat{b}, \hat{L}),$$

and hence

$$L_J \cap L_3 = \bigcup_{b \in J^0 \cap K_1^0} \mathrm{st}(b, L_J) = N(L_J \cap L_1, L_J),$$
$$\hat{L}_J \cap \hat{L}_3 = \bigcup_{\hat{b} \in \hat{J}^0 \cap \hat{K}_1^0} \mathrm{st}(\hat{b}, \hat{L}_J) = N(\hat{L}_J \cap \hat{L}_1, \hat{L}_J).$$

Recall that we defined L in the proof of I.3.16 by introducing new vertices which are the barycenters of the simplices of $N^\circ(K_1, K)$. Hence $\hat{f}^{-1} \circ f|_{L_J} \colon L_J \to \hat{L}_J$ is a simplicial isomorphism, which carries $L_J \cap L_1$ to $\hat{L}_J \cap \hat{L}_1$, and $(*)$ holds by the last two equalities.

Let $(**)$ denote the equality $\hat{f}^{-1} \circ f = \hat{g}^{-1} \circ g$. This holds true on $|J| - |L_3|$ because

$$g = f \quad \text{on} \quad |J| - |L_3|,$$
$$\hat{g} = \hat{f} \quad \text{on} \quad |\hat{J}| - |\hat{L}_3|, \quad \text{and}$$
$$f(|J| - |L_3|) = \hat{f}(|\hat{J}| - |\hat{L}_3|) \quad \text{(by } (*) \text{ and by} \quad f(|J|) = \hat{f}(|\hat{J}|)).$$

Hence it suffices to prove $(**)$ on $|L_J \cap L_3|$.

If we construct the L'_3 and the $\hat{L}'_3$ so that $\hat{f}^{-1} \circ f|_{K'_J \cap L'_3} \colon K'_J \cap L'_3 \to \hat{K}'_J \cap \hat{L}'_3$ is a simplicial isomorphism, then by the definition of h and φ,

$$\hat{h}^{-1} \circ h = \hat{f}^{-1} \circ f \quad \text{on} \quad |L_J \cap L_3|, \quad \text{and}$$
$$\hat{\varphi} \circ \hat{f}^{-1} \circ f = \varphi \quad \text{on} \quad |L_J \cap L_3|.$$

Hence we have

$$\hat{g} \circ \hat{f}^{-1} \circ f = (\hat{\varphi}\hat{h} + (1 - \hat{\varphi})\hat{f}) \circ \hat{f}^{-1} \circ f$$
$$= (\hat{\varphi} \circ \hat{f}^{-1} \circ f)(\hat{h} \circ \hat{f}^{-1} \circ f) + (1 - \hat{\varphi} \circ \hat{f}^{-1} \circ f)f$$
$$= \varphi h + (1 - \varphi)f = g \quad \text{on} \quad |L_J \cap L_3|.$$

Thus the $(**)$ holds true on $|L_J \cap L_3|$. So we need only a simplicial isomorphism $\hat{f}^{-1} \circ f|_{K'_J \cap L'_3} \colon K'_J \cap L'_3 \to \hat{K}'_J \cap \hat{L}'_3$.

Recall that we can adopt arbitrary L'_3 and $\hat{L}'_3$ if the secant maps $h \colon L'_3 \to \mathbf{R}^n$ and $\hat{h} \colon \hat{L}'_3 \to \mathbf{R}^n$ are δ'- and $\hat{\delta}'$-approximations of $f|_{L_3}$ and $\hat{f}|_{\hat{L}_3}$, respectively. We need to choose L'_3 and $\hat{L}'_3$ correlatively. Identify $L_J \cap L_3$ with $\hat{L}_J \cap \hat{L}_3$ through $\hat{f}^{-1} \circ f$ by the equality $(*)$. The identification space $\tilde{L}_3$ of the disjoint union of L_3, and $\hat{L}_3$ is a simplicial complex because $\hat{f}^{-1} \circ f|_{L_J}$ is a simplicial isomorphism and because $L_J \cap L_3$ and $\hat{L}_J \cap \hat{L}_3$ are full subcomplexes of L_3 and $\hat{L}_3$, respectively. A C^∞ map $\tilde{f} \colon \tilde{L}_3 \to \mathbf{R}^n$ is induced by $f|_{L_3}$ and $\hat{f}|_{\hat{L}_3}$ ($\tilde{f}$ is not necessarily a C^∞ imbedding). Let $\tilde{L}_3$ be realized in some Euclidean space. Consider a secant map $\tilde{h} \colon \tilde{L}'_3 \to \mathbf{R}^n$ induced by $\tilde{f}$ which is also a $\tilde{\delta}$-approximation of $\tilde{f}$ for a sufficiently small number $\tilde{\delta} > 0$. Define $h \colon L'_3 \to \mathbf{R}^n$ and $\hat{h} \colon \hat{L}'_3 \to \mathbf{R}^n$ to be the restrictions of this $\tilde{h}$ to $|L_3|$ and to $|\hat{L}_3|$. These L'_3 and $\hat{L}'_3$ fulfill the requirements. $\qquad\square$

Proof of C^∞ triangulability in I.3.13. Note that X is locally closed in $\mathbf{R}^n$. By assumption there are open subsets $U_i \subset V_i$ of X, $i = 1, 2, \ldots$, and C^∞ diffeomorphisms π_i, $i = 1, 2, \ldots$, of $\mathbf{R}^n$ such that all the closures $\overline{V}_i$ in $\mathbf{R}^n$ are included in X, each V_i is bounded and contains $\overline{U}_i$, $\{U_i\}_{i=1,2,\ldots}$ is a covering of X, $\{V_i\}_{i=1,2,\ldots}$ is locally finite at each point of X, and all $\pi_i(V_i)$ are polyhedra. (If X is compact, all V_i except a finite number are empty.) Here we can assume for each i, $\pi_i = \mathrm{id}$ outside of a small neighborhood of V_i. For proof, it suffices to prove the following statement for each $k \in \mathbf{N}$.

$(*)$ Let $h_{k-1} \colon L_{k-1} \to \mathbf{R}^n$ be a C^∞ imbedding such that $h_{k-1}(|L_{k-1}|)$ is a neighborhood of $\cup_{i=1}^{k-1}\overline{U}_i$ in X. Let J_{k-1} be a subcomplex of L_{k-1} such that $h_{k-1}(|J_{k-1}|)$ and V_k are disjoint. There exist a C^∞ imbedding $h_k \colon L_k \to \mathbf{R}^n$ and a subcomplex J_k of L_k such that $h_k(|L_k|)$ is a neighborhood of $\cup_{i=1}^{k}\overline{U}_i$ in X, $h_k(|J_k|)$ equals $h_{k-1}(|J_{k-1}|)$, and $h_k^{-1} \circ h_{k-1}|_{J_{k-1}} \colon J_{k-1} \to J_k$ is a simplicial isomorphism.

Proof of $(*)$. By transforming $\mathbf{R}^n$ by the π_k, we can assume that V_k is a polyhedron. We want to apply I.3.16 to h_{k-1}. For this purpose we subdivide L_{k-1} as follows. First we subdivide L_{k-1} to the complex consisting of the simplices $\sigma_1 * \Delta v_2 \cdots v_l$, where $\sigma_1 \in J_{k-1}$ (possibly empty) and $v_2, \cdots, v_l$

are the barycenters of $\sigma_2 \subset \cdots \subset \sigma_l \in L_{k-1} - J_{k-1}$ with $\sigma_1 \subset \sigma_2$ (possibly $l = 1$). By repeating this subdividing sufficiently many times, we define a subdivision L'_{k-1} of L_{k-1}. Clearly L'_{k-1} equals L_{k-1} on $|J_{k-1}|$. Moreover, we have the following properties. Set

$$K = L'_{k-1}, \qquad f = h_{k-1}|_K,$$
$$K_1 = \{\sigma \in K : f(\sigma) \cap \cup_{i=1}^{k-1}\overline{U_i} \neq \varnothing, \ f(\sigma) \cap (\overline{U_k} - U_k) \neq \varnothing\},$$
$$K_2 = N(K_1, K), \quad \text{and} \quad \partial K_2 = \partial N(K_1, K).$$

Clearly K_1 is finite. Repeat the above subdividing enough times that $f(|K - J_{k-1}|) \cap V_k$ includes $f(|K_2|)$ and, moreover, is its neighborhood in X. It follows that

$$K_2 \cap J_{k-1} = \varnothing. \tag{$**$}$$

Furthermore, the set $f(|K_2| - |\partial K_2|)$ is a polyhedron. Indeed, it is an open subset of $f(|K|)$. Hence $f(|K_2|-|\partial K_2|)\cap f(|K-J_{k-1}|)\cap V_k$ is an open subset of $f(|K - J_{k-1}|) \cap V_k$ because $f(|K - J_{k-1}|) \cap V_k \subset f(|K|)$. Therefore, as $f(|K - J_{k-1}|) \cap V_k \supset f(|K_2|)$, $f(|K_2| - |\partial K_2|)$ is an open subset of $f(|K - J_{k-1}|)\cap V_k$. It follows from this and the above fact that $f(|K-J_{k-1}|)\cap V_k$ is a neighborhood of $f(|K_2|)$ in X and that $f(|K_2| - |\partial K_2|)$ is open in V_k. Hence by the assumption that V_k is a polyhedron, $f(|K_2| - |\partial K_2|)$ is a polyhedron.

Let us apply I.3.16 to $f \colon K \to \mathbf{R}^n$, K_1 and a sufficiently small number $\delta > 0$. We have a δ-approximation $g \colon K' \to \mathbf{R}^n$ of f which is also a C^∞ imbedding such that

(i) g is linear on each simplex of K'_1;

(ii) g equals f outside of $|K_2|$;

(iii) K' equals K outside of K_2 and

(iv) $g(|K_2|)$ coincides with $f(|K_2|)$

where K'_1 denotes the subdivision of K_1 induced by K'.

Set

$$K'_3 = \{\sigma \in K' : g(\sigma) \cap \overline{U_k} = \varnothing\} \cup K'_1, \quad \text{and}$$
$$g_3 = g|_{K'_3}.$$

We want to extend $g_3 \colon K'_3 \to \mathbf{R}^n$ to the required $h_k \colon L_k \to \mathbf{R}^n$. The union of $g_3(|K'_3|)$ and any neighborhood of $\overline{U_k}$ in X is a neighborhood of $\cup_{i=1}^k \overline{U_i}$ in X. The reason is the following. Since $f(|K|)$ is a neighborhood of $\cup_{i=1}^{k-1}\overline{U_i}$ in X, so is $g(|K|)$ by (ii) and (iv). Set

$$A = \{\sigma \in K' : g(\sigma) \cap \cup_{i=1}^{k-1}\overline{U_i} \neq \varnothing\},$$
$$A_1 = \{\sigma \in A : g(\sigma) \subset U_k\},$$
$$A_2 = \{\sigma \in A : g(\sigma) \cap \overline{U_k} = \varnothing\}, \quad \text{and}$$
$$A_3 = \{\sigma \in A : g(\sigma) \cap (\overline{U_k} - U_k) \neq \varnothing\}.$$

Then $g(|A|)$ is a neighborhood of $\cup_{i=1}^{k-1}\overline{U_i}$ in X, any neighborhood of $\overline{U_k}$ in X contains $g(|A_1|)$, and

$$A = A_1 \cup A_2 \cup A_3, \qquad K_3' \supset A_2 \cup K_1', \qquad |A_3| \subset |K_1|.$$

Hence $K_3' \supset A_2 \cup A_3$, and the union of $g(|A_2 \cup A_3|)$ and any neighborhood of $\overline{U_k}$ is a neighborhood of $\cup_{i=1}^{k-1}\overline{U_i}$ in X, which proves the above property.

By (∗∗), (ii) and (iii) there is a subcomplex J_k of K_3' such that the equality $g_3(|J_k|) = h_{k-1}(|J_{k-1}|)$ holds and $g_3^{-1} \circ h_{k-1}|_{J_{k-1}} : J_{k-1} \to J_k$ is a simplicial isomorphism.

Let K_4 be a simplicial decomposition of a compact polyhedral neighborhood of $\overline{U_k}$ in V_k so small that

$$|K_4| \cap g_3(|K_3'|) \subset g_3(|K_1|).$$

By (i) the set $Y = |K_4| \cap g_3(|K_3'|)$ is a compact polyhedron, and the map $g_3|_{g_3^{-1}(Y)} : g_3^{-1}(Y) \to Y$ is a PL homeomorphism. Identify $g_3^{-1}(Y)$ with Y through this homeomorphism, and consider the identification space of the union of $|K_3'|$ and $|K_4|$. The identification space is a polyhedron and admits a simplicial decomposition L_k which equals K_3' on $|J_k|$ and is subdivisions of K_3' on $|K_3'|$ and of K_4 on $|K_4|$. Then L_k and the C^∞ imbedding $h_k : L_k \to \mathbf{R}^n$, defined to be g_3 on $|K_3'|$ and to be the identity on $|K_4|$, fulfill the requirements.   □

Proof of uniqueness in I.3.13. Let $f : K \to \mathbf{R}^n$ and $\hat{f} : \hat{K} \to \mathbf{R}^n$ be C^∞ triangulations of X. We want to find approximations g of f and $\hat{g}$ of $\hat{f}$ which are also C^∞ triangulations of X such that $\hat{g}^{-1} \circ g$ is PL. For this, it suffices to prove the following statement by the same reason as in the above proof of C^∞ triangulability.

(∗) Let $J_1 \subset J_2 \subset K$ and $\hat{J}_1 \subset \hat{J}_2 \subset \hat{K}$ be subcomplexes, let V be an open polyhedral subset of X, and let Y_1 and Y_2 be compact subsets of X such that $f(|J_i|) = \hat{f}(|\hat{J}_i|)$, $i = 1, 2$, $\hat{f}^{-1} \circ f|_{J_2} : J_2 \to \hat{J}_2$ is a simplicial isomorphism, $f(|J_2|)$ is a neighborhood of Y_1 in X, V is a neighborhood of Y_2 in X, and $f(|J_1|)$ and V are disjoint. Let δ be a small positive number. There exist δ-approximations $g : K' \to \mathbf{R}^n$ of f and $\hat{g} : \hat{K}' \to \mathbf{R}^n$ of $\hat{f}$ and subcomplexes J_3 of K' and $\hat{J}_3$ of $\hat{K}'$ such that

(i) g, $\hat{g}$ are C^∞ triangulations of X;

(ii) $g(|J_2' \cup J_3|) = \hat{g}(|\hat{J}_2' \cup \hat{J}_3|),$

and this set is a neighborhood of $Y_1 \cup Y_2$ in X;

(iii) the map

$$\hat{g}^{-1} \circ g|_{J_2' \cup J_3} : J_2' \cup J_3 \to \hat{J}_2' \cup \hat{J}_3$$

is a simplicial isomorphism;

(iv) g equals f on $|J_1|$, and $\hat{g}$ equals $\hat{f}$ on $|\hat{J}_1|$; and

(v) J_1' equals J_1, and $\hat{J}_1'$ equals $\hat{J}_1$

where for each $i = 1, 2$, J_i' and $\hat{J}_i'$ denote the subdivisions of J_i and $\hat{J}_i$ induced by K' and by $\hat{K}'$, respectively.

Proof of (∗). Subdivide K and $\hat{K}$ so fine around $f^{-1}(Y_2)$ and $\hat{f}^{-1}(Y_2)$, respectively, that the following holds. If we set

$$K_1 = \{\sigma \in K : f(\sigma) \cap Y_2 \neq \varnothing\},$$
$$\hat{K}_1 = \{\hat{\sigma} \in \hat{K} : \hat{f}(\hat{\sigma}) \cap Y_2 \neq \varnothing\},$$
$$K_2 = N(K_1, K), \qquad \partial K_2 = \partial N(K_1, K),$$
$$\hat{K}_2 = N(\hat{K}_1, \hat{K}), \quad \text{and} \quad \partial \hat{K}_2 = \partial N(\hat{K}_1, \hat{K}),$$

then $f(|K_2|)$ and $\hat{f}(|\hat{K}_2|)$ are included in V. It follows that $f(|K_2| - |\partial K_2|)$ and $\hat{f}(|\hat{K}_2| - |\partial \hat{K}_2|)$ are polyhedra because $|K_2| - |\partial K_2|$ and $|\hat{K}_2| - |\partial \hat{K}_2|$ are open subsets of $|K|$ and of $|\hat{K}|$, respectively. Here by using the method of subdivision in the proof of C^∞ triangulability, we leave J_1 and $\hat{J}_1$ undivided, and we have $f(|J_2 \cap K_i|) = \hat{f}(|\hat{J}_2 \cap \hat{K}_i|)$, $i = 1, 2$. Apply I.3.17 to f, K, K_1, K_2, δ, $J = J_2$, $\hat{f}$, $\hat{K}$, $\hat{K}_1$, $\hat{K}_2$, $\hat{\delta} = \delta$ and $\hat{J} = \hat{J}_2$. We have δ-approximations $g \colon K' \to \mathbf{R}^n$ of f and $\hat{g} \colon \hat{K}' \to \mathbf{R}^n$ of $\hat{f}$ which are C^∞ imbeddings such that

(vi) g and $\hat{g}$ are linear on each simplex of K_1' and $\hat{K}_1'$, respectively;

(vii) g equals f outside of $|K_2|$, and $\hat{g}$ equals $\hat{f}$ outside of $|\hat{K}_2|$;

(viii) K' equals K outside of K_2, and $\hat{K}'$ equal $\hat{K}$ outside of $\hat{K}_2$;

(ix)   both $g(|K'|)$ and $\hat{g}(|\hat{K}'|)$ equal X, namely, g and $\hat{g}$ are C^∞ triangulations of X; and

$$\hat{f}^{-1} \circ f = \hat{g}^{-1} \circ g \quad \text{on} \quad |J_2|. \tag{x}$$

We need to subdivide these K' and $\hat{K}'$ again to obtain J_3 and $\hat{J}_3$. Set

$$Y_3 = g(|K_1|) \cap \hat{g}(|\hat{K}_1|).$$

By (vi) the sets Y_3, $g^{-1}(Y_3)$ and $\hat{g}^{-1}(Y_3)$ are compact polyhedra, and the map $\hat{g}^{-1} \circ g|_{g^{-1}(Y_3)}$ is PL. Moreover, the map $\hat{g}^{-1} \circ g|_{|J_2| \cup g^{-1}(Y_3)} \colon |J_2| \cup g^{-1}(Y_3) \to |\hat{J}_2| \cup \hat{g}^{-1}(Y_3)$ is a PL homeomorphism because $\hat{g}^{-1} \circ g|_{|J_2|} \colon |J_2| \to |\hat{J}_2|$ is a PL homeomorphism by the assumption on f and $\hat{f}$ and by (x). Through this PL homeomorphism, let us identify $|J_2| \cup g^{-1}(Y_3)$ with $|\hat{J}_2| \cup \hat{g}^{-1}(Y_3)$, and choose a simplicial decomposition of the identification space of the union of $|K|$ and $|\hat{K}|$ in the same way as in the proof of C^∞ triangulability. We can assume that $g^{-1}(Y_3)$ and $\hat{g}^{-1}(Y_3)$ are the underlying polyhedra of subcomplexes J_3 of K' and $\hat{J}_3$ of $\hat{K}'$, respectively, and that $\hat{g}^{-1} \circ g|_{J_2' \cup J_3} : J_2' \cup J_3 \to \hat{J}_2' \cup \hat{J}_3$ is a simplicial isomorphism. Here we do not subdivide K' on $|J_1|$ nor $\hat{K}'$ on $|\hat{J}_1|$, which is possible by (vii), (viii), the property $J_1 \cap K_2 = \hat{J}_1 \cap \hat{K}_2 = \varnothing$ and by the assumption that $\hat{f}^{-1} \circ f|_{J_1} : J_1 \to \hat{J}_1$ is a simplicial isomorphism.

Now $g \colon K' \to \mathbf{R}^n$, $\hat{g} \colon \hat{K}' \to \mathbf{R}^n$, J_3 and $\hat{J}_3$ satisfy the conditions (i), ..., (v). Indeed, (i) coincides with (ix). We have already shown (ii) and (iii) except the property that $g(|J_2' \cup J_3|)$ is a neighborhood of $Y_1 \cup Y_2$ in X. But this property is immediate if we choose sufficiently small $\delta > 0$ because $f(|J_2|)$ is a neighborhood of Y_1 in X and because both $f(|K_1|)$ and $\hat{f}(|\hat{K}_1|)$ are neighborhoods of Y_2 in X. Finally, the conditions (iv) and (v) are consequences of (vii), (viii) and of the property $J_1 \cap K_2 = \hat{J}_1 \cap \hat{K}_2 = \varnothing$. $\qquad\square$

Remark I.3.18. We can generalize I.3.13 to the case of a locally finite family as follows. Let $\{X_i\}$ be a family of closed subsets of $\mathbf{R}^n$ locally finite at each point of $\mathbf{R}^n$. Assume there is a C^∞ diffeomorphism germ φ of $\mathbf{R}^n$ at each point of $\mathbf{R}^n$ such that all $\varphi(X_i)$ are polyhedron germs. There exist a C^∞ triangulation $f \colon K \to \mathbf{R}^n$ of $\mathbf{R}^n$ and subcomplexes K_i of K uniquely up to PL homeomorphism such that for each i, $f|_{K_i}$ is a C^∞ triangulation of X_i. We call such a C^∞ triangulation *compatible* with $\{X_i\}$. Here we can replace $\mathbf{R}^n$ with an open subset of it. We prove this in the same way as I.3.13.

Conjecture I.3.19. *A locally C^∞ triangulable set X in $\mathbf{R}^n$ (i.e., each point of X has a neighborhood U in $\mathbf{R}^n$ such that $X \cap U$ is C^∞ triangulable) is globally C^∞ triangulable.*

Proposition I.3.20. *Let $\{X_i\}_{i=1,\dots,k}$ be a Whitney stratification of a compact set in $\mathbf{R}^n$, and let $\{T_i = (|T_i|, \pi_i, \rho_i)\}_{i=1,\dots,k}$ be a controlled tube system for $\{X_i\}$ such that $\dim X_1 < \cdots < \dim X_k$. For some removal data δ of $\{X_i\}$ and for any sequence of positive numbers $\varepsilon = \{\varepsilon_i\}_{i=1,\dots,k-1} \leq \delta$, there exist C^∞ triangulations (K_i, f_i) of the sets:*

$$X_i - \bigcup_{j=1}^{i-1} \rho_j^{-1}([0, \varepsilon_j[), \quad i = 1, \dots, k,$$

such that for each $1 \leq i_1 < i_2 \leq k$, the map

$$f_{i_1}^{-1} \circ \pi_{i_1} \circ (f_{i_2}|_{f_{i_2}^{-1}(\rho_{i_1}^{-1}(\varepsilon_{i_1}))}): K_{i_2}|_{f_{i_2}^{-1}(\rho_{i_1}^{-1}(\varepsilon_{i_1}))} \longrightarrow K_{i_1}$$

is simplicial.

Proof. First we will require $f_{i_1}^{-1} \circ \pi_{i_1} \circ (f_{i_2}|_{f_{i_2}^{-1}(\rho_{i_1}^{-1}(\varepsilon_{i_1}))})$ to be only PL. We can choose a removal data δ so that for any sequences of positive numbers $\varepsilon = \{\varepsilon_i\}_{i=1,\ldots,k-1}$ and $\varepsilon' = \{\varepsilon_i'\}_{i=1,\ldots,k-1}$ with $\varepsilon \le \delta$ and $\varepsilon_i \le \varepsilon_i' \le 2\varepsilon_i$, $i = 1,\ldots,k-1$, and for each pair $1 \le i_1 < i_2 \le k$, the set

$$Y_{i_2}(\varepsilon') = X_{i_2} - \bigcup_{j=1}^{i_2-1} \rho_j^{-1}([0, \varepsilon_j'[)$$

is a compact C^∞ manifold possibly with boundary and corners =

$$\partial Y_{i_2}(\varepsilon') = Y_{i_2}(\varepsilon') \cap \bigcup_{j=1}^{i_2-1} \rho_j^{-1}(\varepsilon_j').$$

Furthermore, the set $Y_{i_2}(\varepsilon') \cap \rho_{i_1}^{-1}(\varepsilon_{i_1}')$ is a compact C^∞ manifold possible with boundary and corners,

$$\pi_{i_1}(Y_{i_2}(\varepsilon') \cap \rho_{i_1}^{-1}(\varepsilon_{i_1}')) \subset Y_{i_1}(\varepsilon'),$$

and the map $\pi_{i_1}|_{Y_{i_2}(\varepsilon') \cap \rho_{i_1}^{-1}(\varepsilon_{i_1}')}$ is a C^∞ submersion. We fix such δ and ε. We will construct a C^∞ triangulation (K_i, f_i) of each $Y_i(\varepsilon)$ such that for each pair $1 \le i_1 < i_2 \le k$, the map

$$f_{i_1}^{-1} \circ \pi_{i_1} \circ (f_{i_2}|_{f_{i_2}^{-1}(\rho_{i_1}^{-1}([\varepsilon_{i_1}, \varepsilon_{i_1}'[))}): f_{i_2}^{-1}(\rho_{i_1}^{-1}([\varepsilon_{i_1}, \varepsilon_{i_1}'[)) \longrightarrow |K_{i_1}|$$

is PL for some ε_{i_1}' with $\varepsilon_{i_1} < \varepsilon_{i_1}' \le 2\varepsilon_{i_1}$. We call the last property $A(i_1, i_2)$. We will accomplish the construction by triple induction. First assume there exist (K_i, f_i), $i = 1,\ldots,k-1$, with $A(i_1, i_2)$ for all $1 \le i_1 < i_2 \le k-1$. It suffices to find (K_k, f_k) with $A(i, k)$ for any $1 \le i < k$. We carry this out by downward induction as follows. Let $1 \le l < k$ be an integer. Second, assume there exists a C^∞ triangulation $(K_{k,l+1}, f_{k,l+1})$ of a neighborhood of $Y_k(\varepsilon) \cap \cup_{i=l+1}^{k-1} \rho_i^{-1}(\varepsilon_i)$ in $Y_k(\varepsilon)$ such that for each $l < i < k$, $A(i, k)$ is satisfied for some ε_i'. We need only find $(K_{k,l}, f_{k,l})$ with $A(i, k)$ for each $l \le i < k$ and for some ε_i', because if we have $(K_{k,1}, f_{k,1})$ which is a C^∞ triangulation of a neighborhood of $\partial Y_k(\varepsilon)$ in $Y_k(\varepsilon)$ with $A(i, k)$ for any $1 \le i < k$, then by using I.3.16 as in the proof of I.3.13, we can construct a C^∞ triangulation (K_k, f_k) with $A(i, k)$ for each $1 \le i < k$ and for some smaller ε_i'.

Fix ε'_i, $l < i < k$, so that for each i,

$$Y_k(\varepsilon) \cap \rho_i^{-1}([\varepsilon_i, \varepsilon'_i[) \subset \operatorname{Im} f_{k,l+1},$$

and $A(i, k)$ is satisfied for ε'_i. Choose ε'_l so that all $A(l, i)$, $l < i < k$, are satisfied for ε'_l. For each $l < i < k$, $f_l^{-1} \circ \pi_l \circ f_{k,l+1}$ is PL on $f_{k,l+1}^{-1} \{\rho_l^{-1}([\varepsilon_l, \varepsilon'_l[) \cap \rho_i^{-1}([\varepsilon_i, \varepsilon'_i[)\}$, because

$$f_l^{-1} \circ \pi_l \circ f_{k,l+1} = f_l^{-1} \circ \pi_l \circ \pi_i \circ f_{k,l+1} = f_l^{-1} \circ \pi_l \circ f_i \circ f_i^{-1} \circ \pi_i \circ f_{k,l+1}$$

$$\text{on the domain} \quad \text{by} \quad \operatorname{ct}(T_l, T_i),$$

$f_l^{-1} \circ \pi_l \circ f_i$ and $f_i^{-1} \circ \pi_i \circ f_{k,l+1}$ are PL on $f_i^{-1}(\rho_l^{-1}([\varepsilon_l, \varepsilon'_l[))$ and $f_{k,l+1}^{-1}(\rho_i^{-1}([\varepsilon_i, \varepsilon'_i[))$, respectively, and

$$\pi_i^{-1}(\rho_l^{-1}([\varepsilon_l, \varepsilon'_l[)) \cap \rho_i^{-1}([\varepsilon_i, \varepsilon'_i[) = \rho_l^{-1}([\varepsilon_l, \varepsilon'_l[) \cap \rho_i^{-1}([\varepsilon_i, \varepsilon'_i[) \quad \text{by} \quad \operatorname{ct}(T_l, T_i).$$

Hence it suffices to extend $(K_{k,l+1}, f_{k,l+1})$ to a neighborhood of $Y_k(\varepsilon) \cap \rho_l^{-1}(\varepsilon_l)$ in $Y_k(\varepsilon)$. For this we use the third induction. Set

$$Z_{k,l} = Y_k(\varepsilon) \cap \rho_l^{-1}(\varepsilon_l) - \bigcup_{i=l+1}^{k-1} \rho_i^{-1}([\varepsilon_i, \varepsilon'_i[).$$

Let $\{C_i\}_{i=1,\dots,k'}$ be a finite fine covering of $Z_{k,l}$ by compact sets which we will define later so that certain conditions are satisfied. Let $1 \leq l' \leq k'$ be an integer. Assume we have already constructed a C^∞ triangulation $(L_{k,l'-1}, g_{k,l'-1})$ of a small neighborhood of $(Y_k(\varepsilon) \cap \rho_l^{-1}(\varepsilon_l) - Z_{k,l}) \cup (\cup_{i=1}^{l'-1} C_i)$ in $Y_k(\varepsilon)$ so that $f_l^{-1} \circ \pi_l \circ g_{k,l'-1}$ is PL on $|L_{k,l'-1}|$, and

$$(L_{k,l'-1}, g_{k,l'-1})|_{g_{k,l'-1}^{-1}(U)} = (K_{k,l+1}, f_{k,l+1})|_{f_{k,l+1}^{-1}(U)}$$

for some small neighborhood U of $Y_k(\varepsilon) \cap \rho_l^{-1}(\varepsilon_l) \cap \cup_{i=l+1}^{k-1} \rho_i^{-1}(\varepsilon_i)$ in $Y_k(\varepsilon)$. It suffices to obtain $(L_{k,l'}, g_{k,l'})$ with the corresponding properties.

Choose $\{C_i\}$ so that for each i, there exist an open neighborhood V_i of C_i in $Y_k(\varepsilon)$ and open C^∞ imbeddings $\tau_i : V_i \to \mathbf{R}_+^{m_k}$ and $\theta_i : \pi_l(V_i) \to \mathbf{R}_+^{m_l}$, where

$$m_l = \dim X_l, \qquad m_k = \dim X_k \quad \text{and} \quad \mathbf{R}_+ = [0, \infty[,$$

such that $V_i \cap U = \varnothing$, and the composite $\theta_i \circ \pi_l \circ \tau_i^{-1} : \tau_i(V_i) \to \mathbf{R}_+^{m_l}$ is the restriction of the projection of $\mathbf{R}_+^{m_k}$ onto the first m_l-factors. We can reduce the problem to the following assertion.

Assertion. Let $m' > m''$ be non-negative integers, let $p\colon \mathbf{R}_+^{m'} \to \mathbf{R}_+^{m''}$ be the projection onto the first m''-factors, let $\alpha\colon A \to \mathbf{R}_+^{m'}$ be a C^∞ imbedding of a finite simplicial complex A, let (B, β) be a C^∞ triangulation of $\mathbf{R}_+^{m''}$, and let C be a compact subset of $\mathbf{R}_+^{m'}$ such that $\beta^{-1} \circ p \circ \alpha$ is PL. There exist a simplicial complex A_0 and a C^∞ imbedding $\alpha_0\colon A_0 \to \mathbf{R}_+^{m'}$ such that $|A_0| \supset |A|$, a subdivision of A is a subcomplex of A_0, $\alpha_0|_{|A|}\colon A_0|_{|A|} \to \mathbf{R}_+^{m'}$ is a strong approximation of α,

$$A_0 \supset \{\sigma \in A\colon \alpha(\sigma) \cap C = \varnothing\} \quad (\text{set} = A_1),$$

$$\alpha_0|_{|A_1|} = \alpha|_{|A_1|}, \qquad \alpha_0(|A_0|) \supset C, \qquad (\overline{|A_0| - |A|}) \cap |A_1| = \varnothing,$$

and $\beta^{-1} \circ p \circ \alpha_0$ is PL.

Proof of Assertion. By subdividing A outside of A_1 as in the proof of I.3.16, we assume there exists a subcomplex A_2 of A with

$$N(A_1, A) \cap A_2 = \varnothing, \quad \text{and} \quad \alpha(\overline{|A| - |A_2|}) \cap C = \varnothing.$$

(Then the above definition of A_1 is false, and it holds only that A_1 is a subcomplex of A such that for each $\sigma \in A_1$, $\alpha(\sigma) \cap C = \varnothing$.) Moreover, we can suppose that $\beta^{-1} \circ p \circ \alpha\colon A \to B$ is a simplicial map for the following reason. If we subdivide A and B so that $\beta^{-1} \circ p \circ \alpha$ is simplicial and if we prove the assertion for the subdivisions, then the only problem is that the resulting A_0 does not necessarily satisfy the inclusion $A_0 \supset A_1$. We solve the problem as follows. Let A' and B' be subdivisions of A and B respectively, such that $\beta^{-1} \circ p \circ \alpha\colon A' \to B'$ is simplicial. Assume we have proved the assertion for these (A', α) and (B', β), and let $\alpha_0\colon A_0' \to \mathbf{R}_+^{m'}$ be a solution. Then

$$A_0' \supset \{\sigma \in A'\colon \alpha(\sigma) \cap C = \varnothing\} \quad (\text{set} = \hat{A}_1),$$

$$\alpha_0|_{|\hat{A}_1|} = \alpha|_{|\hat{A}_1|}, \quad \text{and} \quad (\overline{|A_0'| - |A|}) \cap |\hat{A}_1| = \varnothing.$$

Since $|\hat{A}_1| \supset |N(A_1, A)|$, we have $(\overline{|A_0'| - |A|}) \cap |N(A_1, A)| = \varnothing$. It is easy to modify such an A_0' so that $A_0' \supset A_1$. Thus we can suppose that $\beta^{-1} \circ p \circ \alpha\colon A \to B$ is simplicial.

Define a homeomorphism $\xi\colon \mathbf{R}_+^{m'} \to \mathbf{R}_+^{m'}$ by

$$\xi(x', x'') = (\beta^{-1}(x'), x'') \quad \text{for} \quad (x', x'') \in \mathbf{R}_+^{m''} \times \mathbf{R}_+^{m'-m''} = \mathbf{R}_+^{m'}.$$

Note that $\xi \circ \alpha\colon A \to \mathbf{R}_+^{m'}$ is a C^∞ imbedding because $\beta^{-1} \circ p \circ \alpha\colon A \to B$ is simplicial. By the latter half of I.3.16 there exists a strong approximation

$\alpha': A' \to \mathbf{R}^{m'}$ of $\xi \circ \alpha$, which is a C^{∞} imbedding by I.3.14, such that α' is linear on each simplex of A'_2, α' equals $\xi \circ \alpha$ on $|A_1|$, and A' equals A on $|A_1|$, where $A'_2 = A'|_{|A_2|}$. Moreover, as we constructed the approximation in the proof of I.3.16 by using a secant map, we can choose (A', α') so that

$$\alpha'(|A'|) \subset \mathbf{R}^{m'}_+ \quad \text{and} \quad p \circ \alpha' = p \circ \xi \circ \alpha = \beta^{-1} \circ p \circ \alpha,$$

because $p \circ \xi \circ \alpha = \beta^{-1} \circ p \circ \alpha$ is simplicial. Then for each $\sigma \in A'$, $p \circ \alpha'(\sigma)$ is contained in some simplex of B, and $p \circ \alpha'|_{\sigma}$ is linear. Hence if $q: \mathbf{R}^{m'} \to \mathbf{R}^{m'-m''}$ denotes the projection onto the last $(m' - m'')$-factors, then $\xi^{-1} \circ \alpha' = (\beta \circ p \circ \alpha', q \circ \alpha'): A' \to \mathbf{R}^{m'}_+$ is a strong approximation of $\alpha: A \to \mathbf{R}^{m'}_+$ such that $\xi^{-1} \circ \alpha'$ equals α on $|A_1|$ and $\beta^{-1} \circ p \circ \xi^{-1} \circ \alpha'$ is PL by the equality $\beta^{-1} \circ p \circ \xi^{-1} \circ \alpha' = p \circ \alpha' = \beta^{-1} \circ p \circ \alpha$.

We will subdivide the A' and extend the $(A', \xi^{-1} \circ \alpha')$ to the required (A_0, α_0). Let H be a simplicial decomposition of a compact polyhedral neighborhood of $\xi(C)$ in $\mathbf{R}^{m'}_+$ so small that

$$\alpha'(\overline{|A| - |A_2|}) \cap |H| = \varnothing.$$

Subdivide H and A' outside of A_1, and keep the same notation. We can assume that for each $\sigma \in H$, $p(\sigma)$ is contained in some simplex of B and there are full subcomplexes A'_3 of A' and H_3 of H such that

$$\alpha'^{-1}(|H|) = |A'_3|, \qquad |H| \cap \alpha'(|A|) = |H_3|, \qquad \{\alpha'(\sigma): \sigma \in A'_3\} = H_3,$$

and $\alpha'|_{|A'_3|}: A'_3 \to H_3$ is a simplicial isomorphism. This is possible because $\alpha'^{-1}(|H|) \subset |A_2|$ and $\alpha'|_{|A_2|}$ is PL. Let A_0 denote the identification space of the disjoint union of A' and H by the isomorphism $\alpha'|_{|A'_3|}: A'_3 \to H_3$. Let us give naturally a simplicial complex structure to A_0. Define a map $\alpha_0: |A_0| \to \mathbf{R}^{m'}_+$ by

$$\alpha_0 = \begin{cases} \xi^{-1} \circ \alpha' & \text{on} \quad |A'| \\ \xi^{-1} & \text{on} \quad |H|. \end{cases}$$

Then $\alpha_0: A_0 \to \mathbf{R}^{m'}_+$ is a C^{∞} imbedding for the following reason. Since $\xi^{-1} \circ \alpha'$ is a C^{∞} imbedding and $|A_2|$ is a neighborhood of $|A'_3|$ in $|A|$, it suffices to prove that $\xi^{-1}|_{H \cup \{\alpha'(\sigma): \sigma \in A'_2\}}$ is a C^{∞} imbedding. But this is easy by the definition of ξ because $H \cup \{\alpha'(\sigma): \sigma \in A'_2\}$ is a simplicial complex in $\mathbf{R}^{m'}_+$ and because for each $\sigma \in H$, $p(\sigma) \subset \sigma'$ for some $\sigma' \in B$. Clearly (A_0, α_0) fulfills the other requirements in the assertion, i.e., there exist C^{∞} triangulations (K_i, f_i) such that all the maps $f_{i_1}^{-1} \circ \pi_{i_1} \circ (f_{i_2}|_{f_{i_2}^{-1}(\rho_{i_1}^{-1}(\varepsilon_{i_1}))})$ are PL.

It remains to subdivide K_i, $i = 1, \ldots, k$, so that the maps

$$f_{i_1}^{-1} \circ \pi_{i_1} \circ (f_{i_2}|_{f_{i_2}^{-1}(\rho_{i_1}^{-1}(\varepsilon_{i_1}))}): K_{i_2}|_{f_{i_2}^{-1}(\rho_{i_1}^{-1}(\varepsilon_{i_1}))} \longrightarrow K_{i_1}, \quad 1 \le i_1 < i_2 \le k,$$

are simplicial. We call these maps $F(f_{i_1}, K_{i_1}, f_{i_2}, K_{i_2})$. Set

$$Z_{i_2, i_1} = f_{i_2}^{-1}(\rho_{i_1}^{-1}(\varepsilon_{i_1})) \subset |K_{i_2}|.$$

We proceed by induction on k. Assume we can subdivide K_i, $i = 1, \ldots, k-1$, so that for any pair $1 \le i_1 < i_2 < k$, $F(f_{i_1}, K_{i_1}, f_{i_2}, K_{i_2})$ is simplicial. First subdivide K_k so that for each $1 \le i < k$, $Z_{k,i}$ is the underlying polyhedron of a subcomplex of K_k, and for each $\sigma \in K_k|_{Z_{k,i}}$, $f_i^{-1} \circ \pi_i \circ (f_k|_\sigma)$ is a linear map into the ambient Euclidean space of K_i. Second, subdivide all K_i, $1 \le i < k$, so that K_i are compatible with

$$\{f_{i_1}^{-1} \circ \pi_{i_1} \circ f_k(\sigma): \sigma \in K_k|_{Z_{k,i}}\}.$$

Third, apply the induction hypothesis to such K_i, $1 \le i < k$. We can assume that for each pair $1 \le i_1 < i_2 < k$, $F(f_{i_1}, K_{i_1}, f_{i_2}, K_{i_2})$ is simplicial, and each K_i, $1 \le i < k$, has the above property of compatibility.

We want to find a subdivision K_k' of K_k such that for each $1 \le i < k$, $F(f_i, K_i, f_k, K_k')$ is simplicial. First we subdivide K_k to the family:

$$L = (K_k - K_k|_{\bigcup_{i=1}^{k-1} Z_{k,i}}) \cup \bigcup_{i=1}^{k-1} \{\sigma_1 \cap (\pi_i \circ f_k)^{-1}(f_i(\sigma_2)):$$

$$\sigma_1 \in (K_k|_{Z_{k,i}} - K_k|_{\bigcup_{i'>i} Z_{k,i'}}), \ \sigma_2 \in K_i, \ \sigma_1^\circ \cap (\pi_i \circ f_k)^{-1}(f_i(\sigma_2^\circ)) \neq \varnothing\}.$$

(Here the condition $\sigma_1^\circ \cap (\pi_i \circ f_k)^{-1}(f_i(\sigma_2^\circ)) \neq \varnothing$ is equivalent to the one $\pi_i \circ f_k(\sigma_1^\circ) \cap f_i(\sigma_2^\circ) \neq \varnothing$, and if this holds, then by the above compatibility property, we have $\pi_i \circ f_k(\sigma_1^\circ) \supset f_i(\sigma_2^\circ)$.) Then L has the following properties. Each simplex of K_k is a union of elements of L. Each element of L is a cell. A face of a cell σ of L is a union of cells of L. Only the last property is not clear. So we prove it. If σ is a simplex of $K_k - K_k|_{\bigcup_{i=1}^{k-1} Z_{k,i}}$, it is clear. Hence assume

$$\sigma = \sigma_1 \cap (\pi_i \circ f_k)^{-1}(f_i(\sigma_2)) \quad \text{for} \quad \sigma_1 \in K_k|_{Z_{k,i}} - K_k|_{\bigcup_{i'>i} Z_{k,i'}},$$

$$\sigma_2 \in K_i \quad \text{with} \quad \sigma_1^\circ \cap (\pi_i \circ f_k)^{-1}(f_i(\sigma_2^\circ)) \neq \varnothing.$$

A face σ_3 of σ is of the form:

$$\sigma_1' \cap (\pi_i \circ f_k)^{-1}(f_i(\sigma_2')),$$

where σ_1' and σ_2' are faces of σ_1 and σ_2, respectively, such that

$$\sigma_1'^{\,o} \cap (\pi_i \circ f_k)^{-1}(f_i(\sigma_2'^{\,o})) \neq \varnothing.$$

We show by induction on $\dim \sigma_1'$ that σ_3 is a union of cells of L. If σ_1' is not contained in $K_k|_{\cup_{i'>i} Z_{k,i'}}$, then by the definition of L, σ_3 is a cell of L. Assume σ_1' is contained in $K_k|_{Z_{k,i'}} - K_k|_{\cup_{i''>i'} Z_{k,i''}}$ for some $i' > i$. By the condition $\mathrm{ct}(T_i, T_{i'})$ and the simplicial property of $F(f_i, K_i, f_{i'}, K_{i'})$, we have

$$\sigma_3 = \sigma_1' \cap (\pi_i \circ f_k)^{-1}(f_i(\sigma_2')) = \sigma_1' \cap (\pi_{i'} \circ f_k)^{-1}(f_{i'}\{(\pi_i \circ f_{i'})^{-1}(f_i(\sigma_2'))\}),$$

and $(\pi_i \circ f_{i'})^{-1}(f_i(\sigma_2'))$ is the union of the simplices of a subcomplex of $K_{i'}$. Let $L_{i'}$ denote the restriction of this subcomplex to $(\pi_i \circ f_{i'})^{-1}(f_i(\sigma_2')) \cap f_{i'}^{-1}(\pi_i \circ f_k(\sigma_1'))$. Then σ_3 is the union of $\sigma_1' \cap (\pi_{i'} \circ f_k)^{-1}(f_{i'}(\sigma_4))$, $\sigma_4 \in L_{i'}$. Let $\sigma_4 \in L_{i'}$. If the intersection $\sigma_1'^{\,o} \cap (\pi_{i'} \circ f_k)^{-1}(f_{i'}(\sigma_4^{\,o}))$ is not empty, then by the definition of L, $\sigma_1' \cap (\pi_{i'} \circ f_k)^{-1}(f_{i'}(\sigma_4))$ is a cell of L. If the intersection is empty, then for some proper face σ_1'' of σ_1' and for a face σ_4' of σ_4,

$$\sigma_1' \cap (\pi_{i'} \circ f_k)^{-1}(f_{i'}(\sigma_4)) = \sigma_1'' \cap (\pi_{i'} \circ f_k)^{-1}(f_{i'}(\sigma_4')),$$

$$\text{with} \quad \sigma_1''^{\,o} \cap (\pi_{i'} \circ f_k)^{-1}(f_{i'}(\sigma_4'^{\,o})) \neq \varnothing.$$

(Here there is no case where $\sigma_1'' = \sigma_1'$ and σ_4' is a proper face of σ_4 because $\pi_{i'} \circ f_k(\sigma_1') \supset f_{i'}(|L_{i'}|)$.) Hence by the induction hypothesis, $\sigma_1' \cap (\pi_{i'} \circ f_k)^{-1} f_{i'}(\sigma_4)$ is a union of cells of L. Thus σ_3 is a union of cells of L.

In the same way we prove that the intersection of two cells of L is a union of cells of L, and if $\sigma_1 \subsetneqq \sigma_2$ are cells of L, then σ_1 is included in a proper face of σ_2.

For such L also we can prove I.3.12 (see 2.9 in [R-S]). Namely, we can subdivide L to a simplicial complex without introducing new vertices. Let K_k' denote such a simplicial subdivision. Then all $F(f_i, K_i, f_k, K_k')$, $1 \leq i < k$, are simplicial for the following reason. It suffices to prove that for each cell σ of L in $Z_{k,i}$, $f_i^{-1} \circ \pi_i \circ (f_k|_{Z_{k,i}})$ carries σ linearly onto a simplex of K_i. Linearity is immediate by the definition of L. Let us prove that the image of σ is a simplex of K_1. Let σ be of the form $\sigma_1 \cap (\pi_{i_1} \circ f_k)^{-1}(f_{i_1}(\sigma_2))$ for $\sigma_1 \in K_k|_{Z_{k,i_1}} - K_k|_{\cup_{i'>i_1} Z_{k,i'}}$ and $\sigma_2 \in K_{i_1}$ with

$$\sigma_1^{\,o} \cap (\pi_{i_1} \circ f_k)^{-1}(f_{i_1}(\sigma_2^{\,o})) \neq \varnothing.$$

It follows from the last nonempty condition that σ is not included in $Z_{k,i'}$ for any $i' > i_1$. Hence we have $i \leq i_1$. If $i = i_1$, $f_i^{-1} \circ \pi_i \circ f_k(\sigma) = \sigma_2$ because $\sigma_2^{\,o}$ is included in $f_{i_1}^{-1} \circ \pi_{i_1} \circ f_k(\sigma_1^{\,o})$. If $i < i_1$,

$$f_i^{-1} \circ \pi_i \circ f_k(\sigma) = f_i^{-1} \circ \pi_i \circ f_{i_1} \circ f_{i_1}^{-1} \circ \pi_{i_1} \circ f_k(\sigma) = f_i^{-1} \circ \pi_i \circ f_{i_1}(\sigma_2),$$

which is a simplex of K_i because $F(f_i, K_i, f_{i_1}, K_{i_1})$ is simplicial. Hence in any case, $f_i^{-1} \circ \pi_i \circ f_k(\sigma)$ is map to a simplex of K_i. Thus we complete the proof. □

Corollary I.3.21 (Triangulation of a Whitney stratification). *For the same $\{X_i\}$, $\{T_i\}$, δ and ε as in I.3.20, there exist finite simplicial complexes $L_1 \subset \cdots \subset L_k$ and a homeomorphism $\tau \colon |L_k| \to X = \cup_{i=1}^k X_i$ such that for each $1 \leq i \leq k$ and for each $\sigma \in L_i - L_{i-1}$, $\tau(|L_i| - |L_{i-1}|) = X_i$, $\tau|_{\sigma - |L_{i-1}|}$ is a C^∞ imbedding, and the map:*

$$\tau^{-1} \circ \pi_i \circ (\tau|_{\tau^{-1}(|T_i|)}) \colon \tau^{-1}(|T_i|) \longrightarrow |L_i|$$

is PL. Here we shrink $|T_i|$ if necessary.

The following proof needs the method of integration of vector fields. But if $\{X_i\}$ and $\{T_i\}$ are of class C^r $\mathfrak{X}$, we can proceed in the $\mathfrak{X}$-category without the method as in the proof of II.6.1′.

In the proof we use the concept of the mapping cylinder of a simplicial map. Let $f \colon K_1 \to K_2$ be a simplicial map between finite simplicial complexes in $\mathbf{R}^n$. By induction on $\dim K_1$ we define the *mapping cylinder* $C_f(K_1, K_2)$ of f, which is a simplicial complex in $\mathbf{R}^n \times \mathbf{R}^n \times \mathbf{R}$ and whose underlying polyhedron equals the mapping cylinder $C_f(|K_1|, |K_2|)$ of the topological map $f \colon |K_1| \to |K_2|$. Let K_1 and K_2 be given in $\mathbf{R}^n \times 0 \times 0 \subset \mathbf{R}^n \times \mathbf{R}^n \times \mathbf{R}$ and $0 \times \mathbf{R}^n \times 1 \subset \mathbf{R}^n \times \mathbf{R}^n \times \mathbf{R}$, respectively, and let K_1' and K_2' denote the barycentric subdivision of K_1 and K_2, respectively. If $\dim K_1 = -1$, i.e., $K_1 = \varnothing$, then set $C_f(K_1, K_2) = K_2'$. Let $\dim K_2 = k$ and assume we have already defined the mapping cylinder $C_f(K_1^{k-1}, K_2)$. For $\sigma \in K_1 - K_1^{k-1}$, let a_σ denote the middle point of the barycenter of σ and of $f(\sigma)$ in $\mathbf{R}^n \times \mathbf{R}^n \times 1/2$. We set

$$C_f(K_1, K_2) = C_f(K_1^{k-1}, K_2)$$
$$\cup \bigcup_{\sigma \in K_1 - K_1^{k-1}} \{a_\sigma, \ \sigma_1, \ a_\sigma * \sigma_1 \colon \sigma_1 \in K_1'|_\sigma \cup K_2'|_{f(\sigma)} \cup C_{f|_{\partial\sigma}}(K_1|_{\partial\sigma}, K_2|_{f(\partial\sigma)})\}.$$

We show some good properties of $C_f(K_1, K_2)$. Clearly it is a simplicial complex in $\mathbf{R}^n \times \mathbf{R}^n \times [0, 1]$, K_1' and K_2' are subcomplexes of $C_f(K_1, K_2)$, and there is a natural simplicial map $p_f \colon C_f(K_1, K_2) \to K_2'$, which is a retraction and carries the barycenter of a simplex σ of K_1 and the above-mentioned a_σ to the barycenter of $f(\sigma)$. Moreover, we can naturally regard $|C_f(K_1, K_2)|$ as the mapping cylinder of $f \colon |K_1| \to |K_2|$ as follows. For a commutative diagram of simplicial maps:

$$
\begin{array}{ccc}
L_1 & \xrightarrow{\ g\ } & L_2 \\
{\scriptstyle \varphi_1}\Big\downarrow & & \Big\downarrow{\scriptstyle \varphi_2} \\
K_1 & \xrightarrow{\ f\ } & K_2,
\end{array}
$$

there exists a natural simplicial map $\varphi\colon C_g(L_1, L_2) \to C_f(K_1, K_2)$. On the other hand, $C_{\mathrm{id}}(K_1, K_1)$ is naturally and simplicially isomorphic to a derived simplicial subdivision L of the usual cell complex $K_1 \times \{0, 1, [0, 1]\}$. Hence we have a natural simplicial map $\eta\colon L \to C_f(K_1, K_2)$, which equals the identity map on $|K_1| \times 0$ and f on $|K_1| \times 1$. By this map $|C_f(K_1, K_2)|$ becomes the mapping cylinder of the topological map f.

Note. Assume $f(|K_1|) = |K_2|$. A continuous map g from $|C_f(K_1, K_2)|$ to a polyhedron is PL if and only if is $g \circ \eta$ is also, because η is PL and surjective.

Proof of I.3.21. For simplicity of notation we assume $X_i \subset \overline{X_{i+1}}$ for each $1 \le i < k$, which is possible because $\{X_i\}_i$ satisfies the weak frontier condition. For a simplicial complex K in a Euclidean space, K' denotes the barycentric subdivision of K in this proof. Let (K_i, f_i), $i = 1, \ldots, k$, denote the C^∞ triangulations of $X_i - \cup_{j=1}^{i-1}\rho_j^{-1}([0, \varepsilon_j[)$ obtained in I.3.20. Let $0 < i \le k$ be an integer. By downward induction on i we shall find finite simplicial complexes $\varnothing = L_{i-1,i} \subset L_{i,i} \subset \cdots \subset L_{k,i}$ and a homeomorphism $\tau_i\colon |L_{k,i}| \to X - \cup_{j=1}^{i-1}\rho_j^{-1}([0, \varepsilon_j[)$ such that for each $i \le i' \le k$ and for each $\sigma \in L_{i',i} - L_{i'-1,i}$,

$$
\tau_i(|L_{i',i}| - |L_{i'-1,i}|) = X_{i'} - \bigcup_{j=1}^{i-1}\rho_j^{-1}([0, \varepsilon_j[),
$$

$\tau_i|_{\sigma - |L_{i'-1,i}|}$ is a C^∞ imbedding, the map

$$
\tau_i^{-1} \circ \pi_{i'} \circ (\tau_i|_{\tau_i^{-1}(|T_{i'}|)})\colon \tau_i^{-1}(|T_{i'}|) \longrightarrow |L_{i'}|
$$

is PL, for each $0 < j < i$, the set $\tau_i^{-1}(\rho_j^{-1}(\varepsilon_j))$ is the underlying polyhedron of a subcomplex of $L_{k,i}$, and the map

$$
f_j^{-1} \circ \pi_j \circ (\tau_i|_{\tau_i^{-1}(\rho_j^{-1}(\varepsilon_j))})\colon \tau_i^{-1}(\rho_j^{-1}(\varepsilon_j)) \longrightarrow |K_j|
$$

is PL. Here we shrink $|T_{i'}|$'s if necessary. This is trivial for $i = k$, and $L_{1,1} \subset \cdots \subset L_{k,1}$ and τ_1 are what we want if they exist. Hence assuming $i < k$, $L_{i+1,i+1} \subset \cdots \subset L_{k,i+1}$ and τ_{i+1}, we will construct $L_{i,i} \subset \cdots \subset L_{k,i}$ and τ_i.

Set
$$X(i) = X \cap \rho_i^{-1}(\varepsilon_i) - \bigcup_{j=1}^{i-1} \rho_j^{-1}([0, \varepsilon_j[).$$

By subdividing $L_{k,i+1}$ and K_i, we suppose that the map

$$f_i^{-1} \circ \pi_i \circ (\tau_{i+1}|_{\tau_{i+1}^{-1}(X(i))}) \colon L_{k,i+1}|_{\tau_{i+1}^{-1}(X(i))} \longrightarrow K_i$$

is simplicial. Let $\tilde{\pi}_i$ denote this map, and set

$$L_{k,i} = L'_{k,i+1} \cup C_{\tilde{\pi}_i}(L_{k,i+1}|_{\tau_{i+1}^{-1}(X(i))}, K_i),$$

$$\tau_i = \tau_{i+1} \quad \text{on} \quad |L'_{k,i+1}|.$$

By the definition of a mapping cylinder, $L_{k,i}$ is a simplicial complex and contains $L'_{k,i+1}$. To extend τ_i to $|L_{k,i}|$, consider the ρ_i on $X - \cup_{j=1}^{i-1} \rho_j^{-1}([0, \varepsilon_j[)$. Shrink the δ if necessary. By the definition of the Whitney condition it is easy to construct a C^∞ vector field $\xi = \{\xi_{i'}\}_{i'=i+1,\ldots,k}$ on $\{X_{i'} \cap \rho_i^{-1}([0, \varepsilon_i]) - \cup_{j=1}^{i-1} \rho_j^{-1}([0, \varepsilon_j[)\}_{i'=i+1,\ldots,k}$ such that

$$\xi_{i'}\rho_i = 1 \quad \text{and} \quad d\pi_i \xi_{i'} = 0, \quad i' = i+1, \ldots, k.$$

Moreover, as in the proof of I.1.5, we modify ξ so as to be a controlled vector field. Note that it follows from the property $d\pi_i \xi_{i'} = 0$ and from the conditions $\mathrm{ct}(T_j, T_i)$, $0 < j < i$, that each $\xi_{i'}$ is tangent to the manifold possibly with boundary $X_{i'} \cap \rho_i^{-1}([0, \varepsilon_i]) \cap \rho_j^{-1}(\varepsilon_j) - \cup_{j'=1}^{i-1} \rho^{-1}([0, \varepsilon_{j'}[)$ at each point for each $0 < j < i$.

By I.1.6, the property $\xi_{i'}\rho_i = 1$ and by the above note, the flow $\theta = \{\theta_{i'}\}_{i'=i+1,\ldots,k}$ of ξ is well-defined and continuous on the set $X(i) \times]-\varepsilon_i, 0]$. Furthermore,

$$\pi_{i'} \circ \theta(x, t) = \theta(\pi_{i'}(x), t) \quad \text{on a neighborhood of } (X(i) \cap X_{i'}) \times]-\varepsilon_i, 0]$$
$$\text{in } X(i) \times]-\varepsilon_i, 0], \quad i' = i+1, \ldots, k; \tag{1}$$

$$\pi_i \circ \theta(x, t) = \pi_i(x) \quad \text{on} \quad X(i) \times]-\varepsilon_i, 0]; \quad \text{and} \tag{2}$$

$$\varepsilon_i + t = \rho_i \circ \theta(x, t) \quad \text{on} \quad X(i) \times]-\varepsilon_i, 0]. \tag{3}$$

θ carries $X(i) \times]-\varepsilon_i, 0]$ bijectively onto $X \cap \rho_i^{-1}(]0, \varepsilon_i]) - \cup_{j=1}^{i-1} \rho_j^{-1}([0, \varepsilon_j[)$. By (2), for each $x \in X(i)$, $\theta(x, t)$ converges to $\pi_i(x)$ as $t \to -\varepsilon_i$. Extend the τ_i to a map:

$$|C_{\tilde{\pi}_i}(L_{k,i+1}|_{\tau_{i+1}^{-1}(X(i))}, K_i)| \longrightarrow X \cap \rho_i^{-1}([0, \varepsilon_i]) - \bigcup_{j=1}^{i-1} \rho_j^{-1}([0, \varepsilon_j[)$$

so that if we let γ denote its composite with the above-mentioned natural map:

$$\eta_i \colon \tau_{i+1}^{-1}(X(i)) \times [0,1] \longrightarrow |C_{\tilde{\pi}_i}(L_{k,i+1}|_{\tau_{i+1}^{-1}(X(i))}, K_i)|,$$

then we also have

$$\gamma(y,t) = \theta(\tau_{i+1}(y), -\varepsilon_i t) \quad \text{for} \quad (y,t) \in \tau_{i+1}^{-1}(X(i)) \times [0,1[. \tag{4}$$

Keep the notation τ_i for the extension. We have

$$\rho_i \circ \gamma(y,t) = (1-t)\varepsilon_i \quad \text{for} \quad (y,t) \in \tau_{i+1}^{-1}(X(i)) \times [0,1] \quad \text{by (3);} \tag{5}$$

$$\gamma(\tau_{i+1}^{-1}(\pi_{i'}(x)), t) = \pi_{i'} \circ \gamma(\tau_{i+1}^{-1}(x), t) \quad \text{on a neighborhood of} \tag{6}$$
$$(X(i) \cap X_{i'}) \times [0,1[\text{ in } X(i) \times [0,1[, \ i' = i+1, \dots, k, \quad \text{by (1);}$$

$$\pi_i \circ \gamma(y \times [0,1]) = \pi_i \circ \tau_{i+1}(y) \quad \text{for} \quad y \in \tau_{i+1}^{-1}(X(i)) \quad \text{by (2); and} \tag{7}$$

$$\gamma(\tau_{i+1}^{-1}(X(i) \cap \rho_j^{-1}(\varepsilon_j)) \times [0,1])$$
$$= X \cap \rho_i^{-1}([0,\varepsilon_i]) \cap \rho_j^{-1}(\varepsilon_j) - \bigcup_{j'=1}^{i-1} \rho_{j'}^{-1}([0,\varepsilon_{j'}[), \tag{8}$$
$$0 < j < i, \quad \text{by (7) and by } \operatorname{ct}(T_j, T_i).$$

Note also that

$$\eta_i(\tau_{i+1}^{-1}(X(i) \cap \rho_j^{-1}(\varepsilon_j)) \times [0,1])$$
$$= |C_{\tilde{\pi}_i}(L_{k,i+1}|_{\tau_{i+1}^{-1}(X(i)\cap\rho_j^{-1}(\varepsilon_j))}, K_i|_{f_i^{-1}(\rho_j^{-1}(\varepsilon_j))})|, \quad 0 < j < i. \tag{9}$$

We will see that the τ_i satisfies the required conditions. First there are subcomplexes $\varnothing = L_{i-1,i} \subset L_{i,i} \subset \cdots \subset L_{k-1,i}$ of $L_{k,i}$ such that

$$\tau_i(|L_{i',i}| - |L_{i'-1,i}|) = X_{i'} - \bigcup_{j=1}^{i-1} \rho_j^{-1}([0,\varepsilon_j[), \ i' = i, \dots, k.$$

Indeed, by (6), if we set

$$L_{i',i} = \begin{cases} K_i', & i' = i \\ L_{i',i+1}' \cup C_{\tilde{\pi}_i}(L_{k,i+1}|_{\tau_{i+1}^{-1}(X(i)\cap\bigcup_{i''=i+1}^{i'} X_{i''})}, K_i), & i' > i, \end{cases}$$

they fulfill the requirements.

Second, for each $i' \geq i$ and for each $\sigma \in L_{i',i} - L_{i'-1,i}$, $\tau_i|_{\sigma - |L_{i'-1,i}|}$ is a C^∞ imbedding for the following reason. If $\sigma \in L'_{k,i+1}$, this follows from the induction hypothesis. The case $\sigma \in K'_i$ also is trivial. So assume

$$\sigma \in C_{\tilde{\pi}_i}(L_{k,i+1}|_{\tau_{i+1}^{-1}(X(i))}, K_i) - L'_{k,i+1} - K'_i.$$

Choose $\sigma_1 \in L_{k,i+1}|_{\tau_{i+1}^{-1}(X(i))}$ so that $\sigma^\circ \subset \eta_i(\sigma_1^\circ \times [0, 1[)$. Then $\sigma_1 \in L_{i',i+1} - L_{i'-1,i+1}$. By the induction hypothesis, $\tau_i|_{\sigma_1 - |L_{i'-1,i+1}|}$ is a C^∞ imbedding. On the other hand, the flow $\theta_{i'}$ is of class C^∞ on $(X(i) \cap X_{i'}) \times]-\varepsilon_i, 0]$. Hence the composite $\gamma|_{(\sigma_1 - |L_{i'-1,i+1}|) \times [0,1[}$ is a C^∞ imbedding. Since η_i is a simplicial map from a derived simplicial subdivision of $L_{k,i+1}|_{\tau_{i+1}^{-1}(X(i))} \times \{0, 1, [0, 1]\}$ to $C_{\tilde{k}_i}(L_{k,i+1}|_{\tau_{i+1}^{-1}(X(i))}, K_i)$, we have a linear imbedding η_σ of σ into $\sigma_1 \times [0, 1]$ such that $\eta_i \circ \eta_\sigma = \mathrm{id}$. Therefore, $\tau_i|_{\sigma - |L_{i'-1,i}|}$ is a C^∞ imbedding.

Third, we prove that for each $i \leq i' \leq k$, the map $\tau_i^{-1} \circ \pi_{i'} \circ (\tau_i|_{\tau_i^{-1}(|T_{i'}|)})$ is PL as follows. For $i' = i$ we choose $|T_{i'}|$ so that

$$|T_{i'}| - \bigcup_{j=1}^{i'-1} \rho_j^{-1}([0, \varepsilon_j[) = \rho_{i'}^{-1}([0, \varepsilon_{i'}]) - \bigcup_{j=1}^{i'-1} \rho_j^{-1}([0, \varepsilon_j[).$$

As in the above second argument, the PL property follows from the equality (7) and from the induction hypothesis that $f_i^{-1} \circ \pi_i \circ (\tau_{i+1}|_{\tau_{i+1}^{-1}(\rho_i^{-1}(\varepsilon_i))})$ is PL. Assume $i' > i$ and choose $|T_i|$ small enough. By the induction hypothesis $\tau_i^{-1} \circ \pi_{i'} \circ \tau_i$ is PL on $\tau_i^{-1}(|T_{i'}|) \cap |L_{k,i+1}|$. Hence it suffices to consider $\tau_i^{-1} \circ \pi_{i'} \circ \tau_i$ on $\tau_i^{-1}(|T_{i'}|) \cap |C_{\tilde{\pi}_i}(L_{k,i+1}|_{\tau_{i+1}^{-1}(X(i))}, K_i)|$. Moreover, we need only prove that $\tau_i^{-1} \circ \pi_{i'} \circ \gamma$ is PL on a neighborhood of $\tau_{i+1}^{-1}(X(i) \cap X_{i'}) \times [0, 1[$ in $\tau_{i+1}^{-1}(X(i)) \times [0, 1[$ because $\eta_i|_{\tau_{i+1}^{-1}(X(i)) \times [0,1[}$ is a PL homeomorphism to its image. By (6) we have

$$\tau_i^{-1} \circ \pi_{i'} \circ \gamma(y, t) = \eta_i \circ \gamma^{-1} \circ \pi_{i'} \circ \gamma(y, t)$$
$$= \eta_i \circ \gamma^{-1} \circ \gamma(\tau_{i+1}^{-1} \circ \pi_{i'} \circ \tau_{i+1}(y), t) = \eta_i(\tau_{i+1}^{-1} \circ \pi_{i'} \circ \tau_{i+1}(y), t)$$

on a neighborhood of $\tau_{i+1}^{-1}(X(i) \cap X_{i'}) \times [0, 1[$ in $\tau_{i+1}^{-1}(X(i)) \times [0, 1[$.

Now by the induction hypothesis $\tau_{i+1}^{-1} \circ \pi_{i'} \circ \tau_{i+1}$ is PL on a neighborhood of $\tau_{i+1}^{-1}(X(i) \cap X_{i'})$ in $\tau_{i+1}^{-1}(X(i))$. By the definition of a mapping cylinder, η_i is PL. Hence the PL property of $\tau_i^{-1} \circ \pi_{i'} \circ \gamma$ on some domain follows. Thus $\tau_i^{-1} \circ \pi_{i'} \circ \tau_i$ is PL on $\tau_i^{-1}(|T_{i'}|)$ for small $|T_{i'}|$.

Finally, we will prove that for each $0 < j < i$, the set $\tau_i^{-1}(\rho_j^{-1}(\varepsilon_j))$ is the underlying polyhedron of a subcomplex of $L_{k,i}$ and the map $f_j^{-1} \circ \pi_j \circ \tau_i$

is PL on the polyhedron. By the definitions of $L_{k,i}$, τ_i and by the equalities (8), (9), we have

$$\tau_i^{-1}(\rho_j^{-1}(\varepsilon_j)) = \tau_{i+1}^{-1}(\rho_j^{-1}(\varepsilon_j)) \cup |C_{\tilde{\pi}_i}(L_{k,i+1}|_{\tau_{i+1}^{-1}(X(i)\cap\rho_j^{-1}(\varepsilon_j))}, K_i|_{f_i^{-1}(\rho_j^{-1}(\varepsilon_j))})|.$$

Hence $\tau_i^{-1}(\rho_j^{-1}(\varepsilon_j))$ is the underlying polyhedron of a subcomplex of $L_{k,i}$ by the induction hypothesis and by the definition of a mapping cylinder. Moreover, the induction hypothesis implies that $f_j^{-1}\circ\pi_j\circ\tau_i$ is PL on $\tau_{i+1}^{-1}(\rho_j^{-1}(\varepsilon_j))$. Thus it remains to prove that the map is PL on $|C_{\tilde{\pi}_i}(L_{k,i+1}|_{\tau_{i+1}^{-1}(X(i)\cap\rho_j^{-1}(\varepsilon_j))},$ $K_i|_{f_i^{-1}(\rho_j^{-1}(\varepsilon_j))})|$. By the note and (9), it suffices to show that $f_j^{-1}\circ\pi_j\circ\gamma$ is PL on $\tau_{i+1}^{-1}(X(i)\cap\rho_j^{-1}(\varepsilon_j))\times[0,1]$. By (7) and the condition $\mathrm{ct}(T_j,T_i)$ we have

$$\pi_j\circ\gamma(y,t) = \pi_j\circ\pi_i\circ\gamma(y,t) = \pi_j\circ\tau_{i+1}(y)$$
$$\text{for}\quad (y,t)\in\tau_{i+1}^{-1}(X(i)\cap\rho_j^{-1}(\varepsilon_j))\times[0,1].$$

Hence we need only prove that $f_j^{-1}\circ\pi_j\circ\tau_{i+1}$ is PL on $\tau_{i+1}^{-1}(X(i)\cap\rho_j^{-1}(\varepsilon_j))$. But this is contained in the induction hypothesis. Thus we complete the proof.   □

Remark I.3.22. The C^r versions of I.3.13, I.3.20 and I.3.21 hold true for any positive integer r. We prove the version of I.3.13 in the same way as in the C^∞ case. By I.1.15, we can reduce the other versions to the C^∞ cases.

We can prove also the C^ω version and the following semialgebraic C^r version of I.3.13, $r = 1,\ldots,\omega$. A compact set $X\subset\mathbf{R}^n$, which is *locally semialgebraically C^r equivalent* to a polyhedron at each point of X, is *semialgebraically C^r triangulable* uniquely up to PL homeomorphism. Here we define naturally the italic words.

CHAPTER II. 𝔛-SETS

In this chapter, r always denotes a positive integer, and smoothness means C^r smoothness. (However, the theorems II and II′ hold for $r = \infty$ and ω if we assume the property (II.1.8) for such r, which is clear by their proofs.) For its axiomatic treatment, the following definition of a subanalytic set, which is equivalent to the definition in §I.2, is adequate. The family of all subanalytic sets in Euclidean spaces is the smallest family $\mathfrak{S}$ of subsets of Euclidean spaces which satisfies the following axioms.

(i)$_{\mathfrak{S}}$ Every analytic set in any Euclidean space is an element of $\mathfrak{S}$.

(ii)$_{\mathfrak{S}}$ If $X_1 \subset \mathbf{R}^n$ and $X_2 \subset \mathbf{R}^n$ are elements of $\mathfrak{S}$, then $X_1 \cap X_2$ and $X_1 - X_2$ are elements of $\mathfrak{S}$.

(iii)$_{\mathfrak{S}}$ If $p \colon \mathbf{R}^m \to \mathbf{R}^n$ is any linear map and $X \subset \mathbf{R}^m$ is an element of $\mathfrak{S}$ such that the restriction of p to the closure of X is proper, then $p(X)$ is an element of $\mathfrak{S}$.

By (I.2.1.1), (I.2.1.4) and by the Hironaka's desingularization theorem, the family of all subanalytic sets in the sense of §I.2 satisfies these axioms. Moreover, by the following fact, we easily prove that this family is the smallest family which satisfies the axioms. Hence the two definitions of a subanalytic set are equivalent.

Let $X \subset \mathbf{R}^n$ be a subanalytic set in the sense of §I.2. There are a finite number of analytic manifolds M_i and proper analytic maps $f_i \colon M_i \to \mathbf{R}^n$ such that $X = \cup_i (f_{2i-1}(M_{2i-1}) - f_{2i}(M_{2i}))$. We prove this as follows. Set $X_1 = \overline{X}$, and

$$X_{i+1} = \overline{X_i - (X_{i-1} - \cdots - (X_1 - X) \cdots)}, \quad i = 1, 2, \ldots .$$

By (I.2.1.1) and (I.2.1.2) $X_1, X_2, \ldots$ are decreasing subanalytic sets closed in $\mathbf{R}^n$ and empty except for a finite number of members, and we have $X = \cup_{i=1,2,\ldots}(X_{2i-1} - X_{2i})$. Apply (I.2.1.12) to each X_i. We obtain M_i and f_i.

We will consider a family of subsets of Euclidean spaces which satisfies certain axioms. When we take the axioms to be only (i)$_{\mathfrak{S}}$, (ii)$_{\mathfrak{S}}$ and (iii)$_{\mathfrak{S}}$, we also need to treat the family of all subsets of Euclidean spaces. But such a family is too large for us to be interested. We need another axiom. Let $\mathfrak{X}$ be a family of subsets of Euclidean spaces which satisfies the following axioms.

Axiom (i) Every algebraic set in any Euclidean space is an element of $\mathfrak{X}$.

Axiom (ii) If $X_1 \subset \mathbf{R}^n$ and $X_2 \subset \mathbf{R}^n$ are elements of $\mathfrak{X}$, then $X_1 \cap X_2$, $X_1 - X_2$ and $X_1 \times X_2$ are elements of $\mathfrak{X}$.

Axiom (iii) If $X \subset \mathbf{R}^n$ is an element of $\mathfrak{X}$ and $p : \mathbf{R}^n \to \mathbf{R}^m$ is a linear map such that the restriction of p to $\overline{X}$ is proper, then $p(X)$ is an element of $\mathfrak{X}$.

Axiom (iv) If $X \subset \mathbf{R}$ and $X \in \mathfrak{X}$, then each point of X has a neighborhood in X which is a finite union of points and intervals.

In Axiom (iv), the family of all the connected components of X is locally finite at each point of $\mathbf{R}$, because for every $x \in \mathbf{R}$, $x \cup X$ is an element of $\mathfrak{X}$ by Axioms (i) and (ii). By (I.2.1.1), (I.2.1.3) and (I.2.1.4) the family of all subanalytic sets in Euclidean spaces is an example of $\mathfrak{X}$. By (I.2.9.1), (I.2.9.3), (I.2.9.4) and (I.2.9.12) the smallest example of $\mathfrak{X}$ is the family of semialgebraic sets in Euclidean spaces. Hence a compact polyhedron in a Euclidean space is an element of any $\mathfrak{X}$. Sometimes we do not mention the ambient Euclidean space of an element of $\mathfrak{X}$.

An $\mathfrak{X}$-*set* is an element of $\mathfrak{X}$, and an $\mathfrak{X}$-*map* is a continuous map between $\mathfrak{X}$-sets whose graph is an $\mathfrak{X}$-set. A C^r $\mathfrak{X}$-*submanifold* of $\mathbf{R}^n$ is an $\mathfrak{X}$-set which is a C^r submanifold of $\mathbf{R}^n$. A C^r $\mathfrak{X}$-*stratification* of an $\mathfrak{X}$-set $\subset \mathbf{R}^n$ is a C^r stratification locally finite at each point of $\mathbf{R}^n$ (not only of the $\mathfrak{X}$-set), whose strata are $\mathfrak{X}$-sets. We define an $\mathfrak{X}$-*function*, an $\mathfrak{X}$-*homeomorphism*, and $\mathfrak{X}$-*homotopy*, an $\mathfrak{X}$-*isotopy*, a (*weak* or *strong*) C^r $\mathfrak{X}$-*isomorphism* between C^r *Whitney* $\mathfrak{X}$-*stratifications*, and a C^r (*Whitney*) $\mathfrak{X}$-*stratification* of an $\mathfrak{X}$-map in the same way as the subanalytic case. Let $X \subset \mathbf{R}^n$ be an $\mathfrak{X}$-set locally closed in $\mathbf{R}^n$, and let $\{Y_i\}$ be a family of $\mathfrak{X}$-sets in $\mathbf{R}^n$ locally finite at each point of X. An $\mathfrak{X}$-*triangulation* of X compatible with $\{Y_i\}$ is a pair consisting of a simplicial complex K in some $\mathbf{R}^{n'}$ and an $\mathfrak{X}$-homeomorphism $\pi : |K| \to X$ such that $|K|$ is an $\mathfrak{X}$-set, for each $\sigma \in K$, $\pi|_{\sigma^\circ}$ is a C^r diffeomorphism onto the image, and the family $\{\pi(\sigma^\circ) : \sigma \in K\}$ is compatible with $\{Y_i\}$. Let f be an $\mathfrak{X}$-function on the X. An $\mathfrak{X}$-*triangulation* of f compatible with $\{Y_i\}$ is an $\mathfrak{X}$-triangulation (K, π) of X compatible with $\{Y_i\}$ such that for each $\sigma \in K$, $f \circ \pi|_\sigma$ is linear.

In this chapter, we show properties of $\mathfrak{X}$-sets, $\mathfrak{X}$-functions and $\mathfrak{X}$-manifolds. The following theorems and the $\mathfrak{X}$-versions of Thom's first and second isotopy lemmata are of capital importance.

Theorem II (Unique $\mathfrak{X}$-triangulation). *Given a finite family $\{Y_i\}$ of $\mathfrak{X}$-sets in $\mathbf{R}^n$, a compact $\mathfrak{X}$-set X in $\mathbf{R}^n$ admits a unique $\mathfrak{X}$-triangulation compatible with $\{Y_i\}$. Moreover, an $\mathfrak{X}$-function f on X admits a unique $\mathfrak{X}$-triangulation compatible with $\{Y_i\}$. Here uniqueness means that for two $\mathfrak{X}$-triangulations (K_j, π_j), $j = 1, 2$, of X compatible with $\{Y_i\}$ there exists a PL homeomorphism τ from $|K_1|$ to $|K_2|$ such that*

$$\tau(\pi_1^{-1}(Y_i)) = \pi_2^{-1}(Y_i).$$

If (K_j, π_j) are $\mathfrak{X}$-triangulations of f compatible with $\{Y_i\}$, then τ satisfies, in addition, the condition $f \circ \pi_2 \circ \tau = f \circ \pi_1$.

Axiom (v) If a subset X of $\mathbf{R}^n$ is an $\mathfrak{X}$-set locally at each point of $\mathbf{R}^n$ (i.e., each point of $\mathbf{R}^n$ has a neighborhood U such that $X \cap U$ is an $\mathfrak{X}$-set), then X is an $\mathfrak{X}$-set.

Theorem II′. *If $\mathfrak{X}$ satisfies Axiom* (v), II *holds true for a locally closed $\mathfrak{X}$-set $X \subset \mathbf{R}^n$ and for a family $\{Y_i\}$ of $\mathfrak{X}$-sets in X locally finite at each point of X.*

Since the family of subanalytic sets is an example of $\mathfrak{X}$, and satisfies Axiom (v) and (II.1.8) for $r = \omega$ (I.2.1.5), we have

Corollary II″ (Unique subanalytic triangulation)[S-Y]. *Let X be a subanalytic set contained and locally closed in $\mathbf{R}^n$, and let $\{Y_i\}$ be a family of subanalytic sets in X locally finite at each point of X. Then X and a subanalytic function on X admit unique subanalytic triangulations compatible with $\{Y_i\}$.*

Note the following four facts: The family of semialgebraic subsets of Euclidean spaces does not satisfy Axiom (v), any abstract simplicial complex can be imbedded in some Euclidean space so that the image is closed, the underlying polyhedron of such an imbedded simplicial complex is an element of $\mathfrak{X}$ if $\mathfrak{X}$ satisfies (v), and a PL map from such a polyhedron to a Euclidean space is of class $\mathfrak{X}$. Here the second fact is shown as follows. Let a simplicial complex K be imbedded in $\mathbf{R}^n$, and let us order all vertices of K as v_1, $v_2, \ldots$. Define a C^0 function α on $|K|$ so that $\alpha(v_i) = i$ for all v_i, and α is linear on each simplex of K. The graph of α fulfills the requirement.

We divide Theorem II into Theorems II.2.1, II.3.1 and Corollary III.1.4, where II.2.1 will show existence of $\mathfrak{X}$-triangulations of $\mathfrak{X}$-sets, its uniqueness will be shown in III.1.4, and II.3.1 will treat the case of $\mathfrak{X}$-functions.

§II.1. $\mathfrak{X}$-sets

In this section, we explain elementary properties of $\mathfrak{X}$-sets.

(II.1.1) Let $X \subset \mathbf{R}^l$, $Y \subset \mathbf{R}^m$ and $Z \subset \mathbf{R}^n$ be $\mathfrak{X}$-sets, let f_1 and f_2 be $\mathfrak{X}$-functions on X, and let $f \colon X \to Y$ and $g \colon Y \to Z$ be $\mathfrak{X}$-maps. If f_1 and f_2 satisfy the first boundedness condition, that for any bounded set $B \subset \mathbf{R}^l$, the images $f_1(X \cap B)$ and $f_2(X \cap B)$ are bounded, then the product $f_1 f_2$ is an $\mathfrak{X}$-function. If at least one of f_1 and f_2 satisfies the first boundedness condition or if f_1 and f_2 are non-negative, the sum $f_1 + f_2$ is of class $\mathfrak{X}$. If f satisfies the first boundedness condition or if g satisfies the second condition that for any bounded set $C \subset \mathbf{R}^n$, $g^{-1}(C)$ is bounded, then the composite $g \circ f$ is an $\mathfrak{X}$-map.

Let $h = (h_1, \ldots, h_m) \colon X \to \mathbf{R}^m$ be a map. If every h_i is an X-function, h is an X-map. If h is an X-map and, in addition, satisfies the first boundedness condition, then every h_i is an X-function.

Proof. We can prove these facts in the same way as the subanalytic case. $\square$

(II.1.2) Given a nonempty X-set $X \subset \mathbf{R}^n$, the closure $\overline{X}$ is an X-set, and the function g which measures distance from X in $\mathbf{R}^n$ is an X-function.

Proof. Set

$$Y = \{(y, t) \in \mathbf{R}^n \times \mathbf{R} \colon g(y) < t\}.$$

By Axiom (iii), Y is an X-set, because it is the image of the X-set:

$$\{(x, y, t) \in X \times \mathbf{R}^n \times \mathbf{R} \colon |x - y| < t\}$$

under the projection $X \times \mathbf{R}^n \times \mathbf{R} \to \mathbf{R}^n \times \mathbf{R}$. Hence by Axiom (ii), the set

$$Y^c \cap \mathbf{R}^n \times (]0, \infty[) = \{(y, t) \in \mathbf{R}^n \times \mathbf{R} \colon 0 < t \le g(y)\}$$

is an X-set. The image of this set under the projection $\mathbf{R}^n \times \mathbf{R} \to \mathbf{R}^n$ equals $\overline{X}^c$. Hence it follows from Axioms (ii) and (iii) that $\overline{X}$ is an X-set.

By the above arguments, the Y and the complement

$$Y^c = \{(y, t) \in \mathbf{R}^n \times \mathbf{R} \colon g(y) \ge t\}$$

are X-sets. Since g is continuous, $\overline{Y}$ equals the set:

$$\{(y, t) \in \mathbf{R}^n \times \mathbf{R} \colon g(y) \le t\}.$$

Hence, graph $g = \overline{Y} \cap Y^c$. Thus g is an X-function. $\square$

(II.1.3) Let f be a C^1 X-function on an open X-subset U of $\mathbf{R}^n$ such that for any bounded set $B \subset \mathbf{R}^n$, $f(B \cap U)$ is bounded. The partial derivatives $\frac{\partial f}{\partial x_1}, \ldots, \frac{\partial f}{\partial x_n}$ are X-functions on U.

Without the boundedness condition, the map $(f, \frac{\partial f}{\partial x_1}, \ldots, \frac{\partial f}{\partial x_n})$ is of class X.

For a C^1 X-map g from a C^1 X-submanifold X of $\mathbf{R}^n$ to another Y, the differential $dg \colon TX \to TY$ is of class X.

Proof. By using (II.1.5), we prove this in the same way as the subanalytic case. $\square$

(II.1.4) Given a C^1 $\mathfrak{X}$-submanifold X of $\mathbf{R}^n$, the tangent bundle $TX(\subset X \times \mathbf{R}^n) \to X$ is an $\mathfrak{X}$-vector bundle (i.e., TX is an $\mathfrak{X}$-set and the projection is an $\mathfrak{X}$-map).

Proof. It suffices to prove that TX is an $\mathfrak{X}$-set. Consider the set:

$$Y = \overline{\{(x, y, t) \in X \times \mathbf{R}^n \times \mathbf{R} : t > 0, \ |y| = 1, \ x + ty \in X\}} \cap X \times \mathbf{R}^n \times 0.$$

Regard Y as a subset of $X \times \mathbf{R}^n$. By (II.1.2), it is an $\mathfrak{X}$-set, and we have

$$TX = \{(x, ty) \in X \times \mathbf{R}^n : (x, y) \in Y, \ t \in \mathbf{R}\}.$$

Hence by Axiom (iii) TX is an $\mathfrak{X}$-set because TX is the image of $Y \times \mathbf{R}$ under the proper $\mathfrak{X}$-map:

$$\mathbf{R}^n \times \{y \in \mathbf{R}^n : |y| = 1\} \times \mathbf{R} \ni (x, y, t) \to (x, ty) \in \mathbf{R}^n \times \mathbf{R}^n. \qquad \square$$

(II.1.5) Given a C^1 $\mathfrak{X}$-submanifold X of $\mathbf{R}^n$ of dimension m, the set $\{(x, T_x X) \in X \times G_{n,m}\}$ is an $\mathfrak{X}$-set.

Proof. Let $e_1, \dots, e_n$ denote the unit vectors subordinate to the coordinate system of $\mathbf{R}^n$. By the definition of the algebraic structure of $G_{n,m}$ it suffices to prove that the map $X \ni x \to (T_x X(e_1), \dots, T_x X(e_n)) \in \mathbf{R}^n \times \cdots \times \mathbf{R}^n$ is an $\mathfrak{X}$-map. Here we regard $T_x X$ as the orthogonal projection of $\mathbf{R}^n$ onto $T_x X$. Moreover, by (II.1.1) we need only prove that for a vector e of $\mathbf{R}^n$, the set $\{(x, T_x X(e)) \in X \times \mathbf{R}^n\}$ is an $\mathfrak{X}$-set. But this follows from Axioms (i), (ii), (iii), (II.1.4) and the fact that $\{(x, T_x X(e)) \in X \times \mathbf{R}^n\}$ is the image of the $\mathfrak{X}$-set:

$$\{(x, v_1, v_2) \in X \times \mathbf{R}^n \times \mathbf{R}^n : v_1 + v_2 = e, \ v_1 \perp v_2, \ v_1 \in T_x X\}$$

under the projection $X \times \mathbf{R}^n \times \mathbf{R}^n \to X \times \mathbf{R}^n$ which forgets the last factor. $\qquad \square$

(II.1.6) Let $X, Y \subset \mathbf{R}^n$ be $\mathfrak{X}$-sets, and let $f : X \to \mathbf{R}^n$ be an $\mathfrak{X}$-map. If $f^{-1}(B)$ is bounded for any bounded set B in $\mathbf{R}^n$, $f(X)$ is an $\mathfrak{X}$-set. If $f(X \cap B)$ is bounded for the same B, $f^{-1}(Y)$ is an $\mathfrak{X}$-set.

Proof. Consider the graph of f and apply Axiom (iii). $\qquad \square$

(II.1.7) (Singular point set of an 𝔛-map). Let r, k be positive and non-negative integers, respectively. Let X and Y be C^r 𝔛-submanifolds of $\mathbf{R}^n$ and let $f \colon X \to Y$ be a C^r 𝔛-map such that $f(X \cap B)$ is bounded for any bounded set $B \subset \mathbf{R}^n$. The set of the points of X where the differential df has rank k is an 𝔛-set.

Proof. We prove this in the same way as the subanalytic case I.2.5.   □

(II.1.8) Let r be a positive integer. A bounded 𝔛-set $X \subset \mathbf{R}^n$ admits a finite C^r 𝔛-stratification with connected strata. (If X is not bounded, by (II.1.10), it admits a finite C^r 𝔛-stratification whose strata are not necessarily connected.)

We call the maximal dimension of the strata the *dimension* of X.

Proof. Without loss of generality, we assume $r > 1$. We proceed by induction on n. If $n = 0$, (II.1.8) is trivial. Hence we assume this for an 𝔛-set in $\mathbf{R}^{n-1}$. Let $p \colon \mathbf{R}^n \to \mathbf{R}^{n-1}$ denote the projection which forgets the last factor. Set

$$p(X) = Y, \qquad Y_1 = \{y \in Y \colon \dim(X \cap p^{-1}(y)) = 0\},$$
$$Y_2 = Y - Y_1, \quad \text{and} \quad X_i = X \cap p^{-1}(Y_i), \ \ i = 1, 2.$$

By Axiom (iv) $p|_{X_1}$ is a finite-to-one map, for each $y \in Y_2$, $X \cap p^{-1}(y)$ is a finite union of points and non-empty open intervals of $p^{-1}(y)$, and X_i and Y_i, $i = 1, 2$, are 𝔛-sets for the following reason. Consider the image Z under the projection $X \times X \times \mathbf{R} \to X \times \mathbf{R}$, which forgets the second factor of the 𝔛-set:

$$\{(x, x', t) \in X \times X \times \mathbf{R} \colon x \neq x', \ p(x) = p(x'), \ |x - x'| \leq t\}.$$

Z is an 𝔛-set. Let Z' denote the image of the 𝔛-set $X \times (]0, \infty[) - Z$ under the projection $X \times \mathbf{R} \to X$. We have $p(X - Z') = Y_2$. Hence Y_2 is an 𝔛-set.

We can consider the problem on X_1 and on X_2 separately. By the induction hypothesis we assume Y_1 and Y_2 are connected 𝔛-submanifolds of $\mathbf{R}^{n-1}$ of dimension m_1 and m_2, respectively. (II.1.8) follows from the following Claims 4 and 5.

Claim 1. There exists an 𝔛-subset Y_1' of Y_1 of dimension $< m_1$ such that $p|_{X_1 - p^{-1}(Y_1')} \colon X_1 - p^{-1}(Y_1') \to Y_1 - Y_1'$ is a local homeomorphism.

Proof of Claim 1. Let X_1'' denote the subset of X_1 where $p|_{X_1}$ is not locally surjective, and let X_1''' denote the subset of X_1 where $p|_{X_1}$ is not locally injective. It suffices to show that X_1'' and X_1''' are 𝔛-sets, and $p(X_1'')$ and

$p(X_1''')$ are of dimension $< m_1$. First we show that X_1'' is an $\mathfrak{X}$-set. Let Z_1 denote the image of the $\mathfrak{X}$-set:

$$\{(x, x', y, t) \in X_1 \times X_1 \times Y_1 \times \mathbf{R} : |x - x'| \le t, \ p(x') = y\}$$

under the projection $X_1 \times X_1 \times Y_1 \times \mathbf{R} \to X_1 \times Y_1 \times \mathbf{R}$ which forgets the second factor. Z_1 is an $\mathfrak{X}$-set and we have

$$Z_1 = \{(x, y, t) \in X_1 \times Y_1 \times \mathbf{R} : t \ge \varphi(x, y)\},$$

where $\varphi(x, y) = \mathrm{dis}(x, p^{-1}(y) \cap X_1)$ (here define $\mathrm{dis}(x, \varnothing) = \infty$). Set $Z_2 = X_1 \times Y_1 \times \mathbf{R} - Z_1$. The X_1'' is the image of the $\mathfrak{X}$-set:

$$\overline{Z_2} \cap \{(x, y, t) \in X_1 \times Y_1 \times \mathbf{R} : y = p(x), \ t > 0\}$$

under the projection $X_1 \times Y_1 \times \mathbf{R} \to X_1$. Hence X_1'' is an $\mathfrak{X}$-set.

Second, we show $\dim p(X_1'') < m_1$, by reduction to absurdity. Assume $\dim = m_1$. Then by shrinking Y_1, we can suppose $Y_1 = p(X_1'')$. Let us define a function ψ on Y_1 by

$$\psi(y) = \max\{x_n \in \mathbf{R} : (y, x_n) \in X_1''\} \quad \text{for} \quad y \in Y_1.$$

The graph of ψ is an $\mathfrak{X}$-set for the following reason. Let $q : Y_1 \times \mathbf{R} \times \mathbf{R} \to Y_1 \times \mathbf{R}$ denote the projection which forgets the second factor. We have

$$\mathrm{graph}\,\psi = q\{(y, x_n, t) \in X_1 \times \mathbf{R} : x_n \ge t\} - q\{(y, x_n, t) \in X_1 \times \mathbf{R} : x_n > t\}.$$

Hence $\mathrm{graph}\,\psi$ is an $\mathfrak{X}$-set. By Claim 2, if we shrink Y_1 once more, ψ is continuous, which means that $p|_{X_1''}$ is locally surjective on $\mathrm{graph}\,\psi$. That contradicts the definition of X_1''.

Third, it follows that X_1''' is an $\mathfrak{X}$-set from the fact that X_1''' is the image of the $\mathfrak{X}$-set:

$$\overline{\{(x, x', t) \in X_1 \times X_1 \times \mathbf{R} : x \ne x', \ p(x) = p(x'), \ |x - x'| = t\}} \cap X_1 \times X_1 \times 0$$

under the projection of $X_1 \times X_1 \times \mathbf{R}$ onto the first factor.

Finally, we show $\dim p(X_1''') < m_1$ by reduction to absurdity. Suppose $\dim = m_1$. In the same way as in the second step, we can assume there exists an $\mathfrak{X}$-function ψ on Y_1 whose graph is included in X_1'''. Apply the same arguments to the positive $\mathfrak{X}$-function ψ_1 on Y_1 defined by

$$\psi_1(y) = \mathrm{dis}((y, \psi(y)), (X_1 - \mathrm{graph}\,\psi) \cap p^{-1}(y)) \quad \text{for} \quad y \in Y_1.$$

We can suppose that ψ_1 is a continuous function on Y_1, which implies that $p|_{X_1}$ is injective locally at each point of graph ψ. That is a contradiction.

Claim 2. Let f be a bounded function on the Y with $\mathfrak{X}$-graph. There exists an $\mathfrak{X}$-subset Y' of Y of dimension $< \dim Y$ such that $Y - Y'$ is a finite disjoint union of connected C^r $\mathfrak{X}$-submanifolds of $\mathbf{R}^{n-1}$ and f is of class C^r on each submanifold.

Proof of Claim 2. Set $\dim Y = m$. By the induction hypothesis of the proof of (II.1.8), we can suppose that Y is a connected C^r $\mathfrak{X}$-submanifold of $\mathbf{R}^{n-1}$. First we will find an $\mathfrak{X}$-subset U of Y of dimension $< m$ so that $f|_{Y-U}$ is continuous. Let g_1 and g_2 be the functions on Y defined by

$$g_1(y) = \overline{\lim_{y' \to y}} \, f(y') \quad \text{and} \quad g_2(y) = \underline{\lim_{y' \to y}} \, f(y') \quad \text{for} \quad y \in Y.$$

In the same way as in the proof of Claim 1, we prove that their graphs are $\mathfrak{X}$-sets. Hence the set

$$U = \{y \in Y : g_1(y) \neq g_2(y)\}$$

is an $\mathfrak{X}$-set. Clearly it coincides with the non-continuous point set of f. We prove $\dim U < m$ by reduction to absurdity. Assume $\dim U = m$. Set

$$U_i = \{u \in U : g_1(u) \geq g_2(u) + 1/i\}, \quad i = 1, 2, \ldots \, .$$

Then $U = \cup_{i=1}^{\infty} U_i$ and each U_i is an $\mathfrak{X}$-set. Hence some U_i's are of dimension m. Consequently, by the induction hypothesis, by shrinking Y, we suppose $g_1 \geq g_2 + \varepsilon$ on Y for some positive number ε. Divide Y into the $\mathfrak{X}$-sets:

$$\{y \in Y : f(y) \geq (g_1(y) + g_2(y))/2\} \quad \text{and} \quad \{y \in Y : f(y) < (g_1(y) + g_2(y))/2\}.$$

By the same reason as above we can assume Y coincides with one of them, say, the latter. Then $f < (g_1 + g_2)/2$ on Y and hence $f < g_1 - \varepsilon/2$ on Y. Let $a_1 \in Y$. The last inequality implies existence of a point a_2 of Y such that $f(a_2) > f(a_1) + \varepsilon/2$. Hence there exists a sequence $a_1, a_2, \ldots$ in Y such that $f(a_i) \to \infty$ as $i \to \infty$, which contradicts the boundedness of f. Thus we have seen $\dim U < m$. Therefore, we assume f is continuous on Y by the induction hypothesis.

Next we will choose a closed $\mathfrak{X}$-subset V of Y of dimension $< m$ so that f is of class C^1 on $Y - V$. Let $v_1, \ldots, v_{n-1}$ be C^{r-1} $\mathfrak{X}$-vector fields on Y which span the tangent space of Y at each point. (For example, we define $v_1, \ldots, v_{n-1}$ so that for each $y \in Y$, $v_{1y}, \ldots, v_{n-1y}$ are the orthogonal projections of $\frac{\partial}{\partial x_1}, \ldots, \frac{\partial}{\partial x_{n-1}}$ onto $T_y Y_1$.) Let $\omega \colon \Omega \to Y$ be a C^{r-1} $\mathfrak{X}$-tubular neighborhood of Y in $\mathbf{R}^{n-1}$ (whose existence is easy to show). The

f is of class C^1 at a point y of Y if and only if for each $1 \leq i \leq n-1$, the limit

$$\lim_{t \to 0}(f \circ \omega(y + tv_{iy}) - f(y))/t$$

exists and is continuous on a neighborhood of y. Set

$$F_i(y,t) = h\{(f \circ \omega(y + tv_{iy}) - f(y)/t\} \quad \text{for} \quad (y,t) \in \Omega' - Y \times 0,$$

where h is a semialgebraic homeomorphism from $\mathbf{R}$ to $]-1,1[$, and Ω' is a small $\mathfrak{X}$-neighborhood of $Y \times 0$ in $Y \times \mathbf{R}$. Each F_i is an $\mathfrak{X}$-function, and by Axiom (iv), we see that for each $y \in Y$, $F_i(y, \cdot)$ is either constant or strictly monotone on the intervals $[-\varepsilon, 0[$ and on $]0, \varepsilon]$ for some number $\varepsilon > 0$. Hence the functions

$$G_{i+}(y) = \lim_{t>0 \to 0} F_i(y,t) \quad \text{and} \quad G_{i-}(y) = \lim_{t<0 \to 0} F_i(y,t)$$

are well-defined, and their graphs are $\mathfrak{X}$-sets.

By the above result on continuity of f we can assume $G_{i\pm}$ are continuous on Y. On the other hand, f is of class C^1 at a point y of Y if and only if $G_{i+} = G_{i-}$ around y, and $G_{i\pm}(y) \neq \pm 1$ for all i because $G_{i\pm}(y) = \pm 1$ means

$$\lim_{t \to 0}(f \circ \omega(y + tv_{iy}) - f(y))/t = \pm\infty.$$

Hence we need only prove that for each i, the sets $\{y \in Y \colon G_{i+}(y) \neq G_{i-}(y)\}$ and $\{y \in Y \colon G_{i\pm}(y) = \pm 1\}$ are $\mathfrak{X}$-sets of dimension $< m$. These sets are $\mathfrak{X}$-sets because $G_{i\pm}$ are $\mathfrak{X}$-functions, and we can easily prove $\dim < m$ by reduction to absurdity, in the same way as in the proof of continuity of $f|_{Y-U}$.

After continuing the same arguments for $v_i f$, $v_i v_j f$ and so on, we prove Claim 2.

Claim 3. There exists an $\mathfrak{X}$-subset Y_1' of Y_1 of dimension $< m_1$ such that the map

$$p|_{\overline{X_1} \cap p^{-1}(Y_1 - Y_1')} \colon \overline{X_1} \cap p^{-1}(Y_1 - Y_1') \longrightarrow Y_1 - Y_1'$$

is a local homeomorphism.

Proof of Claim 3. By Claim 1 we suppose that $p|_{X_1} \colon X_1 \to Y_1$ is a local homeomorphism, and it suffices to prove

$$\dim\{y \in Y_1 \colon \dim(\overline{X_1} \cap p^{-1}(y)) = 1\} < m_1.$$

Assume $\dim = m_1$. By shrinking Y_1, we can suppose $\dim \overline{X_1} \cap p^{-1}(y) = 1$ for all $y \in Y_1$. Remove the isolated points of $\overline{X_1} \cap p^{-1}(y)$ in $p^{-1}(y)$ for all $y \in Y_1$

(we already saw that this point set is an $\mathfrak{X}$-set), and apply the arguments on ψ in the proof of Claim 1 to $\overline{X_1} \cap p^{-1}(Y_1)$. Then we can reduce the problem to the case where there exist two continuous functions $\psi_1 > \psi_2$ on Y_1 such that $\overline{X_1}$ includes

$$O = \{(y, t) \in Y_1 \times \mathbf{R} : \psi_1(y) > t > \psi_2(y)\}.$$

Since O is open in $\overline{X_1} \cap p^{-1}(Y_1)$, O contains a point x of X_1. However, as noted above, X_1 is the graph of some continuous function on Y_1 locally at x, which implies that X_1 is closed locally at x. That is a contradiction.

Claim 4. There exists an $\mathfrak{X}$-subset Y_1' of Y_1 of dimension $< m_1$ such that $X_1 - p^{-1}(Y_1')$ is a finite disjoint union of connected C^r $\mathfrak{X}$-submanifolds of $\mathbf{R}^n$.

Proof of Claim 4. By Claim 3 we suppose that $p|_{\overline{X_1} \cap p^{-1}(Y_1)} \colon \overline{X_1} \cap p^{-1}(Y_1) \to Y_1$ is a local homeomorphism and hence a covering map because $\overline{X_1} \cap p^{-1}(Y_1)$ is closed in $p^{-1}(Y_1)$. Therefore, the number of the elements of $\overline{X_1} \cap p^{-1}(y)$ is constant, say, k. It follows that $\sharp(X_1 \cap p^{-1}(y)) \le k$.

In the same way as the construction of ψ in Claim 1 we obtain $\mathfrak{X}$-sets $Y_{1,1} = Y_1 \supset \cdots \supset Y_{1,k+1} = \varnothing$ and functions ψ_1 on $Y_{1,1}, \dots, \psi_k$ on $Y_{1,k}$ such that the graphs of $\psi_1, \dots, \psi_k$ are $\mathfrak{X}$-sets, $\psi_1 > \cdots > \psi_k$ and X_1 is the union of graph ψ_i, $i = 1, \dots, k$. Apply the induction hypothesis to $Y_{1,1} - Y_{1,2}, \dots, Y_{1,k} - Y_{1,k-1}$. We have a finite C^r $\mathfrak{X}$-stratification of Y_1 with connected strata compatible with $\{Y_{1,i}\}$. Next apply Claim 2 to each ψ_i on each stratum of dimension m included in $Y_{1,i}$. Then we obtain the required Y_1'.

Claim 5. There exist $\mathfrak{X}$-subsets Y_2' of Y_2 of dimension $< m_2$ and X_2' of X_2 such that X_2' admits a finite C^r $\mathfrak{X}$-stratification with connected strata, and $X_2 - X_2' - p^{-1}(Y_2')$ is a finite disjoint union of connected C^r $\mathfrak{X}$-submanifolds of $\mathbf{R}^n$.

Proof of Claim 5. We have already seen the following. Let X_2' denote the union of the points of $X_2 \cap p^{-1}(y)$ which are boundary points of $X_2 \cap p^{-1}(y)$ in $p^{-1}(y)$ for all $y \in Y_2$. By Claim 4, X_2' admits a finite C^r $\mathfrak{X}$-stratification with connected strata, and for each $y \in Y_2$, $(X_2 - X_2') \cap p^{-1}(y)$ is a finite union of open intervals of $p^{-1}(y)$. Let $\tilde{X}_2$ denote the union of the boundary points of $(X_2 - X_2') \cap p^{-1}(y)$ in $p^{-1}(y)$ for all $y \in Y_2$. By Claim 4, we have a closed $\mathfrak{X}$-subset Y_2' of Y_2 of dimension $< m_2$ such that the map $p|_{\tilde{X}_2 - p^{-1}(Y_2')} \colon \tilde{X}_2 - p^{-1}(Y_2') \to Y_2 - Y_2'$ is a local homeomorphism. Since $X_2 - X_2' - p^{-1}(Y_2')$ lies between $\tilde{X}_2 - p^{-1}(Y_2')$, $X_2 - X_2' - p^{-1}(Y_2')$ is a C^r $\mathfrak{X}$-submanifold of $\mathbf{R}^n$. Moreover, the number of connected components of

$X_2 - X_2' - p^{-1}(Y_2')$ is finite, and each of them is an 𝔛-set by the above arguments. Hence Claim 5 follows. □

(II.1.9) Given a nonempty 𝔛-set $X \subset \mathbf{R}^n$,

$$\dim(\overline{X} - X) < \dim X \quad (\text{define } \dim \varnothing = -1).$$

Proof. Since the problem is local, we consider it around $0 \in \mathbf{R}^n$. We proceed by reduction to absurdity. So assume $m' = \dim(\overline{X} - X) \geq \dim X = m$ as germs at 0. By (II.1.2) $\overline{X} - X$ is an 𝔛-set. Apply (II.1.8) to $\overline{X} - X$. There is a C^r 𝔛-submanifold Y in $\overline{X} - X$ of $\mathbf{R}^n$ of dimension m'. We can suppose that Y is a bounded open subset of $\mathbf{R}^{m'} \times 0$ ($\subset \mathbf{R}^n$) for the following reason. By linearly changing the coordinate system of $\mathbf{R}^n$ and by shrinking Y, we assume Y is the graph of a bounded C^r 𝔛-map $\varphi \colon U \to \mathbf{R}^{n-m'}$, U being a bounded open 𝔛-subset of $\mathbf{R}^{m'}$. Define a C^r diffeomorphism Φ of $U \times \mathbf{R}^{n-m'}$ by

$$\Phi(x, y) = (x, y - \varphi(x)) \quad \text{for} \quad (x, y) \in U \times \mathbf{R}^{n-m'}.$$

We have $\Phi(Y) = U \times 0$, and by (II.1.1), we see that Φ is an 𝔛-map. Hence by (II.1.6) $\Phi(X \cap (U \times \mathbf{R}^{n-m'}))$ is an 𝔛-set in $\mathbf{R}^n$. Thus we can suppose $Y = U \times 0$, and also $X \subset U \times \mathbf{R}^{n-m'}$ and $0 \in Y$.

By (II.1.8) X admits a C^r 𝔛-stratification $\{X_i\}$. Since $\{X_i\}$ is locally finite at 0, for some i, $\overline{X_i - 0} - (X_i - 0)$ includes an open subset of Y. Hence we assume X is a bounded C^r 𝔛-submanifold of $\mathbf{R}^n$.

Let $p \colon \mathbf{R}^n \to \mathbf{R}^{m'} \times 0 \subset \mathbf{R}^n$ denote the orthogonal projection. Let S denote the set of the points of X where $d(p|_X)$ is not of maximal rank. By Axiom (iii), (II.1.7) and Sard's Theorem, S is an 𝔛-set, and $p(S)$ is an 𝔛-set of dimension $< m'$. Hence $Y - p(S)$ is an 𝔛-set of dimension m'. By (II.1.8) $Y - p(S)$ includes a C^r 𝔛-submanifold of $\mathbf{R}^{m'} \times 0$ of dimension m', which is, of course, an open subset of $\mathbf{R}^{m'} \times 0$. So by shrinking Y and X, we can assume that $d(p|_X)$ is of constant rank l, and hence $p|_X$ is locally trivial (i.e., each point of X has a neighborhood U in X such that $p(U)$ is a C^r submanifold of $\mathbf{R}^{m'} \times 0$ and $p|_U$ is a C^r submersion onto $p(U)$).

We claim $l = m = m'$. Clearly $l \leq m \leq m'$. Since $d(p|_X)$ is everywhere of rank l, $p(X)$ is a countable or finite union of C^r submanifolds of Y of dimension l, and hence it is of dimension l. Hence if $l < m'$, then by (II.1.8), $Y - p(X)$ includes an open subset Y' of $\mathbf{R}^{m'} \times 0$. Since $X \subset p^{-1}(Y)$, we conclude $\overline{X} \cap Y' = \varnothing$, which is a contradiction. Thus $l = m = m'$. Therefore, we can assume $Y = p(X)$.

It follows from $l = m = m'$ that $p|_X$ is a local C^r diffeomorphism. Therefore, for each $y \in Y$, $p^{-1}(y) \cap X$ is of dimension 0 and hence a finite set. Let ψ denote the function of Y defined by

$$\psi(y) = \mathrm{dis}(y, p^{-1}(y) \cap X) \quad \text{for} \quad y \in Y.$$

It suffices to prove that graph ψ is an X-set for the following reason. Assume graph ψ is an X-set. The set

$$Z = \{(y, t) \in Y \times \mathbf{R} \colon 0 < t < \psi(y)\}$$

is an X-set. Since ψ is positive, we have $p_1(Z) = Y$, where $p_1 \colon Y \times \mathbf{R} \to Y$ denotes the projection. Hence Z is of dimension $m'+1$. By (II.1.8) Z includes a submanifold Z_1 of $Y \times \mathbf{R}$ of dimension $m' + 1$, which is an open subset of $Y \times \mathbf{R}$. Choose a sufficiently small open neighborhood U_0 of a point of Z_1. The image $p_1(U_0)$ is open in Y and $\mathrm{dis}(Y, X \cap p^{-1}(p_1(U_0)))$ is positive. That contradicts the fact $Y \subset \overline{X}$.

It remains to prove that graph ψ is an X-set. Let W_1 and W_2 denote the images of the sets:

$$\{(y, x, t) \in Y \times X \times \mathbf{R} \colon p(x) = y, \ |x| < t\} \quad \text{and}$$
$$\{(y, x, t) \in Y \times X \times \mathbf{R} \colon p(x) = y, \ |x| \le t\}$$

respectively, under the projection $Y \times X \times \mathbf{R} \to Y \times \mathbf{R}$. By Axioms (i), (ii) and (iii) they are X-sets and we have graph $\psi = W_2 - W_1$. Hence by Axiom (ii), graph ψ is an X-set, which completes the proof. $\square$

(II.1.10) (Singular point set of an X-set). Let r be a positive integer. Let $X \subset \mathbf{R}^n$ be a nonempty X-set. The C^r *singular point* set $\Sigma_r X$ of X is an X-set of dimension $< \dim X$.

Proof. Set $m = \dim X$. First we will prove $\dim \Sigma_r X < m$. Since the problem is local, we can assume X is bounded. By (II.1.8) we have a finite C^r X-stratification $\{X_i\}$ of X. Let X_i be of dimension $= m$ for $i = 1, \ldots, i_0$ and of dimension $< m$ for $i = i_0 + 1, \ldots$. By (II.1.9),

$$\dim(\overline{X_i} - X_i) < m, \ \ i = 1, \ldots, i_0 \ \ \text{and}$$
$$\dim \overline{X_i} < m, \ \ i = i_0 + 1, \ldots.$$

Clearly the union $(\cup_{i=1}^{i_0}(\overline{X_i} - X_i)) \cup (\cup_{i=i_0+1,\ldots}\overline{X_i})$ includes $\Sigma_r X$. Hence $\dim \Sigma_r X < m$.

Next we need to show that $\Sigma_r X$ is an X-set. Consider the case $r = 1$. Let Λ denote the linear m-subspaces of $\mathbf{R}^n$ spanned by m-axes of the x_i-axes, $i = 1, \ldots, n$, where $(x_1, \ldots, x_n)$ is the coordinate system of $\mathbf{R}^n$. For each $\lambda \in \Lambda$, let p_λ denote the orthogonal projection of $\mathbf{R}^n$ onto λ. Let X_λ denote the subset of X where $p_\lambda|_X$ is a local C^1 diffeomorphism. We have $\Sigma_1 X = X - \cup_{\lambda \in \Lambda} X_\lambda$. Hence it suffices to show that each X_λ is an X-set. Fix one λ. In the proof of Claim 1 in the proof of (II.1.8), we already proved that the subset of X where $p_\lambda|_X$ is not a local homeomorphism is an X-set. Hence removing this set from X, we assume that $p_\lambda|_X$ is a local homeomorphism and hence X is a C^0 X-submanifold of $\mathbf{R}^n$.

For each $x \in X$, let $TC_x X \subset \mathbf{R}^n$ denote the *tangent cone* of X at x, i.e.,

$$TC_x X = \{ty \in \mathbf{R}^n : (y,0) \in Y, \ t \in \mathbf{R}\},$$

where

$$Y = \overline{\{(y,t) \in \mathbf{R}^n \times \mathbf{R} : t > 0, \ |y| = 1, \ x + ty \in X\}} \cap \mathbf{R}^n \times 0.$$

Set

$$TCX = \{(x,y) \in X \times \mathbf{R}^n : y \in TC_x X\}.$$

(We call $TCX \to X$ the *tangent cone bundle* of X.) As shown in the proof of (II.1.4), TCX is an X-set and the map $TCX \to X$ is an X-map.

We can assume that for each $x \in X$, $TC_x X$ is linear for the following reason. Consider the X-set:

$$\{(x,y_1,y_2,y_3) \in X \times \mathbf{R}^n \times \mathbf{R}^n \times \mathbf{R}^n : y_1, y_2 \in TC_x X, \ y_3$$
$$= y_1 + y_2, \ |y_1| = |y_2| = 1\},$$

and the image of this set under the projection $X \times \mathbf{R}^n \times \mathbf{R}^n \times \mathbf{R}^n \ni (x,y_1,y_2,y_3) \to (x,y_3) \in X \times \mathbf{R}^n$, which is an X-set. For a point x of X, $TC_x X$ is not linear if and only if the intersection of $x \times \mathbf{R}^n$ with this image is not included in $x \times TC_x X$. Hence the set of the points x with nonlinear $TC_x X$ is an X-set. On the other hand, if $TC_x X$ is not linear, $x \in \Sigma_1 X$. Hence we can remove the nonlinear points from X. Thus we assume $TC_x X$ are linear for all $x \in X$.

Moreover, by the same method as in the above reduction to the case where $p_\lambda|_X$ is a local homeomorphism, we can suppose that for any $x \in X$, $p_\lambda|_{TC_x X}$ is a linear isomorphism onto λ. From this it follows that each $TC_x X$ is of dimension m.

Let φ denote the map from X to $G_{n,m}$ defined by

$$\varphi(x) = TC_x X \quad \text{for} \quad x \in X.$$

In the same way as in the proof of (II.1.5) we can prove that the graph of φ is an X-set. It remains to show that the set of the points of X where φ

is non-continuous is an $\mathfrak{X}$-set. Here by (II.1.1) we can replace φ by a finite number of bounded functions on X with $\mathfrak{X}$-graph, because $G_{n,m}$ is compact and included in some Euclidean space. In this case, we can prove in the same way as Claim 2 in the proof of (II.1.8) that the non-continuous point set is an $\mathfrak{X}$-set. Thus the case $r = 1$ is proved.

For the case $r = 2$, apply the same arguments as above to $\{(x, T_x(X - \Sigma_1 X)) \in (X - \Sigma_1 X) \times G_{n,m}\}$. We see that $\Sigma_2 X$ is an $\mathfrak{X}$-set of dimension $< m$. By repeating these arguments, we prove (II.1.10) for general r.   □

(II.1.10)′ (C^0 singular point set of an $\mathfrak{X}$-set). Let $X \subset \mathbf{R}^n$ be a nonempty $\mathfrak{X}$-set. Assume Axiom (v) if X is not bounded. We can define four kinds of C^0 singular points of X as in (I.2.1.5). Any kind of singular point set is an $\mathfrak{X}$-set of dimension $< \dim X$.

Proof. Using Theorems II and II′, we prove this as in the proof of (I.2.1.5).   □

(II.1.11) For a bounded $\mathfrak{X}$-set $X \subset \mathbf{R}^n$, the number of connected components of X is finite, and each connected component is an $\mathfrak{X}$-set.

Proof. Trivial by (II.1.8).   □

(II.1.12) (Curve selection lemma). Given an $\mathfrak{X}$-set $X \subset \mathbf{R}^n$ with $\overline{X} - X \ni 0$, there exists an $\mathfrak{X}$-curve $\varphi \colon [0,1] \to \overline{X}$ such that $\varphi(0) = 0$ and $\varphi(]0,1[) \subset X$.

Proof. We prove this by induction on n. The case $n = 0$ is trivial. Hence assume (II.1.12) for $n - 1$. Set $\dim X = m$. Let $p \colon \mathbf{R}^n \to \mathbf{R}^{n-1}$ denote the projection which forgets the last factor. We can suppose that for any sequence $a_1, a_2, \ldots$ in X, if $p(a_1), p(a_2), \ldots$ converges to 0, then $a_1, a_2, \ldots$ converges to 0 for the following reason. Let $(0, \lambda) \in \mathbf{R}^n \times G_{n,1}$ be an element of $\overline{\{(x, \overrightarrow{0x}) \in X \times G_{n,1} \colon x \neq 0\}}$. By linearly changing the coordinate system of $\mathbf{R}^n$, we assume $\lambda = $ the x_1-axis. Then 0 is adherent to the set:

$$X \cap \{(x_1, \ldots, x_n) \in \mathbf{R}^n \colon |x_n| \leq |x_1|\}.$$

Replace X with this set. We have the above convergence property.

Set $Y = p(X)$. By the proof of (II.1.8) we have $\mathfrak{X}$-sets X_1, $X_2 \supset X_2'$ in $\mathbf{R}^n$, $Y_1 \supset Y_1'$, and $Y_2 \supset Y_2'$ in $\mathbf{R}^{n-1}$ which satisfy the following conditions:

$$X = X_1 \cup X_2, \qquad Y = Y_1 \cup Y_2, \qquad X_1 \cap X_2 = Y_1 \cap Y_2 = \varnothing,$$

$$\dim X_2' < m, \qquad \dim Y_1' < m, \qquad \dim Y_2' < m - 1, \quad \text{and}$$

$$p^{-1}(Y_i) = X_i, \quad i = 1, 2.$$

Furthermore, $Y_1 - Y_1'$ and $Y_2 - Y_2'$ are C^r $\mathfrak{X}$-submanifolds of $\mathbf{R}^{n-1}$ of dimension m and $m-1$, respectively, $X_2 - X_2' - p^{-1}(Y_2')$ is a C^r $\mathfrak{X}$-submanifold of $\mathbf{R}^n$ of dimension m, for each connected component C_1 of $Y_1 - Y_1'$, $X_1 \cap p^{-1}(C_1)$ is a union of graphs of C^r $\mathfrak{X}$-functions on C_1, and each connected component of $X_2 - X_2' - p^{-1}(Y_2')$ lies between the graphs of some two C^r $\mathfrak{X}$-functions on a connected component of $Y_2 - Y_2'$.

Assume $Y_1 - Y_1' \neq \varnothing$ as germs at 0. Let C_1 be a connected component of $Y_1 - Y_1'$ whose closure contains 0, and let ψ be a C^r $\mathfrak{X}$-function on C_1 whose graph is included in X_1. Apply the induction hypothesis to C_1. We have an $\mathfrak{X}$-curve $\varphi' \colon [0,1] \to \overline{C_1}$ such that $\varphi'(0) = 0$ and $\varphi'(]0,1]) \subset C_1$. Set $\varphi = (\varphi', \psi \circ \varphi')$. Then φ fulfills the requirements.

If $Y_1 - Y_1' = \varnothing$ as germs at 0, then $Y_2 - Y_2' \neq \varnothing$ as germs at 0. Let C_2 be a connected component of $Y_2 - Y_2'$ whose closure contains 0, and let ψ_1 and ψ_2 be C^r $\mathfrak{X}$-functions on C_2 such that a connected component of $(X_2 - X_2') \cap p^{-1}(C_2)$ lies between their graphs. Define $\varphi' \colon [0,1] \to \overline{C_2}$ in the same way as above. The curve $\varphi = (\varphi', [(\psi_1 + \psi_2)/2] \circ \varphi')$ is what we want. $\qquad \square$

(II.1.13) Let r be a positive integer. Let X and Y be nonempty C^r $\mathfrak{X}$-submanifolds of $\mathbf{R}^n$. Assume $X \cap Y = \varnothing$ and $\overline{X} \supset Y$. Let Y' denote the set of points of Y where the pair of X and Y does not satisfy the Whitney condition. Then Y' is an $\mathfrak{X}$-set of dimension $< \dim Y$.

Proof. In the same way as the proof of (I.2.1.6), we can show that Y' is an $\mathfrak{X}$-set. Hence it suffices to prove $\dim Y' < m$, where $m = \dim Y$. We assume Y is a neighborhood of 0 in $\mathbf{R}^m \times 0$ ($\subset \mathbf{R}^n$) because the Whitney condition is invariant under a C^1 diffeomorphism of $\mathbf{R}^n$ and the problem is local. Let $p \colon \mathbf{R}^n \to \mathbf{R}^m \times 0$ denote the orthogonal projection. By the definition of the Whitney condition, X and Y satisfy the Whitney condition at a point b of Y if and only if the following conditions are satisfied.

(1) For any sequence $\{a_i\}$ in X converging to b, $\{|T_{a_i}X(a_i - p(a_i))|/|a_i - p(a_i)|\}$ converges to 1. Here we regard $T_{a_i}X$ as the orthogonal projection of $\mathbf{R}^n$ onto $T_{a_i}X$.

(2) For the same $\{a_i\}$ as above, if $\{T_{a_i}X\}$ converges to a subspace $T \subset \mathbf{R}^n$ in $G_{n,m}$, then $T \supset \mathbf{R}^m \times 0$.

Let Y_1 be the set of points b of y such that for some sequence $\{a_i\}$ in X converging to b, (j) does not hold. Similarly, let Y_2 be the set of points where (2) fails. We will show that $\dim Y_j < m$, $j = 1, 2$.

Proof that $\dim Y_1 < m$. First consider the case of $Y = \{0\}$ and $\dim X = 1$. By (II.1.10) and (II.1.11) we assume X is a simple open curve such that $\overline{X} - X$ consists of 0 and another point. By (II.1.2), (II.1.5) and (II.1.9) $T_x X$

converges to a line (say, the x_1-axis of $\mathbf{R}^n$) in $G_{n,1}$ as $x \in X \to 0$. Then X around 0 is the graph of a C^r X-map $f = (f_1, \ldots, f_{n-1})$ from $]-\varepsilon, 0[$ or from $]0, \varepsilon[$ to $\mathbf{R}^{n-1}$ for some $\varepsilon > 0$. Apply the mean value theorem to each f_i. We see that f_i and f can be extended to $]-\varepsilon, 0]$ or to $[0, \varepsilon[$ as a C^1 map. Thus the special case is clear.

Let us consider the general case. We proceed by reduction to absurdity. Assume $\dim Y_1 = m$. By (II.1.8) Y_1 includes an m-dimensional X-submanifold of Y. Hence shrinking Y and Y_1 we suppose $Y_1 = Y$. Define a function φ on X by

$$\varphi(x) = |T_x X(x - p(x))|/|x - p(x)| \quad \text{for} \quad x \in X.$$

By (II.1.1) and (II.1.5) φ is an X-function, and we can assume $\varphi < \varepsilon$ for some number $\varepsilon < 1$ for the following reason. By (II.1.2), the graph of the function Φ on Y, defined by

$$\Phi(y) = \lim_{x \in X \to y} \varphi(x) \quad \text{for} \quad y \in Y,$$

is also an X-set. By hypothesis, Φ is smaller than 1. Apply (II.1.8) to graph Φ. We see that Φ is continuous on an open X-subset of Y. Hence by shrinking Y, we suppose $\Phi < \varepsilon$ on Y for some number $\varepsilon < 1$. Remove from X the closed X-subset of X consisting of points x such that $\varphi(x) \geq \varepsilon$. X is still a C^r X-manifold with the property $Y \subset \overline{X}$. Thus we can assume $\varphi < \varepsilon$ on X.

Now we reduce the problem to the case of $Y = 0$ and $\dim X = 1$ as follows. Let $\psi \colon Y \to \mathbf{R}$ be the function defined by

$$\psi(y) = \mathrm{dis}(y, p^{-1}(y) \cap X) \quad \text{for} \quad y \in Y.$$

We can prove in the same way as (II.1.2) that the graph of ψ is an X-set. By the same arguments as on Φ, we suppose that ψ is continuous on Y. The inclusion $\overline{X} \supset Y$ implies that $\psi = 0$ on Y, that is, each point y of Y is adherent to $X \cap p^{-1}(y)$. By (II.1.12) we have an X-curve C in $X \cap p^{-1}(0)$ such that $0 \in \overline{C}$. Replace X and Y with C and 0, respectively. The inequality $\varphi < \varepsilon$ on X implies that

$$|T_x C(x - p(x))|/|x - p(x)| < \varepsilon \quad \text{for} \quad x \in C.$$

Hence C and 0 do not satisfy the condition (1) for $b = 0$, which contradicts the fact $\dim Y_1 < m$ in the above special case. Thus we have proved $\dim Y_1 < m$.

Proof that $\dim Y_2 < m$. We proceed by induction on $m' - m$, where $m' = \dim X$. There is no case of $m = m'$. So assume $\dim Y_2 < m$ for lower difference of the dimensions. In the same way as above, supposing $Y_2 = Y$, we will arrive at a contradiction.

Let us consider the special case where $m = n - 2$, $m' = n - 1$, and X is the graph of a C^r $\mathfrak{X}$-function g defined on an open $\mathfrak{X}$-set U of $\mathbf{R}^m \times (]0, \infty[) \subset \mathbf{R}^{n-1}$. First by the above arguments concerning ψ we can assume $U = Y \times (]0, \delta[)$ for some positive number δ. (Here we regard $\mathbf{R}^m = \mathbf{R}^m \times 0$ and $Y = Y \times 0$.) Let $L \colon X \to G_{n,m}$ denote the map defined by

$$L(x) = T_x(X \cap (x + \mathbf{R}^m \times 0 \times \mathbf{R})) \quad \text{for} \quad x \in X.$$

Then $L(x)$ is the linear subspace of $\mathbf{R}^n$ spanned by $(1, 0, \dots, 0, \frac{\partial g}{\partial x_1}(x)), \dots$, $(0, \dots, 0, 1, 0, \frac{\partial g}{\partial x_m}(x))$, and we see that L is an $\mathfrak{X}$-map. It follows from the equality $Y_2 = Y$ that for each point b of Y, there is a sequence $\{a_i\}$ in X converging to b such that $\{L(a_i)\}$ converges to some $L \neq \mathbf{R}^m \times 0$ of $G_{n,m}$. Give the metric on $G_{n,m}$ that is induced by the inclusion of $G_{n,m} \subset \mathbf{R}^{n^2}$ and set

$$\varphi'(x) = \operatorname{dis}(L(x), \mathbf{R}^m \times 0) \quad \text{for} \quad x \in X.$$

Then φ' is an $\mathfrak{X}$-function on X, and for the above $\{a_i\}$ and b, $\{\varphi'(a_i)\}$ converges to a positive number. By the same arguments as we used on φ in the proof of $\dim Y_1 < m$, we can assume $\varphi' > \varepsilon$ on X for some positive number ε. (Here we need to shrink Y once more in order to keep the property $U = Y \times (]0, \delta[)$.) Moreover, we suppose that for any sequence $\{a_i\}$ in X converging to a point b of Y, $\{L(a_i)\}$ converges for the following reason. By (II.1.2) and (II.1.12) $\overline{\operatorname{graph} L} \cap (Y \times G_{n,m})$ is of dimension m. Hence by shrinking Y if necessary, we assume $\overline{\operatorname{graph} L} \cap (b \times G_{n,m})$ is of dimension 0 for any $b \in Y$. Then $\overline{\operatorname{graph} L} \cap (b \times G_{n,m})$ consists of one point. If it were to contain two points $b \times L_1$ and $b \times L_2$, there would exist two sequences $\{a_{1,i}\}$ and $\{a_{2,i}\}$ in X converging to b such that $\{L(a_{1,i})\}$ and $\{L(a_{2,i})\}$ converge to L_1 and L_2, respectively, and hence the set

$$\{(x, L(x)) \in X \times G_{n,m} : x \in \bigcup_{i=1}^{\infty} C_i\} \cap (b \times G_{n,m})$$

would not be a finite set, which is a contradiction. Here C_i are curves joining $a_{1,i}$ with $a_{2,i}$ in X such that $\cap_{k=1}^{\infty} \overline{\cup_{i=k}^{\infty} C_i} = b$. (Existence of $\{C_i\}$ follows from the assumption $U = Y \times (]0, \delta[)$, e.g., $C_i = X \cap (\Delta a_{1,i} a_{2,i} + 0 \times \mathbf{R})$.) Thus $\{L(a_i)\}$ converges to some element of $G_{n,m}$. Define $L(b) =$ the element. We have extended L to $X \cup Y$ as an $\mathfrak{X}$-function. Note that $L(y) \neq \mathbf{R}^m \times 0$ for any $y \in Y$ because $\varphi' > \varepsilon$ on X.

Next we want to reduce the problem to the case $\dim Y = 1$. For simplicity of notation, we assume $L(0) \cap 1 \times 0 \times \cdots \times 0 \times \mathbf{R} \neq 1 \times 0 \times \cdots \times 0$.

Set $Z = \mathbf{R} \times 0 \times \cdots \times 0 \times \mathbf{R} \times \mathbf{R} \subset \mathbf{R}^n$ and consider $Y \cap Z$ and $X \cap Z$ in place of Y and X. Then $Y \cap Z$ is an open $\mathfrak{X}$-set in $\mathbf{R} \times 0 \times \cdots \times 0$, $X \cap Z$ is a C^r $\mathfrak{X}$-submanifold of Z because $X \cap Z = \operatorname{graph} g|_{U \cap \mathbf{R} \times 0 \times \cdots \times 0 \times \mathbf{R}}$, and $T_x(X \cap Z \cap (x + \mathbf{R} \times 0 \times \cdots \times 0 \times \mathbf{R}))$ converges to $L(0) \cap \mathbf{R} \times 0 \times \cdots \times 0 \times \mathbf{R} \neq \mathbf{R} \times 0 \times \cdots \times 0$ as $x \to 0$. Hence by replacing Y and X with $Y \cap Z$ and $X \cap Z$ respectively, we reduce the problem to the case $\dim Y = 1$. As already noted, $L(x)$ is spanned by $(1, 0, \ldots, 0, \frac{\partial g}{\partial x_1}(x))$. Hence we can assume $\frac{\partial g}{\partial x_1} > \varepsilon'$ on X around 0 for some positive number ε', which implies

$$g(x_1, x_2) > \varepsilon' x_1 + g(0, x_2) \quad \text{for } (x_1, x_2) \in U \text{ near } 0 \text{ with } x_1 > 0, \; x_2 > 0.$$

Therefore, for fixed small $x_1 > 0$, there is no case where both $g(0, x_2)$ and $g(x_1, x_2)$ converge to 0 as $x_2 \to 0$. That contradicts the fact $Y \subset \overline{X}$. Thus we have proved the special case where $m = n - 2$ and X is the graph of a C^r function.

It remains to reduce the general case to this special case. Let $q \colon \mathbf{R}^n \to \mathbf{R}^{m'} \times 0 \subset \mathbf{R}^n$ denote the orthogonal projection. We assume $q|_X$ is C^r regular and $q(X) \cap Y = \varnothing$. This is possible if we shrink X and Y and linearly change the coordinate system of $\mathbf{R}^n$.

Let us introduce a distance between $L \in G_{n,l}$ and $L' \in G_{n,l'}$ for positive integers $l < n$ and $l' < n$. If $l = l'$, we already did. If $l > l'$, let L'' denote the image of L' under the orthogonal projection onto L and set

$$\operatorname{dis}(L, L') = \begin{cases} 1 & \text{if} \quad \dim L'' < \dim L' \\ \operatorname{dis}(L'', L') & \text{if} \quad \dim L'' = \dim L'. \end{cases}$$

We define similarly $\operatorname{dis}(L, L')$ in the case of $l < l'$. The graph of the map $G_{n,l} \times G_{n,l'} \ni (L, L') \to \operatorname{dis}(L, L') \in \mathbf{R}$ is an $\mathfrak{X}$-set. By the same reason as in the special case, by shrinking X and Y, we assume

$$\operatorname{dis}(T_x X, \mathbf{R}^m \times 0) > \varepsilon \quad \text{for} \quad x \in X, \tag{3}$$

for some positive number ε.

We can suppose $\overline{\overline{X} - X - Y} \cap Y = \varnothing$ (i.e., there exists a neighborhood V of Y in $\mathbf{R}^n$ such that $\overline{X} \cap V = (X \cup Y) \cap V$) for the following reason. If $\dim(\overline{\overline{X} - X - Y} \cap Y) < m$, then it suffices to remove $\overline{\overline{X} - X - Y} \cap Y$ from Y. So assume $\dim(\overline{\overline{X} - X - Y} \cap Y) = m$. By (II.1.8) there is an C^r $\mathfrak{X}$-submanifold X_1 of $\mathbf{R}^n$ such that $X_1 \subset \overline{X} - X - Y$ and $(\overline{X_1} - X_1) \cap Y$ is of dimension m. By (II.1.9), $m < \dim X_1 < m'$. Apply the induction hypothesis to the pairs (X, X_1) and (X_1, Y), and shrink X_1 and Y. We can assume that both (X, X_1) and (X_1, Y) satisfy the condition (2). Hence there is a sequence $\{b_i\}$ in X_1 converging to a point c of Y, and for each i, a sequence $\{a_{i,j}\}_{j=1,2,...}$ in X converging to b_i, such that $\{T_{b_i}X_1\}$ converges to a subspace T of $\mathbf{R}^n$ which includes $\mathbf{R}^m \times 0$ and for each i, $\{T_{c_{i,j}}X\}_{j=1,2,...}$ converges to a subspace T_i of $\mathbf{R}^n$ which includes $T_{b_i}X_1$. This contradicts (3). Thus we can suppose $\overline{X-X-Y} \cap Y = \varnothing$.

If $m' > m+1$, by replacing X with $X_2 = X \cap q^{-1}(\mathbf{R}^{m+1} \times 0)$, we can reduce the problem to the case of $m' = m+1$ for the following reason. Clearly X_2 is an C^r $\mathcal{X}$-submanifold of $\mathbf{R}^n$ because $q|_X$ is C^r regular. It follows from (3) and the inclusion $T_x X_2 \subset T_x X$ for $x \in X_2$ that

$$\mathrm{dis}(T_x X_2, \mathbf{R}^m \times 0) > \varepsilon \quad \text{for} \quad x \in X_2.$$

Hence it suffices to prove $\overline{X_2} \supset Y$ as germs at 0. For that we claim that for an arbitrarily small open neighborhood V of 0 in $\mathbf{R}^n$, $Y \cup q(X \cap V)$ is a neighborhood of 0 in $\mathbf{R}^{m'} \times 0$. For the moment, we assume this claim. Then $Y \cup q(X_2 \cap V)$ is a neighborhood of 0 in $\mathbf{R}^{m+1} \times 0$. Choose V so small that $\overline{X} \cap \overline{V} = (X \cup Y) \cap \overline{V}$, which is possible by the property $\overline{X-X-Y} \cap Y = \varnothing$. It follows that $\overline{X_2} \cap \overline{V} \subset (X_2 \cup Y) \cap \overline{V}$. Suppose $\overline{X_2} \not\supset Y$ as germs at 0. Let $a \in (Y - \overline{X_2}) \cap V$ be such that $q(a) = a$ is adherent to $q(X_2 \cap V)$, and let $\{a_i\}$ be a sequence in $X_2 \cap V$ such that the sequence $\{q(a_i)\}$ converges to a. Here we can assume $\{a_i\}$ converges to a point b of $\overline{X_2} \cap \overline{V}$. Then by the property $\overline{X_2} \cap \overline{V} \subset (X_2 \cup Y) \cap \overline{V}$ we have $b \in X_2 \cup Y$. But $b \notin X_2$ because $q(b) = a$ and $Y \cap q(X) = \varnothing$. Hence $b \in Y$. Moreover, since $q|_Y = \mathrm{id}$, $b = a$. Therefore, $a \in \overline{X_2}$, which is a contradiction.

Proof of the claim. Note that $q(X \cap V)$ is an open subset of $\mathbf{R}^{m'} \times 0$ and does not contain 0 because $q|_X$ is C^r regular and $q(x) \cap Y = \varnothing$. Set

$$W_1 = q(X \cap V),$$
$$W_2 = \mathrm{bdry}\, q(X \cap V) \quad \text{as a subset of} \quad \mathbf{R}^{m'} \times 0, \quad \text{and}$$
$$W_3 = \mathbf{R}^{m'} \times 0 - \overline{W_1}.$$

Since $0 \notin W_1$ and $0 \in W_2$, there are two cases: (i) $0 \in \overline{W_3}$ or (ii) $0 \notin \overline{W_3}$. If (i), W_2 is of dimension $\geq m' - 1$ locally at 0 because $0 \in \overline{W_1}$. Hence if we prove $W_2 \subset Y$ around 0, case (i) does not exist because $m' - 1 > m = \dim Y$. Even supposing case (ii), it suffices to prove $W_2 \subset Y$ around 0 to prove the claim. Choose V so small that $\overline{V}$ is compact and $\overline{X} \cap \overline{V} = (X \cup Y) \cap \overline{V}$. Then

$$\overline{X \cap V} - X \cap V - Y \subset \overline{X} \cap \overline{V} - X \cap V - Y = X \cap (\overline{V} - V).$$

Hence

$$W_2 - Y = \overline{q(X \cap V)} - q(X \cap V) - Y$$
$$= q(\overline{X \cap V}) - q(X \cap V) - Y \subset q(X \cap (\overline{V} - V)).$$

Therefore, we need only show that $0 \notin \overline{q(X \cap (\overline{V} - V))}$. Assume that $0 \in \overline{q(X \cap (\overline{V} - V))}$. There would exist a sequence $\{a_i\}$ in $X \cap (\overline{V} - V)$ such that $q(a_i) \to 0$ as $i \to \infty$. Since $\overline{V} - V$ is compact, we suppose that $\{a_i\}$ converges to a point a of $\mathbf{R}^n$. Then $a \in X$ or $a \in Y$ because $\overline{X} \cap \overline{V} \subset X \cup Y$. If $a \in X$, then $q(a) = 0$, which contradicts the property $q(X) \cap Y = \varnothing$. If $a \in Y$, then $q(a) = a = 0$ and hence $a_i \to 0$ as $i \to 0$, which contradicts $\{a_i\} \subset \overline{V} - V$. Thus we have proved the claim.

Case of $m' = m + 1$. By shrinking X and Y, we suppose that Y is simply connected and $q(X) = Y \times (]0, \delta[)$ for some positive number δ. If we lessen δ, $q|_X : X \to Y \times (]0, \delta[)$ is a finite covering map, which we can prove using the property $\overline{X} - X - Y \cap Y = \varnothing$ in the same way as above. Furthermore, since $Y \times (]0, \delta[)$ is simply connected, the covering map is trivial. On the other hand by (II.1.11), we can replace X with a connected component of X. Hence we assume that X is the graph of a C^r $\mathfrak{X}$-map $\xi = (\xi_1, \dots, \xi_{n-m-1}) : Y \times (]0, \delta[) \to \mathbf{R}^{n-m-1}$. It suffices to consider graph ξ_i, $i = 1, \dots, n - m - 1$, in place of ξ. Therefore, we have reduced the problem to the special case already proved, which completes the proof. □

(II.1.14) (Existence of a canonical C^r Whitney $\mathfrak{X}$-stratification). Let r be a positive integer. Let $\{A_\nu\}$ be a finite family of $\mathfrak{X}$-sets in $\mathbf{R}^n$. An $\mathfrak{X}$-set $X \subset \mathbf{R}^n$ admits the canonical finite C^r Whitney $\mathfrak{X}$-stratification compatible with $\{A_\nu\}$.

Proof. We can prove this in the same way as I.2.2. □

(II.1.15) Let r be a positive integer. Let X, $Y \subset \mathbf{R}^n$ be $\mathfrak{X}$-sets, let $\{A_\nu\}$ and $\{A_{\nu'}\}$ be finite families of $\mathfrak{X}$-sets in $\mathbf{R}^n$, and let $f : X \to Y$ be an $\mathfrak{X}$-homeomorphism such that for a bounded set $B \subset \mathbf{R}^n$, $f(X \cap B)$ and $f^{-1}(B)$ are bounded. There exist finite C^r Whitney $\mathfrak{X}$-stratifications $\{X_i\}$ of X and $\{Y_i\}$ of Y compatible with $\{A_\nu\}$ and $\{A_{\nu'}\}$, respectively, such that $f : \{X_i\} \to \{Y_i\}$ is the canonical C^r $\mathfrak{X}$-isomorphism.

Proof. Same as the proof of I.2.3. □

(II.1.16) Let X, $Y \subset \mathbf{R}^n$ be compact $\mathfrak{X}$-sets, and let a and b be points in $\mathbf{R}^n$ such that the cones $a * X$ and $b * Y$ in $\mathbf{R}^n$ are well-defined.

(i) $a * X$ and $b * Y$ are $\mathfrak{X}$-sets.

Moreover, let $f : X \to Y$ and $F : a * X \to b * Y$ be $\mathfrak{X}$-homeomorphisms such that $F(X) = Y$, and let $g_t : X \to Y$, $t \in [0, 1]$, be an $\mathfrak{X}$-isotopy of $F|_X$.

(ii) The cone extension homeomorphism $h : a * X \to b * Y$ of f is an $\mathfrak{X}$-map.

(iii) There exists an $\mathfrak{X}$-isotopy F_t, $t \in [0, 1]$, of F such that for each $t \in [0, 1]$, $F_t = F$ on X, and F_1 is the cone extension of $F|_X$.

(iv) There exists also an $\mathfrak{X}$-isotopy $G_t : a * X \to b * Y$, $t \in [0, 1]$, of F such that for each $t \in [0, 1]$, $G_t = g_t$ on X, and G_1 is the cone extension of g_1.

We can prove these and construct F_t and G_t in the same way as I.2.4. In the $\mathfrak{X}$-case also, we call the method of construction of F_t and G_t the *Alexander trick* if X and Y are polyhedra.

Let (K, π) be an $\mathfrak{X}$-triangulation of a compact $\mathfrak{X}$-set Y. Let $a * K$ denote the *cone on K* (i.e. $\{a, a * \sigma, \sigma : \sigma \in K\}$). Then $(a * K, a$ cone extension of $\pi)$ is an $\mathfrak{X}$-triangulation of a cone $b * Y$, which we call a *cone extension* of (K, π).

(II.1.17) (The Canonical stratification of an $\mathfrak{X}$-map). Let r be a positive integer. Let X and Y be $\mathfrak{X}$-sets in $\mathbf{R}^n$, let $\{A_\nu\}$ and $\{A_{\nu'}\}$ be finite families of $\mathfrak{X}$-sets in $\mathbf{R}^n$, and let $f : X \to Y$ be an $\mathfrak{X}$-map such that for a bounded set $B \subset \mathbf{R}^n$, $f(X \cap B)$ and $f^{-1}(B)$ are bounded. Then f admits the canonical C^r (Whitney) $\mathfrak{X}$-stratification (with the weak frontier condition) $(\{X_i\}, \{Y_j\})$ such that $\{X_i\}$ and $\{Y_j\}$ are compatible with $\{A_\nu\}$ and $\{A_{\nu'}\}$, respectively. Even if $f^{-1}(B)$ is not always bounded, there exists a finite C^r Whitney $\mathfrak{X}$-stratification $\{X_i\}$ of X compatible with $\{A_\nu\}$ such that $f|_{X_i}$ are of class C^r.

Proof. We prove the first half in the same way as Lemma I.2.6. For the latter half, it suffices to apply the first half to the projection graph $f \to X$. $\quad\square$

(II.1.18) (Friendliness of $\mathfrak{X}$-functions). Let r be a positive integer. Let $\{X_1, X_2\}$ be a C^r $\mathfrak{X}$-stratification of an $\mathfrak{X}$-set $X \subset \mathbf{R}^n$, and let f_1 and f_2 be $\mathfrak{X}$-functions on X. Assume that $f_1^{-1}(0) = f_2^{-1}(0) = X_1$, and the restrictions of f_1 and f_2 to X_2 are positive and of class C^r. Then $f_1|_{X_2 \cap U}$ and $f_2|_{X_2 \cap U}$ are friendly for some open $\mathfrak{X}$-neighborhood U of X_1 in $\mathbf{R}^n$. Here we define *friendliness* of two C^r functions in the same way as the C^∞ case.

Proof. See the proof of I.2.7. $\quad\square$

(II.1.19) Let $X, Y \subset \mathbf{R}^n$ be compact $\mathfrak{X}$-sets, let $\{X_i\}$ and $\{Y_i\}$ be C^r Whitney $\mathfrak{X}$-stratifications of X and of Y, respectively, and let $f \colon \{X_i\} \to \{Y_i\}$ be a C^r $\mathfrak{X}$-isomorphism. Then f is a strong C^r $\mathfrak{X}$-isomorphism for any C^r $\mathfrak{X}$-tube systems for $\{X_i\}$, $\{Y_i\}$ and $\{Z_i = \operatorname{graph} f|_{X_i}\}$.

Proof. Trivial by (II.1.18). $\qquad\qquad\qquad\qquad\qquad\qquad\qquad\qquad\qquad$ $\square$

§II.2. Triangulations of $\mathfrak{X}$-sets

Theorem II.2.1 ($\mathfrak{X}$-triangulation of an $\mathfrak{X}$-set). *For a finite family $\{Y_i\}$ of $\mathfrak{X}$-sets in $\mathbf{R}^n$, a compact $\mathfrak{X}$-set X in $\mathbf{R}^n$ admits an $\mathfrak{X}$-triangulation compatible with $\{Y_i\}$.*

Theorem II.2.1'. *If $\mathfrak{X}$ satisfies Axiom* (v), *II.2.1 holds true for an $\mathfrak{X}$-set $X \subset \mathbf{R}^n$ locally closed in $\mathbf{R}^n$ and for a family $\{Y_i\}$ of $\mathfrak{X}$-sets in X locally finite at each point of X.*

In this section, we prove these theorems. The idea of proof of II.2.1 is similar to that used by [L₁], which proved triangulations of semialgebraic and semianalytic sets. First we will find a good projection of $\mathbf{R}^n$ onto a hyperplane, next triangulate the image of X under this projection and then triangulate X.

Let A be a subset of $\mathbf{R}^n$. Set

$$S^{n-1} = \{\lambda \in \mathbf{R}^n \colon |\lambda| = 1\}.$$

We call a point λ of S^{n-1} a *singular direction* for A at a point a of A if $A \cap (a + \mathbf{R}\lambda)$ has interior points in the line $a + \mathbf{R}\lambda$. A *singular direction* for A is a singular direction for A at some point of A.

Lemma II.2.2 (Singular directions for an $\mathfrak{X}$-set). *Let A be a bounded $\mathfrak{X}$-set in $\mathbf{R}^n$ of dimension $< n$. The set of singular directions for A is an $\mathfrak{X}$-set of dimension $< n - 1$.*

Proof. Set
$$B = \{(x, t, \lambda) \in A \times \mathbf{R} \times S^{n-1} \colon x + t\lambda \in A\},$$

and let p denote the restriction to B of the projection of $A \times \mathbf{R} \times S^{n-1}$ onto $A \times S^{n-1}$. By Axioms (i) and (ii), B and p are of class $\mathfrak{X}$. A point λ of S^{n-1} is a singular direction for A at a point a of A if and only if $p^{-1}(a, \lambda)$ is of dimension 1. Let S denote the set of such points (a, λ) in $A \times S^{n-1}$. Let $q \colon A \times S^{n-1} \to S^{n-1}$ denote the projection. Then $q(S)$ coincides with the set of singular directions for A. Hence by Axiom (iii) it suffices to prove that S is an $\mathfrak{X}$-set and $q(S)$ is of dimension $< n - 1$.

Proof that S is an $\mathfrak{X}$-set. By (II.1.17) p admits a C^r $\mathfrak{X}$-stratification ($\{B_i\}$, $\{C_j\}$). Since each $p|_{B_i}$ is a C^r submersion to some C_j, S is the image under p of the union of B_i's such that

$$\operatorname{rank} d(p|_{B_i}) = \dim B_i - 1.$$

Hence S is a union of $p(B_i)$'s and, consequently, an $\mathfrak{X}$-set.

Proof that $\dim q(S) < n - 1$. By choosing $\{C_j\}$ to be compatible with $\{S\}$ (II.1.17), we assume that S is the union of some C_j. We prove $\dim q(S) < n - 1$ by reduction to absurdity. Assume $\dim q(S) = n - 1$. Let C_j be a stratum included in S such that $\dim q(C_j) = n - 1$. Let (x_0, λ_0) be a point of C_j where $d(q|_{C_j})$ is of rank $n - 1$. Since $\dim(p|_{B_i})^{-1}(x_0, \lambda_0) = 1$ for some B_i, there exist an open neighborhood U of (x_0, λ_0) in C_j and an open interval $I \subset \mathbf{R}$ such that $q|_U$ is a submersion and

$$x + t\lambda \in A \quad \text{for} \quad (x, \lambda) \in U, \ t \in I.$$

This implies the following assertion.

There is an open set U' in some $\mathbf{R}^{n'}$ and C^1 maps $\varphi \colon U' \to \mathbf{R}^n$ and $\psi \colon U' \to S^{n-1}$ such that ψ is a C^1 submersion and the image of $U' \times I$ under the map

$$\Phi \colon U' \times I \ni (y, t) \longrightarrow \varphi(y) + t\psi(y) \in \mathbf{R}^n$$

is of dimension $< n$.

We want to arrive at the contradiction that $\dim \Phi(U' \times I) = n$. Fix a point y in U'. By assumption $d\psi_y$ is of rank $n - 1$, and $d\Phi_{y,t}$ is of rank $< n$ for any $t \in I$. We have

$$d\Phi_{y,t} = d\varphi_y + dt \cdot \psi(y) + t d\psi_y.$$

Regard $d\Phi_{y,t}$ as the map:

$$\mathbf{R}^{n'} \times \mathbf{R} \times I \ni (v, s, t) \longrightarrow d\varphi_y v + s\psi(y) + t d\psi_y v \in \mathbf{R}^n.$$

Then $d\Phi_{y,t}$ is a polynomial map such that for each $t \in I$, the image of $\mathbf{R}^{n'} \times \mathbf{R} \times t$ is a linear space of dimension $< n$. This means that for any $(v_i, s_i) \in \mathbf{R}^{n'} \times \mathbf{R}$, $i = 1, \ldots, n$, and for any $t \in I$, the vectors $d\varphi_y v_i + s_i \psi(y) + t d\psi_y v_i$, $i = 1, \ldots, n$, are linearly dependent. Hence the polynomial extension of $d\Phi_{y,t}$ to $\mathbf{R}^{n'} \times \mathbf{R} \times \mathbf{R}$ keeps the above property. Keep the same notation $d\Phi_{y,t}$ for the extension. Define a rational map $F \colon \mathbf{R}^{n'} \times \mathbf{R} \times (\mathbf{R} - 0) \to \mathbf{R}^n \times \mathbf{R}$ by

$$F(v, s, t) = (d\Phi_{y, 1/t}(v, s), t).$$

Let D denote the image of F. Then D is a semialgebraic set, and we have seen that for each $t \in \mathbf{R}-0$, $D_t \times t = D \cap \mathbf{R}^n \times t$ is a linear space of dimension $< n$. Hence $\overline{D} - D \subset \mathbf{R}^n \times 0$ is a linear space of dimension $< n$ by (I.2.9.2). Define a set $D_0 \subset \mathbf{R}^n$ so that $D_0 \times 0 = \overline{D} - D$.

Now D_0 contains $\psi(y)$ and $d\psi_y \mathbf{R}^{n'}$ for the following reason. For any $(v, s, t) \in \mathbf{R}^{n'} \times \mathbf{R} \times \mathbf{R}$ with $t \neq 0$, it follows that

$$d\varphi_y v/t + s\psi(y)/t + d\psi_y v \in D_{1/t}.$$

Fix v and s, and let t diverge to ∞. We have $d\psi_y v \in D_0$. In the same way we prove $\psi(y) \in D_0$. Hence D_0 includes the linear space spanned by $\psi(y)$ and $d\psi_y \mathbf{R}^{n'}$. However, $d\psi_y \mathbf{R}^{n'}$ is the tangent space of S^{n-1} at $\psi(y)$, and hence, together with $\psi(y)$ it spans $\mathbf{R}^n$, which implies that D_0 is of dimension n. That is a contradiction. $\square$

In II.2.2, let us consider the case where A is parameterized, which will be necessary in the next section. Let A_t, $t \in \mathbf{R}$, be a family of ℐ-sets in $\mathbf{R}^n$ of dimension $< n$. Assume that the set

$$A = \{(a, t) \in \mathbf{R}^n \times \mathbf{R} \colon a \in A_t\}$$

is a bounded ℐ-set in $\mathbf{R}^{n+1}$. For every $t \in \mathbf{R}$, let $T_t \subset S^{n-1}$ denote the set of singular directions for A_t. Set

$$T = \{(\lambda, t) \in S^{n-1} \times \mathbf{R} \colon \lambda \in T_t\}.$$

Lemma II.2.2′. *T is an ℐ-set in $\mathbf{R}^{n+1}$, and for every $t \in \mathbf{R}$, $\overline{T} \cap (S^{n-1} \times t)$ is of dimension $< n - 1$.*

Proof. Replace B and $q \colon A \times S^{n-1} \to S^{n-1}$ in the proof of II.2.2 with

$$\{(x, t, s, \lambda) \in A \times \mathbf{R} \times S^{n-1} \subset \mathbf{R}^n \times \mathbf{R} \times \mathbf{R}^n \times S^{n-1} \colon (x + s\lambda, t) \in A\} \text{ and}$$
$$A \times S^{n-1} \ni (x, t, \lambda) \longrightarrow (\lambda, t) \in S^{n-1} \times \mathbf{R},$$

respectively. Keep the notation B and q, and define $p \colon B \to A \times S^{n-1}$ and $S \subset A \times S^{n-1}$ in the same way. The proof of II.2.2 says that $q(S) = T$, T is an ℐ-set, for each t, T_t is of dimension $< n - 1$, and hence T is of dimension $< n$. By II.1.9, if $T \neq \varnothing$, then

$$\dim(\overline{T} - T) < \dim T < n.$$

Hence

$$\dim \overline{T} \cap (S^{n-1} \times t) \leq \dim(T_t \times t) \cup (\overline{T} - T)$$
$$= \max\{\dim T_t, \dim(\overline{T} - T)\} < n - 1. \qquad \square$$

Remark II.2.3. In the above lemma, fix t and let (λ, t) be a point of $S^{n-1} \times t - \overline{T}$. There exists a small number $\varepsilon > 0$ such that for any $s \in [t-\varepsilon, t+\varepsilon]$, λ is a *non-singular direction* (i.e., not a singular direction) for A_s. Hence II.2.2' implies that A_t, $t \in \mathbf{R}$, locally admits a common non-singular direction.

Remark II.2.4. In II.2.1 we can construct an $\mathfrak{X}$-triangulation (K, τ) of X compatible with $\{Y_j\}$ so that K lies in $\mathbf{R}^n$, $\tau = \mathrm{id}$ on K^0, and τ is extended to an $\mathfrak{X}$-homeomorphism of $\mathbf{R}^n$ which is the identity outside of a bounded set in $\mathbf{R}^n$.

If (K, τ) has these properties and if X is a polyhedron, then $|K| = X$.

Proof of II.2.1 and II.2.4. We proceed by induction on n. If $n = 0$, II.2.1 and II.2.4 are trivial. So assume there exists such an $\mathfrak{X}$-triangulation of any $\mathfrak{X}$-set in $\mathbf{R}^{n-1}$. By replacing X and $\{Y_i\}$ with $\{x \in \mathbf{R}^n : |x| \leq r_0\}$ for a large number r_0 and $\{X\} \cup \{Y_i\}$, respectively, we assume $X = \{|x| \leq r_0\}$. By (II.1.14) there is a finite C^r $\mathfrak{X}$-stratification $\{X_j\}_{j \in J}$ of X with frontier condition which is compatible with $\{Y_i\}$. Set

$$J_1 = \{j \in J : \dim X_j < n\} \quad \text{and} \quad J_2 = J - J_1.$$

By II.2.2 there exists a non-singular direction for all X_j, $j \in J_1$ (say, $(0, \ldots, 0, 1)$ for simplicity of notation). Let $p: \mathbf{R}^n \to \mathbf{R}^{n-1}$ denote the projection which forgets the last factor. Denote a point of $\mathbf{R}^n$ by $x = (x_1, \ldots, x_n) = (x', x_n)$. Set $Z = \cup_{j \in J_1} X_j$. Then Z includes the sphere $\{|x| = r_0\}$ and hence $p(Z) = p(X)$. We will find an $\mathfrak{X}$-triangulation of Z and then extend it to X. For now, we do not require the homeomorphism of this extended triangulation to be the identity on the vertices of the simplicial complex.

By (II.1.17) $p|_Z : Z \to p(Z)$ admits a C^r $\mathfrak{X}$-stratification $(\{X'_{j'}\}, \{W_k\})$ such that $\{X'_{j'}\}$ is compatible with $\{X_j\}_{j \in J_1}$ and satisfies the frontier condition. By replacing $\{X_j\}_{j \in J_1}$ with $\{X'_{j'}\}$, we assume $(\{X_j\}_{j \in J_1}, \{W_k\})$ is a C^r $\mathfrak{X}$-stratification of $p|_Z$. Next, by the induction hypothesis, we can translate $\mathbf{R}^{n-1}$ so that $p(Z)$ is the underlying polyhedron of a finite simplicial complex, L and each W_k is some union of open simplices of L. We assume this. Note that the property $X = \{|x| \leq r_0\}$ now fails and we have $X \subset p(X) \times [-r_0, r_0]$. By a technical reason (which will be clear in a moment) we need a derived subdivision L' of L.

We can assume $\{W_k\}$ coincides with the family of open simplices of L for the following reason. As noted in the definition of a stratification of a map, the image under p of each connected component of each X_j, $j \in J_1$, coincides with some connected component of some W_k. That proof shows, moreover, that for each $\sigma \in L$, the map $p|_{p^{-1}(\sigma^\circ) \cap Z} \colon p^{-1}(\sigma^\circ) \cap Z \to \sigma^\circ$ is a finite (of course, trivial) covering map because $p|_Z$ is a finite-to-one map. We claim that for each connected component A of $p^{-1}(\sigma^\circ) \cap Z$, $\overline{A}$ is X-homeomorphically carried by p onto σ. Note that A is an X-set by (II.1.11). Since $p(\overline{A}) = \sigma$, it suffices to prove that for each $x' \in \partial\sigma$, $p^{-1}(x') \cap \overline{A}$ consists of one point. We can prove this in the same way as proved that $\overline{\operatorname{graph} L} \cap (b \times G_{n,m})$ consists of one point in the proof of $\dim Y_2 < m$ of (II.1.13). We omit the proof. By this claim we easily prove that the family {connected components of $X_j \cap p^{-1}(\sigma^\circ) \colon j \in J_1,\ \sigma \in L$} is a C^r X-stratification of Z with frontier condition such that p homeomorphically carries the closure of each stratum onto some simplex of L. Thus by replacing $\{X_j\}$ with this family of connected components and $\{W_k\}$ with the family of open simplices of L, we assume that $\{W_k\} = \{\sigma^\circ \colon \sigma \in L\}$ and for each $j \in J_1$, $p|_{\overline{X_j}}$ is a homeomorphism onto some $\sigma \in L$.

Note that $p(Z)$ is a compact polyhedral C^0 manifold with boundary such that

$$p^{-1}(\partial p(Z)) \cap Z = \partial p(Z) \times 0, \qquad (*)$$

and $\partial p(Z)$ separates the set $\mathbf{R}^n - \partial p(Z)$ into two connected components: $p(Z)^\circ$ and $\mathbf{R}^n - p(Z)$. In order to easily extend τ to $\mathbf{R}^n$ when we construct (K, τ), we add

$$\{W_k \times (\pm 2r_0)\}_k \quad \text{to} \quad \{X_j\}_{j \in J_1}. \qquad (**)$$

Let us define simplicial complexes Γ and Γ' in $\mathbf{R}^n$ such that $p(\Gamma^0) = L^0$, $p(\Gamma'^0) = L'^0$, and Γ' is a "refinement" of Γ and the required simplicial complex of an X-triangulation of Z, where the symbol ^{0} denotes the 0-skeleton (i.e., the vertices). For each $j \in J_1$ we call a point of $\overline{X_j}$ a vertex of $\overline{X_j}$ if its image under p is a vertex of the simplex $p(\overline{X_j})$. Let $\overline{X_j}^0$ denote the set of vertices of $\overline{X_j}$. Then $\Delta(\overline{X_j}^0)$ is a well-defined simplex by the fact that $p|_{\overline{X_j}} \colon \overline{X_j} \to p(\overline{X_j})$ is a homeomorphism. Also, the map

$$\gamma_j = (p|_{\overline{X_j}})^{-1} \circ (p|_{\Delta \overline{X_j}^0}) \colon \Delta\overline{X_j}^0 \to \overline{X_j}$$

is an X-homeomorphism whose restriction to $(\Delta\overline{X_j}^0)^\circ$ is a C^r diffeomorphism onto the image X_j. Moreover, $\Gamma = \{\Delta\overline{X_j}^0\}_{j \in J_1}$ is a simplicial complex for the following reason.

Clearly any face of $\Delta\overline{X_j}^0$ is one element of Γ. Hence it suffices to prove that $\Delta\overline{X_j}^0 \cap \Delta\overline{X_{j'}}^0$ is a common face of $\Delta\overline{X_j}^0$ and $\Delta\overline{X_{j'}}^0$ for each pair of

j and j' in J_1. We can reduce the problem to the case $p(X_j) = p(X_{j'})$ by replacing X_j and $X_{j'}$ with

$$\overline{X_j} \cap p^{-1}((p(\overline{X_j}) \cap p(\overline{X_{j'}}))^\circ), \quad \text{and} \quad \overline{X_{j'}} \cap p^{-1}((p(\overline{X_j}) \cap p(\overline{X_{j'}}))^\circ)$$

respectively, which are some strata of $\{X_j\}_{j \in J_1}$. Note that if $X_j \neq X_{j'}$ and if $p(X_j) = p(X_{j'})$, then either $X_j < X_{j'}$ or $X_{j'} < X_j$. Here $X_j < X_{j'}$ means that if $x = (x', x_n) \in X_j$ and $y = (x', y_n) \in X_{j'}$, then $x_n < y_n$. Assume $X_j < X_{j'}$. It suffices to prove

$$\Delta \overline{X_j}^0 \cap \Delta \overline{X_{j'}}^0 = \Delta(\overline{X_j}^0 \cap \overline{X_{j'}}^0).$$

The inclusion

$$\Delta \overline{X_j}^0 \cap \Delta \overline{X_{j'}}^0 \supset \Delta(\overline{X_j}^0 \cap \overline{X_{j'}}^0)$$

is trivial, and the reverse inclusion follows from the fact that if v is a vertex of $p(\overline{X_j})$ outside of $p(\overline{X_j}^0 \cap \overline{X_{j'}}^0)$, then

$$p^{-1}(v) \cap \overline{X_j} < p^{-1}(v) \cap \overline{X_{j'}}.$$

Thus Γ is a simplicial complex.

By setting $\gamma = \gamma_j$ on each $\Delta \overline{X_j}^0$, $j \in J_1$, one might expect (Γ, γ) to be an $\mathcal{X}$-triangulation of Z. There may be, however, the case of $\Delta \overline{X_j}^0 = \Delta \overline{X_{j'}}^0$ for some $j \neq j'$. Then we cannot define γ on $\Delta \overline{X_j}^0 = \Delta \overline{X_{j'}}^0$. To solve this problem, we define Γ' and γ' by L' as follows. By the same reason as above, we can replace $\{X_j\}_{j \in J_1}$ and $\{W_k\}$ with $\{Z_l\} = \{X_j \cap p^{-1}(\sigma^\circ): j \in J_1, \sigma \in L'\}$ and $\{W_{k'}'\} = \{\sigma^\circ: \sigma \in L'\}$. Define a simplicial complex $\Gamma' = \{\Delta \overline{Z_l}^0\}$ and an $\mathcal{X}$-homeomorphism $\gamma_l': \Delta \overline{Z_l}^0 \to \overline{Z_l}$ in the same way as Γ and γ_j. Then $Z_l \neq Z_{l'}$ implies $\Delta \overline{Z_l}^0 \neq \Delta \overline{Z_{l'}}^0$. Hence γ', defined to be γ_l' on each $\Delta \overline{Z_l}^0$, is a well-defined $\mathcal{X}$-homeomorphism from $|\Gamma'|$ to Z such that for each $\sigma \in \Gamma'$, $\gamma'|_{\sigma^\circ}$ is a C^r diffeomorphism onto the image. Therefore, (Γ', γ') is an $\mathcal{X}$-triangulation of Z.

Let us extend (Γ', γ') to an $\mathcal{X}$-triangulation (Λ, λ) of X. For $\{X_j\}_{j \in J_2}$ also, substratify it to $\{$connected components of $X_j \cap p^{-1}(\sigma^\circ): j \in J_2, \sigma \in L'\}$, and denote this by $\{A_\alpha\}$. By (II.1.11), each A_α is an $\mathcal{X}$-set. Then $(\{Z_l\} \cup \{A_\alpha\}, \{W_{k'}'\})$ is a C^r $\mathcal{X}$-stratification of $p|_X$. We shall need to extend τ of (K, τ) to $\mathbf{R}^n$. To easily do this, we add $\{$connected components of $W_{k'}' \times (]{-2r_0}, 2r_0[) - X\}$ to $\{A_\alpha\}$ such as $(**)$. Then

$$X = p(X) \times [-2r_0, 2r_0], \quad \text{and} \qquad (***)$$

$$\{A_\alpha\}|_{\partial p(X) \times [-2r_0, 2r_0]} = \{\sigma^\circ \times (]{-2r_0}, 0[), \ \sigma^\circ \times (]0, 2r_0[) : \sigma \in L', \ \sigma \subset \partial p(X)\}.$$

For each α there exist uniquely Z_l and $Z_{l'}$ such that

$$p(A_\alpha) = p(Z_l) = p(Z_{l'}), \qquad Z_l < Z_{l'}, \quad \text{and}$$
$$A_\alpha = \{(x_1, \dots, x_n) \in \mathbf{R}^n : y_n < x_n < z_n, \ (x_1, \dots, x_{n-1}, y_n) \in Z_l,$$
$$(x_1, \dots, x_{n-1}, z_n) \in Z_{l'}\}$$

(in a word, A_α lies between Z_l and Z_l'). For such α, l and l', let B_α denote the usual cell lying between $\Delta \overline{Z_l}^0$ and $\Delta \overline{Z_{l'}}^0$, and let $\theta_\alpha \colon B_\alpha \to \overline{A_\alpha}$ denote the 𝔛-extension of γ_l and $\gamma_{l'}$ defined by

$$\theta_\alpha(x_1, \dots, x_n) = s\gamma_l(x_1, \dots, x_{n-1}, y_n) + (1 - s)\gamma_{l'}(x_1, \dots, x_{n-1}, z_n)$$
$$\text{for} \quad (x_1, \dots, x_n) \in B_\alpha, \ (x_1, \dots, x_{n-1}, y_n) \in Z_l,$$
$$(x_1, \dots, x_{n-1}, z_n) \in Z_{l'} \quad \text{with} \quad x_n = sy_n + (1 - s)z_n.$$

In other words, θ_α carries linearly each segment $B_\alpha \cap p^{-1}(x')$ to $\overline{A_\alpha} \cap p^{-1}(x')$. If the family $\Gamma' \cup \{B_\alpha\}$ were a usual cell complex, $(\Gamma' \cup \{B_\alpha\}, \gamma' \cup \theta_\alpha)$ would be a usual cell 𝔛-triangulation of X. But $\Gamma' \cup \{B_\alpha\}$ is not necessarily a usual cell complex because a face of B_α does not always belong to $\Gamma' \cup \{B_\alpha\}$. So we will subdivide $\Gamma' \cup \{B_\alpha\}$ into a simplicial complex Λ as follows, in the same way as the proof of I.3.12.

For each B_α such that $p(B_\alpha)$ is a vertex in $L'^0 - L^0$, we choose a point b_α in B_α°. For each non-negative integer m, let us define a simplicial complex Λ_m compatible with $\Gamma' \cup \{B_\alpha\}$ such that

$$|\Lambda_m| = |\Gamma'| \cup (X \cap p^{-1}(|L^m|)), \qquad \Lambda_m|_{|\Gamma'|} = \Gamma', \quad \text{and}$$
$$\Lambda_m^0 = \Gamma'^0 \cup \{b_\alpha : p(B_\alpha) \in (L'^0 - L^0) \cap |L^m|\}$$

by induction on m. Set $\Lambda_0 = \Gamma' \cup \{B_\alpha \cap p^{-1}(v) : v \in L^0, \ \alpha\}$, which is clearly a simplicial complex. Assume we have defined Λ_{m-1}. We define Λ_m as follows. For each B_α with $p(B_\alpha)^\circ \subset |L^m| - |L^{m-1}|$, there exists uniquely $B_\beta \subset B_\alpha$ such that $p(B_\beta)$ is a point of $(L'^0 - L^0) \cap (|L^m| - |L^{m-1}|)$. Let C_α denote the union of faces of B_α which do not contain b_β. Then $b_\beta * C_\alpha = B_\alpha$, and C_α is included in $|\Gamma'| \cup (X \cap p^{-1}(|L^{m-1}|))$. Hence C_α is already divided into a simplicial complex D_α. Take the simplicial decomposition of B_α to be $\{b_\beta, b_\beta * \sigma, \sigma : \sigma \in D_\alpha\}$, and set $\Lambda_m = \Lambda_{m-1} \cup \{\text{these simplicial complexes}\}$. It is easy to show that Λ_m is a simplicial complex fulfilling the requirements. Thus we obtain $\Lambda = \Lambda_{n-1}$.

Next we will define an 𝔛-homeomorphism $\lambda \colon |\Lambda| \to X$ so that (Λ, λ) is an 𝔛-triangulation. Set $\lambda = \gamma'$ on $|\Gamma'|$. We cannot set $\lambda = \theta_\alpha$ on each B_α because the inclusion $B_{\alpha'} \subset B_\alpha$ does not imply $\theta_\alpha|_{B_{\alpha'}} = \theta_{\alpha'}$. We need to modify θ_α. Let m be a non-negative integer. We define a modification λ_α of θ_α for each $B_\alpha \subset p^{-1}(|L^m|)$ so that $\lambda_\alpha(B_\alpha) = \overline{A_\alpha}$, if $B_{\alpha'} \subset B_\alpha$, then $\lambda_\alpha|_{B_{\alpha'}} = \lambda_{\alpha'}$, and if $\sigma \subset B_\alpha$ for $\sigma \in \Gamma'$, then $\lambda_\alpha|_\sigma = \gamma'|_\sigma$ by induction on m. If $B_\alpha \subset p^{-1}(L^0)$, set $\lambda_\alpha = \theta_\alpha$. Assume we have already defined λ_α for all $B_\alpha \subset p^{-1}(|L^{m-1}|)$. Let α be such that $B_\alpha^\circ \subset p^{-1}(|L^m| - |L^{m-1}|)$ and let B_β and C_α be given from B_α in the same way as above. Let an $\mathfrak{X}$-homeomorphism $\theta_\alpha' \colon C_\alpha \to \theta_\alpha(C_\alpha)$ be defined by

$$\theta_\alpha' = \begin{cases} \gamma' & \text{on} \quad C_\alpha \cap |\Gamma'| \\ \lambda_{\alpha'} & \text{on} \quad B_{\alpha'} \quad \text{with} \quad B_{\alpha'} \subset C_\alpha. \end{cases}$$

(This is well-defined because $C_\alpha \subset |\Gamma'| \cup p^{-1}(|L^{m-1}|)$.) Extend $\theta_\alpha^{-1} \circ \theta_\alpha'$ to a homeomorphism $\theta_\alpha'' \colon B_\alpha \to B_\alpha$ by cone extension with vertex b_β (the Alexander trick). Then $\lambda_\alpha = \theta_\alpha \circ \theta_\alpha''$ is what we want. Indeed, by (II.1.16) λ_α is an $\mathfrak{X}$-homeomorphism. Clearly $\lambda_\alpha(B_\alpha) = \overline{A_\alpha}$. $B_{\alpha'} \subset B_\alpha$ implies $\lambda_\alpha|_{B_{\alpha'}} = \lambda_{\alpha'}$ because if $B_{\alpha'}^\circ \subset p^{-1}(|L^m| - |L^{m-1}|)$, then $B_\beta = B_{\beta'}$ (where B_β and $B_{\beta'}$ are defined by B_α and $B_{\alpha'}$, respectively, as above). $\theta_\alpha|_{B_{\alpha'}} = \theta_{\alpha'}$ by the definition of θ_α, $\theta_\alpha''|_{B_{\alpha'}} = \theta_{\alpha'}''$ by the definition of a cone extension and because if $B_{\alpha'} \subset p^{-1}(|L^{m-1}|)$, then λ_α is an extension of $\lambda_{\alpha'}$ by the definition of λ_α. For each $\sigma \subset B_\alpha$ with $\sigma \in \Gamma'$, clearly $\lambda_\alpha|_\sigma = \gamma'|_\sigma$. Thus we obtain $\{\lambda_\alpha\}$. For each α, $\lambda_\alpha|_{B_\alpha^\circ}$ is not necessarily a C^r diffeomorphism onto the image because we used a cone extension. But for $\sigma \in \Lambda$ with $\sigma \subset B_\alpha$, by the definition of a cone extension, $\lambda_\alpha|_{\sigma^\circ}$ is a C^r diffeomorphism onto the image. Hence if we define a homeomorphism $\lambda \colon |\Lambda| \to X$ by $\lambda = \lambda_\alpha$ on each B_α and $\lambda = \gamma'$ on $|\Gamma'|$, (Λ, λ) is an $\mathfrak{X}$-triangulation of X.

Since we cannot require λ to be the identity on Λ^0, we modify (Λ, λ) as follows. We need to apply the induction hypothesis carefully when we construct L. We had an $\mathfrak{X}$-triangulation (L, π) of $p(Z)$ compatible with $\{W_k\}$, a derived subdivision L' of L and an $\mathfrak{X}$-triangulation (Λ, λ) of X such that $p \circ \lambda(\Lambda^0) = \pi(L'^0)$. Once more apply, the induction hypothesis to $p(Z)$ and $\pi(L') = \{\pi(\sigma) \colon \sigma \in L'\}$, and let (L_1, π_1) be an $\mathfrak{X}$-triangulation of $p(Z)$ compatible with $\pi(L')$ such that L_1 lies in $\mathbf{R}^{n-1}$, $\pi_1 = \mathrm{id}$ on L_1^0, and π_1 can be extended to an $\mathfrak{X}$-homeomorphism of $\mathbf{R}^{n-1}$ which is the identity outside of a bounded set. Translate $\mathbf{R}^{n-1}$ by the inverse of such an extension of π_1). (This causes no problems because we shall construct the (K, τ) so that $p(K^0) = L_1^0$ and τ is of the form:

$$(\pi_1(x'), \tau_n(x', x_n)) \quad \text{for} \quad (x', x_n) \in |K|.)$$

For each $\sigma \in L'$, the image $\pi(\sigma)$ is some union of simplices of L_1. Let us consider a C^r $\mathfrak{X}$-stratification $\{E_\varepsilon\} = \{\lambda(\sigma^\circ) \cap p^{-1}(\sigma'^\circ) \colon \sigma \in \Lambda, \ \sigma' \in L_1\}$ of X. The pair $(\{E_\varepsilon\}, \{\sigma^\circ \colon \sigma \in L_1\})$ is a C^r $\mathfrak{X}$-stratification of $p|_X$ with the same properties as $(\{Z_l\} \cup \{A_\alpha\}, \{W'_{k'}\})$. Moreover, this stratification has the following good property. Let K_1 be the family of usual cells induced from $\{E_\varepsilon\}$ in the same way as we defined $\Gamma' \cup \{B_\alpha\}$ from $\{Z_l\} \cup \{A_\alpha\}$ (that is, $K_1 = \{\Delta\overline{E_\varepsilon}^0 : p|_{E_\varepsilon}$ is a imbedding$\} \cup \{$usual cells lying between $\Delta\overline{E_{\varepsilon'}}^0$ and $\Delta\overline{E_{\varepsilon''}}^0 :$ some E_ε lies between $E_{\varepsilon'}$ and $E_{\varepsilon''}\})$. This K_1 is a usual cell complex because Λ is a simplicial complex. Let us define a homeomorphism $\tau \colon |K_1| \to X$ as follows. If $\sigma \in K_1$ is induced from E_ε, let $\tau_\sigma \colon \sigma \to \overline{E_\varepsilon}$ be the homeomorphism defined in the same way as γ_j and θ_α. Set $\tau = \tau_\sigma$ on $\sigma \in K_1$. Then (K_1, τ) is a usual cell $\mathfrak{X}$-triangulation of X compatible with $\{Y_i\}$ such that K_1 lies in $\mathbf{R}^n$ and $\tau_1 = \mathrm{id}$ on K_1^0. Let K be a simplicial subdivision of K_1 without new vertices (I.3.12). Then (K, τ) is an $\mathfrak{X}$-triangulation of X with the same properties as (K_1, τ).

We want to extend τ to an $\mathfrak{X}$-homeomorphism of $\mathbf{R}^n$. By $(*)$, $(**)$ and $(***)$, if $E_\varepsilon \subset \partial X$, then $\Delta\overline{E_\varepsilon}^0 = \overline{E_\varepsilon}$. Hence by the method of construction of τ, $\tau = \mathrm{id}$ on ∂X. Therefore, by setting $\tau = \mathrm{id}$ outside of X, we can extend τ to $\mathbf{R}^n$.

It remains to prove the last statement of the remark. Assume $X \subset \mathbf{R}^n$ is a compact polyhedron. Let (K, τ) be an $\mathfrak{X}$-triangulation of X such that K lies in $\mathbf{R}^n$ and $\tau = \mathrm{id}$ on K^0. We prove $\tau^{-1}(X) = X$ by induction on $\dim X$. Let $\Sigma_r X$ denote the C^r singular point set of X, and let X_0 be the closure of a connected component of $X - \Sigma_r X$. It suffices to prove $\tau^{-1}(\Sigma_r X) = \Sigma_r X$ and $\tau^{-1}(X_0) = X_0$. It is shown in the proof of I.3.4 that $\Sigma_r X$ is a compact polyhedron of dimension $< \dim X$, (K, τ) is compatible with $\Sigma_r X$ and X_0, and for some linear subspace Π of $\mathbf{R}^n$ of dimension $= \dim X$, X_0 is the closure of a connected component of $\Pi - \Sigma_r X$. Hence by the induction hypothesis we have $\tau^{-1}(\Sigma_r X) = \Sigma_r X$, and by the hypothesis that $\tau = \mathrm{id}$ on K^0, we see that $\tau^{-1}(X_0) \subset \Pi$. Therefore, by replacing X with X_0 and K with $K|_{\tau^{-1}(X_0)}$, we assume that $\dim X = n$. Moreover, $\overline{X - \Sigma_r X} = X$, and $X - \Sigma_r X$ is connected. In this case, $\Sigma_r X = \mathrm{bdry}\, X$, and by Brouwer's theorem on the invariance of domain, $\tau^{-1}(\Sigma_r X) = \mathrm{bdry}\,|K|$. Hence it suffices to prove

$$(X - \Sigma_r X) \cap (|K| - \mathrm{bdry}\,|K|) \neq \varnothing \tag{0}$$

because $\Sigma_r X = \tau^{-1}(\Sigma_r X)$. Since X is bounded, $\dim \Sigma_r X = n - 1$ and there exists a point x in $(\Sigma_r X - \Sigma_r(\Sigma_r X)) \cap ($the closure of the unbounded connected component of $\mathbf{R}^n - \Sigma_r X)$. Let U be a small neighborhood of x in $\mathbf{R}^n$ such that $(U, U \cap \Sigma_r X)$ is homeomorphic to $(\mathbf{R}^n, 0 \times \mathbf{R}^{n-1})$. Then $U \cap X$ and $U \cap |L|$ are the closures in U of connected components of $U - \Sigma_r X$. These components do not intersect with the unbounded component of $\mathbf{R}^n - \Sigma_r X$, and hence they coincide with each other, which proves (0). $\square$

Remark II.2.5. In I.2.1, assume only that X is bounded in $\mathbf{R}^n$. There exist a finite simplicial complex K in $\mathbf{R}^n$, some union Y of open simplices of K, and an $\mathfrak{X}$-homeomorphism τ from Y to X such that Y is dense in $|K|$ and for each $\sigma \in K$, $|\tau|_{\sigma^\circ \cap Y}$ is a C^r diffeomorphism onto the image. This is immediate by II.2.1.

Here $|K|$ is not unique up to PL homeomorphism. For example, let X be $\partial \sigma -$ (a vertex) for a 2-simplex σ. Then $K_1 = \{$proper faces of $\sigma\}$, $Y_1 = X$ and $\tau_1 = $ id satisfy the above properties. Another triple K_2, Y_2 and τ_2 is defined by $K_2 = \{0,\ 1,\ 2,\ 3,\ [0,1],\ [1,2],\ [2,3]\}$, $Y_2 =]0,3[$ and $\tau_2 = $ the restriction to Y_2 of a surjective simplicial map $K_2 \to K_1$ which carries 0 and 3 to the point $|K_1| - X$. Clearly $|K_1|$ and $|K_2|$ are not PL homeomorphic.

Proof of Theorem II.2.1'. We assume $X = \mathbf{R}^n$ for the following reason. By (II.1.2), $\overline{X}$ is an $\mathfrak{X}$-set in $\mathbf{R}^n$. Hence $\overline{X} - X$ is a closed $\mathfrak{X}$-set. Let φ denote the function which measures distance on X from $\overline{X} - X$, which is an $\mathfrak{X}$-function by (II.1.2). We can replace X with the graph of $1/\varphi$, which is a closed $\mathfrak{X}$-set in $\mathbf{R}^{n+1}$. Therefore, we assume X is closed in $\mathbf{R}^n$. Moreover, replace X and $\{Y_i\}$ with $\mathbf{R}^n$ and $\{X\} \cup \{Y_i\}$, respectively. We can assume $X = \mathbf{R}^n$.

Let $\{\sigma_j\}_{j=1,2,\dots}$ be a simplicial decomposition of $\mathbf{R}^n$ such that for each positive integer k, $\Lambda_k = \{\sigma_j\}_{j=1,\dots,k}$ is a simplicial complex. We will construct an $\mathfrak{X}$-triangulation (K_k, π_k) of $|\Lambda_k|$ compatible with $\{Y_i\}$ inductively on k, so that K_k is a subdivision of Λ_k, $\pi_k(\sigma_j)$ equals σ_j for each $1 \leq j \leq k$, and the limit (K, π) of (K_k, π_k) as $k \to \infty$ is well-defined and fulfills the requirements. Clearly set $(K_1, \pi_1) = (\Lambda_1, \text{id})$ because $|\Lambda_1|$ is a point. By assuming (K_{k-1}, π_{k-1}) to be given, we construct (K_k, π_k) as follows.

For a point $b_k \in \sigma_k^\circ$, let us regard σ_k as a cone with base $\partial \sigma_k$ and vertex b_k, and extend π_{k-1} to σ_k by the cone extension (the Alexander trick). Call the extension $\tilde{\pi}_{k-1}$. Let $\tilde{K}_{k-1}$ denote the subdivision of Λ_k such that $\tilde{K}_{k-1} = K_{k-1}$ on $|\Lambda_{k-1}|$ and

$$\tilde{K}_{k-1}|_{\sigma_k} = (K_{k-1}|_{\partial \sigma_k}) * b_k \ (= \{b_k,\ \sigma,\ \sigma * b_k : \sigma \in K_{k-1}|_{\partial \sigma_k}\}).$$

Then $(\tilde{K}_{k-1}, \tilde{\pi}_{k-1})$ is an $\mathfrak{X}$-triangulation of $|\Lambda_k|$ compatible with $\{Y_i - \sigma_k^\circ\}_i$ such that $\tilde{\pi}_{k-1}(\sigma_j) = \sigma_j$ for $1 \leq j \leq k$. Hence it suffices to modify $(\tilde{K}_{k-1}, \tilde{\pi}_{k-1})$ so that it is compatible with $\{Y_i\}$. By II.2.1 and II.2.4 there exists an $\mathfrak{X}$-triangulation (L, τ) of σ_k compatible with $\{\tilde{\pi}_{k-1}^{-1}(Y_i)\}_i$ such that L is a subdivision of $\tilde{K}_{k-1}|_{\sigma_k}$ and $\tau(\sigma) = \sigma$ for $\sigma \in \tilde{K}_{k-1}|_{\sigma_k}$. We will modify $(\tilde{K}_{k-1}, \tilde{\pi}_{k-1})$ by using (L, τ) as follows.

First we will extend (L, τ) to $|N(\tilde{K}_{k-1}|_{\sigma_k}, \tilde{K}_{k-1})|$. Let $\sigma \in \tilde{K}_{k-1}|_{\sigma_k}$ and $\sigma' \in |\partial N(\tilde{K}_{k-1}|_{\sigma_k}, \tilde{K}_{k-1})|$ with $\sigma * \sigma' \in \tilde{K}_{k-1}$. Set $L_\sigma = L|_\sigma$ and $\tau_\sigma = \tau|_\sigma$. Let us define the extension $(L_\sigma * \sigma', \tau_\sigma * \text{id})$ of (L_σ, τ_σ) to $\sigma * \sigma'$ by

$$L_\sigma * \sigma' = \{\alpha,\ \beta,\ \alpha * \beta : \alpha \in L_\sigma,\ \beta \text{ faces of } \sigma'\}, \quad \text{and}$$

$$\tau_\sigma * \text{id}(sx + (1-s)y) = s\tau_\sigma(x) + (1-s)y \quad \text{for} \quad x \in \sigma,\ y \in \sigma',\ s \in [0,1].$$

The family of $(L_\sigma * \sigma', \tau_\sigma * \mathrm{id})$ for all σ and σ' defines the extension $(\tilde{L}, \tilde{\tau})$ of (L, τ) to $|N(\tilde{K}_{k-1}|_{\sigma_k}, \tilde{K}_{k-1})|$. Clearly $\tilde{\tau}(\sigma) = \sigma$ for $\sigma \in N(\tilde{K}_{k-1}|_{\sigma_k}, \tilde{K}_{k-1})$ and $(\tilde{L}, \tilde{\tau})$ is compatible with $\{\tilde{\pi}_{k-1}^{-1}(Y_i)\}_i$ because each $\tilde{\pi}_{k-1}^{-1}(Y_i) - \sigma_k^\circ$ is some union of open simplices of K_{k-1}.

Next let us extend $(\tilde{L}, \tilde{\tau})$ to $|\Lambda_k|$. By the definition of $(\tilde{L}, \tilde{\tau})$,

$$(\tilde{L}, \tilde{\tau})|_{|\partial N(\tilde{K}_{k-1}|_{\sigma_k}, \tilde{K}_{k-1})|} = (\partial N(\tilde{K}_{k-1}|_{\sigma_k}, \tilde{K}_{k-1}), \mathrm{id}).$$

Hence by setting $(\tilde{L}, \tilde{\tau}) = (K_{k-1}, \mathrm{id})$ outside of $|N(\tilde{K}_{k-1}|_{\sigma_k}, \tilde{K}_{k-1})|$, we can extend $(\tilde{L}, \tilde{\tau})$ to $|\Lambda_k|$. We keep the same notation $(\tilde{L}, \tilde{\tau})$ for the extension. Then $\tilde{\tau}(\sigma) = \sigma$ for $\sigma \in \tilde{K}_{k-1}$, $(\tilde{L}, \tilde{\tau}) = (K_{k-1}, \mathrm{id})$ on $|\Lambda_k| - |N(\tilde{K}_{k-1}|_{\sigma_k}, \tilde{K}_{k-1})|$, and $(\tilde{L}, \tilde{\tau})$ is compatible with $\{\tilde{\pi}_{k-1}^{-1}(Y_i)\}_i$.

Finally, we replace $(\tilde{K}_{k-1}, \tilde{\pi}_{k-1})$ with $(\tilde{L}, \tilde{\pi}_{k-1} \circ \tilde{\tau})$. By the above arguments $(\tilde{L}, \tilde{\pi}_{k-1} \circ \tilde{\tau})$ is an X-triangulation of $|\Lambda_k|$ compatible with $\{Y_i\}$ such that $\tilde{L}$ is a subdivision of Λ_k, $\tilde{\pi}_{k-1} \circ \tilde{\tau}(\sigma_j) = \sigma_j$ for $1 \le j \le k$, and $(\tilde{L}, \tilde{\pi}_{k-1} \circ \tilde{\tau}) = (K_{k-1}, \pi_{k-1})$ on $|\Lambda_k - N(\Lambda_k|_{\sigma_k}, \Lambda_k)|$. Thus $(\tilde{L}, \tilde{\pi}_{k-1} \circ \tilde{\tau}) = (K_k, \pi_k)$ is what we wanted. $\square$

Remark II.2.6. The above proof shows the following facts. In II.2.1′, if X is closed in $\mathbf{R}^n$, we can choose an X-triangulation (K, π) of X so that K lies in $\mathbf{R}^n$ and π can be extended to an X-homeomorphism of $\mathbf{R}^n$. Moreover, for an X-set Y in $\mathbf{R}^n$, there exists an X-triangulation (L, τ) of $\mathbf{R}^n$ such that $|L|$ equals $\mathbf{R}^n$ and $\tau^{-1}(Y)$ is some union of open simplices of L.

In Chapter IV, we shall use the following lemma.

Lemma II.2.7. *Let X be a compact X-set in $\mathbf{R}^{n_1} \times \mathbf{R}^{n_2}$. Let K and $L \supset \tilde{L}$ be finite simplicial complexes in X and $\mathbf{R}^{n_2}$, respectively, let $p \colon \mathbf{R}^{n_1} \times \mathbf{R}^{n_2} \to \mathbf{R}^{n_2}$ denote the projection, and let U be an open neighborhood of X in $\mathbf{R}^{n_1} \times \mathbf{R}^{n_2}$. Assume that $p|_X$ is a finite-to-one-map, $|L|$ includes $p(X)$ and*

$$X \cap p^{-1}(|N(\tilde{L}, L)|) = |K| \cap p^{-1}(|N(\tilde{L}, L)|). \tag{a}$$

There exists an X-isotopy π_t, $0 \le t \le 1$, of $\mathbf{R}^{n_1} \times \mathbf{R}^{n_2}$ of the form:

$$\pi_t(x, y) = (\pi_t'(x, y), \pi_t''(y)) \quad \text{for} \quad (x, y) \in \mathbf{R}^{n_1} \times \mathbf{R}^{n_2}$$

such that $\pi_0 = \mathrm{id}$, $\pi_1(X)$ is a polyhedron, for each $t \in [0, 1]$, $\pi_t = \mathrm{id}$ on $p^{-1}(|\tilde{L}|)$, π_t is invariant on each simplex of K, π_t'' is invariant on each simplex of L, and

$$\pi_t'(x, y) = x \quad \text{for} \quad (x, y) \in \mathbf{R}^{n_1} \times \mathbf{R}^{n_2} - U.$$

Proof. It suffices to find an $\mathfrak{X}$-homeomorphism π of $\mathbf{R}^{n_1} \times \mathbf{R}^{n_2}$ of the form:

$$\pi(x, y) = (\pi'(x, y), \pi''(y)) \quad \text{for} \quad (x, y) \in \mathbf{R}^{n_1} \times \mathbf{R}^{n_2}$$

such that $\pi(X)$ is a polyhedron, $\pi = \mathrm{id}$ on $p^{-1}(|\tilde{L}|)$, π is invariant on each simplex of K, π'' is invariant on each simplex of L,

$$\pi'(x, y) = x \quad \text{for} \quad (x, y) \in \mathbf{R}^{n_1} \times \mathbf{R}^{n_2} - U, \tag{b}$$

and

(c) the linearly defined homotopy from the identity map to π is an isotopy, i.e., for each $t \in [0, 1]$, the map

$$\pi_t \colon \mathbf{R}^{n_1} \times \mathbf{R}^{n_2} \ni (x, y) \longrightarrow t\pi(x, y) + (1 - t)(x, y) \in \mathbf{R}^{n_1} \times \mathbf{R}^{n_2}$$

is a homeomorphism. (For a C^0 transformation f of a Euclidean space, let f_t, $0 \le t \le 1$, denote the homotopy defined in this way.) Indeed, the isotopy π_t, $0 \le t \le 1$, fulfills the requirements in II.2.7.

Let

$$q_j \colon \mathbf{R}^{n_1-j+1} \times \mathbf{R}^{n_2} \longrightarrow \mathbf{R}^{n_1-j} \times \mathbf{R}^{n_2}, \quad j = 1, \dots, n_1$$

denote the projections which forget the first factors. Subdivide K and L. As in the proof of II.2.1 and II.2.4, we obtain simplicial complexes K_j in $\mathbf{R}^{n_1-j} \times \mathbf{R}^{n_2}$, $j = 0, \dots, n_1$, and C^r $\mathfrak{X}$-stratifications with the frontier condition $\{X_i^j\}_i$, $j = 0, \dots, n_1 - 1$, of $q_j \circ \cdots \circ q_1(X)$ and $\{X_i^{n_1}\}_i$ of $|L|$ such that

$$K_0 = K, \qquad |K_j| = q_j \circ \cdots \circ q_1(|K|), \quad j = 1, \dots, n_1 - 1, \qquad K_{n_1} = L,$$

$q_j \colon K_{j-1} \to K_j$, $j = 1, \dots, n_1$, are simplicial maps, each $\{X_i^j\}_i$ is compatible with K_j, and the restriction of q_{j+1} to each X_i^j, $j < n_1$, is a C^r diffeomorphism onto same $X_{i'}^{j+1}$. Here each X_i^j and $q_{j+1}|_{X_i^j}$ are of class C^r. But the class C^0 suffices for the following arguments, and indeed, X_i^j becomes of class C^0 after its modification.

Moreover, since $|L|$ is bounded, by substratifying $\{X_i^j\}_i$, $j = 0, \dots, n_1$, as in the proof of II.2.1 and II.2.4, we find an $\mathfrak{X}$-homeomorphism π'' of $\mathbf{R}^{n_2}$ such that $K_{n_1}^* = \{\pi''(\overline{X_i^{n_1}})\}_i$ is a subdivision of L, $\pi'' = \mathrm{id}$ on $K_{n_1}^{*0}$, π'' is invariant on each simplex of L, and π'' satisfies the condition (c). Here we can assume

$$\{X_i^{n_1}\}_i|_{|\tilde{L}|} = \{\sigma^\circ \colon \sigma \in \tilde{L}\} \quad \text{and} \quad \pi'' = \mathrm{id} \ \text{on} \ |\tilde{L}| \tag{d}$$

for the following reason.

Let $\sigma = \Delta a_1 \cdots a_k b_1 \cdots b_{k'}$ be a simplex of $N(\tilde{L}, L)$ with

$$\Delta a_1 \cdots a_k \in \partial N(\tilde{L}, L) \quad \text{and} \quad \Delta b_1 \cdots b_{k'} \in \tilde{L}.$$

Replace $\{X_i^{n_1}\}_i|_{\sigma^\circ}$ with the family:

$$\left\{ \{ t x_1 + (1-t) x_2 : t \in \,]0, 1[\, , \; x_1 \in X_i^{n_1}, \; x_2 \in (\Delta b_1 \cdots b_{k'})^\circ \} : X_i^{n_1} \subset (\Delta a_1 \cdots a_k)^\circ \right\}$$

and $\pi''|_\sigma$ with the map:

$$t x_1 + (1-t) x_2 \longrightarrow t \pi''(x_1) + (1-t) x_2$$
$$\text{for} \quad t \in [0,1], \; x_1 \in \Delta a_1 \cdots a_k, \; x_2 \in \Delta b_1 \cdots b_{k'}.$$

(Here we also need to modify $\{X_i^j\}_i$, $j = 0, \dots, n_1 - 1$, so that the property that each $q_{j+1}|_{X_i^j}$ is a diffeomorphism onto the image continues to hold. But by (a) this is clearly possible.) Then (d) holds.

For each $j < n_1$, let K_j^* denote the family of all simplices $\Delta(\overline{X_i^j}^0)$, where $\overline{X_i^j}^0 = \overline{X_i^j} \cap (\pi'' \circ q_{n_1} \circ \cdots \circ q_{j+1})^{-1}(K_{n_1}^{*0})$. By the same reason as in the proof of II.2.1 and II.2.4, we can assume each K_j^* is a simplicial complex which is clearly compatible with K_j. We can construct X-functions τ_j' on $\mathbf{R}^{n_1 - j + 1} \times \mathbf{R}^{n_2}$, $j = 1, \dots, n_1$, by downward induction on j such that for each j, the transformation π_j of $\mathbf{R}^{n_1 - j + 1} \times \mathbf{R}^{n_2}$ which is defined by

$$\pi_j(x_j, \dots, x_{n_1}, y) = (\tau_j'(x_j, \dots, x_{n_1}, y), \dots, \tau_{n_1}'(x_{n_1}, y), \pi''(y))$$
$$\text{for} \quad (x_j, \dots, x_{n_1}, y) \in \mathbf{R}^{n_1 - j + 1} \times \mathbf{R}^{n_2},$$

is a homeomorphism,

$$\pi_j = \mathrm{id} \quad \text{on} \quad K_{j-1}^{*0} \cup (q_{n_1} \circ \cdots \circ q_{j+1})^{-1}(|\tilde{L}|),$$
$$\{\pi_j(\overline{X_i^{j-1}})\}_i = K_{j-1}^*,$$

π_j is invariant on each simplex of K_{j-1}, and π_j satisfies the condition (c). Hence $\pi = \pi_1$ fulfills the requirements except (b). Write $\pi_j = (\pi_j', \pi'')$. We need to modify π_1' outside of X so that (b) is satisfied.

Since we do not change π'', we translate X by the homeomorphism (id, π'') of $\mathbf{R}^{n_1} \times \mathbf{R}^{n_2}$. We can assume $\pi'' = \mathrm{id}$, which implies

$$\{\overline{X_i^{n_1}}\}_i = K_{n_1}^*.$$

From now on, we proceed in the C^0 category.

First assume

$$U =]a_1, b_1[\times \cdots \times]a_{n_1}, b_{n_1}[\times \mathbf{R}^{n_2}$$

for some numbers $a_1, \ldots, b_{n_1}$. We modify π_1' by induction on n_1. The case $n_1 = 0$ is trivial. Hence assume

$$\pi_2'(x', y) = x' \quad \text{for} \quad (x', y) \in (\mathbf{R}^{n_1-1} -]a_2, b_2[\times \cdots \times]a_{n_1}, b_{n_1}[) \times \mathbf{R}^{n_2}.$$

By the method of construction of τ_1' (see the proof of II.2.1 and II.2.4), we suppose

$$\tau_1'(x_1, \ldots, x_{n_1}, y) = x_1$$
$$\text{for} \quad (x_1, \ldots, x_{n_1}, y) \in (\mathbf{R}^{n_1} -]a_1, b_1[\times \cdots \times]a_{n_1}, b_{n_1}[) \times \mathbf{R}^{n_2}.$$

Hence

$$\pi_1'(x, y) = x \quad \text{for} \quad (x, y) \in \mathbf{R} \times (\mathbf{R}^{n_1-1} -]a_2, b_2[\times \cdots \times]a_{n_1}, b_{n_1}[) \times \mathbf{R}^{n_2},$$

and we need to modify π_2' on the set:

$$(]-\infty, a_1 + \varepsilon] \cup [b_1 - \varepsilon, \infty[) \times]a_2, b_2[\times \cdots \times]a_{n_1}, b_{n_1}[\times \mathbf{R}^{n_2}.$$

Here we regard π_2' as a map from $\mathbf{R}^{n_1} \times \mathbf{R}^{n_2}$ to $\mathbf{R}^{n_1-1} \times \mathbf{R}^{n_2}$, and ε is a small positive number such that

$$]a_1 + \varepsilon, b_1 - \varepsilon[\times]a_2, b_2[\times \cdots \times]a_{n_1}, b_{n_1}[\times \mathbf{R}^{n_2} \supset X.$$

Let α be a PL function on $\mathbf{R}$ such that

$$0 \leq \alpha \leq 1, \quad \alpha = 0 \text{ on }]-\infty, a_1] \cup [b_1, \infty[\text{ and } \alpha = 1 \text{ on } [a_1 + \varepsilon, b_1 - \varepsilon].$$

Replace π_1 with the map:

$$\pi \colon (x, y) = (x_1, \ldots, x_{n_1}, y) \longrightarrow (\tau_1'(x, y), \pi_{2,\alpha(x_1)}(x_2, \ldots, x_{n_1}, y)).$$

(For the definition of the homotopy $\pi_{2,t}$, $0 \leq t \leq 1$, see the condition (c).) This π satisfies the condition (b). Note that π is equal to π_1 on X and fulfills the requirements, but it does not have the same form as π_1.

Clearly the above proof works in the second case where U is a finite union:

$$\bigcup_k]a_1^k, b_1^k[\times \cdots \times]a_{n_1}^k, b_{n_1}^k[\times \mathbf{R}^{n_2}$$

such that for any i and $k \neq k'$,

$$[a_i^k, b_i^k] \cap [a_i^{k'}, b_i^{k'}] = \varnothing \quad \text{or} \quad [a_i^k, b_i^k] = [a_i^{k'}, b_i^{k'}].$$

Finally, consider the general case of U. At the beginning of the proof, finely subdivide K and L. We can assume the following. For each simplex σ of L there exist a finite number of numbers $a_1^k, b_1^k, \ldots, a_{n_1}^k, b_{n_1}^k$, $k = 1, 2, \ldots,$ such that

$$U \supset \bigcup_k [a_1^k, b_1^k] \times \cdots \times [a_{n_1}^k, b_{n_1}^k] \times \sigma, \tag{σ}$$

$$\bigcup_k]a_1^k, b_1^k[\times \cdots \times]a_{n_1}^k, b_{n_1}^k[\times \sigma \supset X \cap p^{-1}(\sigma),$$

and for any i and $k \neq k'$,

$$[a_i^k, b_i^k] \cap [a_i^{k'}, b_i^{k'}] = \varnothing \quad \text{or} \quad [a_i^k, b_i^k] = [a_i^{k'}, b_i^{k'}].$$

We reduce the problem to the case:

$$\pi_1 = \mathrm{id} \quad \text{on} \quad p^{-1}(|L^l \cup \tilde{L}|), \ \ l = 0, \ldots, n_2, \tag{l}$$

by induction on l. Note that (n_2) is what we want and (l) is equivalent to the condition that each stratum X_i^0 included in $p^{-1}(|L^l \cup \tilde{L}|)$ is an open simplex. Since (0) always holds, we assume $(l-1)$ for a positive number l. Let σ be an l-simplex of L, and let $a_1^k, \ldots, b_{n_1}^k$ be given so that the condition (σ) holds. Let V be a polyhedral closed neighborhood of σ in $\mathbf{R}^{n_2}$ such that the condition (V) holds. Here we define (V) by replacing σ with V in (σ). As in the second case, modify π_1 outside of

$$[a_1^k + \varepsilon, b_1^k - \varepsilon] \times \cdots \times [a_{n_1}^k + \varepsilon, b_{n_1}^k - \varepsilon] \times \mathbf{R}^{n_2}$$

for a sufficiently small positive number ε. Let

$$\pi_\sigma(x, y) = (\pi_\sigma'(x, y), y)$$

denote the modified homeomorphism. Then

$$\pi_\sigma = \mathrm{id} \quad \text{on} \quad K_0^{*0} \cup p^{-1}(|L^{l-1} \cup \tilde{L}|),$$

for a stratum X_i^0 included in $p^{-1}(\sigma)$, $\pi_\sigma(X_i^0)$ is an open simplex, π_σ is invariant on each simplex of K,

$$\pi_\sigma'(x, y) = x \quad \text{for} \quad (x, y) \in \mathbf{R}^{n_1} \times \mathbf{R}^{n_2} - \bigcup_k]a_1^k, b_1^k[\times \cdots \times]a_{n_1}^k, b_{n_1}^k[\times \mathbf{R}^{n_2},$$

and π_σ satisfies condition (c).

Since (b) is not always satisfied for π_σ, we modify π_σ once more, as follows. Let β be a PL function on $\mathbf{R}^{n_2}$ such that

$$0 \le \beta \le 1, \qquad \beta = 0 \text{ on } \mathbf{R}^{n_2} - V \quad \text{and} \quad \beta = 1 \text{ on } \sigma.$$

Replace π_σ with the transformation of $\mathbf{R}^{n_1} \times \mathbf{R}^{n_2}$: $(x,y) \longrightarrow (\pi'_{\sigma,\beta(y)}(x,y),y)$, and keep the notation π_σ. Then π_σ is an $\mathfrak{X}$ homeomorphism, π_σ on X is of the form:

$$\pi_\sigma(x,y) = (\pi'_{\sigma 1}(x,y),\dots,\pi'_{\sigma n_1}(x_{n_1},y),y) \quad \text{for} \quad (x,y) = (x_1,\dots,x_{n_1},y) \in X,$$

π_σ is globally of the form:

$$\pi_\sigma(x,y) = (\pi'_\sigma(x,y),y) \quad \text{for} \quad (x,y) \in \mathbf{R}^{n_1} \times \mathbf{R}^{n_2}, \quad \text{and}$$
$$\pi_\sigma = \mathrm{id} \quad \text{on} \quad K_0^{*0} \cup p^{-1}(|L^{l-1},\tilde{L}|).$$

Furthermore, for a stratum X_i^0 included in $p^{-1}(\sigma)$, $\pi_\sigma(X_i^0)$ is an open simplex (which implies that $\pi_\sigma(X) \cap p^{-1}(\sigma)$ is a polyhedron), π_σ is invariant on each simplex of K, and π_σ satisfies the conditions (b) and (c).

Moreover, after we transform X to $\pi_\sigma(X)$, the condition (σ') continues to hold for any $\sigma' \in K_{n_1}$, because if the closure of a stratum X_i^0 is included in the set:

$$A =]a'_1,b'_1[\times \cdots \times]a'_{n_1},b'_{n_1}[\times p(\overline{X_i^1}),$$

then $\Delta(\overline{X_i^0})$ and hence $\pi_\sigma(\overline{X_i^0})$ are included in A. Hence we can replace X with $\pi_\sigma(X)$. Each stratum X_i^0 included in $p^{-1}(\sigma)$ is an open simplex, and if we define π_1 for this new X, then

$$\pi_1 = \mathrm{id} \quad \text{on} \quad p^{-1}(|L^{l-1} \cup \tilde{L}| \cup \sigma).$$

Pursue the same arguments for all l-simplices of K_{n_1}. We obtain (l), which completes the proof of the lemma. $\qquad\qquad\square$

§II.3. Triangulations of $\mathfrak{X}$-functions

Assuming uniqueness of $\mathfrak{X}$-triangulations of $\mathfrak{X}$-sets(III.1.4), we will prove:

Theorem II.3.1 ($\mathfrak{X}$-triangulation of an $\mathfrak{X}$-function). *For a finite family $\{Y_i\}$ of $\mathfrak{X}$-sets in $\mathbf{R}^n$, an $\mathfrak{X}$-function f on a compact $\mathfrak{X}$-set X admits a unique $\mathfrak{X}$-triangulation compatible with $\{Y_i\}$.*

Theorem II.3.1'. *If* $\mathfrak{X}$ *satisfies Axiom* (v), II.3.1 *holds true for any* $\mathfrak{X}$-*set* $X \subset \mathbf{R}^n$ *locally closed in* $\mathbf{R}^n$ *and for any family* $\{Y_i\}$ *of* $\mathfrak{X}$-*sets in* X *locally finite at each point of* X.

We will prove only II.3.1' because II.3.1 becomes clear midway. Before beginning the proof, we show some remarks.

Remark II.3.2. In II.3.1, we can choose an $\mathfrak{X}$-triangulation (K, π) of f so that K lies in $\mathbf{R}^n$ and π can be extended to an $\mathfrak{X}$-homeomorphism of $\mathbf{R}^n$. If X is the underlying polyhedron of a simplicial complex L, then we can furthermore require K to be a subdivision of L and $\pi(\sigma) = \sigma$ for $\sigma \in L$. For II.3.1', the same statements as above hold true if X is closed in $\mathbf{R}^n$.

Proof. By II.2.4 and II.2.7 we assume X is the underlying polyhedron of a simplicial complex L in $\mathbf{R}^n$. Hence it suffices to consider the second statement. Let (K_1, π_1) be an $\mathfrak{X}$-triangulation of f compatible with $L \cup \{Y_i\}$. Regard (L, id) and (K_1, π_1) as $\mathfrak{X}$-triangulations of the $\mathfrak{X}$-set X compatible with L. By the $\mathfrak{X}$-Hauptvermutung(III.1.4) we have a PL homeomorphism $\tau \colon X \to |K_1|$ such that $\tau(\sigma) = \pi_1^{-1}(\sigma)$ for each $\sigma \in L$. Let K_1' and L_1' be subdivisions of K_1 and of L, respectively, such that $\tau \colon L' \to K_1'$ is a simplicial isomorphism. Then $(K, \pi) = (L', \pi_1 \circ \tau)$ fulfills the requirements. The third statement follows in the same way. $\qquad\square$

Remark II.3.3. Let us assume in II.3.1' that X is not locally closed in $\mathbf{R}^n$. There exist a simplicial complex K, a union Y of open simplices of K and an $\mathfrak{X}$-homeomorphism τ from Y to X such that for each $\sigma \in K$, $\tau|_{\sigma \circ \cap Y}$ is a C^r diffeomorphism onto the image, r a positive integer, and $f \circ \tau$ can be extended to $|K|$ so that the extension is linear on each simplex of K. In II.3.1, a similar modification is possible if f and X are bounded (here K is finite).

Proof. Let $X_1 \subset \mathbf{R}^n \times \mathbf{R}$ denote the graph of f, let $\tilde{X}$ denote the closure of X_1 in $\mathbf{R}^n \times \mathbf{R}$, and let $\tilde{f}$ denote the restriction to $\tilde{X}$ of the projection $\mathbf{R}^n \times \mathbf{R} \to \mathbf{R}$. We can replace f by $\tilde{f}|_{X_1}$. Apply II.3.1 and II.3.1' to $\tilde{f}$ and X_1. We have an $\mathfrak{X}$-triangulation (K, π) of $\tilde{f}$ such that the set $Y = \pi^{-1}(X_1)$ is the union of some open simplices of K. Clearly K, Y and the restriction $\tau = \pi|_Y$ fulfill the requirements. $\qquad\square$

Remark II.3.4. The next fact shows one case where $\mathfrak{X}$-triangulability of an $\mathfrak{X}$-function clarifies the structure of the domain where the function is defined. Let $(W; M_0, M_1)$ be a PL h-cobordism where W is connected and of dimension > 5. Then $W - M_0$ is PL homeomorphic to $M_1 \times]0, 1]$ (the weak h-cobordism theorem, see [R-S]). Hence there exists a C^0 function f on W

such that $f = 0$ on M_0, $f = 1$ on M_1, and $f \circ \varphi(x, t) = t$ for some homeomorphism $\varphi \colon M \times]0, 1[\to W - M_0 - M_1$, where M is a PL manifold. Assume f is an $\mathfrak{X}$-function. Then $(W; M_0, M_1)$ is trivial (i.e., PL homeomorphic to $(M_0 \times [0, 1]; M_0 \times 0, M_0 \times 1))$ as shown below. If $(W; M_0, M_1)$ is not trivial, the level $f^{-1}(x)$ "turns" infinity many times as x converges to either 0 or 1 (in other words, for any segment J in W with $J \cap \partial W =$ (an end point of J), $f|_J$ oscillates infinitely many times near the end point).

Proof of triviality of $(W; M_0, M_1)$. By the s-cobordism theorem (see [R-S]) it suffices to show that $(W; M_0, M_1)$ is topologically trivial (i.e., homeomorphic to $(M_0 \times [0, 1]; M_0 \times 0, M_0 \times 1))$ because the Whitehead torsion is a topological invariant. By II.3.2 we have an $\mathfrak{X}$-homeomorphism π of W such that $\pi(M_i) = M_i$, $i = 0, 1$, and $f \circ \pi$ is PL. Hence we assume f is PL on W. Let K be a simplicial decomposition of W such that for each $\sigma \in K$, $f|_\sigma$ is linear, and let K' be a derived subdivision of K such that for all $\sigma \in K$ with $\sigma \cap M_0 \neq \varnothing$ and $\sigma \not\subset M_0$, the values of f at the new vertices v_σ coincide with one another. Let ε denote the value. Then $f^{-1}([0, \varepsilon])$ is a regular neighborhood of M_0 in W. Hence $f^{-1}([0, \varepsilon])$ is PL homeomorphic to $M_0 \times [0, \varepsilon]$ by uniqueness of regular neighborhoods (see the proof of I.3.8).

In the same way we prove that $f^{-1}([\varepsilon', 1])$ is PL homeomorphic to $M_1 \times [\varepsilon', 1]$ for some $0 < \varepsilon' < 1$. Of course, $f^{-1}([\varepsilon, \varepsilon'])$ is homeomorphic to $f^{-1}(\varepsilon) \times [\varepsilon, \varepsilon']$ by assumption. In conclusion, W is homeomorphic to $M_0 \times [0, 1]$. $\square$

We divide proof of II.3.1$'$ into three steps. First we show existence of an $\mathfrak{X}$-triangulation of $f|_{f^{-1}([-\varepsilon, \varepsilon])}$ for a small positive number ε in the case of compact X. Second, we prove uniqueness of $\mathfrak{X}$-triangulations of f in the compact case, i.e., II.3.1. Finally, we prove the general case. In the first step, we do not use the uniqueness theorem. On the other hand, we can say that in the second and last steps, we need only this if we assume the first step and the Alexander trick(II.1.16). For the first step, the following lemma will be sufficient.

Lemma II.3.5. *Let V be a compact $\mathfrak{X}$-set in $\mathbf{R}^n \times \mathbf{R}$, and let $\{W_i\}$ be a finite family of $\mathfrak{X}$-sets in $\mathbf{R}^n \times \mathbf{R}$. There exist a positive number ε and an $\mathfrak{X}$-triangulation (K, τ) of $V(\varepsilon) = V \cap (\mathbf{R}^n \times [-\varepsilon, \varepsilon])$ compatible with $\{W_i\}$ such that K lies in $\mathbf{R}^n \times \mathbf{R}$ and τ is of the form:*

$$\tau(x, t) = (\tau'(x, t), t) \quad for \quad (x, t) \in |K| \subset \mathbf{R}^n \times \mathbf{R}.$$

Proof. We prove the lemma by induction on n. If $n = 0$, this is trivial. So assume the lemma for $n - 1$. Let p_1 denote the projection $\mathbf{R}^n \times \mathbf{R} \to \mathbf{R}$. Apply (II.1.17) to $p_1|_V$. We have a finite C^r $\mathfrak{X}$-stratification $\{V_j\}$ of V compatible with $\{W_i\}$, which satisfies the frontier condition such that each

V_j is connected and $p_1|_{V_j}$ is either C^r regular or constant. Let $\varepsilon > 0$ be a small number such that for each V_j, where p_1 is constant,

$$p_1(V_j) \cap ([-\varepsilon, \varepsilon] - 0) = \varnothing.$$

Let J_1 denote the set of indices of V_j of dimension $\leq n$ with

$$\dim V_j \cap (\mathbf{R}^n \times 0) < n.$$

Set
$$A = \bigcup_{j \in J_1} V_j \quad \text{and} \quad A_t = \{x \in \mathbf{R}^n : (x, t) \in A\} \quad \text{for} \quad t \in \mathbf{R}.$$

Then A is an $\mathfrak{X}$-set, and for each $-\varepsilon \leq t \leq \varepsilon$, A_t is of dimension $< n$. Hence, by II.2.2' and II.2.3, after changing linearly the coordinate system of $\mathbf{R}^n$, we can assume that $(1, 0, \dots, 0) \in S^{n-1}$ is a non-singular direction for all A_t, $t \in [-\varepsilon, \varepsilon]$. Here we shrink ε if necessary. We replace A, V and $\{V_j\}$ with $A \cap (\mathbf{R}^n \times [-\varepsilon, \varepsilon])$, $V(\varepsilon)$ and $\{V_j \cap p^{-1}(I) : I = -\varepsilon, 0, \varepsilon, \,]-\varepsilon, 0[$ or $]0, \varepsilon[\}$, respectively, and keep the notation A, V and $\{V_j\}$.

Let p denote the projection $\mathbf{R}^n \times \mathbf{R} \to \mathbf{R}^{n-1} \times \mathbf{R}$ which forgets the first factor. The above nonsingularity of $(1, 0, \dots, 0)$ means that $p|_A \colon A \to p(A)$ is a finite-to-one map. Clearly we have $p(A) = p(V)$. We proceed with proof in the same way as the proof of II.2.1. Let us sketch the proof. First, by using the induction hypothesis, we reduce the problem to the case where there exists a simplicial complex L with underlying polyhedron $= p(A)$ such that for each $j \in J_1$, $p|_{\overline{V_j}}$ is an $\mathfrak{X}$-homeomorphism onto a simplex of L. Here we shrink ε once more if necessary. Second, by subdividing L, we construct an $\mathfrak{X}$-triangulation (K_1, τ_1) of A so that $|K_1| \subset \mathbf{R}^n \times \mathbf{R}$, τ_1 is of the form:

$$\tau_1(x, t) = (\tau_1'(x, t), x_2, \dots, x_n, t) \quad \text{for} \quad (x, t) = (x_1, \dots, x_n, t) \in |K_1|,$$

and for each $\sigma \in K$, $p \circ \tau_1|_\sigma$ is a linear homeomorphism onto a simplex of L. Finally, we extend (K_1, τ_1) to an $\mathfrak{X}$-triangulation of V. We omit the details. □

Remark II.3.6. Let (K, τ) be an $\mathfrak{X}$-triangulation of $V(\varepsilon)$ in II.3.5. For any positive number $\varepsilon' < \varepsilon$, $(K, \tau)|_{\tau^{-1}(V(\varepsilon'))}$ is a usual cell $\mathfrak{X}$-triangulation of $V(\varepsilon')$. Here $K|_{\tau^{-1}(V(\varepsilon'))}$ is the usual cell complex generated by $\sigma \cap \tau^{-1}(V(\varepsilon'))$, $\sigma \in K$. Hence if we define K' to be a simplicial subdivision of $K|_{\tau^{-1}(V(\varepsilon'))}$ and τ' to be $\tau|_{\tau^{-1}(V(\varepsilon'))}$, then (K', τ') is an $\mathfrak{X}$-triangulation of $V(\varepsilon')$.

We will apply the uniqueness theorem of $\mathfrak{X}$-triangulations of $\mathfrak{X}$-sets in the following form.

Lemma II.3.7. *Let X and Y be compact polyhedra in $\mathbf{R}^n$, let K be a simplicial complex with underlying polyhedron X, and let $f: X \to Y$ be an $\mathfrak{X}$-homeomorphism such that for each $\sigma \in K$, $f(\sigma)$ is a polyhedron. There exists an $\mathfrak{X}$-isotopy $f_t: X \to Y$, $t \in [0,1]$, of f such that f_1 is PL and*

$$f_t(\sigma) = f(\sigma) \quad for \quad t \in [0,1], \ \sigma \in K.$$

Proof. We construct f_t on $|K^k|$ by induction on k. For $k = 0$, we set $f_t = f$ on $|K^0|$. Assume there exists an $\mathfrak{X}$-isotopy $f_t^k: |K^k| \to f(|K^k|)$, $t \in [0,1]$, of $f^k = f|_{|K^k|}: |K^k| \to f(|K^k|)$ such that f_1^k is PL and

$$f_t^k(\sigma) = f(\sigma) \quad for \quad t \in [0,1], \ \sigma \in K^k.$$

We want to define f_t^{k+1}. For that, it suffices to construct an $\mathfrak{X}$-isotopy $f_t^\sigma: \sigma \to f(\sigma)$, $t \in [0,1]$, of $f^\sigma = f|_\sigma: \sigma \to f(\sigma)$ for each $\sigma \in K^{k+1} - K^k$ such that f_1^σ is PL and

$$f_t^\sigma(\sigma) = f(\sigma) \quad for \quad t \in [0,1],$$
$$f_t^\sigma|_{\partial\sigma} = f_t^k|_{\partial\sigma} \quad for \quad t \in [0,1].$$

Fix $\sigma \in K^{k+1} - K^k$. By the $\mathfrak{X}$-Hauptvermutung(III.1.4), there is a PL homeomorphism $\varphi: f(\sigma) \to \sigma$. The image $\varphi \circ f(\partial\sigma)$ coincides with $\partial\sigma$, which is clear if we regard σ and $f(\sigma)$ as topological manifolds with boundary. Set

$$F = \varphi \circ f: \sigma \to \sigma \quad and$$
$$g_t = \varphi \circ (f_t^k|_{\partial\sigma}): \partial\sigma \to \partial\sigma \quad for \quad t \in [0,1].$$

Then g_t is an $\mathfrak{X}$-isotopy of $F|_{\partial\sigma}$, and g_1 is PL. Hence by the Alexander trick (II.1.16) we have an $\mathfrak{X}$-isotopy $G_t: \sigma \to \sigma$, $t \in [0,1]$ of F such that for each $t \in [0,1]$, $G_t = g_t$ on $\partial\sigma$, and G_1 is the cone extension of g_1 and hence is PL. Therefore, $f_t^\sigma = \varphi^{-1} \circ G_t$, $t \in [0,1]$, is what we want. $\qquad\square$

We subdivide the second step of the proof of II.3.1' into proof of existence and proof of uniqueness.

Lemma II.3.8 (Existence of an $\mathfrak{X}$-triangulation in the compact case). *If X is compact, there exists an $\mathfrak{X}$-triangulation of f compatible with $\{Y_i\}$.*

Proof. For $a < b \in \mathbf{R}$, set

$$X[a,b] = f^{-1}([a,b]) \quad and \quad X[a] = f^{-1}(a).$$

Apply II.3.5 and II.3.6 to the graph of f. There exist numbers $a_0 < \cdots < a_m \in \mathbf{R}$ and $\mathfrak{X}$-triangulations (K_j, π_j) of $f|_{X[a_{j-1}, a_j]}$, $j = 1, \ldots, m$, compatible with $\{Y_i\} \cup \{X[a_{j-1}], X[a_j]\}$ such that $[a_0, a_m]$ includes the image of f. Using II.3.7, we want to paste these $\mathfrak{X}$-triangulations together. We can assume $m = 2$. By II.2.1 we have an $\mathfrak{X}$-triangulation (L, τ) of $X[a_1]$ compatible with the family:

$$\pi_1(K_1) \cup \pi_2(K_2) = \{\pi_j(\sigma) \colon \sigma \in K_j, \ j = 1, 2\}.$$

Set
$$Z_j = \pi_j^{-1}(X[a_1]), \ j = 1, 2.$$

If the homeomorphisms

$$\varphi_1 = \tau^{-1} \circ (\pi_1|_{Z_1}) \colon Z_1 \longrightarrow |L| \quad \text{and}$$
$$\varphi_2 = \tau^{-1} \circ (\pi_2|_{Z_2}) \colon Z_2 \longrightarrow |L|$$

are PL, so is the homeomorphism:

$$\pi_2^{-1} \circ (\pi_1|_{Z_1}) \colon Z_1 \longrightarrow Z_2.$$

Hence the identification space of $|K_1|$ and $|K_2|$ by $\pi_2^{-1} \circ (\pi_1|_{Z_1})$ is a polyhedron and admits a simplicial decomposition K which is a subdivision of K_1 on $|K_1|$ and a subdivision of K_2 on $|K_2|$. If we set $\pi = \pi_1$ on $|K_1|$ and $\pi = \pi_2$ on $|K_2|$, (K, π) is an $\mathfrak{X}$-triangulation of f compatible with $\{Y_i\}$.

We want to modify (K_1, π_1) so that φ_1 becomes PL. It suffices to find an $\mathfrak{X}$-triangulation (K_1', ρ) of $f \circ \pi_1 \colon |K_1| \to \mathbf{R}$ compatible with K_1 such that $\varphi_1 \circ (\rho|_{\rho^{-1}(Z_1)})$ is PL, because the $\mathfrak{X}$-triangulation $(K_1', \pi_1' = \pi_1 \circ \rho)$ of $f|_{\pi_1(|K_1|)}$ satisfies the condition that the map

$$\tau^{-1} \circ (\pi_1'|_{\pi_1'^{-1}(X[a_1])}) \colon \pi_1'^{-1}(X[a_1]) \longrightarrow |L|$$

is PL. Moreover, we require the conditions:

$$|K_1'| = |K_1| \quad \text{and} \quad \rho(\sigma) = \sigma \quad \text{for} \quad \sigma \in K_1$$

for simplicity of the construction of (K_1', ρ).

First we define an $\mathfrak{X}$-triangulation $(K_{1,Z}', \rho_Z)$ of Z_1 compatible with $K_{1,Z} = K_1|_{Z_1}$ such that $\varphi_1 \circ \rho_Z$ is PL,

$$|K_{1,Z}'| = Z_1 \quad \text{and} \quad \rho_Z(\sigma) = \sigma \quad \text{for} \quad \sigma \in K_{1,Z}$$

as follows. Note that $\varphi_1(K_{1,Z})$ is a family of polyhedra. Apply II.3.7 to $\varphi_1 \colon Z_1 \to |L|$ and $K_{1,Z}$. We have a PL homeomorphism $\Phi_1 \colon Z_1 \to |L|$ such that

$$\Phi_1(\sigma) = \varphi_1(\sigma) \quad \text{for} \quad \sigma \in K_{1,Z}.$$

Let $K'_{1,Z}$ be a subdivision of $K_{1,Z}$ such that for each $\sigma \in K'_{1,Z}$, $\Phi_1|_\sigma$ is a linear map into some simplex of L, and set $\rho_Z = \varphi_1^{-1} \circ \Phi_1$. Then $(K'_{1,Z}, \rho_Z)$ is what we want because

$$\varphi_1 \circ \rho_Z = \Phi_1, \quad \text{and}$$
$$\rho_Z(\sigma) = \varphi_1^{-1} \circ \Phi_1(\sigma) = \varphi_1^{-1}(\varphi_1(\sigma)) = \sigma \quad \text{for} \quad \sigma \in K_{1,Z}.$$

Next we extend $(K'_{1,Z}, \rho_Z)$ to (K'_1, ρ) as follows. It suffices to canonically define an $\mathfrak{X}$-triangulation $(K'_{1,\sigma}, \rho_\sigma)$ of $f \circ \pi_1|_\sigma$ for each $\sigma \in K_1$ so that if $\sigma \in K_1$ is in Z_1, then $(K'_{1,\sigma}, \rho_\sigma) = (K'_{1,Z}, \rho_Z)|_\sigma$, and if $\sigma \subset \sigma' \in K_1$, then $(K'_{1,\sigma}, \rho_\sigma) = (K'_{1,\sigma'}, \rho_{\sigma'})|_\sigma$. Indeed, the union K'_1 of $K'_{1,\sigma}$, $\sigma \in K_1$, and ρ, which equals ρ_σ on each $\sigma \in K_1$, fulfill the requirements. For each $\sigma \in K_1$ there are uniquely $\sigma_1 \in K_1$ and $\sigma_2 \in K_1$ (one of them may be empty) such that

$$\sigma = \sigma_1 * \sigma_2, \quad \sigma_1 \in K_{1,Z}, \quad \text{and} \quad \sigma_2 \notin N^\circ(K_{1,Z}, K_1).$$

Set $\rho_{\sigma_1} = \rho_Z|_{\sigma_1}$ and extend ρ_{σ_1} to σ by

$$\rho_\sigma = \mathrm{id} \quad \text{if} \quad \sigma_1 = \varnothing, \quad \text{and}$$

$$\rho_\sigma(tx + (1-t)y) = t\rho_{\sigma_1}(x) + (1-t)y$$
$$\text{for} \quad x \in \sigma_1, \ y \in \sigma_2, \ 0 \le t \le 1, \quad \text{if} \quad \sigma_1 \ne \varnothing, \ \sigma_2 \ne \varnothing.$$

Note that ρ_σ is constructed by a finite repetition of cone extensions. Set

$$K'_{1,\sigma} = \{\sigma'_1, \ \sigma'_1 * \sigma'_2, \ \sigma'_2 \colon \sigma'_1 \in K'_{1,Z}, \ \sigma'_2 \in K_1, \ \sigma'_1 \subset \sigma_1, \ \sigma'_2 \subset \sigma_2\}.$$

Then $(K'_{1,\sigma}, \rho_\sigma)$ is an $\mathfrak{X}$-triangulation of $f \circ \pi_1|_\sigma$ which is an extension of $(K'_{1,Z}, \rho_Z)|_{\sigma_1}$ and compatible with K_1 such that $|K'_{1,\sigma}| = \sigma$. Moreover, if $\sigma \subset \sigma' \in K_1$, then $(K_{1,\sigma}, \rho_\sigma) = (K_{1,\sigma'}, \rho_{\sigma'})|_\sigma$. $\qquad\square$

For the uniqueness in II.3.1$'$ in the compact case, we need the following improvement of II.3.7.

Lemma II.3.9. *Let V and V' be compact polyhedra, let $\{W_i\}$ and $\{W_i'\}$ be finite families of closed subpolyhedra of V and V', respectively, and let $\varphi \colon V \to V'$ be an 𝔛-homeomorphism such that $\varphi(W_i) = W_i'$. There exists an 𝔛-isotopy $\varphi_t \colon V \to V'$, $t \in [0,1]$, of φ such that φ_1 is PL and*

$$\varphi_t(W_i) = W_i' \quad for \quad t \in [0,1].$$

Proof. Let L and L' be simplicial decomposition of V and V' compatible with $\{W_i\}$ and $\{W_i'\}$, respectively. By II.2.1 we have an 𝔛-triangulation (L'', τ) of V, compatible with $L \cup \varphi^{-1}(L') = L \cup \{\varphi^{-1}(\sigma) \colon \sigma \in L'\}$. Define 𝔛-homeomorphisms $\psi \colon V \to |L''|$ and $\psi' \colon V' \to |L''|$ by

$$\psi = \tau^{-1}, \quad and \quad \psi' = \tau^{-1} \circ \varphi^{-1}.$$

For any $\sigma \in L$ and $\sigma' \in L'$, $\psi(\sigma)$ and $\psi'(\sigma')$ are polyhedra. By II.3.7 there are 𝔛-isotopies $\psi_t \colon V \to |L''|$ of ψ and $\psi_t' \colon V' \to |L''|$ of ψ', $t \in [0,1]$, such that ψ_1 and ψ_1' are PL and

$$\psi_t(\sigma) = \psi(\sigma), \qquad \psi_t'(\sigma') = \psi'(\sigma') \quad for \quad t \in [0,1], \ \sigma \in L, \ \sigma' \in L'.$$

Then $\varphi_t = \psi_t'^{-1} \circ \psi_t$, $t \in [0,1]$, fulfills the requirements.   □

Lemma II.3.10 (Uniqueness of an 𝔛-triangulation in the compact case). *If X is compact, an 𝔛-triangulation of f compatible with $\{Y_i\}$ is unique.*

Proof. We shall reduce the problem to a trivial case by stages. Let (K_j, π_j), $j = 1, 2$, be an 𝔛-triangulation of f compatible with $\{Y_i\}$. Apply II.3.8 to f and

$$\pi_1(K_1) \cup \pi_2(K_2) = \{\pi_j(\sigma) \colon \sigma \in K_j, \ j = 1, 2\}.$$

There exists a third 𝔛-triangulation (K_3, π_3) of f compatible with $\pi_1(K_1) \cup \pi_2(K_2)$. Note that for each $\sigma \in K_j$, $j = 1, 2$, $\pi_3^{-1} \circ \pi_j(\sigma)$ is the union of some simplices of K_3. For the proof, it suffices to find PL homeomorphisms:

$$\tau_1 \colon |K_1| \longrightarrow |K_3| \quad and \quad \tau_2 \colon |K_2| \longrightarrow |K_3|$$

such that

$$\tau_1(\sigma) = \pi_3^{-1} \circ \pi_1(\sigma) \quad for \quad \sigma \in K_1,$$
$$\tau_2(\sigma) = \pi_3^{-1} \circ \pi_2(\sigma) \quad for \quad \sigma \in K_2,$$
$$f \circ \pi_3 \circ \tau_1 = f \circ \pi_1 \quad on \quad |K_1|, \quad and$$
$$f \circ \pi_3 \circ \tau_2 = f \circ \pi_2 \quad on \quad |K_2|.$$

Hence we reduce the problem to the case where

$$X = |K_2|, \qquad \pi_2 = \mathrm{id},$$

for each $\sigma \in K_2$, $f|_\sigma$ is linear, $\pi_1 \colon |K_1| \to |K_2|$ is an $\mathfrak{X}$-homeomorphism, for each $\sigma \in K_1$, $\pi_1(\sigma)$ is a polyhedron, $f \circ \pi_1|_\sigma$ is linear, and K_2 is compatible with $\pi_1(K_1)$. We need only choose a PL homeomorphism τ from $|K_1|$ to $|K_2|$ so that

$$\tau(\sigma) = \pi_1(\sigma) \quad \text{for} \quad \sigma \in K_1 \quad \text{and}$$
$$f \circ \tau = f \circ \pi_1 \quad \text{on} \quad |K_1|.$$

Let $a_0 < a_1 < \cdots$ be the numbers of the image of K_2^0 under f, and for each positive integer k, set

$$U_{2,k} = f^{-1}([a_{k-1}, a_k]), \qquad V_{2,k} = f^{-1}(a_k), \quad \text{and}$$
$$U_{1,k} = \pi_1^{-1}(U_{2,k}), \qquad V_{1,k} = \pi_1^{-1}(V_{2,k}).$$

Then $U_{1,k}$, $Y_{2,k}$, $V_{1,k}$ and $V_{2,k}$ are polyhedra. Apply II.3.9 to

$$\pi_1 \colon |K_1| \to |K_2|, \quad \{U_{1,k}, V_{1,k}\} \cup K_1, \quad \text{and} \quad \{U_{2,k}, V_{2,k}\} \cup \pi_1(K_1).$$

We have a PL homeomorphism μ from $|K_1|$ to $|K_2|$ such that

$$\mu(U_{1,k}) = U_{2,k}, \qquad \mu(V_{1,k}) = V_{2,k} \quad \text{for all} \quad k, \quad \text{and}$$
$$\mu(\sigma) = \pi_1(\sigma) \quad \text{for} \quad \sigma \in K_1.$$

Note that $f \circ \mu = f \circ \pi_1$ on $\cup_k V_{1,k}$. But this equality does not hold globally. We need to modify μ so that it holds.

Compare $f \circ \mu$ and $f \circ \pi_1$. By I.3.10 we can modify μ in a neighborhood of $\cup_k V_{1,k}$. To be precise, there exists a PL homeomorphism ν from a closed polyhedral neighborhood W_1 of $\cup_k V_{1,k}$ in $|K_1|$ to another W_2 such that

$$\nu = \mathrm{id} \quad \text{on} \quad \bigcup_k V_{1,k},$$
$$f \circ \mu \circ \nu = f \circ \pi_1 \quad \text{on} \quad W_1,$$
$$\nu(U_{1,k} \cap W_1) = U_{1,k} \cap W_2 \quad \text{for all} \quad k, \quad \text{and}$$
$$\nu(\sigma \cap W_1) = \sigma \cap W_2 \quad \text{for} \quad \sigma \in K_1.$$

We want to extend ν to $|K_1^l| \cup W_1$ by induction on l. The case $l = 0$ is trivial because $|K_1^0| \subset W_1$. By using the induction hypothesis in the same way as in the proof of II.3.8, we reduce the problem to the case where K_1 is generated by a simplex σ, and ν is already extended to $\partial\sigma$. Note that for any $a_{k-1} < c < d < a_k$, the restrictions of $f \circ \mu$ to $\sigma_\mu[c,d] = \sigma \cap (f \circ \mu)^{-1}([c,d])$ and of $f \circ \pi_1$ to $\sigma_\pi[c,d] = \sigma \cap (f \circ \pi_1)^{-1}([c,d])$ are PL trivial by I.3.9, namely, there are PL homeomorphisms

$$\nu_\mu: \sigma_\mu(c) \times [c,d] \longrightarrow \sigma_\mu[c,d] \quad \text{and}$$
$$\nu_\pi: \sigma_\pi(c) \times [c,d] \longrightarrow \sigma_\pi[c,d],$$

where

$$\sigma_\mu(c) = \sigma \cap (f \circ \mu)^{-1}(c) \quad \text{and} \quad \sigma_\pi(c) = \sigma \cap (f \circ \pi_1)^{-1}(c),$$

such that

$$f \circ \mu \circ \nu_\mu(x,t) = t \quad \text{for} \quad (x,t) \in \sigma_\mu(c) \times [c,d], \quad \text{and}$$
$$f \circ \pi_1 \circ \nu_\pi(x,t) = t \quad \text{for} \quad (x,t) \in \sigma_\pi(c) \times [c,d].$$

Hence for each k, by setting

$$Z = \sigma_\pi[a_{k-1}, a_k], \quad f_1 = f \circ \pi|_Z, \quad \text{and} \quad f_2 = f \circ \mu|_Z,$$

and assuming

$$a_{k-1} = 0, \quad a_k = 3 \text{ and } W_1 \cap \sigma_\pi[0,3] = \sigma_\pi[0,1] \cup \sigma_\pi[2,3],$$

we need only prove the following assertion.

Let Z be a usual cell, and let f_1 and f_2 be PL functions on Z such that

$$0 \leq f_1 \leq 3, \quad 0 \leq f_2 \leq 3, \quad f_1^{-1}(0) = f_2^{-1}(0), \quad \text{and} \quad f_1^{-1}(3) = f_2^{-1}(3).$$

Set

$$Z_j[a,b] = f_j^{-1}([a,b]), \quad \text{and} \quad Z_j(a) = f_j^{-1}(a) \quad \text{for} \quad a < b \in \mathbf{R}, \ j = 1,2.$$

Assume that $Z_1(1)$ is a usual cell and there are PL homeomorphisms:

$$\nu: \partial Z \cup Z_1[0,1] \cup Z_1[2,3] \longrightarrow \partial Z \cup Z_2[0,1] \cup Z_2[2,3] \quad \text{and}$$
$$\xi_j: Z_j(1) \times [1,2] \longrightarrow Z_j[1,2], \ j = 1,2,$$

such that

$$f_2 \circ \nu = f_1 \quad \text{on} \quad \partial Z \cup Z_1[0,1] \cup Z_2[2,3] \quad \text{and}$$
$$f_j \circ \xi_j(x,t) = t \quad \text{for} \quad (x,t) \in Z_j(1) \times [1,2], \ j=1,2.$$

We can extend ν to a PL homeomorphism τ of Z so that $f_2 \circ \tau = f_1$.

Consider this assertion on only $Z_1[1,2]$ and $Z_2[1,2]$, and translate it into the following assertion on $Z_1(1) \times [1,2]$ and $Z_2(1) \times [1,2]$ through ξ_1 and ξ_2. There $Z_1 = Z_1(1)$, $Z_2 = Z_2(1)$.

Let Z_1 and Z_2 be PL balls. Let ν be a PL homeomorphism from $\partial(Z_1 \times [0,1])$ to $\partial(Z_2 \times [0,1])$ of the form:

$$\nu(x,t) = (\nu_1(x,t), t) \quad \text{for} \quad (x,t) \in \partial(Z_1 \times [0,1]).$$

Then ν can be extended to a PL homeomorphism from $Z_1 \times [0,1]$ to $Z_2 \times [0,1]$ keeping the property of form.

Trivially, a cone extension of ν is a solution. $\qquad\square$

Proof of existence in II.3.1'. Assume X is not compact. Let $\{X_j\}$ be a family of compact $\mathfrak{X}$-subsets of X locally finite at each point of X whose union equals X. By II.3.8, we can obtain $\mathfrak{X}$-triangulations of f on X_j. Combining these triangulations, we want to construct a global $\mathfrak{X}$-triangulation of f on X. For that, it suffices to prove the following assertion.

Assertion (i). Let X_1 and X_2 be compact $\mathfrak{X}$-subsets of X. Set $X_3 = X_1 \cap X_2$. For each $j = 1,2$, let (K_j, π_j) be an $\mathfrak{X}$-triangulation of $f|_{X_j}$ compatible with $\{Y_i\} \cup \{X_3\}$. Assume that the subcomplex $L_j = K_j|_{\pi_j^{-1}(X_3)}$ of K_j is full. There exists an $\mathfrak{X}$-triangulation (K_j', π_j') of $f|_{X_j}$ compatible with $\{Y_i\} \cup \{X_3\}$ such that K_j' is a subdivision of K_j,

$$(K_j', \pi_j') = (K_j, \pi_j) \quad \text{outside of} \quad |N(L_j, K_j)|,$$

and $\pi_2'^{-1} \circ (\pi_1'|_{|L_1'|}) \colon L_1' \to L_2'$ is a simplicial isomorphism, where $L_j' = K_j'|_{|L_j|}$. We reduce this assertion to an easier one.

Assertion (ii). Under the same conditions as Assertion (i), there exist two $\mathfrak{X}$-triangulations (L_j', τ_j), $j = 1,2$, of $f|_{X_3}$ such that for each j, L_j' is a subdivision of L_j,

$$\tau_j(\sigma_j) = \pi_j(\sigma_j) \quad \text{for} \quad \sigma_j \in L_j,$$
$$f \circ \tau_j = f \circ \pi_j \quad \text{on} \quad |L_j|,$$

and $\tau_2^{-1} \circ \tau_1 \colon |L_1'| \to |L_2'|$ is a PL homeomorphism.

Assume (ii). By subdividing both L_j' if necessary, we can assume $\tau_2^{-1} \circ \tau_1 \colon L_1' \to L_2'$ is a simplicial isomorphism. Set $\rho_j = \pi_j^{-1} \circ \tau_j$. Note that each ρ_j is an X-homeomorphism of $|L_j|$ such that

$$\rho_j(\sigma_j) = \sigma_j \quad \text{for} \quad \sigma_j \in L_j,$$
$$f \circ \pi_j \circ \rho_j = f \circ \pi_j \quad \text{on} \quad |L_j|,$$

and (L_j', ρ_j) is an X-triangulation of $|L_j|$ (i.e., for each $\sigma_j \in L_j'$, $\rho_j|_{\sigma_j^\circ}$, is a C^r diffeomorphism onto the image). By the cone extension in the proof of II.3.8, we can extend L_j' to a subdivision K_j' of K_j and ρ_j to an X-homeomorphism P_j of $|K_j|$ so that

$$P_j(\sigma_j) = \sigma_j \quad \text{for} \quad \sigma_j \in K_j,$$

(K_j', P_j) is an X-triangulation of $|K_j|$, and

$$(K_j', P_j) = (K_j, \mathrm{id}) \quad \text{outside of} \quad |N(L_j, K_j)|.$$

Moreover, since P_j is constructed by finite repetitions of cone extensions of ρ_j and since $f \circ \pi_j$ are linear on each $\sigma_j \in K_j$, we have

$$f \circ \pi_j \circ P_j = f \circ \pi_j \quad \text{on} \quad |K_j|.$$

Hence $(K_j', \pi_j' = \pi_j \circ P_j)$ fulfills the requirements of Assertion (i).

Proof of Assertion (ii). By II.3.8 we have an X-triangulation (L_3, τ_3) of $f|_{X_3}$ compatible with $\pi_1(L_1) \cup \pi_2(L_2)$. For each $j = 1, 2$, apply II.3.10 to the pair of $(L_j, \pi_j|_{|L_j|})$ and (L_3, τ_3). There exists a PL homeomorphism φ_j from $|L_j|$ to $|L_3|$ such that

$$\varphi_j(\sigma_j) = \tau_3^{-1} \circ \pi_j(\sigma_j) \quad \text{for} \quad \sigma_j \in L_j, \quad \text{and}$$
$$f \circ \pi_j = f \circ \tau_3 \circ \varphi_j \quad \text{on} \quad |L_j|.$$

Let L_j' be a subdivision of L_j such that φ_j is linear on each $\sigma_j \in L_j'$ and (L_j', φ_j) is compatible with L_3, and set $\tau_j = \tau_3 \circ \varphi_j$. Then (L_j', τ_j), $j = 1, 2$, are the required X-triangulations of $f|_{X_3}$. □

Proof of uniqueness in II.3.1'. Assume X is not compact. Let (K, π) and (L, τ) be X-triangulations of f compatible with $\{Y_i\}$. We want a PL homeomorphism μ from $|K|$ to $|L|$ such that

$$f \circ \tau \circ \mu = f \circ \pi \quad \text{and} \quad \mu(\pi^{-1}(Y_i)) = \tau^{-1}(Y_i).$$

By the same reason as in the proof of II.3.10 we can assume that (L, τ) is compatible with $\pi(K)$. So we replace the above conditions by

$$f \circ \tau \circ \mu = f \circ \pi, \quad \text{and} \quad \mu(\sigma) = \tau^{-1} \circ \pi(\sigma) \quad \text{for} \quad \sigma \in K.$$

Note that if X is compact, by II.3.10, there is such a PL homeomorphism. Hence it suffices to prove the following assertion as in the proof of II.3.9.

Assertion. Let K_j and L_j, $j = 1, 2, 3$, be finite full subcomplexes of K and L, respectively, such that

$$K_1 \cap K_2 = K_3, \quad L_1 \cap L_2 = L_3 \text{ and}$$
$$\pi(|K_j|) = \tau(|L_j|), \quad j = 1, 2, 3.$$

For each $j = 1, 2$, let μ_j be a PL homeomorphism from $|K_j|$ to $|L_j|$ such that

$$f \circ \tau \circ \mu_j = f \circ \pi \quad \text{on} \quad |K_j| \text{ and} \qquad (1.|K_j|)$$
$$\mu_j(\sigma) = \tau^{-1} \circ \pi(\sigma) \quad \text{for} \quad \sigma \in K_j. \qquad (2.K_j)$$

Keeping these properties, we can modify μ_2 so that $\mu_2 = \mu_1$ on $|K_3|$.

Proof of Assertion. We carry this out in the same way as the proof of Assertion (i) in the proof of existence. Note that

$$\mu_j(|K_3|) = |L_3|, \quad j = 1, 2,$$

and set $\rho = \mu_2^{-1} \circ \mu_1$ on $|K_3|$. Then

$$f \circ \pi \circ \rho = f \circ \pi \quad \text{on} \quad |K_3| \text{ and}$$
$$\rho(\sigma) = \sigma \quad \text{for} \quad \sigma \in K_3.$$

Let P denote the PL extension: $|K_2| \to |K_2|$ of ρ constructed by cone extensions in the same way as in the proofs of II.3.8 and of existence. Then

$$f \circ \pi \circ P = f \circ \pi \quad \text{on} \quad |K_2| \text{ and}$$
$$P(\sigma) = \sigma \quad \text{for} \quad \sigma \in K_2,$$

because $f \circ \pi$ is linear on each $\sigma \in K_2$. Hence if we replace μ_2 by $\mu_2 \circ P$, the conditions $(1.|K_2|)$ and $(2.K_2)$ continue to be true. Indeed,

$$f \circ \tau \circ \mu_2 \circ P = f \circ \pi \circ P = f \circ \pi \quad \text{on} \quad |K_2|, \text{ and}$$
$$\mu_2 \circ P(\sigma) = \mu_2(\sigma) = \tau^{-1} \circ \pi(\sigma) \quad \text{for} \quad \sigma \in K_2.$$

Clearly $\mu_2 = \mu_1$ on $|K_3|$. $\qquad \qquad \qquad \square$

In Chapter IV, we shall need the following lemma and remark, which are refinements of the uniqueness theorem of $\mathfrak{X}$-triangulations of $\mathfrak{X}$-functions.

Lemma II.3.11. *Let f be an $\mathfrak{X}$-function on an $\mathfrak{X}$-set X contained and locally closed in $\mathbf{R}^n$, and let $\{Y_i\}$ be a family of $\mathfrak{X}$-sets in $\mathbf{R}^n$ locally finite at each point of X. Let (K_i, π_i), $i = 1, 2$, be $\mathfrak{X}$-triangulations of f compatible with $\{Y_i\}$. Assume Axiom (v) and $|K_1|$ and $|K_2|$ are closed in their ambient Euclidean spaces. Then there exists an $\mathfrak{X}$-isotopy τ_t, $0 \le t \le 1$, from $|K_1|$ to $|K_2|$ such that for each Y_i and for each $0 \le t \le 1$, τ_1 is PL,*

$$\tau_0 = \pi_2^{-1} \circ \pi_1, \qquad f \circ \pi_2 \circ \tau_t = f \circ \pi_1, \quad and \quad \tau_t(\pi_1^{-1}(Y_i)) = \pi_2^{-1}(Y_i).$$

Proof. As usual, we can reduce the problem to the case where

$$X = |K_2|, \quad \pi_2 = \mathrm{id}, \quad \text{and} \quad \{Y_i\} = \{\pi_1(\sigma^\circ) \colon \sigma \in K_1\}.$$

Note that for each $\sigma \in K_1$, $\pi_1(\sigma)$ is a polyhedron. Then we need only find an $\mathfrak{X}$-isotopy τ_t, $0 \le t \le 1$, from $|K_1|$ to $|K_2|$ such that for each $\sigma \in K_1$ and for each $0 \le t \le 1$, τ_1 is PL,

$$\tau_0 = \pi_1, \qquad f \circ \tau_t = f \circ \pi_1, \quad \text{and} \quad \tau_t(\sigma) = \pi_1(\sigma).$$

By uniqueness of $\mathfrak{X}$-triangulations of f we have a PL homeomorphism τ from $|K_1|$ to $|K_2|$ such that

$$\tau(\sigma) = \pi_1(\sigma) \quad \text{for} \quad \sigma \in K_1 \quad \text{and} \quad f \circ \tau = f \circ \pi_1.$$

We will construct τ_t so that $\tau_1 = \tau$. Hence it suffices to find an $\mathfrak{X}$-isotopy ρ_t, $0 \le t \le 1$, of $|K_1|$ such that for each $\sigma \in K_1$ and for each $0 \le t \le 1$,

$$\rho_0 = \mathrm{id} \qquad \rho_1 = \pi_1^{-1} \circ \tau,$$
$$f \circ \pi_1 \circ \rho_t = f \circ \pi_1, \quad \text{and} \quad \rho_t(\sigma) = \sigma.$$

Note that $f \circ \pi_1$ is a simplicial function on K_1. Set $P(x, t) = (\rho_t(x), t)$ for $x \in |K_1|$, $t = 0, 1$. Then P is an $\mathfrak{X}$-homeomorphism of $|K_1| \times \{0, 1\}$, and we need to extend P and ρ_t to $|K_1| \times [0, 1]$, keeping the form.

Order all the vertices $a_0, a_1, \ldots$ of K_1. Let $\sigma \in K_1$. For simplicity of notation we assume $\sigma = \Delta a_0 \cdots a_k$. Set

$$\sigma_i = \Delta(a_0, 0) \cdots (a_i, 0)(a_i, 1) \cdots (a_k, 1) \subset \sigma \times [0, 1], \quad i = 0, \ldots, k.$$

Then $\sigma \times [0, 1]$ is the union of σ_i, and $|K_1| \times [0, 1]$ is the union of all σ_i for all σ. On each σ_i we define P as follows. By the Alexander trick we extend $P|_{\Delta(a_i, 1) \cdots (a_k, 1)}$ to $\Delta(a_i, 0)(a_i, 1) \cdots (a_k, 1)$, then to $\Delta(a_{i-1}, 0)(a_i, 0)(a_i, 1) \cdots (a_k, 1) \cdots$ $(a_k, 1), \ldots$, and finally, to σ_i. The extension coincides with P on $\sigma_i \cap (\sigma \times 0)$

because $P = \mathrm{id}$ on $\sigma \times 0$. Thus we define P on σ_i. The value of P at a point (x, t) of σ_i does not depend on σ nor on i. To be precise, the values $P(x, t)$ coincide with each other when two such extensions are given by σ, i and by σ', i' so that $\sigma_i \cap \sigma'_{i'}$ contains (x, t), because the order of the vertices of K_1 are fixed and the Alexander trick decides a unique extension. Hence P and ρ_t are well-defined on $|K_1| \times [0, 1]$. Clearly the extended ρ_t is invariant on each $\sigma \in K_1$ for each $0 \leq t \leq 1$, and the equality $f \circ \pi_1 \circ \rho_t = f \circ \pi_1$ follows from the definition of the Alexander trick because $f \circ \pi_1$ is simplicial. □

Remark II.3.12. Theorem II.3.1, II.3.1' and Lemma II.3.11 hold true also in the case where f is an $\mathfrak{X}$-map from X to an $\mathfrak{X}$-polyhedron of dimension 1 as follows. There exists an $\mathfrak{X}$-triangulation (K, π) of X compatible with $\{Y_i\}$ such that $f \circ \pi$ is linear on each simplex of K. For two such $\mathfrak{X}$-triangulations (K_i, π_i), $i = 1, 2$, there exists an $\mathfrak{X}$-isotopy τ_t, $0 \leq t \leq 1$, from $|K_1|$ to $|K_2|$ such that for each Y_i and for each $0 \leq t \leq 1$, τ_1 is PL,

$$\tau_0 = \pi_2^{-1} \circ \pi_1, \qquad f \circ \pi_2 \circ \tau_t = f \circ \pi_1, \quad \text{and}$$
$$\tau_t(\pi_1^{-1}(Y_i)) = \pi_2^{-1}(Y_i).$$

This is immediate by the above proof.

Remark II.3.13. Let $f \colon X \to Y$ be an $\mathfrak{X}$-map between $\mathfrak{X}$-sets. An $\mathfrak{X}$-*triangulation* of f is a quadruplet (K, L, π, τ), where (K, π) and (L, τ) are $\mathfrak{X}$-triangulations of X and Y, respectively, such that $\tau \circ f \circ \pi^{-1} \colon |K| \to |L|$ is PL. An $\mathfrak{X}$-map is not always $\mathfrak{X}$-triangulable. But we can prove the following statement.

(∗) Assume f is proper. There exists an $\mathfrak{X}$-subset Y' of Y of dimension $< \dim Y - 1$ (not only $< \dim Y$) such that $Y - Y'$ is locally closed in the ambient Euclidean space, and for an $\mathfrak{X}$-neighborhood U of each point of $Y - Y'$ in $Y - Y'$, $f|_{f^{-1}(U)} \colon f^{-1}(U) \to U$ is $\mathfrak{X}$-triangulable.

Moreover, for proper f, it seems very possible that if Y is bounded in the ambient Euclidean space, then $f|_{f^{-1}(Y-Y')}$ is $\mathfrak{X}$-triangulable in the sense of II.3.3 (see IV.1.5). We obtain a more general result in the semialgebraic case (II.4.10). If X and Y are bounded and if we allow $\dim Y' < \dim Y$, then $\mathfrak{X}$-triviality of f on each connected component of $Y - Y'$ [Ha] follows immediately, i.e., there exists an $\mathfrak{X}$-homeomorphism $\alpha \colon f^{-1}(y) \times C \to f^{-1}(C)$ for each connected component C of $Y - Y'$ and a point y of C such that $f \circ \alpha$ is the projection.

The statement (∗) is proved in the same way as II.3.1 using the following generalization of II.2.2', which also is proved in the same way as II.2.2'.

(∗∗) Let $A \subset \mathbf{R}^n \times Y$ be an $\mathfrak{X}$-set such that for each $y \in Y$, $\dim(A \cap (\mathbf{R}^n \times y)) < n$. Set

$$T = \{(\lambda, y) \in S^{n-1} \times Y \colon \lambda \text{ is a singular direction for } A \cap (\mathbf{R}^n \times y)\}.$$

Then T is an $\mathfrak{X}$-set, and there exists an $\mathfrak{X}$-subset Y' of Y of dimension $< \dim Y - 1$ such that for every $y \in Y - Y'$, $T \cap (S^{n-1} \times y)$ is of dimension $< n - 1$.

§II.4. Triangulations of semialgebraic and $\mathfrak{X}_0$ sets and functions

There are two kinds of natural generalizations of a semialgebraic set. One is a semialgebraic set over a real closed field, and the other is an $\mathfrak{X}_0$-set, where $\mathfrak{X}_0$ is a family $\mathfrak{X}$ which satisfies the following axioms stronger than Axioms (iii) and (iv), respectively.

Axiom (iii)$_0$. If $X \subset \mathbf{R}^n$ is an element of $\mathfrak{X}$, and if $p \colon \mathbf{R}^n \to \mathbf{R}^{n-1}$ is a linear map, then $p(X)$ is an element of $\mathfrak{X}$.

Axiom (iv)$_0$. If $X \subset \mathbf{R}$ and $X \in \mathfrak{X}$, then X is a finite union of intervals and points.

We define $\mathfrak{X}_0$-*cell triangulations* of an $\mathfrak{X}_0$-set and an $\mathfrak{X}_0$-function in the same way as semialgebraic cell triangulations. Let R always denote a real closed field. This section treats semialgebraic cell triangulations of semialgebraic sets and functions over R and the $\mathfrak{X}_0$-case. In the semialgebraic case, $r = \infty$ (ω if $R = \mathbf{R}$). For the proof we can not use the pasting method because R might not have the Lindelöf property. We need an effective construction of semialgebraic triangulations.

Remark II.4.1. Let $X \subset R^n$ and $X_i \subset R^n$, $i = 1, 2, \ldots$, be a finite number of semialgebraic sets. The arguments in §II.2 work on semialgebraic triangulations of semialgebraic sets over R. Hence if X is bounded and closed, it admits a semialgebraic triangulation compatible with $\{X_i\}$ [L$_1$]. We can regard general X as a bounded semialgebraic set in R^n through the map $R^n \ni x \to x(1 + |x|^2)^{-1/2} \in R^n$. Therefore, for general X there exist a finite simplicial complex K in R^n, some union Y of open simplices of K and a semialgebraic homeomorphism τ from Y to X such that for each $\sigma \in K$, $\tau|_{\sigma^\circ \cap Y}$ is a C^∞ (C^ω if $R = \mathbf{R}$) diffeomorphism onto the image and $\{\tau(\sigma^\circ \cap Y) \colon \sigma \in K\}$ is compatible with $\{X_i\}$. If we can choose K so that $|K| = Y$, then X is bounded and closed.

Assume X is bounded and closed. Uniqueness of semialgebraic triangulations of X is immediate in the case $R = \mathbf{R}$ by III.1.4. The general case of R follows from it and the Tarski-Seidenberg principle. We sketch the proof. See [Co] for the details.

First we describe a semialgebraic set in R^n as a point of $R^{n'}$ for some n' as follows. Let r, s and t be non-negative integers, and $f_{i,j} = \Sigma_{|\alpha| \le t} a_{i,j,\alpha} x^\alpha$, $i = 1, \ldots, r$, $j = 1, \ldots, 2s$, be polynomial functions on R^n. Let us regard $(f_{i,j})$ as the point $a = (a_{i,j,\alpha})$ of $R^{n'(n,r,s,t)}$, $n'(n,r,s,t) = \sharp\{(i,j,\alpha)\}$. Let a

correspond to the semialgebraic set:

$$S(a) = \bigcup_{i=1}^{r} \{x \in R^n : f_{i,j}(x) = 0, \ f_{i,s+j}(x) > 0, \ j = 1, \dots, s\}.$$

Note that the correspondence S is not injective and any semialgebraic set in R^n is in the image for large r, s and t. For non-negative integers r_1, s_1 and t_1, set

$\mathfrak{A}(n, r, s, t, r_1, s_1, t_1)$

$\quad = \{(a, a_1, a_2) \in R^{n'(2n,r,s,t)} \times R^{n'(n,r_1,s_1,t_1)} \times R^{n'(n,r_1,s_1,t_1)} :$

$S(a_1)$ and $S(a_2)$ are bounded closed polyhedra,

$p_1 S(a) = S(a_1),\ p_2 S(a) = S(a_2),\ $ and $p_1|_{S(a)}$ and $p_2|_{S(a)}$ are injective$\}$,

where p_1 and p_2 are the projections of $R^n \times R^n$ onto the first and latter factors, respectively. Set

$\mathfrak{A}'(n, r, s, t, r_1, s_1, t_1) = \{(a, a_1, a_2) \in \mathfrak{A}(n, r, s, t, r_1, s_1, t_1) : S(a) \text{ polyhedron}\}.$

Note for $(a, a_1, a_2) \in \mathfrak{A}$, $p_2 \circ (p_1|_{S(a)})^{-1}$ is a semialgebraic homeomorphism from $S(a_1)$ to $S(a_2)$, and if $(a, a_1, a_2) \in \mathfrak{A}'$, then $p_2 \circ (p_1|_{S(a)})^{-1}$ is a PL homeomorphism. Let $\mathfrak{B}(n, r, s, t, r_1, s_1, t_1)$ and $\mathfrak{B}'(n, r, s, t, r_1, s_1, t_1)$ denote the respective images of $\mathfrak{A}(n, r, s, t, r_1, s_1, t_1)$ and $\mathfrak{A}'(n, r, s, t, r_1, s_1, t_1)$ under the projection of $R^{n'(2n,r,s,t)} \times R^{n'(n,r_1,s_1,t_1)} \times R^{n'(n,r_1,s_1,t_1)}$ onto the last two factors. Then $\mathfrak{B}$ and $\mathfrak{B}'$ are semialgebraic sets and mean pairs of bounded closed polyhedra in R^n which are semialgebraically and PL homeomorphic, respectively. Now we can easily prove (Lemma 2.21 in [A-B-B]) that for any non-negative integers r_1, s_1 and t_1, there exist non-negative integers r_2, s_2 and t_2 such that for points a_1 and a_2 of $R^{n'(n,r_1,s_1,t_1)}$ with bounded closed polyhedra $S(a_1)$ and $S(a_2)$, $S(a_1)$ and $S(a_2)$ are PL homeomorphic if and only if $(a_1, a_2) \in \mathfrak{B}'(n, r_2, s_2, t_2, r_1, s_1, t_1)$. Consequently, for uniqueness, it suffices to show that

$$\mathfrak{B}(n, r, s, t, r_1, s_1, t_1) \subset \mathfrak{B}'(n, r_2, s_2, t_2, r_1, s_1, t_1)$$

The definition of these sets does not depend on special R, and in the case $R = \mathbf{R}$, the inclusion holds as before. Hence the general case follows from the Tarski-Seidenberg principle (see, e.g., [B-C-R]).

Moreover, we can choose as r_2, s_2 and t_2 effective functions in variables n, r, s, t, r_1, s_1 and t_1. This is proved partially by [A-B-B] and completely by [Co].

To investigate closed semialgebraic subsets of R^n, semialgebraic cell triangulations are useful by the following theorem.

Theorem II.4.2 (Cell triangulation of a semialgebraic or $\mathfrak{X}_0$-set). *Let X be a closed semialgebraic set in R^n, and let $\{X_i\}$ be a finite family of semialgebraic sets in R^n. Then X admits a semialgebraic cell triangulation (C, π) compatible with $\{X_i\}$ such that $|C|$ lies in R^n and if X_i is a closed polyhedron in X, then $\pi^{-1}(X_i) = X_i$.*

The same statement holds for a closed $\mathfrak{X}_0$-set.

Remark II.4.3. In the above theorem (C, π) is not unique unless X is bounded as follows. (Here uniqueness means that for two triangulations (C_j, π_j), $j = 1, 2$, there exist cell subdivisions C_j' of C_j, $j = 1, 2$, which are cell isomorphic.)

Let C_1 and C_2 denote the cell complexes generated by

$$\{(x_1, x_2) \in R^2 : 0 \le x_1,\ 0 \le x_2 \le 1\} \quad \text{and}$$

$$\{(x_1, x_2) \in R^2 : 0 \le x_1,\ 0 \le x_2 \le 1 + x_1\}$$

respectively. Then (C_1, id) and (C_2, π) are semialgebraic cell triangulations of $|C_1|$, where

$$\pi(x_1, x_2) = (x_1, x_2/(1 + x_1)) \quad \text{for} \quad (x_1, x_2) \in |C_2|.$$

However, there are no cell subdivisions C_1' of C_1 and C_2' of C_2 which are isomorphic. Indeed, if C_1' and C_2' are respective cell subdivisions of C_1 and C_2, then for any $\sigma_1 = \sigma(\sigma_1', \sigma_1'') \in C_1'$, σ_1'' is of dimension 0. On the other hand, for some $\sigma_2 = \sigma(\sigma_2', \sigma_2'') \in C_2'$, σ_2'' is of dimension 1. (See §I.3 for the definition of $\sigma(\sigma_1', \sigma_1'')$.)

The following lemma reduces II.4.2 to the case where X is bounded.

Lemma II.4.4. *Let C_1 be a cell decomposition of R^n. There exist a cell subdivision C_2 of C_1 and a semialgebraic imbedding τ of R^n into R^n such that for each $\sigma \in C_2$, $\tau(\sigma) \subset \sigma$, $\tau|_{\sigma^\circ}$ is a C^∞ diffeomorphism onto the image, $\overline{\tau(\sigma)}$ is a usual cell, if σ is bounded, then $\tau = \mathrm{id}$ on σ, and the following condition is satisfied. Set $A = \tau(R^n)$. Let C_3 be a usual cell decomposition of $\overline{A}$ compatible with $\{\tau(\sigma) : \sigma \in C_2\}$. There exist a usual cell subdivision C_4 of C_3 and a semialgebraic homeomorphism $\chi \colon A \to R^n$ such that for each $\sigma \in C_4$, $\chi(\sigma \cap A)$ is a cell, $\chi|_{\sigma^\circ \cap A}$ is a C^∞ diffeomorphism onto the image, and for each $\sigma_1 \in C_2$, $\chi \circ \tau(\sigma_1) = \sigma_1$.*

Proof. By I.3.2, which holds for R, and by the definition of a derived subdivision, we have a cell subdivision C_2 of C_1 compatible with the hyperplanes $\{x_1 = 0\}, \ldots, \{x_n = 0\}$ such that each unbounded $\sigma \in C_2$ is of the form $\sigma = \sigma(\sigma', \sigma'')$, where σ' is a simplex and σ'' is a usual cell, and each point of σ is described (uniquely if $t > 0$) as

$$a + tb \quad \text{for} \quad a \in \sigma',\ 0 \le t,\ b \in \sigma''.$$

Let $\varphi\colon [0,\infty[\to [0,1[$ be a semialgebraic C^∞ diffeomorphism. Define the imbedding τ by $\tau = \mathrm{id}$ on bounded $\sigma \in C_2$ and by

$$\tau(a+tb) = a + \varphi(t)b \text{ for } a \in \sigma',\ t \geq 0,\ b \in \sigma'', \text{ unbounded } \sigma = \sigma(\sigma',\sigma'') \in C_2.$$

It is clear that (C_2, τ) fulfills all but the last requirement in the lemma.

     Proof of the last requirement. We can assume that C_3 is a simplicial complex. For a number $0 \leq \varepsilon \leq 1$, let A_ε denote the union of all bounded $\sigma \in C_2$ and of points $a + tb$ for $a \in \sigma'$, $0 \leq t \leq \varepsilon$, $b \in \sigma''$, where $\sigma(\sigma',\sigma'')$ are unbounded cells in C_2. Then $A = \mathrm{Int}\, A_1$, and for each $0 < \varepsilon \leq 1$, $\mathrm{Int}\, A_\varepsilon$ is semialgebraically homeomorphic to R^n. Fix $0 < \varepsilon < 1$ so that $\mathrm{Int}\, A_\varepsilon \cup \mathrm{bdry}\, A_1$ contains all vertices of C_3.

     Let C_4 denote the usual cell complex consisting of $\sigma \cap A_\varepsilon$, $\sigma \cap \mathrm{bdry}\, A_\varepsilon$ and $\sigma - \mathrm{Int}\, A_\varepsilon$ for $\sigma \in C_3$. Set $\chi = \mathrm{id}$ on A_ε. We want to define $\chi\colon A - A_\varepsilon \to R^n - A_\varepsilon$. First let a semialgebraic homeomorphism $\psi\colon A - \mathrm{Int}\, A_\varepsilon \to \mathrm{bdry}\, A_\varepsilon \times [\varepsilon, 1[$ be defined so that $\psi(\cdot) = (\cdot, \varepsilon)$ on $\mathrm{bdry}\, A_\varepsilon$, for each $\sigma \in C_4$ with $\sigma^\circ \subset A - A_\varepsilon$,

$$\psi(\sigma \cap A) = (\sigma \cap \mathrm{bdry}\, A_\varepsilon) \times [\varepsilon, 1[,$$

and $\psi|_{\sigma^\circ}$ is a C^∞ diffeomorphism onto the images as follows. Let $\sigma \in C_3$ with $\sigma \cap (A - \mathrm{Int}\, A_\varepsilon) \neq \varnothing$. There exist σ' and σ'' in C_3 such that $\sigma' * \sigma'' = \sigma$, $\sigma' \subset \mathrm{Int}\, A_\varepsilon$ and $\sigma'' \subset \mathrm{bdry}\, A$. Each point x of $\sigma \cap (A - \mathrm{Int}\, A_\varepsilon)$ is uniquely described as $tc + (1-t)d$ for $0 < t < 1$, $c \in \sigma'$ and $d \in \sigma''$. Let t_0 be the number such that $t_0 c + (1 - t_0)d \in \mathrm{bdry}\, A_\varepsilon$. We set

$$\psi(x) = (t_0 c + (1 - t_0)d,\ (t - t_0 - \varepsilon t + \varepsilon)/(1 - t_0)).$$

This ψ fulfills the requirements. Through χ, we can regard $A - \mathrm{Int}\, A_\varepsilon$ as $\mathrm{bdry}\, A_\varepsilon \times [\varepsilon, 1[$. Hence for construction of χ, it suffices to find a semialgebraic homeomorphism $\chi_0\colon \mathrm{bdry}\, A_\varepsilon \times [\varepsilon, 1[\to R^n - \mathrm{Int}\, A_\varepsilon$ such that $\chi_0(\cdot, \varepsilon) = \mathrm{id}$, for each $\sigma \in C_4$ with $\sigma \subset \mathrm{bdry}\, A_\varepsilon$, $\chi_0|_{\sigma^\circ \times [\varepsilon, 1[}$ is a C^∞ diffeomorphism onto the image, and for each $\sigma_1 \in C_2$,

$$\chi_0\{(\sigma_1 \cap \mathrm{bdry}\, A_\varepsilon) \times [\varepsilon, 1[\} = \sigma_1 - \mathrm{Int}\, A_\varepsilon.$$

For a point $(x, t) \in \mathrm{bdry}\, A_\varepsilon \times [\varepsilon, 1[$, x is described as $a + \varepsilon b$ for some $a \in \sigma'$ and $b \in \sigma''$ with $\sigma = \sigma(\sigma', \sigma'') \in C_2$. Set

$$\chi_0(x, t) = x + \varphi_1(t)b,$$

where $\varphi_1\colon [\varepsilon, 1[\to [0, \infty[$ is a semialgebraic C^∞ diffeomorphism. Then χ_0 satisfies the required properties. $\qquad\square$

Proof of II.4.2. We prove only the former statement. The latter is proved in the same way. By I.3.2 there is a cell decomposition C_1 of R^n such that if X_i is a closed polyhedron in R^n, then X_i is a union of cells of C_1. We can replace X and $\{X_i\}$ with R^n and $\{X, X_i\}$ respectively. Hence it suffices to prove the following assertion.

Let C_1 be a cell decomposition of R^n. There exists a semialgebraic cell triangulation (C, π) of R^n compatible with $\{X_i\}$ such that $|C| = R^n$ and $\pi(\sigma) = \sigma$ for $\sigma \in C_1$.

Let us prove this. By the first half of II.4.4 we have a cell subdivision C_2 of C_1 and a semialgebraic imbedding τ of R^n into R^n such that for each $\sigma \in C_2$, $\tau(\sigma) \subset \sigma$, $\tau|_{\sigma^\circ}$ is a C^∞ diffeomorphism onto the image, $\overline{\tau(\sigma)}$ is a usual cell, and if σ is bounded, then $\tau = \text{id}$ on σ. Apply II.2.1 and II.2.4 to $\overline{\tau(R^n)}$ and $\{\tau(X_i)\} \cup \{\tau(\sigma) : \sigma \in C_2\}$. There exists a semialgebraic triangulation (C_3, ρ) of $\overline{\tau(R^n)}$ compatible with $\{\tau(X_i)\} \cup \{\tau(\sigma) : \sigma \in C_2\}$ such that $|C_3| = \overline{\tau(R^n)}$ and for each $\sigma \in C_2$, $\rho \circ \tau(\sigma) = \tau(\sigma)$. By the latter half of II.4.4 there exist a usual cell subdivision C_4 of C_3 and a semialgebraic homeomorphism $\chi \colon \overline{\tau(R^n)} \to R^n$ such that for each $\sigma \in C_4$, $\chi(\sigma \cap \tau(R^n))$ is a cell, $\chi|_{\sigma^\circ \cap \tau(R^n)}$ is a C^∞ diffeomorphism onto the image, and for each $\sigma_1 \in C_2$, $\chi \circ \tau(\sigma_1) = \sigma_1$. Let C be a cell decomposition of R^n compatible with $\{\chi(\sigma \cap \tau(R^n)) \colon \sigma \in C_4\}$, and set $\pi = \tau^{-1} \circ \rho \circ \chi^{-1}$. Then (C, τ) is what we want. $\qquad\square$

Remark II.4.5. This remark holds also for an X_0-function. Let f be a semialgebraic function on a semialgebraic set $X \subset R^n$, and let $\{X_i\}$ be a finite family of semialgebraic sets in R^n. By II.3.1, if $R = \mathbf{R}$ and if X is compact, then f admits a unique semialgebraic triangulation compatible with $\{X_i\}$. If X is bounded and closed in R^n, we can prove in the same way the uniqueness of semialgebraic triangulations of f compatible with $\{X_i\}$. For its existence we need to modify the proof in the case of $R = \mathbf{R}$ because we used local compactness of $\mathbf{R}$. We will show the modification in II.4.10. If f is bounded, then by using Remark II.3.3 and the map $R^n \ni x \to x(1 + |x|^2)^{-1/2} \in R^n$, we obtain a finite simplicial complex K in R^{n+1}, a union Y of open simplices of K, and a semialgebraic homeomorphism τ from Y to X such that for each $\sigma \in K$, $\tau|_{\sigma^\circ \cap Y}$ is a C^∞ (C^ω if $R = \mathbf{R}$) diffeomorphism onto the image, $\{\tau(\sigma^\circ \cap Y) \colon \sigma \in K\}$ is compatible with $\{X_i\}$ and $f \circ \tau$ can be extended to $|K|$ so that the extension is linear on each simplex of K.

It remains to consider unbounded semialgebraic functions. We need to treat semialgebraic or X_0- cell triangulations of semialgebraic or X_0- functions for the following reason.

Let g be a continuous function on a topological space Z. Assume there exist a finite simplicial complex K, a union Y of open simplices of K and a homeomorphism τ from Y to Z such that $g \circ \tau$ is linear on $\sigma^\circ \cap Y$ for each $\sigma \in K$. Then g is bounded because a linear function on an open simplex is bounded.

Let $f: X \to Y$ be a semialgebraic map between closed semialgebraic sets $X, Y \subset R^n$. We call f *proper* if the inverse image of a bounded set in R^n under f is bounded in R^m.

Theorem II.4.6 (Cell triangulation of a semialgebraic or $\mathfrak{X}_0$-function) (cf. [S$_5$]). *Let X be a closed semialgebraic set in R^n, let $\{X_i\}$ be a finite family of semialgebraic sets in R^n, and let f be a proper semialgebraic function on X. Then f admits a semialgebraic cell triangulation (C, π) compatible with $\{X_i\}$ such that C lies in R^n and if X_i is a closed polyhedron in X, then $\pi^{-1}(X_i) = X_i$.*

The same statement holds for an $\mathfrak{X}_0$-function.

Remark II.4.7 (Non-proper case of II.4.6). Let X be a semialgebraic set in R^n, let $\{X_i\}$ be a finite family of semialgebraic sets in R^n, and let f be a semialgebraic function on X. There exist a cell complex C in R^{n+1}, a union Y of open cells of C and a semialgebraic homeomorphism $\pi: Y \to X$ such that Y is dense in $|C|$. Furthermore, for each $\sigma \in C$, $\pi|_{\sigma^\circ \cap Y}$ is a C^∞ (C^ω if $R = \mathbf{R}$) diffeomorphism onto the image, $\{\pi(\sigma^\circ \cap Y): \sigma \in C\}$ is compatible with $\{X_i\}$, and $f \circ \pi$ is extensible to $|C|$ so that the extension is linear on each cell of C.

The $\mathfrak{X}_0$-case also holds true.

Proof of II.4.7. By imbedding R^n into R^n by the map $R^n \ni x \to x(1 + |x|^2)^{-1/2} \in R^n$, we assume that X is bounded. Replace X, f and $\{X_i\}$ with the closure of graph f in $R^n \times R$, the restriction of the projection $R^n \times R \to R$ onto the closure, and $\{\text{graph } f, X_i \times R\}$, respectively. Then the problem is reduced to II.4.6. □

Remark II.4.8. (i) II.4.6 does not necessarily hold unless f is proper as shown below. Set

$$X = \mathbf{R}, \quad \text{and} \quad f(x) = 1/(1 + x^2) \quad \text{for} \quad x \in \mathbf{R}.$$

If there were a semialgebraic cell triangulation (C, π) of f, then C would contain a non-compact cell σ and $f \circ \pi|_\sigma$ would be either constant or unbounded, because a linear function on $\mathbf{R}_+ = \{\text{non-negative reals}\}$ is always constant or unbounded. That is impossible because f is bounded and not constant locally at each point of $\mathbf{R}$.

(ii) In II.4.7, we can not necessarily choose C in R^n as shown below. Set

$$X = \mathbf{R} - 0, \quad \text{and} \quad f(x) = x + 1/x \quad \text{for} \quad x \in \mathbf{R}.$$

Let C, Y and π be a solution in II.4.7. Assume $|C| \subset \mathbf{R}$. Clearly $C \neq \{\mathbf{R}\}$. There are at most two cells of C which are of the form $[a, \infty[$ or $]-\infty, a]$. On the interior of such a cell, $f \circ \pi$ is bounded above or below, and on another open cell included in Y, $f \circ \pi$ is bounded. Hence for a sufficiently large number $b \in \mathbf{R}$, $(f \circ \pi)^{-1}(\pm b)$ consists of at most two points. However, if $b > 2$, $f^{-1}(\pm b)$ consists of four points by the definition of f.

Proof of II.4.6. We prove only the former statement because the latter is easier. By II.4.2 we can assume that X is the underlying polyhedron of a cell complex C_1 and $C_1 = \{X_i\}$. Let C_2, τ, A and A_ε be defined from C_1 in the same way as in the proof of II.4.4. Let $\theta: R \to]-1, 1[$ be a semialgebraic C^∞ diffeomorphism. Since f is proper, $\theta \circ f \circ \tau^{-1}$ can be extended to a semialgebraic function on $\overline{A}$. Let F denote the extension. Note $F = \pm 1$ on bdry A and $-1 < F < 1$ on A. We assume $F^{-1}(1) \neq \varnothing$ and $F^{-1}(-1) \neq \varnothing$ without loss of generality.

For a while, suppose there exists a semialgebraic triangulation (C_3, ρ) of F compatible with $\{\tau(\sigma) : \sigma \in C_2\}$ such that $|C_3| = \overline{A}$ and $\rho(\tau(\sigma)) = \tau(\sigma)$ for $\sigma \in C_2$(II.4.10). Let us subdivide C_3 as follows. Let $-1 = v_1 < \cdots < v_m = 1$ denote the values of $F \circ \rho$ on the vertices of C_3. Set

$$V_i = (F \circ \rho)^{-1}(v_i), \quad i = 1, \ldots, m \quad \text{and}$$
$$\tilde{V}_i = (F \circ \rho)^{-1}([v_i, v_{i+1}]), \quad i = 1, \ldots, m - 1.$$

We require each of V_i and $\tilde{V}_i$ to be unions of cells of C_3. Note that for all $\sigma \in C_3$, V_i and $\tilde{V}_i$, $\sigma \cap V_i$ and $\sigma \cap \tilde{V}_i$ are usual cells, and such cells generate a usual cell complex. By I.3.12, which also holds for R, there exists a simplicial subdivision of this complex without new vertices. Replace C_3 with this simplicial complex, and keep the same notation C_3 for it. Then V_i and $\tilde{V}_i$ are unions of simplices of C_3.

We define a cell subdivision C_4 of C_3 as follows, which is similar to the definition of C_4 in the proof of II.4.4. Let δ be a positive number with $-1 < -\delta < v_2$ and $v_{m-1} < \delta < 1$. Set

$$B = (F \circ \rho)^{-1}([-\delta, \delta]).$$

Then Int $B \cup$ bdry A contains all vertices of C_3. Let C_4 denote the usual cell complex consisting of $\sigma \cap B$, $\sigma \cap$ bdry B and $\sigma - $ Int B for all $\sigma \in C_3$.

Let us define a semialgebraic homeomorphism $\chi: A \to X$, which is similar to χ in the proof of II.4.4. Set $\chi = $ id on B. Let a semialgebraic homeomorphism $\psi: A - $ Int $B \to$ bdry $B \times [\delta, 1[$ be defined in exactly the same way as ψ in the proof of II.4.4. Then

$$F \circ \rho \circ \psi^{-1}(x, t) = \pm t \quad \text{for} \quad (x, t) \in \text{bdry } B \times [\delta, 1[,$$

because for each $x \in \text{bdry } B$, $F \circ \rho$ is linear on $\psi^{-1}(x \times [\delta, 1[)$. By ψ, it suffices to construct a semialgebraic homeomorphism $\chi_0 \colon \text{bdry } B \times [\delta, 1[\to X - \text{Int } B$. For each $(x, t) \in \text{bdry } B \times [\delta, 1[$, x is described as $a + t_0 b$ for some $a \in \sigma'$, $t_0 > 0$, $b \in \sigma''$ and $\sigma = \sigma(\sigma', \sigma'') \in C_2$. Here a, t_0 and b are uniquely determined by x. Set

$$\chi_0(x, t) = x + \varphi_1(t)b,$$

where $\varphi_1 \colon [\delta, 1[\to [0, \infty[$ is a semialgebraic C^∞ diffeomorphism. Then χ_0 is semialgebraic, and its bijectivity follows from the uniqueness property in the description of elements of $\sigma \in C_2$. We define χ to be $\chi_0 \circ \psi$ on $A - B$.
  Set

$$C = \{\chi(\sigma \cap A) \colon \sigma \in C_4\} \quad \text{and} \quad \pi_0 = \tau^{-1} \circ \rho \circ \chi^{-1}.$$

By the above method of construction of χ, (C, π_0) is a semialgebraic cell triangulation of X compatible with $C_1 = \{X_i\}$ such that $|C| = X$, $\pi_0(X_i) = X_i$, and for each cell $\sigma \in C$, $\theta \circ f \circ \pi_0 \circ \chi$ is linear on $\chi^{-1}(\sigma)$. But $f \circ \pi_0$ is not always linear on each $\sigma \in C$. We want to replace π_0 with a semialgebraic homeomorphism π so that (C, π) is the required cell triangulation. For that, it suffices to find a semialgebraic homeomorphism α of X such that for each $\sigma \in C$, $\alpha|_{\sigma^\circ}$ is a C^∞ diffeomorphism of σ° and $f \circ \pi_0 \circ \alpha$ is linear on σ.
  Set $\alpha = \text{id}$ on $\cup_{i=2}^{m-1} V_i \cup \text{bdry } B$ because $f \circ \pi_0$ is locally constant there. We need to define α on (i) $\tilde{V}_i - V_i - V_{i+1}$ for $1 < i < m-1$, on (ii) $(\tilde{V}_1 - V_2) \cap \text{Int } B$, on (iii) $(\tilde{V}_{m-1} - V_{m-1}) \cap \text{Int } B$ and on (iv) $X - B$. Recall that $\chi = \text{id}$ on B, $\theta \circ f \circ \pi_0 = F \circ \rho$ on B and that $C|_{\tilde{V}_i}$ for all $1 < i < m - 1$ are simplicial complexes. Consider (i). Fix a point $v \in]v_i, v_{i+1}[$. In the same way as ψ in the proof of II.4.4, we can construct a semialgebraic homeomorphism:

$$\beta_i = (\beta_i', \beta_i'') \colon \tilde{V}_i - V_i - V_{i+1} \longrightarrow (\theta \circ f \circ \pi_0)^{-1}(v) \times]v_i, v_{i+1}[$$

such that for each $\sigma \in C$, $\beta_i|_{\sigma^\circ \cap (\tilde{V}_i - V_i - V_{i+1})}$ is a C^∞ diffeomorphism onto $(\sigma^\circ \cap (\theta \circ f \circ \pi_0)^{-1}(v)) \times]v_i, v_{i+1}[$, $\beta_i'' = \theta \circ f \circ \pi_0$, and for each $x \in (\theta \circ f \circ \pi_0)^{-1}(v)$, $\beta_i^{-1}(x \times]v_i, v_{i+1}[)$ is a segment. Note that $F \circ \pi_0$ is strictly monotone on the segment $\beta^{-1}(x \times]v_i, v_{i+1}[)$. We will require α to be invariant on such segments. Let $\tilde{f}_i$ denote the function on $\tilde{V}_i$ such that

$$\tilde{f}_i = f \circ \pi_0 \quad \text{on} \quad V_i \cup V_{i+1},$$

and for each $\sigma \in C$ with $\sigma \subset \tilde{V}_i$, $\tilde{f}_i$ is linear on σ. Compare $f \circ \pi_0$ and $\tilde{f}_i$. They are strictly monotone on each segment $\beta_i^{-1}(x \times]v_i, v_{i+1}[)$ and take the same values on $V_i \cup V_{i+1}$. Hence there exists uniquely a homeomorphism α_i of $\tilde{V}_i$ such that

$$\alpha_i = \text{id} \quad \text{on} \quad V_i \cup V_{i+1}, \qquad \tilde{f}_i = f \circ \pi_0 \circ \alpha_i,$$
$$\alpha_i(\sigma) = \sigma \quad \text{for} \quad \sigma \in C \text{ with } \sigma \subset \tilde{V}_i,$$

and α_i is invariant on each segment $\beta_i^{-1}(x \times]v_i, v_{i+1}[)$. Clearly α_i is semial-gebraic. We define α on $\tilde{V}_i$ to be α_i. Then $f \circ \pi_0 \circ \alpha_i$ is linear on each $\sigma \in C$ with $\sigma \subset \tilde{V}_i$.

On (ii) $(\tilde{V}_1 - V_2) \cap \operatorname{Int} B$ and on (iii) $(\tilde{V}_{m-1} - V_{m-1}) \cap \operatorname{Int} B$, we define α in the same way.

Consider (iv). Let $\tilde{f}_0$ denote the PL function on $X - \operatorname{Int} B$ defined by $\tilde{f}_0 = f \circ \pi_0$ on bdry B and by

$$\tilde{f}_0(a + tb) = t + g(a, b)$$
$$\text{for} \quad a + tb \in X - B, \ a \in \sigma', \ t > 0, \ b \in \sigma'', \ \text{unbounded } \sigma(\sigma', \sigma'') \in C_2,$$

where g is some PL function on $\sigma' \times \sigma''$. (Unique existence of g and $\tilde{f}_0$ is easy to show because bdry B is a polyhedron and $f \circ \pi_0 = \text{const}$ on each connected component of bdry B.) Then $\tilde{f}_0$ is linear on each $\sigma \in C$ with $\sigma \subset X - \operatorname{Int} B$. By the same reason as above there exists a unique semialgebraic homeomorphism α_0 of $X - \operatorname{Int} B$ such that

$$\alpha_0 = \operatorname{id} \quad \text{on} \quad \operatorname{bdry} B, \qquad \tilde{f}_0 = f \circ \pi_0 \circ \alpha_0,$$
$$\alpha_0(\sigma) = \sigma \quad \text{for} \quad \sigma \in C \ \text{with} \ \sigma \subset X - \operatorname{Int} B,$$

and α_0 is invariant on each half line $\{a + tb \in X - \operatorname{Int} B : t \geq 0\}$ for $a \in \sigma'$ and $b \in \sigma''$ with unbounded $\sigma(\sigma', \sigma'') \in C_2$. Setting $\alpha = \alpha_0$ on $X - B$, we complete the construction of α.

It remains to prove the theorem under the assumption that X is bounded and closed in R^n. In this case, uniqueness of semialgebraic triangulations of X holds (see II.4.1). Hence it suffices to construct a semialgebraic triangulation (C, π) of f compatible with $\{X_i\}$, because by II.3.2, we can modify (C, π) so that $|C| = X$ and $\pi^{-1}(X_i) = X_i$. That follows from the following lemma. (The proof of II.3.1 does not work for R.)   □

Remark II.4.9. Let $f \colon X \to Y$ be a semialgebraic map between semialgebraic sets in $\mathbf{R}^n$ (not in R^n). Then there exists a semialgebraic subset Y' of Y of dimension $< \dim Y - 1$ such that $f|_{f^{-1}(Y-Y')}$ is semialgebraically cell triangulable in the following sense. There exist cell complexes C and D, unions A and B of open cells of C and D, respectively, and semialgebraic homeomorphisms $\pi \colon A \to f^{-1}(Y - Y')$ and $\tau \colon B \to Y - Y'$ such that A and B are dense in $|C|$ and $|D|$, respectively, and for each $\sigma \in C$, $\pi|_{\sigma \circ \cap A}$ is a C^∞ diffeomorphism onto the image. A similar statement for D holds, and $\tau \circ f \circ \pi$ is extensible to a cell map from C to D.

We can state the same remark for an $\mathfrak{X}_0$ map.

This follows from (**) in II.3.13 as in the proof of II.4.6.

Lemma II.4.10. *Let X be a bounded closed semialgebraic set in $R^n \times R$, and let $\{X_i\}$ be a finite family of semialgebraic sets in $R^n \times R$. There exists a semialgebraic triangulation (C, π) of X compatible with $\{X_i\}$ such that C lies in $R^n \times R$ and π is of the form:*

$$\pi(x,t) = (\pi'(x,t), t) \quad for \quad (x,t) \in |C|.$$

We prove this in the same way as II.3.5. Since II.2.2' does not hold globally, we replace it by the following:

Lemma II.4.11. *Let Y be an algebraic set in $R^n \times R$ which does not include any algebraic set of the form $R^n \times c$, $c \in R$. There exists a bipolynomial diffeomorphism τ (i.e. τ and τ^{-1} are both polynomial maps) of $R^n \times R$ of the form:*

$$\tau(x,t) = (x_1, x' + \tau'(x_1), t) \quad for \quad (x,t) = (x_1, x', t) = (x_1, \ldots, x_n, t) \in R^n \times R$$

such that $(1, 0, \ldots, 0)$ is a non-singular direction for $\tau(Y)$.

Proof of II.4.11. Let g be a polynomial function on $R^n \times R$ with zero set Y, let $\tau' : R \to R^{n-1}$ be a polynomial map, and let τ be the bipolynomial diffeomorphism of $R^n \times R$ defined by $\tau(x,t) = (x_1, x' + \tau'(x_1), t)$ for $(x,t) = (x_1, x', t) \in R^n \times R$. Then

$$\tau^{-1}(x,t) = (x_1, x' - \tau'(x_1), t).$$

Since $\tau(Y)$ is an algebraic set, $(1, 0, \ldots, 0)$ is a non-singular direction for $\tau(Y)$ if and only if the common zero set in $R^{n-1} \times R$ of the (x', t)-functions $g \circ \tau^{-1}(1, x', t), g \circ \tau^{-1}(2, x', t), \ldots$ is empty. We want to choose C' so that this condition is satisfied. For each positive integer l, let $W_l \subset R^{n-1} \times R$ denote the common zero set of $g \circ \tau^{-1}(1, x', t), \ldots, g \circ \tau^{-1}(l, x', t)$. For proof of the lemma, it suffices to prove the following statement because the polynomial ring $R[x', t]$ is Noetherian and because W_l is determined only by the values $\tau'(i)$, $i = 1, \ldots, l$.

($*$) For $a_1, \ldots, a_l \in R^{n-1}$ there exists an integer $l' > l$ and a point $a_{l'} \in R^{n-1}$ such that for any polynomial map $\tau' : R \to R^{n-1}$ with $\tau'(i) = a_i$, $i = 1, \ldots, l, l'$, $W_{l'}$ is a proper subset of W_l if $W_l \neq \varnothing$.

Proof of ($*$). Let (b, c) be a point of $W_l \subset R^{n-1} \times R$. By assumption $R^n \times c$ is not included in Y. Hence we have an integer $l' > l$ such that $l' \times R^{n-1} \times c$ is not included in Y, i.e., the x'-function $g(l', x', c)$ is not zero. Choose $a_{l'}$ so that $g(l', b - a_{l'}, c) \neq 0$. Then $W_{l'}$ does not contain (b, c) and hence is a proper subset of W_l. $\qquad\square$

Proof of II.4.10. We proceed by induction on n. The case $n = 0$ is trivial. Hence assume the lemma for $n - 1$. Let X and $\{X_i\}$ be given by a finite number of non-zero polynomials $g_j(x, t)$, $j = 1, \ldots, m - 1$, on $R^n \times R$ (i.e., X and X_i are finite unions of finite intersections of the sets $\{g_j = 0\}$ or $\{g_j > 0\}$). Note that if a stratification in $R^n \times R$ is compatible with $\{g_j^{-1}(0)\}$ and if its strata are semialgebraically connected, then the stratification is compatible with X and $\{X_i\}$. Hence we can replace X and $\{X_i\}$ with $Z = \{|x|^2 + t^2 \le c\}$ and $\{g_j^{-1}(0)\}$, respectively, for a number c so large that $Z \supset X$.

Set $g_m(x, t) = |x|^2 + t^2 - c$. We can assume that g_j are all irreducible, $g_1, \ldots, g_{m'}$ are of the form $t - d$, $d \in R$, where t is the last coordinate function of $R^n \times R$, and $g_{m'+1} \ldots g_m$ are not divisible by polynomials of this form.

Let Y denote the zero set of $g_{m'+1} \cdots g_m$. By II.4.11 we can suppose that $(1, 0, \ldots, 0)$ is a non-singular direction for Y. Let p denote the restriction to Z of the projection $R^n \times R \to R^{n-1} \times R$ which forgets the first factor. In the same way as in the proof of II.3.5, applying the induction hypothesis to $p(Z)$, we can construct a semialgebraic triangulation of Z compatible with $\{g_j = 0\}_{j=1,\ldots,m}$. We omit the details. □

§II.5. C^r 𝒳-manifolds

This section introduces a topology of the C^r 𝒳-map space and investigates it and structures of C^r 𝒳-manifolds, $r > 0$. The main results are the following. II.5.4 is an 𝒳-version of Thom's transversality theorem, which is fundamental for further investigations. A C^1 𝒳-manifold admits a unique C^r 𝒳-manifold structure (II.5.6). By pasting a boundary, we can compactify a uniquely bounded non-compact C^r 𝒳-manifold (II.5.7). Any bounded C^r 𝒳-manifold is C^r 𝒳-diffeomorphic to some affine non-singular algebraic variety (II.5.10). Note that the boundedness condition in these results is not necessary if 𝒳 $=$ 𝒳$_0$ (II.5.11). In this section, r' always denotes a non-negative integer $\le r$. A C^0 *𝒳-manifold* in $\mathbf{R}^n$ of dimension m is an 𝒳-set which is locally 𝒳-homeomorphic to $\mathbf{R}^m$.

In the case where 𝒳 is semialgebraic, a $C^{r'}$ 𝒳-manifold is called a $C^{r'}$ *Nash manifold* and a *Nash manifold* is a semialgebraic C^ω manifold. Similarly we also define a $C^{r'}$ *Nash map* and a *Nash map*.

Let X and Y be $C^{r'}$ 𝒳-submanifolds of $\mathbf{R}^n$ and $\mathbf{R}^m$, respectively. We want to define a topology on the space $C_𝒳^{r'}(X, Y)$ of $C^{r'}$ 𝒳-maps from X to Y. It is called the $C^{r'}$ *Nash topology* if 𝒳 is semialgebraic (see [S$_3$]), and it is the $C^{r'}$ *Whitney topology* if 𝒳 satisfies Axiom (v) and if X is closed in $\mathbf{R}^n$.

Consider the case $r' = 0$ and $Y = \mathbf{R}$. For each bounded positive 𝒳-function ε on X, set

$$U_\varepsilon = \{g \in C^0_{\mathfrak{X}}(X, \mathbf{R}) \colon |g| < \varepsilon\}.$$

For each $g \in C^0_{\mathfrak{X}}(X, \mathbf{R})$, $g + U_\varepsilon$ are $\mathfrak{X}$-functions on X(II.1.1). By choosing $\{g + U_\varepsilon\}_\varepsilon$ as a fundamental neighborhood system of g in $C^0_{\mathfrak{X}}(X, \mathbf{R})$, we define the topology on $C^0_{\mathfrak{X}}(X, \mathbf{R})$. The topology on $C^0_{\mathfrak{X}}(X, \mathbf{R}^m)$ is similarly defined. In the case $r' = 0$ and general Y, regard an $\mathfrak{X}$-map from X to Y as an $\mathfrak{X}$-map from X to $\mathbf{R}^m$. Then $C^0_{\mathfrak{X}}(X, Y)$ becomes a subset of $C^0_{\mathfrak{X}}(X, \mathbf{R}^m)$. The topology on $C^0_{\mathfrak{X}}(X, Y)$ is the relativization of the topology on $C^0_{\mathfrak{X}}(X, \mathbf{R}^m)$.

Example 1. If $\mathfrak{X}$ is semialgebraic, $r' = 0$, $X = \mathbf{R}^n$ and $Y = \mathbf{R}$, then to define the above fundamental neighborhood system, we can choose $\{\varepsilon\} = \{1/(n + |x|^n)\}_{n=1,2,\dots}$ (see [S$_3$]).

By this we see the following. Let $L_{n,m}$ denote the space of all linear maps from $\mathbf{R}^n$ to $\mathbf{R}^m$, and regard it as a subspace of $C^0_{\mathfrak{X}}(\mathbf{R}^n, \mathbf{R}^m)$. The relative topology on $L_{n,m}$ is discrete.

Example 2. If $\mathfrak{X}$ satisfies Axiom (v), and if X is closed in $\mathbf{R}^n$, then the $\{\varepsilon\}$ has to contain arbitrarily small positive C^0 functions.

If $r' = 1$, $X = \mathbf{R}^n$ and $Y = \mathbf{R}$, then the topology on $C^1_{\mathfrak{X}}(\mathbf{R}^n, \mathbf{R})$ is by definition the one induced from the topology on $C^0_{\mathfrak{X}}(\mathbf{R}^n, \mathbf{R}^{n+1})$ by the map:

$$C^1_{\mathfrak{X}}(\mathbf{R}^n, \mathbf{R}) \ni f \to (f, \frac{\partial f}{\partial x_1}, \dots, \frac{\partial f}{\partial x_n}) \in C^0_{\mathfrak{X}}(\mathbf{R}^n, \mathbf{R}^{n+1}).$$

In the same way we define the topology on $C^{r'}_{\mathfrak{X}}(\mathbf{R}^n, Y)$ for any $r' \geq 1$ and Y.

For $r' \geq 1$ and general X we define the topology as follows. For $f \in C^{r'}_{\mathfrak{X}}(X, Y)$ and $x \in X$, the differential df_x of f at x means a linear map from the tangent space $T_x X$ of X at x to $\mathbf{R}^m$. Extend df_x to a map $\mathbf{R}^n \to \mathbf{R}^m$ by composing it with the orthogonal projection $\mathbf{R}^n \to T_x X$ and let Df denote the map : $X \to L_{n,m} = \mathbf{R}^{nm}$. Then $(f, Df) : X \to \mathbf{R}^m \times \mathbf{R}^{nm}$ is a $C^{r'-1}$ $\mathfrak{X}$-map as shown below. Assume this. For each $1 \leq k \leq r'$ we inductively define a $C^{r'-k}$ map $D^k f : X \to \mathbf{R}^{n^k m}$ by $D^k f = D(D^{k-1}f)$. Let $\|f\|_{r'}$ denote the function on X defined by

$$\|f\|_{r'}(x) = |f(x)| + |Df(x)| + \dots + |D^{r'}f(x)| \quad \text{for} \quad x \in X.$$

We see that $\|f\|_{r'}$ is an $\mathfrak{X}$-function. If we replace $|g| < \varepsilon$ in the definition of the topology on $C^0_{\mathfrak{X}}(X, \mathbf{R})$ with $\|g\|_{r'} < \varepsilon$, then we define the topology on $C^{r'}_{\mathfrak{X}}(X, Y)$. We call the topology the $C^{r'}$ $\mathfrak{X}$-*topology*. We also can define this topology in the case of $C^{r'}$ $\mathfrak{X}$-manifolds with boundary and corners.

Note that for f and g in $C_{\mathfrak{X}}^{r'}(X, \mathbf{R}^m)$ with $f + g \in C_{\mathfrak{X}}^{r'}(X, \mathbf{R}^m)$ and for $c \in \mathbf{R}$,

$$D^k(f + g) = D^k f + D^k g \quad \text{and} \quad D^k(cf) = cD^k f, \ k = 1, \ldots, r'.$$

Proof that (f, Df) is of class $\mathfrak{X}$. Let $e \in \mathbf{R}^n$, and for each $x \in X$, let $T_x X(e)$ denote the image of e under the orthogonal projection onto $T_x X$. It suffices to prove that the map $X \ni x \to (f(x), df_x(T_x X(e))) \in \mathbf{R}^m \times \mathbf{R}^m$ is of class $\mathfrak{X}$. As shown in the proof of II.1.5, the set $\{(x, T_x X(e) : x \in X\}$ is an $\mathfrak{X}$-set, and by II.1.3, so is the set:

$$\{(x, v, f(x), df_x v) : x \in X, \ v \in T_x X\}.$$

Hence the set:

$$\{(x, T_x X(e), f(x), df_x(T_x X(e))) : x \in X\}$$

is also an $\mathfrak{X}$-set. The graph of the above map is the image of this set under the projection which forgets the second factor, and the map $x \to T_x X(e)$ is bounded. Therefore, the graph and hence the map are of class $\mathfrak{X}$. $\qquad \square$

Note. If we define a topology on $C_{\mathfrak{X}}^1(X, Y)$ to be the induced one from $C_{\mathfrak{X}}^0(TX, TY)$ by the map $C_{\mathfrak{X}}^1(X, Y) \ni f \to df \in C_{\mathfrak{X}}^0(TX, TY)$, and if $\dim X > 0$, then $C_{\mathfrak{X}}^1(X, Y)$ becomes a discrete space. This follows from the fact that $L_{n,m}$ with the relative topology is a discrete space.

Attending to the above note, we give another equivalent definition of the topology. This definition will become useful. Set

$$\tilde{T}X = \{(x, v) \in X \times \mathbf{R}^n : v \in T_x X, \ |v| \leq 1\},$$
$$\tilde{T}^2 X = \{(x, v, u, w) \in X \times \mathbf{R}^n \times \mathbf{R}^n \times \mathbf{R}^n :$$
$$v \in T_x X, \ (u, w) \in T_{x,v}(TX), \ |v| \leq 1, \ |u| \leq 1, \ |w| \leq 1\},$$
$$\vdots$$

Then $\tilde{T}^{r'} X$ is a C^0 $\mathfrak{X}$-manifold possibly with boundary and corners, and each $f \in C_{\mathfrak{X}}^{r'}(X, Y)$ defines an $\mathfrak{X}$-map from $\tilde{T}^{r'} X$ to $T^{r'} Y$, where $T^{r'} Y = T(T(\cdots(TY)\cdots))$. Let $\tilde{d}^{r'} f$ denote the element of $C_{\mathfrak{X}}^0(\tilde{T}^{r'} X, T^{r'} Y)$. An equivalent definition of the topology on $C_{\mathfrak{X}}^{r'}(X, Y)$ is the reduced one by the map:

$$C_{\mathfrak{X}}^{r'}(X, Y) \ni f \to \tilde{d}^{r'} f \in C_{\mathfrak{X}}^0(\tilde{T}^{r'} X, T^{r'} Y).$$

It is clear that these two definitions are equivalent.

Let $B^{r'}(X,Y)$ denote $C^{r'}$ $\mathfrak{X}$-maps from X to Y which carry bounded sets in $\mathbf{R}^n$ to bounded sets in $\mathbf{R}^m$. Let Z be another $C^{r'}$ $\mathfrak{X}$-submanifold of $\mathbf{R}^l$ and let $f \in B^{r'}(X,Y)$ and $h \in B^{r'}(Y,Z)$.

(i) The map $h_* : B^{r'}(X,Y) \to B^{r'}(X,Z)$, induced by h, is continuous.

(ii) Assume that for any bounded set $C \subset Z$, $h^{-1}(C)$ is bounded. Then
$h_* : C_{\mathfrak{X}}^{r'}(X,Y) \to C_{\mathfrak{X}}^{r'}(X,Z)$ is continuous.

(iii) Assume the same condition as in (ii) on f. The map $f^* : C_{\mathfrak{X}}^{r'}(Y,Z) \to C_{\mathfrak{X}}^{r'}(X,Z)$, induced by f, is continuous if and only if f is proper (i.e., the inverse image of any compact set in Z under f is compact).

Proof of (i). By (II.1.1), for any $g \in B^{r'}(X,Y)$, we have $h \circ g \in B^{r'}(X,Z)$. Let $r' = 0$, $g \in B^0(X,Y)$ and $\varepsilon > 0 \in B^0(X,\mathbf{R})$. By the definition of the topology it suffices to prove that the set

$$\{g' \in B^0(X,Y) : |h \circ g' - h \circ g| < \varepsilon\}$$

is an open neighborhood of g in $B^0(X,Y)$. Let $g' \in B^0(X,Y)$. The condition $|h \circ g' - h \circ g| < \varepsilon$ is satisfied if and only if graph g' is contained in the set:

$$G = \{(x,y) \in X \times Y : y \in h^{-1}\{z \in Z : |z - h \circ g(x)| < \varepsilon(x)\}\}.$$

Hence we need only show that G is a neighborhood of graph g in $X \times Y$ and also an $\mathfrak{X}$-set. Clearly G is a neighborhood, and it is an $\mathfrak{X}$-set by the following two facts. First G is the image under the projection of $X \times Y \times Z$ onto $X \times Y$ of the $\mathfrak{X}$-set:

$$\{(x,y,z) \in X \times Y \times Z : |z - h \circ g(x)| < \varepsilon(x), \; z = h(y)\}.$$

Second, the restriction of the projection to the closure of this $\mathfrak{X}$-set is proper, which follows from the assumption $h \in B^0(Y,Z)$.

Let $r' = 1$ and $g \in B^1(X,Y)$. If

$$\tilde{d}g \in B^0(\tilde{T}X, TY), \qquad \tilde{d}h \in B^0(\tilde{T}Y, TZ), \quad \text{and} \quad \tilde{d}g(\tilde{T}X) \subset (\tilde{T}Y)^\circ,$$

then continuity of h_* follows from the same arguments applied to the map:

$$(\tilde{d}h)_* : B^0(\tilde{T}X, TY) \to B^0(\tilde{T}X, TZ).$$

But these conditions are not always satisfied. Keeping the topology, we need to change the definition of $\tilde{T}$. Define an $\mathfrak{X}$-function α on Y by

$$\alpha(y) = \max_{(y,v) \in \tilde{T}Y} |dh_y v| + 1 \quad \text{for} \quad y \in Y,$$

and set

$$\hat{T}Y = \{(y,v) \in TY : |v| \le \alpha(y)\} \quad \text{and} \quad \hat{d}h = dh|_{\hat{T}X}.$$

In the second definition of the topology we can replace $C^0_{\mathfrak{X}}(\tilde{T}Y, TZ)$ with $C^0_{\mathfrak{X}}(\hat{T}Y, TZ)$. If we replace this, then $\hat{d}h \in B^0(\hat{T}Y, TZ)$. Shrink $\tilde{T}X$ more. Then we can define a similar $\mathfrak{X}$-function β on X, $\hat{T}X$ by β, and $\hat{d}g$ for any $g \in C^1_{\mathfrak{X}}(X, Y)$ so that

$$\hat{d}g \in B^0(\hat{T}X, TY) \quad \text{and} \quad \hat{d}g(\hat{T}X) \subset (\hat{T}Y)^\circ.$$

Hence we can apply the arguments, and we obtain (i) in the case of $r' = 1$.
    In the same way we prove the general case of r'. $\square$

Proof of (ii). By (II.1.1) and by the assumption in (ii), $h \circ g \in C^{r'}_{\mathfrak{X}}(X, Z)$ for any $g \in C^{r'}_{\mathfrak{X}}(X, Y)$. The rest of the proof is the same as in the proof of (i).$\square$

Proof of (iii). (iii) is immediate by the above proof and by the following easy fact.
    Let A, B and C be closed subsets of Euclidean spaces. Let $\varphi \in C^0(A, B)$, and let $\varphi^* : C^0(B, C) \to C^0(A, C)$ denote the map induced by φ. Define topologies on $C^0(A, C)$ and on $C^0(B, C)$ in the same way as we defined the topology on $C^0_{\mathfrak{X}}(X, Y)$, by replacing the ε with a positive C^0 function. Then φ^* is continuous if and only if φ is proper. $\square$

Lemma II.5.1 (X-tube). *Let r be a positive integer. Let X be a C^r X-submanifold of $\mathbf{R}^n$. If $r > 1$, there exists an X-tubular neighborhood U of X in $\mathbf{R}^n$, and the tube $T = (|T|, \pi, \rho)$ at X, defined by $|T| = U$, $\pi = $ the projection and $\rho(x) = |x - \pi(x)|^2$ for $x \in U$, is of class C^{r-1} and X. For general r there exists a C^r X-tube at X.*

Proof. Assume $r > 1$. Let $p \colon N \to X$ ($N \subset X \times \mathbf{R}^n$) denote the normal bundle of X in $\mathbf{R}^n$, which is clearly an X-vector bundle by (II.1.4). Let an X-map $\varphi \colon N \to \mathbf{R}^n$ be defined by

$$\varphi(x, y) = x + y \quad \text{for} \quad (x, y) \in N \subset X \times \mathbf{R}^n.$$

By (II.1.7) the subset U_1 of N where $d\varphi$ has rank n is an X-neighborhood of $X = X \times 0$ in N. Let U_2 denote the subset of U_1 consisting of points (x, y) with $\text{dis}(\varphi(x, y), X) = |y|$, and set

$$U_3 = \{(x, y) \in U_2 \colon \sharp(\varphi^{-1}(\varphi(x, y)) \cap U_2) = 1\}.$$

Then U_3 is an $\mathfrak{X}$-neighborhood of X in N, $\varphi|_{U_3}$ is an imbedding, and $U = \varphi(U_3)$ is an $\mathfrak{X}$-tubular neighborhood of X in $\mathbf{R}^n$. By (II.1.1) and (II.1.2), the tube $T = (|T|, \pi, \rho)$ at X, defined in II.5.1, is of class $\mathfrak{X}$, and clearly it is of class C^{r-1}.

Let r be general. By the following II.5.2 we have a C^r $\mathfrak{X}$-approximation $\psi \colon X \to G_{n, \text{codim } X}$ of the characteristic map of the normal bundle of X in $\mathbf{R}^n$ in the C^0 $\mathfrak{X}$-topology. Choose the approximation so strong that for each $x \in X$, $\mathbf{R}^n$ is the direct sum of $\psi(x)$ and $T_x X$, which is possible by II.5.3. We obtain a C^r tube at X from ψ by the method in examples after the definition of a C^r tube. We can prove that the tube is of class $\mathfrak{X}$ in the same way as in the case $r > 1$. $\qquad\square$

Theorem II.5.2 (Approximation of an $\mathfrak{X}$-map). *Let $r > r'$ be positive and non-negative integers, respectively. Let X and Y be C^r $\mathfrak{X}$-submanifolds of $\mathbf{R}^n$, and let $f \colon X \to Y$ be a $C^{r'}$ $\mathfrak{X}$-map such that for a bounded set $C \subset \mathbf{R}^n$, $f(X \cap C)$ is bounded. We can approximate f by a C^r $\mathfrak{X}$-map in the $C^{r'}$ $\mathfrak{X}$-topology.*

Proof. Assume Y is a C^{r+1} $\mathfrak{X}$-submanifold of $\mathbf{R}^n$. It suffices to prove the case $Y = \mathbf{R}$ for the following reason. By the first half of II.5.1, whose proof did not use II.5.2, there exists an $\mathfrak{X}$-tubular neighborhood U of Y in $\mathbf{R}^n$ and the orthogonal projection $p \colon U \to Y$ is a C^r $\mathfrak{X}$-map. Regard f as a $C^{r'}$ $\mathfrak{X}$-map from X to $\mathbf{R}^n$. If we have a C^r $\mathfrak{X}$-approximation $g \colon X \to \mathbf{R}^n$ of f, $p \circ g \colon X \to Y$ is a C^r $\mathfrak{X}$-approximation of f because $p \circ f = f$ and because the induced map p_* is continuous. Thus we can assume $Y = \mathbf{R}^n$ and hence by (II.1.1), $Y = \mathbf{R}$. Here we use the boundedness assumption.

In the case where Y is merely of class C^r also, we reduce the problem to the function case. Assume the theorem in the function case. Then the latter half of II.5.1 holds true because we used II.5.2 in the proof of the latter half of II.5.1 only to approximate an $\mathfrak{X}$-map from X to $G_{n, \text{codim } Y}$, and because we can approximate the map if II.5.2 holds in the function case as shown above. Hence there exists a C^r $\mathfrak{X}$-tube at Y, and it suffices to approximate f by a C^r $\mathfrak{X}$-map $g \colon X \to \mathbf{R}^n$ by the same reason as above. Therefore, we assume $Y = \mathbf{R}$.

Case $r'=0$. For a positive $\mathfrak{X}$-function ε on X we will find a C^r $\mathfrak{X}$-function g on X such that $|f - g| \leq \varepsilon$. By (II.1.17) there exists a finite C^{r+1} $\mathfrak{X}$-stratification $\{X_i\}$ of X with weak frontier condition such that for each i $f|_{X_i}$ is of class C^{r+1}. By the first half of II.5.1 for each i, we have an $\mathfrak{X}$-tubular neighborhood U_i of X_i in $\mathbf{R}^n$ and the orthogonal projection $p_i \colon U_i \to X_i$ is of class C^r and $\mathfrak{X}$. Note that $p_i|_{U_i \cap X}$ is of class C^r. Define a C^r $\mathfrak{X}$-function g_i on $U_i \cap X$ by $g_i = f \circ p_i$. After shrinking U_i, we can assume $|g_i - f|_{U_i \cap X}| \leq \varepsilon|_{U_i \cap X}$. Hence if there exists a partition of unity $\{\varphi_i\}$ of class C^r and $\mathfrak{X}$ subordinate to the covering $\{U_i\}$, then $g = \sum \varphi_i g_i$ satisfies the condition $|f - g| \leq \varepsilon$. (Such g is of class $\mathfrak{X}$ by the boundedness assumption in II.5.2(II.1.1).)

We construct $\{\varphi_i\}$ as follows. Without loss of generality, we assume $\dim X_i = i$. For each i, let χ_i denote the function on X_i defined by

$$\chi_i(x) = \min\{1, \operatorname{dis}(x, E_i(x) - p_i^{-1}(x))\} \quad \text{for} \quad x \in X_i,$$

where $\operatorname{dis}(x, \varnothing)$ means $\varnothing$, and $E_i(x)$ denotes the $(\operatorname{codim} X_i)$-dimensional linear space including $p_i^{-1}(x)$. The graph of χ_i is an $\mathfrak{X}$-set, and χ_i is locally larger than a positive constant. Apply (II.1.17) to the projection graph $\chi_i \to X_i$, and stratify X_i. (Here we use downward induction on i to substratify $\{X_i\}$.) Then we can assume χ_i is a positive C^r $\mathfrak{X}$-function. Let ψ be a semialgebraic (and hence $\mathfrak{X}$) C^r function on $[0,1]$ such that $0 \leq \psi \leq 1$, $\psi = 1$ on $[0, 1/3]$ and $\psi = 0$ on $[1/2, 1]$. Define an $\mathfrak{X}$-function φ_i' on $\mathbf{R}^n$ by

$$\varphi_i'(x) = \begin{cases} 0 & \text{for} \quad x \notin U_i \\ \psi(|x - p_i(x)|/\chi_i \circ p_i(x)) & \text{for} \quad x \in U_i. \end{cases}$$

It is of class C^r on $\mathbf{R}^n - \partial X_i$, and $\{\varphi_i = \varphi_i' \prod_{j<i}(1 - \varphi_j')\}$ fulfills the requirements. Here ∂X_i denotes the boundary $\overline{X_i} - X_i$.

Note. From now on we can use the latter half of II.5.1. Indeed, in its proof, we used II.5.2 only in the case $r' = 0$.

Case $r' > 0$. Let ε be a positive $\mathfrak{X}$-function on X. We need to find a C^r $\mathfrak{X}$-function g on X with $\|f - g\|_{r'} \leq \varepsilon$. For that we will construct a finite C^{r+1} $\mathfrak{X}$-stratification $\{X_i\}$ of X with frontier condition, and a C^r $\mathfrak{X}$-function f_i on a small open $\mathfrak{X}$-neighborhood O_i of each X_i in X such that

$$D^j(f - f_i) = 0 \quad \text{on} \quad X_i, \quad j = 0, \dots, r', \tag{1}$$

(2) (X_i, O_i) is C^r $\mathfrak{X}$-diffeomorphic to $(X_i \times 0, X_i \times I^{m_i})$, where

$$I = \,]-1, 1[, \quad m_i = n' - \dim X_i, \quad n' = \dim X,$$

(3) $f|_{X_i}, \dots, D^{r'} f|_{X_i}$ are C^r $\mathfrak{X}$-functions.

First we obtain the $\{X_i\}$ with property (3) as follows. Let $\{X_i\}$ be a finite C^{r+1} $\mathfrak{X}$-stratification of X with weak frontier condition such that all $f|_{X_i}$ are of class C^r. Note that we use here the assumption that $f(C \cap X)$ is bounded for a bounded set $C \subset \mathbf{R}^n$ and that it follows from the assumption that Df is of class $\mathfrak{X}$. We require, moreover, $Df|_{X_i}$ to be of class C^r. We cannot apply (II.1.17) to Df because $Df(C \cap X)$ is not always bounded for a bounded set $C \subset \mathbf{R}^n$. Hence we need to modify Df. Define a $C^{r'-1}$ $\mathfrak{X}$-map $\partial f \colon X \to G_{n+1,n'}$ by

$$\partial f(x) = T_{x,f(x)}\, \mathrm{graph}\, f \quad \text{for} \quad x \in X.$$

Then $Df|_{X_i}$ is of class C^r if and only if $\partial f|_{X_i}$ is of class C^r. Since ∂f is bounded, we can apply (II.1.17) to ∂f. Hence by substratifying $\{X_i\}$, we assume that for each i, $Df|_{X_i}$ is of class C^r. For $D^2 f|_{X_i}$, imbed $G_{n+1,n'}$ in $\mathbf{R}^{(n+1)^2}$, replace ∂f with its graph $\subset \mathbf{R}^{n+(n+1)^2}$, and apply the same process as above. Then we can assume that for each i, $D^2 f|_{X_i}$ is of class C^r if $r' > 1$. By repeating these arguments, we obtain (3).

The properties (2) and (3) imply existence of $\{f_i\}$ with (1). Indeed, assume (2) and (3). For construction of $\{f_i\}$ we can suppose $(X_i, O_i) = (X_i \times 0, X_i \times I^{m_i})$. For each i, let $y = (y_1, \ldots, y_{m_i})$ denote the coordinate system of $\mathbf{R}^{m_i}$, and set

$$A = \{\alpha = (\alpha_1, \ldots, \alpha_{m_i}) \in \mathbf{N}^{m_i} : |\alpha| = \alpha_1 + \cdots + \alpha_{m_i} \leq r'\},$$

$$D_y^\alpha = \frac{\partial^{|\alpha|}}{\partial y_1^{\alpha_1} \cdots \partial y_{m_i}^{\alpha_{m_i}}} \quad \text{for} \quad \alpha = (\alpha_1, \ldots, \alpha_{m_i}) \in A, \quad \text{and}$$

$$f_i(x,y) = \sum_{\alpha \in A} D_y^\alpha f(x,0) y^\alpha / \alpha! \quad \text{for} \quad (x,y) \in X_i \times I^{m_i}.$$

Then f_i is a C^r $\mathfrak{X}$-function on O_i and (1) is satisfied.

It remains to obtain (2). Let $\{X_i\}$ be a finite C^{r+1} $\mathfrak{X}$-stratification of X with (3). For each i, let Λ_i denote the subset of G_{n,m_i} consisting of the linear spaces spanned by m_i axes of the x_j-axes, $j = 1, \ldots, n$, where $(x_1, \ldots, x_n)$ is the coordinate system of $\mathbf{R}^n$. For each $\lambda \in \Lambda_i$, let $Q_{i\lambda}$ denote the set of points $x \in X_i$ such that the restriction to $T_x X$ of the orthogonal projection $u_x \colon \mathbf{R}^n \to \lambda + T_x X_i$ is bijective. Then $\{Q_{i\lambda}\}_{\lambda \in \Lambda_i}$ is a finite open $\mathfrak{X}$-covering of X_i. By (II.1.14) we can stratify each X_i into a finite C^{r+1} $\mathfrak{X}$-stratification $\{X_{ij}\}_j$ with frontier condition so that each X_{ij} is included in some $Q_{i\lambda}$. Hence by induction on m_i, we can assume $X_i = Q_{i\lambda}$ for some λ. For some open $\mathfrak{X}$-neighborhood O_i of X_i in X, (X_i, O_i) is C^r $\mathfrak{X}$-diffeomorphic to $(X_i \times 0, X_i \times I^{m_i})$ for the following reason.

We can assume $m_i > 0$. Let $q \colon N \to X_i$ be the orthogonal projection of an $\mathfrak{X}$-tubular neighborhood of X_i in $\mathbf{R}^n$. Choose O_i so small that $O_i \subset N$ and for each $x \in X_i$, $u_x|_{q^{-1}(x) \cap O_i}$ is a C^r imbedding. For each $x \in O_i$, let $y \in \lambda$ denote the λ-factor of $u_{q(x)}(x - q(x))$ in $\lambda + T_x X_i$. We easily prove that the map $O_i \ni x \to (q(x), y) \in X_i \times \lambda$ is a C^r $\mathfrak{X}$-imbedding. Therefore, we can regard O_i as an $\mathfrak{X}$-neighborhood of $X_i \times 0$ in $X_i \times \mathbf{R}^{m_i}$. Let χ_i be the function on X_i defined in the same way as in the case $r' = 0$, i.e.,

$$\chi_i(x) = \min\{1, \mathrm{dis}((x,0), x \times \mathbf{R}^{m_i} - O_i)\} \quad \text{for} \quad x \in X_i.$$

Approximate χ_i by a positive C^r $\mathfrak{X}$-function in the C^0 $\mathfrak{X}$-topology, keep the notation χ_i for the approximation, and consider the C^r $\mathfrak{X}$-diffeomorphism:

$$X_i \times \mathbf{R}^{m_i} \ni (x,y) \longrightarrow (x, (m_i+1)y/\chi_i(x)) \in X_i \times \mathbf{R}^{m_i}.$$

The image of O_i under this diffeomorphism includes $X_i \times I^{m_i}$. Thus if we replace O_i with the inverse image of $X_i \times I^{m_i}$, the property (2) follows.

We assume $\dim X_i = i$ and $O_i \cap X_j = \varnothing$ for all $i > j$ for simplicity of notation. For proof of the theorem, it suffices to find open $\mathfrak{X}$-neighborhoods $W_i \subset W_i'$ of X_i in O_i and C^r $\mathfrak{X}$-functions φ_i on $X - \partial X_i$ for all $0 \le i < n'$ such that

$$\overline{W_i'} \subset O_i \cap \partial X_i, \qquad 0 \le \varphi_i \le 1, \qquad \varphi_i = 1 \quad \text{on} \ \ W_i,$$

$$\varphi_i = 0 \quad \text{outside of} \ \ W_i' \quad \text{and}$$

$$\|(f - f_i)(1 - \varphi_0) \cdots (1 - \varphi_{i-1})\varphi_i\|_{r'} \le \varepsilon/n' \quad \text{on} \quad O_i$$

for the following reason. Let $\tilde{f}_i$ and $\tilde{\varphi}_i$ denote the natural extensions of f_i and φ_i to X, respectively, (i.e., $\tilde{f}_i = 0$ on O_i^c and $\tilde{\varphi}_i = 0$ on ∂X_i). Then $(1 - \tilde{\varphi}_0) \cdots (1 - \tilde{\varphi}_{i-1})\tilde{\varphi}_i$ and $\tilde{f}_i(1 - \tilde{\varphi}_0) \cdots (1 - \tilde{\varphi}_{i-1})\tilde{\varphi}_i$ are C^r $\mathfrak{X}$-functions on X because

$$(1 - \varphi_0) \cdots (1 - \varphi_{i-1})\varphi_i = 0 \quad \text{on} \quad O_i \cap (W_0 \cup \cdots \cup W_{i-1} \cup (O_i - W_i'))$$

and because $O_i^c \cup W_0 \cup \cdots \cup W_{i-1} \cup (O_i - W_i')$ is a neighborhood of O_i^c in X. The function

$$g = \tilde{f}_0\tilde{\varphi}_0 + \cdots + \tilde{f}_{n'-1}(1-\tilde{\varphi}_0) \cdots (1-\tilde{\varphi}_{n'-2})\tilde{\varphi}_{n'-1} + f_{n'}(1-\tilde{\varphi}_0) \cdots (1-\tilde{\varphi}_{n'-1})$$

is what we wanted, because

$$f = f\tilde{\varphi}_0 + \cdots + f(1-\tilde{\varphi}_0) \cdots (1-\tilde{\varphi}_{n'-2})\tilde{\varphi}_{n'-1} + f(1-\tilde{\varphi}_0) \cdots (1-\tilde{\varphi}_{n'-1}).$$

Here note that $f_{n'} = f|_{X_{n'}}$. We want to construct $W_i \subset W_i'$ and φ_i for each $i < n'$. We proceed by induction on i. Assume we have already constructed $W_0, W_0', \varphi_0, \ldots, W_{i-1}, W_{i-1}', \varphi_{i-1}$. Consider $(f - f_i)(1 - \tilde{\varphi}_0) \cdots (1 - \tilde{\varphi}_{i-1})$, and shrink O_i so that $\overline{O_i} - X = \overline{X_i} - X$. We can restate the problem as follows. Here we replace X_i with X and write $m = m_i$ for simplicity of notation.

Assertion. Let m be a positive integer, let h be a $C^{r'}$ $\mathfrak{X}$-function on $X \times I^m$ and let ε be a positive $\mathfrak{X}$-function on $X \times I^m$ such that $D_y^\alpha h = 0$ on $X \times 0$ for all $\alpha \in A$, where

$$y \in I^m \quad \text{and} \quad A = \{\alpha = (\alpha_1, \ldots, \alpha_m) \in \mathbf{N}^m : |\alpha| \le r'\}.$$

There exist open $\mathfrak{X}$-neighborhoods $\Omega \subset \Omega'$ of $X \times 0$ in $\mathbf{R}^n \times \mathbf{R}^m$ and a C^r $\mathfrak{X}$-function ω on $X \times I^m$ such that

$$\overline{\Omega'} \subset X \times I^m \cup \partial X \times 0, \qquad 0 \leq \omega \leq 1,$$
$$\omega = 1 \quad \text{on } \Omega, \qquad \omega = 0 \quad \text{on } X \times I^m - \Omega', \quad \text{and}$$
$$\|h\omega\|_{r'} \leq \varepsilon. \tag{4}$$

Proof of Assertion. We proceed by induction on $n' = \dim X$. Assume the assertion holds for X of dimension $< n'$. By using a C^r $\mathfrak{X}$-tube at $X \times I^m$ in $\mathbf{R}^n \times \mathbf{R}^m$ as usual, we can extend h and ε to $(\mathbf{R}^n - Z) \times I^m$, where $Z = \partial X$. Hence we assume in the assertion that $X = \mathbf{R}^n - Z$. Note that $\dim Z < n'$. Set $\dim Z = n''$. Let χ be a C^r $\mathfrak{X}$-function on X such that $0 < \chi \leq 1$, and let ψ be a C^r $\mathfrak{X}$-function on $[0, \infty[$ such that $0 \leq \psi \leq 1$, $\psi = 1$ on $[0, 1/2]$, and $\psi = 0$ on $[1, \infty[$. Set

$$\omega = \psi(|y|^2/\chi(x)) \quad \text{for} \quad (x, y) \in X \times I^m.$$

For some Ω and Ω'', ω satisfies the conditions except (4) in the assertion. Fixing ψ, we will choose χ so that (4) is satisfied. We want to replace (4) with an easy equivalent condition on χ. Set

$$A' = \{\alpha = (\alpha_1, \ldots, \alpha_{n+m}) \in \mathbf{N}^{n+m} : |\alpha| \leq r'\}.$$

The condition (4) is equivalent to

$$|D^\alpha h D^\beta \omega| \leq \varepsilon \quad \text{for} \quad \alpha \in A' \text{ and } \beta \in A' \text{ with } \alpha + \beta \in A', \tag{4'}$$

where $D^\alpha = \partial^{|\alpha|}/\partial x_1^{\alpha_1} \cdots \partial x_n^{\alpha_n} \partial y_1^{\alpha_{n+1}} \cdots \partial y_m^{\alpha_{n+m}}$. Let h_α denote the non-negative function on $X \times I^m$ defined so that $h_\alpha = 0$ on $X \times 0$ and

$$|D^\alpha h(x, y)| = |y|^{r' - |\alpha|} h_\alpha(x, y) \quad \text{for} \quad (x, y) \in X \times I^m. \tag{5}$$

It is easy to show that h_α is of class $\mathfrak{X}$. By (5), for each $\alpha \in A'$ and $\beta \in A'$ with $\alpha + \beta \in A'$, we have

$$|D^\alpha h(x, y) D^\beta \omega(x, y)| = |y|^{r' - |\alpha|} h_\alpha(x, y) |D^\beta \omega(x, y)|$$
$$\leq |y|^\beta h_\alpha(x, y) |D^\beta \omega(x, y)| \leq \chi^{|\beta|/2}(x) \tilde{h}_\alpha(x) |D^\beta \omega(x, y)| \quad \text{for} \quad (x, y) \in X \times I^m,$$

where

$$\tilde{h}_\alpha(x) = \max_{|y| \leq \sqrt{\chi(x)}} h_\alpha(x, y).$$

Here the last inequality follows from the fact that $D^\beta \omega(x, y) \neq 0$ only for (x, y) with $|y|^2/\chi(x) \leq 1$. Hence for (4'), it suffices to choose χ so that for each $\alpha \in A'$ and $\beta \in A'$,

$$\chi^{|\beta|/2}(x)|D^\beta \omega(x, y)| \leq c \quad \text{on} \quad X \times I^m, \quad \text{and} \tag{6}$$

$$\tilde{h}_\alpha \leq \varepsilon/c \quad \text{on} \quad \mathbf{R}^n - Z \tag{7}$$

for some constant c.

By easy calculations of partial derivatives of ω we obtain (6) for some constant c which does not depend on χ, if χ satisfies the condition:

$$\left|\frac{\partial \chi}{\partial x_i}\right| \leq \sqrt{\chi}, \ i = 1, \ldots, n, \quad \text{and} \quad |D^\beta \chi| \leq 1 \quad \text{for} \ \beta \in B \text{ with } |\beta| > 1, \tag{8}$$

where

$$B = \{\beta = (\beta_1, \ldots, \beta_n) \in \mathbf{N}^n : |\beta| \leq r'\}.$$

Fix such c. Since $h_\alpha = 0$ on $X \times 0$ for any $\alpha \in A'$, we have a positive $\mathfrak{X}$-function χ_c on X such that for any χ with $\chi \leq \chi_c$, (7) is satisfied. Thus we require χ to satisfy (8) and the condition:

$$\chi \leq \chi_c. \tag{9}$$

The condition (8) is, moreover, replaced with the condition:

$$|D^\beta \chi| \leq 1 \quad \text{for} \quad \beta \in B \tag{8'}$$

because if χ satisfies (8'), then $c'\chi^2$ satisfies (8) for a small positive number c'.

Existence of such a χ is not clear because Z is not smooth. We need to stratify Z. We will construct a finite C^r $\mathfrak{X}$-stratification $\{Z_i\}_{i=0,\ldots,n''}$ of Z with weak frontier condition, $\mathfrak{X}$-tubes $T_i = (|T_i|, \pi_i, \rho_i)$ of class C^r at Z_i, $i = 0, \ldots, n''$, and C^r $\mathfrak{X}$-functions ψ_i on $\mathbf{R}^n - \partial Z_i$, κ_i on $|T_i|$, $i = 0, \ldots, n''$, and κ_{-1} on $\mathbf{R}^n$ with the following properties. For each i,

$$\dim Z_i = i, \ \ 0 \leq \psi_i \leq 1,$$

$$\psi_i = 1 \ \text{ on an open } \mathfrak{X}\text{-neighborhood } V_i \text{ of } Z_i \text{ in } |T_i|,$$

$$\psi_i = 0 \ \text{ outside of another } V_i' \text{ with } \overline{V_i'} \subset |T_i| \cup \partial Z_i,$$

$$\kappa_i \geq 0, \ \ \kappa_i > 0 \ \ \text{ on } |T_i| - Z_i, \ \ \kappa_{-1} > 0,$$

and $\tilde{\kappa}_{n''}\psi_{n''}, \tilde{\kappa}_{n''-1}(1-\psi_{n''})\psi_{n''-1}, \ldots, \tilde{\kappa}_0(1-\psi_{n''})\cdots(1-\psi_1)\psi_0$ and $\kappa_{-1}(1-\psi_{n''})\cdots(1-\psi_0)$ on X satisfy (8') and (9), where $\tilde{\kappa}_i$ denotes the natural extensions of κ_i to X. Note that the functions $\tilde{\kappa}_{n''}\psi_{n''},\ldots$ are of class C^r and $\mathfrak{X}$ on X but not necessarily positive on X. If we succeed in construction, then the function

$$\chi = (\tilde{\kappa}_{n''}\psi_{n''} + \cdots + \tilde{\kappa}_{-1}(1-\psi_{n''})\cdots(1-\psi_0))/(n''+2)$$

is what we want because it is positive on X.

We will construct $\{Z_i, T_i, \psi_i, \kappa_i, V_i, V_i'\}_{i=0,\ldots,n''}$ by downward induction on i. We postpone construction of κ_{-1}. Let $0 \le i \le n''$. Assuming there exists $\{Z_j, T_j, \psi_j, \kappa_j, V_j, V_j'\}_{j=i+1,\ldots,n''}$ with $\dim(Z - \cup_{j=i+1}^{n''} Z_j) \le j$, we will construct Z_i, T_i, ψ_i, κ_i, V_i and V_i'. Set $Z' = Z - \cup_{j=i+1}^{n''} Z_j$ and suppose $\dim Z' = i$ (if $< i$, then set $Z_i = \varnothing$). By (II.1.14) we have a C^r $\mathfrak{X}$-manifold Z_i of dimension i contained in Z' such that $Z' - Z_i$ is of dimension $< i$ and $\{Z_j\}_{j=i,\ldots,n''}$ satisfies the weak frontier condition. By the Note, there is a C^r $\mathfrak{X}$-tube $T_i = (|T_i|, \pi_i, \rho_i)$ at Z_i. Let $\tilde{\psi}_j$ denote the natural extension of ψ_j to $\mathbf{R}^n$ for each $i+1 \le j \le n''$. Then $\psi' = (1-\tilde{\psi}_{i+1})\cdots(1-\tilde{\psi}_{n''})$ is of class C^r and $\mathfrak{X}$ on $\mathbf{R}^n - Z'$. Hence we can replace the requirement that $\tilde{\kappa}_i(1-\psi_{n''})\cdots(1-\psi_{i+1})\psi_i$ on X satisfies $(8')$ and (9) with the condition:

$$\|\kappa_i\psi_i\|_{r'} \le \delta \quad \text{on} \quad |T_i| \tag{8i}$$

for some non-negative $\mathfrak{X}$-function δ on $|T_i|$ with $\delta^{-1}(0) = Z_i$.

Moreover, we reduce this condition as follows. First it suffices to consider the conditions

$$\kappa_i\psi_i \le \delta \quad \text{on} \quad |T_i| \quad \text{and} \tag{8i1}$$

$$\|\kappa_i\psi_i\|_{r'} \le 1 \quad \text{on} \quad |T_i| \tag{8i2}$$

because for such κ_i and ψ_i, $c'\kappa_i^{r'+1}\psi_i^{r'+1}$ satisfies (8i) for some small positive number c'. Second, since $0 \le \psi_i \le 1$, we can simplify (8i1) as

$$\kappa_i \le \delta \quad \text{on} \quad |T_i|. \tag{8i1$'$}$$

Furthermore, if we can construct κ_i so that

$$\|\kappa_i\|_{r'} = 0 \quad \text{on} \quad Z_i, \tag{10}$$

the existence of ψ_i with (8i2) follows from the induction hypothesis of the proof of Assertion. Thus (8i1$'$) and (10) are sufficient.

We will construct κ_i of the form $\zeta \circ P$ where P is the map from $|T_i|$ to $Z_i \times [0, \infty[$ given by

$$P(x) = (\pi_i(x), \rho(x)) \quad \text{for} \quad x \in |T_i|,$$

and ζ is a certain C^r $\mathfrak{X}$-function on $Z_i \times [0, \infty[$ which is positive on $Z_i \times]0, \infty[$. The conditions $(8i1')$ and (10) are induced from

$$\zeta \leq \delta_0 \quad \text{on} \quad Z_i \times]0, \infty[, \quad \text{and} \tag{$8i1''$}$$

$$\|\zeta\|_r = 0 \quad \text{on} \quad Z_i \times 0, \tag{$10'$}$$

where δ_0 is some positive $\mathfrak{X}$-function on $Z_i \times]0, \infty[$. (Here we replace r' with r for simplicity of notation.) We obtain such ζ as follows. Approximate the $\mathfrak{X}$-function $\min\{t^{2r}/2, \delta_0(x,t)/2\}$ on $Z_i \times]0, \infty[$ by a positive C^r $\mathfrak{X}$-function ζ' on $Z_i \times]0, \infty[$ in the C^0 $\mathfrak{X}$-topology so that

$$\zeta'(x,t) < t^{2r} \quad \text{and} \quad \zeta'(x,t) < \delta_0(x,t) \quad \text{for} \quad (x,t) \in Z_i \times]0, \infty[.$$

Let ζ denote the natural extension of ζ' to $Z_i \times 0$. Since ζ is not always of class C^r or does not always satisfy $(10')$ globally on $Z_i \times 0$, we will find a closed $\mathfrak{X}$-subset Z'' of Z_i of dimension $< i$ and an $\mathfrak{X}$-neighborhood V_i of $(Z_i - Z'') \times 0$ in $Z_i \times [0, \infty[$ such that $\zeta|_{V_i}$ is of class C^r and $(10')$ holds on $(Z_i - Z'') \times 0$. If there exist such Z'' and V_i, then we complete the construction of $Z_i, T_i, \psi_i, \kappa_i, V_i$ and V_i'.

We define $D_x \zeta$ in the same way as $D\zeta$ when we fix $t \in]0, \infty[$, and regard $\zeta(x,t)$ as a function on Z_i. Similarly we define $\partial_x \zeta$, $D_t \zeta$ and $\partial_t \zeta$. Existence of such Z'' and V_i follows from the following assertions 1 and 2.

Assertion 1. For some Z'' and for each $k = 0, \ldots, r$, $|D_t^k \zeta'|$ can be extended continuously to $(Z_i - Z'') \times 0$ so that the extension attains 0 at $(Z_i - Z'') \times 0$, and, moreover, for some V_i,

$$|D_t^k \zeta'|(x,t) < t^{2r-2k} \quad \text{for} \quad (x,t) \in V_i - Z_i \times 0.$$

Proof of Assertion 1. By induction on k it suffices to prove this for $k = 1$. Since $|\zeta'(x,t)| < t^{2r}$, $D_t \zeta$ exists and equals to 0 on $Z_i \times 0$. But $D_t \zeta$ is not always continuous at $Z_i \times 0$. Hence we need to remove some Z'' from Z_i. By the definition of ∂_t and D_t, $\partial_t \zeta'$ is a C^{2r-1} $\mathfrak{X}$-map from $Z_i \times]0, \infty[$ to $G_{2,1}$, and $D_t \zeta$ is continuous at $(x,0) \in Z_i \times 0$ if and only if $\partial_t \zeta$ is also. Let Z''' denote the set of the points x of Z_i such that

$$\sharp(x \times 0 \times G_{2,1} \cap \overline{\text{graph } \partial_t \zeta'}) > 1.$$

We will prove that if we set $Z'' = \overline{Z'''}$, the first half of Assertion 1 holds true. Clearly Z''' is an $\mathfrak{X}$-set. First $\dim Z''' < i$ for the following reason.

Let $x \in Z'''$. Then $x \times 0 \times G_{2,1} \cap \overline{\text{graph } \partial_t \zeta'}$ is connected because for a fundamental system $\{B_k\}_{k=1,2,\ldots}$ of neighborhoods of $(x,0)$ in $Z_i \times [0, \infty[$

such that $B_k - Z_i \times 0$ are all connected, $\overline{\mathrm{graph}(\partial_t \zeta' |_{B_k - Z_i \times 0})}$ are connected, and

$$x \times 0 \times G_{2,1} \cap \overline{\mathrm{graph}\,\partial_t \zeta'} = \bigcap_{k=1}^{\infty} \overline{\mathrm{graph}(\partial_t \zeta' |_{B_k - Z_i \times 0})}.$$

Hence $x \times 0 \times G_{2,1} \cap \overline{\mathrm{graph}\,\partial_t \zeta'}$ is of dimension > 0. Therefore, if Z''' were of dimension i, $Z''' \times 0 \times G_{2,1} \cap \overline{\mathrm{graph}\,\partial_t \zeta'}$ would be of dimension $> i$. But by (II.1.9) $\overline{\mathrm{graph}\,\partial_t \zeta'} - \mathrm{graph}\,\partial_t \zeta'$ $(= Z_i \times 0 \times G_{2,1} \cap \overline{\mathrm{graph}\,\partial_t \zeta'})$ is of dimension $\leq i$. Thus $\dim Z''' < i$.

Let $x \in Z_i - \overline{Z'''}$. Next we prove that $|D_t \zeta|$ is continuous at $(x, 0)$. By the definition of Z''' there exists $d \in \mathbf{R} \cup \{\pm\infty\}$ such that for any sequence $\{(a_k, t_k)\}$ in $Z_i \times]0, \infty[$ converging to $(x, 0)$, $\{|D_t \zeta'|(a_k, t_k)\}_k$ converges or diverges to d. But by the Mean Value Theorem and the property $|\zeta'(x, t)| < t^{2r}$ there exists a sequence $\{(a_k, t_k)\}$ convergent to $(x, 0)$ such that $\{|D_t \zeta'|(a_k, t_k)\}_k$ converges to 0. Hence d must be 0, which shows the continuity of $|D_t \zeta|$ at $(x, 0)$.

It remains to prove that $|D_t \zeta'|(x, t) < t^{2r-2}$ on $V_i - Z_i \times 0$ for some V_i. Let S_i denote the set of the points (x, t) of $(Z_i - \overline{Z'''}) \times]0, \infty[$ such that $|D_t \zeta'|(x, t) \geq t^{2r-2}$. Clearly S_i is a closed $\mathfrak{X}$-subset of $(Z_i - \overline{Z'''}) \times]0, \infty[$. We will see that $V_i = (Z_i - \overline{Z'''}) \times]0, \infty[- S_i$ and Z'', defined so that $Z'' \times 0 = \overline{Z'''} \times 0 \cup (\overline{S_i} \cap Z_i \times 0)$, fulfill the requirements. For that, it suffices to show that $\dim \overline{S_i} \cap Z_i \times 0 < i$. Fix a point x_0 of $Z_i - \overline{Z'''}$. Then $S_i \cap x_0 \times]0, t_0] = \varnothing$ for some positive number t_0 for the following reason. If not, then $S_i \supset x_0 \times]0, t_0]$ for some t_0 because $S_i \cap x_0 \times]0, \infty[$ is an $\mathfrak{X}$-set in the line $x_0 \times]0, \infty[$. Hence $\zeta(x_0, t) \geq t^{2r-1}/(2r-1)$ on $]0, t_0[$ because $\zeta'(x_0, t)$ is the integration of $|D_t \zeta'|(x_0, t)$, which is a contradiction. Hence either

$$S_i \cap x_0 \times]0, \infty[= \varnothing,$$

or there exists $t_0 \in]0, \infty[$ such that

$$(x_0, t_0) \in S_i \quad \text{and} \quad S_i \cap x_0 \times]0, \infty[\subset x_0 \times [t_0, \infty[.$$

Let S_i' denote the set of such points (x_0, t_0). Then S_i' is an $\mathfrak{X}$-set of dimension $\leq i$ and

$$\overline{S_i} \cap Z_i \times 0 = \overline{S_i'} \cap Z_i \times 0.$$

By (II.1.9), $\overline{S_i'} \cap Z_i \times 0$ is of dimension $< i$. Thus we prove Assertion 1.

Assertion 2. Assume ζ is continuous at $Z_i \times 0$. For some Z'', V_i, and for each $k = 0, \ldots, r$, $D_x^k \zeta' |_{V_i}$ can be extended continuously to $(Z_i - Z'') \times 0$ so that the extension attains 0 at $(Z_i - Z'') \times 0$.

Proof of Assertion 2. By induction on k it suffices to prove Assertion 2 in the case $k = 1$. By the same arguments as in the proof of Assertion 1, we prove that there exist Z'' and V_i such that the X-map $\partial_x \zeta'|_{V_i} : V_i \to G_{n+1,i}$ can be extended continuously to $(Z_i - Z'') \times 0$. We need only prove that the extension attains $T_x Z_i \times 0 \subset \mathbf{R}^n \times \mathbf{R}$ at $(x, 0) \in (Z_i - Z'') \times 0$. Since the problem is local, we can assume $Z_i = \mathbf{R}^i = \mathbf{R}^n$ and $(x, 0) = (0, 0)$. What we prove is $\frac{\partial \zeta'}{\partial x_j}(x, t) \to 0$, $j = 1, \dots, i$, as $(x, t) \to (0, 0)$ with $t > 0$. We prove also this by reduction to absurdity. Assume for some j, $\frac{\partial \zeta'}{\partial x_j}(x, t) \to d \in \mathbf{R} \cup \{\pm\infty\} - 0$ as $(x, t) \to (0, 0)$. Let $d' \neq 0$ be a number with $0 < d' < d$ or $0 > d' > d$. For sufficiently small $t > 0$ and $x_j > 0$,

$$|\zeta'(0, t) - \zeta'(0, \dots, 0, x_j, 0, \dots, 0, t)| \geq |d'| x_j,$$

which implies

$$|\zeta(0, 0) - \zeta(0, \dots, 0, x_j, 0, \dots, 0, 0)| \geq |d'| x_j.$$

That contradicts $\zeta|_{Z_i \times 0} = 0$ and the continuity of ζ. Thus we complete the proof of Assertion 2 and hence the induction step of the construction of $\{Z_i, T_i, \psi_i, \kappa_i, V_i, V_i'\}$.

For proof of the assertion, it remains only to construct κ_{-1} so that $\kappa_{-1}(1 - \psi_{n''}) \cdots (1 - \psi_0)$ satisfies the conditions (8') and (9). As above, we reduce the problem to a simple case. Since $(1 - \psi_{n''}) \cdots (1 - \psi_0)$ vanishes on an X-neighborhood of Z in $\mathbf{R}^n$, we can replace the χ_c in (9) with a positive X-function on $\mathbf{R}^n$. Hence there exists a positive X-function γ on $\mathbf{R}^n$ such that if κ_{-1} satisfies the condition:

$$\|\kappa_{-1}\|_{r'} \leq \gamma \quad \text{on} \quad \mathbf{R}^n, \tag{11}$$

then (8') and (9) hold for $\kappa_{-1}(1 - \psi_{n''}) \cdots (1 - \psi_0)$.

We will choose κ_{-1} of the form $\zeta_{-1} \circ \eta$, where $\eta(x) = |x|^2$ for $x \in \mathbf{R}^n$, and ζ_{-1} is a positive C^r X-function on $[0, \infty[$. It is easy to find a positive X-function ξ on $[0, \infty[$ such that if ζ_{-1} satisfies the condition:

$$\|\zeta_{-1}\|_{r'} \leq \xi,$$

then $\zeta_{-1} \circ \eta$ satisfies (11). Moreover, we can replace this condition with

$$\zeta_{-1} \leq \xi \quad \text{and} \quad |\zeta_{-1}^{(k)}| \leq 1, \quad k = 1, \dots, r', \tag{12}$$

because if ζ_{-1} satisfies (12) and if $\xi \leq 1$, then for a small positive number c'', $\|c' \zeta_{-1}^{r'+1}\|_{r'} \leq \xi$. Thus we have reduced the problem to the following assertion.

Assertion'. Let ξ be a positive $\mathfrak{X}$-function on $[0, \infty[$. There exists a positive C^r $\mathfrak{X}$-function ζ_{-1} on $[0, \infty[$ with (12).

Proof of Assertion'. We can assume ξ is of class C^{r+1} by the case $r' = 0$. Set $H = (\prod_{k=1}^{r'+1} \xi^{(k)})^{-1}(0)$. There are three cases to consider.

(i) The case where H is bounded. Let H' be an interval $[0, h]$ containing H. Clearly $\xi, \dots, \xi^{(r')}$ are strictly monotone outside of H'. If $\xi^{(k)}$ for some $k = 0, \dots, r'$ is positive and monotone increasing outside H', so are $\xi, \dots, \xi^{(k-1)}$. Hence $\zeta_{-1} =$ (a small positive constant) satisfies (12). If one of $\xi, \dots, \xi^{(r')}$ is negative and monotone decreasing outside of H', so is ξ, which is impossible. If every $\xi^{(k)}$ is either positive and monotone decreasing, or negative and monotone increasing outside of H', then (12) holds for $\zeta_{-1} =$ (a small positive number)$\cdot\xi$.

(ii) The case where $[0, \infty[- H$ is bounded. Let H' be an interval $[h, \infty[$ contained in H. In this case, ξ is a polynomial function on H'. Hence it suffices to set $\zeta_{-1} =$ a small positive constant.

(iii) The case where H and $[0, \infty[- H$ are unbounded. Set $H' = H \cap ([0, \infty[- H)$, which is an infinite discrete $\mathfrak{X}$-set by Axiom (iv). Let φ be a C^r $\mathfrak{X}$-function on $[0, \infty[$ such that $0 \le \varphi \le 1$, $\varphi = 1$ on an $\mathfrak{X}$-neighborhood of H', and $W^+ = \operatorname{supp} \varphi$ has an infinite number of connected components $W_1^+, W_2^+, \dots$. (Existence of φ is shown in the same way as the construction of a partition of unity in the proof of the case $r' = 0$.) Set

$$W^- = \operatorname{supp}(1 - \varphi), \qquad \xi^+ = \xi\varphi \quad \text{and} \quad \xi^- = \xi \cdot (1 - \varphi).$$

Then $[0, \infty[= W^+ \cup W^-$, and W^- also has an infinite number of connected components $W_1^-, W_2^-, \dots$. Define functions $\theta^{\pm}$ on $W^{\pm}$ by

$$\theta^{\pm}(x) = \max_{y \in W_l^{\pm}, \ k=1, \dots, r'} \{1, |\xi^{\pm(k)}(y)|\} \quad \text{if} \quad x \in W_l^{\pm}, \ l = 1, 2, \dots .$$

We see that $\theta^{\pm}$ and $1/\theta^{\pm}$ are locally constant $\mathfrak{X}$-functions, and $\zeta_{-1} = \widetilde{\xi^+/2\theta^+} + \widetilde{\xi^-/2\theta^-}$ fulfills the requirements, where the symbol $\tilde{}$ denotes the natural extension. We complete the proof. $\qquad\square$

In the case of $\mathfrak{X} =$ semialgebraic, we can approximate a $C^{r'}$ Nash map between Nash manifolds by a Nash map in the $C^{r'}$ Nash topology (see [S_3]).

Let $\varphi : L_1 \to L_2$ be a linear map between linear spaces in Euclidean spaces such that $0 \in L_1$, $0 \in L_2$ and $\varphi(0) = 0$. Set

$$|\varphi|^{\max} = \max_{v \in L_1, |v|=1} |\varphi(v)| \quad \text{and}$$

$$|\varphi|^{\min} = \min_{v \in L_1, |v|=1} |\varphi(v)|.$$

Lemma II.5.3. *Let r be a positive integer. Let X and Y be C^r $\mathfrak{X}$-subman-ifolds of $\mathbf{R}^n$, and let $f\colon X \to Y$ be a C^r $\mathfrak{X}$-map such that for a bounded set $B \subset \mathbf{R}^n$, $f(X \cap B)$ is bounded. If f is an immersion(a diffeomorphism or a diffeomorphism onto its image), a strong C^r $\mathfrak{X}$-approximation of f in the C^r $\mathfrak{X}$-topology is an immersion(a diffeomorphism or a diffeomorphism onto its image, respectively).*

If f is a diffeomorphism and, moreover, for a bounded set $B \subset \mathbf{R}^n$, $f^{-1}(B)$ is bounded, then $g^{-1} \to f^{-1}$ as $g \to f$.

Proof. We prove only the case $r = 1$. The general case follows in the same way. Let $f \in B^1(X, Y)$ and let $|df|^{\max}$ and $|df|^{\min}$ denote the functions defined by

$$|df|^{\max}(x) = |df_x|^{\max} \quad \text{and} \quad |df|^{\min}(x) = |df_x|^{\min} \quad \text{for} \quad x \in X.$$

In the case $r = 1$, it is convenient to consider these functions in place of the map $Df\colon X \to \mathbf{R}^{n^2}$. It is easy to prove that these functions are of class $\mathfrak{X}$,

$$|Df|/n \le |df|^{\max} \le |Df|,$$

and hence the topology on $B^1(X, Y)$ defined by the norm $|f| + |df|^{\max}$ is the C^1 $\mathfrak{X}$-topology. Clearly f is an immersion if and only if $|df|^{\min}$ is positive.

Assume f is an immersion. Let g be a C^1 $\mathfrak{X}$-approximation of f in the C^1 $\mathfrak{X}$-topology so strong that

$$|df_x - dg_x|^{\max} < |df_x|^{\min}/2 \quad \text{for} \quad x \in X.$$

Then

$$|dg_x|^{\min} \ge |df_x|^{\min} - |df_x - dg_x|^{\max} > |df_x|^{min}/2 > 0.$$

Hence g is an immersion.

Assume f is a diffeomorphism onto the image. We will find positive $\mathfrak{X}$-functions ε_1 and ε_2 on X so that if a C^1 $\mathfrak{X}$-map g satisfies the conditions $|f - g| < \varepsilon_1$ and $|df - dg|^{\max} < \varepsilon_2$, then g is a diffeomorphism onto the image.

Set $\varepsilon_2 = |df|^{\min}/10$. There exists an $\mathfrak{X}$-neighborhood U of the diagonal in $X \times X$ such that if $|df - dg|^{\max} < \varepsilon_2$, then for each $x \in X$, $g|_{U_x}$ is a diffeomorphism onto its image for the following reason, where U_x is the $\mathfrak{X}$-neighborhood of x in X given by $x \times U_x = x \times X \cap U$.

Let $p\colon TX \to X$ $(TX \subset X \times \mathbf{R}^n)$ denote the tangent bundle of X. For each $x \in X$, let $q_x\colon U_x \to x + T_xX$ denote the restriction to U_x of the orthogonal projection. Let us choose the U so that for each $x \in X$, q_x is a C^1 diffeomorphism onto a solid sphere B_x with center x in $x + T_xX$,

$$||df|^{\min}(x') - |df|^{\min}(x)| < |df|^{\min}(x)/2 \quad \text{for} \quad x' \in U_x,$$

$$|d(q_x^{-1})|^{\max}(y) < 2 \qquad \qquad \text{for} \quad y \in B_x,$$

$$\text{and} \quad |d(f \circ q_x^{-1})_y - df_x|^{\max} < |df|^{\min}(x)/5 \quad \text{for} \quad y \in B_x,$$

which is possible because

$$d(q_x)_x = \mathrm{id}, \qquad d(f \circ q_x^{-1})_x = df_x,$$

and because $d(f \circ q_x^{-1})_y$ is a linear map from $T_x X$ to $\mathbf{R}^n$. It follows that for $x \in X$ and $y \in B_x$,

$$|d(g \circ q_x^{-1})_y - df_x|^{\max}$$
$$\leq |d(g \circ q_x^{-1})_y - d(f \circ q_x^{-1})_y|^{\max} + |d(f \circ q_x^{-1})_y - df_x|^{\max}$$
$$< |dg_{q_x^{-1}(y)} - df_{q_x^{-1}(y)}|^{\max} \cdot |d(q_x^{-1})_y|^{\max} + |df_x|^{\min}/5$$
$$< |df_{q_x^{-1}(y)}|^{\min}/5 + |df_x|^{\min}/5$$
$$< |df_x|^{\min}/2. \tag{$*$}$$

It suffices to prove that for each x, $g|_{U_x}$ is injective because $g|_{U_x}$ is an immersion and U_x is compact. We prove this by reduction to absurdity. Assume $g(x') = g(x'')$ for some $x' \neq x''$ in U_x. Set

$$\varphi(t) = t q_x(x'') + (1-t) q_x(x') \quad \text{for} \quad 0 \leq t \leq 1 \quad \text{and}$$
$$\Phi = g \circ q_x^{-1} \circ \varphi.$$

Let μ be the linear map from $\mathbf{R}$ to $\mathbf{R}^n$ given by

$$\mu(1) = df_x(q_x(x'') - q_x(x')).$$

Then Φ is a C^1 $\mathfrak{X}$-curve in $\mathbf{R}^n$, $\Phi(0) = \Phi(1)$, and by $(*)$, we have

$$|d\Phi_t - \mu|^{\max} = |d\Phi_t(1) - \mu(1)|$$
$$= |d(g \circ q_x^{-1})_{\varphi(t)}(d\varphi_t(1)) - df_x(q_x(x'') - q_x(x'))|$$
$$= |(d(g \circ q_x^{-1})_{\varphi(t)} - df_x)(q_x(x'') - q_x(x'))|$$
$$\leq |d(g \circ q_x^{-1})_{\varphi(t)} - df_x|^{\max} \cdot |q_x(x'') - q_x(x')|$$
$$< |df_x|^{\min} \cdot |q_x(x'') - q_x(x')|/2 \leq |\mu|/2 \quad \text{for} \quad t \in [0,1].$$

Hence for each $t \in [0,1]$ the angle constituted by the vectors $\mu(1)$ and $d\Phi_t(1)$ is smaller that $\pi/6$. Therefore, the composite of Φ and the orthogonal projection of $\mathbf{R}^n$ onto the line $\mu(\mathbf{R})$ is a diffeomorphism onto the image, which contradicts $\Phi(0) = \Phi(1)$. Thus $g|_{U_x}$ is a diffeomorphism onto the image.

Next we need to define ε_1. For each $x \in X$, let V_x denote the neighborhood of x in X defined by $V_x \times x = X \times x \cap U$. Set

$$\delta_1(x) = \inf_{x' \in X - U_x} |f(x) - f(x')|, \qquad \delta_2(x) = \inf_{x' \in X - V_x} |f(x) - f(x')|, \quad \text{and}$$
$$\varepsilon_1(x) = \min\{\delta_1(x), \delta_2(x)\}/4 \quad \text{for} \quad x \in X.$$

Then δ_1, δ_2 and ε_1 are positive $\mathfrak{X}$-functions on X such that

$$|f(x) - f(x')| \geq \delta_1(x) \quad \text{and} \quad |f(x) - f(x')| \geq \delta_2(x) \quad \text{for} \quad (x, x') \in X \times X - U.$$

Hence if $|f - g| < \varepsilon_1$, then for $(x, x') \in X \times X - U$,

$$\begin{aligned}
|g(x) - g(x')| &\geq |f(x) - f(x')| - |f(x) - g(x)| - |f(x') - g(x')| \\
&> (\delta_1(x) + \delta_2(x'))/2 - \varepsilon_1(x) - \varepsilon_1(x') \\
&> \varepsilon_1(x) + \varepsilon_1(x') > 0.
\end{aligned}$$

Therefore, for each $x \in X$, $g(x) \notin g(X - U_x)$ and, moreover, there is no sequence of points in $g(X - U_x)$ converging to $g(x)$. Consequently, if $|f - g| < \varepsilon_1$ and $|df - dg|^{\max} < \varepsilon_2$, then g is a diffeomorphism onto the image.

Let f be a diffeomorphism, and let g be an approximation as above. Then g can be proper for the following reason. Let C be a compact set in Y. Since f is a diffeomorphism, $f^{-1}(C)$ is compact. Let C' denote the union of U_x, $x \in f^{-1}(C)$. Then C' includes $g^{-1}(C)$ because if there were $x' \in g^{-1}(C) - C'$, then for $x = f^{-1} \circ g(x')$,

$$\begin{aligned}
0 = |g(x') - g(x')| &= |f(x) - g(x')| \\
&\geq |f(x) - f(x')| - |f(x') - g(x')| \geq \delta_2(x') - \delta_2(x')/4 > 0
\end{aligned}$$

because $(x, x') \in X \times X - U$. In the above definition of U, choose B_x, $x \in X$, so that the function $X \ni x \rightarrow \text{radius} B_x \in \mathbf{R}$ is continuous. Then C' is compact because each B_x and hence U_x are compact. Hence $g^{-1}(C)$ is compact, i.e., g is proper.

It follows from properness that $g(X)$ is closed in Y. On the other hand, it is open in Y because g is a diffeomorphism onto $g(X)$. Hence $g(X)$ is the union of some connected components of Y. However, by using a C^1 $\mathfrak{X}$-tube at Y, we can construct a homotopy from f to g. Therefore, $g(X) = Y$,

It remains to prove the last statement in the lemma. We can replace f and g with id and $f^{-1} \circ g$, respectively, for the following reason. Let $g \rightarrow f$. Then $f^{-1} \circ g \rightarrow f^{-1} \circ f = \text{id}$. Indeed, since $f^{-1} \in B^1(Y, X)$, the induced map $(f^{-1})_* : C^1_{\mathfrak{X}}(X, Y) \rightarrow C^1_{\mathfrak{X}}(X, X)$ is continuous. Assume we have already proved $(f^{-1} \circ g)^{-1} = g^{-1} \circ f \rightarrow \text{id}$. Then $g^{-1} = g^{-1} \circ f \circ f^{-1} \rightarrow f^{-1}$ because the induced map $(f^{-1})^* : C^1_{\mathfrak{X}}(Y, X) \rightarrow C^1_{\mathfrak{X}}(X, X)$ is continuous, which follows from the facts that f^{-1} is proper and $f^{-1} \in B^1(Y, X)$. Thus we assume $f = \text{id}$.

For a small positive $\mathfrak{X}$-function ε on X, it suffices to define positive $\mathfrak{X}$-functions ε_1 and ε_2 so that if $|g - \text{id}| < \varepsilon_1$ and $|dg - \text{id}|^{\max} < \varepsilon_2$, then $|g^{-1} - \text{id}| < \varepsilon$ and $|d(g^{-1}) - \text{id}|^{\max} < \varepsilon$. Set

$$\varepsilon_1(x) = \inf_{|x' - x| \leq \varepsilon(x)} \varepsilon(x') \quad \text{for} \quad x \in X.$$

Let ε be sufficiently small. Then ε_1 is a positive $\mathfrak{X}$-function on X. If $|g-\mathrm{id}| < \varepsilon_1$,

$$|x - g^{-1}(x)| < \varepsilon_1(g^{-1}(x)) \leq \varepsilon(x) \quad \text{for} \quad x \in X,$$

because $\varepsilon_1(x') \leq \varepsilon(g(x'))$ for $x' \in X$. Similarly set

$$\varepsilon_2(x) = \inf_{|x'-x| \leq \varepsilon(x)} \{\varepsilon(x'), 1\}/2 \quad \text{for} \quad x \in X.$$

Assume $|g - \mathrm{id}| < \varepsilon_1$ and $|dg - \mathrm{id}|^{\max} < \varepsilon_2$. Let $x \in X$. Then

$$
\begin{aligned}
|\mathrm{id} - d(g^{-1})|^{\max}(x) &= |d(g \circ g^{-1}) - d(g^{-1})|^{\max}(x) \\
&= |dg_{g^{-1}(x)} \circ d(g^{-1})_x - d(g^{-1})_x|^{\max} \\
&\leq |dg - \mathrm{id}|^{\max}(g^{-1}(x)) \cdot |d(g^{-1})|^{\max}(x) \\
&< \varepsilon(x)|d(g^{-1})|^{\max}(x)/2.
\end{aligned}
$$

On the other hand,

$$
\begin{aligned}
1 = |\mathrm{id}|^{\max}(x) &= |d(g \circ g^{-1})|^{\max}(x) \\
&\geq |dg|^{\min}(g^{-1}(x)) \cdot |d(g^{-1})|^{\max}(x), \quad \text{and} \\
1 = |\mathrm{id}|^{\min}(g^{-1}(x)) &\leq |\mathrm{id} - dg|^{\max}(g^{-1}(x)) + |dg|^{\min}(g^{-1}(x)) \\
&\leq 1/2 + |dg|^{\min}(g^{-1}(x)),
\end{aligned}
$$

which imply

$$|d(g^{-1})|^{\max}(x) \leq 1/|dg|^{\min}(g^{-1}(x)) \leq 2.$$

Hence

$$|\mathrm{id} - d(g^{-1})|^{\max} < \varepsilon. \qquad \square$$

We shall need the $\mathfrak{X}$-version of Thom's transversality theorem [Th$_1$]. Before stating it, we give some definitions. Let X and Y be C^r $\mathfrak{X}$-submanifolds of $\mathbf{R}^n$ and $\mathbf{R}^m$, respectively, with $\dim X = n'$. Set

$$
\begin{aligned}
C_{\mathfrak{X}}^r(X, x; Y, y) &= \{f \in C_{\mathfrak{X}}^r(X, Y) \colon f(x) = y\} \quad \text{for} \quad (x, y) \in X \times Y, \\
J_{x,y}^{r'}(X, Y) &= C_{\mathfrak{X}}^r(X, x; Y, y)/ \sim \ (= \text{jet space}), \quad \text{and} \\
J^{r'}(X, Y) &= \bigcup_{(x,y) \in X \times Y} J_{x,y}^{r'}(X, Y),
\end{aligned}
$$

where for f and f' in $C_{\mathfrak{X}}^r(X, x; Y, y)$, the equivalence $f \sim f'$ means $Df(x) = Df'(x), \ldots, D^{r'}f(x) = D^{r'}f'(x)$. For $f \in C_{\mathfrak{X}}^r(X, x; Y, y)$ the class represented by f in $J_{x,y}^{r'}(X, Y)$ is called the r'-jet of f at x and is written

$J^{r'}f(x)$. Hence $J^{r'}f$ is a map from X to $J^{r'}(X,Y)$. We imbed $J^{r'}_{x,y}(X,Y)$ for each $(x,y) \in X \times Y$ in a Euclidean space so that $J^{r'}(X,Y)$ is a $C^{r-r'}$ X-submanifold of the Euclidean space and $J^{r'}f$ is a $C^{r-r'}$ X-map as follows.

If $X = \mathbf{R}^n$ and $Y = \mathbf{R}^m = \mathbf{R}$, then for a point $(x_0,y_0) \in \mathbf{R}^n \times \mathbf{R}$, a natural representation of $J^{r'}_{x_0,y_0}(\mathbf{R}^n,\mathbf{R})$ is $x_0 \times y_0 \times P_{x_0,y_0}$, where

$$P_{x_0,y_0} = \{\sum_{\alpha \in A} a_\alpha(x-x_0)^\alpha + y_0 : a_\alpha \in \mathbf{R}\}, \quad \text{and}$$

$$A = \{\alpha = (\alpha_1,\dots,\alpha_n) \in \mathbf{N}^n : 1 \le |\alpha| \le r'\}.$$

We identify $J^{r'}(\mathbf{R}^n,\mathbf{R})$ with $\mathbf{R}^n \times \mathbf{R}^{m'}$, where $m' = \sharp A + 1$, by the correspondence

$$J^{r'}(\mathbf{R}^n,\mathbf{R}) \ni (x_0,y_0,\sum_{\alpha \in A} a_\alpha(x-x_0)^\alpha + y_0) \longrightarrow (x_0,y_0,a_\alpha)_{\alpha \in A} \in \mathbf{R}^n \times \mathbf{R}^{m'}.$$

For $f \in C^r_X(\mathbf{R}^n,\mathbf{R})$ we have

$$J^r f(x_0) = (x_0, D^\alpha f(x_0)/\alpha!)_{\alpha \in \mathbf{N}^n, |\alpha| \le r'} \quad \text{for} \quad x_0 \in \mathbf{R}^n,$$

where

$$D^\alpha = \partial^{|\alpha|}/\partial x_1^{\alpha_1} \cdots \partial x_n^{\alpha_n} \quad \text{for} \quad \alpha = (\alpha_1,\dots,\alpha_n) \in \mathbf{N}^n.$$

Note that P_{x_0,y_0} becomes a commutative $\mathbf{R}$-ring without unit if we define ring operations by

$$c\left(\sum_{\alpha \in A} a_\alpha(x-x_0)^\alpha + y_0\right) = \sum_{\alpha \in A} c a_\alpha(x-x_0)^\alpha + y_0,$$

$$\sum_{\alpha \in A} a_\alpha(x-x_0)^\alpha + y_0 + \sum_{\alpha \in A} b_\alpha(x-x_0)^\alpha + y_0 = \sum_{\alpha \in A}(a_\alpha+b_\alpha)(x-x_0)^\alpha + y_0,$$

$$\left(\sum_{\alpha \in A} a_\alpha(x-x_0)^\alpha + y_0\right)\left(\sum_{\beta \in A} b_\beta(x-x_0)^\beta + y_0\right) = \sum_{\alpha+\beta \in A} a_\alpha b_\beta(x-x_0)^{\alpha+\beta} + y_0,$$

$$\text{for} \quad c, a_\alpha, b_\alpha, b_\beta \in \mathbf{R}.$$

Let $X \subset \mathbf{R}^n$ be general and $Y = \mathbf{R}$. We give naturally a commutative $\mathbf{R}$-ring structure to $J^{r'}_{x_0,y_0}(X,\mathbf{R})$. The canonical map from P_{x_0,y_0} to $J^{r'}_{x_0,y_0}(X,\mathbf{R})$ is a homomorphism, where P_{x_0,y_0} is defined by $\mathbf{R}^n, \mathbf{R}, x_0$ and y_0 as above. Let $\mathfrak{m}_{x_0,y_0}$ denote the kernel. Then

$$\mathfrak{m}_{x_0,y_0} = \{f \in P_{x_0,y_0} : D(f|_X) = \cdots = D^{r'}(f|_X) = 0 \text{ at } x_0\}.$$

Hence $\cup_{(x,y)\in X\times \mathbf{R}} x \times y \times \mathfrak{m}_{x,y}$ is an $\mathfrak{X}$-subset of $\mathbf{R}^n \times \mathbf{R}^{m'}$. Moreover, by using a local $\mathfrak{X}$-coordinate system of X, we easily prove that it is a $C^{r-r'}$ $\mathfrak{X}$-submanifold. Since $\mathfrak{m}_{x_0,y_0}$ is a linear subspace of P_{x_0,y_0}, the linear subspace $\mathfrak{m}_{x_0,y_0}^{\perp}$ of P_{x_0,y_0} consisting of the vectors normal to $\mathfrak{m}_{x_0,y_0}$ is mapped bijectively onto $J_{x_0,y_0}^{r'}(X,\mathbf{R})$ by the canonical map. Consider $\cup_{(x,y)\in X\times \mathbf{R}} x \times y \times \mathfrak{m}_{x,y}^{\perp}$ and identify it with $J^{r'}(X,\mathbf{R})$. Then $J^{r'}(X,\mathbf{R})$ becomes a $C^{r-r'}$ $\mathfrak{X}$-submanifold of $\mathbf{R}^n \times \mathbf{R}^{m'}$ such that for each $f \in C_{\mathfrak{X}}^r(X,\mathbf{R})$, $J^{r'}f$ is a $C^{r-r'}$ $\mathfrak{X}$-map.

If X is general and if $Y = \mathbf{R}^m$, we regard $J^{r'}(X,\mathbf{R}^m)$ as a $C^{r-r'}$ $\mathfrak{X}$-submanifold of $\mathbf{R}^n \times \mathbf{R}^{mm'}$ by

$$J^{r'}(X,\mathbf{R}^m) = \bigcup_{(x,y_1,\dots,y_m)\in X\times \mathbf{R}^m} J_{x,y_1}^{r'}(X,\mathbf{R}) \times \cdots \times J_{x,y_m}^{r'}(X,\mathbf{R}).$$

Clearly for each $f \in C_{\mathfrak{X}}^r(X,\mathbf{R}^m)$, $J^{r'}f$ is a $C^{r-r'}$ $\mathfrak{X}$-map.

Let X and Y be general. Let $T = (|T|,\pi,\rho)$ be a C^r $\mathfrak{X}$-tube at Y. Regard π as a C^r map from $|T|$ to $\mathbf{R}^m$. Then

$$J^{r'}(X,Y) = \{f \in J_{x,y}^{r'}(X,\mathbf{R}^m): x \in X,\ y \in Y,\ (J^{r'}\pi(y)) \circ f = f\}.$$

Hence $J^{r'}(X,Y)$ is an $\mathfrak{X}$-subset of $J^{r'}(X,\mathbf{R}^m)$. We see that it is also a $C^{r-r'}$ $\mathfrak{X}$-submanifold, and hence for $f \in C_{\mathfrak{X}}^r(X,Y)$, $J^{r'}f$ is a $C^{r-r'}$ $\mathfrak{X}$-map from X to $J^{r'}(X,Y)$. Thus we give a $C^{r-r'}$ $\mathfrak{X}$-manifold structure to $J^{r'}(X,Y)$. If $r = r'$, $J^{r'}(X,Y)$ is *locally flat* in $\mathbf{R}^n \times \mathbf{R}^{mm'}$, i.e., the germ of $(\mathbf{R}^n \times \mathbf{R}^{mm'}, J^{r'}(X,Y))$ at each point of $J^{r'}(X,Y)$ is $\mathfrak{X}$-homeomorphic to the germ of $(\mathbf{R}^{n+mm'}, \mathbf{R}^{m''} \times 0)$ at 0, $m'' = \dim J^{r'}(X,Y)$. This is immediate by the above arguments.

Theorem II.5.4 (Transversality theorem). *Let r,r' be positive and non-negative integers, respectively. Let X and Y be C^r $\mathfrak{X}$-submanifolds of $\mathbf{R}^n$ and $\mathbf{R}^m$ respectively, with $\dim X = n'$. Let $\{X_i\}$ be a finite $C^{r-r'}$ Whitney $\mathfrak{X}$-stratification in $J^{r'}(X,Y)$ with $\operatorname{codim}\cup X_i = n''$. Assume $\cup X_i$ is closed in $J^{r'}(X,Y)$, and $r - r' > \max(n' - n'', 0)$. The subset of $B^r(X,Y)$ consisting of f such that $J^{r'}f$ are transversal to each X_i is open and dense in $B^r(X,Y)$ in the C^r $\mathfrak{X}$-topology.*

Proof. Assume $\dim X_1 < \dim X_2 < \cdots$. We easily show openness as follows. Let $f \in B^r(X,Y)$ be such that $J^{r'}f$ is transversal to each X_i. In the same way as in the proof of II.5.3, we find an open neighborhood U_1 of f in $B^r(X,Y)$ so that for $f' \in U_1$, $J^{r'}f'$ is transversal to X_1. Shrink U_1. By the Whitney condition there exists an open $\mathfrak{X}$-neighborhood O_1 of X_1 in $J^{r'}(X,Y)$ such that for $f' \in U_1$, $J^{r'}f'$ is transversal to each $O_1 \cap X_i$. Next find smaller $U_2 \subset U_1$ so that for $f' \in U_2$, $J^{r'} f'$ is transversal to $X_2 - O_1$, and so on. Thus we prove openness.

Density is not trivial. We can proceed with its proof in the same way as above. Hence we assume $\{X_i\} = \{X_1\}$. Moreover, we can suppose $Y = \mathbf{R}^m$ for the following reason. Let $T = (|T|, \pi, \rho)$ be a C^r $\mathfrak{X}$-tube at Y. Let π_* denote the map from $J^{r'}(X, |T|)$ to $J^{r'}(X, Y)$ induced by π. It is a surjective $C^{r-r'}$ $\mathfrak{X}$-submersion. Let $f \in B^r(X, Y)$. Regard f as a map to $|T|$ and assume there exists a C^r $\mathfrak{X}$-approximation f' of f such that $J^{r'} f'$ is transversal to $\pi_*^{-1}(X_1)$. Then $J^{r'}(\pi \circ f)$ is transversal to X_1, and, by continuity of the map:

$$B^r(X, |T|) \ni f' \longrightarrow \pi \circ f' \in B^r(X, Y),$$

$\pi \circ f'$ is a C^r $\mathfrak{X}$-approximation of f.

Next we reduce the problem to the case where X is open in $\mathbf{R}^n$ as follows. Let $T' = (|T'|, \pi', \rho')$ be a C^r $\mathfrak{X}$-tube at X. Let X'_1 denote the subset of $J^{r'}(|T'|, \mathbf{R}^m)$ such that for each $(x_0, y_0) \in X \times \mathbf{R}^m$, $X'_1 \cap J^{r'}_{x_0, y_0}(|T'|, \mathbf{R}^m)$ is the inverse image of $X_1 \cap J^{r'}_{x_0, y_0}(X, \mathbf{R}^m)$ under the canonical map $J^{r'}_{x_0, y_0}(|T'|, \mathbf{R}^m) \to J^{r'}_{x_0, y_0}(X, \mathbf{R}^m)$, and for each $(x_0, y_0) \in |T'| \times \mathbf{R}^m$, $X'_1 \cap J^{r'}_{x_0, y_0}(|T'|, \mathbf{R}^m)$ is the image of $X'_1 \cap J^{r'}_{\pi'(x_0), y_0}(|T'|, \mathbf{R}^m)$ under the map $J^{r'}_{\pi'(x_0), y_0}(|T'|, \mathbf{R}^m) \to J^{r'}_{x_0, y_0}(|T'|, \mathbf{R}^m)$ induced by the parallel transformation $\mathbf{R}^n \ni x \to x - x_0 + \pi'(x_0) \in \mathbf{R}^m$. Then X'_1 is a $C^{r-r'}$ $\mathfrak{X}$-submanifold of $J^{r'}(|T'|, \mathbf{R}^m)$ of codimension n''. Let $f \in B^r(X, \mathbf{R}^m)$. Clearly $f \circ \pi' \in B^r(|T'|, \mathbf{R}^m)$. Assume there exists a C^r $\mathfrak{X}$-approximation g of $f \circ \pi'$ such that $J^{r'} g$ is transversal to X'_1. Then $g|_X$ is a C^r $\mathfrak{X}$-approximation of f, and $J^{r'}(g|_X)$ is transversal to X_1. Thus we assume X is open in $\mathbf{R}^n$.

Let $f = (f_1, \ldots, f_m) \in B^r(X, \mathbf{R}^m)$, and let ε be a small positive C^r $\mathfrak{X}$-function on X. We give an order to the set $\{\alpha^1, \ldots, \alpha^{m'}\} = \{\alpha \in \mathbf{N}^n : |\alpha| \leq r'\}$ so that $|\alpha^1| \leq |\alpha^2| \leq \cdots$. Define C^r $\mathfrak{X}$-maps $\theta: X \times I^n \to \mathbf{R}^n$, $\lambda_j: X \to \mathbf{R}$, $j = 1, \ldots, m'$, $\tilde{f}: X \times I^{mm'} \to \mathbf{R}^m$ and $\hat{f}: X \times I^n \times I^{mm'} \to \mathbf{R}^m$, where $I = [-1.1]$, by

$$\theta(x, a) = (x_1 + a_1 \varepsilon(x), \ldots, x_n + a_n \varepsilon(x))$$
$$\text{for} \quad (x, a) = (x_1, \ldots, x_n, a_1, \ldots, a_n) \in X \times I^n,$$

$$\lambda_j(x) = x^{\alpha^j} \varepsilon(x) \quad \text{for} \quad x \in X,$$

$$\tilde{f}(x, \beta) = \left(f_1(x) + \sum_{j=1}^{m'} \beta_{1j} \lambda_j(x), \ldots, f_m(x) + \sum_{j=1}^{m'} \beta_{mj} \lambda_j(x) \right)$$
$$\text{for} \quad x \in X \quad \text{and} \quad \beta = (\beta_{ij}) \in I^{mm'}, \quad \text{and}$$

$$\hat{f}(x, a, \beta) = \tilde{f}(\theta^{-1}(x, a), \beta) \quad \text{for} \quad (x, a, \beta) \in X \times I^n \times I^{mm'}.$$

Here θ^{-1} denotes the inverse map of the x-map $\theta(x, a)$ for fixed a. Choose ε so small in the C^1 $\mathfrak{X}$-topology that for any $a \in I^n$ and $\beta \in I^{mm'}$, $\theta(\cdot, a)$ is a diffeomorphism of X and $\hat{f}(\cdot, a, \beta)$ is a strong approximation of f, which is possible by II.5.3 We shall find $a \in I^n$ and $\beta \in I^{mm'}$ so that $J^{r'}\hat{f}(\cdot, a, \beta)$ is transversal to X_1. For that, it suffices to prove the following statement, because a countable intersection of open and dense subsets in $I^n \times I^{mm'}$ is dense.

(1) Each point of X has a compact neighborhood N such that the set

$$\{(a, \beta) \in I^n \times I^{mm'} : J^{r'}\hat{f}(\cdot, a, \beta) \text{ is transversal to } X_1 \text{ at } N\}$$

is open and dense in $I^n \times I^{mm'}$. Here the term "at N" means "at each point of N".

Since openness is clear, we need only require density. We reduce it to an easier statement. Regard $\hat{f}$ as an x-map and set

$$F(x, a, \beta) = (J^{r'}\hat{f})(\theta(x, a), a, \beta) \quad \text{for} \quad (x, a, \beta) \in X \times I^n \times I^{mm'}.$$

Then F is a $C^{r-r'}$ map from $X \times I^n \times I^{mm'}$ to $J^{r'}(X, \mathbf{R}^m)$ such that

$$F(x, 0, 0) = J^{r'}\hat{f}(x, 0, 0) = J^{r'}f(x).$$

Consider the following statement.

(2) For each point $(x_0, a_0, \beta_0) \in X \times I^n \times I^{mm'}$, x_0 has a compact neighborhood N in X such that (a_0, β_0) is adherent to the set:

$$Z(N) = \{(a, \beta) \in I^n \times I^{mm'} : F(\cdot, a, \beta) \text{ is transversal to } X_1 \text{ at } N\}.$$

The reason why we consider $F(\cdot, a, \beta)$ in place of $J^{r'}\hat{f}(\cdot, a, \beta)$ is that calculations of F are easier than those of $J^{r'}\hat{f}$. It suffices to prove (2) for the following reason. Assume (2). Fix x_0. For each $(a, \beta) \in I^n \times I^{mm'}$ there exists a closed neighborhood $N_{a,\beta}$ of x_0 in X such that (a, β) is adherent to $Z(N_{a,\beta})$. As already noted, $Z(N_{a,\beta})$ is open in $I^n \times I^{mm'}$. Hence $\cup_{(a,\beta) \in I^n \times I^{mm'}} Z(N_{a,\beta})$ is open and dense in $I^n \times I^{mm'}$. For $N_{a,\beta}$, choose the set $\{x \in X : |x - x_0| \le 1/k\}$ for some positive integer k. Then we easily show by reduction to absurdity that for some N, $Z(N)$ is dense in $I^n \times I^{mm'}$. Clearly $Z(N)$ is open there. It follows from these that for some $(a, \beta) \in I^n \times I^{mm'}$, $F(\cdot, a, \beta)$ is transversal to X_1. Consequently, $F(\theta^{-1}(\cdot, a), a, \beta)$ is transversal to X_1. Now by the definition of F this new map equals $J^{r'} \hat{f}$. Hence $J^{r'} \hat{f}(\cdot, a, \beta)$ is transversal to X_1. Thus it suffices to prove (2).

To prove (2) we use only the property of F that for each $(x_0, a_0, \beta_0) \in X \times I^n \times I^{mm'}$, $F|_{x_0 \times I^n \times I^{mm'}}$ is locally a diffeomorphism at (x_0, a_0, β_0). We prove this property as follows. Let $(z, y) = (z_1, \dots, z_n, y_{ij})_{\substack{i=1,\dots,m \\ j=1,\dots,m'}}$ denote the coordinate system of $J^{r'}(X, \mathbf{R}^m)$ such that for a point $x = (x_1, \dots, x_n)$ of X, the coordinate of the point $J^{r'} f(x)$ is given by

$$z_i = x_i \quad \text{and} \quad y_{i,j} = D^{\alpha^j} f_i(x)/\alpha^j!.$$

Write

$$F = (F_1, \dots, F_n, F_{ij})_{\substack{i=1,\dots,m \\ j=1,\dots,m'}}, \qquad \theta = (\theta_1, \dots, \theta_n) \quad \text{and} \quad \hat{f} = (\hat{f}_1, \dots, \hat{f}_m).$$

Then

$$\frac{\partial F_k}{\partial a_{k'}}(x, a, \beta) = \frac{\partial \theta_k}{\partial a_{k'}}(x, a) = \begin{cases} \varepsilon(x) & \text{if} \quad k = k' \\ 0 & \text{if} \quad k \neq k', \end{cases}$$

$$\frac{\partial F_k}{\partial \beta_{i,j}}(x, a, \beta) = \frac{\partial \theta_k}{\partial \beta_{i,j}}(x, a) = 0, \quad \text{and}$$

$$\begin{aligned} \frac{\partial F_{i',j'}}{\partial \beta_{i,j}}(x, a, \beta) &= \left. \frac{\partial D^{\alpha^{j'}} \hat{f}_{i'}(x, a, \beta)}{\partial \beta_{i,j}} \right|_{(x,a,\beta)=(\theta(x,a),a,\beta)} /\alpha^{j'}! \\ &= \left. D^{\alpha^{j'}} \left(\frac{\partial \hat{f}_{i'}(x, a, \beta)}{\partial \beta_{i,j}} \right) \right|_{(x,a,\beta)=(\theta(x,a),a,\beta)} /\alpha^{j'}! \\ &= \begin{cases} D^{\alpha^{j'}} (\lambda_j \circ \theta^{-1}(x, a))|_{(x,a)=(\theta(a,a),a)} /\alpha^{j'}! & \text{if} \quad i' = i \\ 0 & \text{if} \quad i' \neq i. \end{cases} \end{aligned}$$

By these first and second equalities it suffices to prove that the $mm' \times mm'$-matrix $\left(\frac{\partial F_{i',j'}}{\partial \beta_{i,j}} \right)$ is of rank mm' at each point of $X \times I^n \times I^{mm'}$. Moreover, by the last equality we need only show that the $m' \times m'$-matrix $\left(D^{\alpha^{j'}} (\lambda_j \circ \theta^{-1}) \right)$ is of rank m' at each point of $X \times I^n$. To prove this, fix $a \in I^n$, set $\xi = (\xi_1, \dots, \xi_n) = \theta^{-1}(x, a)$, and regard $(\xi_1, \dots, \xi_n)$ as a new coordinate system of X. Let D_ξ^α denote $\partial^{|\alpha|}/\partial \xi_1^{\alpha_1} \cdots \partial \xi_n^{\alpha_n}$. Each D^{α^j} at each point of X is described as a linear combination of $D_\xi^{\alpha^{j'}}$, $j' = 1, \dots, m'$, with coefficients in $\mathbf{R}$. Hence it suffices to prove that the matrix $\left(D_\xi^{\alpha^{j'}}(\lambda_j \circ \xi)\right)$ is of rank m' at each point of $X \times I^n$. But this is equivalent to the fact that the matrix $\left(D^{\alpha^{j'}}\lambda_j\right)$ is of rank m' at each point of X.

Assume $x = 0$. Then

$$D^{\alpha^{j'}}\lambda_j(0)/\alpha^{j'}! = \begin{cases} \varepsilon(0) & \text{if } j' = j \\ 0 & \text{if } j' < j \end{cases}$$

because $|\alpha^{j'}| \le |\alpha^j|$ if $j' \le j$. Hence $\left(D^{\alpha^{j'}}\lambda_j(0)\right)$ is of rank m'. For general $x = x_0$ we can replace λ_j with $(x - x_0)^{\alpha^j}\varepsilon(x)$ because the linear space spanned by λ_j, $j = 1, \ldots, m'$, equals the one spanned by $(x - x_0)^{\alpha^j}\varepsilon(x)$, $j = 1, \ldots, m'$. Hence we can reduce this general case to the case $x = 0$. Thus F has the required property.

Since F has many properties that are not needed for proof of (2), we restate (2) in the following more general and simpler form. Here we change notation.

(3) Let $g : \mathbf{R}^n \times \mathbf{R}^m \to \mathbf{R}^m$ be a C^r map such that $g(0) = 0$ and $g|_{0 \times \mathbf{R}^m}$ is a local diffeomorphism at 0×0. Let $X_1 \subset \mathbf{R}^m$ be a C^r submanifold of codimension n' such that $0 \in X_1$ and $r > n - n'$. There exist an open neighborhood N of 0 in $\mathbf{R}^n$ and a subset Z of $\mathbf{R}^m$ such that $0 \in \overline{Z}$, and for each $a \in Z$, $g|_{N \times a}$ is transversal to X_1.

Reduction of (3) to the case $X_1 = l\{0\}$. Since the problem is local at 0×0 in $\mathbf{R}^n \times \mathbf{R}^m$ and 0 in $\mathbf{R}^m$, we assume there exists a C^r submersion φ from $\mathbf{R}^m$ to $\mathbf{R}^{n'}$ such that $\varphi^{-1}(0) = X_1$. The restriction $g|_{N \times a}$ is transversal to X_1 if and only if 0 is a regular value of $\varphi \circ g|_{N \times a}$. Hence we can replace $g : \mathbf{R}^n \times \mathbf{R}^m \to \mathbf{R}^m$ and $X_1 \subset \mathbf{R}^m$ with $\varphi \circ g : \mathbf{R}^n \times \mathbf{R}^m \to \mathbf{R}^{n'}$ and $0 \in \mathbf{R}^{n'}$, respectively. Moreover, since $\varphi \circ g|_{0 \times \mathbf{R}^m}$ is locally a submersion at 0×0, we can assume that $d(\varphi \circ g|_{0 \times \mathbf{R}^m})_{0,0}$ carries the subspace $0 \times \mathbf{R}^{n'} \times 0$ of $0 \times \mathbf{R}^m$ onto $\mathbf{R}^{n'}$. Hence if we choose a in (3) of the form $(a', 0) \in \mathbf{R}^{n'} \times \mathbf{R}^{m-n'}$, we can replace $\varphi \circ g$ with $\varphi \circ g|_{\mathbf{R}^n \times \mathbf{R}^{n'} \times 0}$. Thus we assume $X_1 = \{0\}$ and hence $n' = m$ in (3).

Proof of (3) in the case $X_1 = \{0\}$. Let $(x, y) = (x_1, \ldots, x_n, y_1, \ldots, y_m)$ denote the coordinate system of $\mathbf{R}^n \times \mathbf{R}^m$, and write $g = (g_1, \ldots, g_m)$. The local diffeomorphism condition in (3) means that the matrix $\left(\frac{\partial g_k}{\partial y_j}\right)$ is of rank m at $(0, 0)$. Let $N \times M \subset \mathbf{R}^n \times \mathbf{R}^m$ be a small open neighborhood of $(0, 0)$ in $\mathbf{R}^n \times \mathbf{R}^m$ where $\left(\frac{\partial g_k}{\partial y_j}\right)$ is of rank m. By the implicit function theorem we assume there exists a C^r map $\psi = (\psi_1, \ldots, \psi_m)$ from N to M such that

$$\text{graph } \psi = N \times M \cap g^{-1}(0).$$

It suffices to find a dense subset Z of M such that if $\psi(x_0) \in Z$, the matrix whose (i, k)-element is $\frac{\partial g_k}{\partial x_i}(x_0, \psi(x_0))$ is of rank m. Consider the equation:

$$g(x, \psi(x)) = 0.$$

We have an equation of matrices:

$$0 = \left(\frac{\partial g_k(x, \psi(x))}{\partial x_i} \right) = \left(\frac{\partial g_k}{\partial x_i}(x, \psi(x)) \right) + \left(\frac{\partial \psi_j}{\partial x_i}(x) \right) \left(\frac{\partial g_k}{\partial y_j}(x, \psi(x)) \right).$$

Let Z denote the complement in M of the critical value set of ψ. By Sard's theorem, Z is dense in M. It follows that if $\psi(x_0) \in Z$, $\left(\frac{\partial \psi_j}{\partial x_i}(x_0) \right)$ is of rank m and hence $\left(\frac{\partial g_k}{\partial x_i}(x_0, \psi(x_0)) \right)$ is of rank m. $\qquad \square$

Lemma II.5.5. *Let r be a positive integer. Let X and Y be C^r X-submanifolds of $\mathbf{R}^n$, let Z be a C^r X-submanifold of Y, and let $f \in B^r(X, Y)$. Assume that Z is closed in Y and f is transversal to Z. For a strong C^r X-approximation f' of f in the C^r X-topology, there exists a C^r X-diffeomorphism τ of X such that $\tau(f^{-1}(Z)) = f'^{-1}(Z)$. Moreover, we can choose τ so that $\tau \to \mathrm{id}$ as $f' \to f$.*

Proof. Set $Z_1 = f^{-1}(Z)$, and let $T = (|T|, \pi, \rho)$ and $T_1 = (|T_1|, \pi_1, \rho_1)$ be C^r X-tubes at Z and Z_1, respectively, such that $f(X \cap |T_1|) \subset |T|$ and for a bounded set $B \subset \mathbf{R}^n$, $\pi_1^{-1}(B)$ is bounded. Let W be an open X-neighborhood of Z in Y whose closure is included in $|T|$, and let U be a small open X-neighborhood of the diagonal in $Z \times Z$. For each $z \in Z$, define a set $U_z \subset Z$ so that $U = \cup_{z \in Z} z \times U_z$, and set

$$V_z = \pi^{-1}(U_z) \cap W \quad \text{and} \quad V = \bigcup_{z \in Z} z \times V_z.$$

Choose U so small that there exists a C^r X-map $p \colon V \to Y \cap |T|$ such that for each $z \in Z$, $p|_{z \times V_z}$ is a projection into $Y \cap \pi^{-1}(z)$. Let a C^r X-map $\psi \colon V \to Y \cap |T|$ be defined by

$$\psi(z, y) = p(z, y - \pi(y) + z) \quad \text{for} \quad (z, y) \in V.$$

Then

$$\psi(z \times V_z) \subset \pi^{-1}(z) \quad \text{and} \quad \psi^{-1}(z) = z \times (V_z \cap Z) \quad \text{for} \quad (z, y) \in V.$$

Since we shall choose τ to be the identity outside of a small neighborhood of Z_1 in $X \cap |T_1|$, and since we shall modify f' to coincide with f outside the neighborhood without changing $f'^{-1}(Z)$, we can forget the outside and suppose $X \subset |T_1|$. Let f' be a strong C^r $\mathfrak{X}$-approximation of f, and define C^r $\mathfrak{X}$-maps F and F' from X to $Y \cap |T|$ by

$$F(x) = \psi(f \circ \pi_1(x), f(x)) \quad \text{and} \quad F'(x) = \psi(f \circ \pi_1(x), f'(x)) \quad \text{for} \quad x \in X.$$

Here we assume that $|T_1|$ is so small and f' is so close to f that F and F' are well-defined, and for each $x \in Z_1$, $F|_{X \cap \pi_1^{-1}(x)}$ and $F'|_{X \cap \pi_1^{-1}(x)}$ are C^r imbeddings. Clearly

$$F^{-1}(Z) = Z_1, \qquad F'^{-1}(Z) = f'^{-1}(Z),$$

and F' is an C^r $\mathfrak{X}$-approximation of F. Hence we can replace f and f' with F and F', respectively. The advantage of replacement is that for each $x \in Z_1$, $F(X \cap \pi_1^{-1}(x))$ equals $F'(X \cap \pi_1^{-1}(x))$. By this property there exists a unique C^r $\mathfrak{X}$-diffeomorphism τ_x of $X \cap \pi_1^{-1}(x)$ such that $F' \circ \tau_x = F$ on $\pi_1^{-1}(x)$. Define the bijective map $\tau: |T_1| \to |T_1|$ by

$$\tau = \tau_x \quad \text{on} \quad X \cap \pi_1^{-1}(x) \quad \text{for} \quad x \in Z_1.$$

Then τ is a C^r $\mathfrak{X}$-diffeomorphism because (F, π_1) and (F', π_1) are C^r imbeddings of X into $|T| \times Z_1$ with the same images, and hence we have

$$(F', \pi_1) \circ \tilde{\tau} = (F, \pi_1)$$

for some C^r $\mathfrak{X}$-diffeomorphism $\tilde{\tau}$ of X, which, consequently, equals τ. Clearly we have $F' \circ \tau = F$ and hence $\tau(F^{-1}(Z)) = F'^{-1}(Z)$. By II.5.3, $\tilde{\tau} = \tau \to \text{id}$ as $f' \to f$ because $\tilde{\tau} = (F', \pi_1)^{-1} \circ (F, \pi_1)$.

We modify τ and f' so that $\tau = \text{id}$ and $f' = f$ outside of a small neighborhood of Z_1 as follows. Let φ be a C^r $\mathfrak{X}$-function on $|T_1|$ such that $0 \le \varphi \le 1$, $\psi = 0$ outside of a small neighborhood of Z_1, and $\varphi = 1$ on a smaller one. We can replace f' with $q(\varphi f' + (1 - \varphi)f)$ without changing the set $f'^{-1}(Z)$, where q is the projection of a C^r tubular $\mathfrak{X}$-neighborhood of Y in $\mathbf{R}^n$. Then τ is automatically equal to the identity outside of $\text{supp}\, \varphi$. $\quad\square$

Theorem II.5.6 (Approximation of an $\mathfrak{X}$-manifold). *Let $r' < r$ be positive integers. Let X be a $C^{r'}$ $\mathfrak{X}$-submanifold of $\mathbf{R}^n$. We can approximate the inclusion map of X into $\mathbf{R}^n$ by a $C^{r'}$ $\mathfrak{X}$-imbedding in the $C^{r'}$ topology so that the image is a C^r $\mathfrak{X}$-submanifold of $\mathbf{R}^n$.*

Proof. Let $\text{codim}\, X = n'$, and let $(E_{n,n'}, \pi_G, G_{n,n'})$ denote the universal vector bundle over $G_{n,n'}$. By the proof of II.5.1 there exist a C^r $\mathfrak{X}$-tube $T = (|T|, \pi, \rho)$ at X and a $C^{r'}$ $\mathfrak{X}$-map $\Phi: |T| \to E_{n,n'}$ such that

$$\Phi^{-1}(G_{n,n'}) = X, \qquad \Phi(\pi^{-1}(x)) \subset \pi_G^{-1}(\Phi(x)) \quad \text{for} \quad x \in X,$$

and Φ is transversal to $G_{n,n'}$. The theorem follows if we apply II.5.4 and II.5.5 to Φ. $\square$

Theorem II.5.7 (Compactification of an X-manifold). *Let r be a positive integer. For a non-compact bounded C^r X-submanifold X of $\mathbf{R}^n$, there exists a unique compact C^r X-submanifold Y of $\mathbf{R}^n$ with boundary such that X is C^r X-diffeomorphism to the interior Y°. Here uniqueness means that for another Y_1 with the same property, Y_1 is C^r X-diffeomorphic to Y.*

Before beginning the proof, we make some remarks. A manifold always means a manifold without boundary. We call Y in the above theorem a C^r *X-compactification* of X. We define also a C^r *compactification* Y of X if Y is a compact C^r manifold with boundary such that X is C^r diffeomorphic to Y°.

Remark II.5.8. As noted in VI.2.6 in [S₃], there exists a bounded Nash (and hence C^r X-)submanifold of $\mathbf{R}^n$ whose two distinct C^r compactifications exist.

Remark II.5.9 (see VI.2.5 in [S₃]). Let r be a positive integer. Let A_1, A_2 and A_3 denote the quotient spaces {compact C^r X-submanifolds of Euclidean spaces possibly with boundary}/C^r X-diffeomorphisms, {compact C^1 manifolds possibly with boundary}/C^1 diffeomorphisms and {bounded C^r X-submanifolds of Euclidean spaces}/C^r X-diffeomorphisms, respectively. By II.5.7 the map : $A_1 \ni Y \to Y^\circ \in A_3$ is bijective. Moreover, the natural map $A_1 \to A_2$ also is bijective as shown below.

Surjectivity easily follows from a result of Nash which states that a compact C^r manifold admits a Nash manifold structure (see I.3.7 in [S₃] for the details).

Proof of injectivity of $A_1 \to A_2$. Let X and Y be C^1 diffeomorphic compact C^r X-submanifolds of $\mathbf{R}^n$ possibly with boundary. We will prove that they are C^r X-diffeomorphic. First assume that X and Y are without boundary. Let $f: X \to Y$ be a C^1 diffeomorphism, and let $T = (|T|, \pi, \rho)$ be a C^r X-tube at Y. Regard f as a map from X to $\mathbf{R}^n$, and approximate f by the restriction f a polynomial map φ to X o in the C^1 topology. Then $f' = \pi \circ \varphi$ is a C^r X-approximation of f in the C^1 topology and hence is a diffeomorphism. Therefore, X and Y are C^r X-diffeomorphic.

Next assume X and Y have boundary. Set

$$X_1 = \{x \in X: \operatorname{dis}(x, \partial X) \geq \varepsilon\}, \quad \text{and}$$
$$Y_1 = \{y \in Y: \operatorname{dis}(y, \partial Y) \geq \varepsilon\},$$

where ε is a small positive number. Then X_1 and Y_1 are C^r X-manifolds with boundary, and by using X-tubes at ∂X and ∂Y, we easily prove that X_1

and Y_1 are C^r $\mathfrak{X}$-diffeomorphic to X and Y, respectively. Hence it suffices to find a C^r $\mathfrak{X}$-diffeomorphism from X_1 to Y_1. It is also easy to construct a C^1 diffeomorphism $f \colon (X, X_1) \to (Y, Y_1)$. Let $T_{\partial X_1} = \{|T_{\partial X_1}|, \pi_{\partial X_1}, \rho_{\partial X_1}\}$, $T_{\partial Y_1} = \{|T_{\partial Y_1}|, \pi_{\partial Y_1}, \rho_{\partial Y_1}\}$ and $T_Y = \{|T_Y|, \pi_Y, \rho_Y\}$ be C^r $\mathfrak{X}$-tubes at ∂X_1, ∂Y_1 and Y, respectively, and let φ be a C^r $\mathfrak{X}$-function on $|T_{\partial X_1}|$ such that $0 \le \varphi \le 1$, $\varphi = 0$ outside of a small neighborhood of ∂X_1, and $\varphi = 1$ on a smaller one. Let $f' \colon X \to \mathbf{R}^n$ be a strong C^r $\mathfrak{X}$-approximation of f in the C^1 topology. The C^r $\mathfrak{X}$-map $f'' \colon X_1 \to Y$, defined by

$$
f''(x) = \begin{cases} \pi_Y\{\varphi(x)(f'(x) + \pi_{\partial Y_1} \circ f' \circ \pi_{\partial X_1}(x) \\ \quad -f' \circ \pi_{\partial X_1}(x)) + (1 - \varphi(x))f'(x)\} & \text{for} \quad x \in X_1 \cap |T_{\partial X_1}| \\ \pi_Y \circ f'(x) & \text{for} \quad x \in X_1 - |T_{\partial X_1}|, \end{cases}
$$

is a C^r $\mathfrak{X}$-diffeomorphism onto Y_1. $\qquad\square$

Remark II.5.10. Let r be a positive integer. A bounded C^r $\mathfrak{X}$-submanifold X of $\mathbf{R}^n$ is C^r $\mathfrak{X}$-diffeomorphic to an affine non-singular algebraic variety. I do not know whether the variety and the diffeomorphism can be constructed by the elementary method in Preface.

Proof. If X is compact, this follows from a result of Tognoli [To] that a compact C^1 manifold is C^1 diffeomorphic to an affine non-singular variety. Assume X is not compact. By II.5.7 we can regard X as the interior of some compact C^r $\mathfrak{X}$-submanifold Y of $\mathbf{R}^n$ with boundary. By VI.2.5 in [S$_3$] there exists a Nash submanifold $\tilde{Y}$ of some $\mathbf{R}^{n'}$ with boundary which is C^r diffeomorphic to Y. Since $\tilde{Y}$ is a C^r $\mathfrak{X}$-submanifold of $\mathbf{R}^{n'}$, it follows from II.5.9 that Y and $\tilde{Y}$ are C^r $\mathfrak{X}$-diffeomorphic. Moreover, by VI.2.11 in [S$_3$], $\tilde{Y}^\circ$ is Nash diffeomorphic to an affine non-singular algebraic variety. Thus the remark holds. $\qquad\square$

Proof of II.5.7. Existence. Apply II.5.2 to the function which measures distance from $\overline{X} - X$ defined on X. We have a positive C^r $\mathfrak{X}$-function φ on X such that $\varphi(x) \to 0$ as $x \in X$ converges to a point of $\overline{X} - X$. By (II.1.7) the critical value set of φ is a bounded $\mathfrak{X}$-set of dimension 0 and hence finite. Hence for a small positive number ε, $]0, \varepsilon[$ does not contain critical values of φ. Lessen ε. By the following theorem II.6.5, $\varphi|_{\varphi^{-1}(]0,\varepsilon[)}$ is C^r $\mathfrak{X}$-trivial (see below for the definition of this term). Hence $\varphi^{-1}(]0, \varepsilon[)$ is C^r $\mathfrak{X}$-diffeomorphic to $\varphi^{-1}(]\varepsilon/2, \varepsilon[)$. Moreover, we can easily extend this diffeomorphism to a diffeomorphism $\varphi^{-1}(]0, \infty[) \to \varphi^{-1}(]\varepsilon/2, \infty[)$. Clearly $\varphi^{-1}(]\varepsilon/2, \infty[)$ is the interior of the compact C^r $\mathfrak{X}$-manifold $\varphi^{-1}([\varepsilon/2, \infty[)$ with boundary. Thus it suffices to prove II.6.5.

Uniqueness. Let Y_i, $i = 1, 2$, be C^r $\mathfrak{X}$-compactifications of X. There exists a C^r $\mathfrak{X}$-diffeomorphism $\tau\colon Y_1^\circ \to Y_2^\circ$. Under this condition, it suffices to find a C^r $\mathfrak{X}$-diffeomorphism from Y_1 to Y_2. Moreover, by II.5.9 we need only construct a C^1 diffeomorphism from Y_1 to Y_2. We assume $r > 1$ by II.5.6 for compact $C^{r'}$ $\mathfrak{X}$-manifolds with boundary, which we can prove in the same way as in the proof of II.5.9.

For each $i = 1, 2$ it is easy to construct a non-negative C^r $\mathfrak{X}$-function φ_i on Y_i such that $\varphi_i^{-1}(0) = \partial Y_i$, and φ_i is C^r regular on ∂Y_i. Then Y_i is C^r diffeomorphic to $\varphi_i^{-1}([\varepsilon, \infty[)$ for a small enough positive number ε. Hence we will find small ε and a C^1 diffeomorphism from $\varphi_1^{-1}([\varepsilon, \infty[)$ to $\varphi_2^{-1}([\varepsilon, \infty[)$. By (II.1.18), φ_1 and $\varphi_2 \circ \tau$ are friendly on $U \cap Y_1^\circ$ for some neighborhood U of ∂Y_1 in Y_1. Therefore, what we prove is the following statement.

Let M be a compact C^r manifold, and let φ be a non-negative function on $M \times [0, 3]$ such that $\varphi^{-1}(0) = M \times 0$, $\varphi|_{M \times]0,3]}$ is C^r regular, $\varphi^{-1}(1) \subset M \times]0, 1[$, $\varphi^{-1}(2) \subset M \times]1, 2[$, and φ on $M \times]0, 3]$ and the projection $p\colon M \times]0, 3] \to]0, 3]$ are friendly. Then $\varphi^{-1}([1, \infty[)$ is C^r diffeomorphic to $M \times [1, 3]$.

Let v be a C^r function on $\mathbf{R}$ such that $v(1) = 0$, $v = 1$ on $[2, \infty[$ and v is C^r regular on $]-\infty, 2[$. Consider the C^r function $\psi = p \cdot v \circ \varphi$ on $M \times]0, 3]$. We have

$$\psi^{-1}(0) = \varphi^{-1}(1) \quad \text{and} \quad \psi^{-1}(t) = p^{-1}(t) = M \times t \quad \text{for} \quad t \in [2, 3],$$

and ψ is C^r regular on $\psi^{-1}([0, 3])$. Then $\psi^{-1}([0, 2])$ is C^r diffeomorphic to $\psi^{-1}(2) \times [0, 2q]$. Hence the above statement is clear. $\square$

Theorem II.5.11 (Case of $\mathfrak{X}_0$). *If* $\mathfrak{X} = \mathfrak{X}_0$, *all the results in this section hold without the boundedness condition.*

Proof. Since $\mathbf{R}^n$ is Nash diffeomorphic to an open ball in $\mathbf{R}^n$, we can reduce the problems to the bounded case. $\square$

§II.6. $\mathfrak{X}$-triviality of $\mathfrak{X}$-maps

Let $X \subset \mathbf{R}^n$ and $Y \subset \mathbf{R}^m$ be $\mathfrak{X}$-sets, and let $f\colon X \to Y$ be an $\mathfrak{X}$-map. We say that f satisfies the *first (second) boundedness condition* if for any bounded set $B \subset \mathbf{R}^n$ ($\subset \mathbf{R}^m$, resp.), $f(X \cap B)$ ($f^{-1}(B)$, resp.,) is bounded in $\mathbf{R}^m$ ($\mathbf{R}^n$, resp.). We call f $\mathfrak{X}$-*trivial* if there exists an $\mathfrak{X}$-homeomorphism with the first and second boundedness conditions $g\colon f^{-1}(y) \times Y \to X$ for some $y \in Y$ such that $f \circ g\colon f^{-1}(y) \times Y \to Y$ is the projection. Note that this definition does not depend on y, namely, if there exists such a

g for some y, then there exists g for any y. For an $\mathfrak{X}$-subset Y' of Y we call f $\mathfrak{X}$-*trivial* over Y' if $f|_{f^{-1}(Y')}$ is $\mathfrak{X}$-trivial. Let X and Y be C^r $\mathfrak{X}$-submanifolds of $\mathbf{R}^n$, $r > 0$. We call a C^r $\mathfrak{X}$-submersion $f : X \to Y$ C^r $\mathfrak{X}$-*trivial* if there exists a C^r $\mathfrak{X}$-diffeomorphism $g : f^{-1}(y) \times Y \to X$, $y \in Y$ with the first and second boundedness conditions, such that $f \circ g$ is the projection. Let $X \xrightarrow{f} Y \xrightarrow{p} P$ be a diagram of $\mathfrak{X}$-sets and $\mathfrak{X}$-maps. We call (f, p) $\mathfrak{X}$-*trivial* if there exist $\mathfrak{X}$-homeomorphisms with the first and second boundedness conditions $g : (p \circ f)^{-1}(a) \times P \to X$ and $h : p^{-1}(a) \times P \to Y$ for $a \in P$ such that the diagram

$$
\begin{array}{ccccc}
(p \circ f)^{-1}(a) \times P & \xrightarrow{\ f \times \mathrm{id}\ } & p^{-1}(a) \times P & \xrightarrow{\ \mathrm{proj}\ } & P \\
\downarrow{\scriptstyle g} & & \downarrow{\scriptstyle h} & & \downarrow{\scriptstyle \mathrm{id}} \\
X & \xrightarrow{\quad f \quad} & Y & \xrightarrow{\quad p \quad} & P
\end{array}
$$

commutes. If X, Y and P are C^r $\mathfrak{X}$-manifolds, $r > 0$, f and p are C^r $\mathfrak{X}$-maps, and g and h are C^r $\mathfrak{X}$-diffeomorphisms, then we call (f, p) C^r $\mathfrak{X}$-*trivial*. We define naturally (C^r) *triviality*. In this section, we consider when f and (f, p) are (C^r) $\mathfrak{X}$-trivial. Note that the boundedness condition in the following theorems is not necessary in the case $\mathfrak{X} = \mathfrak{X}_0$(II.6.12). First we have the following.

Theorem II.6.1 ($\mathfrak{X}$-version of Thom's first isotopy lemma). *Let $X \subset \mathbf{R}^n$, $Y \subset \mathbf{R}^m$ be $\mathfrak{X}$-sets, and let $f : X \to Y$ be a proper $\mathfrak{X}$-map with the second boundedness condition. Assume that f is a C^1 map, i.e., f can be extended to a C^1 map from a neighborhood of X in $\mathbf{R}^n$ to $\mathbf{R}^m$. Assume further that Y is C^1 smooth, and X admits a finite C^1 Whitney $\mathfrak{X}$-stratification $\{X_i\}$ such that for each i, $f|_{X_i}$ is a C^1 submersion onto Y. Then there exists an open $\mathfrak{X}$-covering $\{U_i\}$ of Y locally finite at each point of not only Y but also of $\mathbf{R}^m$, such that for each i, $f|_{f^{-1}(U_i)} : f^{-1}(U_i) \to U_i$ is $\mathfrak{X}$-trivial.*

Complement. *We can choose the $\mathfrak{X}$-homeomorphism of triviality $f^{-1}(y) \times U_i \to f^{-1}(U_i)$ so that it induces a C^1 $\mathfrak{X}$-diffeomorphism $(f^{-1}(y) \cap X_j) \times U_i \to X_j \cap f^{-1}(U_i)$ for each j.*

We can weaken the assumption that X admits a C^1 Whitney $\mathfrak{X}$-stratification $\{X_i\}$ as follows.

There exist a finite C^1 $\mathfrak{X}$-stratification $\{X_i\}$ of X and its controlled C^1 $\mathfrak{X}$-tube system $\{T_i = (|T_i|, \pi_i, \rho_i)\}$ such that for each i, $f|_{X_i}$ is a C^1 submersion onto Y, $f \circ \pi_i = f$ on $X \cap |T_i|$, and for any i' with $X_i \cap (\overline{X_{i'}} - X_{i'}) \neq \varnothing$, $(\pi_i, \rho_i)|_{X_{i'} \cap |T_i|} : X_{i'} \cap |T_i| \to X_i \times \mathbf{R}$ is a C^1 submersion. (II.6.10 shows that these assumptions are weaker than the assumption on a Whitney stratification.)

These assertions will be clear by the proof of II.6.1.

Theorem II.6.1' ($\mathfrak{X}$-version of Thom's second isotopy lemma). *Let $X \subset \mathbf{R}^n, Y \subset \mathbf{R}^m$ and $P \subset \mathbf{R}^{m'}$ be $\mathfrak{X}$-sets, and let $f: X \to Y$ and $p: Y \to P$ be proper $\mathfrak{X}$-maps with the second boundedness condition. Assume that f and p are C^1 maps, P is C^1 smooth, and f admits a finite C^1 Whitney $\mathfrak{X}$-stratification sans éclatement ($\{X_i\}, \{Y_j\}$) such that for each j, $p|_{Y_j}$ is a C^1 submersion onto P. Then there exists an open $\mathfrak{X}$-covering $\{U_i\}$ of P locally finite at each point of $\mathbf{R}^{m'}$ such that for each i, $(f|_{(p\circ f)^{-1}(U_i)}, p|_{p^{-1}(U_i)})$ is $\mathfrak{X}$-trivial.*

Complement. *As in the preceding complement, we can choose the $\mathfrak{X}$-homeomorphisms of triviality so that they induce C^1 $\mathfrak{X}$-diffeomorphisms.*

We can replace the assumptions that p is proper and satisfies the second boundedness condition with ones that Y is of the form $Y_0 \times P$, Y_0 is locally closed in $\mathbf{R}^{m-m'}$, p is the projection and $\{Y_j\}$ is of the form $\{Y_{0j} \times P\}$.

The assumption that f admits a finite C^1 Whitney $\mathfrak{X}$-stratification sans éclatement can be weakened as follows.

There exist a finite C^1 $\mathfrak{X}$-stratification ($\{X_i\}, \{Y_j\}$) of f, a controlled C^1 $\mathfrak{X}$-tube system $\{T_j^Y = (|T_j^Y|, \pi_j^Y, \rho_j^Y)\}$ for $\{Y_j\}$ and a C^1 $\mathfrak{X}$-tube system $\{T_i^X = (|T_i^X|, \pi_i^X, \rho_i^X)\}$ for $\{X_i\}$ controlled over $\{T_j^Y\}$ such that the following hold. For each j, $p|_{Y_j}$ is a C^1 submersion onto P, and $p \circ \pi_j^Y = p$ on $Y \cap |T_j^Y|$. For any j' with $Y_j \cap (\overline{Y_{j'}} - Y_{j'}) \neq \varnothing$, (1) $(\pi_j^Y, \rho_j^Y)|_{Y_{j'} \cap |T_j^Y|}$ is a C^1 submersion to $Y_j \times \mathbf{R}$. For any i and i' with $f(X_i) \subset Y_j$, $f(X_{i'}) \subset Y_{j'}$ and $X_i \cap (\overline{X_{i'}} - X_{i'}) \neq \varnothing$, if $Y_j = Y_{j'}$ then (2) $(\pi_i^X, \rho_i^X)|_{X_{i'} \cap |T_i^X|}$ is a C^1 submersion to $X_i \times \mathbf{R}$, and if $Y_j \neq Y_{j'}$ then (3) $(\pi_i^X, f)|_{X_{i'} \cap |T_i^X|}$ is a C^1 submersion to $\{(x, y) \in X_i \times (Y_{j'} \cap |T_j^Y|): f(x) = \pi_j^Y(y)\}$.

These also will be clear. The last assumption is convenient for applications. Note that the above conditions (1) and (2) are satisfied if $\{X_i\}$ and $\{Y_i\}$ are Whitney stratifications respectively, and (3) is equivalent to that for $a \in X_i$ and $b \in X_{i'} \cap |T_i^X|$ with $X_i \cap \overline{X_{i'}} - X_{i'} \neq \varnothing$ and $\pi_i^X(b) = a$, the tangent map $d(\pi_i)_b: T_b(f|_{X_{i'}})^{-1} f(b) \to T_a(f|_{X_i})^{-1} f(a)$ is surjective.

If we forget $\mathfrak{X}$, these theorems are called the Thom's first and second isotopy lemmata, which follow from I.1.6 easily (see [G-al]).

Theorem II.6.2 ($\mathfrak{X}$-triviality). *Let $X \subset \mathbf{R}^n$, $Y' \subset Y \subset \mathbf{R}^m$ be $\mathfrak{X}$-sets and let $f: X \to Y$ be an $\mathfrak{X}$-map with the first boundedness condition. Assume Axiom (v) unless Y is bounded, that Y' is closed in Y and a deformation retract of Y, $f|_{f^{-1}(Y')}$ is $\mathfrak{X}$-trivial, and there exists an open $\mathfrak{X}$-covering $\{U_i\}$ of Y locally finite at each point of $\mathbf{R}^m$ such that for each i, $f|_{f^{-1}(U_i)}$ is $\mathfrak{X}$-trivial. Then f is $\mathfrak{X}$-trivial.*

Theorem II.6.2'. *Let $X \subset \mathbf{R}^n, Y \subset \mathbf{R}^m$ and $P' \subset P \subset \mathbf{R}^{m'}$ be $\mathfrak{X}$-sets, and let $f: X \to Y$ and $p: Y \to P$ be $\mathfrak{X}$-maps. Assume Axiom (v) unless*

P is bounded, that $p \circ f$ and p satisfy the first boundedness condition, P' is closed in P and a deformation retract of P, $(f|_{(p \circ f)^{-1}(P')}, p|_{p^{-1}(P')})$ is $\mathfrak{X}$-trivial, and there exists an open $\mathfrak{X}$-covering $\{U_i\}$ of P locally finite at each point of $\mathbf{R}^{m'}$ such that for each i, $(f|_{(p \circ f)^{-1}(U_i)}, p|_{p^{-1}(U_i)})$ is $\mathfrak{X}$-trivial. Then (f, p) is $\mathfrak{X}$-trivial.

Theorem II.6.3. *Let r be a positive integer. In II.6.2, if X and Y are C^r $\mathfrak{X}$-manifolds, f is a C^r $\mathfrak{X}$-map with the first boundedness condition, and $f|_{f^{-1}(U)}$ and $f|_{f^{-1}(U_i)}$ are C^r $\mathfrak{X}$-trivial for an open $\mathfrak{X}$-neighborhood U of Y' in Y, then f is C^r $\mathfrak{X}$-trivial. A similar statement for II.6.2' holds true.*

Remark II.6.4. In II.6.2 and II.6.3, the assumptions that Y' is closed in Y and a deformation retract of Y and $f|_{f^{-1}(Y')}$ is (C^r) $\mathfrak{X}$-trivial are satisfied if Y is contractible and Y' is any point of Y. Similarly, in II.6.2' and II.6.3, P' is closed and a deformation retract of P and $(f|_{(p \circ f)^{-1}(P')}, p|_{p^{-1}(P')})$ is (C^r) $\mathfrak{X}$-trivial if P is contractible and P' is any point of P.

Theorem II.6.5 (C^r $\mathfrak{X}$-triviality). *Let r be a positive integer. Let X and B be C^r $\mathfrak{X}$-submanifolds of $\mathbf{R}^n$ and let $p: X \to B$ be a proper C^r $\mathfrak{X}$-submersion with the second boundedness condition. Assume Axiom (v) unless B is bounded. There exists an open $\mathfrak{X}$-covering $\{U_i\}$ of B locally finite at each point of $\mathbf{R}^n$ such that for each i, $p|_{p^{-1}(U_i)}: p^{-1}(U_i) \to U_i$ is C^r $\mathfrak{X}$-trivial.*

Theorem II.6.5'. *Let $r > 0$ be an integer. Let X, Y and P be C^r $\mathfrak{X}$-submanifolds of $\mathbf{R}^n$, and let $f: X \to Y$ and $p: Y \to P$ be proper C^r $\mathfrak{X}$-submersions with the second boundedness condition. Assume Axiom (v) unless P is bounded. There exists an open $\mathfrak{X}$-covering $\{U_i\}$ of P locally finite at each point of $\mathbf{R}^n$ such that for each i, $(f|_{(p \circ f)^{-1}(U_i)}, p|_{p^{-1}(U_i)})$ is C^r $\mathfrak{X}$-trivial.*

Corollary II.6.6. *In II.6.5 and II.6.5', assume p and (f, p) are C^r trivial (e.g., B and P are contractible) and satisfy the first boundedness condition. Then they are C^r $\mathfrak{X}$-trivial.*

We postpone proof of II.6.1 and II.6.1' because in the proof we need II.6.5, II.6.5' and their proof. II.6.2 of the C^0 category and II.6.3 of the C^r category both without $\mathfrak{X}$ are well-known [St]. We use the same idea as in [St] to prove II.6.2 and II.6.3 in the case where Y is closed in $\mathbf{R}^m$, and the idea of a collapse (see [R-S]) to reduce the general case to this special one. In II.6.5, the properness condition is necessary. For example, $X = \mathbf{R}^2 - \mathbf{R}_+ \times 0$, where $\mathbf{R}_+ = \{\text{nonnegative reals}\}$, $B = \mathbf{R}$ and $p = $ the projection to the second factor. Conjectures are that $\mathfrak{X}$-triviality in II.6.5, II.6.5' and II.6.6 holds without C^r differentiability and that II.6.6 holds without the properness condition. In the Nash case, II.6.5 holds on an arbitrary real closed field and implies an important corollary [C-S$_1$].

Proof of II.6.2 and II.6.2′. We prove only II.6.2 because we can do II.6.2′ in the same way but with more complicated notation. Let (K, π) be an 𝔛-triangulation of $\overline{Y}$ compatible with Y, Y' and $\{U_i\}$ such that $|K|$ is closed in $\mathbf{R}^n$ and hence π satisfies the first and second boundedness conditions. Since $\{U_i\}$ is an open covering of Y, for each $\sigma \in K$, $\sigma \cap Y$ is included in some U_i and, hence, f is 𝔛-trivial over $\sigma \cap Y$. As we consider II.6.2 in the C^0 𝔛-category, we can translate the problem on Y to the one on $\pi^{-1}(Y)$. Hence we assume Y is a union of open simplexes of K. Let K' be a derived subdivision of K, and let L denote its subcomplex consisting of the simplexes contained in Y. Note that for each $\sigma \in K' - L$, $\sigma - Y$ is included in a face of σ of the same dimension because K' is a derived subdivision. The face is $\overline{\sigma - Y}$.

First we want to reduce the problem to the case where Y is closed in $\mathbf{R}^m$. This is because we require a deformation retraction of Y to Y' to be not only of class 𝔛 but also to satisfy the first boundedness condition. If Y is closed and the deformation retraction is PL, then these conditions are satisfied. It suffices to find commutative 𝔛-maps:

$$
\begin{array}{ccc}
X & \xrightarrow{\;\Phi\;} & f^{-1}(|L|) \\
{\scriptstyle f}\downarrow & & \downarrow{\scriptstyle f} \\
Y & \xrightarrow[\;\varphi\;]{} & |L|
\end{array}
$$

such that φ and Φ satisfy the first and second boundedness conditions, and for each $y \in Y$, $\Phi|_{f^{-1}(y)}$ is an 𝔛-homeomorphism onto $f^{-1}(\varphi(y))$ because $|L|$ is closed in $\mathbf{R}^m$, the inclusion $|L| \cap Y' \to |L|$ is weakly homotopy equivalent (see the following construction of φ), and hence $|L| \cap Y'$ is a deformation retract of $|L|$. Let k be a positive integer. Let $\sigma \in K' - L$ of dimension k. We shall construct commutative 𝔛-maps:

$$
\begin{array}{ccc}
f^{-1}(Y \cap \sigma) & \xrightarrow{\;\Phi_\sigma\;} & f^{-1}(Y \cap \partial\sigma) \\
{\scriptstyle f}\downarrow & & \downarrow{\scriptstyle f} \\
Y \cap \sigma & \xrightarrow[\;\varphi_\sigma\;]{} & Y \cap \partial\sigma
\end{array}
$$

such that φ_σ and Φ_σ are 𝔛-retractions and for each $y \in Y \cap \sigma$, $\Phi_\sigma|_{f^{-1}(y)}$ is an 𝔛-homeomorphism onto $f^{-1}(\varphi_\sigma(y))$. Assume we have such φ_σ and Φ_σ, set $\varphi_k = \varphi_\sigma$ and $\Phi_k = \Phi_\sigma$ for all $\sigma \in K'' - L$ of dimension k. Then the maps

$$
\varphi = \begin{cases} \varphi_1 \circ \cdots \circ \varphi_{\dim Y} & \text{on} \quad Y - |L| \\ \mathrm{id} & \text{on} \quad |L| \text{ and} \end{cases}
$$

$$
\Phi = \begin{cases} \Phi_1 \circ \cdots \circ \Phi_{\dim Y} & \text{on} \quad f^{-1}(Y - |L|) \\ \mathrm{id} & \text{on} \quad f^{-1}(|L|) \end{cases}
$$

fulfill the requirements. Hence we need only construct φ_σ and Φ_σ. Existence of the φ_σ is shown as follows. Clearly $(Y \cap \sigma, Y \cap \partial\sigma)$ is $\mathfrak{X}$-homeomorphic to $(I^k - J \times 1, \partial(I^k) - J \times 1)$, where $I = [0,1]$ and J is an $\mathfrak{X}$-polyhedron such that $(I^{k-1})^\circ \subset J \subset I^{k-1}$, and there exists an $\mathfrak{X}$-retraction τ of $I^k - (I^{k-1})^\circ \times 1$ to $\partial(I^k) - (I^{k-1})^\circ \times 1$ such that $\tau^{-1}(\partial(I^{k-1}) \times 1) = \partial(I^{k-1}) \times 1$. The τ induces φ_σ. Existence of the Φ_σ follows from triviality of f over $\sigma \cap Y$. Thus we can assume Y is closed in $\mathbf{R}^m$.

Next, we reduce the problem to the case where $Y = Y \times [0,1]$, $Y' = Y \times 1$. Let $\tau_t \colon Y \to Y$, $t \in [0,1]$, be a deformation retraction to Y', i.e., $\tau_0 = \mathrm{id}$, $\tau_1(Y) = Y'$ and $\tau_1|_{Y'} = \mathrm{id}$. We can assume τ_t is PL and hence of class $\mathfrak{X}$. Note that we can replace f with its graph because f satisfies the first boundedness condition. Set

$$X^I = \{(x, y, t) \in X \times Y \times I : f(x) = \tau_t(y)\} \quad \text{and}$$
$$f^I(x, y, t) = (y, t) \quad \text{for} \quad (x, y, t) \in X^I.$$

Then $f^I \colon X^I \to Y \times I$ is an $\mathfrak{X}$-map and satisfies the first boundedness condition, $X^I \cap (X \times Y \times 0)$ is the graph of f, $f^I|_{X^I \cap (X \times Y \times 0)}$ is the projection to Y, f^I is $\mathfrak{X}$-trivial locally at each point of $Y \times I$, and f^I is $\mathfrak{X}$-trivial over $Y \times 1$ because τ_1 satisfies the first boundedness condition. As we want $\mathfrak{X}$-triviality of f^I over $Y \times 0$, we can assume $Y = Y \times [0,1]$ and $Y' = Y \times 1$. For simplicity of notation, we write $f \colon X \to Y \times I$ for $f^I \colon X^I \to Y \times I$.

Choose a simplicial decomposition K_0 of Y so fine that for each $\sigma \in K_0$, there exist numbers $0 = a_0 < \cdots < a_l = 1$ such that f is $\mathfrak{X}$-trivial over each $\sigma \times [a_i, a_{i+1}]$. Let k be a non-negative integer. We show $\mathfrak{X}$-triviality of f over $|K_0^k| \times I \cup Y \times 1$ by induction on k. Assume $\mathfrak{X}$-triviality of f over $|K_0^{k-1}| \times I \cup Y \times 1$. Let $\sigma \in K_0$ be of dimension k. Then it suffices to extend the $\mathfrak{X}$-triviality of f over $\partial\sigma \times I \cup \sigma \times 1$ to $\sigma \times 1$. Let $0 = a_0 < \cdots < a_l = 1$ be given as above for σ. There exists an $\mathfrak{X}$-retraction of $\sigma \times [a_{l-1}, 1]$ to $\partial\sigma \times [a_{l-1}, 1] \cup \sigma \times 1$. Then by the same way as the above reduction to the case of closed Y, we can extend the $\mathfrak{X}$-triviality of f over $\partial\sigma \times [a_{l-1}, 1] \cup \sigma \times 1$ to all of $\sigma \times [a_{l-1}, 1]$. Repeating this argument, we obtain extended $\mathfrak{X}$-triviality of f over $\sigma \times [0, 1]$. $\qquad\square$

Proof of II.6.3. Here we prove only the first half. By the above proof we have an $\mathfrak{X}$-map $\tau \colon Y \times I \to Y$ with the first boundedness condition, $I = [0,1]$, an open $\mathfrak{X}$-covering $\{V_i\}$ of Y, and for each i, numbers $0 = a_{i0} < \cdots < a_{il_i} = 1$ such that $\tau(\cdot, 0) = \mathrm{id}$, $\tau(Y \times 1) \subset Y'$, $\{V_i\}$ is locally finite at each point of $\mathbf{R}^m$, and f is C^r $\mathfrak{X}$-trivial over an open $\mathfrak{X}$-neighborhood of each $\tau(V_i \times [a_{ij}, a_{ij+1}])$. Let $T \colon Y \times I \to Y$ be a C^r $\mathfrak{X}$-approximation of τ with the same properties except that $T(Y \times 1) \subset Y'$. We replace it with $T(Y \times 1) \subset U$.

Using T as above, we replace the problem with the following statement. Let $X \subset \mathbf{R}^n$, $Y \subset \mathbf{R}^m$ be C^r $\mathfrak{X}$-manifolds, let $f\colon X \to Y \times I$, $I = [0,1]$, be a C^r $\mathfrak{X}$-map with the first boundedness condition, and let $\{V_i\}$ be an open $\mathfrak{X}$-covering of Y locally finite at each point of $\mathbf{R}^m$. Assume f is C^r $\mathfrak{X}$-trivial over $Y \times [1/2, 1]$, and for each i, there exist numbers $0 = a_0 < \cdots < a_l = 1$ such that f is C^r $\mathfrak{X}$-trivial over $V_i \times [a_j, a_{j+2}]$, $j = 0, \dots, l-2$. Then f is C^r $\mathfrak{X}$-trivial.

We can prove this by the same idea as before. But we need to change two points. First, since we cannot replace the problem of C^r $\mathfrak{X}$-triviality of f over $Y \times I$ to the one over a polyhedron, we regard the simplicial complex K_0 in the above proof as the family of the images of simplexes of K_0 in Y. Second, since there is not a C^r $\mathfrak{X}$-deformation retraction except in special cases, we always consider C^r $\mathfrak{X}$-triviality not over unions of elements of K_0 but over their open $\mathfrak{X}$-neighborhoods. Since the modification is easy, we omit details.   □

For proof of II.6.5 we shall use an induction method. In the induction step, we meet a more general and complicated situation than in the theorem. So we modify the theorem as follows.

Let B be a bounded open $\mathfrak{X}$-set in $\mathbf{R}^d$, and let S, X and M be bounded $\mathfrak{X}$-sets in $\mathbf{R}^{n-d} \times B \subset \mathbf{R}^n$ with $S \subset X$ and $S \subset M$. Let $q\colon \mathbf{R}^{n-d} \times B \to B$ denote the projection and set

$$q_X = q|_X, \qquad q_M = q|_M \quad \text{and} \quad q_S = q|_S.$$

For each $b \in B$, set

$$X_b = q_X^{-1}(b), \qquad M_b = q_M^{-1}(b) \quad \text{and} \quad S_b = q_S^{-1}(b).$$

Assume that $X - S$ is a C^r $\mathfrak{X}$-submanifold of $\mathbf{R}^n$ of dimension $k+d$, M is a C^r $\mathfrak{X}$-submanifold of $\mathbf{R}^n$ of dimension $k+d+1$, M includes a neighborhood of S in X, q_X is proper, $q|_{X-S}$ is a C^r submersion, for each $b \in B$, M_b is C^r smooth and of dimension $k+1$, and $\sharp S_b = 0$ or $= 1$. Furthermore, assume that there exists a C^r $\mathfrak{X}$-function f on M such that $f^{-1}(0) = X \cap M$, and for each $s \in S$, $f|_{M_{p(s)}}$ takes the *Morse type* at s (i.e., $f|_{M_{p(s)}}$ is of the form $\sum_{i=1}^{k+1} \pm x_i^2 + \text{const}$ locally at s for some C^r $\mathfrak{X}$-local coordinate system $(x_1, \dots, x_{k+1})$ of $M_{p(s)}$ at s).

A *Morse function* on a C^r manifold is a C^r function which takes the Morse type at each critical point. We call an $\mathfrak{X}$-homeomorphism between $\mathfrak{X}$-sets a C^r $\mathfrak{X}$-*diffeomorphism* if we can extend it to a C^r $\mathfrak{X}$-diffeomorphism between open $\mathfrak{X}$-neighborhoods in the ambient spaces. We define a C^r $\mathfrak{X}$-*function* on an $\mathfrak{X}$-set in the same way.

Theorem II.6.7. *Assume r is larger than a certain number which depends on n, and shrink M. There exists a positive integer r', a finite $C^{r'}$ $\mathfrak{X}$-stratification $\{B_i\}$ of B, and for each B_i, and for each $b_i \in B_i$, a $C^{r'}$ $\mathfrak{X}$-imbedding,*

$$\pi_i \colon q_X^{-1}(B_i) \cup q_M^{-1}(B_i) \longrightarrow (X_{b_i} \cup M_{b_i}) \times B_i$$

of the form $\pi_i = (\pi_i', q)$ such that

$$\pi_i(q_X^{-1}(B_i)) = X_{b_i} \times B_i, \qquad \pi_i(q_M^{-1}(B_i)) \subset M_{b_i} \times B_i,$$
$$\pi_i'|_{X_{b_i} \cup M_{b_i}} = \mathrm{id} \quad and \quad f \circ \pi_i' = f \quad on \quad q_M^{-1}(B_i).$$

Moreover, if π_i is first defined on $q_M^{-1}(B_i)$ so that

$$\pi_i(q_M^{-1}(B_i)) \subset M_{b_i} \times B_i, \qquad \pi_i'|_{M_{b_i}} = \mathrm{id} \quad and \quad f \circ \pi_i' = f \quad on \quad q_M^{-1}(B_i),$$

by shrinking M and substratifying B_i, we can extend π_i to $q_X^{-1}(B_i)$.

Proof that II.6.7 implies II.6.5. Assume II.6.7 and consider II.6.5. We can easily reduce II.6.5 to the case where B is bounded in $\mathbf{R}^n$. We omit the proof and suppose the boundedness.

We shall prove that there exists a C^r $\mathfrak{X}$-stratification $\{B_i\}$ of B locally finite at each point of $\mathbf{R}^n$ such that for each i, $p|_{p^{-1}(B_i)} \colon p^{-1}(B_i) \to B_i$ is C^r $\mathfrak{X}$-trivial. This suffices for the proof for the following reason. Let $\{B_i\}$ be such a C^r $\mathfrak{X}$-stratification of B. For each i, let $b_i \in B_i$, and let $g_i \colon f^{-1}(b_i) \times B_i \to f^{-1}(B_i)$ be a C^r $\mathfrak{X}$-trivialization of $f|_{f^{-1}(B_i)}$, i.e., a C^r $\mathfrak{X}$-diffeomorphism such that $p \circ g_i \colon f^{-1}(b_i) \times B_i \to B_i$ is the projection. Let $p_i \colon U_i \to B_i$ be a small open C^r tubular $\mathfrak{X}$-neighborhood of B_i in B. By II.6.3 it suffices to extend g_i to $p^{-1}(b_i) \times U_i \to f^{-1}(U_i)$ keeping the above property. There exists a C^r submersive $\mathfrak{X}$-retraction $q_i \colon p^{-1}(U_i) \to p^{-1}(B_i)$ such that $p \circ q_i = p_i \circ p$ because p is submersive and proper. By p_i and q_i we can define an extension $\tilde{g}_i$ uniquely so that

$$q_i \circ \tilde{g}_i(x, y) = g_i(x, p_i(y)) \quad and \quad p \circ \tilde{g}_i(x, y) = y \quad for \quad (x, y) \in p^{-1}(y_i) \times U_i.$$

By Theorem II we assume B is open in $\mathbf{R}^d$. Replace X and p with graph p and the restriction to graph p of the projection $X \times B \to B$. If r is large enough, then we can apply II.6.7 to $X = \mathrm{graph}\, p$ and $M = S = \varnothing$, and we obtain $C^{r'}$ $\mathfrak{X}$-triviality of p over each stratum of a finite $C^{r'}$ $\mathfrak{X}$-stratification of B. Here by substratifying, we can choose a C^r $\mathfrak{X}$-stratification of B. Hence it suffices to prove the following two facts. Let r_1 be an integer larger than r. First for $p \colon X \to B$, there exists a bounded C^{r_1} $\mathfrak{X}$-submanifold Y of $\mathbf{R}^n$ and a C^r X-diffeomorphism $\varphi\colon Y \to X$ such that $p \circ \varphi$ is of class C^{r_1}. Second, if a C^r X-map $X \to B$ is $C^{r'}$ X-trivial, $r' < r$, then it is C^r X-trivial.

We prove the first fact as follows. By II.5.6 there exist a bounded C^{r_1} X-submanifold Y of $\mathbf{R}^n$ and a C^r X-diffeomorphism $\varphi\colon Y \to X$. We need to modify φ so that $p \circ \varphi$ is of class C^{r_1}. By II.5.2 we have a strong C^{r_1} X-approximation p_1 of $p \circ \varphi\colon Y \to B$ which is proper and C^{r_1} submersive. As in the proof of II.5.5, we construct an open X-neighborhood U of the set $\cup_{b \in B} b \times p^{-1}(b)$ in $B \times X$ and a C^r X-map $\theta\colon U \to X$ such that for each $b \in B$, $\theta(b, \cdot)\colon U_b \to p^{-1}(b)$ is a C^r submersive retraction, where U_b is defined by $b \times U_b = b \times X \cap U$. Set

$$\varphi_1(y) = \theta(p_1(y), \varphi(y)) \quad \text{for} \quad y \in Y.$$

Then φ_1 is a C^r X-diffeomorphism from Y to X, and $p \circ \varphi_1$ is of class C^{r_1} because $p \circ \varphi_1 = p_1$.

The second fact is immediate by II.5.2. $\qquad\qquad\qquad\qquad\qquad\square$

Proof of II.6.7. We proceed by induction on the k. The case $k = 0$ is immediate by the undermentioned Claim 1. So we assume that $k > 0$ and the theorem for $k - 1$ holds true.

Claim 1. Shrink M, stratify B and transform M by some C^{r-k-4} X-diffeomorphism of $\mathbf{R}^{n-d} \times B$ which preserves each fibre of q. We can assume

$$S = 0^{n-d} \times B \subset \mathbf{R}^{n-d} \times B,$$

M is an open neighborhood of S in $\mathbf{R}^{k+1} \times 0^{n-d-k-1} \times B$, and

$$f(x_1, \dots, x_{k+1}, 0, x_{n-d+1}, \dots, x_n) = \sum_{i=1}^{k+1} \pm x_i^2$$

$$\text{for} \quad (x_1, \dots, x_{k+1}, 0, x_{n-d+1}, \dots, x_n) \in M.$$

Proof of Claim 1. By (II.1.17) q_S admits a finite C^r X-stratification. Hence, by the assumption that $\sharp S_b = 0$ or 1, there exists a finite C^r X-stratification $\{B_i\}$ of B such that for each i, either $q_S|_{q_S^{-1}(B_i)}\colon q_S^{-1}(B_i) \to B_i$ is a C^r diffeomorphism or $q_S^{-1}(B_i) = \varnothing$. Thus we assume q_S is a C^r diffeomorphism (the case $S = \varnothing$ is easier to prove and hence omitted). Shrink M if necessary. Then q_M is a C^r submersion because q_S is a diffeomorphism. The equality $S = 0^{n-d} \times B$ follows after the transformation of $\mathbf{R}^{n-d} \times B$ by the C^r X-diffeomorphism:

$$\mathbf{R}^{n-d} \times B \ni (x, b) \longrightarrow (x - q_S^{-1}(b), b) \in \mathbf{R}^{n-d} \times B.$$

For each $s \in S$, let $v(s) \in G_{n,k+1}$ denote the tangent space of $M_{q(s)}$ at s. Then $v \colon S \to G_{n,k+1}$ is a C^{r-1} X-map. We reduce the problem to the case where for each $s \in S$, $v(s)$ equals $\mathbf{R}^{k+1} \times 0^{n-k-1}$ as follows. By stratifying B, we assume there exists a linear $(k+1)$-subspace of $\mathbf{R}^n$, spanned by some of the x_i-axes, $i = 1, \ldots, n-d$, (say, $x_1, \ldots, x_{k+1}$) of $\mathbf{R}^n$ such that for each $s \in S$, the restriction to $v(s)$ of the orthogonal projection of $\mathbf{R}^n$ onto the space is bijective. For each $s \in S$, set $q(s) = b$, and let ρ_b denote the linear isomorphism of $\mathbf{R}^{n-d} \times 0^d$ such that $\rho_b|_{v(s)}$ is the above bijection and $\rho_b = \mathrm{id}$ on $0^{k+1} \times \mathbf{R}^{n-d-k-1} \times 0^d$. The C^{r-1} X-diffeomorphism ρ of $\mathbf{R}^{n-d} \times B$, given by

$$\rho(x,b) = \rho_b(x,0) + (0,b) \quad \text{for} \quad (x,b) \in \mathbf{R}^{n-d} \times B,$$

carries each $v(s) + (0,b)$ to $\mathbf{R}^{k+1} \times 0^{n-d-k-1} \times b$. Hence we may suppose $v(s) = \mathbf{R}^{k+1} \times 0^{n-k-1}$.

Moreover, we can suppose $M \subset \mathbf{R}^{k+1} \times 0^{n-d-k-1} \times B$ for the following reason. Let M' denote the image of M under the projection $\mathbf{R}^{k+1} \times \mathbf{R}^{n-d-k-1} \times B \to \mathbf{R}^{k+1} \times B$. Shrink M. Then M' is open in $\mathbf{R}^{k+1} \times B$ and there exists a C^{r-1} X-map $\xi \colon M' \to \mathbf{R}^{n-d-k-1}$ such that

$$M = \{(x, \xi(x,b), b) \colon (x,b) \in M'\}.$$

Let η be the C^{r-1} X-function on $\mathbf{R}^{k+1} \times B$ which is equal to 0 outside of a small X-neighborhood of $0^{k+1} \times B$ in $\mathbf{R}^{k+1} \times B$ and equal to 1 on a smaller one. Move $\mathbf{R}^{n-d} \times B$ by the C^{r-1} X-diffeomorphism:

$$\mathbf{R}^{k+1} \times \mathbf{R}^{n-d-k-1} \times B \ni (x,y,b) \longrightarrow$$

$$\begin{cases} (x, y - \eta(x,b)\xi(x,b), b) \in \mathbf{R}^{k+1} \times \mathbf{R}^{n-d-k-1} \times B & \text{if } (x,b) \in M' \\ (x, y, b) \in \mathbf{R}^{k+1} \times \mathbf{R}^{n-d-k-1} \times B & \text{if } (x,b) \notin M', \end{cases}$$

and shrink M once more. Then M becomes included in $\mathbf{R}^{k+1} \times 0^{n-d-k-1} \times B$. It remains to reduce the problem to the case where $f = \sum_{i=1}^{k+1} \pm x_i^2$. Set

$$F_1(x_1, \ldots, x_{k+1}, 0, x_{n-d+1}, \ldots, x_n)$$

$$= \sum_{1 \le i,j \le k+1} \frac{\partial^2 f}{\partial x_i \partial x_j}(0, \ldots, 0, x_{n-d+1}, \ldots, x_n) x_i x_j / 2 \quad \text{and}$$

$$F_2(x_1, \ldots, x_{k+1}, 0, x_{n-d+1}, \ldots, x_n)$$

$$= \sum_{\substack{\alpha = (\alpha_1, \ldots, \alpha_{k+1}, 0, \ldots, 0) \in \mathbf{N}^{k+d+1} \\ |\alpha| \le k+1}} D^\alpha f(0, \ldots, 0, x_{n-d+1}, \ldots, x_n) x^\alpha / \alpha!$$

$$\text{for} \quad (x_1, \ldots, 0, \ldots x_n) \in \mathbf{R}^{k+1} \times 0^{n-d-k-1} \times B.$$

Then F_1 and F_2 are C^{r-3} and C^{r-k-2} 𝔛-functions on $\mathbf{R}^{k+1} \times 0^{n-d-k-1} \times B$ respectively. Note that if we fix $(x_{n-d+1}, \dots, x_n)$, the $(x_1, \dots, x_{k+1})$-functions F_1 and F_2 are nondegenerate. First we can assume F_1 is of the form $\sum_{i=1}^{k+1} \pm x_i^2$ because by Assertion 1 below, for some C^{r-3} 𝔛-stratification $\{B_i\}$ of B, there exists a C^{r-3} 𝔛-diffeomorphism τ of each $\mathbf{R}^{k+1} \times 0^{n-d-k-1} \times B_i$ of the form:

$$\tau(x_1, \dots, x_{k+1}, 0, x_{n-d+1}, \dots, x_n) = (\tau'(x_1, \dots, x_n), 0, x_{n-d+1}, \dots, x_n)$$

such that $\tau = \mathrm{id}$ on S and $F_1 \circ \tau$ is of the form $\sum_{i=1}^{k+1} \pm x_i^2$.

Assertion 1. Set

$$Z = \{a = (a_{ij})_{1 \le i,j \le n} \in \mathbf{R}^{n^2} : \sum a_{i,j} t_i t_j \text{ nondegenerate}\}.$$

Let a function g on $\mathbf{R}^n \times Z$ be given by

$$g(t, a) = \sum a_{i,j} t_i t_j \quad \text{for} \quad (t, a) = (t_i, a_{i,j}) \in \mathbf{R}^n \times Z.$$

Then Z admits a finite Nash ($= C^\omega$ semialgebraic) stratification $\{Z_{i'}\}$ such that for each i', there exists a Nash diffeomorphism $\tau_{i'}$ of $\mathbf{R}^n \times Z_{i'}$ of the form $\tau_{i'}(t, a) = (\tau'_{i'}(t, a), a)$ such that $\tau_{i'} = \mathrm{id}$ on $0^n \times Z_{i'}$ and $g \circ \tau_{i'}$ is an x-function (i.e., does not depend on $Z_{i'}$).

Proof of Assertion 1. By some stratification $\{Z_{i'}\}$ of Z and by some diffeomorphism of each $\mathbf{R}^n \times Z_{i'}$ of the same form as the above $\tau_{i'}$, we can assume $a_{1,1} = \pm 1$. We have

$$\sum a_{i,j} t_i t_j = \pm(t_1 \pm \sum_{i \ne 1} (a_{1,i} + a_{i,1}) t_i/2)^2 + \sum_{2 \le i,j \le n} b_{i,j} t_i t_j$$

for some $(a_{i,j})_{1 \le i,j \le n}$-polynomials $b_{i,j}$, $2 \le i, j \le n$. The assertion follows easily by induction on n.

Continued proof of Claim 1. Next, by Assertion 2 below we can assume F_2 is of the form:

$$\sum_{I=1}^{k+1} \pm x_i^2 + \sum_{i=1}^{k+1} f_i x_i^2$$

for some C^{r-k-2} 𝔛-functions f_i on M which vanish on S.

Assertion 2. Let n' be a positive integer. Set

$$A = \{\alpha \in \mathbf{N}^n : 3 \le |\alpha| \le n'\} \quad \text{and} \quad \sharp A = n''.$$

Let a function h on $\mathbf{R}^n \times \mathbf{R}^{n''}$ be given by

$$h(x, a_\alpha) = \sum_{i=1}^n \pm x_i^2 + \sum_{\alpha \in A} a_\alpha x^\alpha.$$

There exists a C^ω semialgebraic diffeomorphism τ of $\mathbf{R}^n \times \mathbf{R}^{n''}$ of the form $\tau(x, a_\alpha) = (\tau'(x, a_\alpha), a_\alpha)$ such that $\tau = \mathrm{id}$ on $0^n \times \mathbf{R}^{n''}$ and

$$h \circ \tau(x, a_\alpha) = \sum_{i=1}^n \pm x_i^2 + \sum_{i=1}^n f_i x_i^2$$

for some C^ω semialgebraic functions f_i on $\mathbf{R}^n \times \mathbf{R}^{n''}$ which vanish on $0^n \times \mathbf{R}^{n''}$.

Proof of Assertion 2. It suffices to construct a Nash diffeomorphism τ_1 of $\mathbf{R}^n \times \mathbf{R}^{n''}$ of the form $\tau_1(x, a_\alpha) = (\tau_1'(x, a_\alpha), a_\alpha)$ so that $\tau_1 = \mathrm{id}$ on $0^n \times \mathbf{R}^{n''}$ and

$$h \circ \tau_1 = \sum_{i=1}^n \pm x_i^2 + \sum_{i=1}^n f_i x_i^2 + h_1, \tag{$*$}$$

where f_i are the same as above and h_1 is a Nash function on $\mathbf{R}^n \times \mathbf{R}^{n''}$ which is $\max\{2, n\}$-*flat* at $0^n \times \mathbf{R}^{n''}$ (i.e., $D^\beta h_1 = 0$ on $0^n \times \mathbf{R}^{n''}$ for all $\beta \in \mathbf{N}^{n+n''}$ with $|\beta| \le \max\{2, n\}$), because such h_1 can be described as $\sum_{i=1}^n f_i' x_i^2$ for some f_i' with the same properties as f_i.

First we deal with only the variable x_1. Let h be of the form:

$$\varphi_1 x_1^2 \pm x_1^2 + 2\varphi_2 x_1 + \varphi_3, \tag{$**$}$$

where φ_i are Nash functions on $\mathbf{R}^n \times \mathbf{R}^{n''}$ such that $\varphi_1 = \varphi_2 = \varphi_3 = 0$ on $0^n \times \mathbf{R}^{n''}$ and φ_2 and φ_3 do not depend on x_1. We can assume φ_2 is $\max\{1, n-1\}$-flat at $0^n \times \mathbf{R}^{n''}$. We prove this by an induction method as follows. By the definition of h, φ_2 is 1-flat at $0^n \times \mathbf{R}^{n''}$. Assume φ_2 is m-flat at $0^n \times \mathbf{R}^{n''}$ for some integer m. We want to reduce the problem to the case where φ_2 is $(m+1)$-flat there. Define a Nash diffeomorphism τ_2 of $\mathbf{R}^n \times \mathbf{R}^{n''}$ by

$$\tau_2(x, a_\alpha) = (x_1 \mp \varphi_2, x_2, \ldots, x_n, a_\alpha).$$

Then

$$h \circ \tau_2 = \varphi_1(x_1 \mp \varphi_2, x_2, \ldots)(x_1 \mp \varphi_2)^2 \pm x_1^2 + \varphi_4 = \varphi_5 x_1^2 \pm x_1^2 + 2\varphi_6 x_1 + \varphi_7,$$

where $\varphi_4, \ldots, \varphi_7$ are C^ω semialgebraic functions on $\mathbf{R}^n \times \mathbf{R}^{n''}$ such that $\varphi_4 = \cdots = \varphi_7 = 0$ on $0^n \times \mathbf{R}^{n''}$ and φ_4, φ_6 and φ_7 do not depend on x_1. It is easy to calculate that φ_6 is $(m+1)$-flat at $0 \times \mathbf{R}^{n''}$. Hence we assume the φ_2 in $(**)$ is $\max\{1, n-1\}$-flat at $0^n \times \mathbf{R}^{n''}$. If $n = 1$, then $(*)$ holds. So assume $n > 1$.

Next we consider the variable x_2. We modify φ_3 in $(**)$ by a Nash diffeomorphism τ_3 of $\mathbf{R}^{n-1} \times \mathbf{R}^{n''}$ with $\tau_3 = \mathrm{id}$ on $0^{n-1} \times \mathbf{R}^{n''}$ in the same way, so that

$$\varphi_3 \circ \tau_3 = \varphi_8 x_2^2 \pm x_2^2 + 2\varphi_9 x_2 + \varphi_{10},$$

where φ_8, φ_9 and φ_{10} are Nash functions on $\mathbf{R}^{n-1} \times \mathbf{R}^{n''}$, $\varphi_8 = \varphi_9 = \varphi_{10} = 0$ on $0^{n-1} \times \mathbf{R}^{n''}$, φ_9 and φ_{10} do not depend on x_2, and φ_9 is $\{1, n-1\}$-flat at $0^{n-1} \times \mathbf{R}^{n''}$. Then

$$h \circ \tilde{\tau}_3 = \varphi_1 \circ \tilde{\tau}_3 \cdot x_1^2 \pm x_1^2 + 2\varphi_2 \circ \tilde{\tau}_3 \cdot x_1 + \varphi_3 \circ \tilde{\tau}_3,$$

where

$$\tilde{\tau}_3(x_1, \ldots, x_n, a_\alpha) = (x_1, \tau_3(x_2, \ldots, x_n, a_\alpha)) \text{ for } (x_1, \ldots, x_n, a_\alpha) \in \mathbf{R}^n \times \mathbf{R}^{n''},$$

and $\varphi_2 \circ \tilde{\tau}_3$ continues to be $\max\{1, n-1\}$-flat at $0^{n-1} \times \mathbf{R}^{n''}$. In this way, by induction on n, we reduce h to the required form $(*)$, and Assertion 2 holds.

Continued proof of Claim 1. Assume F_2 is of the above form. Then

$$f = \sum_{i=1}^{k+1} \pm x_i^2 + \sum_{i=1}^{k+1} f_i x_i^2 + \sum_{\substack{\alpha=(\alpha_1, \ldots, \alpha_{k+1}, 0, \ldots, 0) \in \mathbf{N}^{k+d+1} \\ |\alpha| = k+2}} f_\alpha x^\alpha$$

and hence

$$= \sum_{i=1}^{k+1} \pm x_i^2 + \sum_{i=1}^{k+1} f_i' x_i^2$$

for some C^{r-k-3} X-functions f_α and f_i' on M with $f_i' = 0$ on S. Then

$$\sum_{i=1}^{k+1} \pm x_i^2 + \sum_{i=1}^{k+1} f_i' x_i^2 = \sum \pm \left[\sqrt{1 \mp f_i'} x_i \right]^2.$$

Thus if we shrink M, there exists a C^{r-k-3} X-imbedding τ of M into $\mathbf{R}^{k+1} \times 0^{n-d-k-1} \times B$ of the form $\tau(x, 0, b) = (\tau'(x, b), 0, b)$ such that $f \circ \tau = \sum_{i=1}^{k+1} \pm x_i^2$. Shrink M once more. By Assertion 3 below we can extend τ to $\mathbf{R}^{k+1} \times 0^{n-d-k-1} \times B$. Hence Claim 1 follows.

Assertion 3. Let M be an open $\mathfrak{X}$-neighborhood of $0^{k+1} \times 0^{n-d-k-1} \times B$ in $\mathbf{R}^{k+1} \times 0^{n-d-k-1} \times B$ and let τ be a C^r $\mathfrak{X}$-imbedding of M into $\mathbf{R}^{k+1} \times 0^{n-d-k-1} \times B$ such that $\tau = $ id on $0^{k+1} \times 0^{n-d-k-1} \times B$ and $\tau(M_b) \subset \mathbf{R}^{k+1} \times 0^{n-d-k-1} \times b$ for $b \in B$. Keeping the last property and shrinking M, we can extend τ to a C^{r-1} $\mathfrak{X}$-diffeomorphism of $\mathbf{R}^{k+1} \times 0^{n-d-k-1} \times B$.

Proof of Assertion 3. Define a C^{r-1} $\mathfrak{X}$-diffeomorphism τ_0 of $\mathbf{R}^{k+1} \times 0^{n-d-k-1} \times B$ by

$$\tau_0(x_1, \ldots, x_{k+1}, 0, b) = \left(\sum_{i=1}^{k+1} \frac{\partial \tau}{\partial x_i}(0, 0, b) x_i, 0, b \right)$$

for $(x_1, \ldots, x_{k+1}, 0, b) \in \mathbf{R}^{k+1} \times 0^{n-d-k-1} \times B$.

Let θ be a non-negative C^{r-1} $\mathfrak{X}$-function on $\mathbf{R}$ which is equal to 0 outside of a neighborhood of 0 and equal to 1 on a smaller one. For each $t > 0$, set

$$\tau_t(x, 0, b) = \begin{cases} \theta(|x|/t)\tau(x, 0, b) + (1 - \theta(|x|/t))\tau_0(x, 0, b) & \text{for} \quad (x, 0, b) \in M \\ \tau_0(x, 0, b) & \text{for} \quad (x, 0, b) \in \mathbf{R}^{k+1} \times 0^{n-d-k-1} \times B - M. \end{cases}$$

For each $b \in B$ there exists $t > 0$ such that $\tau_t(\cdot, 0, b)$ is a well-defined C^{r-1} $\mathfrak{X}$-diffeomorphism from $\mathbf{R}^{k+1}$ to $\mathbf{R}^{k+1} \times 0^{n-d-k-1} \times b$. Let T_b denote the set of such t. It includes some open interval $]0, \varepsilon[$, and the set $T = \cup_{b \in B} b \times T_b \subset B \times \mathbf{R}$ is an $\mathfrak{X}$-set. There exists a C^{r-1} $\mathfrak{X}$-function ξ on B whose graph is included in T. Then $\tau_{\xi(b)}(x, 0, b)$ is the required extension of τ.

Note. In Claim 1, we have transformed $\mathbf{R}^{n-d} \times B$ by a C^{r-k-4} diffeomorphism. Hence $X - S$ is now of class C^{r-k-4}. For simplicity of notation, we assume it is of class C^r.

Claim 2. We can suppose there exists a bounded C^r $\mathfrak{X}$-function ψ on X such that $\psi = x_1$ on $M \cap X$, and for each $b \in B$, $\psi|_{X_b - S}$ is a Morse function.

Proof of Claim 2. Let ψ be defined to be x_1 on $M \cap X$ and a bounded C^r $\mathfrak{X}$-function on $X - S$. It suffices to approximate ψ on $X - \overline{M}$ by a C^r $\mathfrak{X}$-function so that the Morse condition is satisfied, because $\psi|_{X_b - S}$ is C^r regular at each point of $\overline{M_b} - S$ and any strong approximation keeps the above property. Hence we assume $S = M = \varnothing$, but here we cannot require $q_X : X \to B$ to be proper. We will apply the transversality theorem II.5.4. In II.5.4, set $Y = \mathbf{R}$ and $r' = 2$ and define $\{X_i\}$ to be a finite C^{r-2} Whitney $\mathfrak{X}$-stratification of the set $Q \subset J^2(X, \mathbf{R})$, given by

$$Q \cap J^2_{x,y}(X, \mathbf{R}) = \{J^2\lambda(x) : \lambda \in C^r_{\mathfrak{X}}(X, \mathbf{R}), \; \lambda(x) = y,$$

$$x \text{ is a degenerate critical point of } \lambda|_{X_{q(x)}}\} \quad \text{for} \quad (x, y) \in X \times \mathbf{R}.$$

We see that Q is an $\mathfrak{X}$-set of codimension $k + 1$. Choose r so that $r > k + d - (k + 1) + 2 = d + 1$. By II.5.4 we can assume $(J^2\psi)^{-1}Q$ is of dimension $\leq k + d - \mathrm{codim}\, Q = d - 1 < \dim B$. Hence by stratifying B, we suppose $(J^2\psi)^{-1}Q = \varnothing$, which implies that $\psi|_{X_b - S}$ is a Morse function.

Claim 3. In Claim 2, we can assume that for each $b \in B$, 0 is not a critical value of $\psi|_{X_b - S}$.

Proof of Claim 3. Add the set:

$$\bigcup_{x \in X} \{\lambda \in J^2_{x,0}(X, \mathbf{R}) \colon \lambda \text{ are critical at } x\}$$

to Q in the proof of Claim 2. Claim 3 follows because this set is an $\mathfrak{X}$-set of dimension $k + 1$.

Note. In Claim 2, $\psi = 0$ on S.

Claim 4. For each $b \in B$, let Z_b denote the critical point set of $\psi|_{X_b - S}$. Set $Z = \cup_{b \in B} Z_b$. We can assume that ψ is constant on each connected component of Z and the values are distinct from each other.

Proof of Claim 4. Clearly Z is an $\mathfrak{X}$-set and $q|_Z \colon Z \to B$ is a finite-to-one map. Hence by stratifying B, we assume $q|_Z$ is a C^r $\mathfrak{X}$-covering map. Let $Y \subset B \times \mathbf{R}$ denote the image of Z under the map $(q_X, \psi) \colon X \to B \times \mathbf{R}$. Then Y is an $\mathfrak{X}$-set and the restriction to Y of the projection $B \times \mathbf{R} \to B$ can be assumed to be a C^r covering map. Under this assumption we modify ψ so that it is constant on each connected component of Z as follows. There exists a C^r $\mathfrak{X}$-function η on $B \times \mathbf{R}$ such that $\eta(b, t) = t$ on a neighborhood of $B \times 0$, for each $b \in B$, $\eta|_{b \times \mathbf{R}}$ is a C^r $\mathfrak{X}$-diffeomorphism onto $\mathbf{R}$, and η is constant on each connected component of Y. The function $\eta \circ (q_X, \psi)$ on X keeps the properties of ψ and, moreover, $\eta \circ (q_X, \psi)$ is constant on each connected component of Z. Hence it suffices to replace ψ with $\eta \circ (q_X, \psi)$.

It remains to reduce the problem to the case where for any two connected components Z' and Z'' of Z, $\psi(Z') \neq \psi(Z'')$. Assume $\psi(Z') = \psi(Z'')$. We will modify ψ in a neighborhood of Z'. Let $U_1 \subset U_2$ be small open $\mathfrak{X}$-neighborhoods of Z' in X such that $\overline{U_1} \cap X \subset U_2$ and $\overline{U_2} \cap (Z - Z') = \varnothing$. Let φ be a non-negative C^r $\mathfrak{X}$-function on X such that $\varphi = 0$ outside of U_2 and $\varphi = 1$ on U_1. Let φ_1 be a positive C^r $\mathfrak{X}$-function on X so small that $(J^2(\psi + \varphi_1\varphi))^{-1}Q = \varnothing$, where Q is given in the proof of Claims 2 and 3. (This is possible by the openness property in II.5.4.) Then $\psi + \varphi_1\varphi$ keeps the same properties as ψ except that $\psi(Z')$ consists of one point and $\psi(Z') = \psi(Z'')$, and it has the property

$$(\psi + \varphi_1\varphi)(Z') \cap (\psi + \varphi_1\varphi)(Z - Z') = \varnothing.$$

Hence by replacing ψ with $\psi + \varphi_1\varphi$ and by repeating the above arguments, we reduce the problem to the case where $\psi(Z')$ is a point and is distinct from $\psi(Z'')$. Thus we obtain Claim 4.

Claim 5. Let Y be a connected component of Z. Set $\tilde{Y} = \psi^{-1}(\psi(Y))$, and let b_0 be a point of B. We can assume there exist an open $\mathfrak{X}$-neighborhood U of $\tilde{Y}$ in X and a $C^{r'}$ $\mathfrak{X}$-imbedding $u\colon U \to U_0 \times B$ of the form $u = (u', q)$ such that $u'|_{U_0} = \mathrm{id}$ and $\psi \circ u' = \psi|_U$, where $U_0 = U \cap X_{b_0}$ and r' is some positive integer $< r$.

Proof of Claim 5. Note that $\dim \tilde{Y}_0 = k-1$, where $\tilde{Y}_0 = \tilde{Y} \cap X_{b_0}$, and that B, Y, $\tilde{Y}$, $X - S$ and $(\psi - \psi(Y))|_{X-S}$ satisfy the conditions of B, S, X, M and f in II.6.7. Hence by the induction hypothesis we assume there exist an open $\mathfrak{X}$-neighborhood V of Y in X and a $C^{r'}$ $\mathfrak{X}$-imbedding $v\colon \tilde{Y} \cup V \to (\tilde{Y}_0 \cup V_0) \times B$ of the form $v = (v', q)$ such that

$$v(\tilde{Y}) = \tilde{Y}_0 \times B, \qquad v(V) \subset V_0 \times B,$$
$$v'|_{\tilde{Y}_0 \cup V_0} = \mathrm{id} \quad \text{and} \quad \psi \circ v' = \psi \quad \text{on} \quad V,$$

where $V_0 = V \cap X_{b_0}$. It suffices to extend v to an $\mathfrak{X}$-neighborhood of $\tilde{Y} - Y$ in X. First we will find an open $\mathfrak{X}$-neighborhood W of $\tilde{Y} - Y$ in X and a $C^{r'}$ $\mathfrak{X}$-imbedding $\eta\colon W \to W_0 \times B$ with the required properties of u, where $W_0 = W \cap X_{b_0}$. For this we identify an $\mathfrak{X}$-neighborhood of $\tilde{Y} - Y$ in X with a part of the trivial line bundle over $\tilde{Y} - Y$ by the following assertion.

Assertion 4. There exist an open $\mathfrak{X}$-neighborhood W of $\tilde{Y} - Y$ in X and a C^{r-1} $\mathfrak{X}$-imbedding $w\colon W \to (\tilde{Y} - Y) \times \mathbf{R}$ of the form $w = (w', \psi)$ such that $w'|_{\tilde{Y}-Y} = \mathrm{id}$ and $q \circ w' = q$ on W.

Proof of Assertion 4. Choose W so small that for each $x \in W$, there exists uniquely $y \in \tilde{Y}_{q(x)} - Y$ such that

$$\mathrm{dis}(x, \tilde{Y}_{q(x)} - Y) = |x - y|$$

and the correspondence $W \ni x \to y \in \tilde{Y} - Y$ is a C^{r-1} imbedding, where $\tilde{Y}_b = \tilde{Y} \cap X_b$ for $b \in B$. Set $w'(x) = y$. We can shrink W so that w is a C^{r-1} imbedding. The equalities $w'|_{\tilde{Y}-Y} = \mathrm{id}$ and $q \circ w' = q$ on W are clear, which proves the assertion.

Continued proof of Claim 5. Let us regard W as a subset of $(\tilde{Y} - Y) \times \mathbf{R}$ through w. Note that $\psi(y, t) = t$ and $q(y, t) = q(y)$ for $(y, t) \in W$. Shrink W so that if $(y, t) \in W$, then $(v'(y), t) \in W$. We define naturally the $C^{r'}$ $\mathfrak{X}$-imbedding $\eta\colon W \to W_0 \times B$ by

$$\eta(y, t) = (v'(y), t, q(y)) \quad \text{for} \quad (y, t) \in W.$$

We cannot expect $\eta = v$ on $W \cap V$. Hence we modify η and v as follows.

Let O be a small open $\mathfrak{X}$-neighborhood of $\tilde{Y} \cap V - Y$ in X such that

$$v'(O) \subset O_0 = O \cap X_{b_0} \quad \text{and} \quad O \subset V \cap W.$$

By also regarding O as a subset of $(\tilde{Y} - Y) \times \mathbf{R}$ through the ω, we describe $v|_O$ as

$$v(y,t) = (v''(y,t), t, q(y)) \quad \text{for} \quad (y,t) \in O \subset (\tilde{Y} - Y) \times \mathbf{R}.$$

Here $v'' \colon O \to \tilde{Y}_0 - Y$ is a $C^{r'}$ $\mathfrak{X}$-map such that

$$v''(y,t) = y \qquad \text{for} \quad (y,t) \in O \cap (\tilde{Y}_0 - Y) \times \mathbf{R} \ \text{ and}$$
$$v''(y, \psi(Y)) = v'(y) \quad \text{for} \quad y \in \tilde{Y} \cap V - Y.$$

So it suffices to modify v'' on $O - V'$ so that

$$v''(y,t) = v'(y) \quad \text{for} \quad (y,t) \in O - V'$$

for some small $\mathfrak{X}$-neighborhood V' of Y in X with $\overline{V'} \subset V$.

We modify v'' on $O - V'$ as follows. Let $\varepsilon \colon E \to \tilde{Y}_0 - Y$ be a C^{r-1} $\mathfrak{X}$-tubular neighborhood in $X_0 - Y$. Let $\{\xi_V, \xi_W\}$ be a C^r $\mathfrak{X}$-partition of unity subordinate to the covering $\{V, W\}$ of $V \cup W$. Set

$$\theta(y,t) = \varepsilon(\xi_W(y,t)v'(y) + \xi_V(y,t)v''(y,t)) \quad \text{for} \quad (y,t) \in O.$$

Choose sufficiently small O. Then θ is a well-defined C^{r-1} $\mathfrak{X}$-map from O to $\tilde{Y}_0 - Y$ such that

$$\theta(y,t) = v'(y) \quad \text{for} \quad (y,t) \in O - V',$$
$$\theta = v'' \quad \text{on} \quad O \cap V'' \ \text{ and}$$
$$\theta(y, \psi(Y)) = v'(y) = v''(y, \psi(Y)) \quad \text{for} \quad y \in \tilde{Y} \cap V - Y,$$

where V' and V'' are some open $\mathfrak{X}$-neighborhoods of Y in X with $\overline{V''} \cap X \subset V'$ and $\overline{V'} \cap X \subset V$. Replace v'' on $O - V'$ with θ. To be precise, set

$$U = V'' \cup O \cup (W - \overline{V'}) \quad \text{and} \quad u(y,t) = \begin{cases} v(y,t) & \text{on} \quad V'' \\ (\theta(y,t), t, q(y)) & \text{on} \quad O \\ \eta(y,t) & \text{on} \quad W - \overline{V'}, \end{cases}$$

and shrink U so that $u'(U) \subset U_0$ and $u: U \to U_0 \times B$ is a $C^{r'}$ imbedding (which is easy to prove by the equality $v''(y, \psi(Y)) = v'(y)$ for $y \in \tilde{Y} \cap V - Y$). Then we obtain Claim 5.

Claim 6. Set $\tilde{S} = \psi^{-1}(0)$ and let b_0 be a point of B. We can suppose there exist an open $\mathfrak{X}$-neighborhood Σ of $\tilde{S}$ in X and a $C^{r'}$ $\mathfrak{X}$-imbedding $\sigma: \Sigma \cup M \to (\Sigma_0 \cup M_0) \times B$ of the form $\sigma = (\sigma', q)$ such that

$$\sigma(\Sigma) \subset \Sigma_0 \times B, \qquad \sigma(M) \subset M_0 \times B,$$
$$\sigma'|_{\Sigma_0 \cup M_0} = \mathrm{id}, \qquad \psi \circ \sigma' = \psi|_{\Sigma \cup M} \quad \text{and} \quad f \circ \sigma' = f \quad \text{on } M,$$

where $\Sigma_0 = \Sigma \cap X_{b_0}$, $M_0 = M_{b_0}$, and r' is some positive integer $< r$.

Proof of Claim 6. We proceed as in the proof of Claim 5. Apply the induction hypothesis to B, S, $\tilde{S}$, $M' = M \cap \{x_1 = 0\}$ and $f|_{M'}$. We can assume there exists a $C^{r'}$ $\mathfrak{X}$-imbedding $\chi: \tilde{S} \cup M' \to (\tilde{S}_0 \cup M'_0) \times B$ of the form $\chi = (\chi', q)$ such that

$$\chi(\tilde{S}) = \tilde{S}_0 \times B, \qquad \chi(M') \subset M'_0 \times B,$$
$$\chi'|_{\tilde{S}_0 \cup M'_0} = \mathrm{id} \quad \text{and} \quad f \circ \chi' = f \quad \text{on } M',$$

where $\tilde{S}_0 = \tilde{S} \cap X_{b_0}$ and $M'_0 = M' \cap M_{b_0}$. Here, if χ is first defined on M' so that

$$\chi(M') \subset M'_0 \times B, \qquad \chi'|_{M'_0} = \mathrm{id} \quad \text{and} \quad f \circ \chi' = f \quad \text{on } M',$$

then by shrinking M' and by stratifying B, we can extend χ to $\tilde{S}$. Hence it is possible to let

$$\chi'(0, x_2, \dots, x_{k+1}, 0, b) = (0, x_2, \dots, x_{k+1}, 0, b_0)$$
$$\text{for} \quad (0, x_2, \dots, x_{k+1}, 0, b) \in M' \subset 0 \times \mathbf{R}^k \times 0^{n-d-k-1} \times B.$$

We can extend χ to an imbedding $M \to M_0 \times B$, $M_0 = M_{b_0}$, by setting

$$\chi'(x, 0, b) = (x, 0, b_0) \quad \text{for} \quad (x, 0, b) \in M \subset \mathbf{R}^{k+1} \times 0^{n-d-k-1} \times B.$$

Furthermore, we extend χ to an open $\mathfrak{X}$-neighborhood Σ of $\tilde{S}$ in X as in the proof of Claim 5, so that the required properties are satisfied. We do not repeat that proof.

Claim 7. Let $I =]i_1, i_2[$ be a bounded connected component of $\mathbf{R} - 0 - \{\text{critical values of } \psi|_{X-S}\}$, and (b_0, t_0) a point of $B \times I$. Set $\Lambda = \psi^{-1}(I)$,

$\Lambda_{0,0} = X_{b_0} \cap \psi^{-1}(t_0)$. We can assume there exists a $C^{r'}$ $\mathfrak{X}$-diffeomorphism $\lambda \colon \Lambda \to \Lambda_{0,0} \times B \times I$ of the form (λ', q, ψ) such that $\lambda'|_{\Lambda_{0,0}} = \mathrm{id}$, where r' is some positive integer $< r$.

Proof of Claim 7. Let an $\mathfrak{X}$-map $p \colon \Lambda \to B \times I$ be given by $p = (q, \psi)$. Then p is a proper C^r submersion, $p^{-1}(b_0, t_0) = \Lambda_{0,0}$ and $\dim \Lambda_{0,0} = k - 1$. Hence, as in the proof that II.6.7 implies II.6.5, Claim 7 follows from II.6.7 for $k - 1$.

By assuming Claims 5, 6 and 7, we will define the $C^{r'}$ $\mathfrak{X}$-imbedding $\pi \colon X \cup M \to (X_0 \cup M_0) \times B$ with the required properties in II.6.7, where $X_0 = X_{b_0}$, $M_0 = M_{b_0}$ and b_0 is a point of B. In Claim 6, the π was already defined on M. Hence it remains to define it on X. Let $I =]i_1, i_2[$, $t_0 \in I$, Λ, $\Lambda_{0,0}$ and $\lambda = (\lambda', q, \psi)$ be the same as in Claim 7. By Claims 5 and 6 we have an $\mathfrak{X}$-neighborhood Ω of $\psi^{-1}\{i_1, i_2\}$ in X and a $C^{r'}$ $\mathfrak{X}$-imbedding $\omega \colon \Omega \to \Omega_0 \times B$ of the form $\omega = (\omega', q)$ such that $\omega'|_{\Omega_0} = \mathrm{id}$ and $\psi \circ \omega' = \psi|_{\Omega}$, where $\Omega_0 = \Omega \cap X_{b_0}$. Set $\Delta = \Omega \cap \Lambda$, $\Delta_0 = \Omega_0 \cap \Lambda$ and $\delta = \omega|_\Delta$. Then δ is a $C^{r'}$ $\mathfrak{X}$-imbedding of Δ into $\Delta_0 \times B$ of the form $\delta = (\delta', q)$ such that $\delta'|_{\Delta_0} = \mathrm{id}$ and $\psi \circ \delta' = \psi|_\Delta$. We need to extend δ to an imbedding $\Lambda \to \Lambda_0 \times B$ keeping these properties, where $\Lambda_0 = \Lambda \cap X_{b_0}$.

Identify Λ with $\Lambda_{0,0} \times B \times I$ through λ. Since $\Lambda_{0,0}$ is compact, by shrinking Ω, we can assume $\Delta = \Lambda_{0,0} \times L$ for

$$L = \{(b, t) \in B \times I \colon t \in]i_1, i_1 + h(b)[\cup]i_2 - h(b), i_2[\},$$

where h is a positive $\mathfrak{X}$-function on B. Describe δ with the coordinate system of $\Lambda_{0,0} \times B \times I$. Then $\delta(x, b, t)$ for $(x, b, t) \in \Delta$ is of the form $(\delta''(x, b, t), b, t)$, where δ'' is a $C^{r'}$ $\mathfrak{X}$-map from Δ to $\Lambda_{0,0}$ such that $\delta''(\cdot, b_0, t) = \mathrm{id}$ for each $(b_0, t) \in L$ and $\delta''(\cdot, b, t)$ is a $C^{r'}$ $\mathfrak{X}$-diffeomorphism of $\Lambda_{0,0}$ for each $(b, t) \in L$.

It suffices to extend $\delta''|_{\Lambda_{0,0} \times L/2}$ to $\Lambda_{0,0} \times B \times I$ so that $\delta''(\cdot, b_0, t) = \mathrm{id}$ for each $t \in I$, and $\delta''(\cdot, b, t)$ is a $C^{r'}$ $\mathfrak{X}$-diffeomorphism of $\Lambda_{0,0}$ for each $(b, t) \in B \times I$, where $L/2$ is defined by $h/2$ in the same way as L was defined by h. Obviously we set

$$\delta''(x, b_0, t) = x \quad \text{for} \quad (x, t) \in \Lambda_{0,0} \times I.$$

By Theorem II we can assume B is $\mathfrak{X}$-diffeomorphic to an open simplex. It is easy to construct a $C^{r'}$ $\mathfrak{X}$-map $\theta \colon B \times I \to B \times I$ such that

$$\mathrm{Im}\,\theta \subset L \cup b_0 \times I \quad \text{and} \quad \theta = \mathrm{id} \quad \text{on} \quad L/2 \cup b_0 \times I.$$

Define δ'' on $\Lambda_{0,0} \times B \times I$ by setting

$$\delta''(x, b, t) = \delta''(x, \theta(b, t)) \quad \text{for} \quad (x, b, t) \in \Lambda_{0,0} \times B \times I.$$

Then δ'' fulfills the requirements. By repeating these arguments for each connected component of $\mathbf{R} - 0 - \{\text{critical values of } \psi|_{X-S}\}$, we construct the $C^{r'}$ $\mathfrak{X}$-imbedding $\pi \colon X \cup M \to (X_0 \cup M_0) \times B$.

It remains only to consider the case where π is first given on M so that

$$\pi(M) \subset M_0 \times B, \qquad \pi'|_{M_0} = \text{id} \quad \text{and} \quad f \circ \pi' = f \quad \text{on} \quad M.$$

The reduction of f to the form $\sum_{i=1}^{k+1} \pm x_i^2$ (Claim 1) is easy to prove as follows. As in the proof of Claim 1, we can assume $S = 0^{n-d} \times B$ and $M \subset \mathbf{R}^{k+1} \times 0^{n-d-k-1} \times B$. By Assertion 3, by shrinking M, we can extend π to a C^r $\mathfrak{X}$-diffeomorphism $\hat{\pi}$ from $\mathbf{R}^{k+1} \times 0^{n-d-k-1} \times B$ to $(\mathbf{R}^{k+1} \times 0^{n-d-k-1} \times b_0) \times B$ which keeps the properties that $\hat{\pi}$ is of the form $(\hat{\pi}', q)$ and $\hat{\pi}'|_{\mathbf{R}^{k+1} \times 0 \times b_0} = \text{id}$. Transform $\mathbf{R}^{n-d} \times B$ by the composite of

$$\mathbf{R}^{k+1} \times \mathbf{R}^{n-d-k-1} \times B \ni (x', x'', b)$$
$$\longrightarrow \hat{\pi}(x', 0, b) + (0, x'', 0, 0) \in \mathbf{R}^{k+1} \times \mathbf{R}^{n-d-k-1} \times b_0 \times B$$

and the projection $\mathbf{R}^{k+1} \times \mathbf{R}^{n-d-k-1} \times b_0 \times B \to \mathbf{R}^{k+1} \times \mathbf{R}^{n-d-k-1} \times B$. We can assume that for each $(x, 0, b) \in M$, $\pi'(x, 0, b) = (x, 0, b_0)$ and the value $f(x, 0, b)$ does not depend on b. It is easy to shrink M and find a $C^{r'}$ $\mathfrak{X}$-diffeomorphism of $\mathbf{R}^{k+1} \times 0^{n-d-k-1} \times b_0$ whose composite with f is of the form $\sum_{i=1}^{k+1} \pm x_i^2$ on M. Extend it naturally to $\mathbf{R}^{k+1} \times \mathbf{R}^{n-d-k-1} \times B$ as above, and transform $\mathbf{R}^{n-d} \times B$ by its inverse map. Then we obtain Claim 1 keeping the equality $\pi'(x, 0, b) = (x, 0, b_0)$ for $(x, 0, b) \in M$. The rest of the proof proceeds in the same way as before. Thus we complete the proof of II.6.7. $\qquad\square$

Proof of II.6.5'. As the problem is local in $\mathbf{R}^n$ with respect to P, by II.2.1 we can assume that P is bounded and contractible. We want to prove that (f, p) is C^r $\mathfrak{X}$-trivial. By II.6.3 and II.6.5, p is C^r $\mathfrak{X}$-trivial. Hence we assume Y is the product of P and a C^r $\mathfrak{X}$-manifold and p is the projection. The theorem follows from the following lemma (the case where $V = W = \varnothing$). $\qquad\square$

Lemma II.6.8. *Let r be a positive integer. Let A, B, C, $V \subset A$, $W \subset B \times C$ be bounded C^r $\mathfrak{X}$-submanifolds of $\mathbf{R}^n$ or of $\mathbf{R}^n \times \mathbf{R}^n$, let $\widehat{W}$ be a locally closed $\mathfrak{X}$-subset of $(\overline{B} - B) \times C$ of the product form $\widehat{W}_1 \times C$, and let $\Psi = (\psi, \psi') \colon A \to B \times C$ be a surjective C^r $\mathfrak{X}$-submersion. Assume that $0 \in C$, C is contractible, V is open in A, W is the intersection of $B \times C$ with an open $\mathfrak{X}$-neighborhood of $\widehat{W}$ in $\overline{B} \times C$, and $\Psi|_{A-V}$ is proper. Also assume that for any contractible open $\mathfrak{X}$-subset B' of B, $\psi|_{\psi^{-1}(B')_0} \colon \psi^{-1}(B')_0 \to B'$ is C^r*

𝔛*-trivial, and there exists a* C^r 𝔛*-map* $\chi\colon V \cup \Psi^{-1}(W) \to (V \cup \Psi^{-1}(W))_0$
such that $\chi(V) = V_0$, $\chi(\Psi^{-1}(W)) = \Psi^{-1}(W)_0$, $(\chi, \psi')\colon V \cup \Psi^{-1}(W) \to$
$(V \cup \Psi^{-1}(W))_0 \times C$ *is a* C^r *imbedding,* $\chi = \mathrm{id}$ *on* $(V \cup \Psi^{-1}(W))_0$ *and*
$\psi \circ \chi = \psi$. *Here for a subset* Z *of* A *and* $c \in C$, Z_c *denotes* $Z \cap \psi'^{-1}(c)$.
Keeping all these properties, we shrink V *and* W *sufficiently. Then we can*
extend χ *to* $A \to A_0$ *so that* $(\chi, \psi')\colon A \to A_0 \times C$ *is a* C^r *diffeomorphism*
and $\psi \circ \chi = \psi$.

Proof. **Case where** B **is a point, that is to say, we ignore** B, W
and ψ**.** By the following Note we can assume $(\chi, \psi')\colon V \to V_0 \times C$ is a C^r
diffeomorphism. We extend χ to $\psi'^{-1}(U_i) \to A_0$ and then to $A \to A_0$ for a
finite open 𝔛-covering $\{U_i\}$ of C. To extend χ to $\psi'^{-1}(U_i)$, we proceed as
in the proof of II.6.7. We can assume A is the interior of a bounded C^r 𝔛-
manifold possibly with boundary $\tilde{A}$ such that ψ', χ are extended to $\tilde{A}$, $V \cup \partial\tilde{A}$
respectively, so that $(\chi, \psi')|_{\partial\tilde{A}}$ is a C^r diffeomorphism onto $\partial\tilde{A}_0 \times C$, and $\tilde{A}_0$
is compact. (First we construct $\tilde{A}_0$ by II.5.7, then, $V \cup \partial\tilde{A}$ by triviality of V,
and finally, $\tilde{A}$ by pasting $V \cup \partial\tilde{A}$ with A.) Shrink V so that we have a non-
singular non-negative C^r 𝔛-function φ on $\overline{V_0}$ with zero set $= \partial\tilde{A}_0$. Extend
φ to be $\varphi \circ \chi$ on $V \cup \partial\tilde{A}$ and $\varphi \circ \chi$ to $\tilde{A}$ so that the extension on A satisfies
the conditions in Claims 2, 3 and 4 in the proof of II.6.7. Then as in Claims
5, 6 and 7 we obtain a finite C^r 𝔛-stratification $\{C_i\}$ of C, and for each i,
a C^r 𝔛-map $\chi_i\colon \psi'^{-1}(C_i) \to \tilde{A}_0$ such that $(\chi_i, \psi')\colon \psi'^{-1}(C_i) \to \tilde{A}_0 \times C_i$ is
a diffeomorphism, and $\chi_i = \chi$ on $V \cap \psi'^{-1}(C_i)$. Here we shrink V and we
use the property that $\tilde{A}_0$ is compact. Next, as in the beginning of the proof
that II.6.7 implies II.6.5, we extend χ_i to a C^r 𝔛-map $\psi'^{-1}(U_i) \to A_0$ for a
small open 𝔛-neighborhood U_i of C_i in C so that $\chi_i = \chi$ on $V \cap \psi'^{-1}(U_i)$.
Here we use the assumption that ψ' is a submersion and $\psi'|_{A-V}$ is proper.
Finally, in the same way as the proof of II.6.3, we paste the extensions χ_i
and obtain a global C^r 𝔛-extension of χ. Thus the case is proved.

Note. There exist open 𝔛-subsets $V_1 \subset V_2 \subset V$ and a C^r 𝔛-map $\chi'\colon V_2 \to$
V_{20} such that $\chi' = \chi$ on V_1, $\psi'|_{A-V_1}$ is proper, and $(\chi', \psi')\colon V_2 \to V_{20} \times C$ is
a C^r diffeomorphism.

Proof of Note. Let A_1 denote the union of $\overline{A_c}$ for all $c \in C$. Here by replacing
A with graph ψ' if necessary, we assume

$$\overline{A_c} \cap \overline{A_{c'}} = \varnothing \quad \text{for} \quad c \neq c' \in C.$$

Since $\psi'|_{A-V}$ is proper, $V \cup (A_1 - A)$ is a neighborhood of $A_1 - A$ in A_1, and
for each $c \in C$, $\chi(V_c) \cup (\overline{A_0} - A_0)$ is a neighborhood of $\overline{A_0} - A_0$ in $\overline{A_0}$. Hence
as above, we regard A as the interior of a C^r 𝔛-manifold with boundary
$\tilde{A}$, and ψ' and χ are extended to $\psi'\colon \tilde{A} \to C$ and $\chi\colon V \cup \partial\tilde{A} \to (V \cup \partial\tilde{A})_0$
respectively, so that ψ' is proper, $\psi'|_{\partial\tilde{A}}$ is C^r 𝔛-trivial and $(\chi, \psi')\colon V \cup \partial\tilde{A} \to$

$(V \cup \partial \tilde{A})_0 \times C$ is a C^r imbedding. Let $\alpha \colon \partial \tilde{A}_0 \times [0,1] \to \tilde{A}_0$ be a C^r $\mathfrak{X}$-collar, i.e., a C^r $\mathfrak{X}$-imbedding such that $\alpha(x,0) = x$ for $x \in \partial \tilde{A}_0$. Then we can assume $V_0 = \alpha(\partial \tilde{A}_0 \times]0, 1/2[)$, and it is easy to find a positive C^r $\mathfrak{X}$-function β on C such that $0 < \beta < 1/2$, and

$$\chi(V_y) \supset \alpha(\partial \tilde{A}_0 \times]0, \beta(y)[) \quad \text{for} \quad y \in C.$$

Let $\gamma = (\gamma_1, \gamma_2)$ be a C^r $\mathfrak{X}$-diffeomorphism of $\alpha(\partial \tilde{A}_0 \times]0, 1[) \times C$ such that

$$\gamma_2(x,y) = y,$$
$$\gamma_1(x,y) = x \quad \text{if} \quad x \in \alpha(\partial \tilde{A}_0 \times (]0, \beta(y)/2] \cup [2/3, 1[)), \quad \text{and}$$
$$\gamma_1(\alpha(\partial X_0 \times]0, \beta(y)[), y) = \alpha(\partial \tilde{A}_0 \times]0, 1/2[) \quad \text{for} \quad y \in C.$$

Replace χ with $\gamma_1 \circ (\chi, \mathrm{id})$. Then for the new χ, we have

$$\chi(V_y) \supset \alpha(\partial \tilde{A}_0 \times]0, 1/2[) \quad \text{for} \quad y \in C,$$

and the Note follows when we shrink V to be $\chi^{-1}(\alpha(\partial \tilde{A}_0 \times]0, 1/2[))$.

Case where B is contractible and $W = \widehat{W} = \varnothing$. Assume $0 \in B$. By hypothesis $\psi|_{A_0}$ is C^r $\mathfrak{X}$-trivial and we have a C^r $\mathfrak{X}$-map $\chi' \colon A_0 \to \Psi^{-1}(0,0)$ such that $(\chi', \psi) \colon A_0 \to \Psi^{-1}(0,0) \times B$ is a diffeomorphism. We suppose $\chi'(V_0) = V \cap \Psi^{-1}(0,0)$ by shrinking V if necessary. Apply the lemma in the previous case to Ψ and $\chi' \circ \chi \colon V \to V \cap \Psi^{-1}(0,0)$. By shrinking V, we have a C^r $\mathfrak{X}$-extension $\nu \colon A \to \Psi^{-1}(0,0)$ of $\chi' \circ \chi$ such that $(\nu, \psi, \psi') \colon A \to \Psi^{-1}(0,0) \times B \times C$ is a diffeomorphism. Then the C^r $\mathfrak{X}$-map $(\chi', \psi)^{-1} \circ (\nu, \psi) \colon A \to A_0$ is an extension of χ such that (the extended $\chi, \psi') \colon A \to A_0 \times C$ is a diffeomorphism. The condition $\chi = \mathrm{id}$ on A_0 is satisfied when we replace χ with $(\chi|_{A_0})^{-1} \circ \chi$.

Case where B is contractible and W is the interior of a C^r $\mathfrak{X}$-submanifold with boundary of $B \times C$ which is closed in $B \times C$ and whose boundary is contractible. Define a C^r $\mathfrak{X}$-map $\nu_W \colon \Psi^{-1}(W) \to \Psi^{-1}(0,0)$ to be $\chi' \circ \chi|_{\Psi^{-1}(W)}$ for the above χ'. Let $\nu \colon A \to \Psi^{-1}(0,0)$ be the above C^r $\mathfrak{X}$-map constructed without W. By the above proof it suffices to shrink W and modify ν so that $\nu_W = \nu$ on $\Psi^{-1}(W)$. Let $\widehat{W}$ denote the C^r $\mathfrak{X}$-manifold with boundary whose interior is W. Let U be a C^r $\mathfrak{X}$-collar of $\widehat{W}$ such that $W - U$ is also the intersection of $B \times C$ with an open $\mathfrak{X}$-neighborhood of $\widehat{W}$ in $\overline{B} \times C$, which is possible if we choose U so small that $\overline{U} - U = \partial \widehat{W} - \partial \widehat{W}$. We shall replace W with $W - U$. Regard $(U, \partial \widehat{W})$ as $(\partial \widehat{W} \times [-1,1], \partial \widehat{W} \times -1)$. We shall modify ν on U so that it equals ν_W on

$\partial \widetilde{W} \times [1/2, 1]$. Set $\partial \widetilde{W} = D$. Assume $0 \in D$. Then what we need to prove is the following.

Let $(\chi_i, \psi_1, \psi_2) \colon A \to E \times D \times\,]-1, 1[$, $i = 1, 2$, be C^r 𝔛-diffeomorphisms such that $\chi_1 = \chi_2$ on V, $E = (\psi_1, \psi_2)^{-1}(0, 0)$, $\chi_1 = \chi_2 = \mathrm{id}$ on E and $(\psi_1, \psi_2)|_{A-V}$ is proper. Shrink V enough keeping the properness property. Then there exists a C^r 𝔛-map $\chi \colon A \to E$ such that $(\chi, \psi_1, \psi_2) \colon A \to E \times D \times\,]-1, 1[$ is a diffeomorphism, $\chi = \chi_1$ on V, $\chi = \mathrm{id}$ on E, and

$$\chi = \begin{cases} \chi_1 & \text{on} \quad \psi^{-1}(]-1, -1/2[) \\ \chi_2 & \text{on} \quad \psi^{-1}(]1/2, 1[). \end{cases}$$

We simplify the problem. Define a C^r 𝔛-map $\alpha \colon E \times D \times\,]-1, 1[\to E$ to be $\chi_1 \circ (\chi_2, \psi_1, \psi_2)^{-1}$. Then $\alpha(\cdot, y, z)$ is a diffeomorphism of E for each $(y, z) \in D \times\,]-1, 1[$,

$$\alpha(x, y, z) = x \quad \text{if} \quad (x, y, z) \in (\chi_1, \psi_1, \psi_2)(V) \quad \text{or} \quad (y, z) = (0, 0),$$

and we need only construct a C^r 𝔛-map $\beta \colon E \times D \times\,]-1, 1[\to E$ such that $\beta(\cdot, y, z)$ is a diffeomorphism of E for each $(y, z) \in D \times\,]-1, 1[$ and

$$\beta(x, y, z) = \begin{cases} x & \begin{array}{l} \text{for} \quad (x, y, z) \in \psi_2^{-1}(]1/2, 1[) \cup (\chi_1, \psi_1, \psi_2)(V) \\ \text{or if} \quad (y, z) = (0, 0), \end{array} \\ \alpha(x, y, z) & \text{for} \quad (x, y, z) \in \psi_2^{-1}(]-1, -1/2[). \end{cases}$$

Let γ be a C^r 𝔛-map $D \times\,]-1, 1[^2 \to D \times\,]-1, 1[$ such that

$$\gamma(0 \times 0 \times\,]-1, 1[) = (0, 0) \quad \text{and}$$
$$\gamma(y, z, t) = \begin{cases} (y, z) & \text{if} \quad t \in\,]-1, -1/2[\\ (0, 0) & \text{if} \quad t \in\,]1/2, 1[. \end{cases}$$

The existence of such a γ follows from contractibility of $D \times\,]-1, 1[$ as shown in the proof of II.6.1. Set

$$\beta(x, y, z) = \alpha(x, \gamma(y, z, z)) \quad \text{for} \quad (x, y, z) \in E \times D \times\,]-1, 1[,$$

which satisfies the above requirements. Here we shrink V.

General case. We assume B is noncompact and $\widehat{W}_1$ is nonempty because the other case is easy to prove. First we simplify the problem so that $\overline{B}$ is a C^r 𝔛-manifold with boundary whose interior is B. We have a compact C^r 𝔛-manifold with boundary $\tilde{B} \subset \mathbf{R}^{n'}$ and a C^r 𝔛-diffeomorphism $\mu \colon \tilde{B}^\circ \to B$. We want to replace B with $\tilde{B}^\circ$. For that we also need to replace $\widehat{W}_1$ with some set. The new $\widehat{W}_1$ should be the image under the projection $\mathbf{R}^{n'} \times \mathbf{R}^n \to \mathbf{R}^{n'}$ of the set $\overline{\operatorname{graph} \mu} \cap \mathbf{R}^{n'} \times \widehat{W}_1$. We need the property that an open $\mathfrak{X}$-subset of $\tilde{B}^\circ \times C$ is the intersection of $\tilde{B}^\circ \times C$ with an open $\mathfrak{X}$-neighborhood of (the new $\widehat{W}_1) \times C$ in $\tilde{B} \times C$ if and only if the corresponding open subset of $B \times C$ is the intersection of $B \times C$ with an open $\mathfrak{X}$-neighborhood of (the original $\widehat{W}_1) \times C$ in $\overline{B} \times C$. We choose $\tilde{B}$ and μ with this property as follows.

We can assume $\widehat{W}_1 \neq \overline{B} - B$. First we consider the case where $\widehat{W}_1$ is closed. Let α and α_W be non-negative $\mathfrak{X}$-functions on $\overline{B}$ such that they are of class C^r on B, $\alpha^{-1}(0) = \overline{B} - B$ and $\alpha_W^{-1}(0) = \overline{W}_1$. Let a and a_W be small positive numbers such that α and $\alpha|_{\alpha_W^{-1}(a_W)}$ are C^1 regular on $\alpha^{-1}(]0, 2a[)$ and $\alpha^{-1}(]0, 2a[) \cap \alpha_W^{-1}(a_W)$ respectively, and $\alpha_W^{-1}(]0, 2a_W[)$ admits a C^r Whitney $\mathfrak{X}$-stratification with a stratum $B \cap \alpha_W^{-1}(]0, 2a_W[)$ from which to $]0, 2a_W[$, α_W is a C^r Whitney stratified map. Then α and $\alpha|_{\alpha_W^{-1}(a_W)}$ are C^r $\mathfrak{X}$-trivial over $]0, 2a[$ by II.6.5, and $\alpha_W|_B$ is so over $]0, 2a_W[$ by II.6.1 and its remark. (To prove II.6.1 we shall use this lemma without $\widehat{W}_1$. In the present proof, we do not need II.6.1 when $\widehat{W}_1$ is empty.)

By the triviality of α and $\alpha|_{\alpha_W^{-1}(a_W)}$ and by this lemma in the case $B = a$ point, $\alpha|_{B - \alpha_W^{-1}(]0, a_W[)}$ is C^r $\mathfrak{X}$-trivial over $]0, 2a[$. Indeed, we can extend the C^r $\mathfrak{X}$-map of triviality $\chi: \alpha^{-1}(]0, 2a[) \cap \alpha_W^{-1}(a_W) \to \alpha^{-1}(a) \cap \alpha_W^{-1}(a_W)$ to a small open $\mathfrak{X}$-neighborhood in $\alpha^{-1}(]0, 2a[) \cap \alpha_W^{-1}(]0, a_W[)$ as usual and then to $\alpha^{-1}(]0, 2a[) - \alpha_W^{-1}(]0, a_W[)$ by the lemma. Hence $(B - \alpha_W^{-1}(]0, a_W[), B \cap \alpha_W^{-1}(a_W))$ is C^r $\mathfrak{X}$-diffeomorphic to $(B - \alpha^{-1}(]0, a[)) - \alpha_W^{-1}(]0, a_W[), \alpha_W^{-1}(a_W) - \alpha^{-1}([0, a]))$.

By the triviality of $\alpha_W|_B$, $(B \cap \alpha_W^{-1}(]0, a_W]), B \cap \alpha_W^{-1}(a_W))$ is C^r $\mathfrak{X}$-diffeomorphic to $((B \cap \alpha_W^{-1}(a_W)) \times]0, a_W], (B \cap \alpha_W^{-1}(a_W)) \times a_W)$, and the new $\widehat{W}_1$ in $\overline{(B \cap \alpha_W^{-1}(a_W)) \times]0, a_W]} = \alpha_W^{-1}(a_W) \times [0, a_W]$ is $\alpha_W^{-1}(a_W) \times 0$ and has the above required property. Moreover, by the above first diffeomorphism, $((B \cap \alpha_W^{-1}(a_W)) \times]0, a_W], (B \cap \alpha_W^{-1}(a_W)) \times a_W)$ is C^r $\mathfrak{X}$-diffeomorphic to $((\alpha_W^{-1}(a_W) - \alpha^{-1}([0, a])) \times]0, a_W], (\alpha_W^{-1}(a_W) - \alpha^{-1}([0, a])) \times a_W)$, and the $\widehat{W}_1$ becomes $(\alpha_W^{-1}(a_W) - \alpha^{-1}([0, a])) \times 0$ and keeps the property. Paste $B - \alpha^{-1}(]0, a]) - \alpha_W^{-1}(]0, a_W[)$ and $(\alpha_W^{-1}(a_W) - \alpha^{-1}([0, a]) \times]0, a_W]$ at $\alpha_W^{-1}(a_W) - \alpha^{-1}([0, a])$ and $(\alpha_W^{-1}(a_W) - \alpha^{-1}([0, a])) \times a_W$ respectively. Then we obtain a C^r $\mathfrak{X}$-diffeomorphism from B to $B - \alpha^{-1}(]0, a]) - \alpha_W^{-1}(]0, a_W])$ such that the new $\widehat{W}_1$ is $\alpha_W^{-1}(a_W) - \alpha^{-1}([0, a[)$ and has the property.

$B - \alpha^{-1}(]0, a]) - \alpha_W^{-1}(]0, a_W])$ is not yet what we want because its closure has corners. We need to smooth them. The smoothing is always possible, to be precise, for a compact C^r $\mathfrak{X}$-manifold with corners M, there exist a compact C^r $\mathfrak{X}$-manifold with boundary M' and an $\mathfrak{X}$-homeomorphism $\tau: M \to M'$ such that $\tau|_{M^\circ}$ is a C^r diffeomorphism. But in the present case, smoothing is trivial because the manifold with corners around the corners is C^r X-diffeomorphic to $[0,1[^2 \times (\alpha^{-1}(a) \cap \alpha_W^{-1}(a_W))$. The closed case is proved.

If $\widehat{W}_1$ is not closed, we modify the above proof. Let Q be an open X-neighborhood of $\widehat{W}_1$ in $\overline{B} - B$ where $\widehat{W}_1$ is closed. Define a non-negative X-function $\alpha_{\overline{B}-B-Q}$ on $\overline{B}$ as α_W, i.e., it is of class C^r on B, the zero set equals $\overline{B} - B - Q$, and for a positive number ε, $\alpha_{\overline{B}-B-Q}^{-1}(]0,2\varepsilon[)$ admits a C^r Whitney X-stratification with a stratum $B \cap \alpha_{\overline{B}-B-Q}^{-1}(]0,2\varepsilon[)$ compatible with $\widehat{W}_1$, from which to $]0,2\varepsilon[$, $\alpha_{\overline{B}-B-Q}$ is a C^r Whitney X-stratified map. By the triviality, we can assume B is the union of a C^r X-manifold with boundary B_1 and $\partial B_1 \times]0,1]$, where ∂B_1 and $\partial B_1 \times 1$ are naturally pasted,

$$Q = (\overline{B_1} - B_1) \cup (\overline{\partial B_1} - \partial B_1) \times]0,1] = \overline{B} - B - \overline{\partial B_1} \times 0,$$

and $\widehat{W}_1 \cap (\overline{\partial B_1} - \partial B_1) \times]0,1]$ is of the form $\widehat{W}_2 \times]0,1]$ for a closed X-subset $\widehat{W}_2$ of $\overline{\partial B_1} - \partial B_1$. We make the same arguments as above for B_1 and $\widehat{W}_1 \cap (\overline{B_1} - B_1)$. (This is the case with boundary, but we can do it in the same way.) Then we obtain a C^r X-manifold with corners B_2 such that its proper faces of dimension $= \dim B - 1$ consists of two C^r X-manifolds with boundary B_3 and B_4, $(B_2^{\circ} \cup B_3^{\circ}, B_3^{\circ})$ is C^r X-diffeomorphic to $(B_1, \partial B_1)$, and the new $\widehat{W}_1 \cap (\overline{B_1} - B_1)$ in B_4 through this diffeomorphism has the property in question. Extend naturally this diffeomorphism to (the C^r X-manifold which we construct by pasting naturally $B_2 - B_4$ with $B_3^{\circ} \times]0,1]$) $\to B$, and smooth the closure of the pasted manifold. Thus we can assume $\overline{B}$ is a C^r X-manifold with boundary whose interior is B.

We want to show that if $\widehat{W}_1$ is contractible, then we can shrink W so that it satisfies the conditions in the previous case.

Let O be an open X-neighborhood of $\widehat{W}$ in $\partial\overline{B} \times C$ such that $O \cup W$ is open in $\overline{B} \times C$. By II.2.1 we have a contractible open X-neighborhood O' of $\widehat{W}$ in O. We want to shrink W so that $\overline{W} \cap B \times C$ is a C^r X-manifold with boundary and $\overline{W} \cap B \times C - W$ is diffeomorphic to O'. We can assume $(\overline{B} \times C, \partial\overline{B} \times C) = (B^* \times [0,1[, B^* \times 0)$ for a C^r X-manifold B^* because the problem of shrinking is considered around $\partial\overline{B} \times C$. Since $O \cup W$ is an X-neighborhood of O', there exists a non-negative C^r X-function $\nu < 1$ on $\overline{O'}$ such that

$$\nu^{-1}(0) = \overline{O'} - O' \quad \text{and} \quad \{(x,t) \in O' \times [0,1[: t < \nu(x)\} \subset O \cup W.$$

(This set) $-B^* \times 0$ satisfies the conditions.

There exists a finite open X-covering $\{B_i\}$ of B such that for each i, the restriction of χ to $\psi^{-1}(B_i) \cap (V \cup \Psi^{-1}(W))$ can be extended to $\psi^{-1}(B_i) \to$

$\psi^{-1}(B_i)_0$ as required, for the following reason. We can regard $(\overline{B}, \partial \overline{B})$ as $(B^* \times [0, 1[, B^* \times 0)$ because the inside of a collar of $\overline{B}$ has no problem as shown already. By II.2.1 it is easy to find a finite open $\mathfrak{X}$-covering $\{\Lambda_i\}$ of B^* such that each Λ_i and $\overline{\Lambda_i} \times 0 \cap \widehat{W}_1$ are contractible and $\Lambda_i \times 0 \cap \widehat{W}_1 = \overline{\Lambda_i} \times 0 \cap \widehat{W}_1$. Then $\{B_i = \Lambda_i \times]0, 1[\}$ fulfills the requirements by this lemma in the previous case because for each i, $B_i \times C \cap W$ can be shrunk so that it is the intersection of $B_i \times C$ with an open $\mathfrak{X}$-neighborhood of $(\overline{B_i} \cap \widehat{W}_1) \times C$ in $\overline{B_i} \times C$, and its closure in $B_i \times C$ is a C^r $\mathfrak{X}$-manifold with contractible boundary.

By induction on the minimal number of the elements of such a covering of B we can assume B is the union of open $\mathfrak{X}$-subsets B_1 and B_2 and for each $i = 1, 2$, $\chi|_{(V \cup \Psi^{-1}(W)) \cap \psi^{-1}(B_i)}$ is extended to $\chi_i \colon \psi^{-1}(B_i) \to \psi^{-1}(B_i)_0$ as required. It suffices to paste χ_1 and χ_2. We proceed as in the previous case. Shrink B_1 so that $B \cap \overline{B_1}$ is a C^r $\mathfrak{X}$-manifold with boundary. Let F denote the boundary, and let $\kappa \colon F \times [-1, 1] \to U$ be a C^r $\mathfrak{X}$-collar of $B \cap \overline{B_1}$ such that $U \subset B_2$ and $\kappa(\cdot, -1) = \mathrm{id}$. Choose U small enough. We can choose κ so that κ can be extended to an $\mathfrak{X}$-map $\overline{\kappa} \colon \overline{F} \times [-1, 1] \to \overline{B}$ as $\overline{\kappa}(x \times [-1, 1]) = x$ for $x \in \overline{F} - F$. Then we can reduce the problem to the following one.

Assume B is of the form $F \times]-1, 1[$ for a C^r $\mathfrak{X}$-manifold F. Let χ_1 and χ_2 be C^r $\mathfrak{X}$-maps $A \to A_0$ such that $(\chi_i, \psi') \colon A \to A_0 \times C$, $i = 1, 2$, are diffeomorphisms, $\chi_1 = \chi_2$ on $V \cap \Psi^{-1}(W)$, $\chi_1 = \chi_2 = \mathrm{id}$ on A_0, and $\psi \circ \chi_1 = \psi \circ \chi_2 = \psi$. Shrink V and W enough. Then there exists another C^r $\mathfrak{X}$-map $\chi \colon A \to A_0$ such that $(\chi, \psi') \colon A \to A_0 \times C$ is a diffeomorphism, $\chi = \chi_1$ on $V \cap \Psi^{-1}(W)$, $\chi = \mathrm{id}$ on A_0, $\psi \circ \chi = \psi$, and

$$\chi = \begin{cases} \chi_1 & \text{on } \psi^{-1}(F \times]-1, -1/2[) \\ \chi_2 & \text{on } \psi^{-1}(F \times]1/2, 1[). \end{cases}$$

We prove this in the same way as in the previous case. Note that F is not necessarily contractible, but contractibility is not needed. We need only be careful with the shrinking of W. First we must ensure that $\widehat{W}_1$ is contractible after replacing B with $F \times]-1, 1[$. By the above special choice of κ this requirement is fulfilled. Then, as shown above, the conditions on W in the previous case are satisfied. Second, as we use a C^r $\mathfrak{X}$-map $\gamma \colon C \times I \to C$, $I = [0, 1]$, such that $\gamma(C \times 1) = 0$ and $\gamma(\cdot, 0) = \mathrm{id}$, we have to shrink W to W^* so that for each $c \in C$ and for each $t \in I$, the image of $W^* \cap B \times c$ under the projection $B \times C \to B$ is included in the projection image of $W \cap B \times \gamma(c, t)$. This is possible because I is compact and $\widehat{W}$ is of the form $\widehat{W}_1 \times C$ and does not depend on a point of C. We omit the details. $\square$

We can generalize II.6.5 as the following theorem.

Theorem II.6.9. *Let X and B be $\mathfrak{X}$-sets in $\mathbf{R}^n$, and let $p\colon X \to B$ be an $\mathfrak{X}$-map with the second boundedness condition. Assume that for each point $b \in B$, $p^{-1}(b)$ is a compact C^1 $\mathfrak{X}$-submanifold of $\mathbf{R}^n$. There exists a C^1 $\mathfrak{X}$-stratification $\{B_i\}$ of B locally finite at each point of $\mathbf{R}^n$ such that for each i, $p|_{p^{-1}(B_i)}\colon p^{-1}(B_i) \to B_i$ is C^1 $\mathfrak{X}$-trivial.*

Proof. As before we can suppose B and hence X are bounded. We proceed by induction on $\dim B$. By stratifying B, we assume that B is a C^1 $\mathfrak{X}$-submanifold of $\mathbf{R}^n$ and p is surjective. By II.6.5 it suffices to find a closed $\mathfrak{X}$-subset B' of B, of dimension smaller than $k_1 = \dim B$, such that $X - p^{-1}(B')$ is a C^1 $\mathfrak{X}$-submanifold of $\mathbf{R}^n$ and $p|_{X-p^{-1}(B')}$ is a proper C^1 submersion onto some union of connected components of $B - B'$.

First there exists a closed $\mathfrak{X}$-subset B_1 of B of dimension $< k_1$ such that $p|_{X-p^{-1}(B_1)}$ is proper for the following reason. Set $\tilde{X} = \operatorname{graph} p$, and let $\tilde{p}$ denote the projection of $\mathbf{R}^n \times \overline{B}$ onto $\overline{B}$. Since $\tilde{X}$ is bounded in $\mathbf{R}^n \times \mathbf{R}^n$, we need only prove

$$\dim \tilde{p}(\overline{\tilde{X}} - \tilde{X}) < k_1.$$

Assume $\dim \tilde{p}(\overline{\tilde{X}} - \tilde{X}) = k_1$. Apply (II.1.17) to $\tilde{p}|_{\overline{\tilde{X}} \cap \tilde{p}^{-1}(B)}$. There exist a nonempty open $\mathfrak{X}$-subset B_2 of B and a C^1 Whitney $\mathfrak{X}$-stratification $\{\tilde{X}_i\}$ of $\overline{\tilde{X}} \cap \tilde{p}^{-1}(B_2)$ with the frontier condition such that $\tilde{p}(\overline{\tilde{X}} - \tilde{X}) \supset B_2$, $\{\tilde{X}_i\}$ is compatible with $\{\tilde{X}\}$, and for each i, $\tilde{p}|_{\tilde{X}_i}$ is a C^1 submersion onto B_2. Note that $\tilde{p}|_{\overline{\tilde{X}} \cap \tilde{p}^{-1}(B_2)}\colon \overline{\tilde{X}} \cap \tilde{p}^{-1}(B_2) \to B_2$ is proper. Hence by the Thom's first isotopy lemma (see [G-al]), by shrinking B_2, we see that $\{\tilde{X}_i\}$ is weakly C^1 isomorphic to $\{(\tilde{X}_i \cap \tilde{p}^{-1}(b)) \times B_2\}$ for a point b of B_2. Therefore, $\tilde{X} \cap \tilde{p}^{-1}(B_2)$ is closed in $\mathbf{R}^n \times B_2$ because $\tilde{X} \cap \tilde{p}^{-1}(b) = p^{-1}(b) \times b$ is compact and $\{\tilde{X}_i\}$ is compatible with $\{\tilde{X}\}$. It follows that $\overline{\tilde{X}} \cap \tilde{p}^{-1}(B_2) = \tilde{X} \cap \tilde{p}^{-1}(B_2)$, which contradicts the above condition $\tilde{p}(\overline{\tilde{X}} - \tilde{X}) \supset B_2$.

The above proof shows, moreover, that there is a closed $\mathfrak{X}$-subset B_3 ($\supset B_1$) of B of dimension $< k_1$ such that $p|_{X-p^{-1}(B_3)}\colon X - p^{-1}(B_3) \to B - B_3$ is C^0 trivial over each connected component of $B - B_3$. We generalize this fact as follows. Set

$$\hat{X} = \{(x, \lambda) \in X \times G_{n,k_x}\colon \lambda = T_x p^{-1}(p(x))\} \subset \mathbf{R}^n \times \mathbf{R}^{n^2},$$

where $k_x = \dim p^{-1}(p(x))$, and let $\hat{p}\colon \hat{X} \to B$ denote the composite of the projection of $\tilde{X}$ onto X and p. Then $\hat{X}$ is bounded in $\mathbf{R}^n \times \mathbf{R}^{n^2}$ and for each point $b \in B$, $\hat{p}^{-1}(b)$ is a compact C^0 $\mathfrak{X}$-submanifold of $\mathbf{R}^n \times \mathbf{R}^{n^2}$. Hence by the same reason as above we assume $\hat{p}|_{\hat{X}-\hat{p}^{-1}(B_3)}\colon \hat{X} - \hat{p}^{-1}(B_3) \to B - B_3$ is C^0 trivial over each connected component of $B - B_3$.

We want to show

$$\dim p(\Sigma_1(X - p^{-1}(B_3))) < k_1, \tag{1}$$

where Σ_1 denotes the C^1 singular point set of an $\mathfrak{X}$-set. If this inequality holds, we obtain the required B' as follows. Set

$$B_4 = p(\Sigma_1(X - p^{-1}(B_3))) \cup B_3.$$

Apply (II.1.17) to $p|_{X-p^{-1}(B_4)}$. We have a closed $\mathfrak{X}$-subset B_5 ($\supset B_4$) of B of dimension $< k_1$ and a C^1 Whitney $\mathfrak{X}$-stratification $\{X_i\}$ of $X - p^{-1}(B_5)$ such that for each i, $p|_{X_i}$ is a C^1 submersion onto some connected component of $B - B_5$. Define $B' = B_5$. This B' satisfies the required conditions that $X - p^{-1}(B')$ is a C^1 $\mathfrak{X}$-manifold and $p|_{X-p^{-1}(B')}$ is a proper C^1 submersion. It remains to show that $p|_{X-p^{-1}(B')}$ is a C^1 submersion. This is clear because for each point x of $X - p^{-1}(B')$, x belongs to some X_i and

$$\text{rank } d(p|_{X-p^{-1}(B')}) \geq \text{rank } d(p|_{X_i}) \quad \text{at} \quad x.$$

Thus it suffices to prove (1).

We can assume $k_x = \text{const}$, say, $= k_2$. Set

$$A = \{\alpha = (\alpha_1, \dots, \alpha_{k_2}) \in \mathbf{Z}^{k_2} : 1 \leq \alpha_1 < \cdots < \alpha_{k_2} \leq n\} \quad \text{and}$$
$$X_\alpha = \{x \in X - p^{-1}(B_3) : p_\alpha(T_x p^{-1}(p(x))) = \mathbf{R}^{k_2}\} \quad \text{for} \quad \alpha \in A,$$

where $p_\alpha : \mathbf{R}^n \to \mathbf{R}^{k_2}$ are the projections defined by

$$p_\alpha(x_1, \dots, x_n) = (x_{\alpha_1}, \dots, x_{\alpha_{k_2}}) \quad \text{for} \quad (x_1, \dots, x_n) \in \mathbf{R}^n.$$

Then $\{X_\alpha\}_{\alpha \in A}$ is an open $\mathfrak{X}$-covering of $X - p^{-1}(B_3)$. Let $\{X'_\alpha\}_{\alpha \in A}$ be an open $\mathfrak{X}$-covering of $X - p^{-1}(B_3)$ such that for each $\alpha \in A$,

$$\overline{X'_\alpha} \cap X - p^{-1}(B_3) \subset X_\alpha.$$

For each $\alpha \in A$, let Y_α denote the subset of X'_α consisting of points where the map $(p, p_\alpha) : X'_\alpha \to B \times \mathbf{R}^{k_2}$ is not locally injective. Clearly

$$\bigcup_{\alpha \in A} (X'_\alpha - Y_\alpha) \supset X - p^{-1}(B_3) - \Sigma_1(X - p^{-1}(B_3)).$$

We prove

$$\dim p(Y_\alpha) < k_1. \tag{2}$$

Assume (2) does not hold. There exists an open $\mathfrak{X}$-subset B_α of $B - B_3$ such that $B_\alpha \subset p(Y_\alpha)$. Set

$$Z_\alpha = \{(x, x') \in X'_\alpha \times X'_\alpha - \Delta_X : p(x) = p(x'),\ p_\alpha(x) = p_\alpha(x')\},$$

where Δ_X is the diagonal in $X \times X$. Then

$$\Delta_{Y_\alpha} = \overline{Z_\alpha} \cap \Delta_X, \tag{3}$$

and we can assume the image of Z_α under the map $X \times X \ni (x_1, x_2) \to p(x_1) \in B$ contains B_α. Let a function with $\mathfrak{X}$-graph ρ on B_α be defined so that for each $b \in B_\alpha$, $\rho(b)$ is the infimum of $|x - x'|$ as (x, x') runs through the nonempty set $Z_\alpha \cap (p^{-1}(b) \times p^{-1}(b))$. By the hypothesis that $p^{-1}(b)$ is a C^1 $\mathfrak{X}$-manifold and by the definition of X_α, $p_\alpha|_{X_\alpha \cap p^{-1}(b)}$ is a C^1 $\mathfrak{X}$-immersion. Hence by the property $\overline{X'_\alpha} \cap X - p^{-1}(B_3) \subset X_\alpha$, ρ is a positive function. Apply (II.1.17) to ρ. Then ρ is continuous on an open $\mathfrak{X}$-subset of B_α. However, (3) implies that for each $b \in B_\alpha$, there is a sequence $\{b_i\}$ in B_α convergent to b such that $\{\rho(b_i)\}$ converges to 0, which is a contradiction. Thus we see (2).

We now prove that $X'_\alpha - Y_\alpha$ is a C^1 $\mathfrak{X}$-submanifold of $\mathbf{R}^n$. First we want to prove that the tangent space of X'_α at each point x_0 of $X'_\alpha - Y_\alpha$ exists. We assume that $x_0 = 0$, 0 is contained in a stratum, say, X_1 of the stratification $\{X_i\}$, and $\dim X_1 = k_1$ (which is possible if we substratify X_1). Let L denote the linear subspace of $\mathbf{R}^n$ spanned by $T_0 p^{-1}(p(0))$ and $T_0 X_1$, and let $l\colon \mathbf{R}^n \to L$ denote the orthogonal projection. By Brouwer's invariance theorem of domain, (p, p_α) is a local homeomorphism at 0. Hence it suffices to prove

$$|l(x) - x|/|x| \longrightarrow 0 \quad \text{as} \quad x \longrightarrow 0 \quad \text{in} \quad X'_\alpha.$$

Without loss of generality, we can assume

$$\mathbf{R}^n = \mathbf{R}^{k_1} \times \mathbf{R}^{k_2} \times \mathbf{R}^{k_3},$$

$$X_1 = \mathbf{R}^{k_1} \times 0^{k_2} \times 0^{k_3}, \quad \text{and} \quad p^{-1}(p(0)) = 0^{k_1} \times \mathbf{R}^{k_2} \times 0^{k_3} \quad \text{around } 0.$$

Then

$$L = \mathbf{R}^{k_1} \times \mathbf{R}^{k_2} \times 0^{k_3}, \qquad l(x_1, x_2, x_3) = (x_1, x_2, 0), \quad \text{and}$$

$$|l(x) - x|/|x| = |x_3|/|x| \quad \text{for} \quad x = (x_1, x_2, x_3) \in \mathbf{R}^{k_1} \times \mathbf{R}^{k_2} \times \mathbf{R}^{k_3}.$$

For each small positive number δ, set

$$U_\delta = \{(x_1, x_2, x_3) \in X'_\alpha \subset \mathbf{R}^{k_1} \times \mathbf{R}^{k_2} \times \mathbf{R}^{k_3} : |x_i| < \delta,\ i = 1, 2, 3\} \quad \text{and}$$

$$V_\delta = U_\delta \cap (0^{k_1} \times \mathbf{R}^{k_2} \times 0^{k_3}) = \{(0, x_2, 0) \in \mathbf{R}^{k_1} \times \mathbf{R}^{k_2} \times \mathbf{R}^{k_3} : |x_2| < \delta\}.$$

Let ε be a positive number. By the local C^0 triviality of $\hat{p}|_{\hat{X}-\hat{p}^{-1}(B_3)}$ there exists a small number δ such that for each $x \in \overline{U_\delta}$ and for each $y = (y_1, y_2, y_3) \in T_x p^{-1}(p(x)) \subset \mathbf{R}^n$, we have $\overline{U_\delta} \subset X'_\alpha - Y_\alpha$ and

$$|(y_1, y_3)| \le \varepsilon |y_2|. \tag{4}$$

Let l_2 denote the orthogonal projection of $\mathbf{R}^{k_1} \times \mathbf{R}^{k_2} \times \mathbf{R}^{k_3}$ onto $0^{k_1} \times \mathbf{R}^{k_2} \times 0^{k_3}$. For each $x \in \overline{U_\delta}$, $l_2|_{U_\delta \cap p^{-1}(p(x))}$ is a C^1 immersion. Moreover, we can choose sufficiently small δ' so that for any $x \in U_{\delta'}$, the map

$$l_2|_{U_\delta \cap p^{-1}(p(x))} : U_\delta \cap p^{-1}(p(x)) \longrightarrow V_\delta$$

is proper. For each $x = (x_1, x_2, x_3) \in U_{\delta'}$, consider the segment s joining $l_2(x)$ with 0 and the closed C^1 curve $\sigma = l_2^{-1}(s) \cap U_\delta \cap p^{-1}(p(x))$. There uniquely exists a C^1 path $\varphi \colon [-1, 1] \to \sigma$ such that $\varphi(1) = x$ and for each $t \in [-1, 1]$, $\varphi(t)$ is of the form $(\varphi_1(t), tx_2, \varphi_3(t))$. Then $\varphi(0) \in X_1$, and $\varphi_1 \colon [-1, 1] \to \mathbf{R}^{k_1}$ and $\varphi_3 \colon [-1, 1] \to \mathbf{R}^{k_3}$ are C^1 paths. It follows from (4) that

$$\left| \left(\frac{d\varphi_1}{dt}, \frac{d\varphi_3}{dt} \right) \right| \le \varepsilon |x_2|.$$

Hence

$$|(\varphi_1(1), \varphi_3(1)) - (\varphi_1(0), \varphi_3(0))| \le \varepsilon |x_2|,$$

which implies

$$|x_3|/|x| = |\varphi_3(1) - \varphi_3(0)|/|x| \le \varepsilon |x_2|/|x| \le \varepsilon.$$

Thus the tangent space $T_{x_0} X'_\alpha$ exists.

For proof of C^1 smoothness of $X'_\alpha - Y_\alpha$, we need to show that the map $X'_\alpha - Y_\alpha \ni x \to T_x X'_\alpha \in \mathbf{R}^{n^2}$ is continuous. But this follows from the property:

$$T_x X'_\alpha = T_x X_i + T_x p^{-1}(p(x)) \quad \text{for} \quad x \in X_i \cap X'_\alpha,$$

where $\{X_i\}$ is the above C^1 Whitney stratification of $p|_{X-p^{-1}(B_5)}$. $\qquad\square$

The C^r case, $r > 0$, of this theorem is an open problem.

Proof of II.6.6. We consider only C^r $\mathfrak{X}$-triviality of p of II.6.5. We can treat (f, p) of II.6.5$'$ in the same way. By II.6.3 and II.6.5, it suffices to find a closed $\mathfrak{X}$-subset B' of B and an open $\mathfrak{X}$-neighborhood U of B' in B such that B' is a deformation retract of B and p is C^r $\mathfrak{X}$-trivial over U. As in the proof of II.6.2, we have a closed $\mathfrak{X}$-subset B' of B which is closed also in $\mathbf{R}^n$

and a deformation retract of B. Let U be a small open $\mathfrak{X}$-neighborhood of B' in B whose closure in B is closed in $\mathbf{R}^n$. We want to show that p is C^r $\mathfrak{X}$-trivial over U.

First assume U is bounded. Let $0 \in U$. By hypothesis there exists a C^r map $h: p^{-1}(\overline{U}) \to p^{-1}(0)$ such that $(h,p): p^{-1}(\overline{U}) \to p^{-1}(0) \times \overline{U}$ is a C^r diffeomorphism. Approximate h by a C^r $\mathfrak{X}$-map $\tilde{h}$ in the usual C^r topology on the C^r map space $C^r(p^{-1}(\overline{U}), p^{-1}(0))$. This is possible by the polynomial approximation theorem and II.5.1 because $p^{-1}(\overline{U})$ and $p^{-1}(0)$ are compact. Then, since $p^{-1}(\overline{U})$ is compact, the map $(\tilde{h}, p): p^{-1}(\overline{U}) \to p^{-1}(0) \times \overline{U}$ is a C^r $\mathfrak{X}$-diffeomorphism. Hence p is C^r $\mathfrak{X}$-trivial over U.

Note that this $\tilde{h}$ is C^r homotopic to h, and hence there exists a C^r map $H: p^{-1}(\overline{U}) \times I \to p^{-1}(0)$, $I = [0,1]$, such that $H(\cdot,0) = h(\cdot)$, $H(\cdot,1) = \tilde{h}(\cdot)$, and for each $t \in I$, $(H(\cdot,t), p(\cdot)): p^{-1}(\overline{U}) \to p^{-1}(0) \times \overline{U}$ is a C^r diffeomorphism.

Assume U is unbounded. Set

$$U_i = \{y \in U : i - 1 < |y| < i+1\}, \quad i = 0,1,\ldots .$$

Since each U_i is bounded, p is C^r $\mathfrak{X}$-trivial over an open $\mathfrak{X}$-neighborhood of $\overline{U}_i$ in B. We paste their C^r $\mathfrak{X}$-triviality. Let $0 \in U$. Let $h: p^{-1}(\overline{U}) \to p^{-1}(0)$ be a C^r map such that $(h,p): p^{-1}(\overline{U}) \to p^{-1}(0) \times \overline{U}$ is a C^r diffeomorphism. As shown above, for each i, there exists a C^∞ map $H_i: p^{-1}(\overline{U}_i) \times I \to p^{-1}(0)$ such that $H_i(\cdot,0) = h(\cdot)$, $H_i(\cdot,1)$ is of class $\mathfrak{X}$, and for each $t \in I$, $(H_i(\cdot,t), p(\cdot)): p^{-1}(\overline{U}_i) \to p^{-1}(0) \times \overline{U}_i$ is a C^r diffeomorphism. It suffices to construct a C^r $\mathfrak{X}$-map $h_i: p^{-1}(\overline{U}_i) \to p^{-1}(0)$ such that $(h_i, p): p^{-1}(\overline{U}_i) \to p^{-1}(0) \times \overline{U}_i$ is a C^r diffeomorphism, and for $x \in \overline{U}_i$,

$$h_i(x) = \begin{cases} H_i(x,1) & \text{if } i - 1 < |p(x)| < i - 1/3 \\ H_{i+1}(x,1) & \text{if } i + 1/3 < |p(x)| < i + 1. \end{cases}$$

Let α be a C^r $\mathfrak{X}$-function on $]i - 1, i+1[$ such that $0 \le \alpha \le 1$ and

$$\alpha = \begin{cases} 0 & \text{on }]i - 1/5, i + 1/5[\\ 1 & \text{on }]i - 1, i - 1/4[\, \cup \,]i + 1/4, i + 1[. \end{cases}$$

Define a map $\beta: p^{-1}(\overline{U}_i) \to p^{-1}(0)$ by

$$\beta(x) = \begin{cases} H_i(x, \alpha(|p(x)|)) & \text{if } i - 1 < |p(x)| \le i \\ H_{i+1}(x, \alpha(|p(x)|)) & \text{if } i < |p(x)| < i + 1. \end{cases}$$

Then β is of class C^r, $(\beta, p): p^{-1}(\overline{U}_i) \to p^{-1}(0) \times \overline{U}_i$ is a C^r diffeomorphism, and

$$\beta(x) = \begin{cases} H_i(x,1) & \text{if } i - 1 < |p(x)| < i - 1/4 \\ H_{i+1}(x,1) & \text{if } i + 1/4 < |p(x)| < i + 1. \end{cases}$$

Approximate β by a C^r $\mathfrak{X}$-map without changing it on $\{x \in \overline{U_i}: |p(x)| \in$ $]i-1, i-1/3[\,\cup\,]i+1/3, i+1[\}$. The approximation is the required h_i. $\square$

Proof of II.6.1. As usual, we assume $0 \in Y$, Y is bounded and contractible, $X \subset \overline{X_1}$, each X_i is connected, and $\dim X_1 > \dim X_2 \geq \cdots$, and we set $Z_0 = f^{-1}(0) \cap Z$ for any subset Z of X. By the forthcoming lemma II.6.10 we have a controlled C^1 $\mathfrak{X}$-tube system $\{T_i = (|T_i|, \pi_i, \rho_i)\}$ for $\{X_i\}$ such that $f \circ \pi_i = f$ on $X \cap |T_i|$. By its proof we can assume that for each $i > i'$, the map $(\pi_i, \rho_i)|_{X_{i'} \cap |T_i|}$ is a C^1 submersion to $X_i \times]0,1[$ and $(\pi_i, \rho_i)|_{\cup_{j=i'}^{i-1} X_j \cap |T_i|}$ is proper to $X_i \times]0,1[$.

Set $f_i = f|_{X_i}$. For each i we will construct a C^1 $\mathfrak{X}$-map $h_i : X_i \to X_{i0}$ such that $(h_i, f_i): X_i \to X_{i0} \times Y$ is a C^1 diffeomorphism and the following four conditions are satisfied by induction on the dimension of X_0. Let $i > i'$, and let $|T_i|'$ be some open $\mathfrak{X}$-neighborhoods of X_i smaller than $|T_i|$.

$$h_{i'}(X_{i'} \cap |T_i|') = (X_{i'} \cap |T_i|')_0. \tag{i}$$

$$h_i \circ \pi_i = \pi_i \circ h_{i'} \quad \text{on} \quad X_{i'} \cap |T_i|'. \tag{ii}$$

$$\rho_i \circ h_{i'} = \rho_i \quad \text{on} \quad X_{i'} \cap |T_i|'. \tag{iii}$$

$$h_{i'} = \text{id} \quad \text{on} \quad X_{i'0}. \tag{iv}$$

If there exist such h_i's and we set $h = h_i$ on X_i, $i = 1, 2, \ldots$, then h is continuous by (ii) and (iii) and hence f is $\mathfrak{X}$-trivial. If $\dim X_0 = 0$, then $X = X_1$ and existence of h_1 is clear. Hence assuming such h_i, $i = 2, 3, \ldots$, we construct h_1.

We use another induction. Let $1 < l$ be an integer. By induction on l we will construct a C^1 $\mathfrak{X}$-map $h_1^l : X_1 \cap \cup_{i=2}^l |T_i|'' \to (X_1 \cap \cup_{i=2}^l |T_i|'')_0$ such that (h_1^l, f_1) is a C^1 imbedding and the conditions (i), (ii), (iii) for $1 = i' < i \leq l$ and (iv) on $(X_1 \cap \cup_{i=2}^l |T_i|)_0$ for $i' = 1$ are satisfied. Assume such a h_1^{l-1} (nothing if $l = 2$). Here $|T_i|''$ is an open $\mathfrak{X}$-neighborhood of X_i smaller than $|T_i|'$ and $|T_i|'$ in (i), (ii) and (iii) are replaced with $|T_i|''$.

For construction of h_1^l we want to apply II.6.8 in the case $W = \widehat{W} = \varnothing$. We try to apply this to

$$A = X_1 \cap |T_l|, \quad B = X_{l0} \times]0,1[, \quad C = Y,$$

$$V = A \cap \bigcup_{i=2}^{l-1} |T_i|'', \quad \psi = (h_l \circ \pi_l, \rho_l), \quad \text{and} \quad \psi' = f_1.$$

We check whether these data satisfy the conditions in II.6.8. The non-trivial conditions are only C^1 $\mathfrak{X}$-triviality of $\psi|_{\psi^{-1}(B')_0}$ and existence of

χ. The triviality follows when we apply the first induction hypothesis to $(\pi_l, \rho_l)\colon (\cup_{i=1}^{l-1} X_i \cap |T_l|)_0 \to X_{l0} \times {]}0,1{[}$ because the dimension of its fibre is smaller than $\dim X_0$. It is natural to hope $h_1^{l-1}|_V$ to be the required map χ. But the equality $\psi \circ \chi = \psi$ does not necessarily hold. We modify the above data so that it holds.

By the first induction hypothesis, for any contractible open $\mathfrak{X}$-subset D of X_{l0}, the map $(\pi_l, \rho_l)\colon (X \cap |T_l|)_0 \to X_{l0} \times {]}0,1{[}$ is $\mathfrak{X}$-trivial over $D \times {]}0,1{[}$. Then by the pasting method in the proof of II.6.8 we have an $\mathfrak{X}$-map $\gamma\colon (X \cap |T_l|)_0 \to (X \cap |T_l| \cap \rho_l^{-1}(1/2))_0$ such that (γ, ρ_l) is a homeomorphism onto $(X \cap |T_l| \cap \rho_l^{-1}(1/2))_0 \times {]}0,1{[}$, and $\pi_l \circ \gamma = \pi_l$. Moreover, by the induction process and the proof of II.6.8, for each $i = 1, \ldots, l-1$, $(\gamma, \rho_l)|_{(X_i \cap |T_l|)_0}$ can be a C^1 diffeomorphism onto $(X_i \cap |T_l| \cap \rho_l^{-1}(1/2))_0 \times {]}0,1{[}$. There exists a positive small C^1 $\mathfrak{X}$-function α on X_l such that conditions (ii) and (iii) on $\{x \in X_{i'} \cap |T_l|\colon \rho_l(x) < \alpha \circ \pi_l(x)\}$ for $i = l$, $i' = 2, \ldots, l-1$ hold. Let β be a C^1 $\mathfrak{X}$-function on $X_l \times \mathbf{R}$ such that for each $x \in X_l$, $\beta(x \times [0, \alpha(x)]) = [0,1]$, and $\beta(x, \cdot)$ is a diffeomorphism of $\mathbf{R}$ equal to the identity on $[0, \alpha(x)/2]$. Replace ρ_l with $\tilde{\rho}_l(x) = \beta(\pi_l(x), \rho_l(x))$ for $x \in |T_l|$. Set

$$\tilde{A} = X_1 \cap \{\tilde{\rho}_l < 1\}, \quad \tilde{B} = B, \quad \tilde{C} = C,$$
$$\tilde{V} = \tilde{A} \cap V, \quad \tilde{\psi} = (h_l \circ \pi_l, \tilde{\rho}_l), \quad \text{and} \quad \tilde{\psi}' = \psi',$$

and define a C^1 $\mathfrak{X}$-map $\tilde{h}_1^{l-1}\colon \tilde{V} \to \tilde{V}_0$ so that

$$\gamma \circ \tilde{h}_1^{l-1} = \gamma \circ h_1^{l-1}, \quad \text{and} \quad \tilde{\rho}_l \circ \tilde{h}_1^{l-1} = \tilde{\rho}_l \quad \text{on} \quad \tilde{V}.$$

Here $\tilde{h}_1^{l-1}$ is well-defined when we shrink $|T_i|''$, $i = 2, \ldots, l-1$, suitably. Then $(\tilde{h}_1^{l-1}, f_1)\colon \tilde{V} \to \tilde{V}_0 \times C$ is a C^1 $\mathfrak{X}$-imbedding, and we have

$$\tilde{h}_1^{l-1} = \mathrm{id} \quad \text{on} \quad \tilde{V}_0 \quad \text{and because} \quad h_1^{l-1} = \mathrm{id} \quad \text{on} \quad V_0,$$
$$h_l \circ \pi_l \circ \tilde{h}_1^{l-1} = h_l \circ \pi_l \circ \gamma \tilde{h}_1^{l-1} = h_l \circ \pi_l \circ \gamma \circ h_1^{l-1} = h_l \circ \pi_l \circ h_1^{l-1}$$
$$= h_l \circ \pi_l \circ \pi_i \circ h_1^{l-1} = h_l \circ \pi_l \circ h_i \circ \pi_i = h_l \circ h_l \circ \pi_l \circ \pi_i = h_l \circ \pi_l$$
$$\text{on} \quad \tilde{V} \cap |T_i|'', \quad i = 2, \ldots, l-1.$$

Hence $\tilde{\psi} \circ \tilde{h}_1^{l-1} = \tilde{\psi}$ on $\tilde{V}$, and $\tilde{h}_1^{l-1}$ satisfies the requirements on χ.

Now by II.6.8 we can extend $\tilde{h}_1^{l-1}$ to a C^1 $\mathfrak{X}$-map $\chi\colon \tilde{A} \to \tilde{A}_0$ so that $\tilde{\psi} \circ \chi = \tilde{\psi}$. Here we shrink $\tilde{V}$ and hence V, but V does not lose its above properties. The map

$$h_1^l = \begin{cases} h_1^{l-1} & \text{on} \quad X_1 \cap \bigcup_{i=2}^{l-1} |T_i|'' \\ \chi & \text{on} \quad |T_l|' \end{cases}$$

fulfills all the requirements as follows. (Here we need to shrink $|T_i|'''$'s and $|T_l|'$.) By definition of $\tilde{h}_1^{l-1}$, $\tilde{h}_1^{l-1} = h_1^{l-1}$ on $\{x \in \tilde{V} : \rho_l(x) < \alpha \circ \pi_l(x)/2\}$. Condition (iii) holds for $h_1 = \chi$, $i' = 1$ and $i = l$ if we shrink $|T_l|'$ to be $\{x \in |T_l| : \rho_l(x) < \alpha \circ \pi_l(x)/2\}$, because $\tilde{\rho}_l = \rho_l$ there. Conditions (i) and (iv) can be satisfied for $h_1 = \chi$, $i' = 1$ and $i = l$ if we shrink $|T_l|'$, and modify χ as before. By the equality $\tilde{\psi} \circ \chi = \tilde{\psi}$ and (iv) for $i' = l$, we have

$$h_l \circ \pi_l = h_l \circ \pi_l \circ \chi = \pi_l \circ \chi \quad \text{on} \quad \tilde{A}.$$

Hence condition (ii) holds for $h_1 = \chi$, $i' = 1$ and $i = l$.

Thus by induction we obtain a C^1 𝒳-map $h_1 : X_1 \cap \cup_{i>1}|T_i|'' \to (X_1 \cap \cup_{i>1}|T_i|'')_0$ which satisfies (i),...,(iv). Moreover, by II.6.8, by shrinking $\cup_{i>1}|T_i|''$, we can extend it to the required C^1 𝒳-map $h_1 : X_1 \to X_{10}$, which completes the proof. □

Proof of II.6.1'. We proceed as in the proof of II.6.1. We assume that all the X_i, Y_j and P are bounded and connected, $0 \in P$, P is contractible, $\dim X_1 \geq \dim X_2 \geq \cdots$ and $\dim Y_1 \geq \dim Y_2 \geq \cdots$. We prove that (f, p) is 𝒳-trivial and the 𝒳-homeomorphism of triviality induces C^1 𝒳-triviality of $(f|_{X_i}, p|_{f(X_i)})$ for each i. By induction on $\dim X$ we show the 𝒳-triviality as follows. It is clear if $\dim X = 0$. (We shall use a double induction method.) By II.6.10 below and the proof of II.6.1 we have a controlled 𝒳-tube system $\{T_j' = (|T_j'|, \pi_j', \rho_j')\}$ for $\{Y_j\}$ and a tube system $\{T_i = (|T_i|, \pi_i, \rho_i)\}$ for $\{X_i\}$ controlled over $\{T_j'\}$. Furthermore, we have C^1 𝒳-maps $h_j' : Y_j \to Y_{j0}$, $j = 1, 2, \ldots$, such that for each i and for each $j' < j$,

$$\rho_i < 1, \quad \rho_{j'}' < 1, \qquad p \circ \pi_{j'} = p \text{ on } Y \cap |T_{j'}'|,$$

$(h_{j'}', p) : Y_{j'} \to Y_{j'0} \times P$ is a diffeomorphism, and the following are satisfied:

$$h_{j'}'(Y_{j'} \cap |T_j'|) = Y_{j'0} \cap |T_j'|, \tag{i}'$$
$$h_j' \circ \pi_j' = \pi_j' \circ h_{j'}' \quad \text{on} \quad Y_{j'} \cap |T_j'|, \tag{ii}'$$
$$\rho_j' \circ h_{j'}' = \rho_j' \quad \text{on} \quad Y_{j'} \cap |T_j'|, \tag{iii}'$$
$$\text{and} \quad h_{j'}' = \text{id} \quad \text{on} \quad Y_{j'0}. \tag{iv}'$$

Here Y_{j0} denotes $Y_j \cap p^{-1}(0)$, and, in general, Z_a denotes $Z \cap p^{-1}(a)$ for $Z \subset Y$ and $a \in P$. Define a map $h' : Y \to Y_0$ to be h_j' on each Y_j, which is continuous by (ii)' and (iii)'. We want to lift $\{h_j'\}$ to $\{X_i\}$. By induction we can assume

$$\dim X_1 > \dim X_2, \quad \dim Y_1 > \dim Y_2,$$
$$f(X_1) = Y_1, \quad X \subset \overline{X_1}, \quad Y \subset \overline{Y_1},$$

and that there exist C^1 $\mathfrak{X}$-maps $h_i \colon X_i \to X_{i0}$, $i = 2, \ldots$, such that for $1 < i' < i$, $(h_{i'}, p \circ f) \colon X_{i'} \to X_{i'0} \times P$ is a diffeomorphism,

$$f \circ h_{i'} = h'_{j'} \circ f \qquad \text{on } X_{i'}, \qquad \text{where } f(X_{i'}) = Y_{j'}, \qquad (0)$$
$$h_{i'}(X_{i'} \cap |T_i|) = X_{i'0} \cap |T_i|, \tag{i}$$
$$h_i \circ \pi_i = \pi_i \circ h_{i'} \qquad \text{on } X_{i'} \cap |T_i|, \tag{ii}$$
$$\rho_i \circ h_{i'} = \rho_i \qquad \text{on } X_{i'} \cap |T_i| \text{ if } f(X_i) = f(X_{i'}), \tag{iii}$$
$$\text{and} \quad h_{i'} = \text{id} \qquad \text{on } X_{i'0}. \tag{iv}$$

Here $X_{i0} = X_i \cap (p \circ f)^{-1}(0)$. We need to construct h_1 with these properties.

Let $l > 1$ be an integer. As the second induction, assume there exists a C^1 $\mathfrak{X}$-map $h_1^{l-1} \colon X_1 \cap \cup_{i=2}^{l-1} |T_i| \to X_{10}$ such that $(h_1^{l-1}, p \circ f) \colon X_1 \cap \cup_{i=2}^{l-1} |T_i| \to X_{10} \times P$ is a C^1 imbedding, for $1 < i < l$,

$$f \circ h_1^{l-1} = h'_1 \circ f \qquad \text{on } X_1 \cap \bigcup_{i=2}^{l-1} |T_i|, \tag{$0)_1^{l-1}$}$$
$$h_1^{l-1}(X_1 \cap |T_i|) = X_{10} \cap |T_i|, \tag{$(i)_1^{l-1}$}$$
$$h_i \circ \pi_i = \pi_i \circ h_1^{l-1} \qquad \text{on } X_1 \cap |T_i|, \tag{$(ii)_1^{l-1}$}$$
$$\rho_i \circ h_1^{l-1} = \rho_i \qquad \text{on } X_1 \cap |T_i| \text{ if } f(X_i) = Y_1, \tag{$(iii)_1^{l-1}$}$$
$$\text{and} \quad h_1^{l-1} = \text{id} \qquad \text{on } X_{10} \cap \bigcup_{i=2}^{l-1} |T_i|. \tag{$(iv)_1^{l-1}$}$$

We want to construct h_1^l.

Case of $f(X_l) = Y_1$. This case is easy because we can forget Y. Note that (iii) holds for $i = l$ and any $i' \neq 1$ with $X_l \subset \overline{X_{i'}}$. As in the proof of II.6.1, we try to apply II.6.8 with $A = X_1 \cap |T_l|$, $B = X_{l0} \times {]}0, 1{[}$, $C = P$, $V = X_1 \cap |T_l| \cap \cup_{i=2}^{l-1} |T_i|$, $W = \widehat{W} = \varnothing$, $\psi = (h_l \circ \pi_l, \rho_l)$, and $\psi' = p \circ f$. Neither the data in the proof of II.6.1 nor these satisfy the conditions in II.6.8, and we modify these in the same way. Shrink $|T_i|$, $i = 2, \ldots, l$, and modify ρ_l and h_1^{l-1} outside of a small $\mathfrak{X}$-neighborhood of X_l in $|T_l|$ and its intersection with $X_1 \cap \cup_{i=2}^{l-1} |T_i|$ respectively. Let $\tilde{\rho}_l$ and $\tilde{h}_1^{l-1}$ denote the modified maps. Then the conditions in II.6.8 are satisfied, and we have a C^1 $\mathfrak{X}$-map $h_1^l \colon X_1 \cap |T_l| \to X_{10} \cap |T_l|$ such that $(h_1^l, p \circ f) \colon X_1 \cap |T_l| \to (X_{10} \cap |T_l|) \times P$ is a diffeomorphism, $h_1^l = \tilde{h}_1^{l-1}$ on $X_1 \cap |T_l| \cap \cup_{i=2}^{l-1} |T_i|$ for shrunk $|T_i|$'s, $h_l \circ \pi_l = \pi_l \circ h_1^l$, $\tilde{\rho}_l \circ h_1^l = \tilde{\rho}_l$, and $h_1^l = \text{id}$ on $X_{10} \cap |T_l|$. The equality $f \circ h_1^l = h'_1 \circ f$ on $X_1 \cap |T_l|$ is automatically satisfied. Indeed, since $f \circ h_l = h'_1 \circ f$ on X_l and $\{T_i\}$ is controlled over $\{T'_j\}$, we have

$$f \circ h_1^l = \pi'_1 \circ f \circ h_1^l = f \circ \pi_1 \circ h_1^l = f \circ h_l \circ \pi_l = h'_1 \circ f \circ \pi_l = h'_1 \circ \pi'_1 \circ f = h'_1 \circ f$$
$$\text{on } X_1 \cap |T_l|.$$

Shrink $|T_l|$ so that $\tilde{\rho}_l = \rho_l$ on $|T_l|$ and $\tilde{h}_1^{l-1} = h_1^{l-1}$ on $X_1 \cap |T_l| \cap \cup_{i=2}^{l-1}|T_i|$ keeping the property $h_1^l(X_1 \cap |T_l|) = X_{10} \cap |T_l|$, $((h_1^l, p \circ f)$ is no more a diffeomorphism but a C^1 imbedding). Define h_1^l on $\cup_{i=2}^{l-1}|T_i| - |T_l|$ to be h_1^{l-1}. Then h_1^l satisfies conditions $(0)_1^l,\ldots,(iv)_1^l$.

Case where if $X_l \subset \overline{X_i} - X_i$, ***then*** $f(X_l) \cap f(X_i) = \varnothing$. Note that $f(X_l) \cap Y_1 = \varnothing$. The proof is similar to the above case. But now $W \neq \varnothing$, and the arguments become complicated. Let $f(X_l) = Y_{l'}$. As the problem is considered only around X_l, we assume $\overline{X_i} \supset X_l$ and $\overline{Y_j} \supset Y_{l'}$ for all i, j. Then $f^{-1}(f(X_l)) = X_l$ and properness of f fails. To apply II.6.8 we set

$$A = X_1 \cap |T_l|, \qquad B = \{(x,y) \in X_{l0} \times (Y_{10} \cap |T_{l'}'|): f(x) = \pi_{l'}'(y)\},$$

$$C = P, \qquad V = X_1 \cap |T_l| \cap \bigcup_{\substack{1 < i < l \\ f(X_i) = Y_1}} |T_i|,$$

$$W = (h_l \circ \pi_l, h_1' \circ f, p \circ f)(X_1 \cap |T_l| \cap \bigcup_{\substack{1 < i < l \\ f(X_i) \neq Y_1}} |T_i|),$$

$$\widehat{W}_1 = \{(x,y) \in X_{l0} \times (\bigcup_{j \neq 1, l'} Y_{j0} \cap |T_{l'}'|): f(x) = \pi_{l'}'(y)\}, \qquad \widehat{W} = \widehat{W}_1 \times C,$$

$$\Psi = (\psi, \psi') = (h_l \circ \pi_l, h_1' \circ f, p \circ f): A \longrightarrow B \times C, \quad \text{and}$$

$$\chi = h_1^{l-1} \quad \text{on} \quad X_1 \cap |T_l| \cap \bigcup_{i=2}^{l-1} |T_i|.$$

Note that

$$h_l \circ \pi_l = h_l \circ \pi_l \circ \pi_i = \pi_l \circ h_i \circ \pi_i = \pi_l \circ \pi_i \circ h_1^{l-1} = \pi_l \circ h_1^{l-1} \quad \text{and}$$
$$h_1' \circ f = f \circ h_1^{l-1} \quad \text{on} \quad X_1 \cap |T_l| \cap |T_i|, \quad i = 2, \ldots, l-1,$$

which implies

$$\psi = (\pi_l \circ h_1^{l-1}, f \circ h_1^{l-1}) \quad \text{on} \quad X_1 \cap |T_l| \cap \bigcup_{i=2}^{l-1} |T_i|.$$

We check if these data satisfy the conditions in II.6.8.

The B is a C^1 $\mathfrak{X}$-submanifold of $\mathbf{R}^n \times \mathbf{R}^m$ because $f|_{X_{l0}}$ and $\pi_{l'}'|_{Y_{10} \cap |T_{l'}'|}$ are C^1 $\mathfrak{X}$-submersions. The $\widehat{W}_1$ is open in $\overline{B} - B$ because

$$\widehat{W}_1 = \overline{B} - B - (\overline{X_{l0}} - X_{l0}) \times Y_0 - \overline{X_{l0}} \times (|\overline{T_{l'}'}|_0 - |T_{l'}'|_0) - \overline{X_{l0}} \times \overline{Y_{l'0}}.$$

We have the following difficulties (a), (b), (d) and (e). **(a)** W is not necessarily the intersection of $B \times C$ with an open $\mathfrak{X}$-neighborhood of $\widehat{W}$ in $\overline{B} \times C$ because we do not assume

$$h_i(X_{ia} \cap |T_l|) = X_{i0} \cap |T_l| \quad \text{for all } a \in P, \ i = 2, \dots, l-1.$$

(b) $\Psi|_{A-V}$ is not always surjective nor proper by a similar reason.

But Ψ can be a C^1 $\mathfrak{X}$-submersion as follows. For this it suffices to see that for each $a \in P$, $(h_l \circ \pi_l, h_1' \circ f)|_{X_{1a} \cap |T_l|} : X_{1a} \cap |T_l| \to B$ is a C^1 submersion because $p \circ f$ is a C^1 submersion onto P. Moreover, since the map

$$(h_l, h_1') \colon \{(x, y) \in X_{la} \times (Y_{1a} \cap |T_{l'}'|) : f(x) = \pi_l'(y)\} \longrightarrow B$$

is a C^1 imbedding, we need only prove that the map $(\pi_l, f) \colon X_{1a} \cap |T_l| \to$ the above domain of (h_l, h_1') is a C^1 submersion. For simplicity of notation we consider only the case $a = 0$. Then the map is $(\pi_l, f) \colon X_{10} \cap |T_l| \to B$. We reduce the problem further. The composite of this map with the C^1 submersion $B \ni (x, y) \to f(x) \in Y_{l'0}$ is $f \circ \pi_l \colon X_{10} \cap |T_l| \to Y_{l'0}$ and hence is a C^1 submersion. Therefore, we have to prove that for each $b \in Y_{l'0}$, the map

$$(\pi_l, f) \colon X_{10} \cap |T_l| \cap (f \circ \pi_l)^{-1}(b)$$
$$\longrightarrow \{(x, y) \in X_{10} \times (Y_{10} \cap |T_{l'}'|) : f(x) = \pi_{l'}'(y) = b\}$$

is a C^1 submersion. Note that this target space equals the product:

$$\{x \in X_{l0} : f(x) = b\} \times \{y \in Y_{10} \cap |T_{l'}'| : \pi_{l'}'(y) = b\}.$$

By the condition of sans éclatement, for each $x \in X_{10} \cap |T_l|$ near X_l, the tangent spaces $T_x(X_{10} \cap \pi_l^{-1}(\pi_l(x)))$ and $T_x(X_{10} \cap f^{-1}(f(x)))$ intersect transversally in the space $T_x(X_{10} \cap (f \circ \pi_l)^{-1}(b))$. Hence the map is a C^1 submersion because the maps

$$\pi_l \colon X_{10} \cap |T_l| \cap (f \circ \pi_l)^{-1}(b) \longrightarrow X_{10} \cap f^{-1}(b) \quad \text{and}$$
$$f \colon X_{10} \cap |T_l| \cap (f \circ \pi_l)^{-1}(b) \longrightarrow Y_1 \cap |T_{l'}'| \cap \pi_{l'}'^{-1}(b)$$

are C^1 submersions. Thus by shrinking $|T_l|$, we can assume Ψ is a C^1 submersion.

By the same reason we can suppose, moreover, that

(c) for each i and j with $f(X_i) = Y_j$, the map

$$(h_l \circ \pi_l, h_j' \circ f, p \circ f) \colon X_i \cap |T_l| \longrightarrow \{(x, y) \in X_{l0} \times (Y_{j0} \cap |T_{l'}'|) : f(x) = \pi_{l'}'(y)\} \times C$$

is a C^1 submersion.

(d) The map $\psi|_{\psi^{-1}(B')_0}: \psi^{-1}(B')_0 \to B'$ is not always C^1 $\mathfrak{X}$-trivial for a contractible open $\mathfrak{X}$-subset B' of B by a reason similar to (a).

(e) χ is defined on V but not necessarily on $\Psi^{-1}(W)$ because $\Psi^{-1}(W)$ may be larger than $X_1 \cap |T_l| \cap \cup_{i=2}^{l-1} |T_i|$.

It is clear that $(\chi, \psi'): X_1 \cap |T_l| \cap \cup_{i=2}^{l-1} |T_i| \to (X_{10} \cap |T_l| \cap \cup_{i=2}^{l-1} |T_i|) \times C$ is a C^1 imbedding, $\chi = \mathrm{id}$ on $X_{10} \cap |T_l| \cap \cup_{i=2}^{l-1} |T_i|$ and $\psi \circ \chi = \psi$.

We need to modify these data outside of small $\mathfrak{X}$-neighborhoods of X_l and $Y_{l'}$ and to correct (a), (b), (d) and (e). We do not change notation after each step of shrinking and modification. From now on we do not need T_i, T_j' for $i \neq l$, $j \neq l'$, nor we do not ask if their properties, e.g., (ii)$'$ for $j \neq l'$, hold. But we have to keep the properties of $T_l, T_{l'}'$ and $f|_{|T_l|}$. If one fails, we mention and correct later it.

First we shrink $|T_{l'}'|_0$ and modify $\rho_{l'}'|_{|T_{l'}'|_0}$ outside a smaller $\mathfrak{X}$-neighborhood of $Y_{l'0}$ so that the closure of new $|T_{l'}'|_0$ is contained in original $|T_{l'}'|_0$, the map

$$(\pi_{l'}', \rho_{l'}'): Y_0 \cap |T_{l'}'| - Y_{l'} \longrightarrow Y_{l'0} \times]0, 1[$$

is proper and $\mathfrak{X}$-trivial over any contractible open $\mathfrak{X}$-subset of $Y_{l'0} \times]0, 1[$, and the $\mathfrak{X}$-map of triviality induces a C^1 $\mathfrak{X}$-triviality of $(\pi_{l'}', \rho_{l'}')|_{Y_{j0} \cap |T_{l'}'|}$ over such a set for each $j \neq l'$. This is easy by II.6.1. Here (i)$'$, (iii)$'$ for $j = l'$ and the conditions that $f(X \cap |T_l|) \subset |T_{l'}'|$ may fail. Replace $Y \cap |T_{l'}'|$ with h'^{-1}(the new $Y_0 \cap |T_{l'}'|$), and $X \cap |T_l|$ with $(h' \circ f)^{-1}$(the new $Y_0 \cap |T_{l'}'|$). (The new $|T_{l'}'|$ is the union of $|T_{l'}'| - Y$ and the new $Y \cap |T_{l'}'|$. This is important, and we do not mention the outside of Y or of X.) Then (i)$'$ for $j = l'$ and $f(X \cap |T_l|) \subset |T_{l'}'|$ hold. For (iii)$'$ for $j = l'$, it suffices to replace $\rho_{l'}'$ with (the new $\rho_{l'}'$ on $Y_0 \cap |T_{l'}'|$) $\circ h'$.

By the above triviality we have an $\mathfrak{X}$-map $\alpha: Y_0 \cap |T_{l'}'| - Y_{l'} \to Y_0 \cap |T_{l'}'| \cap \rho_{l'}'^{-1}(1/2)$ such that $(\alpha, \rho_{l'}')$ is a homeomorphism,

$$\pi_{l'}' \circ \alpha = \pi_{l'}', \qquad \alpha = \mathrm{id} \quad \text{on} \quad Y_0 \cap |T_{l'}'| \cap \rho_{l'}'^{-1}(1/2),$$

and for each $j \neq l'$, $(\alpha, \rho_{l'}')|_{Y_{j0} \cap |T_{l'}'|}$ is a C^1 diffeomorphism onto $(Y_{j0} \cap |T_{l'}'| \cap \rho_{l'}'^{-1}(1/2)) \times]0, 1[$.

We want to lift α to $X_0 \cap |T_l| - X_l$. Set

$$\hat{B}_j = \{(x, y) \in X_{l0} \times (Y_{j0} \cap |T_{l'}'|): f(x) = \pi_{l'}'(y)\}, \quad j \neq l' \quad \text{and}$$

$$\hat{B} = \bigcup_j \hat{B}_j.$$

Then $\hat{B}_1 = B$, $\{\hat{B}_j\}$ is a C^1 Whitney $\mathfrak{X}$-stratification of $\hat{B}$, and by (c), the map $(\pi_l, f): \{X_{i0} \cap |T_l| - X_l\} \to \{\hat{B}_j\}$ is a C^1 Whitney stratified map sans éclatement. For the lifting we shrink $|T_l|_0$ and $\hat{B}$. There exists a C^1 $\mathfrak{X}$-function ε on X_{l0} such that $0 < \varepsilon < 1$ and

$$\rho'_{l'} \circ f(x) > \varepsilon \circ \pi_l(x) \quad \text{for} \quad x \in X_0 \cap (\overline{|T_l|_0} - |T_l|),$$

because $f^{-1}(Y_{l'}) = X_l$. Replace $X_0 \cap |T_l|$ with $\{x \in X_0 \cap |T_l| : \rho'_{l'} \circ f(x) < \varepsilon \circ \pi_l(x)\}$. Then for each $x \in X_{l0}$, $f|_{X_0 \cap \pi_l^{-1}(x)}$ is surjective and proper onto $\{y \in Y_0 \cap \pi_{l'}^{-1}(f(x)) : \rho'_{l'}(y) < \varepsilon(x)\}$. Hence the map

$$(\pi_l, f) \colon X_0 \cap |T_l| \longrightarrow \{(x, y) \in \hat{B} \subset X_{l0} \times (Y_0 \cap |T'_{l'}| - Y_{l'}) : \rho'_{l'}(y) < \varepsilon(x)\}$$

is surjective and proper. Set $U = $ the target space. Let δ be a C^1 $\mathfrak{X}$-function on $\{(s,t) \in \,]0,1[^2 : t < s\}$ such that for each $s \in \,]0,1[$, $\delta(s, \cdot)$ is a diffeomorphism from $]0, s[$ to $]0, 1[$ equal to the identity on $]0, s/2[$. x The sequence of maps

$$\{X_{i0} \cap |T_l| - X_l\} \xrightarrow{(\pi_l, f)} \{\hat{B}_j \cap U\} \ni (x, y) \longrightarrow \delta(\varepsilon(x), \rho'_{l'}(y)) \in \,]0, 1[$$

satisfies the conditions of II.6.1′, and $\dim \cup_i (X_{i0} \cap |T_l| - X_l) < \dim X$. Hence by the first induction hypothesis, $\mathfrak{X}$-triviality of the map $\{\hat{B}_j \cap U\} \to \,]0, 1[$ can be lifted to the map $\{X_{i0} \cap |T_l| - X_l\} \to \,]0, 1[$. Particularly, take the $\mathfrak{X}$-map of triviality $\hat{B} \cap U \to \hat{B} \cap U \cap \delta(\varepsilon, \rho'_{l'})^{-1}(1/2)$ induced from the α. Then we have an $\mathfrak{X}$-map

$$\beta \colon X_0 \cap |T_l| - X_l \longrightarrow X_0 \cap |T_l| \cap \delta(\varepsilon \circ \pi_l, \rho'_{l'} \circ f)^{-1}(1/2)$$

which is a lift of α, such that $(\beta, \delta(\varepsilon \circ \pi_l, \rho'_{l'} \circ f)) \colon X_0 \cap |T_l| - X_l \to (X_0 \cap |T_l| \cap \delta(\varepsilon \circ \pi_l, \rho'_{l'} \circ f)^{-1}(1/2)) \times \,]0, 1[$ is an $\mathfrak{X}$-homeomorphism, for each $i \neq l$, $(\beta, \delta(\varepsilon \circ \pi_l, \rho'_{l'} \circ f))|_{X_{i0} \cap |T_l|}$ is a C^1 diffeomorphism onto $(X_{i0} \cap |T_l| \cap \delta(\varepsilon \circ \pi_l, \rho'_{l'} \circ f)^{-1}(1/2)) \times \,]0, 1[$,

$$\beta = \mathrm{id} \quad \text{on} \quad X_0 \cap |T_l| \cap \delta(\varepsilon \circ \pi_l, \rho'_{l'} \circ f)^{-1}(1/2),$$
$$\pi_l = \pi_l \circ \beta, \quad \text{and} \quad \alpha \circ f = \alpha \circ f \circ \beta.$$

Second, we shrink $|T'_{l'}|$ outside $|T'_{l'}|_0$ and modify $\rho'_{l'}$ and h' so that the map $(h', p) \colon Y \cap |T'_{l'}| \to (Y_0 \cap |T'_{l'}|) \times P$ is a homeomorphism as follows. There exists a C^1 $\mathfrak{X}$-function θ on $Y_{l'}$ such that

$$0 < \theta \leq 1, \qquad \theta = 1 \quad \text{on} \quad Y_{l'0} \text{ and}$$
$$h'(\pi_{l'}'^{-1}(y)) \supset \{y' \in Y_0 \cap |T'_{l'}| : \pi_{l'}'(y') = h'_{l'}(y), \ \rho'_{l'}(y') < \theta(y)\} \quad \text{for} \quad y \in Y_{l'}.$$

Replace $|T'_{l'}|$ with $|\tilde{T}'_{l'}| = \{y \in |T'_{l'}|: \rho'_{l'}(y) < \theta \circ \pi'_{l'}(y)\}$, $\rho'_{l'}$ with $\tilde{\rho}'_{l'} = \delta(\theta \circ \pi'_{l'}, \rho'_{l'})$ and $h'|_{Y \cap |T'_{l'}|}$ with an 𝔛-map $\tilde{h}'$ such that

$$\tilde{\rho}'_{l'} \circ \tilde{h}' = \tilde{\rho}'_{l'} \qquad \text{on} \quad Y \cap |\tilde{T}'_{l'}|,$$
$$\alpha \circ h' = \alpha \circ \tilde{h}' \qquad \text{on} \quad Y \cap |\tilde{T}'_{l'}| - Y_{l'}, \quad \text{and}$$
$$\tilde{h}' = h' \qquad \text{on} \quad Y_{l'}.$$

Such $\tilde{h}'$ is uniquely decided, and $|\tilde{T}'_{l'}|$, $\tilde{\rho}'_{l'}$ and $\tilde{h}'$ fulfill all the requirements. We keep the notation $|T'_{l'}|$, $\rho'_{l'}$ and h'.

Since we changed $|T'_{l'}|$ and h', the properties (0) for $i' \neq 1, l$, $(0)_1^{l-1}$, and $f(X \cap |T_l|) \subset \overline{|T'_{l'}|}$ may fail. So we shrink $|T_l|$ and modify $h_{i'}$, $i' \neq 1, l$, and h_1^{l-1} as follows. Note that by the definition of $\tilde{h}'$, we have

$$\alpha \circ f \circ h_{i'} = \alpha \circ h' \circ f \quad \text{and} \quad \alpha \circ f \circ h_1^{l-1} = \alpha \circ h' \circ f.$$

The inclusion $f(X \cap |T_l|) \subset |T'_{l'}|$ is satisfied if we shrink $|T_l|$. Replace $h_{i'}$ with a C^1 𝔛-map $\tilde{h}_{i'}$ such that $\beta \circ h_{i'} = \beta \circ \tilde{h}_{i'}$ and $\rho'_{l'} \circ f = \rho'_{l'} \circ f \circ \tilde{h}_{i'}$. This is clearly uniquely decided if we shrink $|T_l|$, further. Note that

$$\pi_l \circ h_{i'} = \pi_l \circ \beta \circ h_{i'} = \pi_l \circ \beta \circ \tilde{h}_{i'} = \pi_l \circ \tilde{h}_{i'}.$$

Set $\tilde{h}_l = h_l$, and define $\tilde{h}_1^{l-1}$ in the same way as $\tilde{h}_{i'}$. Then

$$\alpha \circ f \circ \tilde{h}_{i'} = \alpha \circ f \circ \beta \circ \tilde{h}_{i'} = \alpha \circ f \circ \beta \circ h_{i'} = \alpha \circ f \circ h_{i'} = \alpha \circ h' \circ f \quad \text{and}$$
$$\rho'_{l'} \circ f \circ \tilde{h}_{i'} = \rho'_{l'} \circ f = \rho'_{l'} \circ h' \circ f.$$

Hence (0) $f \circ \tilde{h}_{i'} = h' \circ f$. In the same way we see $(0)_1^{l-1}$ $f \circ \tilde{h}_1^{l-1} = h' \circ f$.

Third, we shrink $X_0 \cap |T_l|$ and modify $f|_{X_0 \cap |T_l|}$ so that (d) is corrected. Since $(\pi_l, f): \{X_{i0} \cap |T_l| - X_l\} \to \{\hat{B}_j\}$ is a C^r Whitney 𝔛-stratified map sans éclatement, keeping this property, we need only the property that $(\pi_l, f)|_{X_0 \cap |T_l|}$ is surjective and proper onto $\hat{B}$. For that it suffices to shrink $X_0 \cap |T_l|$ and modify $f|_{X_0 \cap |T_l|}$ outside of X_l so that for each $x \in X_l$, $f|_{X_0 \cap \pi_l^{-1}(x)}$ is surjective and proper onto $Y_0 \cap \pi'^{-1}_{l'}(f(x))$. Shrink $|T_l|_0$ so that $\overline{|T_l|_0}$ is contained in the original $|T_l|_0$ and hence $\rho'_{l'} \circ f$ and π_l are well-defined on $\overline{|T_l|_0}$. There exists a C^1 𝔛-function θ on X_{l0} such that $0 < \theta < 1$, and

$$\rho'_{l'} \circ f(x) > \theta \circ \pi_l(x) \quad \text{for} \quad x \in X_0 \cap (\overline{|T_l|_0} - |T_l|).$$

Replace $X_0 \cap |T_l|$ with $\{x \in X_0 \cap |T_l|: \rho'_{l'} \circ f(x) < \theta \circ \pi_l(x)\}$. Then for each $x \in X_{l0}$, $f|_{X_0 \cap \pi_l^{-1}(x)}$ is surjective and proper onto $\{y \in Y_0 \cap \pi'^{-1}_{l'}(f(x)): \rho'_{l'}(y) <$

$\theta(x)\}$. Next replace $f|_{X_0 \cap |T_l|}$ with an $\mathfrak{X}$-map $\tilde{f}$ such that $\tilde{f} = f$ on X_l, $\alpha \circ \tilde{f} = \alpha \circ f$ on $X_0 \cap |T_l| - X_l$, and $\rho'_{l'} \circ \tilde{f} = \delta(\theta \circ \pi_l, \rho'_{l'} \circ f)$. Keep the notation f. Then $f|_{X_0 \cap \pi_i^{-1}(x)}$ is surjective and proper onto $Y_0 \cap \pi_{l'}^{\prime -1}(f(x))$. Thus we correct (d). Here $f|_{X_0 \cap |T_l|}$ is only a C^1 $\mathfrak{X}$-stratified map. It may be not of class C^1 nor a stratified map sans éclatement outside of an $\mathfrak{X}$-neighborhood of X_l, but we do not need these properties.

In this step also, we need to shrink $|T_l|$ outside $|T_l|_0$ and modify $f|_{X \cap |T_l|}$ outside $X_l \cup |T_l|_0$ because (0), (i), $(0)_1^{l-1}$ and $(i)_1^{l-1}$ may fail. (i) and $(i)_1^{l-1}$ are easily satisfied. Recall the above modification of $f|_{X_0 \cap |T_l|}$. There exists a family of $\mathfrak{X}$-homeomorphisms:

$$\gamma_t \colon \{y \in Y_0 \cap |T'_{l'}| \colon \rho'_{l'}(y) < t\} \longrightarrow Y_0 \cap |T'_{l'}|, \quad t \in \,]0,1],$$

such that the new f is $\gamma_{\theta \circ \pi_l(x)} \circ f(x)$, for each t,

$$\gamma_1 = \mathrm{id}, \qquad \pi'_{l'} \circ \gamma_t = \pi'_{l'},$$
$$\gamma_t = \mathrm{id} \quad \text{on} \quad \{y \in Y_0 \cap |T'_{l'}| \colon \rho'_{l'}(y) < t/2\},$$
$$\rho'_{l'} \circ \gamma_t = \mathrm{const} \quad \text{on} \quad Y_0 \cap |T'_{l'}| \cap \rho'^{-1}_{l'}(\varepsilon) \quad \text{for each } \varepsilon \in [t/2, t[,$$

for each $j \neq l'$, $\gamma_t|_{\{y \in Y_{j0} \cap |T'_{l'}| \colon \rho'_{l'}(y) < t\}}$ is a C^1 diffeomorphism onto $Y_{j0} \cap |T'_{l'}|$, and the map:

$$\{(y,t) \in (Y_0 \cap |T'_{l'}|) \times \,]0,1] \colon \rho'_{l'}(y) < t\} \ni (y,t) \longrightarrow \gamma_t(y) \in Y_0 \cap |T'_{l'}|$$

and its restriction to the intersection with $(Y_{j0} \cap |T'_{l'}|) \times \,]0,1]$ are of class $\mathfrak{X}$ and C^1 $\mathfrak{X}$ respectively. We define the new $f|_{X \cap |T_l| - |T_l|_0}$ to be $(h'|_{Y_{p \circ f(x)}})^{-1} \circ \gamma_{\theta \circ h_l \circ \pi_l(x)} \circ h' \circ f(x)$. If we shrink $|T_l|$, then this is well-defined and satisfies (0) and $(0)_1^{l-1}$.

Now β is an $\mathfrak{X}$-map from $X_0 \cap |T_l| - X_l$ to $X_0 \cap |T_l| \cap (\rho'_{l'} \circ f)^{-1}(1/2)$ such that $(\beta, \rho'_{l'} \circ f) \colon X_0 \cap |T_l| - X_l \to (X_0 \cap |T_l| \cap (\rho'_{l'} \circ f)^{-1}(1/2)) \times \,]0,1[$ is an $\mathfrak{X}$-homeomorphism, for each $i \neq l$, $(\beta, \rho'_{l'} \circ f)|_{X_{i0} \cap |T_l|}$ is a C^1 diffeomorphism onto $(X_{i0} \cap |T_l| \cap (\rho'_{l'} \circ f)^{-1}(1/2)) \times \,]0,1[$, and

$$\pi_l = \pi_l \circ \beta, \qquad \alpha \circ f = f \circ \beta,$$
$$\beta = \mathrm{id} \quad \text{on} \quad X_0 \cap |T_l| \cap (\rho'_{l'} \circ f)^{-1}(1/2).$$

By β we can lift γ_t to $\mathfrak{X}$-homeomorphisms:

$$\mu_t \colon \{x \in X_0 \cap |T_l| \colon \rho'_{l'} \circ f(x) < t\} \longrightarrow X_0 \cap |T_l|, \quad t \in \,]0,1],$$

so that for each $t \in \,]0,1]$,

$$\gamma_t \circ f = f \circ \mu_t, \qquad \mu_1 = \mathrm{id}, \qquad \pi_l \circ \mu_t = \pi_l,$$
$$\mu_t = \mathrm{id} \quad \text{on} \quad \{x \in X_0 \cap |T_l| : \rho'_{l'} \circ f(x) < t/2\},$$
$$\rho'_{l'} \circ f \circ \mu_t = \mathrm{const} \quad \text{on} \quad X_0 \cap |T_l| \cap (\rho'_{l'} \circ f)^{-1}(\varepsilon) \quad \text{for each } \varepsilon \in [t/2, t[,$$

for each $i \neq l$, $\mu_t|_{\{x \in X_{i0} \cap |T_l| : \rho'_{l'} \circ f(x) < t\}}$ is a C^1 diffeomorphism onto $X_{i0} \cap |T_l|$, and the map

$$\{(x,t) \in (X_0 \cap |T_l|) \times \,]0,1] : \rho'_{l'} \circ f(x) < t\} \ni (x,t) \longrightarrow \mu_t(x) \in X_0 \cap |T_l|$$

and its restriction to the intersection with $(X_{i0} \cap |T_l|) \times \,]0,1]$ are of class $\mathfrak{X}$ and $C^1 \, \mathfrak{X}$ respectively.

Finally, we will shrink $|T_l| - |T_l|_0$ and modify f, h_i and h_1^{l-1} so that for each $a \in P$, $f|_{X_a \cap |T_l|}$ is surjective and proper onto $Y_a \cap |T'_{l'}|$. This is enough for correction of (a) and (b). For (e), it suffices to replace W with $B \times C - \Psi(\Psi^{-1}(B \times C - W))$ because by surjectiveness and properness of $f|_{X_a \cap |T_l|}$, the new W also is the intersection of $B \times C$ with an open $\mathfrak{X}$-neighborhood of $\widehat{W}$ in $\overline{B} \times C$. As before, we have a $C^1 \, \mathfrak{X}$-function ξ on X_l such that $0 < \xi \leq 1$, $\xi = 1$ on X_{l0}, and for $x \in X_l$, $f|_{X \cap \pi_l^{-1}(x) \cap (\rho'_{l'} \circ f)^{-1}([0, \xi(x)[)}$ is surjective and proper onto $Y \cap \pi_{l'}'^{-1}(f(x)) \cap \rho_{l'}'^{-1}([0, \xi(x)[)$. Here the condition $\xi = 1$ on X_{l0} can be satisfied because we already shrank $|T_l|_0$ so that $\overline{|T_l|_0} \subset$ the original $|T_l|_0$. Replace $X \cap |T_l|$ with $\{x \in X \cap |T_l| : \rho'_{l'} \circ f(x) < \xi \circ \pi_l(x)\}$, f with $\gamma_{\pi_l(x)} \circ f(x)$ and h_i, h_1^{l-1} with $\beta_{\pi_l(x)} \circ h_i$ and $\beta_{\pi_l(x)} \circ h_1^{l-1}$ respectively. Here we shrink $|T_i|$, $i = 2, \ldots, l-1$, so that the new h_1^{l-1} carries each $X_1 \cap |T_l| \cap |T_i|$ to $X_{10} \cap |T_l| \cap |T_i|$. But it continues to be an $\mathfrak{X}$-neighborhood of X_i. Then $f|_{X_a \cap |T_l|}$ is surjective and proper onto $Y_a \cap |T'_{l'}|$, and all the conditions are kept.

Now we can apply II.6.8, and we obtain an $\mathfrak{X}$-extension of $h_1^{l-1}|_{X_1 \cap |T_l| \cap \cup_{i=2}^{l-1} |T_i|}$ to $h_1^l : X_1 \cap |T_l| \to X_{10} \cap |T_l|$ such that $(h_1^l, p \circ f) : X_1 \cap |T_l| \to (X_{10} \cap |T_l|) \times P$ is a C^1 diffeomorphism,

$$h_l \circ \pi_l \circ h_1^l = h_l \circ \pi_l, \quad \text{and} \quad h_1' \circ f \circ h_1^l = h_1' \circ f.$$

Since
$$h_l \circ \pi_l \circ h_1^l = \pi_l \circ h_1^l, \quad \text{and} \quad h_1' \circ f \circ h_1^l = f \circ h_1^l,$$

it follows that

$$(0)_1^l \ f \circ h_1^l = h_1' \circ f \quad \text{and} \quad (ii)_1^l \ h_l \circ \pi_l = \pi_l \circ h_1^l, \quad \text{on } X_1 \cap |T_l|.$$

(i)$_1^l$ for $i = l$ is clear, and (iv)$_1^l$ on $X_{10} \cap |T_l|$ can be satisfied. Since h_1^l is an extension of $h_1^{l-1}|_{X_1 \cap |T_l| \cap \cup_{i=2}^{l-1}|T_i|}$ and we did not change h_1^{l-1} on an 𝔛-neighborhood of X_l in $X \cap |T_l|$, h_1^l becomes a real extension of the original h_1^{l-1} if we shrink $|T_l|$ further. Thus we complete the construction of h_1^l in the second case.

Case where there exists i such that $X_l \subset \overline{X_i} - X_i$ and $f(X_l) = f(X_i) \neq Y_1$. Let $f(X_l) = Y_{l'}$. As above we can suppose $\overline{X_i} \supset X_l$ and $\overline{Y_j} \supset Y_{l'}$ for all i, j. Changing order, we assume for some $1 < l^* < l$,

$$f(X_l) = f(X_i), \quad i = 2, \ldots, l^*, \quad \text{and}$$
$$f(X_l) \neq f(X_i), \quad i = l^* + 1, \ldots, l - 1.$$

The reason why the above proof does not work is that the map $(\pi_l, f) \colon X_0 \cap |T_l| \to U$ in the previous case cannot be proper because $f^{-1}(f(X_l)) \neq X_l$. We want to modify $\cup_{i=2}^{l^*} X_i$ so that it is a C^1 𝔛-manifold and $f(\cup_{i=2}^{l^*} X_i) \neq f(X_l)$, and to reduce this case to the case where the above map is proper.

At first, we consider only $f \colon X_0 \to Y_0$. By (c) and the same arguments as above, for each $x \in X_{l0}$ and for each small positive number ε, there exists $\delta > 0$ such that the map

$$f \colon \{x' \in X_0 \cap |T_l| \cap \cup_{i=2}^{l^*}|T_i| \colon \pi_l(x') = x, \ \rho_l(x') = \varepsilon, \ 0 \le \rho_{l'}' \circ f(x') < 2\delta\}$$
$$\longrightarrow \{y \in Y_0 \cap |T_{l'}'| \colon \pi_{l'}'(y) = f(x), \ 0 \le \rho_{l'}'(y) < 2\delta\}$$

is surjective and proper, and its restriction to the intersection of this domain with each X_{i0} is C^1 submersive onto the intersection of the target space with $f(X_{i0})$. Hence we have positive C^1 𝔛-functions ε on X_{l0} and δ on $\{(x, s) \in X_{l0} \times \mathbf{R} \colon 0 < s \le \varepsilon(x)\}$ such that for each $x \in X_{l0}$ and for each $(s, t) \in \mathbf{R}^2$ with $0 < s \le \varepsilon(x)$ and $0 < t \le \delta(x, s)$, the map

$$f \colon \{x' \in X_0 \cap |T_l| \cap \cup_{i=2}^{l^*}|T_i| \colon \pi_l(x') = x, \ \rho_l(x') = s, \ \rho_{l'}' \circ f(x') = t\}$$
$$\longrightarrow \{y \in Y_0 \cap |T_{l'}'| \colon \pi_{l'}'(y) = f(x), \ \rho_{l'}'(y) = t\}$$

is surjective and proper, and the same statement as above about restriction holds. Therefore, if we choose δ so that for each $x \in X_{l0}$, $\delta(x, \cdot)$ is C^1 regular, then for each $x \in X_l$, the map

$$f \colon \{x' \in X_0 \cap |T_l| \cap \cup_{i=2}^{l^*}|T_i|$$
$$: \pi_l(x') = x, \ \rho_l(x') < \varepsilon(x), \ \rho_{l'}' \circ f(x') = \delta(x, \rho_l(x'))\}$$
$$\longrightarrow \{y \in Y_0 \cap |T_{l'}'| \colon \pi_{l'}'(y) = f(x), \ \rho_{l'}'(y') < \delta(x, \varepsilon(x))\}$$

is surjective and proper, and the statement about restriction holds. Assume δ is as above. It follows that the map

$$(\pi_l, f): \{x \in X_0 \cap |T_l| \cap \cup_{i=2}^{l^*}|T_i|$$
$$: \rho_l(x) < \varepsilon \circ \pi_l(x), \; \rho'_{l'} \circ f(x) = \delta(\pi_l(x), \rho_l(x))\}$$
$$\longrightarrow \{(x, y) \in X_{l0} \times (Y_0 \cap |T'_{l'}|): f(x) = \pi'_{l'}(y), \; \rho'_{l'}(y) < \delta(x, \varepsilon(x))\}$$

is surjective and proper, and its restriction to the intersection of the domain with X_i is C^1 submersive onto the intersection of the target space with $X_{l0} \times f(X_{i0})$.

We can easily modify $\rho_l|_{|T_l|_0}$ and then ρ_l so that $\varepsilon = 1$. Set

$$Q = \{x \in |T_l|_0 \cap \cup_{i=2}^{l^*}|T_i|: \rho_l(x) < 1, \; \rho'_{l'} \circ f(x) < \delta(\pi_l(x), \rho_l(x))\} \quad \text{and}$$
$$Z = \{x \in |T_l|_0 \cap \cup_{i=2}^{l^*}|T_i|: \rho_l(x) < 1, \; \rho'_{l'} \circ f(x) = \delta(\pi_l(x), \rho_l(x))\}.$$

Here f is extended to $\mathbf{R}^n \to \mathbf{R}^m$ as follows. By replacing (X, f) with (graph f, the restriction to graph f of the projection $\mathbf{R}^n \times \mathbf{R}^m \to \mathbf{R}^m$), we assume f is extended to a C^1 $\mathfrak{X}$-submersion $q: \mathbf{R}^n \to \mathbf{R}^m$ with the first boundedness condition. Moreover, we extend $f: \{X_i\} \to \{Y_j\}$ to

$$q|_{\mathbf{R}^n - q^{-1}(\overline{Y} - Y)}: \{X_i, q^{-1}(Y_j) - \overline{X}, \mathbf{R}^n - q^{-1}(\overline{Y})\} \longrightarrow \{Y_j, \mathbf{R}^m - \overline{Y}\}.$$

We easily see that this also is a C^1 Whitney $\mathfrak{X}$-stratified map sans éclatement. Choose $\{T'_j\}$ and $\{T_i\}$ so that they can be extended to a controlled tube system for the extension of $\{Y_j\}$ and a weakly controlled tube system for the extension of $\{X_i\}$ controlled over $\{T'_j\}$ respectively, and repeat the above arguments for these tube system. Then $Q \cup Z$ is a C^1 $\mathfrak{X}$-submanifold with boundary $= Z$ of $|T_l|_0 - X_l$, and Z is transversal to each X_i. By shrinking $|T_l|_0$, we assume $Q \cup Z$ and Z are closed in $|T_l|_0 - X_l$. Set

$$X_0^* = X_0 \cap |T_l| - Q - X_l \quad \text{and}$$
$$(X_{i10}^*, X_{i20}^*) = (X_{i0} \cap |T_l| - Q - Z, X_{i0} \cap Z), \quad i = 1, l^* + 1, \ldots, l - 1.$$

Then $f: \{X_{ik0}^*\} \to \{Y_{j0}\}$ is a C^1 Whitney $\mathfrak{X}$-stratified map sans éclatement, which easily follows from the above transversality. We will replace $\{X_{i0}\}$ with $\{X_{ik0}^*\}$. (We do not need the property that $f: \{X_{ik0}^*, X_{l0}\} \to \{Y_{j0}\}$ is a C^1 $\mathfrak{X}$-stratified map sans éclatement. Indeed, it is only the following fact that we used in the second case which follows from the hypothesis about a C^1 Whitney $\mathfrak{X}$-stratified map sans éclatement in II.6.1′: The sequence

$$\{X_{i0} \cap |T_l| - X_l\} \xrightarrow{(\pi_l, f)} \{\hat{B}_j \cap U\} \ni (x, y) \longrightarrow \delta(\varepsilon(x), \rho'_{l'}(y)) \in \,]0, 1[$$

in the proof in the second case satisfies the conditions of II.6.1'. After reduction to the second case, we have the same fact even if $f\colon \{X^*_{ik0}, X_{l0}\} \to \{Y_{j0}\}$ is not a C^1 Whitney $\mathcal{X}$-stratified map sans éclatement.)

Now we modify δ to be C^1 regular as required above. We reduce the problem to the case $X_{l0} = {]}0, 1]$ as follows. Let $0 \le \alpha < 1$ be an $\mathcal{X}$-function on $\overline{X_{l0}}$ whose zero set is $\overline{X_{l0}} - X_{l0}$. Set

$$\beta(u, s) = \min_{\alpha(x)=u} \delta(x, s) \quad \text{for} \quad (u, s) \in {]}0, 1]^2.$$

Then β is a positive $\mathcal{X}$-function such that

$$\beta(\alpha(x), s) \le \delta(x, s) \quad \text{for} \quad (x, s) \in X_{l0} \times {]}0, 1],$$

and we can replace $\delta(x, s)$ with $\beta(\alpha(x), s)$. Hence it suffices to lessen β so that it is of class C^1 and for each $u \in {]}0, 1]$, $\beta(u, \cdot)$ is C^1 regular. A further reduction is to the case $X_{l0} \times {]}0, 1] = {]}0, 1]$. Set

$$\gamma(v) = \min_{us=v} \beta(u, s) \quad \text{for} \quad v \in {]}0, 1].$$

Then γ is a positive $\mathcal{X}$-function such that $\gamma(us) \le \beta(u, s)$. If there exists a positive C^1 regular $\mathcal{X}$-function γ' on ${]}0, 1]$ smaller than γ, then the function ${]}0, 1]^2 \ni (u, s) \to \gamma'(us) \in \mathbf{R}$ is what we want. Construction of γ' is easy. Indeed, there exists a small positive number c such that γ is C^1 regular on ${]}0, c]$ and $\gamma > \gamma(c)$ on ${]}c, 1]$, and hence $\gamma'(v) = \gamma(v/c)$ fulfills the requirements.

We want to extend $\{X^*_{ik0}\}$ to a C^1 $\mathcal{X}$-stratification $\{X^*_{ik}\}$ of $X \cap |T_l| - X_l$ (an $\mathcal{X}$-neighborhood of $\cup^{l^*}_{i=2} X_i$). The following is the only possibility:

$$X^*_{ik} = \begin{cases} h^{-1}_i(X^*_{ik0}), & i = l^* + 1, \dots, l-1 \\ (h^{l-1}_1)^{-1}(X^*_{120}), & i = 1, \ k = 2 \\ X_1 - (h^{l-1}_1)^{-1}(Q \cup Z), & i = 1, \ k = 1. \end{cases}$$

Shrink $|T_l|$ enough. Clearly X^*_{ik}, $i = l^* + 1, \dots, l-1$, and X^*_{12} are C^1 $\mathcal{X}$-manifolds, and $h_i|_{X^*_{ik}}$, $i = l^*+1, \dots, l-1$, and $h^{l-1}_1|_{X^*_{12}}$ are C^1 $\mathcal{X}$-submersions to respective X^*_{ik0}. If for each $x \in X_l$ with $p \circ f(x) = a$,

$$h^{l-1}_1 \Big(X_{1a} \cap \bigcup^{l^*}_{i=2} |T_i| \Big) \supset X_{10} \cap (Q \cup Z) \cap \rho^{-1}_l({]}0, d[) \quad \text{for some} \quad d > 0, \quad (*)$$

then X^*_{11} is a C^1 $\mathcal{X}$-manifold, $\{X^*_{ik}\}$ is a C^1 $\mathcal{X}$-stratification, $f\colon \{X^*_{ik}\} \to \{Y_j\}$ is a C^1 $\mathcal{X}$-stratified map, and we can construct h^l_1 in the same way as in the second case. (h_1^{l-1} is defined on the intersection of X_{11}^* with an ℋ-neighborhood of $X_{12}^* \cup \bigcup_{i=l^*+1}^{l-1} \bigcup_{k=1,2} X_{ik}^*$ in $\mathbf{R}^n$.) We will modify h_1^{l-1} outside of the intersection of X_1 with a small ℋ-neighborhood of $\bigcup_{i=2}^{l^*} X_i$ in $\mathbf{R}^n$ so that condition (∗) is satisfied. Shrinking $|T_l|$, $|T_i|$, $i = 2, \ldots, l^*$, we assume

$$h_1^{l-1}\left(X_1 \cap |T_l| \cap \bigcup_{i=2}^{l^*} |T_i|\right) \subset X_{10} \cap (Q \cup Z).$$

By the same arguments as above we have a positive C^1 ℋ-function δ_1 on $X_l \times \,]0,1]$ such that for each $x \in X_l$, $\delta_1(x, \cdot)$ is C^1 regular, and

$$h_1^{l-1}\left(X_1 \cap \pi_l^{-1}(x) \cap \bigcup_{i=2}^{l^*} |T_i|\right)$$
$$\supset \{x' \in X_{10} \cap |T_l| \cap \cup_{i=2}^{l^*}|T_i|:$$
$$\rho_l(x') < d,\ \pi_l(x') = h_l(x),\ \rho_{l'}' \circ f(x') \le \delta_1(x, \rho_l(x'))\},$$

for some $d > 0$. Set

$$R_1(x) = \{x' \in X_{10} \cap |T_l| \cap \cup_{i=2}^{l^*}|T_i|:\ \rho_l(x') < 1,\ \pi_l(x') = h_l(x),\ \rho_{l'}' \circ f(x') \le \delta_1(x, \rho_l(x'))\}.$$

Let δ_2 be another positive C^1 ℋ-function on $X_l \times \,]0,1]$ smaller than δ_1, and let $R_2(x)$ denote the set defined by δ_2 in the same way as $R_1(x)$ by δ_1. If for each $x \in X_l$ there exists a positive number d such that

$$\delta(h_l(x), s) = \delta_1(x, s) \quad \text{for} \quad s \in \,]0, d],$$

condition (∗) is satisfied. We modify h_1^{l-1} so that this equality holds. For that, it suffices to find a C^1 ℋ-diffeomorphism:

$$\eta_x: R_1(x) \cap (\rho_{l'}' \circ f)^{-1}(\,]0, \delta_2(x,1)[\,) \longrightarrow X_{10} \cap (Q \cup Z) \cap (\rho_{l'}' \circ f)^{-1}(\,]0, \delta_2(x,1)[\,)$$

for each $x \in X_l$ such that

$$\eta_x = \mathrm{id} \quad \text{on} \quad R_2(x), \qquad f \circ \eta_x = f,$$

and the map

$$\bigcup_{x \in X_l} x \times R_1(x) \ni (x, x') \longrightarrow \eta_x(x') \in X_{10} \cap (Q \cup Z)$$

is of class C^1 ℋ. Indeed, $\eta_x(h_1^{l-1}(x))$ is an admitted modification of h_1^{l-1} and satisfies the equality.

In abuse of notation, set

$$\hat{B}_j = \{(x, y, s) \in X_{l0} \times (Y_{j0} \cap |T'_{l'}|) \times \,]0, 1[: f(x) = \pi'_{l'}(y), \; \rho'_{l'}(y) \le \delta(x, s)\},$$

$$\text{and} \quad \hat{B} = \bigcup_j \hat{B}_j.$$

Then the map $(\pi_l, f, \rho_l) \colon \{X_{i0} \cap (Q \cup Z)\} \to \{\hat{B}_j\}$ is a C^1 Whitney X-stratified map sans éclatement, and the map $X_0 \cap (Q \cup Z) \to \hat{B}$ itself is surjective and proper. Here the strata have boundary. We can extend naturally the definition of a C^1 Whitney X-stratified map sans éclatement to this case. (Note that the image of the boundary of a stratum under f is the boundary of some stratum.) We will find an X-function $\theta \colon \hat{B} \to [0, 1[$ so that the sequence of maps

$$\{X_{i0} \cap (Q \cup Z)\} \longrightarrow \{\hat{B}_j\} \xrightarrow{\ \theta\ } [0, 1[$$

satisfies the conditions of II.6.1', $\theta^{-1}(0)$ and $(\theta \circ (\pi_l, f, \rho_l))^{-1}(0)$ are the unions of the boundaries of the respective strata, and we can extend $\theta \circ (\pi_l, f, \rho_l)$ and θ over $\theta^{-1}(0)$ and $(\theta \circ (\pi_l, f, \rho_l))^{-1}(0)$ respectively, as functions to $]d, 1[$ for some $d < 0$ (see the below for the precise meaning). For such a θ, by the first induction hypothesis, we can apply II.6.1' to the sequence because $\dim X_0 \cap (Q \cup Z) < \dim X$.

For each $(x, y) \in X_{l0} \times (Y_0 \cap |T'_{l'}|)$ with $f(x) = \pi'_{l'}(y)$, $\theta|_{\hat{B} \cap x \times y \times \mathbf{R}}$ should be a C^1 diffeomorphism. Note that

$$\hat{B} \cap x \times y \times \mathbf{R} = x \times y \times [v, 1[\quad \text{for} \quad v \in \,]0, 1[\quad \text{with} \quad \rho'_{l'}(y) = \delta(x, v).$$

So a natural construction of θ is as follows. Let $\xi \colon \{(s, v) \in \,]0, 1[^2 : s \ge v\} \to [0, 1[$ be the C^1 X-function defined so that for each $v \in \,]0, 1[$, $\xi(\cdot, v)$ is a linear homeomorphism onto $[0, 1[$. Note that $(\mathbf{R} \times v) \cap (\text{the domain}) = [v, 1[\times v$, and hence $\xi(v, v) = 0$. Define θ by

$$\theta(x, y, s) = \xi(s, v) \quad \text{for} \quad (x, y, s) \in \hat{B},$$

where $v \in \,]0, 1[$ is the number such that $\delta(x, v) = \rho'_{l'}(y)$. Since v is uniquely decided by x and y, θ is well-defined. Moreover, clearly θ fulfills the above requirements.

By the induction hypothesis we obtain the following statement. Let $\chi_2 \colon \hat{B} \to \hat{B} \cap \theta^{-1}(0)$ be an X-map such that $(\chi_2, \theta) \colon \hat{B} \to (\hat{B} \cap \theta^{-1}(0)) \times [0, 1[$ is a homeomorphism and its restriction to each $\hat{B}_j$ is a C^1 diffeomorphism onto $(\hat{B}_j \cap \theta^{-1}(0)) \times [0, 1[$. Then there exists an X-map $\chi_1 \colon X_0 \cap (Q \cup$

$Z) \to X_0 \cap (Q \cup Z) \cap (\theta \circ (\pi_l, f, \rho_l))^{-1}(0)$ such that $(\chi_1, \theta \circ (\pi_l, f, \rho_l))$ is a homeomorphism, its restriction to each $X_{i0} \cap (Q \cup Z)$ is a C^1 diffeomorphism, and $(\pi_l, f, \rho_l) \circ \chi_1 = \chi_2 \circ (\pi_l, f, \rho_l)$. We let χ_2 be defined by $\chi_2(x, y, s) = (x, y, v)$, where v is defined by x and y as above. Then it follows from the equality $(\pi_l, f, \rho_l) \circ \chi_1 = \chi_2 \circ (\pi_l.f, \rho_l)$ that $\pi_l \circ \chi_1 = \pi_l$ and $f \circ \chi_1 = f$.

Using χ_1, we construct η_x as follows. By a C^1 𝔛-partition of unity we have a C^1 𝔛-function $\hat{\theta} \colon \{(s, v_1, v_2) \in \,]0, 1[^3 : s \geq v_1, \; v_1 < v_2\} \to [0, 1[$ such that for each $0 < v_1 < v_2 < 1$, $\hat{\theta}(\cdot, v_1, v_2)$ is a diffeomorphism onto $[0, 1[$ and is the identity on $[v_2, 1[$. For each $x \in X_l$ and for each $x' \in R_1(x) \cap (\rho'_{l'} \circ f)^{-1}(]0, \delta_2(x, 1)[)$, $x'' = \eta_x(x')$ is defined so that

$$\chi_1(x'') = \chi_1(x') \quad \text{and} \quad \hat{\theta}(\theta \circ (\pi_l, f, \rho_l)(x'), v_1, v_2) = \theta \circ (\pi_l, f, \rho_l)(x''),$$

where $0 < v_1 < v_2 < 1$ are given by

$$\theta(h_l(x), f(x'), s_i) = v_i, \quad i = 1, 2,$$

and $0 < s_1 < s_2 < 1$ are given by

$$\rho'_{l'} \circ f(x') = \delta_i(x, s_i), \quad i = 1, 2.$$

It is easy to check that η_x is well-defined and fulfills the requirements. Thus we complete the construction of h_1^l in any case.

As the second induction system works, there exists a C^1 𝔛-map $h_1^k \colon X_1 \cap \cup_{i=2}^k |T_i| \to X_{10}$, where k is the number of the strata X_i, such that $(h_1^k, p \circ f) \colon X_1 \cap \cup_{i=2}^k |T_i| \to X_{10} \times P$ is a C^1 imbedding and all the conditions $(0)_1^k, \ldots,$ $(iv)_1^k$ are satisfied. It remains only to extend h_1^k to X_1. For that we apply II.6.8 to

$$A = X_1, \qquad B = Y_{10}, \qquad C = P, \qquad V = X_1 \cap \bigcup_{i=2}^k |T_i|,$$

$$W = \widehat{W} = \varnothing, \qquad \Psi = (h_1' \circ f, p \circ f), \qquad \chi = h_1^k.$$

Clearly these satisfy the conditions of II.6.8. Hence, by shrinking V, we obtain a C^1 𝔛-extension $h_1 \colon X_1 \to X_{10}$ of h_1^k such that $(h_1, p \circ f)$ is a diffeomorphism and $h_1' \circ f = f \circ h_1$ ((0) for $i = 1$). Conditions (i), (ii) and (iii) for $i' = 1$ are the same as $(i)_1^k$, $(ii)_1^k$ and $(iii)_1^k$ respectively. It is easy to modify h_1 so that (iv) for $i' = 1$ is satisfied. We complete the proof. $\square$

To prove also the 𝔛-Hauptvermutung(III.1.4) we shall use the following lemmas.

Lemma II.6.10 (Existence of a controlled X-tube system). *Let r be a positive integer. Let $\{T_i = (|T_i|, \pi_i, \rho_i)\}_{i=1,2,\ldots}$ be a C^r X-tube system for a finite C^r Whitney X-stratification $\{X_i\}_{i=1,2,\ldots}$ of an X-set in $\mathbf{R}^n$ with $\dim X_1 < \dim X_i$, $i \neq 1$. Assume Axiom (v) unless $\cup_i X_i$ is bounded. There exists a controlled C^r X-tube system $\{T_i' = (|T_i'|, \pi_i', \rho_i')\}$ for $\{X_i\}$ such that for each i,*

$$|T_i'| \subset |T_i|, \qquad \rho_i = \rho_i' \quad \text{on} \quad |T_i'|, \quad \text{and}$$
$$\pi_1 = \pi_1' \quad \text{on} \quad |T_1'|.$$

Let $f: X \to Y$ be a C^r X-map between X-subsets of $\mathbf{R}^n$ with the first boundedness condition, let f admit a C^r Whitney X-stratification sans éclatement $(\{X_i\}, \{Y_j\})$, and let $\{T_j'\}$ be a weakly controlled C^r X-tube system for $\{Y_j\}$. Assume Axiom (v) unless X is bounded. Then there exists a C^r X-tube system $\{T_i\}$ for $\{X_i\}$ controlled over $\{T_j'\}$.

Proof. In exactly the same way as in the proof of I.1.3, we construct a controlled C^{r-1} X-tube system with the required properties. To obtain a system of class C^r we modify the orthogonal projection $\rho_{y,t}$ in the proof of I.1.3 as in the proof of II.5.1. We omit the details because they are easy. $\square$

We define naturally a C^r X-*triangulation* of a compact X-set in $\mathbf{R}^n$ and local C^r X-*equivalence*.

Lemma II.6.11 (C^r X-triangulation). *Let r be a positive integer. Let $X \subset \mathbf{R}^n$ be either a compact C^r X-submanifold possibly with boundary and corners of $\mathbf{R}^n$, or a compact X-set which is locally C^r X-equivalent to a polyhedron at each point of X. Let (K, f) be a C^r triangulation of X. There is an arbitrarily strong approximation (K', f') of (K, f) in the C^1 topology which is a C^r X-triangulation of X.*

Proof. We will prove this by an induction method similar to the proof of C^∞ triangulability in I.3.13. Order the vertices of K as $v_1, \ldots, v_l$. Set

$$C_i = |N(N(\operatorname{st}(v_i, K), K), K)|, \quad i = 1, \ldots, l.$$

We can assume that for each $i = 1, \ldots, l$, there exists a C^r X-diffeomorphism φ_i from a closed neighborhood U_i of $f(C_i)$ in X to a polyhedron in $\mathbf{R}^n$. Let $1 \leq k \leq l$. By induction, assume there exists a strong approximation (K_{k-1}, f_{k-1}) of (K, f) such that (K_{k-1}, f_{k-1}) is a C^r triangulation of X, each U_i is a neighborhood of $f_{k-1}(C_i)$, and f_{k-1} is of class X on $\cup_{i=1}^{k-1} |\operatorname{st}(v_i, K)|$. It suffices to construct (K_k, f_k) with the corresponding properties.

Apply I.3.16 to $\varphi_k \circ f_{k-1}|_{C_k}$ and $|\operatorname{st}(v_k, K)|$. There is a strong approximation (L_k, g_k) of $(K_{k-1}|_{C_k}, \varphi_k \circ f_{k-1}|_{C_k})$ such that g_k equals the secant map induced by $\varphi_k \circ f_{k-1}|_{C_k}$ on $L_k|_{|\operatorname{st}(v_k,K)|}$, g_k equals $\varphi_k \circ f_{k-1}|_{C_k}$ outside of $|N(\operatorname{st}(v_k, K), K)|$, L_k equals $K_{k-1}|_{C_k}$ outside of $|N(\operatorname{st}(v_k, K), K)|$, and $g_k(C_k)$ coincides with $\varphi_k \circ f_{k-1}(C_k)$. In addition, if we construct g_k as in the proof of I.3.16, g_k is of class $\mathfrak{X}$ on an $\mathfrak{X}$-subset of C_k where $\varphi_k \circ f_{k-1}$ is of class $\mathfrak{X}$. Hence we assume g_k is of class $\mathfrak{X}$ on $\cup_{i=1}^k |\operatorname{st}(v_i, K)|$. Set

$$(K_k, f_k) = \begin{cases} (K_{k-1}, f_{k-1}) & \text{outside of } C_k \\ (L_k, \varphi_k^{-1} \circ g_k) & \text{on } C_k. \end{cases}$$

Then (K_k, f_k) is what we want. □

By the same reason as II.5.11 we have the following.

Theorem II.6.12 (Case of $\mathfrak{X}_0$). *If $\mathfrak{X} = \mathfrak{X}_0$, all the results in this section hold without the boundedness condition.*

§II.7. 𝔛-singularity theory

Let $X_1 \subset X_1'$ and $X_2 \subset X_2'$ be $\mathfrak{X}$-sets, and let f_1, f_2 be $\mathfrak{X}$-functions on X_1', X_2' respectively. We call X_1 and X_2 $\mathfrak{X}$-*equivalent* if there exists an $\mathfrak{X}$-homeomorphism τ from an $\mathfrak{X}$-neighborhood of X_1 in X_1' to one of X_2 in X_2' such that $\tau(X_1) = X_2$. The germs of f_1 at X_1 and of f_2 at X_2 are called *locally* $\mathfrak{X}$-*equivalent* or *locally R $\mathfrak{X}$-equivalent* if we can choose the above τ so that $f_1 = f_2 \circ \tau$ as germs at X_1. If τ is an $\mathfrak{X}$-homeomorphism from X_1' to X_2' and $f_1 = f_2 \circ \tau$ holds globally, then f_1 and f_2 are called $\mathfrak{X}$-*equivalent* or *R $\mathfrak{X}$-equivalent*. (*Local*) *R-L $\mathfrak{X}$-equivalence* is defined when $\pi \circ f_1 = f_2 \circ \tau$ for an $\mathfrak{X}$-homeomorphism π of **R**. If τ and π are homeomorphisms but not necessarily of class $\mathfrak{X}$, then (*local*) (*R-L*) C^0 *equivalence* is defined; if they are C^r ($\mathfrak{X}$-)diffeomorphisms, then (*local*) (*R-L*) C^r ($\mathfrak{X}$-)*equivalence* is defined. Here recall the definition of a C^r $\mathfrak{X}$-diffeomorphism. It is an $\mathfrak{X}$-homeomorphism which can be extended to a C^r $\mathfrak{X}$-diffeomorphism between open $\mathfrak{X}$-neighborhoods of the ambient Euclidean spaces. For $\mathfrak{X}$-maps between $\mathfrak{X}$-sets also, we define R and R-L $\mathfrak{X}$-*equivalence* as in the function case.

The main theorem of this section is II.7.3, which shows uniqueness of a controlled C^1 $\mathfrak{X}$-tube system of an $\mathfrak{X}$-equivalence class of C^1 $\mathfrak{X}$-stratifications with open ball strata. By II.7.3, for classification of $\mathfrak{X}$-sets, $\mathfrak{X}$-functions and $\mathfrak{X}$-maps by $\mathfrak{X}$-equivalence relations, we can assume that a controlled $\mathfrak{X}$-tube system is always attached to a C^1 $\mathfrak{X}$-stratification. Research of C^1 $\mathfrak{X}$-stratifications with controlled C^1 $\mathfrak{X}$-tube systems is much more fruitful than that without C^1 X-tube systems. For example, X-equivalence of two C^1 X-stratifications follows from their C^0 equivalence when controlled C^1 X-tube systems are counted (II.7.5). II.7.5 treats, moreover, X-functions and X-maps.

Another purpose of this section is to show some properties of X-equivalence. By II.7.1, 2, 3 and 9 we see the differences from and advantages over properties of C^0 equivalence. First the X-equivalence class of the zero set of an X-function almost determines the local X-equivalence class of the germ of the function at the zero set as shown below. This does not hold for C^0 equivalence of polynomial function germs [Ki]. To be precise, there exist two polynomial function germs on $\mathbf{R}^n$ at 0, $n \geq 7$, with isolated singularities at 0 which are not locally R-L C^0 equivalent but whose zero set germs are locally C^0 equivalent.

Theorem II.7.1 (X-equivalence of X-functions and zero sets). *Let X be an X-subset of $\mathbf{R}^n$, and let f_1, f_2 be X-functions on X. Assume Axiom* (v) *or that X is bounded. If $f_1^{-1}(0)$ and $f_2^{-1}(0)$ are X-equivalent as X-subsets of X, the germs of f_1 at $f_1^{-1}(0)$ and of f_2 at $f_2^{-1}(0)$ are locally X-equivalent up to sign, i.e., the germs of $|f_1|$ and $|f_2|$ are locally X-equivalent.*

Proof. By the above hypothesis we can assume $f_1^{-1}(0) = f_2^{-1}(0)$ and

$$\overline{\operatorname{graph} f_i} \cap \mathbf{R}^n \times 0 = \overline{f_1^{-1}(0)} \times 0, \qquad i = 1, 2,$$

because the problem is local at $f_1^{-1}(0)$ and $f_2^{-1}(0)$. Consider graph$(f_1, f_2) \subset \mathbf{R}^n \times \mathbf{R}^2$ and the restrictions to the graph of the projections of $\mathbf{R}^n \times \mathbf{R}^2$ onto the last two factors in place of X, f_1 and f_2. Then we can extend f_1 and f_2 to $\overline{X}$ so that the zero sets of the extensions coincide with each other. Hence it suffices to prove the following statement.

Let $X_1 \supset X_2$ be X-subsets of $\mathbf{R}^n$, and let f_1, f_2 be X-functions on X_1 with $f_1^{-1}(0) = f_2^{-1}(0)$. Assume X_1 is closed in $\mathbf{R}^n$. There exists an X-homeomorphism τ of X_1 such that $|f_1| = |f_2| \circ \tau$ on a neighborhood of $f_1^{-1}(0)$ and τ is invariant on X_2.

We can replace X_2 with decreasing closed X-subsets $X_3, \dots, X_k$ of X_1 requiring τ to be invariant on each of $X_3, \dots, X_k$. We assume $X_1, X_3, \dots, X_k$ and $f_1^{-1}(0)$ are polyhedra by II.2.1 and II.2.1′, and f_1 and f_2 are PL by II.3.1, II.3.1′ and II.3.2. Then existence of the τ follows from I.3.10.   □

Remark II.7.2. An immediate corollary is the following. Let f_1 and f_2 be X-function germs at 0 in $\mathbf{R}^n$ which are locally R-L X-equivalent. Then f_1 is locally X-equivalent to f_2 or $-f_2$.

This is not the case of C^∞ function germs. For example, the germs at 0 in **R**

$$f_1 = \begin{cases} \sin(1/|x|)\exp(-1/|x|) & \text{for } x \neq 0 \text{ and} \\ 0 & \text{for } x = 0 \end{cases} \qquad f_2 = 2f_1$$

are clearly locally R-L C^∞ equivalent but not locally C^0 equivalent up to sign because the C^0 singular value sets of f_1 and f_2 are the germs at 0 of $\{0, \frac{1}{\sqrt{2}}\exp(3\pi/4 + n\pi): n \in \mathbf{Z}\}$ and $\{0, \sqrt{2}\exp(3\pi/4 + n\pi): n \in \mathbf{Z}\}$ respectively, and the C^0 singular value set of a C^0 function germ is an invariant of its local C^0 equivalence class.

We can construct examples with isolated C^∞ singular values as follows. It is easy to find a C^∞ regular C^∞ function g on $[1/2, 1] \times \mathbf{R}$ such that $g(x,y) = x$ outside $[2/3, 5/6] \times [-1, 1]$ and the restriction to $g^{-1}(3/4)$ of the projection $p: \mathbf{R}^2 \ni (x,y) \to y \in \mathbf{R}$ is a triple covering over $[-1/2, 1/2]$. Extend g to $[1/2^{n+1}, 1/2^n]$ for each $n = 1, 2, \ldots$ by setting $g(x,y) = g(2^n x, y)/2^n$, and outside $]0, 1] \times \mathbf{R}$ by $g(x,y) = x$. Set

$$A = \{3/2^n: n = 2, 3, \ldots\} \quad \text{and} \quad B = \{1/2^n: n = 0, 1, \ldots\}.$$

Then g is a C^0 function on $\mathbf{R}^2$ and C^∞ regular outside its zero set $= 0 \times \mathbf{R}$. For $t \in \mathbf{R}$, $p|_{g^{-1}(t)}$ is a triple covering over $[-1/2, 1/2]$ if $t \in A$ and a one-to-one map if $t \in B$. Let φ_i, $i = 1, 2$, be C^∞ homeomorphisms of **R** such that $\varphi_i^{-1}(0) = 0$, φ_i are C^∞ regular outside 0, $\varphi_1 \circ g$ is of class C^∞ and $\varphi_2 \circ \varphi_1(A) \subset \varphi_1(B)$. These are easy to construct. Define f_1 and f_2 to be the germs at 0 of $\varphi_1 \circ g$ and $\varphi_2 \circ \varphi_1 \circ g$ respectively. Their local R-L C^0 equivalence is clear, and non local C^0 equivalence up to $\pm$ follows from the fact that for each $t \in \varphi_2 \circ \varphi_1(A)$ near 0, $f_1^{-1}(t)$ is connected and $f_2^{-1}(t)$ has 3 connected components. (Here we confuse a set and a germ. But the meaning of connectedness may be clear.)

When (R-L) C^0 equivalence of sets and maps is treated in singularity theory, their Whitney stratifications and then controlled tube systems are used in most cases. Hence uniqueness of a controlled tube system is desirable. To be precise, suppose we are given a stratified homeomorphism $f: \{X_i\} \to \{Y_i\}$ between C^1 stratifications and C^1 controlled tube systems $\{T_i^X = (|T_i^X|, \pi_i^X, \rho_i^X)\}$ for $\{X_i\}$ and $\{T_i^Y = (|T_i^Y|, \pi_i^Y, \rho_i^Y)\}$ for $\{Y_i\}$ such that for each i, $f|_{X_i}$ is a C^1 diffeomorphism onto Y_i and $(\pi_i^X, \rho_i^X)|_{X_j \cap T_i^X}: X_j \cap |T_i^X| \to X_i \times \mathbf{R}$ and $(\pi_i^Y, \rho_i^Y)|_{Y_j \cap T_i^Y}: Y_j \cap |T_i^Y| \to Y_i \times \mathbf{R}$ are C^1 submersions for i and j with $X_i \cap \overline{X_j} - X_j \neq \varnothing$. Then we can shrink $|T_i^X|$ and $|T_i^Y|$ and modify f so that $f(X \cap |T_i^X|) = Y \cap |T_i^Y|$,

$$\pi_i^Y \circ f = f \circ \pi_i^X \quad \text{and} \quad \rho_i^Y \circ f = \rho_i^X \quad \text{on } X \cap |T_i^X| \tag{$*$}$$

for each i, namely, f is *compatible* with $\{T_i^X\}$ and $\{T_i^Y\}$, where $X = \cup_i X_i$ and $Y = \cup_i Y_i$. This is not always the case (e.g., $\{X_1, X_2\}$ and $\{Y_1, Y_2\}$ in I.1.12), and the failure is the largest reason why we cannot expect a general and precise theory of (R-L) C^0 equivalence. (We have a counterexample where X_i and Y_i are diffeomorphic to Euclidean spaces. Set $X_1 = $ a point of S^n, $n \geq 5$, $X_2 = S^n - X_1$. Let $M \subset \mathbf{R}^m \times 1 \subset \mathbf{R}^{m+1}$ be a compact contractible C^∞ manifold of dimension n whose boundary is not simply connected, smooth $(0 * \partial M) \cup M$ at ∂M, and set $Y = $ the set, $Y_1 = 0$ and $Y_2 = Y - Y_1$. Then $\{X_i\}$ and $\{Y_i\}$ are C^∞ Whitney stratifications, and by the weak h-cobordism theorem, there is a stratified homeomorphism $f \colon \{X_1, X_2\} \to \{Y_1, Y_2\}$ such that $f|_{X_2}$ is a diffeomorphism onto Y_2 but f cannot be compatible with any controlled tube systems.) On the other hand, uniqueness of a controlled $\mathfrak{X}$-tube system holds partially as follows.

Theorem II.7.3 (Uniqueness of controlled $\mathfrak{X}$-tube system). *Let $\{X_i\}$ and $\{Y_i\}$ be C^1 (not necessarily Whitney) $\mathfrak{X}$-stratifications of locally closed $\mathfrak{X}$-subsets X and Y of $\mathbf{R}^n$ respectively, such that all X_i and Y_i are C^1 $\mathfrak{X}$-diffeomorphic to open balls. Assume Axiom (v) or that X and Y are bounded in $\mathbf{R}^n$. Let $\{T_i^X\}$ and $\{T_i^Y\}$ be controlled C^1 $\mathfrak{X}$-tube systems for $\{X_i\}$ and $\{Y_i\}$ respectively, such that $(\pi_i^X, \rho_i^X)|_{X_j \cap |T_i^X|} \colon X_j \cap |T_i^X| \to X_i \times \mathbf{R}$ and $(\pi_i^Y, \rho_i^Y)|_{Y_j \cap |T_i^Y|} \colon Y_j \cap |T_i^Y| \to Y_i \times \mathbf{R}$ are C^1 submersions for i and j with $X_i \subset \overline{X_j} - X_j$. Let $f \colon \{X_i\} \to \{Y_i\}$ be an $\mathfrak{X}$-stratified homeomorphism such that each $f|_{X_i}$ is a C^1 diffeomorphism onto Y_i. Then shrinking $|T_i^X|$ and $|T_i^Y|$, we can modify f to be compatible with $\{T_i^X\}$ and $\{T_i^Y\}$.*

Proof. Suppose $\dim X_1 < \dim X_2 < \cdots$ for simplicity of notation. In the following arguments, we frequently shrink $|T_i^X|$ and $|T_i^Y|$ and modify ρ_i^X, ρ_i^Y and ρ_i^V outside small $\mathfrak{X}$-neighborhoods of X_i, Y_i and V_i respectively, for simplicity of notation and for applications of II.6.1, and we always assume $f(X \cap |T_i^X|) = Y \cap |T_i^Y|$. Note that we may make these changes without explicitly telling the reader.

By induction, assuming (*) for all $i > 1$, we need only modify f on an open small $\mathfrak{X}$-neighborhood N of X_1 in X so that the modification $\hat{f}$ satisfies (*) for $i = 1$ and coincides with f outside another $\mathfrak{X}$-neighborhood contained and closed in N. Also, for a sequence $\{a_j\}$ in N converging to a point of $\overline{N} - N$ such that $\{f(a_j)\}$ converges, $\{\hat{f}(a_j)\}$ also converges to the same point. (The second condition is necessary for extension of $\hat{f}$ to X. The last is for the induction step. The stratum X_1 is not always of the smallest dimension. Even if it is, by the last condition, we can extend $\hat{f}$ to $X \cap (\overline{X_1} - X_1)$ by setting $= f$ there.)

First we find an open $\mathfrak{X}$-subset U of X_1 and modify f so that (*) for $i = 1$ holds on $X \cap \pi_1^{X-1}(U)$. Let $k > 1$ be an integer. By induction, suppose

(*) for $i = 1$ holds on $X \cap \pi_1^{X-1}(U) \cap \cup_{1<i<k}|T_i^X|$. We want (*) for $i = 1$ on $X_k \cap \pi_1^{X-1}(U)$ and then on $X \cap \pi_1^{X-1}(U) \cap |T_k^X|$. Set

$$V_i = \text{graph } f|_{X_i \cap \pi_1^{X-1}(U)}, \quad V = \bigcup_{i=1}^{k} V_i,$$

$$W_k = V_k \cap \text{graph } f|_{X \cap \cup_{1<i<k}|T_i^X|}, \quad \text{and} \quad n_1 = \dim X_1.$$

We can assume

$$U = f(U) = \mathbf{R}^{n_1} \times 0 \subset \mathbf{R}^n, \quad \pi_1^{X-1}(U) = \pi_1^{Y-1}(f(U)) = \mathbf{R}^n,$$

$$f|_U = \text{id},$$

$$\pi_1^X(x) = \pi_1^Y(x) = (x_1, \dots, x_{n_1}, 0, \dots 0) \quad \text{for } x = (x_1, \dots, x_n) \in \mathbf{R}^n,$$

$$\rho_1^X(x) = \rho_1^Y(x) = x_{n_1+1}^2 + \dots + x_n^2,$$

and for each $i > 1$, the pair (V_1, V_i) satisfies the Whitney condition. Then V is not necessarily included in $\mathbf{R}^n \times \mathbf{R}^n$. For the inclusion, we need to weaken the assumption $f(U) = \mathbf{R}^{n_1} \times 0$ as $f(U) \subset \mathbf{R}^{n_1} \times 0$. For simplicity of notation, we suppose both $f(U) = \mathbf{R}^{n_1} \times 0$ and $V \subset \mathbf{R}^n \times \mathbf{R}^n$. Let p and q denote the projections of $\mathbf{R}^n \times \mathbf{R}^n$ onto the former and latter factors respectively, and define two $\mathcal{X}$-tubes $(\mathbf{R}^n \times \mathbf{R}^n, \pi_1^V, \rho_1^V)$ and $(\mathbf{R}^n \times \mathbf{R}^n, \pi_1^{V'}, \rho_1^V)$ at V_1 by

$$\pi_1^V(x, y) = (x_1, \dots, x_{n_1}, 0, \dots, 0, x_1, \dots, x_{n_1}, 0, \dots, 0),$$

$$\pi_1^{V'}(x, y) = (y_1, \dots, y_{n_1}, 0, \dots, 0, y_1, \dots, y_{n_1}, 0, \dots, 0), \quad \text{and}$$

$$2\rho_1^V(x, y) = (x_1 - y_1)^2 + \dots + (x_{n_1} - y_{n_1})^2$$

$$+ x_{n_1+1}^2 + \dots + x_n^2 + y_{n_1+1}^2 + \dots + y_n^2$$

$$\text{for } (x, y) = (x_1, \dots, x_n, y_1, \dots, y_n) \in \mathbf{R}^n \times \mathbf{R}^n.$$

Compare (π_1^V, ρ_1^V) with $(\pi_1^V, \rho_1^X \circ p)$ on V_k. There exists a C^1 $\mathcal{X}$-diffeomorphism τ^X of V_k such that τ^X can be extended to V as a homeomorphism, τ^X is invariant on each fibre $V_k \cap \pi_1^{V-1}(z)$, $z \in V_1$,

$$\tau^X = \text{id} \quad \text{on } W_k \cup (V_k - \text{a small } \mathcal{X}\text{-neighborhood of } V_1), \quad \text{and}$$

$$\rho_1^V = \rho_1^X \circ p \circ \tau^X \quad \text{on } V_k \cap (\text{a smaller } \mathcal{X}\text{-neighborhood of } V_1) \quad (1)$$

for the following reason. Note that

$$\pi_1^V = (p|_V)^{-1} \circ \pi_1^X \circ p \quad \text{on } V, \quad \text{and}$$

$$\rho_1^X \circ p(x, y) = x_{n_1+1}^2 + \dots + x_n^2$$

$$\text{for } (x, y) = (x_1, \dots, x_n, y_1, \dots, y_n) \in \mathbf{R}^n \times \mathbf{R}^n.$$

Hence

$$\rho_1^V = \rho_1^X \circ p \quad \text{on} \ \ W_k \tag{2}$$

because of $(*)$ for $i = 1$ on $p(W_k)$. On the other hand, for each $z \in V_1$, $\rho_1^V|_{V_k \cap \pi_1^{V-1}(z)}$ and $\rho_1^X \circ p|_{V_k \cap \pi_1^{V-1}(z)}$ are friendly on $V_k \cap \pi_1^{V-1}(z) \cap (\text{a neigh-}$ borhood of z). Therefore, shrinking U if necessary, we can assume

$$\rho_1^V|_{V_k \cap \pi_1^{V-1}(z)} \quad \text{and} \quad \rho_1^X \circ p|_{V_k \cap \pi_1^{V-1}(z)} \ \ \text{are friendly for any} \ \ z \in V_1.$$

Then there exists a C^1 flow on V_k of class locally $\mathfrak{X}$ (i.e., of class $\mathfrak{X}$ locally at each point of V_k) such that each orbit is contained in a fibre of π_1^V and the restrictions of ρ_1^V and $\rho_1^X \circ p$ to an orbit is C^1 regular. Translate points of V_k along the flow as in the proof of I.1.8. Then we obtain a C^1 diffeomorphism τ' of V_k of class locally $\mathfrak{X}$ such that τ' is invariant on each fibre $V_k \cap \pi_1^{V-1}(z)$, $z \in V_1$,

$$\tau' = \text{id} \quad \text{on} \ \ W_k \cup (V_k \cap \text{a very small } \mathfrak{X}\text{-neighborhood of } V_1),$$

$$\text{and} \ \ \rho_1^V = \rho_1^X \circ p \circ \tau' \quad \text{on} \ \ V_k - \text{a small } \mathfrak{X}\text{-neighborhood of } V_1. \tag{4}$$

Clearly, shrinking U, we can assume τ' is of class $\mathfrak{X}$ and the pair of $\rho_1^V|_{V_k}$ and $\rho_1^X \circ p \circ \tau'$ keeps the above properties (2) and (3). C^1-Whitney-substratify $\{V_i\}_{i<k}$ if necessary, shrink U, choose a C^1 controlled $\mathfrak{X}$-tube system whose tube at V_1 is $(\mathbf{R}^n \times \mathbf{R}^n, \pi_1^V, \rho_1^V)$, and apply II.6.1 to $(\pi_1^V, \rho_1^V)|_{V-V_1} : V - V_1 \to V_1 \times]0, \infty[$ and $(\pi_1^V, \rho_1^X \circ p \circ \tau')|_{V-V_1} : V - V_1 \to V_1 \times]0, \infty[$. (We can assume the conditions in II.6.1 for $\{V_i\}$ are satisfied without loss of generality.) Then, by its proof, we have two C^1 flows of class $\mathfrak{X}$ on V_k such that if $\mathfrak{F}$ and $\mathfrak{F}'$ denote the respective families of orbits, each orbit is contained in a fibre of π_1^V, the restrictions of ρ_1^V to each $\delta \in \mathfrak{F}$ and of $\rho_1^X \circ p \circ \tau'$ to $\delta' \in \mathfrak{F}'$ are C^1 regular, and

$$\mathfrak{F} = \mathfrak{F}' \quad \text{on} \ \ W_k \cup (V_k - \text{a small } \mathfrak{X}\text{-neighborhood of } V_1).$$

By $\mathfrak{F}$ and $\mathfrak{F}'$ we can construct the τ^X as follows. For $\delta \in \mathfrak{F}$ and $\delta' \in \mathfrak{F}'$ with $\delta = \delta'$ outside a small $\mathfrak{X}$-neighborhood of V_1, define $\tau^X|_\delta$ so that

$$\tau^X(\delta) = \delta', \quad \text{and} \quad \rho_1^V = \rho_1^X \circ p \circ \tau' \circ \tau^X \quad \text{on} \ \ \delta.$$

Then $\tau^X|_\delta$ is uniquely determined, τ^X is a C^1 $\mathfrak{X}$-diffeomorphism of V_k, τ^X is invariant on each fibre $V_k \cap \pi_1^{V-1}(z)$, $z \in V_1$,

$$\tau^X = \text{id} \quad \text{on} \ \ W_k \cup (V_k - \text{a small } \mathfrak{X}\text{-neighborhood of } V_1), \quad \text{and}$$

$$\rho_1^V = \rho_1^X \circ p \circ \tau' \circ \tau^X \quad \text{on} \ \ V_k. \tag{5}$$

The equality (1) follows from (4) and (5), and if we set $\tau^X = \mathrm{id}$ on $\cup_{i<k} V_i$ then τ^X is an $\mathfrak{X}$-homeomorphism of V. Thus τ^X is what we wanted.

In the same way, we construct an $\mathfrak{X}$-homeomorphism τ^Y of V such that $\tau^Y|_{V_k}$ is a C^1 diffeomorphism of V_k, τ^Y is invariant on each $V \cap \pi_1^{V'-1}(z)$, $z \in V_1$,

$$\tau^Y = \mathrm{id} \quad \text{on} \quad W_k \cup \bigcup_{i<k} V_i \cup (V_k - \text{a small } \mathfrak{X}\text{-neighborhood of } V_1), \quad \text{and}$$

$$\rho_1^V = \rho_1^Y \circ q \circ \tau^Y \quad \text{on a smaller } \mathfrak{X}\text{-neighborhood of } V_1 \text{ in } V.$$

Next we will construct an $\mathfrak{X}$-homeomorphism τ^V of V such that $\tau^V|_{V_k}$ is a C^1 diffeomorphism of V_k, τ^V is invariant on each $V \cap \rho_1^{V-1}(t)$, $t \in \mathbf{R}$,

$$\tau^V = \mathrm{id} \quad \text{on} \quad W_k \cup \bigcup_{i<k} V_i \cup (V_k - \text{a small } \mathfrak{X}\text{-neighborhood of } V_1), \quad \text{and}$$

$$\pi_1^V = \pi_1^{V'} \circ \tau^V \quad \text{on a smaller } \mathfrak{X}\text{-neighborhood of } V_1 \text{ in } V.$$

As above, we assume $\{V_i\}_{i\leq k}$ is a C^1 Whitney stratification and $(\pi_1^V, \rho_1^V)|_{V-V_1}$ together with $\{V_i\}_{1<i\leq k}$ satisfies the conditions in II.6.1. Hence we have a third C^1 $\mathfrak{X}$-tube $(\mathbf{R}^n \times \mathbf{R}^n, \pi_1^{V''}, \rho_1^V)$ at V_1 such that

$$\pi_1^{V''} = \begin{cases} \pi_1^V & \text{outside a small } \mathfrak{X}\text{-neighborhood of } V_1 \text{ (say, } \rho_1^{V-1}([0,1])) \\ \pi_1^{V'} & \text{on } W_k \cup (\text{a smaller one (say, } \rho_1^{V-1}([0,1/2]))), \end{cases}$$

and $(\pi_1^{V''}, \rho_1^V)|_{V-V_1}$ also satisfies the conditions. (Note that

$$\pi_1^V = \pi_1^{V'} = \pi_1^{V''} \quad \text{on } W_k \quad \text{and}$$

$$\pi_1^{V-1}(0) \cap \rho_1^{V-1}(1) = \pi_1^{V''-1}(0) \cap \rho_1^{V-1}(1),$$

the former of which follows from (∗) for $i = 1$ on $p(W_k)$.) By II.6.1 and its proof, there exist $\mathfrak{X}$-homeomorphisms

$$g, g'' \colon V_1 \times]0, \infty[\times (V \cap \pi_1^{V-1}(0) \cap \rho_1^{V-1}(1)) \longrightarrow V - V_1$$

such that $(\pi_1^V, \rho_1^V) \circ g$ and $(\pi_1^{V''}, \rho_1^V) \circ g''$ are both the projection onto $V_1 \times]0, \infty[$, the restrictions of g and g'' to $V_1 \times]0, \infty[\times (V_i \cap \pi_1^{V-1}(0) \cap \rho_1^{V-1}(1))$ are C^1 diffeomorphisms onto V_i for each $i > 1$, and

$$g = g'' \quad \text{on} \quad V_1 \times [1, \infty[\times (V \cap \pi_1^{V-1}(0) \cap \rho_1^{V-1}(1))$$

$$\cup (\text{an } \mathfrak{X}\text{-neighborhood of } V_1 \times]0, \infty[\times (\bigcup_{1<i<k} V_i \cap \pi_1^{V-1}(0) \cap \rho_1^{V-1}(1))).$$

Set $\tau^V = g'' \circ g^{-1}$ on V_k and $\tau^V = \mathrm{id}$ on $V - V_k$. Then it is clear that all the conditions on τ^V are satisfied except for the last. The last is also easy to show. Indeed,

$$\pi_1^{V'} \circ \tau^V \circ g(z, t, a) = \pi_1^{V''} \circ g''(z, t, a) = z$$

$$\text{for } (z, t, a) \in V_1 \times \,]0, 1/2] \times (V \cap \pi_1^{V-1}(0) \cap \rho_1^{V-1}(1)).$$

Hence $\pi_1^{V'} \circ \tau^V = \pi_1^V$ on the image of this domain under g.

Define an X-homeomorphism τ of V to be $\tau^Y \circ \tau^V \circ \tau^{X-1}$. Then we have

$$\tau = \mathrm{id} \quad \text{on } V_1 \cup (\text{an X-neighborhood of } \bigcup_{1 < i < k} V_i \text{ in } V)$$

$$\cup (V_k - \text{a small X-neighborhood of } V_1 \text{ in } V),$$

$$\rho_1^X \circ p = \rho_1^V \circ \tau^{X-1}$$

$$= \rho_1^V \circ \tau^V \circ \tau^{X-1} = \rho_1^Y \circ q \circ \tau^Y \circ \tau^V \circ \tau^{X-1} = \rho_1^Y \circ q \circ \tau$$

on a smaller X-neighborhood of V_1 in V, and

$$(p|_V)^{-1} \circ \pi_1^X \circ p = \pi_1^V = \pi_1^V \circ \tau^{X-1} = \pi_1^{V'} \circ \tau^V \circ \tau^{X-1}$$

$$= \pi_1^{V'} \circ \tau^Y \circ \tau^V \circ \tau^{X-1} = (q|_V)^{-1} \circ \pi_1^Y \circ q \circ \tau$$

on the smaller one.

Hence, if we replace $f|_{p(V)}$ with $f' = q \circ \tau \circ (q|_V)^{-1} \circ f|_{p(V)}$, then

$$f' = f \quad \text{on } U \cup (\text{an X-neighborhood of } \pi_1^{X-1}(U) \cap \bigcup_{1 < i < k} X_i \text{ in } \bigcup_{1 < i \leq k} X_i)$$

$$\cup (X_k \cap \pi_1^{X-1}(U) - \text{a small X-neighborhood of } U \text{ in } \bigcup_{i \leq k} X_i),$$

$$\pi_1^Y \circ f' = \pi_1^Y \circ q \circ \tau \circ (q|_V)^{-1} \circ f = q \circ (p|_V)^{-1} \circ \pi_1^X \circ p \circ (q|_V)^{-1} \circ f$$

$$= q \circ (p|_V)^{-1} \circ \pi_1^X = \pi_1^X = f' \circ \pi_1^X$$

on a smaller X-neighborhood of U in $\bigcup_{i \leq k} X_i$, and

$$\rho_1^Y \circ f' = \rho_1^Y \circ q \circ \tau \circ (q|_V)^{-1} \circ f = \rho_1^X \circ p \circ (q|_V)^{-1} \circ f = \rho_1^X$$

on the smaller one.

Since f' cannot always be extended to $\cup_{i \leq k} X_i$, we modify it as follows. By the above method of construction of τ^X, τ^Y and τ^V, there exist nonempty, bounded and open X-subsets $V_1^* \subset V_1^{**}$ of V_1 and an X-homeomorphism τ^* of V such that $\tau^*|_{V_k}$ is a C^1 diffeomorphism of V_k and

$$\tau^* = \begin{cases} \tau & \text{on } V \cap \pi_1^{V-1}(V_1^*) \\ \mathrm{id} & \text{on } V \cap \pi_1^{V-1}(V_1 - V_1^{**}) \cup \{z \in V : \tau(z) = z\}. \end{cases}$$

Define an $\mathfrak{X}$-homeomorphism $f^*\colon \pi_1^{X-1}(U)\cap\cup_{i\leq k}X_i \to \pi_1^{Y-1}(f(U))\cap\cup_{i\leq k}Y_i$ by τ^* as we defined f' by τ, and set $U^* = p(V_1^*)$. Then we can extend f^* to an $\mathfrak{X}$-stratified homeomorphism $\{X_i\}_{i\leq k} \to \{Y_i\}_{i\leq k}$ by setting $f^* = f$ outside $\pi_1^{X-1}(U)$, and we have

$$f^* = f \quad \text{on } X_1 \cup (\text{an } \mathfrak{X}\text{-neighborhood of } \bigcup_{1<i<k} X_i \text{ in } \bigcup_{i\leq k} X_i) \text{ and}$$

$$\pi_1^Y \circ f^* = f^* \circ \pi_1^X, \ \rho_1^Y \circ f^* = \rho_1^X \quad \text{on } X_k \cap \pi_1^{X-1}(U^*).$$

Here we shrink $|T_1^X|$.

To extend f^* to X, we use an $\mathfrak{X}$-isotopy from f^* to f, whose existence is clear by the above arguments. To be precise, there exists an $\mathfrak{X}$-stratified isotopy $f_t^*\colon \{X_i\}_{i\leq k} \to \{Y_i\}_{i\leq k}$, $t \in [0,1]$, such that

$$f_t^* = \begin{cases} f & \text{on } X_1 \cup (\text{an } \mathfrak{X}\text{-neighborhood of } \bigcup_{1<i<k} X_i \text{ in } \bigcup_{i\leq k} X_i) \\ f^* & \text{for } t \in [0,1/3] \\ f & \text{for } t \in [2/3,1], \end{cases}$$

and the map $X_k \times [0,1] \ni (x,t) \to (f_t^*(x),t) \in Y_k \times [0,1]$ is a C^1 diffeomorphism.

We need also a precise $\mathfrak{X}$-trivialization of (π_k^X, ρ_k^X) as follows. For simplicity of notation but without loss of generality, we assume that for $i < j$, the map $X_j \cap |T_i^X| \ni x \to (\pi_i^X(x), \rho_i^X(x)) \in X_i \times \mathbf{R}$ is a proper C^1 submersion onto $X_i \times]0,1[$. Choose a point a_i in each X_i. Then there exist $\mathfrak{X}$-maps $\varphi_i\colon X \cap |T_i^X| \to X \cap \pi_i^{X-1}(a_i)$ such that

$$\varphi_i = \mathrm{id} \quad \text{on } X \cap \pi_i^{X-1}(a_i), \qquad \rho_i^X \circ \varphi_i = \rho_i^X, \text{ and}$$

$(\varphi_i, \pi_i^X)\colon X \cap |T_i^X| \to (X \cap \pi_i^{X-1}(a_i)) \times X_i$ is a homeomorphism. Furthermore, for $i < j$, the restriction of (φ_i, π_i^X) to $X_j \cap |T_i^X|$ is a C^1 diffeomorphism onto $(X_j \cap \pi_i^{X-1}(a_i)) \times X_i$,

$$\varphi_i(X \cap |T_i^X| \cap |T_j^X|) \subset |T_j^X|, \text{ and}$$

$$\varphi_j \circ \varphi_i = \varphi_j, \ \pi_j^X \circ \varphi_i = \varphi_i \circ \pi_j^X \quad \text{on } X \cap |T_i^X| \cap |T_j^X|.$$

We construct φ_i by a double induction method. By the first induction, we assume existence of φ_i, $i > 1$, and it suffices to construct φ_1. By the second induction, we assume φ_1 is already defined on $X \cap |T_1^X| \cap \cup_{1<i<k} |T_i^X|$ with the above required properties. We want to extend φ_1 to $X \cap |T_1^X| \cap |T_k|$. If φ_1 is

extended to $|T_1^X| \cap X_k$, it is uniquely extended, moreover, to $X \cap |T_1^X| \cap |T_k^X|$ by

$$\varphi_1|_{X \cap \pi_k^{X-1}(x)} = (\varphi_k|_{X \cap \pi_k^{X-1}(\varphi_1(x))})^{-1} \circ \varphi_k|_{X \cap \pi_k^{X-1}(x)} \quad \text{for } x \in |T_1^X| \cap X_k.$$

Hence we need only extend φ_1 to $|T_1^X| \cap X_k$. Apply II.6.8 to

$$A = |T_1^X| \cap X_k, \quad B = \{0\}, \quad C = X_1 \times]0,1[,$$
$$V = |T_1^X| \cap X_k \cap \bigcup_{1<i<k} |T_i^X|, \quad W = \varnothing,$$
$$\Psi = (\pi_1^X, \rho_1^X), \quad \text{and} \quad \chi = \varphi_1|_{|T_1^X| \cap X_k \cap \cup_{1<i<k}|T_i^X|}.$$

Then we obtain the required extension. (In II.6.8, A and C are assumed to be bounded, and in the present case, they are not so. If they are unbounded, we see existence of the local extension by II.6.8 and then obtain the global extension as in the proof of II.6.3.)

Now we can extend f^* to $X \cap |T_k^X|$ and then to X. For $x \in X \cap |T_k^X|$, set $x_0 = \pi_k^X(x)$ and $t = \rho_k^X(x)$, let $y_0 \in X_k$ be such that $f(y_0) = f_t^*(x_0)$, and set

$$f^*(x) = f \circ (\varphi_k|_{X \cap \pi_k^{X-1}(y_0)})^{-1} \circ \varphi_k(x).$$

Then f^* is an extension to $X \cap |T_k^X| \to Y \cap |T_k^Y|$ and satisfies $(*)$ for $i = 1$ on $X \cap \pi_1^{X-1}(U^*) \cap |T_k^X|$. (Here we replace $|T_k^X|$ with $\rho_k^{X-1}([0,1/3[)$ and keep the notation.) Moreover, since

$$f^* = f \quad \text{on } X_1 \cup (X \cap \rho_k^{X-1}([2/3,1[))$$
$$\cup(\text{an } X\text{-neighborhood of } \bigcup_{1<i<k} X_i \text{ in } \bigcup_{i \leq k} X_i)$$

and we can assume

$$\overline{X \cap \rho_k^{X-1}([0,2/3])} - X \cap \rho_k^{X-1}([0,2/3]) = \partial X_k,$$

the extension of f^* to X, defined to be f outside $|T_k^X|$, is the required one. Note that by the above method of construction, we can keep $(*)$ for $i > 1$ and $i = 1$ on $\pi_1^{X-1}(U) \cap \cup_{1<i<k} |T_i^X|$. Thus we can assume $(*)$ for $i = 1$ holds on $X \cap \pi_1^{X-1}(U)$ for an open X-subset U of X.

To obtain $(*)$ for $i = 1$ on $X \cap |T_1^X|$, we reduce X_1 to U. Without loss of generality, we can assume by II.6.1 and II.6.3

$$X_1 = Y_1 = B' \times 0 \subset \mathbf{R}^{n_1} \times \mathbf{R}^{n-n_1}, \quad U = C' \times 0,$$
$$B' = \{x' \in \mathbf{R}^{n_1} : |x'| < 1\}, \qquad C' = \{|x'| \leq 1/2\},$$

where

$$enskip X_i \cap |T_1^X| = Y_i \cap |T_1^Y| = B' \times Z_i \text{ for some } Z_i \subset \mathbf{R}^{n-n_1}, \ i = 1, 2, \dots,$$
$$\pi_1^X(x) = \pi_1^Y(x) = (x_1, \dots, x_{n_1}, 0 \dots, 0) \quad \text{for } x = (x_1, \dots, x_n) \in \mathbf{R}^n,$$
$$\rho_1^X(x) = \rho_1^Y(x) = x_{n_1+1}^2 + \dots + x_n^2,$$
$$f = \text{id on } X_1 \cup 0 \times Z \text{ where } Z = \bigcup_{i=1,2\dots} Z_i, \tag{6}$$

and Z is contained and closed in $B'' = \{x'' \in \mathbf{R}^{n-n_1} : |x''| < 1\}$. Moreover, by shrinking $|T_1^X|$ and considering graph f, we suppose

$$f(X \cap |T_1^X|) \subset Y \cap |T_1^Y| \quad \text{and}$$

$$f(a_i) \to a_i$$
$$\text{for any } a_i \in X \cap |T_1^X| \text{ with } \pi_1^X(a_i) \to a \in \partial X_1. \tag{7}$$

(Remember $f(a) = a$.)

Let α be a strictly decreasing C^1 𝔛-function on $[0, 1]$ such that $\alpha = 1$ at 0, $\alpha = 0$ at 1 and

$$\{(x', x'') \in B' \times Z : |x''| \leq \alpha(|x'|)\} \subset f(B' \times Z).$$

Let Q denote the former set, and let

$$\xi = (\xi', \xi'') \colon B' \times B'' \to B' \times B'' - \{1/2 \leq |x'| < 1\} \times 0$$

be a C^1 𝔛-diffeomorphism such that for each $(x', x'') \in B' \times B''$, $s, t \in [0, 1[$,

$$\xi'(x', x'') = ux' \quad \text{for some } u > 0,$$
$$\xi'(x', x'') = x' \quad \text{if } |x'| \leq 1/3 \text{ or } (x', x'') \notin Q,$$
$$|\xi'(x', x'')| \leq 1/2 \quad \text{if } 2|x''| \leq \alpha(|x'|),$$
$$\xi''(x', x'') = x'', \quad \text{and}$$
$$\xi\{(y', y'') \in B' \times B'' : |y'| = s, |v''| = t\} = \{|y'| = s', |y''| = t'\} \tag{8}$$
$$\text{for some } s', t' \in [0, 1[.$$

Define the modification $\hat{f}\colon X\cap|T_1^X|\to Y\cap|T_1^Y|$ of $f|_{X\cap|T_1^X|}$ by $\xi\circ\hat{f}=f\circ\xi$.
It is easy to see that $\hat{f}$ is well-defined and is a stratified C^0 imbedding
$\{X_i\cap|T_1^X|\}\to\{Y_i\cap|T_1^Y|\}$; $\hat{f}|_{X_i\cap|T_1^X|}$ is a C^1 imbedding into $Y_i\cap|T_1^Y|$ for
each i; $\hat{f}$ satisfies (*) for $i=1$ on $\{(x',x'')\in X\cap|T_1^X|\colon 2|x''|\le\alpha(|x'|)\}$
because the image of the domain under ξ is contained in $X\cap\pi_1^{X-1}(U)$ and
(*) for $i=1$ holds on $X\cap\pi_1^{X-1}(U)$; $\hat{f}=f$ outside of $Q\cup f^{-1}(Q)$, which
is a closed $\mathfrak{X}$-neighborhood of X_1 in X smaller than $X\cap|T_1^X|$; $\hat{f}$ keeps (*)
for $i>1$ because of (8), controlledness of $\{T_i^X\}$ and $\{T_i^Y\}$ and (*) for f and
$i>1$; and by (6) and (7), for any sequence $\{a_i\}$ in $X\cap|T_1^X|$ converging to a
point of $\overline{X\cap|T_1^X|}-|T_1^X|$ such that $\{f(a_i)\}$ converges, $\{\hat{f}(a_i)\}$ converges to
the same point. Hence $\hat{f}$ fulfills all the requirements, and we complete the
proof. $\square$

Remark II.7.4. It may be possible to weaken the assumption in II.7.3 that
X_i and Y_i are C^1 $\mathfrak{X}$-diffeomorphic to open balls by replacing it with the
assumption that X_i and Y_i are simply connected. Without assumption, the
theorem cannot hold as follows.

By [Si$_1$], there exist a compact C^∞ manifold $M\subset 1\times\mathbf{R}^n\subset\mathbf{R}^{n+1}$
and a C^∞ diffeomorphism $\tau=(\tau_1,\tau_2)\colon[0,1]\times M\to[0,1]\times M$ such that
$\tau_1(0,\cdot)=0$, $\tau_1(1,\cdot)=1$, $\tau_2(0,\cdot)$ is the identity map of M and $\tau_2(1,\cdot)$ is a C^∞
diffeomorphism of M which is not C^∞ isotopic to the identity. Set

$$X=S^1\times(0^{n+1}*M-M),\quad X_1=S^1\times 0^{n+1},\quad\text{and}\quad X_2=X-X_1.$$

Let Z denote the quotient C^∞ manifold $M\times[0,1]/\sim$ by the equivalence
relation defined by $(0,x)\sim(1,\tau_2(1,x))$ for $x\in M$, and let $p\colon Z\to S^1$ denote
the C^∞ submersion induced by the projection $M\times[0,1]\to[0,1]$. Note that
p is a non-trivial fibre bundle. We can assume $Z\subset S^1\times 1\times\mathbf{R}^n$ and p is the
restriction to Z of the projection $S^1\times\mathbf{R}^{n+1}\to S^1$. Set

$$Y=\bigcup_{t\in S^1}(t,0^{n+1})*p^{-1}(t)-Z,\quad Y_1=S^1\times 0^{n+1},\quad\text{and}\quad Y_2=Y-Y_1.$$

By II.5.9, we suppose all the manifolds are of class C^1 $\mathfrak{X}$. Then $\{X_1,X_2\}$ and
$\{Y_1,Y_2\}$ satisfy all the assumptions in II.7.3 except the ones that X_i and Y_i
are C^1 $\mathfrak{X}$-diffeomorphic to open balls, they are C^1 Whitney stratifications,
and any f cannot be compatible with any C^1 $\mathfrak{X}$-tube systems.

*Proof that there exists an $\mathfrak{X}$-stratified homeomorphism $f\colon\{X_i\}\to\{Y_i\}$ such
that $f|_{X_2}$ is a C^1 diffeomorphism onto Y_2.* It suffices to prove the following
statement. There exists an $\mathfrak{X}$-homeomorphism $\theta=(\theta_1,\theta_2)\colon[0,1]\times(0^{n+1}*$

$M) \to [0,1] \times (0^{n+1} * M)$ such that $\theta_1(0,\cdot) = 0$, $\theta_1(1,\cdot) = 1$, $\theta_2(0,\cdot) = \mathrm{id}$, $\theta_2(1,\cdot) =$ the cone extension of $\tau_2(1,\cdot)$, and the restrictions of θ to $[0,1] \times 0^{n+1}$ and $[0,1] \times (0^{n+1} * M - 0^{n+1})$ are C^1 diffeomorphisms of $[0,1] \times 0^{n+1}$ and of $[0,1] \times (0^{n+1} * M - 0^{n+1})$ respectively. Define θ to be τ on $[0,1] \times M$ and the identity on $\{0, 1/2, 1\} \times 0^{n+1}$. Note that if we set

$$A = (0, 0^{n+1}) * (1/2, 0^{n+1}) * (0 \times M),$$
$$B = (1/2, 0^{n+1}) * ([0,1] \times M), \quad \text{and}$$
$$C = (1, 0^{n+1}) * (1/2, 0^{n+1}) * (1 \times M),$$

then we have

$$A \cup B \cup C = [0,1] \times (0^{n+1} * M), \quad A \cap B = (1/2, 0^{n+1}) * (0 \times M),$$
$$A \cap C = (1/2, 0^{n+1}), \quad \text{and} \quad B \cap C = (1/2, 0^{n+1}) * (1 \times M).$$

Hence we can extend θ to $(1/2, 0^{n+1}) * ([0,1] \times M)$ and then to $[0,1] \times (0^{n+1} * M)$ by the Alexander trick. Here the restriction of θ to $[0,1] \times (0^{n+1} * M - 0^{n+1})$ is not always a C^1 diffeomorphism of $[0,1] \times (0^{n+1} * M - 0^{n+1})$. For this problem, it suffices to modify τ so that $\tau(t, x) = (t, x)$ for small $t > 0$ and $\tau(t, x) = (t, \tau_2(1, x))$ for t near 1, which is clearly possible. Then θ fulfills all the requirements.

Proof that f cannot be compatible. Assume some f is compatible with some C^1 X-tubes $T_1^X = (|T_1^X|, \pi_1^X, \rho_1^X)$ at X_1 and $T_1^Y = (|T_1^Y|, \pi_1^Y, \rho_1^Y)$ at Y_1. By the following assertion, we can suppose π_1^X and π_1^Y are the restrictions of the projections $S^1 \times \mathbf{R}^{n+1} \to S^1 \times 0^{n+1}$ and ρ_1^X and ρ_1^Y are the squares of the functions which measure distance from $S^1 \times 0^{n+1}$. Let $\varepsilon > 0$ be a small number. Then the commutative diagram

$$
\begin{array}{ccc}
\rho_1^{X-1}(\varepsilon) & \xrightarrow{\ f\ } & \rho_1^{Y-1}(\varepsilon) \\
\pi_1^X \downarrow & & \pi_1^Y \downarrow \\
S^1 \times 0^{n+1} & \xrightarrow{\ f\ } & S^1 \times 0^{n+1}
\end{array}
$$

shows that $p \colon Z \to S^1$ is trivial, which is a contradiction.

Assertion. Let $\{X_i\}$ be a C^1 Whitney X-stratification of a locally closed X-set $X \subset \mathbf{R}^n$. Assume Axiom (v) or that X is bounded. Let $\{T_i = (|T_i|, \pi_i, \rho_i)\}$ and $\{T_i' = (|T_i'|, \pi_i', \rho_i')\}$ be controlled C^1 X-tube systems for $\{X_i\}$. Shrink $|T_i|$ and $|T_i'|$. Then there exists an X-stratified homeomorphism $f \colon \{X_i\} \to \{X_i\}$ compatible with $\{T_i\}$ and $\{T_i'\}$ such that each $f|_{X_i}$ is a C^1 diffeomorphism of X_i.

Proof of Assertion. As in the proof of II.7.3, we assume $\dim X_1 < \dim X_2 < \cdots$ and each X_i is connected for simplicity of notation. Shrink $|T_i|$ and $|T_i'|$, modify ρ_i and ρ_i' outside a small $\mathfrak{X}$-neighborhood of X_i for each i, and set

$$Y = X \times]-1, 2[\text{ and } Y_i = X_i \times]-1, 2[.$$

Then by II.6.10 and its proof, we can suppose that there exists a controlled C^1 $\mathfrak{X}$-tube system $\{T_i^Y = (|T_i^Y|, \pi_i^Y, \rho_i^Y)\}$ for $\{Y_i\}$ such that for each $i < j$, π_i^Y is commutative with the projection $p \colon Y \to]-1, 2[$. Also,

$$|T_i^Y| \subset |T_i| \times]-1, 2[, \qquad |T_i^Y| \subset |T_i'| \times]-1, 2[,$$

π_i^Y has a C^1 $\mathfrak{X}$-extension $\tilde{\pi}_i^Y$ to $|T_i| \times]-1, 2[\to Y_i$ which also is commutative with p,

$$\pi_i^Y(x, t) = \begin{cases} (\pi_i(x), t) & \text{for } t \in]-1, 0] \\ (\pi_i'(x), t) & \text{for } t \in [1, 2[, \end{cases}$$

$$\rho_i^Y(x, t) = \begin{cases} \rho_i(x) & \text{for } t \in]-1, 0] \\ \rho_i'(x) & \text{for } t \in [1, 2[, \end{cases}$$

$(\pi_i^Y, \rho_i^Y)|_{Y_j \cap |T_i^Y|}$ is a proper C^1 submersion onto $Y_i \times]0, 1[$, and $\tilde{\pi}_i^Y|_{Y_j \cap |T_i| \times]-1, 2[}$ is submersive. Set $Y(0) = X \times 0$ and $Y_i(0) = X_i \times 0$, and let $q \colon Y \to Y(0)$ denote the projection.

It suffices to find an $\mathfrak{X}$-map $\varphi \colon Y \to Y(0)$ such that for each i, $(\varphi, p)|_{Y_i}$ is a C^1 diffeomorphism onto $Y_i(0) \times]-1, 2[$,

$$\varphi = \begin{cases} \text{id} & \text{on } Y(0) \\ q & \text{on } Y_i - \bigcup_{j<i} \rho_j^{Y-1}(]0, 1 - \varepsilon_i[) \text{ for some } 0 < \varepsilon_i < 1/2, \end{cases}$$
$$\text{and } \pi_i^Y \circ \varphi = \varphi \circ \pi_i^Y, \quad \rho_i^Y \circ \varphi = \rho_i^Y \quad \text{on } Y \cap \rho_i^{Y-1}(]0, \varepsilon_i[),$$

where $\varepsilon_1 > \varepsilon_2 > \cdots$. We construct φ by double induction. Define φ on Y_1 to be $q|_{Y_1}$. Let $k \geq 0$ and $l > 1$ be integers with $k < l$. Assuming φ is defined on $(\cup_{i<l} Y_i) \cup (Y_l \cap \cup_{k<i<l} |T_i| \times]-1, 2[)$, we need only extend φ to Y_l if $k = 0$ and to $Y_l \cap |T_k| \times]-1, 2[$ if $k > 0$. If $k = 0$, it suffices to set $\varphi = q$ on $Y_l - \cup_{i<l} |T_i| \times]-1, 2[$. If $k > 0$, apply II.6.8 to

$$A = Y_l \cap \rho_k^{Y-1}(]0, 1/2 + \varepsilon_l[), \qquad B = Y_k(0) \times]0, 1/2 + \varepsilon_l[, \qquad C =]-1, 2[,$$
$$V = Y_l \cap \rho_k^{Y-1}(]0, 1/2 + \varepsilon_l[) \cap \bigcup_{k<i<l} |T_i| \times]-1, 2[, \qquad W = \varnothing$$
$$\Psi = (\psi, \psi') = (\varphi \circ \pi_k^Y, \rho_k^Y, p), \quad \text{and}$$
$$\chi = \varphi \quad \text{on } V.$$

(Though A and B are assumed to be bounded in II.6.8, they are not so in the present case. If they are bounded, by C^1-$\mathfrak{X}$-triangulating $Y_k(0)$ we can easily reduce the problem to the bounded case.) Shrink ε_l. Then we extend φ to $Y_l \cap \rho_k^{Y-1}(]0, 1/2 + \varepsilon_l[)$. Keep the notation φ for the extension. We need to extend it, moreover, to $Y_l \cap |T_k| \times]-1, 2[$. Apply, once more, II.6.8 to

$$A = Y_l \cap |T_k| \times]-1, 2[- \rho_k^{Y-1}(]0, 1/2]), \qquad B = Y_k(0), \qquad C =]-1, 2[,$$
$$V = Y_l \cap (\rho_k^{Y-1}(]1/2, 1/2 + \varepsilon_l[) \cup (|T_k| \times]-1, 2[- \rho_k^{Y-1}(]0, 1 - \varepsilon_l]))$$
$$\cup ((|T_k| \cap \bigcup_{k<i<l} |T_i|) \times]-1, 2[)),$$
$$W = \varnothing, \qquad \Psi = (\psi, \psi') = (\varphi \circ \tilde{\pi}_k^Y, p), \quad \text{and}$$
$$\chi = \begin{cases} \varphi & \text{on } Y_l \cap (\rho_k^{Y-1}(]1/2, 1/2 + \varepsilon_l[) \cup ((|T_k| \cap \bigcup_{k<i<l} |T_i|) \times]-1, 2[)) \\ q & \text{on the complement in } V. \end{cases}$$

Then shrinking ε_l more, we have the required extension of φ to $Y_l \cap |T_k| \times]-1, 2[$. $\qquad\square$

Let X, X', Y and Y' be closed subsets of $\mathbf{R}^n$, and let $f: X \to Y$ and $f': X' \to Y'$ be C^1 maps. Let $\{Y_j\}$ and $\{T_j^Y = (|T_j^Y|, \pi_j^Y, \rho_j^Y)\}$ be a C^1 stratification of Y and its controlled C^1 tube system such that $(\pi_j^Y, \rho_j^Y)|_{Y_{j'} \cap |T_j^Y|}$ is a C^1 submersion to $Y_j \times \mathbf{R}$ for j and j' with $Y_j \cap (\overline{Y_{j'}} - Y_{j'}) \neq \varnothing$. Let $\{X_i\}$ and $\{T_i^X = (|T_i^X|, \pi_i^X, \rho_i^X)\}$ be a C^1 stratification of X and its C^1 tube system controlled over $\{T_j^Y\}$ such that for i, i', j and j' with $f(X_i) \subset Y_j$, $f(X_{i'}) \subset Y_{j'}$ and $X_i \cap (\overline{X_{i'}} - X_{i'}) \neq \varnothing$, if $Y_j = Y_{j'}$ then $(\pi_i^X, \rho_i^X)|_{X_{i'} \cap |T_i^X|}$ is a C^1 submersion to $X_i \times \mathbf{R}$, and if $Y_j \neq Y_{j'}$ then $(\pi_i^X, f)|_{X_{i'} \cap |T_i^X|}$ is a C^1 submersion to $\{(x, y) \in X_i \times (Y_{j'} \cap |T_j^Y|): f(x) = \pi_j^Y(y)\}$. Let $\{X_i'\}$, $\{Y_j'\}$, $\{T_i^{X'}\}$ and $\{T_j^{Y'}\}$ be given for X', f' and Y' in the same way. Let $\tau: \{X_i\} \to \{X_i'\}$ and $\theta: \{Y_j\} \to \{Y_j'\}$ be stratified C^0 maps such that $\tau|_{X_i}$ and $\theta|_{Y_j}$ are C^1 diffeomorphisms onto X_i' and Y_j' respectively, and the following diagram is commutative:

$$\begin{array}{ccc} \{X_i\} & \xrightarrow{\ \tau\ } & \{X_i'\} \\ {\scriptstyle f}\downarrow & & \downarrow{\scriptstyle f'} \\ \{Y_j\} & \xrightarrow{\ \theta\ } & \{Y_j'\}. \end{array}$$

We call (τ, θ) *compatible* with $\{T_i^X, T_j^Y\}$ and $\{T_i^{X'}, T_j^{Y'}\}$ if θ is compatible with $\{T_j^Y\}$ and $\{T_j^{Y'}\}$, for each i,

$$\tau(X \cap |T_i^X|) = X' \cap |T_i^{X'}|,$$

$$\pi_i^{X'} \circ \tau = \tau \circ \pi_i^X \quad \text{on } X \cap |T_i^X|, \text{ and}$$

$$\rho_i^{X'} \circ \tau = \rho_i^X \quad \text{on } X \cap |T_i^X| \cap f^{-1}(f(X_i)).$$

Note that the last equality holds on a neighborhood of $X \cap |T_i^X| \cap f^{-1}(f(X_i)) - X_i$ in X.

For a tube T_k^A, we denote its elements by $|T_k^A|$, π_k^A and ρ_k^A.

Theorem II.7.5 (Compatibility implies X-equivalence). *Let X, $\{X_i\}$, $\{T_i^X\}$, etc. be as above. We suppose they are of class $\mathfrak{X}$ except τ and θ. Assume Axiom* (v) *or that the following $\mathfrak{X}$-sets are all bounded.*

(i) *If θ is compatible with $\{T_j^Y\}$ and $\{T_j^{Y'}\}$, then Y and Y' are $\mathfrak{X}$-homeomorphic.*

(ii) *Let $g: Z \to \mathbf{R}$ be a C^1 $\mathfrak{X}$-function on a closed $\mathfrak{X}$-set and let $\{Z_i\}$ and $\{T_i^Z\}$ be a C^1 $\mathfrak{X}$-stratification of Z and its weak controlled tube system. Assume for each i, $g|_{Z_i}$ is constant or C^1 regular; for i' with $Z_i \cap (\overline{Z_{i'}} - Z_{i'}) \neq \varnothing$, $(\pi_i^Z, \rho_i^Z)|_{Z_{i'} \cap |T_i^Z|}$ is a C^1 submersion to $Z_i \times \mathbf{R}$; if $g|_{Z_i}$ is constant and $g|_{Z_{i'}}$ is C^1 regular then for each $z \in Z_i$ and for each $t > 0 \in \mathbf{R}$, $g|_{\pi_i^{Z-1}(z) \cap Z_{i'}}$ is C^1 regular and $g|_{\pi_i^{Z-1}(z) \cap Z_{i'} \cap \rho_i^{Z-1}(t)}$ is C^1 regular at $\pi_i^{Z-1}(z) \cap Z_{i'} \cap \rho_i^{Z-1}(t) \cap g^{-1}(g(Z_i))$; if $g|_{Z_i}$ is C^1 regular then $g|_{Z_{i'}}$ is C^1 regular, $g|_{\pi_i^{Z-1}(z) \cap Z_{i'}}$ is constant for $z \in Z_i$ and $\rho_i^Z \circ \pi_{i'}^Z = \rho_i^Z$ on $Z \cap |T_i^Z| \cap |T_{i'}^Z|$ for $i \neq i'$; and if $g|_{Z_i}$ and $g|_{Z_{i'}}$ are constant then $\rho_i^Z \circ \pi_{i'}^Z = \rho_i^Z$ on $Z \cap |T_i^Z| \cap |T_{i'}^Z|$ for $i \neq i'$. Let $g': Z' \to \mathbf{R}$, $\{Z_i'\}$ and $\{T_i^{Z'}\}$ be given with the same properties. Let $\xi: \{Z_i\} \to \{Z_i'\}$ be a stratified homeomorphism such that for each i, $\xi|_{Z_i}$ is a C^1 diffeomorphism onto Z_i',*

$$g' \circ \xi = g, \qquad \xi(Z \cap |T_i^Z|) = Z' \cap |T_i^{Z'}|,$$

$$\pi_i^{Z'} \circ \xi = \xi \circ \pi_i^Z \quad \text{on } Z \cap |T_i^Z|, \text{ and}$$

$$\rho_i^{Z'} \circ \xi = \rho_i^Z \quad \text{on a neighborhood of } Z \cap |T_i^Z| \cap g^{-1}(g(Z_i)) - Z_i \text{ in } Z.$$

Then g and g' are $\mathfrak{X}$-equivalent.

(iii) *If (τ, θ) is compatible with $\{T_i^X, T_j^Y\}$ and $\{T_i^{X'}, T_j^{Y'}\}$ and if f and f' are proper, then f and f' are R-L $\mathfrak{X}$-equivalent. If $Y = Y'$ and if (τ, id) is compatible with $\{T_i^X, T_j^Y\}$ and $\{T_j^{X'}, T_j^{Y'}\}$, then f and f' are R $\mathfrak{X}$-equivalent.*

The conditions in (ii) may look complicated. But they are weaker than the conditions in (iii). Indeed, (ii) in the case of a proper function is equivalent to the latter half of (iii) in the case of $Y = Y' = \mathbf{R}$.

We can choose the $\mathfrak{X}$-homeomorphisms (of equivalence) in II.7.5 and (iii) of the next corollary so that they are compatible with the respective tube systems and their restrictions to the strata are C^1 diffeomorphisms onto the strata, which is clear by the forthcoming proofs.

By the following corollary, we know that $\mathfrak{X}$-equivalence is weaker than C^1 equivalence. The former is strictly weaker than the latter. For example, $f_1(x) = x$ and $f_2(x) = x^3$ are semialgebraically equivalent but not C^1 equivalent.

Corollary II.7.6 (C^1 equivalence implies $\mathfrak{X}$-equivalence). *Assume Axiom (v) or that the following $\mathfrak{X}$-sets are all bounded.*

(i) Two C^1 equivalent closed $\mathfrak{X}$-subsets of $\mathbf{R}^n$ are $\mathfrak{X}$-equivalent.

(ii) Two C^1 equivalent C^1 $\mathfrak{X}$-functions on closed $\mathfrak{X}$-subsets of $\mathbf{R}^n$ are $\mathfrak{X}$-equivalent.

(iii) Let X, X', Y and Y' be closed $\mathfrak{X}$-subsets of $\mathbf{R}^n$, and let $f: \{X_i\} \to \{Y_j\}$ and $f': \{X_i'\} \to \{Y_j'\}$ be C^1 Whitney $\mathfrak{X}$-stratifications sans éclatement of proper C^1 $\mathfrak{X}$-maps $X \to Y$ and $X' \to Y'$. Assume that f and f' are R-L C^1 equivalent, i.e., there exist C^1 diffeomorphisms $\tau: X \to X'$ and $\theta: Y \to Y'$ such that $f' \circ \tau = \theta \circ f$, $\tau(X_i) = X_i'$ and $\theta(Y_j) = Y_j'$ for every i and j. Then f and f' are R-L $\mathfrak{X}$-equivalent.

Assume, in addition, that $Y = Y'$, $\{Y_j\} = \{Y_j'\}$ and $\theta = \mathrm{id}$, and remove the properness assumption. Then f and f' are R $\mathfrak{X}$-equivalent.

An open problem is whether they are C^1 (R) (R-L) $\mathfrak{X}$-equivalent.

Proof of II.7.6. Proof of (i) Let $\theta: Y \to Y'$ be a C^1 diffeomorphism between closed $\mathfrak{X}$-subsets of $\mathbf{R}^n$. Let $\{Y_j\}$ denote the connected components of the canonical C^1 Whitney $\mathfrak{X}$-stratification of Y (II.1.15). Since the construction of $\{Y_j\}$ depends only on the C^1 structure of Y but not the $\mathfrak{X}$-structure, $\{Y_j' = \theta(Y_j)\}$ consists of the connected components of the canonical C^1 Whitney $\mathfrak{X}$-stratification of Y'. Let $\{T_j^Y\}$ and $\{T_j^{Y'}\}$ denote controlled C^1 $\mathfrak{X}$-tube systems for $\{Y_j\}$ and $\{Y_j'\}$ respectively (II.6.10). By II.7.5 it suffices to modify θ to be compatible with $\{T_j^Y\}$ and $\{T_j^{Y'}\}$. Define a controlled C^1 (not necessarily $\mathfrak{X}$) tube system $\{T_{0j}^{Y'}\}$ for $\{Y_j'\}$ to be $\{\theta_* T_j^Y = (\theta(|T_j^Y|), \theta \circ \pi_j^Y \circ \theta^{-1}, \rho_j^Y \circ \theta^{-1})\}$. (Here by shrinking $|T_j^Y|$, we assume θ is defined on $\cup |T_j^Y|$ and a C^1 diffeomorphism onto a neighborhood of Y' in $\mathbf{R}^n$.) Then θ is compatible with $\{T_j^Y\}$ and $\{T_{0j}^{Y'}\}$. Hence we need only a stratified homeomorphism $\theta': \{Y_j'\} \to \{Y_j'\}$ compatible with $\{T_{0j}^{Y'}\}$ and $\{T_j^{Y'}\}$ such that $\theta'|_{Y_j'}$ is a C^1 diffeomorphism of Y_j' for each j. But this is just the assertion without $\mathfrak{X}$ in II.7.4, which is proved more easily.

Proof of (ii) Let $g: Z \to \mathbf{R}$ and $g': Z' \to \mathbf{R}$ be C^1 $\mathfrak{X}$-functions on closed $\mathfrak{X}$-subsets of $\mathbf{R}^n$, and let $\xi: Z \to Z'$ be a C^1 diffeomorphism such that $g' \circ \xi =$

g. We want to construct canonically a C^1 Whitney $\mathfrak{X}$-stratification $\{Z_i\}$ of Z and its weak controlled C^1 $\mathfrak{X}$-tube system $\{T_i^Z\}$ so that the conditions in (ii) of II.7.5 are satisfied. Recall that $\Sigma_1 Z$ denotes the C^1 singular point set of Z. Let $\Lambda_1 Z$ denote the subset of $Z - \Sigma_1 Z$ where the germ of g is C^1 regular or constant, let $\Lambda_2 Z$ denote the subset of $\Lambda_1(Z - \Lambda_1 Z)$ consisting of points z such that the two pairs of $\Lambda_1 Z$ and $\Lambda_1(Z - \Lambda_1 Z)$ and of $\Lambda_1 Z \cap g^{-1}(g(z))$ and $\Lambda_1(Z - \Lambda_1 Z) \cap g^{-1}(g(z))$ satisfy the Whitney condition at z, and so on. Then $\{Z_i\}$ is defined to be the connected components of $\Lambda_1 Z, \Lambda_2 Z, \ldots$. It is easy to construct the required $\{T_i^Z\}$ as in the proof of I.1.3. By the same reason as in (i), $\{Z_i' = \xi(Z_i)\}$ coincides with the C^1 Whitney $\mathfrak{X}$-stratification of Z' constructed in the same way, and it suffices to shrink $|T_i^Z|$ and $|T_i^{Z'}|$ and find a stratified homeomorphism $\xi' \colon \{Z_i'\} \to \{Z_i'\}$ such that for each i, $\xi'|_{Z_i'}$ is a C^1 diffeomorphism of Z_i',

$$g' \circ \xi' = g', \qquad \xi'(Z' \cap |T_{0i}^{Z'}|) = Z' \cap |T_i^{Z'}|,$$

$$\pi_i^{Z'} \circ \xi' = \xi' \circ \pi_{0i}^{Z'} \quad \text{on } Z' \cap |T_{0i}^{Z'}|, \quad \text{and}$$

$$\rho_i^{Z'} \circ \xi' = \rho_{0i}^{Z'} \quad \text{on a neighborhood of } Z' \cap |T_{0i}^{Z'}| \cap g'^{-1}(g'(Z_i')) - Z_i' \text{ in } Z',$$

where $\{T_{0i}^{Z'}\} = \{\xi_* T_i^Z\}$. Hence what we prove is the following function case without $\mathfrak{X}$ of the assertion in II.7.4.

For each $k = 0, 1$, let $\{T_{ki}^Z\}_i$ be a weakly controlled C^1 tube system for $\{Z_i\}$ such that the conditions in (ii) of II.7.5 are satisfied for $\{T_{ki}^Z\}_i$. Then there exists a stratified homeomorphism $\eta \colon \{Z_i\} \to \{Z_i\}$ such that for each i, $\eta|_{Z_i}$ is a C^1 diffeomorphism of Z_i,

$$g \circ \eta = g, \qquad \eta(Z \cap |T_{0i}^Z|) = Z \cap |T_{1i}^Z|,$$

$$\pi_{1i}^Z \circ \eta = \eta \circ \pi_{0i}^Z \quad \text{on } Z \cap |T_{0i}^Z|, \quad \text{and}$$

$$\rho_{1i}^Z \circ \eta = \rho_{0i}^Z \quad \text{on a neighborhood of } Z \cap |T_{0i}^Z| \cap g^{-1}(g(Z_i)) - Z_i \text{ in } Z.$$

We prove this in the same way as the assertion. Set

$$\hat{Z} = Z \times {]}-1, 2{[}, \quad \text{and} \quad \hat{Z}_i = Z_i \times {]}-1, 2{[},$$

and define $\hat{g} \colon \hat{Z} \to \mathbf{R}$ by $\hat{g}(z, t) = g(z)$. As in the proof of the assertion, we can assume there exists a weakly controlled C^1 tube system $\{T_i^{\hat{Z}}\}$ for $\{\hat{Z}_i\}$ such that the conditions in (ii) of II.7.5 are satisfied for $\{T_i^{\hat{Z}}\}$. Also, for each

i, $\pi_i^{\hat{Z}}$ is commutative with the projection $p \colon \hat{Z} \to \,]{-1}, 2[$,

$$|T_i^{\hat{Z}}| \subset |T_{ki}^Z| \times \,]{-1}, 2[, \quad k = 0, 1,$$

$$\pi_i^{\hat{Z}}(z, t) = \begin{cases} (\pi_{0i}^Z(z), t) & \text{for } t \in \,]{-1}, 0] \\ (\pi_{1i}^Z(z), t) & \text{for } t \in [1, 2[, \end{cases}$$

$$\rho_i^{\hat{Z}}(z, t) = \begin{cases} \rho_{0i}^Z(z) & \text{for } t \in \,]{-1}, 0] \\ \rho_{1i}^Z(z) & \text{for } t \in [1, 2[, \end{cases}$$

and for i' with $Z_i \subset \overline{Z_{i'}} - Z_{i'}$, $(\pi_i^{\hat{Z}}, \rho_i^{\hat{Z}})|_{\hat{Z}_{i'} \cap |T_i^{\hat{Z}}|}$ is a proper C^1 submersion onto $\hat{Z}_i \times \,]0, 1[$. Then it suffices to find a C^0 map $\varphi \colon \hat{Z} \to Z$ such that for each i, $(\varphi, p)|_{\hat{Z}_i}$ is a C^1 diffeomorphism of $\hat{Z}_i$,

$$g \circ \varphi = \hat{g}, \qquad \varphi(\hat{Z} \cap |T_i^{\hat{Z}}|) \subset |T_{0i}^Z|,$$

$$\pi_{0i}^Z \circ \varphi = \varphi \circ \pi_i^{\hat{Z}} \quad \text{on } \hat{Z} \cap |T_i^{\hat{Z}}|, \quad \text{and}$$

$$\rho_{0i}^Z \circ \varphi = \rho_i^{\hat{Z}} \quad \text{on a neighborhood of } \hat{Z} \cap |T_i^{\hat{Z}}| \cap \hat{g}^{-1}(\hat{g}(\hat{Z}_i)) - \hat{Z}_i \text{ in } \hat{Z}.$$

We construct φ by double induction as in the proof of the assertion. We need only modify the extension of φ to $\hat{Z}_l \cap |T_k^{\hat{Z}}| \times \,]{-1}, 2[$ in the case where $\hat{g}|_{\hat{Z}_k}$ is constant and $\hat{g}|_{\hat{Z}_l}$ is C^1 regular. Then there are two possible cases: $\hat{Z}_l \cap \hat{g}^{-1}(\hat{g}(\hat{Z}_k)) = \varnothing$ or not. For simplicity of notation, we assume each Z_i is connected. In the first case, we replace $\rho_k^{\hat{Z}}$ with $\hat{g}$ and apply II.6.8 without $\mathfrak{X}$, which is easily shown. Then we obtain the required φ. We omit the details. In the second case, we extend φ first to $\hat{Z}_l \cap \hat{g}^{-1}(\hat{g}(\hat{Z}_k)) \cap |T_k^{\hat{Z}}| \times \,]{-1}, 2[$ in the same way as in the proof of the assertion, second, to a neighborhood of $\hat{Z}_l \cap \hat{g}^{-1}(\hat{g}(\hat{Z}_k)) \cap |T_k^{\hat{Z}}| \times \,]{-1}, 2[$ in $\hat{Z}_l$, and then to $\hat{Z}_l \cap |T_k^{\hat{Z}}| \times \,]{-1}, 2[$ as in the first case.

Proof of (iii) By the same reason as above, it suffices to prove the following map case of the assertion without $\mathfrak{X}$.

For each $k = 0, 1$, let $\{T_{kj}^Y\}_j$ be a controlled C^1 tube system for $\{Y_j\}$, and let $\{T_{ki}^X\}_i$ be a C^1 tube system for $\{X_i\}$ controlled over $\{T_{kj}^Y\}$. Shrink $|T_{ki}^X|$ and $|T_{kj}^Y|$. Then there exist stratified homeomorphisms $\lambda^X \colon \{X_i\} \to \{X_i\}$ and $\lambda^Y \colon \{Y_j\} \to \{Y_j\}$ compatible with $\{T_{0i}^X, T_{0j}^Y\}$ and $\{T_{1i}^X, T_{1j}^Y\}$ such that $f \circ \lambda^X = \lambda^Y \circ f$, and $\lambda^X|_{X_i}$ and $\lambda^Y|_{Y_j}$ are C^1 diffeomorphisms of X_i and Y_j. Moreover, if $\{T_{0j}^Y\} = \{T_{1j}^Y\}$, we can choose λ^Y to be the identity. This proof also is similar to that of the assertion. Set

$$\hat{X}_i = X_i \times \,]{-1}, 2[, \quad \hat{Y}_j = Y_j \times \,]{-1}, 2[, \quad \hat{X} = \cup \hat{X}_i, \quad \hat{Y} = \cup \hat{Y}_j,$$

$$\text{and} \quad \hat{f} = f \times \text{id} \colon \hat{X} \longrightarrow \hat{Y}.$$

There exist a controlled C^1 tube system $\{T_j^{\hat{Y}}\}$ for $\{\hat{Y}_j\}$ and a C^1 tube system $\{T_i^{\hat{X}}\}$ for $\{\hat{X}_i\}$ controlled over $\{T_j^{\hat{Y}}\}$ such that for each i, $\pi_i^{\hat{X}}$ is commutative with the projection $p^X : \hat{X} \to]-1, 2[$,

$$|T_i^{\hat{X}}| \subset |T_{ki}^X| \times]-1, 2[, \quad k = 0, 1,$$

$$\pi_i^{\hat{X}}(x, t) = \begin{cases} (\pi_{0i}^X(x), t) & \text{for } t \in]-1, 0] \\ (\pi_{1i}^X(x), t) & \text{for } t \in [1, 2[, \end{cases}$$

$$\rho_i^{\hat{X}}(x, t) = \begin{cases} \rho_{0i}^X(x) & \text{for } t \in]-1, 0] \\ \rho_{1i}^X(x) & \text{for } t \in [1, 2[, \end{cases}$$

and for i' with $X_i \subset \overline{X_{i'}} - X_{i'}$, $(\pi_i^{\hat{X}}, \rho_i^{\hat{X}})|_{\hat{X}_{i'} \cap |T_i^{\hat{X}}|}$ is a proper C^1 submersion onto $\hat{X}_i \times]0, 1[$. Furthermore, the same properties hold for $\pi_j^{\hat{Y}}$ and $\rho_j^{\hat{Y}}$, and if $\{T_{0j}^Y\} = \{T_{1j}^Y\}$ then $\pi_j^{\hat{Y}} = (\pi_j^Y, p^Y)$ and $\rho_j^{\hat{Y}} = \rho_j^Y$, where p^Y denotes the projection $\hat{Y} \to]-1, 2[$. Then, as above, we need only find C^0 maps $\psi : \hat{X} \to X$ and $\varphi : \hat{Y} \to Y$ such that for each i and for each j, $(\psi, p^X)|_{\hat{X}_i}$ and $(\varphi, p^Y)|_{\hat{Y}_j}$ are C^1 diffeomorphisms of $\hat{X}_i$ and $\hat{Y}_j$ respectively,

$$f \circ \psi = \varphi \circ \hat{f},$$

$$\psi(\hat{X} \cap |T_i^{\hat{X}}|) \subset |T_{0i}^X|, \qquad \varphi(\hat{Y} \cap |T_j^{\hat{Y}}|) \subset |T_{0j}^Y|,$$

$$\pi_{0i}^X \circ \psi = \psi \circ \pi_i^{\hat{X}} \quad \text{on } \hat{X} \cap |T_i^{\hat{X}}|,$$

$$\rho_{0i}^X \circ \psi = \rho_i^{\hat{X}} \quad \text{on } \hat{X} \cap |T_i^{\hat{X}}| \cap \hat{f}^{-1}(\hat{f}(\hat{X}_i)),$$

$$\pi_{0j}^Y \circ \varphi = \varphi \circ \pi_j^{\hat{Y}}, \quad \rho_{0j}^Y \circ \varphi = \rho_j^{\hat{Y}} \quad \text{on } \hat{Y} \cap |T_j^{\hat{Y}}|, \text{ and}$$

$$\varphi = \text{id} \quad \text{if } \{T_{0j}^Y\} = \{T_{1j}^Y\}.$$

Existence of φ is already shown in the proof of the assertion, and construction of ψ is the same as that of φ in (ii). We do not repeat the proof. □

Proof of II.7.5. We easily see that $\{X_i\}$, $\{Y_j\}$ and $\{Z_i\}$ satisfy the weak frontier condition. For simplicity of notation, we assume $\dim X_1 < \dim X_2 < \cdots$, $\dim Y_1 < \dim Y_2 < \cdots$, $\dim Z_1 < \dim Z_2 < \cdots$, and X_i, Y_j and Z_i are all connected. Moreover, we suppose Axiom (v) and that X, X',... are unbounded because the bounded case is easy to prove.

(i) Let $k > 0$ be an integer. We modify θ so that for any $j \le k$, the C^1 diffeomorphism $\theta|_{Y_j} : Y_j \to Y_j'$ is of class locally $\mathfrak{X}$, by assuming this for $k-1$. Let θ_k be a strong C^1 locally $\mathfrak{X}$ approximation of $\theta|_{Y_k} : Y_k \to Y_k'$ in the C^1

Whitney topology, whose existence is clear, and extend it to $\theta_k \colon \{Y_j\}_{j\leq k} \to \{Y'_j\}_{j\leq k}$ by setting $\theta_k = \theta$ on $\{Y_j\}_{j<k}$.

First, θ_k can be compatible with $\{T_j^Y\}_{j<k}$ and $\{T_j^{Y'}\}_{j<k}$ for the following reason. Let $0 < l < k$ be an integer, and assume θ_k is compatible with $\{T_j^Y\}_{l<j<k}$ and $\{T_j^{Y'}\}_{l<j<k}$. Then shrinking $|T_l^Y|$ and $|T_l^{Y'}|$, we will modify θ_k only on $Y_k \cap |T_l^Y|$ for compatibility with T_l^Y and $T_l^{Y'}$. Without loss of generality, we assume $\theta_k(Y_k \cap |T_l^Y|) \subset |T_l^{Y'}|$, and $(\pi_l^Y, \rho_l^Y)|_{Y_k \cap |T_l^Y|}$ and $(\pi_l^{Y'}, \rho_l^{Y'})|_{Y_k' \cap |T_l^{Y'}|}$ are proper C^1 submersions onto $Y_l \times \,]0,1[$ and $Y_l' \times \,]0,1[$ respectively. Define a modification $\tilde{\theta}_k$ of θ_k on Y_k by

$$
\tilde{\theta}_k(y) = \begin{cases} \theta_k(y) & \text{for } y \in Y_k - |T_l^Y| \\ \pi_k^{Y'}(\alpha \circ \rho_k^Y(y) \cdot \theta_k(y) + (1 - \alpha \circ \rho_k^Y(y)) p_{\theta \circ \pi_l^Y(y), \rho_l^Y(y)} \circ \theta_k(y)) \\ \qquad\qquad \text{for } y \in Y_k \cap |T_l^Y|, \end{cases}
$$

where α is a C^1 $\mathfrak{X}$-function on $\mathbf{R}$ with $0 \leq \alpha \leq 1$, $\alpha = 0$ on $[0,1/3]$ and $\alpha = 1$ on $[2/3,1]$, and for each $(y,t) \in Y_l' \times \,]0,1[$, $p_{y,t}$ is a C^1 $\mathfrak{X}$-retraction of an $\mathfrak{X}$-neighborhood of $Y_k' \cap \pi_l^{Y'-1}(y) \cap \rho_l^{Y'-1}(t)$ in Y_k' to it such that the map $(y,y',t) \to p_{y',t}(y)$ is defined on an $\mathfrak{X}$-neighborhood of $\mathrm{graph}(\pi_l^{Y'}, \rho_l^{Y'})|_{Y_k' \cap |T_l^{Y'}|}$ in $(Y_k' \cap |T_l^{Y'}|) \times Y_l' \times \,]0,1[$ and is of class C^1 $\mathfrak{X}$. Let the approximation θ_k be very strong. Then $\tilde{\theta}_k$ is well-defined and a strong C^1 locally $\mathfrak{X}$ approximation of $\theta|_{Y_k}$, and its extension $\tilde{\theta}_k \colon \{Y_j\}_{j\leq k} \to \{Y'_j\}_{j\leq k}$ is compatible with T_l^Y and $T_l^{Y'}$. Note that $\tilde{\theta}_k = \theta_k$ on $Y_k \cap \cup_{l<j<k} |T_j^Y|$ by controlledness of $\{T_j^Y\}$ and $\{T_j^{Y'}\}$ and by compatibility of θ_k with $\{T_j^Y\}_{l<j<k}$ and $\{T_j^{Y'}\}_{l<j<k}$. Hence $\tilde{\theta}_k$ is compatible with $\{T_j^Y\}_{l\leq j<k}$ and $\{T_j^{Y'}\}_{l\leq j<k}$, and we can assume θ_k is compatible with $\{T_j^Y\}_{j<k}$ and $\{T_j^{Y'}\}_{j<k}$.

Next we need to extend θ_k to the outside of $\cup_{j\leq k} Y_j$. For that it suffices to prove the following statement.

Let η be a C^1 diffeomorphism of Y_k sufficiently close to the identity such that for each $j < k$,

$$
\eta(Y_k \cap |T_j^Y|) = Y_k \cap |T_j^Y|, \quad \text{and}
$$
$$
\pi_j^Y \circ \eta = \pi_j^Y, \quad \rho_j^Y \circ \eta = \rho_j^Y \quad \text{on } Y_k \cap |T_j^Y|.
$$

Shrink $|T_j^Y|$ for all j. Then there exists a stratified homeomorphism $\tilde{\eta} \colon \{Y_j\} \to \{Y_j\}$ which is an extension of η and is compatible with $\{T_j^Y\}$ and $\{T_j^Y\}$ such that each $\tilde{\eta}|_{Y_j}$ is a C^1 diffeomorphism of Y_j.

Clearly we set $\tilde{\eta} = \mathrm{id}$ on $\cup_{j<k} Y_j$. Shrink $|T_j^Y|$, set

$$
\hat{Y} = Y \times [0,1], \qquad \hat{Y}_j = Y_j \times [0,1], \quad \text{and}
$$
$$
T_j^{\hat{Y}} = (|T_j^Y| \times [0,1], (\pi_j^Y, p), \rho_j^Y),
$$

where p denotes the projection $\hat{Y} \to [0,1]$, and let $\hat{\eta}$ be a strong C^1 approximation of the projection $\hat{Y}_k \to Y_k$ such that for each $j < k$ and $t \in [0,1]$,

$$\hat{\eta}(\cdot, t) = \begin{cases} \mathrm{id} & \text{if} \quad t = 0 \\ \eta & \text{if} \quad t = 1, \end{cases}$$

$$\hat{\eta}(\hat{Y}_k \cap |T_j^{\hat{Y}}|) = Y_k \cap |T_j^Y|,$$

$$\pi_j^Y \circ \hat{\eta}(\cdot, t) = \pi_j^Y \quad \text{on} \quad Y_k \cap |T_j^Y|, \quad \text{and} \quad \rho_j^Y \circ \hat{\eta} = \rho_j^{\hat{Y}} \quad \text{on} \quad \hat{Y}_k \cap |T_j^{\hat{Y}}|,$$

whose existence is shown as in the first step above. Then we need only extend $\hat{\eta}$ to a stratified C^0 map $\hat{\eta} \colon \{\hat{Y}_j\} \to \{Y_j\}$ so that $(\hat{\eta}, p) \colon \{\hat{Y}_j\} \to \{\hat{Y}_j\}$ is a stratified homeomorphism compatible with $\{T_j^{\hat{Y}}\}$ and $\{T_j^{\hat{Y}}\}$, and for each j, $(\hat{\eta}, p)|_{\hat{Y}_j}$ is a C^1 diffeomorphism of $\hat{Y}_j$. This is easy to see as in the proof of the assertion in II.7.4.

In conclusion, we can assume each $\theta|_{Y_j} \colon Y_j \to Y_j'$ is locally $\mathfrak{X}$.

Shrink $|T_j^Y|$ and $|T_j^{Y'}|$, and let h be a proper positive C^1 $\mathfrak{X}$-function on $\mathbf{R}^n$ such that for each $y \in Y_j$, $h = \mathrm{const}$ on $\pi_j^{Y-1}(y)$. Recall the definition of a removal data. Let $I \subset \mathbf{R}$ denote the common C^1 regular values of h, $h|_{Y_1}, h|_{Y_2}, \dots$, let $I' \subset I$ be a closed subset of $\mathbf{R}$ which is not bounded from above and is a finite union of intervals locally at each point of $\mathbf{R}$, and let H denote the family of positive C^1 $\mathfrak{X}$-functions on $\mathbf{R}$ which are locally constant on $\mathbf{R} - I'$. Let $\delta = \{\delta_j \colon H^{j-1} \to H\}_{j=1,2\dots}$ be a removal data of $\{Y_j\}$ for h and $\{T_j^Y\}$, and let $\varepsilon = \{\varepsilon_j\} \le \delta$ be a sequence of elements of H. For each j, set

$$Y_{j,\varepsilon} = Y_j - \bigcup_{i=1}^{j-1} \{\rho_i^Y < \varepsilon_i \circ h\},$$

which is a C^1 $\mathfrak{X}$-manifold possibly with corners.

Set $h' = h \circ \theta^{-1}$ on Y', and extend it to $\cup |T_j^{Y'}|$ so that for each $y \in Y_j'$, $h' = \mathrm{const}$ on $\pi_j^{Y'-1}(y)$, which is possible because θ is compatible with $\{T_j^Y\}$ and $\{T_j^{Y'}\}$. Then h' is a C^1 $\mathfrak{X}$-function because each $\theta|_{Y_j}$ is C^1 and locally $\mathfrak{X}$; δ is also a removal data of $\{Y_j'\}$ for h' and $\{T_j^{Y'}\}$; the set $Y_{j,\varepsilon}'$, which we define from Y_j' as we did $Y_{j,\varepsilon}$ from Y_j, is a C^1 $\mathfrak{X}$-manifold possibly with corners; and $\theta|_{Y_{j,\varepsilon}}$ is a C^1 $\mathfrak{X}$-diffeomorphism onto $Y_{j,\varepsilon}'$.

It remains to modify θ outside of $\cup_j Y_{j,\varepsilon}$. By a downward induction method, it suffices to prove the following statement.

Fix j. Set

$$A = Y \cap \pi_j^{Y-1}(Y_{j,\varepsilon}) \cap \{\rho_j^Y \le \varepsilon_j \circ h\}, \quad \text{and} \quad B = A \cap \{\rho_j^Y = \varepsilon_j \circ h\}, \quad \text{and} \quad C = Y_{j,\varepsilon},$$

and define A', B' and C' by h' and $T_j^{Y'}$ in the same way. Let $\beta\colon B\cup C \to B'\cup C'$ be an $\mathfrak{X}$-homeomorphism such that

$$\beta(B) = B', \quad \beta = \theta \text{ on } C, \text{ and } \pi_j^{Y'}\circ\beta = \beta\circ\pi_j^Y.$$

Then, keeping the last equality, we can extend β to an $\mathfrak{X}$-homeomorphism from A to A' so that $\rho_j^{Y'}\circ\beta = \rho_j^Y$. (We do not need the last equality for the proof of (i), but it will be used for (iii).)

By II.6.1', the sequences of maps

$$A - C \ni y \longrightarrow (\pi_j^Y(y), \rho_j^Y(y)/\varepsilon_j\circ h(y)) \in C\times\,]0,1] \xrightarrow{\text{proj}}\,]0,1]$$

and $A' - C' \to C'\times\,]0,1] \to\,]0,1]$ defined in the same way are $\mathfrak{X}$-trivial. Hence the above statement follows immediately.

(ii) The proof is similar to (i). We check each of the steps in (i). For reduction to the case where each $\xi|_{Z_i}\colon Z_i \to Z_i'$ is of class locally $\mathfrak{X}$, we need only prove the following statements (1) and (2).

(1) Let $0 < l < k$ be integers. Assume $\xi|_{Z_l}\colon Z_l \to Z_l'$ is locally $\mathfrak{X}$. Let $\xi_k\colon Z_k \to Z_k'$ be a C^1 locally $\mathfrak{X}$ strong approximation of $\xi|_{Z_k}$ such that for each $l < i < k$,

$$g'\circ\xi_k = g \quad \text{on } Z_k\cap|T_i^Z|,$$

$$\xi_k(Z_k\cap|T_i^Z|) = Z_k'\cap|T_i^{Z'}|,$$

$$\pi_i^{Z'}\circ\xi_k = \xi\circ\pi_i^Z \quad \text{on } Z_k\cap|T_i^Z|, \text{ and}$$

$$\rho_i^{Z'}\circ\xi_k = \rho_i^Z \quad \text{on a neighborhood of } Z_k\cap|T_i^Z|\cap g^{-1}(g(Z_i)) \text{ in } Z_k.$$

Then we can modify ξ_k so that these equalities hold for $i = l$ also, and if $l = 1$ then $g'\circ\xi_k = g$ on Z_k.

Proof of (1). Assume $l > 1$. If $g|_{Z_k}$ and $g|_{Z_l}$ are C^1 regular or they are constant, the proof is the same as in (i) because we need not consider the equality $g'\circ\xi_k = g$. Assume $g|_{Z_k}$ is C^1 regular and $g|_{Z_l}$ is constant. Then there are two possible cases: $Z_k\cap|T_l^Z|\cap g^{-1}(g(Z_l)) = \varnothing$ or not. In the first case, it suffices to replace $p_{\theta\circ\pi_l^Y(y),\rho_l^Y(y)}$ in (i) with $p_{\xi\circ\pi_l^Z(z),g(z)}$, where $p_{z,t}$, $(z,t) \in Z_l'\times\mathbf{R}$, is a C^1 $\mathfrak{X}$-retraction of an $\mathfrak{X}$-neighborhood of $Z_k'\cap\pi_l^{Z'-1}(z)\cap g'^{-1}(t)$ in Z_k' to it. In the second case, we modify ξ_k first on a small neighborhood of $Z_k\cap|T_l^Z|\cap g^{-1}(g(Z_l))$ in Z_k by using a C^1 $\mathfrak{X}$-retraction $p_{z,s,t}$, $(z,s,t) \in Z_l'\times\mathbf{R}^2$, of an $\mathfrak{X}$-neighborhood of $Z_k'\cap\pi_l^{Z'-1}(z)\cap\rho_l^{Z'-1}(s)\cap g'^{-1}(t)$ in Z_k' in place of the above $p_{z,t}$, and then on $Z_k\cap|T_l^Z|$ as in the first case.

In the case of $l = 1$, it remains to obtain the equality $g' \circ \xi_k = g$ on $Z_k - \cup_{i<k}|T_i^Z|$. But this is clear if, as above, we use a C^1 𝔛-retraction of an 𝔛-neighborhood of $Z_k' \cap g'^{-1}(t)$ in Z_k' to it for each $t \in \mathbf{R}$.

(2) Let k be an integer. Let η be a C^1 diffeomorphism of Z_k sufficiently close to the identity such that for each $i < k$,

$$g \circ \eta = g \quad \text{on } Z_k,$$

$$\eta(Z_k \cap |T_i^Z|) = Z_k \cap |T_i^Z|,$$

$$\pi_i^Z \circ \eta = \pi_i^Z \quad \text{on } Z_k \cap |T_i^Z|, \quad \text{and}$$

$$\rho_i^Z \circ \eta = \rho_i^Z \quad \text{on a neighborhood of } Z_k \cap |T_i^Z| \cap g^{-1}(g(Z_i)) \text{ in } Z_k.$$

Shrink $|T_i^Z|$. Then there exists a stratified homeomorphism $\tilde{\eta}: \{Z_i\} \to \{Z_i\}$ which is an extension of η such that each $\tilde{\eta}|_{Z_i}$ is a C^1 diffeomorphism of Z_i and the above equalities hold for all i and k.

Proof of (2). As in the proof of (1), we can construct a C^1 diffeomorphism isotopy η_t, $t \in [0, 1]$, of η to the identity with the same properties as η. Hence we reduce (2) to a problem of triviality as in (i). We solve the problem in the same way as in (ii) of the proof of II.7.6. We omit the details.

We removed a subset from each Y_j in (i) outside of which θ is of class 𝔛. Here we change the subset because we need to take account of g and g'. Let h, I, I', H and δ be given for $\{Z_i\}$ and $\{T_i^Z\}$ as in (i). Let $\varepsilon^+ = \{\varepsilon_i^+\} \leq \delta$ and $\varepsilon^- = \{\varepsilon_i^-\}$ be sequences of elements of H such that $\varepsilon_1^+ \gg \varepsilon_1^- \gg \varepsilon_2^+ \gg \cdots$. Choose the sequences such that for each $i < i'$ and for each $z_0 \in Z_i$ with $g = \text{const}$ on Z_i and $g \neq \text{const}$ on $Z_{i'}$, the restriction of g to $\{z \in Z_{i'}: \pi_i^Z(z) = z_0, \rho_i^Z(z) = \varepsilon_i^+ \circ h(z_0)\}$ is C^1 regular at $\{z \in Z_{i'}: \pi_i^Z(z) = z_0, \rho_i^Z(z) = \varepsilon_i^+ \circ h(z_0), |g(z) - g(z_0)| \leq \varepsilon_i^- \circ h(z_0)\}$ and $\rho_i^{Z'} \circ \xi = \rho_i^Z$ on the same set. For each i, set

$$Z_{i,\varepsilon^\pm} = Z_i - \bigcup_{j=1}^{i-1}\{\rho_j^Z < \varepsilon_j^+ \circ h, \text{ if } g = \text{const on } Z_j \text{ then } |g - g(Z_j)| < \varepsilon_j^- \circ h\}.$$

We can choose ε^+ and ε^- so that $Z_{i,\varepsilon^\pm}$ is a C^1 𝔛-manifold possibly with corners, which will be clear in the following (iii). Note that if $g = \text{const}$ on Z_i, $Z_{i,\varepsilon^\pm}$ coincides with Z_{i,ε^+} which is defined without ε^-.

Define h' and $Z_{i,\varepsilon^\pm}'$ for $\{Z_i'\}$ and $\{T_i^{Z'}\}$ in the same way. Then each $Z_{i,\varepsilon^\pm}'$ is a C^1 𝔛-manifold possibly with corners, and $\xi|_{Z_{i,\varepsilon^\pm}}$ is a C^1 𝔛-diffeomorphism onto $Z_{i,\varepsilon^\pm}'$. We need to modify ξ outside of $\cup_i Z_{i,\varepsilon^\pm}$. As in (i), it suffices to prove the following statement.

(3) Fix i. Set

$$A = Z \cap \pi_i^{Z-1}(Z_{i,\varepsilon\pm})$$

$$\cap \{\rho_i^Z \le \varepsilon_i^+ \circ h, \text{ if } g = \text{const on } Z_i \text{ then } |g - g(Z_i)| \le \varepsilon_i^- \circ h\}$$

$$B = \begin{cases} A \cap \{\rho_i^Z = \varepsilon_i^+ \circ h\} & \text{if } g \ne \text{const on } Z_i \\ A \cap \{\rho_i^Z = \varepsilon_i^+ \circ h \text{ or } |g - g(Z_i)| = \varepsilon_i^- \circ h\} & \text{if } g = \text{const on } Z_i, \end{cases}$$

$$C = Z_{i,\varepsilon\pm},$$

and define A', B' and C' by h' and $T_i^{Z'}$. Let $\gamma \colon B \cup C \to B' \cup C'$ be an X-homeomorphism such that

$$\gamma(B) = B', \qquad \gamma = \xi \quad \text{on } C, \qquad g = g' \circ \gamma, \text{ and } \pi_i^{Z'} \circ \gamma = \gamma \circ \pi_i^Z.$$

Then, keeping the last two equalities, we can extend γ to an X-homeomorphism $\tilde{\gamma}$ from A to A' so that $\rho_i^{Z'} \circ \tilde{\gamma} = \rho_i^Z$ on a neighborhood of $A \cap g^{-1}(g(Z_i)) - Z_i$ in A.

Proof of (3). If $g \ne$ const on Z_i, (3) is an immediate consequence of (non-proper) II.6.1' as in (i). If not, it is not trivial. Assume $g = 0$ on Z_i. Then, as the problem is local around Z_i, Z_i' and 0, we can assume $g \colon \{Z_k\} \to \{]-\infty, 0[, 0,]0, \infty[\}$ and $g' \colon \{Z_k'\} \to \{]-\infty, 0[, 0,]0, \infty[\}$ are stratified functions, by stratifying $g^{-1}(0)$ and $g'^{-1}(0)$. We will prove (3) for stratified functions in a more general situation (4), which treats stratified maps.

(iii) The first half of (iii) follows from the latter one for the following reason. By (i) and its proof, we have a stratified X-homeomorphism $\theta_1 \colon \{Y_j\} \to \{Y_j'\}$ compatible with $\{T_j^Y\}$ and $\{T_j^{Y'}\}$ such that $\theta_1|_{Y_j}$ is a C^1 diffeomorphism onto Y_j'. Moreover, by the method of construction of θ_1, there exists an isotopy θ_t, $t \in [0, 1]$, from $\theta_0 = \theta$ to θ_1 such that each θ_t has the same properties as θ and the map $Y_j \times [0, 1] \ni (y, t) \to (\theta_t(y), t) \in Y_j' \times [0, 1]$ is a C^1 diffeomorphism for each j. (We choose $\theta_{1/2}$ so that each $\theta_{1/2}|_{Y_j} \colon Y_j \to Y_j'$ is locally X, and define θ_t, $t \in [1/2, 1]$, by shrinking each ε_j to the zero function.) Then it suffices to find an isotopy $\tau_t \colon X \to X'$, $t \in [0, 1]$, such that for each $t \in [0, 1]$, (τ_t, θ_t) has the same properties as (τ, θ). Replace τ_t with $\tau_0 \circ \tau_t$. Then what we prove is the following statement.

Set

$$\hat{X} = X \times [0, 1], \qquad \hat{X}_i = X_i \times [0, 1], \quad \text{and}$$

$$T_i^{\hat{X}} = (|T_i^X| \times [0, 1], (\pi_i^X, p^X), \rho_i^X),$$

where p^X denotes the projection $\hat{X} \to [0, 1]$, and define $\hat{Y}$, $\hat{Y}_j$, $\{T_j^{\hat{Y}}\}$ and $p^Y \colon \hat{Y} \to [0, 1]$ in the same way. Let $\eta^Y \colon \{\hat{Y}_j\} \to \{Y_j\}$ be a stratified map such that $\eta^{\hat{Y}} = (\eta^Y, p^Y)\colon \{\hat{Y}_j\} \to \{\hat{Y}_j\}$ is a stratified homeomorphism compatible with $\{T_j^{\hat{Y}}\}$, and $\{T_j^{\hat{Y}}\}$ and each $\eta^{\hat{Y}}|_{\hat{Y}_j}$ is a C^1 diffeomorphism of $\hat{Y}_j$. Then there exists a stratified map $\eta^{\hat{X}}\colon \{\hat{X}_i\} \to \{X_i\}$ such that $\eta^{\hat{X}} = (\eta^X, p^X)\colon \{\hat{X}_i\} \to \{\hat{X}_i\}$ is a stratified homeomorphism, each $\eta^{\hat{X}}|_{\hat{X}_i}$ is a C^1 diffeomorphism of $\hat{X}_i$, and $(\eta^{\hat{X}}, \eta^{\hat{Y}})$ is compatible with $\{T_i^{\hat{X}}, T_j^{\hat{Y}}\}$ and $\{T_i^{\hat{X}}, T_j^{\hat{Y}}\}$.

We can construct η^X in the same way as φ in the proof of (ii) of II.7.6. We omit the details.

Now we prove the latter half. As in (ii) we can assume each $\tau|_{X_i}\colon X_i \to X_i'$ is locally $\mathfrak{X}$. We need to generalize the rest of (ii) to the map case. Let h, I, I' and δ be given for $\{X_i\}$ and $\{T_i^X\}$ as in (ii). Let $\varepsilon^+ = \{\varepsilon_i^+\} \leq \delta$ and $\varepsilon^- = \{\varepsilon_i^-\}$ be sequences of elements of H given inductively so that $\infty \gg \varepsilon_1^+ \gg \varepsilon_1^- \gg \varepsilon_2^+ \gg \cdots$, and the following conditions (a),...,(e) are satisfied. As usual, we extend each ρ_i^X outside $|T_i^X|$ by setting it equal to 1 there. Fix any i_0. For each $i < i_0$, let $A(i)$ denote the family of the sets:

$$\{x \in X_{i_0}\colon \rho_i^X(x) * \varepsilon_i^+ \circ h(x)\}, \qquad * \in \{=, >\},$$
$$\text{if } f(X_i) \subset Y_j \text{ and } f(X_{i_0}) \subset Y_j \text{ for some } j, \quad \text{and}$$

$$\{x \in X_{i_0}\colon \rho_i^X(x) *_1 \varepsilon_i^+ \circ h(x), *_3 \, \rho_j^Y \circ f(x) *_2 \varepsilon_i^- \circ h(x)\},$$
$$(*_1, *_2, *_3) \in \{(=, <, \text{and}), (=, =, \text{and}), (<, =, \text{and}), (>, >, \text{or})\},$$
$$\text{if } f(X_i) \subset Y_j \text{ and } f(X_{i_0}) \not\subset Y_j \text{ for some } j.$$

Let M_k be any elements of $A(k)$, $k = 1, \ldots i-1$, and let t be any number of I'.

(a) $\cap_{k=1}^{i-1} M_k$ and M_i are C^1 submanifolds of X_{i_0}.

(b) $\cap_{k=1}^{i-1} M_k$, $h^{-1}(t) \cap X_{i_0}$ and M_i are transversal to each other in X_{i_0}.

(c) If $f(X_i) \subset Y_j$ and $f(X_{i_0}) \subset Y_j$ for some j, the restrictions of ρ_i^X to $\{x \in \cap_{k=1}^{i-1} M_k\colon 0 < \rho_i^X(x) \leq \varepsilon_i^+ \circ h(x)\}$ and to $\{x \in \cap_{k=1}^{i-1} M_k\colon 0 < \rho_i^X(x) \leq \varepsilon_i^+ \circ h(x), h(x) = t\}$ are C^1 regular.

(d) If $f(X_i) \subset Y_j$ and $f(X_{i_0}) \not\subset Y_j$ for some j, the restrictions of ρ_i^X and $\rho_j^Y \circ f$ to $\{x \in \cap_{k=1}^{i-1} M_k\colon 0 < \rho_i^X(x) \leq \varepsilon_i^+ \circ h(x), \rho_j^Y \circ f(x) \leq \varepsilon_i^- \circ h(x)\}$ and to $\{x \in \cap_{k=1}^{i-1} M_k\colon 0 < \rho_i^X(x) \leq \varepsilon_i^+ \circ h(x), \rho_j^Y \circ f(x) \leq \varepsilon_i^- \circ h(x), h(x) = t\}$ and the restriction of $\rho_j^Y \circ f$ to $\{x \in \cap_{k=1}^{i-1} M_k\colon \rho_i^X(x) = \varepsilon_i^+ \circ h(x), \rho_j^Y \circ f(x) \leq \varepsilon_i^- \circ h(x)\}$ are C^1 regular.

(e) If $f(X_i) \subset Y_j$ and $f(X_{i_0}) \not\subset Y_j$, $\rho_i^{X'} \circ \tau = \rho_i^X$ on $\{x \in X_{i_0}\colon \rho_i^X(x) = \varepsilon_i^+ \circ h(x), \rho_j^Y \circ f(x) \leq \varepsilon_i^- \circ h(x)\}$.

For each i, set

$$X_{i,\varepsilon\pm} = X_i - \bigcup_{k=1}^{i-1}\{\rho_k^X(x) < \varepsilon_k^+ \circ h, \ \rho_j^Y \circ f < \varepsilon_k^- \circ h \text{ for } j \text{ with } f(X_k) \subset Y_j\},$$

which is C^1 𝔛-manifold possibly with corners. Define $X'_{i,\varepsilon\pm}$ in the same way, which is also a C^1 𝔛-manifold possibly with corners. Then $\tau|_{X_{i,\varepsilon\pm}}$ is a C^1 𝔛-diffeomorphism onto $X'_{i,\varepsilon\pm}$, and it suffices to prove the following statement.

(4) Fix i_0, let j_0 be such that $f(X_{i_0}) \subset Y_{j_0}$, and set

$$A = X \cap \pi_{i_0}^{X-1}(X_{i_0,\varepsilon\pm}) \cap \{\rho_{i_0}^X \le \varepsilon_{i_0}^+ \circ h, \ \rho_{j_0}^Y \circ f \le \varepsilon_{i_0}^- \circ h\},$$
$$B = A \cap \{\rho_{i_0}^X = \varepsilon_{i_0}^+ \circ h \text{ or } \rho_{j_0}^Y \circ f = \varepsilon_{i_0}^- \circ h\}, \ \text{ and } \ C = X_{i_0,\varepsilon\pm},$$

and define A' B' and C' by h', $T_{i_0}^{X'}$ and $T_{j_0}^{Y'}$. Let $\gamma\colon B \cup C \to B' \cup C'$ be an 𝔛-homeomorphism such that

$$\gamma(B) = B', \quad \gamma = \tau \text{ on } C, \qquad f = f' \circ \gamma, \ \text{ and } \ \pi_{i_0}^{X'} \circ \gamma = \gamma \circ \pi_{i_0}^X.$$

Then, keeping the last two equalities, we can extend γ to an 𝔛-homeomorphism $\tilde\gamma$ from A to A' so that $\rho_{i_0}^{X'} \circ \tilde\gamma_{i_0}^X = \rho_{i_0}^X$ on a neighborhood of $A \cap f^{-1}(Y_{j_0}) - X_{i_0}$ in A.

Proof of (4). Let $\varepsilon_{i_0 t}^+$ and $\varepsilon_{i_0 t}^-$, $t \in {]0,2[}$, be elements of H such that $\varepsilon_{i_0 1}^+ = \varepsilon_{i_0}^+$, $\varepsilon_{i_0 1}^- = \varepsilon_{i_0}^-$, $\varepsilon_{i_0 t}^+ \to 0$ and $\varepsilon_{i_0 t}^- \to 0$, as $t \to 0$, the maps $\mathbf{R} \times {]0,2[} \ni (s,t) \to \varepsilon_{i_0 t}^+(s), \ \varepsilon_{i_0 t}^-(s) \in \mathbf{R}$ are of class C^1, for each $s \in \mathbf{R}$, the maps ${]0,2[} \ni t \to \varepsilon_{i_0 t}^+(s), \ \varepsilon_{i_0 t}^-(s) \in \mathbf{R}$ are C^1 regular, and for each $t \in {]0,2[}$, $\varepsilon_{i_0 t}^+$ and $\varepsilon_{i_0 t}^-$ have the same properties as $\varepsilon_{i_0}^+$ and $\varepsilon_{i_0}^-$. Define A_t, B_t and C_t for each $t \in {]0,2[}$ by $\varepsilon_{i_0 t}^+$ and $\varepsilon_{i_0 t}^-$ as we did A, B and C by $\varepsilon_{i_0}^+$ and $\varepsilon_{i_0}^-$. Note that $C_t = C$. Set

$$D_t = \{(x,y) \in C \times (Y \cap |T_{j_0}^Y|) \colon f(x) = \pi_{j_0}^Y(y), \ \rho_{j_0}^Y(y) \le \varepsilon_{i_0 t}^- \circ h(x)\},$$
$$E_t = D_t \cap \{\rho_{j_0}^Y = \varepsilon_{i_0 t}^- \circ h\}, \ \text{ and }$$
$$\hat{A} = \bigcup_{t \in {]0,2[}} A_t \times t \subset X \times {]0,2[}.$$

Define $\hat{B}$, $\hat{C}$, $\hat{D}$, $\hat{E}$ in the same way and define a proper C^1 𝔛-map $\hat{f}\colon \hat{A} \to \hat{D}$ to be $(\pi_{i_0}^X, f) \times \text{id}$. Let $p\colon \hat{D} \to {]0,2[}$ denote the projection. Let $\{\hat{A}_i\}$ denote the family of sets:

$$\{(x,t) \in \hat{B}: x \in X_{i_1},\ \rho_{i_0}^X(x) = \varepsilon_{i_0t}^+ \circ h(x),\ \rho_{j_0}^Y \circ f(x) = \varepsilon_{i_0t}^- \circ h(x)\},$$

$$\{(x,t) \in \hat{B}: x \in X_{i_2},\ \rho_{i_0}^X(x) = \varepsilon_{i_0t}^+ \circ h(x),\ \rho_{j_0}^Y \circ f(x) \neq \varepsilon_{i_0t}^- \circ h(x)\},$$

$$\{(x,t) \in \hat{B}: x \in X_{i_3},\ \rho_{i_0}^X(x) \neq \varepsilon_{i_0t}^+ \circ h(x),\ \rho_{j_0}^Y \circ f(x) = \varepsilon_{i_0t}^- \circ h(x)\},$$

$$\text{and} \quad (\hat{A} - \hat{B}) \cap (X_{i_4} \times \,]0,2[),$$

and let $\{\hat{D}_j\}$ consist of $\hat{E} \cap (Y_{j_1} \times \,]0,2[)$ and $(\hat{D} - \hat{E}) \cap (Y_{j_2} \times \,]0,2[)$. Then $\hat{f}: \{\hat{A}_i\} \to \{\hat{D}_j\}$ is a stratified map, and there are a controlled C^1 $\mathfrak{X}$-tube system $\{T_j^{\hat{D}}\}$ for $\{\hat{D}_j\}$ and a C^1 $\mathfrak{X}$-tube system $\{T_i^{\hat{A}}\}$ for $\{\hat{A}_i\}$ controlled over $\{T_j^{\hat{D}}\}$ such that $\{T_i^{\hat{A}}\}|_{\hat{A}-\hat{B}}$ and $\{T_j^{\hat{D}}\}|_{\hat{D}-\hat{E}}$ are naturally induced from $\{T_i^X\}$ and $\{T_j^Y\}$ respectively. For i and i' with $\hat{A}_i \cap (\overline{\hat{A}_{i'}} - \hat{A}_{i'}) \neq \varnothing$, if $\hat{f}(\hat{A}_i) \cup \hat{f}(\hat{A}_{i'}) \subset \hat{D}_j$ for some j then $(\pi_i^{\hat{A}}, \rho_i^{\hat{A}})|_{\hat{A}_{i'} \cap |T_i^{\hat{A}}|}$ is a C^1 submersion to $\hat{A}_i \times \mathbf{R}$. For the same i and i', if $\hat{f}(\hat{A}_i) \subset \hat{D}_j$ and $\hat{f}(\hat{A}_{i'}) \subset \hat{D}_{j'}$ for some $j \neq j'$ then $(\pi_i^{\hat{A}}, \hat{f})|_{\hat{A}_{i'} \cap |T_i^{\hat{A}}|}$ is a C^1 submersion to $\{(x,t,y,t) \in \hat{A}_i \times (\hat{D}_{j'} \cap |T_j^{\hat{D}}|): \hat{f}(x,t) = \pi_j^{\hat{D}}(y,t)\}$. For j and j' with $\hat{D}_j \cap (\overline{\hat{D}_{j'}} - \hat{D}_{j'}) \neq \varnothing$, $(\pi_j^{\hat{D}}, \rho_j^{\hat{D}})|_{\hat{D}_{j'} \cap |T_j^{\hat{D}}|}$ is a C^1 submersion to $\hat{D}_j \times \mathbf{R}$. Hence by II.6.1$'$ (see its complement), the sequence $\{\hat{B}_i\} \xrightarrow{\hat{f}} \{\hat{D}_j\} \xrightarrow{p} \,]0,2[$ is $\mathfrak{X}$-trivial. In the same way we define a sequence $\{\hat{B}_i'\} \xrightarrow{\hat{f}'} \{\hat{D}_j\} \xrightarrow{p} \,]0,2[$ by $\{T_i^{X'}\}$, $\{T_j^{Y'}\}$ and f', which is also $\mathfrak{X}$-trivial. Remember that we constructed the triviality as a lift of triviality of p. (4) immediately follows from the triviality. $\qquad \square$

The $\mathfrak{X}$-equivalence class of an $\mathfrak{X}$-set or an $\mathfrak{X}$-function does not depend on $\mathfrak{X}$.

Theorem II.7.7 (Independence of $\mathfrak{X}$-equivalence). *Let $\mathfrak{X}_1 \subset \mathfrak{X}_2$ be families of $\mathfrak{X}$. Let $X \subset \mathbf{R}^n$ be an $\mathfrak{X}_1$-set, let X_1 and X_2 be $\mathfrak{X}_1$-subsets of X closed in X, and let f_1 and f_2 be $\mathfrak{X}_1$-functions on X with the first boundedness condition, i.e., for any bounded set $B \subset \mathbf{R}^n$, $f_1(X \cap B)$ and $f_2(X \cap B)$ are bounded. Assume (1) X is bounded and locally closed in $\mathbf{R}^n$ or (2) Axiom (v) and X is closed in $\mathbf{R}^n$. If X_1 and X_2 are $\mathfrak{X}_2$-equivalent as $\mathfrak{X}_2$-subsets of X, then they are $\mathfrak{X}_1$-equivalent. The same statement holds for f_1 and f_2.*

Proof. We prove the former statement assuming X is closed in $\mathbf{R}^n$. There exists an $\mathfrak{X}_2$-homeomorphism τ from a closed $\mathfrak{X}_2$-neighborhood U_1 of X_1 in X to a closed U_2 of X_2 in X such that $\tau(X_1) = X_2$. We suppose X,

X_1 and X_2 are polyhedra by II.1.1 and II.1.1' and, moreover, U_1 and U_2 also are polyhedra because by their proofs, there exist $\mathcal{X}_2$-homeomorphisms π_i, $i = 1, 2$, of X such that $\pi_i(U_i)$ are polyhedra and $\pi_i = $ id on some smaller neighborhoods of X_i. Since (U_1, X_1) and (U_2, X_2) are $\mathcal{X}_2$-homeomorphic, by uniqueness of $\mathcal{X}$-triangulations, they are PL homeomorphic and hence $\mathcal{X}_1$-homeomorphic.

In the same way we can prove the latter statement in the case where X is closed in $\mathbf{R}^n$.

Let us prove the latter when X is bounded and locally closed but not closed. (The former in the same case follows in the same way only more easily.) We will reduce the problem to the closed case. Let τ be an $\mathcal{X}_2$-homeomorphism of X such that $f_1 = f_2 \circ \tau$. By local closedness $\overline{X} - X$ is closed. Let φ be an $\mathcal{X}$-function on $\mathbf{R}^n$ with $\varphi^{-1}(0) = \overline{X} - X$. Replace X with the $\mathcal{X}$-set:

$$\{(\varphi(x)x, \varphi(x)) \in \mathbf{R}^n \times \mathbf{R} : x \in X\}.$$

Then we can assume $\overline{X} - X = \{0\}$ and hence τ can be extended to an $\mathcal{X}_2$-homeomorphism of $\overline{X}$. Set $Y_i = \operatorname{graph} f_i$, $i = 1, 2$, and let g_i denote the restrictions to Y_i of the projection $\mathbf{R}^n \times \mathbf{R} \to \mathbf{R}$. Define an $\mathcal{X}_2$-homeomorphism $\pi \colon Y_1 \to Y_2$ by

$$\pi(x, f_1(x)) = (\tau(x), f_2 \circ \tau(x)) \quad \text{for} \quad x \in X_1.$$

Then $g_1 = g_2 \circ \pi$ and we can consider Y_i, g_i and π, $i = 1, 2$, in place of X, f_i and τ. Let $\overline{g_i}$, $i = 1, 2$, denote the $\mathcal{X}_1$-extension of g_i to $\overline{Y_i}$, which is possible because of the definition of g_i. By the properties $\overline{X} - X = \{0\}$ and $g_1 = g_2 \circ \pi$ we can extend π to an $\mathcal{X}_2$-homeomorphism $\overline{\pi} \colon \overline{Y_1} \to \overline{Y_2}$. Then $\overline{g_1} = \overline{g_2} \circ \overline{\pi}$, and it suffices to find an $\mathcal{X}_1$-homeomorphism $\pi' \colon \overline{Y_1} \to \overline{Y_2}$ such that

$$\pi'(\overline{Y_1} - Y_1) = \overline{Y_2} - Y_2 \text{ and } \overline{g_1} = \overline{g_2} \circ \pi'.$$

We can easily prove this in the same way as the above closed case because $\overline{Y_i}$ and $\overline{Y_i} - Y_i$ are closed. $\qquad\square$

A very possible and important conjecture is the following. If it holds true, then research of $\mathcal{X}$-sets and $\mathcal{X}$-maps does not depend on a special choice of $\mathcal{X}$.

Conjecture II.7.8. *Let $\mathcal{X}_1 \subset \mathcal{X}_2$ be families of $\mathcal{X}$. Let $f \colon X \to Y$ and $f' \colon X' \to Y'$ be $\mathcal{X}_1$-maps between $\mathcal{X}_1$-sets. Assume either* (i) *X, Y, X', Y' are bounded in their respective ambient Euclidean spaces,* (ii) *$\mathcal{X}_1$ and $\mathcal{X}_2$ satisfy Axiom* (v), *or* (iii) *$\mathcal{X}_1$ and $\mathcal{X}_2$ are $\mathcal{X}_0$. If f and f' are R-L $\mathcal{X}_2$-equivalent, then they are R-L $\mathcal{X}_1$-equivalent.*

V.1.4 is a partial answer.

Example II.7.9. (Local) C^0 equivalence of two X-functions (germs) does not imply their (local) X-equivalence. There exist two homogeneous polynomial functions on $\mathbf{R}^7$ with an isolated singularity at 0 which are C^0 equivalent but not X-equivalent and whose germs at 0 are locally C^0 equivalent but not locally X-equivalent.

Construction of polynomials. Let $M_0 \subset \mathbf{R}^6$ be a compact connected C^∞ manifold of dimension 5 with the fundamental group $\mathbf{Z}/5\mathbf{Z}$. Let f_1 be a proper C^∞ function on $\mathbf{R}^6$ such that $f_1^{-1}(0) = M_0$, $\{f_1 < 0\}$ is bounded and f_1 is C^∞ regular on M_0. Let f_2 be a polynomial function on $\mathbf{R}^6$ which is a C^1 approximation of f_1 on the compact set $\{f_1 \le 1\}$. We can replace M_0 with $f_2^{-1}(0) \cap \{f_1 \le 1\}$. But $f_2^{-1}(0)$ may have other connected components. We need to avoid this. Without loss of generality, assume

$$\{f_1 \le 1\} \subset \{x_1^2 + \cdots + x_6^2 < 1\},$$

and replace f_2 with $f_2 + |x|^{2l}/l$ for a sufficiently large integer l. Then $f_2^{-1}(0)$ has no more other connected components. Moreover, the sum of the terms of the maximal degree $2l$ takes positive values except at 0. Set

$$f_3(x_1, \ldots, x_7) = x_7^{2l} f_2(x_1/x_7, \ldots, x_6/x_7),$$

and write $M = S^6 \cap (0 * (f_2^{-1}(0) \times 1)) \subset \mathbf{R}^7$, where $S^6 = \{|x| = 1\}$. Then f_3 is homogeneous, is positive on $\{x_7 = 0\} - 0$, and has an isolated singularity at 0, and $f_3^{-1}(0)$ is the union of the infinite cones $\overrightarrow{0 * M} = \{tx : t \in [0, \infty[,\ x \in M\}$ and $\overrightarrow{0 * (-M)}$.

Let $p: U \to M$ be a sufficiently small closed C^∞ tubular neighborhood of M in S^6 such that the restriction of f_3 to each fibre is C^∞ regular and $|f_3|$ is constant on ∂U. Let M_1 and M_2 denote the connected components of ∂U with $f_3 > 0$ on M_1. Note that M_1 and M_2 are both diffeomorphic to M. It is known that if the fundamental group of a C^∞ manifold is $\mathbf{Z}/5\mathbf{Z}$, the Whitehead group of the manifold does not vanish (see [Hi]). Hence there exists a non-trivial C^∞ h-cobordism one of whose boundaries is M_1. Let $(W; M_1, N)$ be a non-trivial C^∞ h-cobordism such that $W \subset U - M_2$. In the same way as we define f_3 from M_0, we define homogeneous polynomial functions g_3, h on $\mathbf{R}^7$ from the images of $\overrightarrow{0 * N} \cap \mathbf{R}^6 \times 1$ and $\overrightarrow{0 * M_1} \cap \mathbf{R}^6 \times 1$ respectively, under the projection $\mathbf{R}^6 \times 1 \to \mathbf{R}^6$. Set $f = f_3 h$, $g = g_3 h$.

Proof that f and g are C^0 equivalent (cf. [Ki]). For simplicity of notation, we assume

$$\deg g_3 = 2l \ \text{ and } \ g_3^{-1}(0) \cap S^6 = N \cup (-N),$$

and we prove only C^0 equivalence of f_3 and g_3. We can prove C^0 equivalence of f and g in the same way, and local C^0 equivalence of the germs of f and

g at 0 is clear. We will construct a homeomorphism π of $\mathbf{R}^7$ such that $f_3 \circ \pi = g_3$ and $\pi(\overrightarrow{0 * N}) = \overrightarrow{0 * M}$. Let ε be a sufficiently small positive number, and let $q: V \to N$ be a closed tubular neighborhood in S^6 such that the restriction of g_3 to each fibre is C^∞ regular and g_3 takes the values $\pm\varepsilon$ on ∂V. Set $N_1 = \partial V \cap \{g_3 > 0\}$. We can multiply f_3 by a positive number because they are C^0 equivalent. Hence we assume $f_3 = \pm\varepsilon$ on ∂U. Set

$$\tilde{A} = \overrightarrow{0 * A} \cap \{|x| \geq 1\} \qquad \text{for a set } A \subset S^6.$$

Extend $p: U \to M$ to a tubular neighborhood $\tilde{p}: \hat{U} \to \tilde{M}$ in $\{|x| \geq 1\}$ so that

$$\tilde{p}(x) = |x| p(x/|x|) \quad \text{and} \quad f_3 = \pm\varepsilon \text{ on } \partial\hat{U},$$

which is possible because f_3 is homogeneous. There exists a C^∞ imbedding $\tau: W \to \tilde{M}$ with $\tau = p$ on M_1. Set $P = \tau(W)$ and $Q = \tau(N)$. Then $(P; M, Q)$ is a non-trivial h-cobordism.

We now define π. Set $\pi = \tau|_N$ on N, and extend it to $V \to \tilde{p}^{-1}(Q)$ so that

$$\tilde{p} \circ \pi = \pi \circ q \quad \text{and} \quad f_3 \circ \pi = g_3.$$

This is uniquely determined because the restrictions of f_3 and g_3 to each fibre of $\tilde{p}$ and q respectively, are C^∞ regular. Next we want to extend π^{-1} to $\tilde{p}^{-1}(P) \cap \{f_3 = \varepsilon\}$. Modifying τ, we obtain a C^∞ diffeomorphism $\tau': (W - V^\circ; M_1, N_1) \to (\tilde{p}^{-1}(P) \cap \{f_3 = \varepsilon\}; M_1, \tilde{p}^{-1}(Q) \cap \{f_3 = \varepsilon\})$ such that $\tau' = \text{id}$ on M_1 and $\tau' = \pi$ on N_1. Define π^{-1} on $\tilde{p}^{-1}(P) \cap \{f_3 = \varepsilon\}$ so that

$$\tau'^{-1}(x) = \pi^{-1}(x)/|\pi^{-1}(x)| \quad \text{and} \quad g_3 \circ \pi^{-1} = \varepsilon,$$

which also is uniquely determined because g_3 is homogeneous. Note that $f_3 = g_3 \circ \pi^{-1}$ on $\tilde{p}^{-1}(P) \cap \{f_3 = \varepsilon\}$ and its image under π^{-1} is included in $\{|x| \leq 1\}$. Since $\tau' = \text{id}$ on M_1, we have $\pi^{-1}(x)/|\pi^{-1}(x)| = x$ for $x \in M_1$. Hence we can extend π^{-1}, moreover, to the hypersurface $\{x = (x_1, \ldots, x_7) \in \mathbf{R}^7: f_3(x) = \varepsilon, |x| \leq 1, x_7 \geq 0\}$ so that

$$\pi^{-1}(x)/|\pi^{-1}(x)| = x/|x| \quad \text{and} \quad f_3 = g_3 \circ \pi^{-1}.$$

In the same way we define π^{-1} on $(\tilde{p}^{-1}(P) \cap \{f_3 = -\varepsilon\}) \cup \{x = (x_1, \ldots, x_7) \in \mathbf{R}^7: f_3(x) = -\varepsilon, |x| \leq 1, x_7 \geq 0\}$. Let B denote the domain where π is already defined. Then $B \subset \{x_7 \geq 0\}$, and we can define π on $-B$ to be

$$\pi(x) = -\pi(-x) \quad \text{for} \quad x \in -B,$$

because f_3 and g_3 are homogeneous polynomials of even degree,

$$B \cap (-B) = B \cap \{x_7 = 0\} \text{ and}$$

$$\pi(x)/|\pi(x)| = x/|x| \quad \text{and hence} \quad \pi(x) = -\pi(-x) \quad \text{for} \quad x \in B \cap \{x_7 = 0\}.$$

Note that

$$B \cup (-B) = \{|g_3| = \varepsilon, \ |x| \leq 1\} \cup (S^6 \cap \{|g_3| < \varepsilon\}).$$

We extend π to (the inside of $B \cup (-B)$) $\to$ (the inside of $\pi(B \cup (-B))$). Recall that $(P; M, Q)$ is invertible, i.e., there exists a C^∞ h-cobordism $(P'; Q, M)$ such that $P \cup_Q P'$ and $P \cup_M P'$ (the unions of P and P' pasted at Q and M respectively) are C^∞ diffeomorphic to $M \times [0, 1]$ and $Q \times [0, 1]$ respectively. It is easy to construct a C^∞ map $\xi \colon P \times {]0, 1]} \to P \cup 0 * M - 0$ such that $\xi|_{P \times 1} = \mathrm{id}$, $\xi|_{P \times t}$ is a C^∞ imbedding for each $t \in {]0, 1]}$, $\xi(x, t) = tx$ for $(x, t) \in M \times {]0, 1]}$, and $\xi|_{Q \times]0,1]}$ is a C^∞ diffeomorphism onto $P \cup 0 * M - 0$. For each $t \in {]0, 1]}$, set

$$\varepsilon_t = \varepsilon t^{2l}$$
$$p_t(x) = t p(x/t) \quad \text{for} \quad x \in tU,$$
$$q_t(x) = t q(x/t) \quad \text{for} \quad x \in tV,$$
$$\overrightarrow{p}(x) = |x| p(x/|x|) \quad \text{for} \quad x \in \overrightarrow{0 * U} - 0,$$
$$\tilde{U}_t = \overrightarrow{p}^{-1}(t\tilde{M}) \cap \{|f_3| \leq \varepsilon_t\},$$
$$\tilde{p}_t(x) = |x| p(x/|x|) \quad \text{for} \quad x \in \tilde{U}_t,$$
$$\tau_t(x) = \xi(\tau(x/t), t) \quad \text{for} \quad x \in tW \quad \text{and}$$
$$\tau_t'(x) = \overrightarrow{p}^{-1}(\xi(\tilde{p} \circ \tau'(x/t), t)) \cap \{f_3 = \varepsilon_t\} \quad \text{for} \quad x \in t(W - V^\circ).$$

Then $p_t \colon tU \to tM$ and $q_t \colon tV \to tN$ are tubular neighborhoods in tS^6 such that the restrictions of f_3 and g_3 to each fibre of p_t and q_t respectively, are C^∞ regular, and the above arguments work for $\varepsilon = \varepsilon_t$, tS^6, tM, tN, $(\xi(P, t); tM, \xi(Q, t))$, $p_t \colon tU \to tM$, $q_t \colon tV \to tN$, $\tilde{p}_t \colon \tilde{U}_t \to t\tilde{M}$, $\tau_t \colon tW \to \xi(P, t)$ and $\tau_t' \colon t(W - V^\circ) \to \overrightarrow{p}^{-1}(\xi(P, t)) \cap \{f_3 = \varepsilon_t\}$. Construct B and $\pi \colon B \cup (-B) \to \mathbf{R}^7$ from these data as we constructed B and π from ε, S^6, etc., and write them as B_t and π_t. Then for $t \neq t' \in {]0, 1]}$,

$$(B_t \cup (-B_t)) \cap (B_{t'} \cup (-B_{t'})) = \varnothing,$$
$$\pi_t(B_t \cup (-B_t)) \cap \pi_{t'}(B_{t'} \cup (-B_{t'})) = \varnothing,$$

$0 \cup_{t \in]0,1]} B_t \cup (-B_t)$ and $0 \cup_{t \in]0,1]} \pi_t(B_t \cup (-B_t))$ are the insides of $B \cup (-B)$ and $\pi(B \cup (-B))$ respectively, the map π between them defined to be π_t on each $B_t \cup (-B_t)$ and $\pi(0) = 0$ is a homeomorphism, and $f_3 \circ \pi = g_3$.

In the same way as above we can extend π to (the outside of $B \cup (-B)$) $\to$ (the outside of $\pi(B \cup (-B))$).

Proof that the germs of f and g at 0 are not locally ℵ-*equivalent.* Assume they are locally ℵ-equivalent. There is an ℵ-homeomorphism $\pi \colon O_1 \to O_2$

between open $\mathfrak{X}$-neighborhoods of 0 in $\mathbf{R}^7$ such that $\pi(O_1 \cap g^{-1}(0)) = O_2 \cap f^{-1}(0)$. We can assume

$$\pi(O_1 \cap 0 * W) = O_2 \cap 0 * U_+, \qquad \text{where} \quad U_+ = U \cap \{f_3 \geq 0\}.$$

Define $\mathfrak{X}$-functions ρ_1 and ρ_2 on $O_1 \cap 0 * W$ to be $\rho_1(x) = |x|$ and $\rho_2(x) = |\pi(x)|$. Then the levels of ρ_1 and ρ_2 near 0 are all homeomorphic to W and U_+ respectively. On the other hand, by II.7.1 the germs of ρ_1 and ρ_2 at 0 are locally $\mathfrak{X}$-equivalent because $\rho_1^{-1}(0) = \rho_2^{-1}(0) = 0$. Hence W and U_+ are homeomorphic, which contradicts the fact that W is non-trivial and U_+ is trivial. (Here we use the well-known fact that C^0 triviality implies C^∞ triviality (see the proof of II.3.4)).

It follows clearly that f and g are not $\mathfrak{X}$-equivalent. $\qquad\square$

Note that the above homeomorphism π of $\mathbf{R}^7$ can be chosen so that $\pi|_{\mathbf{R}^7-0}$ is a C^∞ diffeomorphism onto $\mathbf{R}^7 - 0$.

The following theorem shows how many R-L $\mathfrak{X}$-equivalence classes of $\mathfrak{X}$-functions exist. This is the same as in the case of R-L C^0 equivalence but not of R-L C^1 equivalence. Indeed, the germs at 0 of any two polynomial functions in x, y variables of $\{xy(x-y)(x-ty): t > 1 \in \mathbf{R}\}$ are not locally R-L C^1 equivalent, and hence the local R-L C^1 equivalence classes of polynomial function germs on $\mathbf{R}^2$ at 0 have the cardinal number of the continuum.

Theorem II.7.10 (Cardinality of $\mathfrak{X}$-equivalence classes). *There are a countable number of R-L $\mathfrak{X}$-equivalence classes of all bounded $\mathfrak{X}$-functions on all bounded $\mathfrak{X}$-sets.*

Proof. For a bounded $\mathfrak{X}$-function f on a bounded $\mathfrak{X}$-set, by II.3.1, there exist a finite simplicial complex K, a union X of open simplexes of K, and a PL function g on $|K|$ such that f is $\mathfrak{X}$-equivalent to $g|_X$. There are only a countable number of finite simplicial complexes up to a simplicial isomorphism, and for each finite simplicial complex, there are a finite number of unions of its open simplexes. Hence it suffices to count the R-L $\mathfrak{X}$-equivalence classes of the restrictions to X of all PL functions on $|K|$. By I.3.11 and its proof, the cardinal number is countable. $\qquad\square$

Remark II.7.11 [B-S]. In the semialgebraic case, we have a more precise result. Let n and d be non-negative integers. Let $S(n, d)$ denote all the semialgebraic functions f on semialgebraic sets in $\mathbf{R}^n$ such that graph f is defined by equalities and inequalities of d' polynomial functions of degree $\leq d$, $d' \leq d$. The R-L semialgebraic equivalence classes of $S(n, d)$ is finite and, moreover, bounded by an effective function in (n, d)-variables.

Consider the above theorems in the case $\mathfrak{X} = \mathfrak{X}_0$. By the same reason as II.5.11, we have the following

Theorem II.7.12 (Case of $\mathfrak{X}_0$). *If $\mathfrak{X}$, $\mathfrak{X}_1$ and $\mathfrak{X}_2$ in II.7.1, 3, 7, 10 are all of class $\mathfrak{X}_0$, then the theorems hold without the boundedness condition.*

But II.7.7 for $\mathfrak{X}_1 = \mathfrak{X}_0$ and $\mathfrak{X}_2 \neq \mathfrak{X}_0$, II.7.5 and II.7.6 are different. Given $\mathfrak{X}_0$, let $\mathfrak{X}_1$ denote the smallest $\mathfrak{X}$ with $\mathfrak{X}_0 \subset \mathfrak{X}$ and Axiom (v). $\mathfrak{X}_0$-equivalence of $\mathfrak{X}_0$-functions is strictly stronger than $\mathfrak{X}_1$-equivalence.

Example II.7.13. There exist two polynomial functions on $\mathbf{R}^8$ which are $\mathfrak{X}_1$-equivalent and C^ω equivalent but not R-L $\mathfrak{X}_0$-equivalent.

Construction of polynomials. As in the above construction, we have a C^∞ trivial h-cobordism $(W_1; M, M_1)$ in $\mathbf{R}^6$, a non-trivial one $(W_2; M, M_2)$ and proper polynomial functions f_1, f_2 on $\mathbf{R}^6$ such that for each i, $\dim W_i = 6$, $W_2 - M \subset W_1^\circ$, M and M_i are connected, $\{f_i \geq 0\} = W_i$ and f_i is C^∞ regular on $f_i^{-1}(0) = M \cup M_i$. Define polynomial functions g_i on $\mathbf{R}^8$ to be

$$g_i(x_1, \ldots, x_8) = f_i(x_1, \ldots, x_6)x_8.$$

Proof that g_1 and g_2 are $\mathfrak{X}_1$-equivalent and C^ω equivalent. We show only that they are C^0 equivalent. Their C^∞ equivalence and $\mathfrak{X}_1$-equivalence are clear by the following proof, and by Theorem 8.4 in [S$_1$]. C^ω equivalence follows from C^∞ equivalence.

Let $q = (q', q'')\colon W_1 \to M_1 \times [0, 1]$ be a C^∞ diffeomorphism such that $q'|_{M_1} = \mathrm{id}$. As in the proof of II.7.9, we have a C^∞ map $\xi\colon W_2 \times \mathbf{R} \to M_1 \times \mathbf{R}$ such that $\xi|_{M \times \mathbf{R}} = (q', \mathrm{id})$, $\xi|_{M_2 \times \mathbf{R}}$ is a diffeomorphism onto $M_1 \times \mathbf{R}$, and for each $t_0 \in \mathbf{R}$, $\xi|_{W_2 \times t_0}$ is a C^∞ imbedding into $M_1 \times \{t \geq t_0\}$. Let $p_i\colon U_i \to M_i$, $i = 1, 2$, be closed C^∞ tubular neighborhoods in $\mathbf{R}^6$, and let ε be a sufficiently small positive increasing C^∞ regular C^∞ function on $\mathbf{R}$. Set

$$U_{it} = U_i \cap \{|f_i| \leq \varepsilon(t)\} \quad \text{for} \quad t \in \mathbf{R},$$

$$\tilde{p}_i = (p_i, \mathrm{id})\colon \tilde{U}_i = U_i \times \mathbf{R} \longrightarrow \tilde{M}_i = M_i \times \mathbf{R},$$

$$\tilde{U}_{it} = U_{it} \times \mathbf{R}, \quad \text{and} \quad \tilde{f}_i(x_1, \ldots, x_7) = f_i(x_1, \ldots, x_6).$$

We have a unique C^∞ diffeomorphism $\pi_t\colon U_{2t} \to \tilde{U}_{1t} \cap \tilde{p}_1^{-1}(\xi(M_2, t))$ for each t such that

$$\pi_t = \xi(\cdot, t) \quad \text{on} \quad M_2, \quad \text{and}$$

$$f_2 = \tilde{f}_1 \circ \pi_t, \quad \pi_t \circ p_2 = \tilde{p}_1 \circ \pi_t \quad \text{on} \quad U_{2t}.$$

Then the map

$$\bigcup_{t\in\mathbf{R}} U_{2t} \times t \ni (x_1,\dots,x_7) \longrightarrow \pi_{x_7}(x_1,\dots,x_6) \in \bigcup_{t\in\mathbf{R}} \tilde{U}_{1t} \cap \tilde{p}_1{}^{-1}(\xi(M_2,t))$$

is a C^∞ diffeomorphism from a neighborhood of $\tilde{U}_2$ to one of $\tilde{U}_1$. Extend each π_t to a P C^∞ diffeomorphism from $\mathbf{R}^6$ to $\pi_t(U_{2t})\cup$(the closed domain in $\partial\tilde{U}_{1t}$ lying between $\partial U_{1t}\times t$ and $\pi_t(\partial U_{2t}))\cup(\mathbf{R}^6-U_{1t})\times t$ so that the map $\pi': \mathbf{R}^6 \times \mathbf{R} \ni (x_1,\dots,x_7) \to \pi_{x_7}(x_1,\dots,x_6) \in \mathbf{R}^7$ is a P C^∞ diffeomorphism. This is possible because for $t \neq t' \in \mathbf{R}$, $\operatorname{Im}\pi_t \cap \operatorname{Im}\pi_{t'} = \varnothing$ and $\operatorname{Im}\pi_t$ can move "differentiably" as t moves in $\mathbf{R}$.

We want to define a homeomorphism π of $\mathbf{R}^8$ such that $g_2 = g_1 \circ \pi$. For each $a \in (\mathbf{R}^6 - M \cup M_1) \times \mathbf{R}$ and $b \in (\mathbf{R}^6 - M \cup M_2) \times \mathbf{R}$ there exists a linear homeomorphism $l_{a,b}$ of $\mathbf{R}$ such that

$$g_2(b, x_8) = g_1(a, l_{a,b}(x_8)),$$

and the map

$$(\mathbf{R}^6 - M \cup M_1) \times \mathbf{R} \times (\mathbf{R}^6 - M \cup M_2) \times \mathbf{R} \times \mathbf{R} \ni (a, b, x_8) \longrightarrow l_{a,b}(x_8) \in \mathbf{R}$$

is of class C^∞. Hence we have a unique C^0 function π'' on $(\mathbf{R}^6 - M\cup M_2)\times\mathbf{R}^2$ such that

$$g_2(x_1,\dots,x_8) = g_1(\pi'(x_1,\dots,x_7), \pi''(x_1,\dots,x_8)) \quad \text{if} \quad f_2(x_1,\dots,x_6) \neq 0.$$

Then the map

$$(\mathbf{R}^6 - M\cup M_2)\times\mathbf{R}^2 \ni (x_1,\dots,x_8) \longrightarrow (\pi'(x_1,\dots,x_7), \pi''(x_1,\dots,x_8)) \in \mathbf{R}^8$$

is a homeomorphism onto $(\mathbf{R}^6 - M \cup M_1) \times \mathbf{R}^2$ and

$$\pi''(x_1,\dots,x_8) = x_8 \quad \text{if} \quad 0 < |f_2(x_1,\dots,x_6)| < \varepsilon(x_7).$$

Therefore, we can extend (π', π'') to a homeomorphism π of $\mathbf{R}^8$ by setting $\pi = (\pi', \text{id})$ on $(M \cup M_2) \times \mathbf{R}^2$. Clearly $g_2 = g_1 \circ \pi$.

Proof that g_1 and g_2 are not R-L X_0-equivalent. Assume they are R-L X_0-equivalent. Since the C^0 critical value sets of g_1 and g_2 are both $\{0\}$, we have an X_0-homeomorphism π from $(\mathbf{R}^8, g_1^{-1}(0))$ to $(\mathbf{R}^8, g_2^{-1}(0))$. Recall

$$g_i^{-1}(0) = \mathbf{R}^7 \times 0 \cup (M \cup M_i) \times \mathbf{R}^2, \quad i = 1, 2.$$

Their C^0 singular point sets are $(M\cup M_i)\times\mathbf{R}\times 0$, and $(M\cup M_i)\times\mathbf{R}\times(\mathbf{R}-0)$ is not homeomorphic to $(\mathbf{R}^6 - M \cup M_{i'}) \times \mathbf{R} \times 0$ for $i' \neq i$. Hence π carries $(\mathbf{R}^7 \times 0, (M\cup M_1) \times \mathbf{R} \times 0)$ to $(\mathbf{R}^7 \times 0, (M\cup M_2) \times \mathbf{R} \times 0)$, which is impossible as we saw in the proof of II.7.9. $\qquad\square$

CHAPTER III.
HAUPTVERMUTUNG FOR POLYHEDRA

By (II.1.15) and (II.1.19), an $\mathcal{X}$-homeomorphism between compact $\mathcal{X}$-subsets of $\mathbf{R}^n$ is a strong C^r isomorphism for some C^r Whitney $\mathcal{X}$-stratifications. This property seems to be the key of the $\mathcal{X}$-Hauptvermutung (uniqueness of $\mathcal{X}$-triangulations of $\mathcal{X}$-sets). Hence we conjecture.

Conjecture. *If there exist strongly isomorphic Whitney stratifications of compact polyhedra X and Y, then X and Y are PL homeomorphic.*

I do not know whether this conjecture is true. In the following sections, we will prove it with conditions attached (III.1.1 and III.1.2).

§III.1. Certain conditions for two polyhedra
to be PL homeomorphic

Let X and Y be compact polyhedra in $\mathbf{R}^n$, let $\{X_i\}_{i=1,\ldots,k}$ and $\{Y_i\}_{i=1,\ldots,k}$ be their respective Whitney stratifications with $\dim X_1 = \dim Y_1 < \cdots < \dim X_k = \dim Y_k$, and let $f\colon \{X_i\} \to \{Y_i\}$ be an isomorphism (i.e., for each i, $f(X_i) = Y_i$, $f|_{X_i}$ is a C^∞ diffeomorphism onto Y_i, and $\{Z_i = \operatorname{graph} f|_{X_i}\}_{i=1,\ldots,k}$ is a Whitney stratification in $\mathbf{R}^n \times \mathbf{R}^n$). Set $Z = \operatorname{graph} f$ and let $p_1\colon Z \to X$ and $p_2\colon Z \to Y$ denote the projections. Let $\{T_i^X = (|T_i^X|, \pi_i^X, \rho_i^X)\}_{i=1,\ldots,k}$, $\{T_i^Y = (|T_i^Y|, \pi_i^Y, \rho_i^Y)\}_{i=1,\ldots,k}$ and $\{T_i^Z = (|T_i^Z|, \pi_i^Z, \rho_i^Z)\}_{i=1,\ldots,k}$ be controlled tube systems for $\{X_i\}$, $\{Y_i\}$ and $\{Z_i\}$ respectively. Note that by I.1.3 we may choose each ρ_i^X, ρ_i^Y and ρ_i^Z to be the squares of the functions which measure distance from X_i, Y_i and Z_i respectively. Let $\delta = \{\delta_i\}_{i=1,\ldots,k-1}$ be a common removal data of $\{X_i\}$, $\{Y_i\}$ and $\{Z_i\}$ for $\{T_i^X\}$, $\{T_i^Y\}$ and $\{T_i^Z\}$ respectively. The main theorems of this chapter are the following two.

Theorem III.1.1. *For any integer $1 \leq i \leq k$, for any sequence of positive numbers $\varepsilon = \{\varepsilon_j\}_{j=1,\ldots,k}$ with $\varepsilon \leq \delta$, and for any C^∞ triangulation (K_i, f_i) of $X_i - \cup_{j=1}^{i-1} \rho_j^{X-1}([0, \varepsilon_j[)$, assume there exist an arbitrarily strong approximation (K_i', f_i') of (K_i, f_i) and a small neighborhood U of X_i in $\mathbf{R}^n$ such that (K_i', f_i') also is a C^∞ triangulation of the same set. Assume also that for each $\sigma \in K_i'$ and for each $i' > i$, the restrictions of $\rho_i^Z \circ p_1^{-1}$ and ρ_i^X to $U \cap X_{i'} \cap \pi_i^{X-1}(f'(\sigma^\circ))$ are friendly. Assume the same statement for $\{Y_i\}$, $\{\rho_i^Y\}$ and $\{\rho_i^Z \circ p_2^{-1}\}$. Then X and Y are PL homeomorphic.*

The conditions in Theorem III.1.1 are stronger than the condition that $f\colon \{X_i\} \to \{Y_i\}$ is a strong isomorphism for $\{T_i^X\}$, $\{T_i^Y\}$ and $\{T_i^Z\}$ (i.e., for each $i < i'$, the restrictions of $\rho_i^Z \circ p_1^{-1}$ and ρ_i^X to $U \cap X_{i'} - \cup_{j=1}^{i-1} \rho_j^{X-1}([0, \varepsilon_j[)$ are friendly, and $\rho_i^Z \circ p_2^{-1}$ are ρ_i^Y have a similar property).

Theorem III.1.2. *Assume that* $f : \{X_i\} \to \{Y_i\}$ *is a strong isomorphism for* $\{T_i^X\}$, $\{T_i^Y\}$ *and* $\{Y_i^Z\}$, *and for each connected component* C *of* X_i, $i = 1, \ldots, k$, $\{X_{i'} \cap \overline{C}\}_{i'}$ *is strongly isomorphic to a Whitney stratification of a solid sphere* $\{x \in \mathbf{R}^{\dim C} : |x| \leq 1\}$ *for some tube systems. Then* X *and* Y *are PL homeomorphic.*

We shall prove these theorems in the next section.

Remark III.1.3. Let r be a positive integer. The C^r versions of the above theorems hold true. They are proved in the same way as I.1.15, I.1.13′ and I.1.16. We omit their proofs.

Corollary III.1.4 ($\mathfrak{X}$-Hauptvermutung). *Two* $\mathfrak{X}$-*homeomorphic compact polyhedra in* $\mathbf{R}^n$ *are PL homeomorphic.*

Proof of III.1.4. Let X and Y be compact polyhedra in $\mathbf{R}^n$, let $f : X \to Y$ be an $\mathfrak{X}$-homeomorphism and let r be a positive integer. We can apply both of the above theorems to the proof. In order to show that the conditions in the theorems are not strange, we apply both.

Application of III.1.1. By (II.1.15) there exist C^r Whitney $\mathfrak{X}$-stratifications $\{X_i\}_{i=1,\ldots,k}$ of X and $\{Y_i\}_{i=1,\ldots,k}$ of Y such that $\dim X_1 = \dim Y_1 < \cdots < \dim X_k = \dim Y_k$ and $f : \{X_i\} \to \{Y_i\}$ is a C^r $\mathfrak{X}$-isomorphism. Set

$$Z = \mathrm{graph}\, f \quad \text{and} \quad Z_i = \mathrm{graph}\, f|_{X_i}, \ i = 1, \ldots, k.$$

By II.5.1 and II.6.9 we have controlled C^r $\mathfrak{X}$-tube systems $\{T_i^X\}$, $\{T_i^Y\}$ and $\{T_i^Z\}$ for $\{X_i\}$, $\{Y_i\}$ and $\{Z_i\}$ respectively. It follows from (II.1.19) that f is a strong C^r $\mathfrak{X}$-isomorphism for $\{T_i^X\}$, $\{T_i^Y\}$ and $\{T_i^Z\}$. Let δ and $\{\varepsilon\}_{i=1,\ldots,k-1}$ be the same as before.

In this situation, we want to apply III.1.1. Let (K_i, f_i) be a C^r triangulation of $X_i - \cup_{j=1}^{i-1} \rho_j^{X-1}([0, \varepsilon_j[)$ as in III.1.1. If f_i is of class $\mathfrak{X}$, the friendliness condition in the theorem is automatically satisfied for $(K_i', f_i') = (K_i, f_i)$ by (II.1.18). If f_i is not so, by II.6.10, we find an arbitrarily strong approximation (K_i', f_i') of (K_i, f_i) which is a C^r $\mathfrak{X}$-triangulation of $X_i - \cup_{j=1}^{i-1} \rho_j^{X-1}([0, \varepsilon_j[)$, and hence the condition in the theorem is satisfied. Therefore, the theorem implies that X and Y are PL homeomorphic.

Application of III.1.2. Since a C^r $\mathfrak{X}$-isomorphism is a strong C^r $\mathfrak{X}$-isomorphism (II.1.19), it is sufficient to find C^r Whitney $\mathfrak{X}$-stratifications $\{X_i\}_{i=1,\ldots,k}$ of X and $\{Y_i\}_{i=1,\ldots,k}$ of Y such that $f : \{X_i\} \to \{Y_i\}$ is a C^r isomorphism, and for each connected component C of X_i, $i = 1, \ldots, k$, $\{X_{i'} \cap \overline{C}\}_{i'}$ is C^r $\mathfrak{X}$-isomorphic to a C^r Whitney $\mathfrak{X}$-stratification of a solid sphere. Here, if we do not require the second condition, (II.1.15) shows existence of such $\{X_i\}$ and $\{Y_i\}$. So we will run over the proof of (II.1.15) so that the second condition is satisfied.

For simplicity of notation, the index of $\{X_i\}$ runs from 0 to $k = \dim X$, and some of X_i may be empty. We use a downward induction. Let l be a nonnegative integer $\leq k$. Assume there exist $\mathfrak{X}$-subsets $X(l)$ of X and $Y(l)$ of Y of dimension $\leq l$ and C^r Whitney $\mathfrak{X}$-stratifications $\{X_i\}_{i=l+1,\ldots,k}$ of $X - X(l)$ and $\{Y_i\}_{i=l+1,\ldots,k}$ of $Y - Y(l)$ such that the following hold:

$$f(X(l)) = Y(l), \qquad f(X_i) = Y_i, \qquad \dim X_i = i, \ \ i = l+1,\ldots,k,$$

$f|_{X-X(l)} \colon \{X_i\}_{i=l+1,\ldots,k} \to \{Y_i\}_{i=l+1,\ldots,k}$ is a C^r $\mathfrak{X}$-isomorphism, and for each connected component C of X_i, $i = l+1,\ldots,k$, the family consisting of the elements of $\{X_{i'} \cap \overline{C}\}_{i'=l+1,\ldots,k}$ and of the elements of a C^r Whitney $\mathfrak{X}$-stratification of $\overline{C} - \cup_{i'=l+1}^{k} X_{i'}$ is C^r $\mathfrak{X}$-isomorphic to a C^r Whitney $\mathfrak{X}$-stratification of a solid sphere. Here we call the last condition $A(C; X_{i'}, \ i' = l+1,\ldots,k)$. We need only find $\mathfrak{X}$-subsets $X(l-1)$ of $X(l)$ and $Y(l-1)$ of $Y(l)$ together with $\{X_i\}_{i=l+1,\ldots,k} \cup \{X_l = X(l) - X(l-1)\}$ and $\{Y_i\}_{i=l+1,\ldots,k} \cup \{Y_l = Y(l) - Y(l-1)\}$ complete the induction step.

If $\dim X(l) < l$, we set $X(l-1) = X(l)$ and $Y(l-1) = Y(l)$. If $\dim X(l) = l = 0$, trivially $X(-1) = Y(-1) = \varnothing$ are sufficient. So assume $\dim X(l) = l > 0$. As in the proof of I.2.3, we obtain $\mathfrak{X}$-subsets $X'(l-1)$ of $X(l)$ and $Y'(l-1)$ of $Y(l)$ such that

$$\dim X'(l-1) < l, \qquad f(X'(l-1)) = Y'(l-1),$$

$f|_{X-X'(l-1)} \colon \{X_i\}_{i=l+1,\ldots,k} \cup \{X(l) - X'(l-1)\} \to \{Y_i\}_{i=l+1,\ldots,k} \cup \{Y(l) - Y'(l-1)\}$ is a C^r $\mathfrak{X}$-isomorphism, and for each connected component C of X_i, $i = l+1,\ldots,k$, condition $A(C; X_{i'}, \ i' = l+1,\ldots,k, \ X(l) - X'(l-1))$ is satisfied. Here if we set $X(l-1) = X'(l-1)$ and $Y(l-1) = Y'(l-1)$, then for a connected component C of $X_l = X(l) - X(l-1)$, condition $A(C; X_i, \ i = l,\ldots,k) = A(C; X_l)$ is not necessarily satisfied. Hence we will enlarge $X'(l-1)$ and $Y'(l-1)$ so that it is so. By II.2.1 there exists an $\mathfrak{X}$-subset $X(l-1)$ of $X(l)$ containing $X'(l-1)$ such that $\dim X(l-1) < l$, for each connected component C of $X(l)-X(l-1)$, $\overline{C}$ is $\mathfrak{X}$-homeomorphic to a simplex and the homeomorphism carries C, C^r diffeomorphically to the interior of the simplex. Here we can replace a simplex by a solid sphere because there is a semialgebraic homeomorphism between them which carries one interior to the other interior C^ω diffeomorphically. Set $Y(l-1) = f(X(l-1))$. For such $X(l-1)$ and $Y(l-1)$ we complete the induction step. Indeed, the above arguments imply that for each connected component C of $X_l = X(l) - X(l-1)$, the family of C and a C^r Whitney $\mathfrak{X}$-stratification of $\overline{C} - C$ is C^r $\mathfrak{X}$-isomorphic to a C^r $\mathfrak{X}$-stratification of a solid sphere, which satisfies

$A(C; X_i, \ i = l, \ldots, k)$. On the other hand, if C is a connected component of $X_i, \ i = l+1, \ldots, k$, $A(C; X_{i'}, \ i' = l, \ldots, k)$ follows from $A(C; X_{i'}, \ i' = l+1, \ldots, k$, $X(l) - X'(l-1))$. Thus the conditions in III.1.2 are satisfied, and hence X and Y are PL homeomorphic. □

Remark III.1.5. Let X and Y be homeomorphic PL manifolds of dimension $\neq 4$. There exist stratifications $\{X_i\}$ of X and $\{Y_i\}$ of Y and a homeomorphism $\pi \colon X \to Y$ such that for each i, $\pi(X_i) = Y_i$, and $\pi|_{X_i}$ is a C^∞ diffeomorphism onto Y_i.

I do not know whether this holds true for polyhedra X and Y nor whether there exist Whitney stratifications $\{X_i\}$ and $\{Y_i\}$ with the same properties.

Proof of III.1.5. If $\dim X \leq 3$, X and Y are PL homeomorphic by uniqueness of a PL structure on a topological manifold of dimension ≤ 3. Hence we assume $\dim X > 4$.

Let K, K' be simplicial complexes with underlying polyhedra X and Y respectively, and let $\pi_0 \colon X \to Y$ be a homeomorphism. Enumerate the vertices $a_1, a_2, \ldots$ of K, and for each a_i, let U_i denote the open star of a_i in K (i.e., the union of open simplexes of K whose closures contain a_i). By subdividing K and K' if necessary, we assume there exist homeomorphisms $\varphi_i \colon U_i \to V_i$ and $\psi_i \colon U'_i \to V'_i$ onto open subsets of a Euclidean space which are linear on $\sigma \cap U_i$ and on $\sigma' \cap U'_i$ respectively, for each $\sigma \in K$ and for each $\sigma' \in K'$, where U'_i are some open sets in Y which contain $\overline{\pi_0(U_i)}$.

We use the following known fact to modify π_0 (see Essays I and IV in [K-S]). Let V and V' be C^∞ submanifolds of $\mathbf{R}^m$ of dimension > 4 whose cohomology groups are 0, let $\tau \colon V \to V'$ be a homeomorphism, and let ε be a positive continuous function on V. We can ε-approximate τ by a C^∞ diffeomorphism τ', i.e.,

$$\mathrm{dis}(\tau(x), \tau'(x)) \leq \varepsilon(x) \quad \text{for} \quad x \in V.$$

Apply this fact to $\psi_1 \circ \pi_0 \circ \varphi_1^{-1} \colon V_1 \to \psi_1 \circ \pi_0(U_0)$. We have a homeomorphism $\pi_1 \colon X \to Y$ such that $\pi_1(U_1) = \pi_0(U_1)$, $\pi_1 = \pi_0$ on U_1^c, for any i, U'_i includes the closure of $\pi_1(U_i)$, and $\psi_1 \circ \pi_1 \circ \varphi_1^{-1} \colon V_1 \to \psi_1 \circ \pi_0(U_1)$ is a C^∞ diffeomorphism. Moreover, by the transversality theorem, we can choose π_1 so that for each pair of open simplexes ρ of K in U_1 and ρ' of K', $\psi_1 \circ \pi_1 \circ \varphi_1^{-1}|_{\varphi_1(\rho)}$ is transversal to $\psi_1(\rho' \cap U'_1)$, which implies that $\rho \cap \pi_1^{-1}(\rho')$ and $\pi_1(\rho) \cap \rho'$ are C^∞ submanifolds of X and Y respectively.

Next we proceed with the same arguments for $\psi_2 \circ \pi_1 \circ \varphi_2^{-1} \colon V_2 \to \psi_2 \circ \pi_1(U_2)$, and we obtain a homeomorphism $\pi_2 \colon X \to Y$ which is a modification of π_1 at U_2. Note that for each pair of open simplexes ρ of K in $U_1 \cup U_2$ and ρ' of K', $\rho \cap \pi_2^{-1}(\rho')$ and $\pi_2(\rho) \cap \rho'$ are C^∞ submanifolds of X and Y respectively, and $\pi_2|_{\rho \cap \pi_2^{-1}(\rho')}$ is a C^∞ diffeomorphism onto $\pi_2(\rho) \cap \rho'$.

In the same way we construct homeomorphisms $\pi_1, \pi_2, \ldots : X \to Y$. Its limit map π is a well-defined homeomorphism from X to Y; for each pair of open simplexes ρ of K and ρ' of K', $\rho \cap \pi^{-1}(\rho')$ and $\pi(\rho) \cap \rho'$ are C^∞ submanifolds of X and Y respectively; and $\pi|_{\rho \cap \pi^{-1}(\rho')}$ is a C^∞ diffeomorphism onto $\pi(\rho) \cap \rho'$. Hence the stratifications

$$\{X_i\} = \{\rho \cap \pi^{-1}(\rho') \colon \rho, \rho' \text{ open simplexes of } K, K' \text{ respectively}\}$$

and $\{Y_i\} = \{\pi(X_i)\}$ are what we wanted. $\square$

§III.2. Proofs of Theorems III.1.1 and III.1.2

Keep the notation X, Y, Z, f, $\{X_i\}$, $\{Y_i\}$, $\{Z_i\}$, $\{T_i^X\}$, $\{T_i^Y\}$, $\{T_i^Z\}$, δ and ε of §III.1. For simplicity of notation, we extend ρ_i^X, ρ_i^Y and ρ_i^Z to $\mathbf{R}^n$, $\mathbf{R}^n$ and $\mathbf{R}^n \times \mathbf{R}^n$ respectively, by setting $\rho_i^X = \rho_i^Y = \rho_i^Z = 1$ outside their respective original domains. We always assume $\delta_i < 1$. Hence the statements in the theorems are not influenced by these extensions. Let Γ denote the set of maps from the set $\{1, \ldots, k-1\}$ to the set $\{=, \geq\}$. For each $\gamma \in \Gamma$ and for each $i = 1, \ldots, k$, set

$$X_{i,\gamma,\varepsilon} = \{x \in X_i \colon \rho_j^X(x)\gamma(j)\varepsilon_j, \ j = 1, \ldots, i-1\},$$

where $\rho_j^X(x)\gamma(j)\varepsilon_j$ means $\rho_j^X(x) = \varepsilon_j$ if $\gamma(j)$ is "=", and $\rho_j^X(x) \geq \varepsilon_j$ if $\gamma(j)$ is "$\geq$". We similarly define $Y_{i,\gamma,\varepsilon}$ and $Z_{i,\gamma,\varepsilon}$. Let γ_0 denote the constant map with the value "$\geq$". Note that $X_{i,\gamma,\varepsilon}$, $Y_{i,\gamma,\varepsilon}$ and $Z_{i,\gamma,\varepsilon}$ are compact C^∞ manifolds possibly with boundary and corners and by I.3.13, they are C^∞ triangulable.

The conditions of a strong isomorphism in III.1.2 imply certain properties of $X_{i,\gamma,\varepsilon}$ and $Y_{i,\gamma,\varepsilon}$ from which it follows that X and Y are PL homeomorphic. First we explain the properties.

(III.2.1) *For some ε, assume there exist C^∞ diffeomorphisms $f_i \colon X_{i,\gamma_0,\varepsilon} \to Y_{i,\gamma_0,\varepsilon}$, $i = 1, \ldots, k$, which carry $X_{i,\gamma,\varepsilon}$ to $Y_{i,\gamma,\varepsilon}$ for each $\gamma \in \Gamma$ such that for any integers $1 \leq i < i' \leq k$ and any connected component C of $X_{i',\gamma_0,\varepsilon} \cap \rho_i^{X-1}(\varepsilon_i)$, $\pi_i^Y \circ f_{i'}(C) = f_i \circ \pi_i^X(C)$. Assume also that C^∞ triangulations of $X_{i,\gamma,\varepsilon}$, $i = 1, \ldots, k$, $\gamma \in \Gamma$, are finite disjoint unions of PL balls. Then X and Y are PL homeomorphic.*

Proof that (III.2.1) implies III.1.2. The conditions in (III.2.1) hold true for any ε if they do for some ε. This is because by the definition of a removal data, for another ε', there are C^∞ diffeomorphisms from $X_{i,\gamma_0,\varepsilon}$ to $X_{i,\gamma_0,\varepsilon'}$ and from $Y_{i,\gamma_0,\varepsilon}$ to $Y_{i,\gamma_0,\varepsilon'}$ which carry $X_{i,\gamma,\varepsilon}$ and $Y_{i,\gamma,\varepsilon}$ to $X_{i,\gamma,\varepsilon'}$ and $Y_{i,\gamma,\varepsilon'}$ respectively, for each $\gamma \in \Gamma$. Hence we shall lessen ε in (III.2.1) without telling. If $\{X_i\}$ and $\{Y_i\}$ are strongly isomorphic for $\{T_i^X\}$, $\{T_i^Y\}$ and $\{T_i^Z\}$, then by I.1.13 and I.1.14, there are C^∞ diffeomorphisms f_i which satisfy the first of the two conditions in (III.2.1) for some ε. Hence we need only prove the following.

Assertion. If for each connected component C of X_i, $i = 1, \dots, k$, $\{X_{i'} \cap \overline{C}\}_{i'=1,\dots,k}$ is strongly isomorphic to a Whitney stratification of a solid sphere for some tube systems, then C^∞ triangulations of $X_{i,\gamma,\varepsilon}$ for all $i = 1, \dots, k$ and $\gamma \in \Gamma$ are finite disjoint union of PL balls.

Proof of Assertion. We proceed by induction on $m = \dim X$. If $m = 0$, the assertion is trivial. So assume the assertion holds for the smaller dimensional case. In particular, we assume the assertion for $\{X_i\}_{i=1,\dots,k-1}$. Hence it suffices to consider only C^∞ triangulations of $X_{k,\gamma,\varepsilon}$. Clearly we can suppose X_k is connected and $\overline{X_k} = X$. (From now on X is not necessarily a polyhedron.) By hypothesis $\{X_i\}_{i=1,\dots,k}$ is strongly isomorphic to a Whitney stratification $\{B_i\}_{i=1,\dots,k}$ of the set $B = \{x \in \mathbf{R}^m : |x| \le 1\}$ for some tube systems. Let $\{T_i^B = (|T_i^B|, \pi_i^B, \rho_i^B)\}_{i=1,\dots,k}$ denote the tube system for $\{B_i\}$ appearing in this strong isomorphism. By I.1.13 and I.1.14, $X_{k,\gamma,\varepsilon}$ are C^∞ diffeomorphic to $B_{k,\gamma,\varepsilon}$, which are defined in the same way as $X_{k,\gamma,\varepsilon}$. Hence assuming that C^∞ triangulations of all $B_{i,\gamma,\varepsilon}$, $i < k$, are finite disjoint unions of PL balls, we need only prove that C^∞ triangulations of $B_{k,\gamma,\varepsilon}$ are PL balls.

To explain plainly its proof, we assume ρ_i^B are the squares of the functions which measure distance from B_i, which is possible by I.1.16. Set

$$B_{i,\gamma,\varepsilon}^+ = \{b \in B : \rho_i^B(b) \le \varepsilon_i, \ \rho_j^B(b)\gamma(j)\varepsilon_j, \ j = 1, \dots, i-1\}, \ i = 1, \dots, k.$$

Then $B_{k,\gamma,\varepsilon}^+ = B_{k,\gamma,\varepsilon}$, and we have natural divisions:

$$B = B_{1,\gamma_0,\varepsilon}^+ \cup \dots \cup B_{k,\gamma_0,\varepsilon}^+ \quad \text{and} \tag{1}$$

$$B_{l,\gamma,\varepsilon}^+ \cap \rho_l^{B-1}(\varepsilon_l) = B_{l+1,\gamma,\varepsilon}^+ \cup \dots \cup B_{k,\gamma,\varepsilon}^+ \quad \text{if} \quad \gamma \ne \gamma_0, \tag{2}$$

where l is the integer such that $\gamma(l) = $ "$=$" and $\gamma(i) = $ "$\ge$" for $i = l+1, \dots, k-1$. Note that $1 \le l < k$. We are concerned with only C^∞ triangulations of $B_{k,\gamma,\varepsilon}$. But to use the divisions (1) and (2) we will triangulate all $B_{i,\gamma,\varepsilon}^+$. Clearly the family $\{B_{i,\gamma,\varepsilon}^+ : i = 1, \dots, k, \ \gamma \in \Gamma\}$ is locally C^∞ equivalent to a family of polyhedra (see I.3.18). (Here ε is chosen to be small enough.) Hence by I.3.13 and I.3.18 there is a C^∞ triangulation (K, f) of B compatible with $\{B_{i,\gamma,\varepsilon}^+ : i = 1, \dots, k, \ \gamma \in \Gamma\}$. Set

$$A_{i,\gamma} = f^{-1}(B_{i,\gamma,\varepsilon}^+), \ i = 1, \dots, k, \ \gamma \in \Gamma.$$

Then (1) and (2) are transformed to the following divisions respectively:

$$|K| = A_{1,\gamma_0} \cup \cdots \cup A_{k,\gamma_0} \text{ and} \tag{1}'$$

$$A_{l,\gamma} \cap f^{-1}(\rho_l^{B-1}(\varepsilon_l)) = A_{l+1,\gamma} \cup \cdots \cup A_{k,\gamma}. \tag{2}'$$

Case $\gamma = \gamma_0$. Let $1 \leq i \leq k$ be an integer. We will prove by induction on i that $A_{i,\gamma_0} \cup \cdots \cup A_{k,\gamma_0}$ is a PL ball. This is trivial for $i = 1$ by uniqueness of C^∞ triangulations (I.3.13). Hence by assuming $i < k$ and the case i, we will prove the case $i + 1$. For simplicity of notation, we assume A_{i,γ_0} is connected (if this is not the case, it suffices to consider each connceted component). By I.3.8 it suffices to show that

(i) A_{i,γ_0} is a PL m-ball,

(ii) $A_{i,\gamma_0} \cap (A_{i+1,\gamma_0} \cup \cdots \cup A_{k,\gamma_0})$ is a PL $(m-1)$-ball,

(iii) $A_{i,\gamma_0} \cap (A_{i+1,\gamma_0} \cup \cdots \cup A_{k,\gamma_0})$ is included in $\partial A_{i,\gamma_0}$, and

(iv) $(A_{i,\gamma_0} \cap (A_{i+1,\gamma_0} \cup \cdots \cup A_{k,\gamma_0}))^\circ$ is included in $(A_{i,\gamma_0} \cup \cdots \cup A_{k,\gamma_0})^\circ$.

By the definition of $B_{i,\gamma_0,\varepsilon}^+$ we have

$$B_{i,\gamma_0,\varepsilon}^+ \cap (B_{i+1,\gamma_0,\varepsilon}^+ \cup \cdots \cup B_{k,\gamma_0,\varepsilon}^+)$$

$$= \{b \in B : \rho_i^B(b) = \varepsilon_i, \ \rho_j^B(b) \geq \varepsilon_j, \ j = 1, \ldots, i-1\} = B_{i,\gamma_0,\varepsilon}^+ \cap \rho_i^{B-1}(\varepsilon_i)$$

$$\text{and } B_{i,\gamma_0,\varepsilon}^+ \cup \cdots \cup B_{k,\gamma_0,\varepsilon}^+ = \{b \in B : \rho_j^B(b) \geq \varepsilon_j, \ j = 1, \ldots, i-1\}.$$

Hence (iii) and (iv) are clear. For proof of (i) and (ii), it suffices to prove that the triple

$$(B_{i,\gamma_0,\varepsilon}^+, B_{i,\gamma_0,\varepsilon}^+ \cap \rho_i^{B-1}(\varepsilon_i), B_{i,\gamma_0,\varepsilon})$$

is C^∞ diffeomorphic to the product:

$$(B^{m'} \cap \{x_1 \geq 0\}, \partial B^{m'} \cap \{x_1 \geq 0\}, 0) \times B_{i,\gamma_0,\varepsilon},$$

where

$$B^{m'} = \{x = (x_1, \ldots, x_{m'}) \in \mathbf{R}^{m'} : |x| \leq 1\}, \ m' = \text{codim } B_i.$$

This is because of uniqueness of C^∞ triangulations and the assumption that a C^∞ triangulation of $B_{i,\gamma_0,\varepsilon}$ is a PL ball.

We find such a diffeomorphism as follows. By the definition of a removal data $\pi_i^B|_{B_{i,\gamma_0,\varepsilon}^+} : B_{i,\gamma_0,\varepsilon}^+ \to B_{i,\gamma_0,\varepsilon}$ is a C^∞ fibre bundle with fibre $B^{m'} \cap \{x_1 \geq 0\}$ (which is easily shown because ρ_j^B are the squares of the functions which measure distance), and $\pi_i^B|_{B_{i,\gamma_0,\varepsilon}^+ \cap \rho_i^{B-1}(\varepsilon_i)} : B_{i,\gamma_0,\varepsilon}^+ \cap \rho_i^{B-1}(\varepsilon_i) \to$

$B_{i,\gamma_0,\varepsilon}$ is its C^∞ subbundle with fibre $\partial B^{m'} \cap \{x_1 \geq 0\}$. By hypothesis the base $B_{i,\gamma_0,\varepsilon}$ is homeomorphic to a PL ball. Hence $\pi_i^B|_{B_{i,\gamma_0,\varepsilon}^+}$ is trivial, and $\pi_i^B|_{B_{i,\gamma_0,\varepsilon}^+ \cap \rho_i^{B-1}(\varepsilon_i)}$ is trivial as a C^∞ subbundle, from which the required diffeomorphism follows. Thus the case γ_0 is proved.

General case of γ. Assume $\gamma \neq \gamma_0$ and let l be as in (2). We proceed as in the case γ_0. Here also we assume that $A_{l+1,\gamma},\ldots,A_{k-1,\gamma}$ are connected for simplicity of notation. Apply inductively I.3.8 to the division (2'). As in the case γ_0, it suffices to prove that for each $i = l+1,\ldots,k-1$,

(v) $A_{l,\gamma} \cap f^{-1}(\rho_l^{B-1}(\varepsilon_l))$ and $A_{i,\gamma}$ are PL m''-balls, where $m'' = \dim B_{k,\gamma,\varepsilon}$,

(vi) $A_{i,\gamma} \cap (A_{i+1,\gamma} \cup \cdots \cup A_{k,\gamma})$ is a PL $(m''-1)$-ball,

(vii) $A_{i,\gamma} \cap (A_{i+1,\gamma} \cup \cdots \cup A_{k,\gamma})$ is included in $\partial A_{i,\gamma}$ and

(viii) $(A_{i,\gamma} \cap (A_{i+1,\gamma} \cup \cdots \cup A_{k,\gamma}))^\circ$ is included in $(A_{i,\gamma} \cup \cdots \cup A_{k,\gamma})^\circ$.

We have equalities:

$$B_{i,\gamma,\varepsilon}^+ \cap (B_{i+1,\gamma,\varepsilon}^+ \cup \cdots \cup B_{k,\gamma,\varepsilon}^+) = B_{i,\gamma,\varepsilon}^+ \cap \rho_i^{B-1}(\varepsilon_i) \quad \text{and}$$
$$B_{i,\gamma,\varepsilon}^+ \cup \cdots \cup B_{k,\gamma,\varepsilon}^+ = \{b \in B : \rho_j^B(b)\gamma(j)\varepsilon_j, \ j = 1,\ldots,i-1\},$$
$$i = l+1,\ldots,k-1.$$

Hence (vii) and (viii) are clear, and (vi) is equivalent to the condition:

(vi)' $A_{i,\gamma} \cap f^{-1}(\rho_i^{B-1}(\varepsilon_i))$ is a PL $(m''-1)$-ball.

To prove (v) and (vi)' we use the trivialization in the case γ_0. Let $l \leq i < k$ be an integer. The diffeomorphism from $(B_{i,\gamma_0,\varepsilon}^+, B_{i,\gamma_0,\varepsilon}^+ \cap \rho_i^{B-1}(\varepsilon_i), B_{i,\gamma_0,\varepsilon})$ to $(B^{m'} \cap \{x_1 \geq 0\}, \partial B^{m'} \cap \{x_1 \geq 0\}, 0) \times B_{i,\gamma_0,\varepsilon}$ given in the case γ_0 can be chosen so as to carry each $B_{i,\gamma,\varepsilon}^+$ to $(B^{m'} \cap \{x_1 \geq 0\}) \times B_{i,\gamma,\varepsilon}$. Hence by the hypothesis that $B_{i,\gamma,\varepsilon}$ is a PL ball, $A_{i,\gamma}$ and $A_{i,\gamma} \cap f^{-1}(\rho_i^{B-1}(\varepsilon_i))$ are PL balls. The dimensional properties in (v) and (vi)' are clearly verified by the definition of a removal data, which completes the proof of the assertion. $\square$

To prove (III.2.1) we need the following lemma.

Lemma III.2.2. *Let ε be sufficiently small. There exists a C^∞ triangulation of X compatible with $\{X_{i,\gamma,\varepsilon}, X_{i,\gamma,\varepsilon}^+\}_{1 \leq i \leq k, \gamma \in \Gamma}$, where*

$$X_{i,\gamma,\varepsilon}^+ = \{x \in X : \rho_i^X(x) \leq \varepsilon_i, \ \rho_j^X(x)\gamma(j)\varepsilon_j, \ j = 1,\ldots,i-1\}.$$

Fix i and γ. Assume that a C^∞ triangulation of $X_{i,\gamma,\varepsilon}$ is a PL ball. Then a C^∞ triangulation of $X_{i,\gamma,\varepsilon}^+$ is PL homeomorphic to a cone with vertex corresponding to any given point of $X_{i,\gamma,\varepsilon}^\circ$ and with base corresponding to the union of $X_{i,\gamma,\varepsilon}^+ \cap \rho_i^{X-1}(\varepsilon_i)$ and all $X_{i,\gamma',\varepsilon}^+$ such that $X_{i,\gamma',\varepsilon} \subset \partial X_{i,\gamma,\varepsilon}$.

If we do not assume that a C^∞ triangulation of $X_{i,\gamma,\varepsilon}$ is a PL ball, replace $X_{i,\gamma,\varepsilon}$ with a C^∞ triangulable subset C of $X_{i,\gamma,\varepsilon}$ whose C^∞ triangulation is a PL ball. Then $X_{i,\gamma,\varepsilon}^+ \cap \pi_i^{X-1}(C)$ is C^∞ triangulable, and its C^∞ triangulation is PL homeomorphic to a cone with vertex corresponding to any given point of C° and with base corresponding to the union of $X_{i,\gamma,\varepsilon}^+ \cap \pi_i^{X-1}(C) \cap \rho_i^{X-1}(\varepsilon_i)$ and $X_{i,\gamma,\varepsilon}^+ \cap \pi_i^{X-1}(\partial C)$.

Proof. For simplicity of notation, we assume X_i and $X_{i,\gamma,\varepsilon}$ are connected and a C^∞ triangulation of $X_{i,\gamma,\varepsilon}$ is a PL ball, and we prove only the second statement. The other statements follow in the same way. By III.2.3 below there exists a Whitney stratification $\{W_j\}_{j=1,\dots,l}$ of X such that each connected component of W_j is an open set of a linear space, its closure is a polyhedron, and {connected components of X_j, $j = 1,\dots,k$} is compatible with $\{W_j\}_{j=1,\dots,l}$. Assume $\dim W_1 < \cdots < \dim W_l$.

Note. For each W_j and for each $x \in W_j$ there exists a neighborhood V of x in X and a linear isometric homeomorphism from $(V \cap W_j, \dots, V \cap W_l)$ to $(L \cap V \cap W_j (= x), \dots, L \cap V \cap W_l) \times (V \cap W_j)$, where L is a linear space in $\mathbf{R}^n$ of codimension $= \dim W_j$ and is transversal to W_j at x.

This is clear. Let i be fixed. We assume $X_i \subset W_{i+1}$ for convenience. Set

$$\tilde{X}_i = X_i, \qquad \tilde{X}_{i+1} = W_{i+1} - \overline{X_i}, \quad \text{and} \quad \tilde{X}_j = W_j, \ j = i+2, \dots, l,$$

and in place of $\{X_j\}_{j=i,\dots,k}$, consider $\{\tilde{X}_j\}_{j=i,\dots,l}$.

First we may assume π_i^X is the orthogonal projection for the following reason. Let $\tilde{\pi}_i^X \colon V \to X_{i,\gamma_0,\varepsilon}$ denote the orthogonal projection of a tubular neighborhood in $\mathbf{R}^n$. Lessen ε_i and define a C^∞ map $q \colon X_{i,\gamma_0,\varepsilon}^+ \to V$ so that for each $x \in X_{i,\gamma_0,\varepsilon}^+$, $q(x)$ is the image of the orthogonal projection of x to $\tilde{\pi}_i^{X-1}(\pi_i^X(x))$. By the Note the image of q is included in X, $q(X_{i,\gamma_0,\varepsilon}^+ \cap \tilde{X}_j) \subset \tilde{X}_j$ for $j = i+1, \dots, l$, q is a diffeomorphism onto the image, the image is a neighborhood of $X_{i,\gamma,\varepsilon}$ in X, and

$$\pi_i^X = \tilde{\pi}_i^X \circ q \quad \text{on} \quad X_{i,\gamma_0,\varepsilon}^+.$$

Hence we can assume π_i^X is the orthogonal projection. (Here we replace ρ_i^X with $\rho_i^X \circ q^{-1}$.)

Let $\tilde{\rho}_i^X$ denote the square of the function which measures distance from X_i. Next, we reduce the problem to the case $\rho_i^X = \tilde{\rho}_i^X$. For that, it suffices to prove that for any small $\varepsilon_i > 0$, there exists a C^∞ diffeomorphism from $(X_{i,\gamma,\varepsilon}^+)_{\gamma \in \Gamma}$ to $(\tilde{X}_{i,\gamma,\varepsilon}^+)_{\gamma \in \Gamma} = ($ the sets similarly defined by $\tilde{\rho}_i^X$ in place of ρ_i^X). Let $x \in X_{i,\gamma_0,\varepsilon}$, and let l be a half line with end x and

$$l \cap |T_i| \subset X \cap \pi_i^{X-1}(x).$$

The correspondence $l \cap \rho_i^{X-1}(\varepsilon_i) \to l \cap \tilde{\rho}_i^{X-1}(\varepsilon_i)$ defines a C^∞ diffeomorphism from $(X_{i,\gamma,\varepsilon}^+ \cap \rho_i^{X-1}(\varepsilon_i))_{\gamma \in \Gamma}$ to $(\tilde{X}_{i,\gamma,\varepsilon}^+ \cap \tilde{\rho}_i^{X-1}(\varepsilon_i))_{\gamma \in \Gamma}$, and the diffeomorphism is easily extended to the required diffeomorphism by l and a partition of unity. We omit the details. We keep the notation ρ_i^X.

Fix γ and set

$$F = \pi_i^{X-1}(x_0) \cap X \cap \{\rho_i^X \le \varepsilon_i\},$$

where x_0 is a point of $X_{i,\gamma,\varepsilon}$, and lessen ε_i. We easily prove that F is C^∞ triangulable, and its C^∞ triangulation is PL homeomorphic to a cone with vertex corresponding to x_0 and with base corresponding to $F \cap \rho_i^{X-1}(\varepsilon_i)$. Clearly $\partial X_{i,\gamma,\varepsilon}$ is the union of some $X_{i,\gamma',\varepsilon}$, $\gamma' \in \Gamma$. Hence it suffices to prove that $X_{i,\gamma,\varepsilon}^+$ is C^∞ diffeomorphic to $X_{i,\gamma,\varepsilon} \times F$.

For that we use the concept of a fibre bundle as in the proof that (III.2.1) implies III.1.2. Let G denote the group of C^∞ diffeomorphisms of F. We will see that the system $\eta = (X_{i,\gamma,\varepsilon}^+, \pi_i^X|_{X_{i,\gamma,\varepsilon}^+}, X_{i,\gamma,\varepsilon}, F, G)$ is a fibre bundle. If this is true, then the bundle is trivial, because by hypothesis, $X_{i,\gamma,\varepsilon}$ is homeomorphic to a PL ball. Moreover, the proof of the covering homotopy theorem (see [St]) shows that $X_{i,\gamma,\varepsilon}^+$ is C^∞ diffeomorphic to $X_{i,\gamma,\varepsilon} \times F$. Thus it suffices to prove that η is a fibre bundle.

Furthermore, we can reduce the problem as follows. (The reason is clear.) There exists a neighborhood U of x_0 in X_i and a C^∞ diffeomorphism τ from $U \times F$ to $\pi_i^{X-1}(U) \cap X \cap \{\rho_i^X \le \varepsilon_i\}$ such that $\pi_i^X \circ \tau$ is the projection onto U.

Since X_i is a C^∞ submanifold of W_{i+1}, we have a C^∞ diffeomorphism $\tau': U \times (\pi_i^{X-1}(x_0) \cap W_{i+1}) \to \pi_i^{X-1}(U) \cap W_{i+1}$ such that $\pi_i^X \circ \tau'$ is the projection onto U and $\tau'|_{x \times (\pi_i^{X-1}(x_0) \cap W_{i+1})}$ is linear and isometric for each $x \in U$. By the note, we can extend τ' to a C^∞ diffeomorphism $\tau'': U \times (\pi_i^{X-1}(x_0) \cap X) \to \pi_i^{X-1}(U) \cap X$ keeping the last two properties. Then $\tau''(U \times F) = \pi_i^{X-1}(U) \cap X \cap \{\rho_i^X \le \varepsilon_i\}$. Hence $\tau = \tau''|_{U \times F}$ is what we wanted. $\qquad\square$

Lemma III.2.3. *There exists a Whitney stratification $\{W_j\}_{j=1,\ldots,l}$ of X such that each connected component of W_j is an open set of a linear space, its closure is a polyhedron, and any Whitney stratification of X with connected strata is compatible with $\{W_j\}$.*

Proof. We construct $\{W_j\}$ by an induction method as in the proof of I.2.2. Let $\Sigma_\infty X$ denote the C^∞ singular point set of X. Then $\Sigma_\infty X$ is a compact polyhedron, and each connected component of $X - \Sigma_\infty X$ is included in a linear space as an open set (see the proof of I.3.4). Set $W_1 = X - \Sigma_\infty X$. Note that the closure of a connected component of W_1 is a polyhedron, which easily follows if we take a simplicial decomposition of $(X, \Sigma_\infty X)$. Let $j > 1$ be an integer. Assume we have defined $W_1, \ldots, W_{j-1}$ so that $\{W_1, \ldots, W_{j-1}\}$ is a Whitney stratification, each connected component of $W_1, \ldots, W_{j-1}$ is included in a linear space as an open set, its closure is a polyhedron, $W_{j,1} = X - W_1 \cup \cdots \cup W_{j-1}$ is a compact polyhedron, and $\dim W_1 > \cdots > \dim W_{j-1} > \dim W_{j,1}$. Then we define W_j as follows. Let $\{W_{j,\alpha}\}_{\alpha \in A}$ denote the smallest family of subsets of $W_{j,1}$ which contains all the sets $W_{j,1} \cap \overline{W_i}$, $i = 1, \ldots, j-1$, and which is closed under the formation of closures, unions, intersections and complements. Set

$$W_j = W_{j,1} - \bigcup_{\alpha \in A} \Sigma_\infty W_{j,\alpha} \cup \bigcup_{\dim W_{j,\alpha} < \dim W_{j,1}} W_{j,\alpha}.$$

Thus we define a stratification $\{W_j\}_{j=1,\ldots,l}$ of X by induction. We change the order of the indices of $\{W_j\}_{j=1,\ldots,l}$ so that $\dim W_1 < \cdots < \dim W_l$. We want to show that $\{W_j\}$ fulfills the requirements. It is only the compatibility property that is not clear.

Let $\{W_\beta\}_{\beta \in B}$ denote the family of connected components of all W_j, $j = 1, \ldots, l$. For each $\beta \in B$, let L_β denote the linear space including W_β as an open set. By the definition of $\{W_j\}$ we have, for each $\beta \in B$,

$$W_\beta = \text{a connected component of the } C^\infty \text{ regular point set of}$$

$$X - \bigcup_{\dim W_\beta < \dim W_{\beta'}} (\text{bdry}(\overline{W_{\beta'}} \cap L_\beta)) \cup W_{\beta'}, \qquad (*)$$

where the symbol bdry denotes the boundary of a set as a topological subset of L_β. As usual, let $\{X_i\}_{1,\ldots,k}$ denote a Whitney stratification of X such that $\dim X_1 < \cdots < \dim X_k$. We will prove that $\{$connected components of $X_i\}_{i=1,\ldots,k}$ is compatible with $\{W_j\}_{j=1,\ldots,l}$. We prove this by reduction to absurdity. Assume that for some integer i_0, a connected component C of X_{i_0} is not included in any one W_j, and each connected component of $X_{i_0+1}, \ldots, X_k$ is included in some W_j. Let j_0 be the integer such that C is included in $W_1 \cup \cdots \cup W_{j_0}$ but not in $W_1 \cup \cdots \cup W_{j_0-1}$. Let W_{β_0} be a connected component of W_{j_0} which intersects with C, and let x_0 be a point of $C - W_{\beta_0}$. For simplicity of notation, we assume $C \cap W_{\beta_0}$ is connected.

First assume $\dim C < \dim W_{\beta_0} = n_0$.

Assertion. The x_0 is an inner point of $\overline{W_{\beta_0}}$ when we regard $\overline{W_{\beta_0}}$ as a subset of L_{β_0}.

We will prove this by carrying a neighborhood of a point a of $C \cap W_{\beta_0}$ in W_{β_0} to a neighborhood of x_0 in $\overline{W_{\beta_0}}$ as follows. Let U be a small neighborhood of a in W_{β_0} which is homeomorphic to $\mathbf{R}^{n_0}$, and does not intersect with

$X_{i_0} - C$ nor with X_i, $i = 1, \ldots, i_0 - 1$. Let $\{\tilde{X}_\lambda\}_{\lambda \in \Lambda}$ denote the family of the connected components of all $X_i \cap U$, $i = i_0 + 1, \ldots, k$. For each $\lambda \in \Lambda$, let X_λ denote the connected component of some X_i which includes $\tilde{X}_\lambda$. (There may be the case where $X_\lambda = X_{\lambda'}$ for $\lambda \neq \lambda'$.) Note that all X_λ are included in W_{β_0} because $\{X_\lambda\}_{\lambda \in \Lambda}$ is compatible with $\{W_j\}_{j=1,\ldots,l}$. Shrink U. By using Thom's first isotopy lemma (see II.6.1 and the note after II.6.1'), we can construct a homeomorphism τ from U to a neighborhood V of x_0 in $C \cup \bigcup_{\lambda \in \Lambda} X_\lambda$ such that $\tau(a) = x_0$. Clearly V is homeomorphic to $\mathbf{R}^{n_0}$ and included in $C \cup W_{\beta_0}$. Hence by the theorem of invariance of dimension of Euclidean spaces, $V \cap C$ and hence V are included in $\overline{W_{\beta_0}} \subset L_{\beta_0}$. It follows from the Brouwer's invariance theorem of domain that V is open in L_{β_0}. Hence x_0 is an inner point of V and then of $\overline{W_{\beta_0}}$, when we regard V and $\overline{W_{\beta_0}}$ as subsets of L_{β_0}. Thus the assertion is proved.

The assertion shall induce a contradiction to $(*)$. By the same arguments as above we can extend τ to a homeomorphism from a small neighborhood $\hat{U}$ of a in X to a neighborhood $\hat{V}$ of x_0 in X which carries $\hat{U} \cap W_\beta$ to $\hat{V} \cap W_\beta$ for any $\beta \in B$ with $\dim W_\beta > n_0$. (Here as above, we choose $\hat{U}$ so small that $\hat{U}$ does not intersect with $X_{i_0} - C$ nor with X_i for $i = 1, \ldots, i_0 - 1$.) Hence x_0 is contained in the interior of $\overline{W_\beta} \cap L_{\beta_0}$ in L_{β_0} for any β such that $\dim W_\beta > n_0$ and $a \in \overline{W_\beta}$. On the other hand, it follows from the assertion that $X - \bigcup_{\dim W_\beta > n_0} W_\beta \supset L_{\beta_0}$ as germs at x_0. But, since $\{X_i\}_i$ is a Whitney stratification, $(X - L_{\beta_0}) \cap \hat{V}$ is included in $\bigcup_{i=i_0+1}^k X_i$ and hence in $\bigcup_{\dim W_\beta > n_0} W_\beta$. Hence $X - \bigcup_{\dim W_\beta > n_0} W_\beta = L_{\beta_0}$ as germs at x_0, which implies that x_0 is a C^∞ regular point of $X - \bigcup_{\dim W_\beta > n_0} W_\beta$. Furthermore, we see these for any point x of C, i.e., x is contained in the interior of $\overline{W_\beta} \cap L_{\beta_0}$ in L_{β_0} if $\dim W_\beta > n_0$ and $a \in \overline{W_\beta}$, and x is a C^∞ regular point of $X - \bigcup_{\dim W_\beta > n_0} W_\beta$. Hence, by $(*)$ and the connectedness of C, C is included in W_{β_0}, which is a contradiction.

Second, assume $\dim C = n_0$. As C is a subset of the polyhedron $\overline{W_{j_0}}$ of maximal dimension, we see that C is contained in L_{β_0}. On the other hand, for any β such that $\dim W_\beta > n_0$ and $C \cap \overline{W_\beta} \neq \varnothing$, C is contained in $\overline{W_\beta}$ for the following reason. Since $\overline{W_\beta}$ is a union of connected components of $\overline{X_i}$, $i = i_0 + 1, \ldots, k$, some connected component intersects with C. Moreover, since C is compatible with $\{\overline{X_i}\}_{i=i_0+1,\ldots,k}$, C is contained in the component. Hence $C \subset \overline{W_\beta}$. By these two facts, if C intersects with $\overline{W_\beta}$ for some β with $\dim W_\beta > n_0$, then $C \subset \overline{W_\beta} \cap L_{\beta_0}$ and hence $C \cap \mathrm{bdry}(\overline{W_\beta} \cap L_{\beta_0}) = \varnothing$. Therefore, C is contained in the set:

$$X - \bigcup_{\dim W_\beta > n_0} (\mathrm{bdry}(\overline{W_\beta} \cap L_{\beta_0})) \cup W_\beta. \qquad (**)$$

If C consists of C^∞ regular points of this set, then we arrive the same contradiction as the first case. This is clear because C consists of C^∞ regular points of $X - \cup_{i=i_0+1}^{k} X_i$ and $X - \cup_{i=i_0+1}^{k} X_i$ contains the set $(**)$. Thus we complete the proof. ∎

The key idea of proof is to consider the divisions $\{X_{i,\gamma_0,\varepsilon}^{+}\}$ and $\{Y_{i,\gamma_0,\varepsilon}^{+}\}$ in place of $\{X_i\}$ and $\{Y_i\}$. This idea works well because of III.2.2. I was inspired by the concept of the dual subdivision of a simplicial complex.

Proof of (III.2.1). By the first statement of III.2.2 there exist C^∞ triangulations (K, f) of X and (L, g) of Y compatible with $\{X_{i,\gamma,\varepsilon}, X_{i,\gamma,\varepsilon}^{+}\}_{1 \leq i \leq k, \gamma \in \Gamma}$ and with $\{Y_{i,\gamma,\varepsilon}, Y_{i,\gamma,\varepsilon}^{+}\}_{1 \leq i \leq k, \gamma \in \Gamma}$ respectively, where $X_{i,\gamma,\varepsilon}^{+}$ and $Y_{i,\gamma,\varepsilon}^{+}$ are defined as in III.2.2. Set

$$\tilde{X} = |K|, \qquad \tilde{X}_{i,\gamma,\varepsilon} = f^{-1}(X_{i,\gamma,\varepsilon}), \qquad \tilde{X}_{i,\gamma,\varepsilon}^{+} = f^{-1}(X_{i,\gamma,\varepsilon}^{+}),$$

$$\tilde{\pi}_i^X = f^{-1} \circ \pi_i^X \circ f, \qquad \tilde{\rho}_i^X = \rho_i^X \circ f,$$

$$\tilde{Y} = |L|, \qquad \tilde{Y}_{i,\gamma,\varepsilon} = g^{-1}(Y_{i,\gamma,\varepsilon}), \qquad \tilde{Y}_{i,\gamma,\varepsilon}^{+} = g^{-1}(Y_{i,\gamma,\varepsilon}^{+}),$$

$$\tilde{\pi}_i^Y = g^{-1} \circ \pi_i^Y \circ g, \text{ and } \tilde{\rho}_i^Y = \rho_i^Y \circ g,$$

$$i = 1, \ldots, k, \quad \gamma \in \Gamma.$$

By uniqueness of C^∞ triangulations (I.3.13, I.3.18) and the first hypothesis in (III.2.1) we have PL homeomorphisms $\tilde{f}_i \colon \tilde{X}_{i,\gamma_0,\varepsilon} \to \tilde{Y}_{i,\gamma_0,\varepsilon}$, $i = 1, \ldots, k$, which carry $\tilde{X}_{i,\gamma,\varepsilon}$ to $\tilde{Y}_{i,\gamma,\varepsilon}$ for each $\gamma \in \Gamma$ such that for each pair of integers $1 \leq i < i' \leq k$ and each connected component C of $\tilde{X}_{i',\gamma_0,\varepsilon} \cap \tilde{\rho}_i^{X-1}(\varepsilon_i)$, we have

$$\tilde{\pi}_i^Y \circ \tilde{f}_{i'}(C) = \tilde{f}_i \circ \tilde{\pi}_i^X(C).$$

Now I.3.13 implies also that $\tilde{X}$ and $\tilde{Y}$ are PL homeomorphic to X and Y respectively. Hence it suffices to prove that $\tilde{X}$ and $\tilde{Y}$ are PL homeomorphic.

We will construct a PL homeomorphism τ from $\tilde{X}$ to $\tilde{Y}$ by double induction. Consider divisions:

$$\tilde{X} = \bigcup_{i=1}^{k} \tilde{X}_{i,\gamma_0,\varepsilon}^{+} \quad \text{and} \quad \tilde{Y} = \bigcup_{i=1}^{k} \tilde{Y}_{i,\gamma_0,\varepsilon}^{+}.$$

By downward induction on $i = 1, \ldots, k$, let us construct PL homeomorphisms:

$$\tau_i \colon \bigcup_{i'=i}^{k} \tilde{X}_{i',\gamma_0,\varepsilon}^{+} \to \bigcup_{i'=i}^{k} \tilde{Y}_{i',\gamma_0,\varepsilon}^{+}$$

so that for each $i' = i, \ldots, k$, $j = 1, \ldots, i-1$, $\gamma \in \Gamma$ and for each connected component C of $\tilde{X}_{i',\gamma_0,\varepsilon}^{+} \cap \tilde{\rho}_j^{X-1}(\varepsilon_j)$, we have

$$\tau_i(\tilde{X}_{i',\gamma,\varepsilon}^{+}) = \tilde{Y}_{i',\gamma,\varepsilon}^{+} \quad \text{and} \quad \tilde{\pi}_j^Y \circ \tau_i(C) = \tilde{f}_j \circ \tilde{\pi}_j^X(C).$$

For $i = k$, set $\tau_k = \tilde{f}_k$. Hence for an integer $1 \le i < k$, by assuming τ_{i+1}, we need only construct τ_i because $\tau = \tau_1$ is what we want.

By the definitions of $X_{i,\gamma_0,\varepsilon}$ and $Y_{i,\gamma_0,\varepsilon}$ we have

$$\tilde{X}^+_{i,\gamma_0,\varepsilon} \cap \Big(\bigcup_{i'=i+1}^{k} \tilde{X}^+_{i',\gamma_0,\varepsilon} \Big) = \tilde{X}^+_{i,\gamma_0,\varepsilon} \cap \tilde{\rho}_i^{X-1}(\varepsilon_i), \qquad \partial \tilde{X}_{i,\gamma_0,\varepsilon} = \bigcup_{\gamma \in \Gamma(i)} \tilde{X}_{i,\gamma,\varepsilon},$$

$$\tilde{Y}^+_{i,\gamma_0,\varepsilon} \cap \Big(\bigcup_{i'=i+1}^{k} \tilde{Y}^+_{i',\gamma_0,\varepsilon} \Big) = \tilde{Y}^+_{i,\gamma_0,\varepsilon} \cap \tilde{\rho}_i^{Y-1}(\varepsilon_i), \quad \text{and} \quad \partial \tilde{Y}_{i,\gamma_0,\varepsilon} = \bigcup_{\gamma \in \Gamma(i)} \tilde{Y}_{i,\gamma,\varepsilon},$$

where $\Gamma(i)$ denotes the subset of Γ consisting of γ such that $\gamma(j) = $ "$=$" for some $1 \le j < i$. On the other hand, by the second hypothesis of (III.2.1) each connected component of $\tilde{X}_{i,\gamma_0,\varepsilon}$ and of $\tilde{Y}_{i,\gamma_0,\varepsilon}$ is a PL ball. Hence by the second statement of III.2.2 and by the Alexander trick, for extension of τ_{i+1} to $\tilde{X}^+_{i,\gamma_0,\varepsilon}$, it suffices to extend $\tau_{i+1}|_{\bigcup_{\gamma \in \Gamma(i)} \tilde{X}^+_{i,\gamma,\varepsilon} \cap \tilde{\rho}_i^{X-1}(\varepsilon_i)}$ to $\tau_i|_{\bigcup_{\gamma \in \Gamma(i)} \tilde{X}^+_{i,\gamma,\varepsilon}}$ so that for each integer $j = 1, \dots, i-1$ and for each connected component C of $\tilde{X}^+_{i,\gamma_0,\varepsilon} \cap \tilde{\rho}_j^{X-1}(\varepsilon_j)$, we have

$$\tilde{\pi}_j^Y \circ \tau_i(C) = \tilde{f}_j \circ \tilde{\pi}_j^X(C).$$

Note that $C \subset \bigcup_{\gamma \in \Gamma(i)} \tilde{X}^+_{i,\gamma,\varepsilon}$.

For that we use another induction. We have sequences of compact polyhedra

$$\varnothing = \bigcup_{\gamma \in \Gamma(1)} \tilde{X}^+_{i,\gamma,\varepsilon} \subset \cdots \subset \bigcup_{\gamma \in \Gamma(i)} \tilde{X}^+_{i,\gamma,\varepsilon} \quad \text{and}$$

$$\varnothing = \bigcup_{\gamma \in \Gamma(1)} \tilde{Y}^+_{i,\gamma,\varepsilon} \subset \cdots \subset \bigcup_{\gamma \in \Gamma(i)} \tilde{Y}^+_{i,\gamma,\varepsilon}.$$

Let $1 < j \le i$ be an integer. Assume there exists a PL homeomorphism:

$$\tau_{i,j-1}: \bigcup_{i'=i+1}^{k} \tilde{X}^+_{i',\gamma_0,\varepsilon} \cup \bigcup_{\gamma \in \Gamma(j-1)} \tilde{X}^+_{i,\gamma,\varepsilon} \to \bigcup_{i'=i+1}^{k} \tilde{Y}^+_{i',\gamma_0,\varepsilon} \cup \bigcup_{\gamma \in \Gamma(j-1)} \tilde{Y}^+_{i,\gamma,\varepsilon}$$

which is an extension of τ_{i+1} such that for each integer $j' = 1, \dots, j-2$, for each connected component C of $\tilde{X}^+_{i,\gamma_0,\varepsilon} \cap \tilde{\rho}_{j'}^{X-1}(\varepsilon_{j'})$ and for each $\gamma \in \Gamma(j-1)$, we have

$$\tau_{i,j-1}(\tilde{X}_{i,\gamma,\varepsilon}) = \tilde{Y}_{i,\gamma,\varepsilon}, \quad \tau_{i,j-1}(\tilde{X}^+_{i,\gamma,\varepsilon}) = \tilde{Y}^+_{i,\gamma,\varepsilon}$$
$$\text{and} \quad \tilde{\pi}_{j'}^Y \circ \tau_{i,j-1}(C) = \tilde{f}_{j'} \circ \tilde{\pi}_{j'}^X(C).$$

We need only construct $\tau_{i,j}$ with the properties corresponding to these. For this we use the same arguments as above. Let γ_j be an element of $\Gamma(j)$ such that $\gamma_j(j') = "\geq"$ for $j' = 1, \ldots, j-2$ and $\gamma_j(j-1) = "="$. We have

$$\bigcup_{\gamma \in \Gamma(j)} \tilde{X}_{i,\gamma,\varepsilon} = \tilde{X}_{i,\gamma_j,\varepsilon} \cup \bigcup_{\gamma \in \Gamma(j-1)} \tilde{X}_{i,\gamma,\varepsilon}, \qquad \tilde{X}_{i,\gamma_j,\varepsilon} \cap \big(\bigcup_{\gamma \in \Gamma(j-1)} \tilde{X}_{i,\gamma,\varepsilon} \big) = \partial \tilde{X}_{i,\gamma_j,\varepsilon},$$

$$\bigcup_{\gamma \in \Gamma(j)} \tilde{Y}_{i,\gamma,\varepsilon} = \tilde{Y}_{i,\gamma_j,\varepsilon} \cup \bigcup_{\gamma \in \Gamma(j-1)} \tilde{Y}_{i,\gamma,\varepsilon}, \text{ and } \quad \tilde{Y}_{i,\gamma_j,\varepsilon} \cap \big(\bigcup_{\gamma \in \Gamma(j-1)} \tilde{Y}_{i,\gamma,\varepsilon} \big) = \partial \tilde{Y}_{i,\gamma_j,\varepsilon}.$$

Hence by using the second hypothesis of (III.2.1), III.2.2 and the Alexander trick in the same way as above, we can define $\tau_{i,j}$ with the required properties, which completes the double induction step. $\qquad \square$

We want to prove III.1.1. This shall follow from the next assertion as III.1.2 follows from (III.2.1). From now on, for C^∞ (cell) triangulations we are interested in only the images of simplexes (cells). Hence we identify a triangulation with the images of the simplexes (cells).

(III.2.4) For some ε, assume there exist C^∞ cell triangulations C_i of $X_{i,\gamma_0,\varepsilon}$ and homeomorphisms $f_i \colon X_{i,\gamma_0,\varepsilon} \to Y_{i,\gamma_0,\varepsilon}$, $i = 1, \ldots, k$, such that for any integers $1 \leq i < i' \leq k$ and for any $\sigma \in C_i$, f_i carries $X_{i,\gamma,\varepsilon}$ to $Y_{i,\gamma,\varepsilon}$ for each $\gamma \in \Gamma$, $f_i|_\sigma$ is a C^∞ diffeomorphism onto the image, $C_{i'}$ and $\{f_{i'}(\sigma') \colon \sigma' \in C_{i'}\}$ are compatible with the families:

$$\{\pi_i^{X-1}(\sigma_i) \cap \rho_i^{X-1}(\varepsilon_i) \colon \sigma_i \in C_i\} \quad \text{and} \quad \{\pi_i^{Y-1}(f_i(\sigma_i)) \cap \rho_i^{Y-1}(\varepsilon_i) \colon \sigma_i \in C_i\}$$

respectively, and for any connected component C of $X_{i',\gamma_0,\varepsilon} \cap \rho_i^{X-1}(\varepsilon_i)$, we have

$$\pi_i^Y \circ f_{i'}(C) = f_i \circ \pi_i^X(C).$$

Then X and Y are PL homeomorphic.

Proof of (III.2.4). We can easily proceed as in the proof of (III.2.1). We omit the details. $\qquad \square$

Proof that (III.2.4) implies III.1.1. We can prove the $\mathfrak{X}$-case by replacing the method of integration in the following arguments with the method of lifting of a flow by II.6.8 as in the proof of II.6.1′, though it is not carried out. First we define C^∞ cell triangulations of $Z_{i,\gamma_0,\varepsilon}$, $i = 1, \ldots, k$, which induce the C_i. As in the proof of I.1.5 for each $j = 1, \ldots, k-1$, we easily find a controlled vector field $\xi^j = \{\xi_i^j\}_{i=j+1,\ldots,k}$ on $\{Z_i\}_{i=j+1,\ldots,k}$ for

$\{T_i^Z\}_{i=j+1,\ldots,k}$ such that

$$\xi_i^j(\rho_j^Z|_{Z_i})(z) = 1 \quad \text{and} \quad d(\pi_j^Z|_{Z_i})\xi_{i,z}^j = 0$$
$$\text{for} \quad z \in Z_i \cap U_j, \quad i = j+1,\ldots,k, \qquad (*)$$

where U_j is a neighborhood of Z_j in $\mathbf{R}^n \times \mathbf{R}^n$. Lessen δ so that for any sequence $\varepsilon = \{\varepsilon_i\}$ of positive numbers with $\varepsilon < \delta$ and for any integers $1 \leq j < i \leq i' \leq k$, I.3.20 holds, U_j includes $Z_{j,\gamma_0,\varepsilon}^+$, and ξ^j satisfies conditions $\mathrm{scv}(T_i, T_{i'})$ and $\mathrm{cv}(T_i, T_{i'})$ on $Z_{i'} \cap Z_{i,\gamma_0,\varepsilon}^+$. (From now on we always assume $\varepsilon < \delta$.)

Let $\varepsilon = \{\varepsilon_i\}$ and $\varepsilon' = \{\varepsilon_i'\}$ be sequences of positive numbers with $\varepsilon \leq \varepsilon' < \delta$. We fix ε' and shall lessen δ and ε later. Then the inequality $\varepsilon' < \delta$ shall not hold. By I.3.20 we have C^∞ triangulations K_i of $Z_{i,\gamma_0,\varepsilon'}$, $i = 1,\ldots,k$, such that for any integers $1 \leq i < i' \leq k$ and for a simplex $\sigma \in K_{i'}$ with $\sigma \subset \rho_i^{Z-1}(\varepsilon_i')$, $\pi_i^Z(\sigma)$ is a simplex of K_i. We construct special C^∞ cell triangulations $C_{i,\varepsilon}^Z$ of $Z_{i,\gamma_0,\varepsilon}$, $i = 1,\ldots,k$, such that each $C_{i,\varepsilon}^Z$ includes K_i, and for any integers $1 \leq i < i' \leq k$ and for any cell $\sigma \in C_{i',\varepsilon}^Z$ with $\sigma_1 = \sigma \cap \rho_i^{Z-1}(\varepsilon_i) \neq \varnothing$, we have

$$\sigma_1 \in C_{i',\varepsilon}^Z, \qquad \pi_i^Z(\sigma_1) \in C_{i,\varepsilon}^Z, \quad \text{and}$$
$$\pi_i^{Z-1}(\pi_i^Z(\sigma_1)) \supset \sigma \quad \text{in a neighborhood of } \sigma_1 \text{ in } \mathbf{R}^n \times \mathbf{R}^n \text{ if} \quad \varepsilon_i < \varepsilon_i'$$

as follows.

We proceed by a downward induction method as in the proof of I.1.13. Let $1 \leq i \leq k$ be an integer. Define a sequence of numbers $\varepsilon(i) = \{\varepsilon_j(i)\}_{j=1,\ldots,k-1}$ by

$$\varepsilon_j(i) = \begin{cases} \varepsilon_j', & j = 1,\ldots,i-1 \\ \varepsilon_j, & j = i,\ldots,k-1. \end{cases}$$

Note that $\varepsilon = \varepsilon(1) \leq \cdots \leq \varepsilon(k) = \varepsilon'$. By assuming $\varepsilon_i < \varepsilon_i'$ and $C_{i',\varepsilon(i+1)}^Z$, $i' = 1,\ldots,k$, we will construct $C_{i',\varepsilon(i)}^Z$. We have

$$Z_{i',\gamma_0,\varepsilon(i)} = Z_{i',\gamma_0,\varepsilon(i+1)} \qquad \text{if } i' \leq i, \text{ and}$$
$$Z_{i',\gamma_0,\varepsilon(i)} - Z_{i',\gamma_0,\varepsilon(i+1)} = \{z \in Z_{i'} : \rho_j^Z(z) \geq \varepsilon_j' \text{ for } j = 1,\ldots,i-1,$$
$$\varepsilon_i \leq \rho_i^Z(z) < \varepsilon_i', \ \rho_j^Z(z) \geq \varepsilon_j \text{ for } j = i+1,\ldots,i'-1\} \qquad \text{if } i' > i.$$

For any i', set

$$C_{i',\varepsilon(i)}^Z|_{Z_{i',\gamma_0,\varepsilon(i+1)}} = C_{i',\varepsilon(i+1)}^Z.$$

It suffices to define $C_{i',\varepsilon(i)}^Z$ on $\overline{Z_{i',\gamma_0,\varepsilon(i)} - Z_{i',\gamma_0,\varepsilon(i+1)}}$ for all $i' = i+1,\dots,k$. By $(*)$ the flows of $\xi_{i'}^i$ induce C^∞ diffeomorphisms:

$$\tau_{i,i',\varepsilon}\colon \{\overline{(Z_{i',\gamma_0,\varepsilon(i)} - Z_{i',\gamma_0,\varepsilon(i+1)})} \cap \rho_i^{Z-1}(\varepsilon_i')\} \times [\varepsilon_i, \varepsilon_i']$$
$$\longrightarrow \overline{Z_{i',\gamma_0,\varepsilon(i)} - Z_{i',\gamma_0,\varepsilon(i+1)}}$$

such that $\rho_i^Z \circ \tau_{i,i',\varepsilon}$ are the projections on $[\varepsilon_i, \varepsilon_i']$, and for integers $i < i' < i'' \le k$, we have

$$\pi_{i'}^Z \circ \tau_{i,i'',\varepsilon}(z,t) = \tau_{i,i',\varepsilon}(\pi_{i'}^Z(z),t) \qquad\qquad (**)$$

for $z \in$ (a neighborhood of $\overline{(Z_{i'',\gamma_0,\varepsilon(i)} - Z_{i'',\gamma_0,\varepsilon(i+1)})} \cap \rho_i^{Z-1}(\varepsilon_i') \cap \rho_{i'}^{Z-1}(\varepsilon_{i'})$

in $\overline{(Z_{i'',\gamma_0,\varepsilon(i)} - Z_{i'',\gamma_0,\varepsilon(i+1)})} \cap \rho_i^{Z-1}(\varepsilon_i'))$

and $t \in [\varepsilon_i, \varepsilon_i']$.

Set

$$C_{i',\varepsilon(i)}^Z \Big|_{\overline{Z_{i',\gamma_0,\varepsilon(i)} - Z_{i',\gamma_0,\varepsilon(i+1)}}}$$
$$= \{\tau_{i,i',\varepsilon}(\sigma \times [\varepsilon_i, \varepsilon_i']),\ \tau_{i,i',\varepsilon}(\sigma \times \varepsilon_i),\ \tau_{i,i',\varepsilon}(\sigma \times \varepsilon_i')\colon$$
$$\sigma \in C_{i',\varepsilon(i+1)}^Z \Big|_{\overline{(Z_{i',\gamma_0,\varepsilon(i)} - Z_{i',\gamma_0,\varepsilon(i+1)})} \cap \rho_i^{Z-1}(\varepsilon_i')}\},\ i' = i+1,\dots,k.$$

Then by $(**)$, the $C_{i',\varepsilon(i)}^Z$ satisfy the required conditions. Thus we define $C_{i,\varepsilon}^Z,\ i = 1,\dots,k$. Recall that each $C_{i,\varepsilon}^Z$ is the family of the images of the cells of a C^∞ cell triangulation.

For proof, it suffices to find homeomorphisms $Z_{i,\gamma_0,\varepsilon} \to X_{i,\gamma_0,\varepsilon}$ and $Z_{i,\gamma_0,\varepsilon} \to Y_{i,\gamma_0,\varepsilon},\ i = 1,\dots,k$, which carry $Z_{i,\gamma,\varepsilon}$ to $X_{i,\gamma,\varepsilon}$ and $Y_{i,\gamma,\varepsilon}$ for each $\gamma \in \Gamma$ respectively, and whose images of the cells of $C_{i,\varepsilon}^Z$ satisfy the conditions in (III.2.4). But both $\{X_i\}$ and $\{Y_i\}$ present the same conditions. Hence it turns out that it suffices to construct homeomorphisms from $Z_{i,\gamma_0,\varepsilon}$ to $Y_{i,\gamma_0,\varepsilon}$. Replace $Y,\ \{Y_i\},\ \{T_i^Y = (|T_i^Y|, \pi_i^Y, \rho_i^Y)\}$ with $Z,\ \{Z_i\}$, $\{p_2^{-1}(|T_i^Y|), (p_2|_{Z_i})^{-1} \circ \pi_i^Y \circ (p_2|_{p_2^{-1}(|T_i^Y|)}), \rho_i^Y \circ (p_2|_{p_2^{-1}(|T_i^Y|)})\}$, where p_2 is the projection $\mathbf{R}^n \times \mathbf{R}^n \to \mathbf{R}^n$ onto the latter factor. The last system is not a tube system for $\{Z_i\}$. Nevertheless, define a family $\{Z_{i,\gamma,\varepsilon}'\}$ by $\{\rho_i^Y \circ (p_2|_{p_2^{-1}(|T_i^Y|)})\}$ in the same way as we defined $\{Z_{i,\gamma,\varepsilon}\}$ by $\{\rho_i^Z\}$. Then it suffices to consider (III.2.4) for $\{Z_{i,\gamma,\varepsilon}\}$ and $\{Z_{i,\gamma,\varepsilon}'\}$. Thus we reduce the problem to the following assertion. Here we change notation to adjust it to the proof of I.1.13.

Assertion. Let $\{X_i\}_{i=1,\dots,k}$ be a Whitney stratification of a compact set $X \subset \mathbf{R}^n$ with $\dim X_1 < \cdots < \dim X_k$. Let $\{T_i = (|T_i|, \pi_i, \rho_i)\}$ be a controlled tube system for $\{X_i\}$, let δ be a removal data of $\{X_i\}$ for $\{T_i\}$ with $\delta_i < 1$ for $i = 1, \ldots, k-1$, and let ε be a sequence of positive numbers ($\varepsilon < \delta$ as stated before). For each $i = 1, \ldots, k$ let $C_{i,\varepsilon}$ be a C^∞ cell triangulation of $X_{i,\gamma_0,\varepsilon}$ defined as above, let U_i be an open neighborhood of X_i in $\mathbf{R}^n$, let π_i' be a submersive C^∞ retraction of U_i to X_i, and let ρ_i' be a non-negative function on $\mathbf{R}^n$ which is of class C^∞ on U_i and equal to 1 outside U_i, and whose zero set in X equals X_i. Let $X_{i,\gamma,\varepsilon}'$ and $X_{i,\gamma,\varepsilon}'^+$, $i = 1, \ldots, k$, $\gamma \in \Gamma$, be defined by ρ_i', $i = 1, \ldots, k$, as $X_{i,\gamma,\varepsilon}$ and $X_{i,\gamma,\varepsilon}^+$. Assume the following conditions (1) and (2) for any ε. Let $1 \leq i < i' \leq k$ be any integers.

$$\pi_i' \circ \pi_{i'}' = \pi_i' \quad \text{and} \quad \rho_i' \circ \pi_{i'}' = \rho_i' \quad \text{on} \quad U_i \cap U_{i'}. \tag{1}$$

(2) For any C^∞ cell triangulation K_i of $X_{i,\gamma_0,\varepsilon}'$ there exists an arbitrarily strong cell approximation K_i' of K_i such that K_i' also is a C^∞ cell triangulation of $X_{i,\gamma_0,\varepsilon}'$ and if we lessen δ_i, then the restrictions of ρ_i and ρ_i' to $X_{i'} \cap \pi_i'^{-1}(\sigma^\circ) \cap \rho_i^{-1}([0, \varepsilon_i[)$ are friendly for any $\sigma \in K_i'$. (To be precise, K_i is the family of images of the cells of a C^∞ cell triangulation of $X_{i,\gamma_0,\varepsilon}'$, and K_i' is the family of the images of the cells of a cell approximation of the C^∞ cell triangulation.)

Shrink δ. Then for any ε there exist homeomorphisms $f_{i,\varepsilon} \colon X_{i,\gamma_0,\varepsilon} \to X_{i,\gamma_0,\varepsilon}'$, $i = 1, \ldots, k$, which carry $X_{i,\gamma,\varepsilon}$ to $X_{i,\gamma,\varepsilon}'$ for any $\gamma \in \Gamma$ such that for any integers $1 \leq i < i' \leq k$ and for any $\sigma \in C_{i',\varepsilon}$, $f_{i',\varepsilon}|_\sigma$ is a C^∞ diffeomorphism onto the image (we say that $f_{i',\varepsilon}$ is a $P\ C^\infty$ *diffeomorphism* with respect to $C_{i',\varepsilon}$), and

$$f_{i,\varepsilon} \circ \pi_i(\sigma) = \pi_i' \circ f_{i',\varepsilon}(\sigma) \quad \text{if} \quad \sigma \subset \rho_i^{-1}(\varepsilon_i).$$

Proof of Assertion.   In the proof of I.1.13, we already constructed $f_{i,\varepsilon}$ without the last property. We will modify the construction so that the last property holds. We proceed by induction. Let $0 \leq i < k$ be an integer. Set

$$X_{j,i,\gamma_0,\varepsilon} = \{x \in X_j : \rho_{j'}(x) \geq \varepsilon_{j'}, \ j' = 1, \ldots, i\}, \quad j = 1, \ldots, k,$$

and define $X_{j,i,\gamma,\varepsilon}$, $X_{j,i,\gamma,\varepsilon}^+$, $X_{j,i,\gamma,\varepsilon}'$ and $X_{j,i,\gamma,\varepsilon}'^+$ in the same way for $\gamma \in \Gamma$. Note that $X_{j,i,\gamma,\varepsilon} = X_{j,\gamma,\varepsilon}$ if $j \leq i+1$. We fix $\varepsilon_1, \ldots, \varepsilon_i$ and vary $\varepsilon_{i+1}, \ldots, \varepsilon_{k-1}$. Assume $i < k-1$ and there exist homeomorphisms $f_{j,i,\varepsilon} \colon X_{j,i,\gamma_0,\varepsilon} \to X_{j,i,\gamma_0,\varepsilon}'$, $j = 1, \ldots, k$, and a controlled tube system $\{T_{j,i} = (|T_{j,i}|, \pi_{j,i}, \rho_{j,i})\}_{j=i+1,\ldots,k}$ for $\{X_{j,i,\gamma_0,\varepsilon}'\}_{j=i+1,\ldots,k}$ with the following conditions. The $f_{j,i,\varepsilon}$ and $T_{j,i}$ depend on $\varepsilon_1, \ldots, \varepsilon_i$ but not on $\varepsilon_{i+1}, \ldots, \varepsilon_{k-1}$. The restrictions $f_{j,i,\varepsilon}|_{X_{j,\gamma_0,\varepsilon}}$ are $P\ C^\infty$ diffeomorphisms with respect to $C_{j,\varepsilon}$. For any $1 \leq j < j' \leq j'' \leq k$, $\gamma \in \Gamma$, $\varepsilon_{i+1}, \ldots, \varepsilon_{k-1}$, $\sigma' \in C_{j',\varepsilon}|_{X_{j',\gamma_0,\varepsilon} \cap \rho_j^{-1}(\varepsilon_j)}$, and $\sigma'' \in C_{j'',\varepsilon}|_{X_{j'',\gamma_0,\varepsilon} \cap \rho_j^{-1}(\varepsilon_j) \cap \rho_{j'}^{-1}(\varepsilon_{j'})}$, we have

$$f_{1,i,\varepsilon} = \mathrm{id}, \tag{3}$$

$$f_{j',i,\varepsilon} \text{ carries } X_{j',i,\gamma,\varepsilon} \text{ to } X'_{j',i,\gamma,\varepsilon}, \tag{4}$$

$$f_{j,i,\varepsilon} \circ \pi_j(\sigma') = \pi'_j \circ f_{j',i,\varepsilon}(\sigma') \quad \text{if} \quad j \le i, \tag{5}$$

$$f_{j',i,\varepsilon}(X_{j',i,\gamma_0,\varepsilon} \cap |T_j|) \supset X'_{j',i,\gamma_0,\varepsilon} \cap |T_{j,i}|$$

$$\text{and} \quad f_{j,i,\varepsilon} \circ \pi_j = \pi_{j,i} \circ f_{j',i,\varepsilon} \quad \text{on} \quad f_{j',i,\varepsilon}^{-1}(X'_{j',i,\gamma_0,\varepsilon} \cap |T_{j,i}|) \quad \text{if} \quad i < j, \tag{6}$$

$$|T_{j,i}| \subset U_j \cap |T_j| \quad \text{and} \quad \rho_{j,i} = \rho_j \quad \text{on} \quad |T_{j,i}| \quad \text{if} \quad i < j, \tag{7}$$

$$\rho_j \circ f_{j',i,\varepsilon} = \rho_j \quad \text{on} \quad X_{j',i,\gamma_0,\varepsilon} \cap X_{j,\gamma_0,\varepsilon}^+ \quad \text{if} \quad i < j, \quad \text{and} \tag{8}$$

$$\pi'_j \circ \pi_{j',i} \circ f_{j'',i,\varepsilon}(\sigma'') = \pi'_j \circ f_{j'',i,\varepsilon}(\sigma'') \quad \text{if} \quad j \le i < j'. \tag{9}$$

Here the manifolds $X'_{j,i,\gamma_0,\varepsilon}$ may have boundary and corners. Until now we did not treat a tube at a manifold with boundary and corners. We need certain additional conditions for its definition as follows. Let M be a C^∞ submanifold with boundary and corners of $\mathbf{R}^n$. We call $T = (|T|, \pi, \rho)$ a *tube* at M if there exist a C^∞ submanifold M_1 of $\mathbf{R}^n$ and a tube $T_1 = (|T_1|, \pi_1, \rho_1)$ at M_1 such that

$$\dim M_1 = \dim M, \qquad M_1 \supset M, \qquad |T| = \pi_1^{-1}(M),$$

$$\pi = \pi_1|_M, \quad \text{and} \quad \rho = \rho_1|_M.$$

In the present case, we assume, moreover, that $|T_{j,i}|$ are neighborhoods of $X'_{j,i,\gamma_0,\varepsilon}$ in $\{x \in \mathbf{R}^n \colon \rho'_{j'}(x) \ge \varepsilon_{j'}, \ j' = 1, \dots, i\}$, which implies by (1) that

$$\pi_{j,i}\{x \in |T_{j,i}| \colon \rho'_{j'}(x)\gamma(j')\varepsilon_{j'}, \ j' = 1, \dots, i\} = X'_{j,i,\gamma,\varepsilon} \quad \text{for} \quad \gamma \in \Gamma, \quad \text{and}$$

$$\pi_{j,i}^{-1}(X'_{j,i,\gamma_0,\varepsilon} \cap \rho'^{-1}_{i'}(\varepsilon_{i'})) = |T_{j,i}| \cap \pi'^{-1}_j(X'_{j,i,\gamma_0,\varepsilon} \cap \rho'^{-1}_{i'}(\varepsilon_{i'})), \ i' = 1, \dots, i. \tag{10}$$

For $i = 0$, $f_{j,i,\varepsilon} = \text{id}$ and $T_{j,i} = T_j$, $j = 1, \dots, k$, satisfy the above conditions, and $f_{j,\varepsilon} = f_{j,k-1,\varepsilon}$, $j = 1, \dots, k$, fulfill the requirements in the assertion. Hence by induction it suffices to lessen $\delta_{i+1}, \dots, \delta_{k-1}$ and to construct $f_{j,i+1,\varepsilon}$ and $\{T_{j,i+1}\}$ for any ε_{i+1}. We set

$$C'_{j,i,\varepsilon} = \{f_{j,i,\varepsilon}(\sigma) \colon \sigma \in C_{j,\varepsilon}\}, \quad j = 1, \dots, k.$$

Note that condition (9) follows from (5) and (6) because

$$\begin{aligned}
\pi'_j \circ \pi_{j',i} \circ f_{j'',i,\varepsilon}(\sigma'') &= \pi'_j \circ f_{j',i,\varepsilon} \circ \pi_{j'}(\sigma'') && \text{by} \quad (6) \\
&= f_{j,i,\varepsilon} \circ \pi_j \circ \pi_{j'}(\sigma'') && \text{by} \quad (5) \\
&= f_{j,i,\varepsilon} \circ \pi_j(\sigma'') && \text{by} \quad \text{ct}(T_j, T_{j'}) \\
&= \pi'_j \circ f_{j'',i,\varepsilon}(\sigma'') && \text{by} \quad (5).
\end{aligned}$$

First we reduce the problem to the case where

$$\pi_{i+1,i} = \pi'_{i+1} \quad \text{on} \quad |T_{i+1,i}|. \tag{11}$$

For that we will find C^∞ diffeomorphisms τ_j of $X'_{j,i,\gamma_0,\varepsilon}$, $j = i+1,\ldots,k$, and a controlled tube system $\{T^1_{j,i} = (|T^1_{j,i}|, \pi^1_{j,i}, \rho^1_{j,i})\}_{j=i+1,\ldots,k}$ for $\{X'_{j,i,\gamma_0,\varepsilon}\}_{j=i+1,\ldots,k}$ which depend on only $\varepsilon_1,\ldots,\varepsilon_i$, such that for any $1 \le i' \le i < j \le j' \le k$ and for any $\sigma \in C'_{j',i,\varepsilon}|_{X'_{j',\gamma_0,\varepsilon} \cap \rho'^{-1}_{i'}(\varepsilon_{i'})}$, we have

$$\tau_{j'}(X'_{j',i,\gamma,\varepsilon}) = X'_{j',i,\gamma,\varepsilon} \quad \text{for} \quad \gamma \in \Gamma,$$

$$|T^1_{j,i}| \subset |T_{j,i}|, \qquad \tau^{-1}_{j'}(X'_{j',i,\gamma_0,\varepsilon} \cap |T^1_{j,i}|) \subset |T_{j,i}|,$$

$$\pi^1_{j,i} \circ \tau_{j'} = \tau_j \circ \pi_{j,i} \quad \text{on} \quad \tau^{-1}_{j'}(X'_{j',i,\gamma_0,\varepsilon} \cap |T^1_{j,i}|),$$

$$\pi^1_{i+1,i} = \pi'_{i+1} \quad \text{on} \quad |T^1_{i+1,i}|',$$

$$\rho^1_{j,i} = \rho_j \quad \text{on} \quad |T^1_{j,i}|,$$

$$\rho_j \circ \tau_{j'} = \rho_j \quad \text{on} \quad \tau^{-1}_{j'}(X'_{j',i,\gamma_0,\varepsilon} \cap |T^1_{j,i}|),$$

$$\tau_{i+1} = \text{id}, \qquad \tau_{j'} = \text{id} \quad \text{on} \quad X'_{j',i,\gamma_0,\varepsilon} - |T^1_{i+1,i}|, \quad \text{and}$$

$$\pi'_{i'}(\sigma) = \pi'_{i'} \circ \tau_{j'}(\sigma),$$

where $|T^1_{i+1,i}|'$ is some neighborhood of $X'_{i+1,i,\gamma_0,\varepsilon}$ in $\{x \in \mathbf{R}^n : \rho'_{i'}(x) \ge \varepsilon_{i'}, \ i' = 1,\ldots,i\}$ smaller than $|T^1_{i+1,i}|$. For such τ_j and $T^1_{j,i}$, replace $f_{j,i,\varepsilon}$ with $\tau_j \circ f_{j,i,\varepsilon}$ for $j = i+1,\ldots,k$ and $T_{j,i}$ with $T^1_{j,i}$ for $j = i+2,\ldots,k$ and with $T^1_{i+1,i}|_{|T^1_{i+1,i}|'}$ for $j = i+1$. Then we can assume (11) because conditions (3),$\ldots$,(9) continue to hold after the replacement.

We shall define τ_j by the flow of a vector field. For that we vary $\pi_{i+1,i}$ to π'_{i+1} continuously. Shrinking $|T_{i+1,i}|$ and lessening $\delta_{i+1},\ldots,\delta_{k-1}$, we want a C^∞ map:

$$|T_{i+1,i}| \times [0,1] \ni (x,t) \longrightarrow \pi^t_{i+1}(x) \in X'_{i+1,\gamma_0,\varepsilon}$$

such that for each $t \in [0,1]$, $\pi^t_{i+1} \colon |T_{i+1,i}| \to X'_{i+1,\gamma_0,\varepsilon}$ is a submersive C^∞ retraction,

$$\pi^0_{i+1} = \pi_{i+1,i} \quad \text{and} \quad \pi^1_{i+1} = \pi'_{i+1} \quad \text{on} \quad |T_{i+1,i}|, \tag{12}$$

$$\pi'_{i'} \circ \pi^t_{i+1}(\sigma) = \pi'_{i'}(\sigma), \ 1 \le i' \le i < j \le k, \tag{13}$$

for $\sigma \in C'_{j,i,\varepsilon}|_{X_{j,\gamma_0,\varepsilon} \cap \rho'^{-1}_{i'}(\varepsilon_{i'}) \cap \rho^{-1}_{i+1}(\varepsilon_{i+1})}$ and for any $\varepsilon_{i+1},\ldots,\varepsilon_{k-1}$, and

$$(\pi^t_{i+1})^{-1}(X'_{i+1,\gamma_0,\varepsilon} \cap \rho'^{-1}_{i'}(\varepsilon_{i'})) \subset \pi^{-1}_{i+1}(X'_{i+1,\gamma_0,\varepsilon} \cap \rho'^{-1}_{i'}(\varepsilon_{i'})), \ i' = 1,\ldots,i. \tag{14}$$

Let us construct π_{i+1}^t by a downward induction method as in the proof of I.1.3. Let $0 \le i' \le i$ be an integer. Assume there exists an open neighborhood $V_{i'+1}$ of $X'_{i+1,\gamma_0,\varepsilon} \cap (\cup_{i''=i'+1}^{i} \rho_{i''}'^{-1}(\varepsilon_{i''}))$ in $|T_{i+1,i}|$ and that π_{i+1}^t is defined on $V_{i'+1} \times [0,1]$ so that for each $t \in [0,1]$, π_{i+1}^t is a submersive C^∞ retraction of $V_{i'+1}$ to $X'_{i+1,\gamma_0,\varepsilon} \cap V_{i'+1}$. Furthermore, assume that condition (12) holds on $V_{i'+1}$,

$$\pi_{i''}' \circ \pi_{i+1}^t(\sigma \cap V_{i'+1}) \subset \pi_{i''}'(\sigma) \quad \text{for} \quad t \in [0,1], \;\; 1 \le i'' \le i < j \le k,$$

$$\sigma \in C'_{j,i,\varepsilon}|_{X'_{j,\gamma_0,\varepsilon} \cap \rho_{i''}'^{-1}(\varepsilon_{i''}) \cap \rho_{i+1}^{-1}(\varepsilon_{i+1})}, \quad (13.i'+1)$$

and

$$(\pi_{i+1}^t)^{-1}(X'_{i+1,\gamma_0,\varepsilon} \cap \rho_{i''}'^{-1}(\varepsilon_{i''}) \cap V_{i'+1}) \subset \pi_{i+1}'^{-1}(X'_{i+1,\gamma_0,\varepsilon} \cap \rho_{i''}'^{-1}(\varepsilon_{i''})),$$

$$i'' = 1, \ldots, i. \quad (14.i'+1)$$

It suffices to define $V_{i'}$ and π_{i+1}^t on $V_{i'} \times [0,1]$ with $(13.i')$ and $(14.i')$. (Here V_0 is a neighborhood of $X'_{i+1,\gamma_0,\varepsilon}$ in $|T_{i+1,i}|$.) This is because by shrinking $|T_{i+1,i}|$, we can suppose $V_0 = |T_{i+1,i}|$ and then (13.0) implies (13), which is easy to show.

Assume $i' = 0$. Set

$$V_0 = V_1 \cup (\text{a small neighborhood of } X'_{i+1,\gamma_0,\varepsilon} - V_1 \text{ in } \mathbf{R}^n) \quad \text{and}$$

$$\pi_{i+1}^t(x) = \begin{cases} \pi_{i+1}'\{\varphi_0(x)\pi_{i+1}^t(x) + (1 - \varphi_0(x))(t\pi_{i+1}'(x) + (1-t)\pi_{i+1,i}(x))\} \\ \qquad\qquad\qquad\qquad\qquad\qquad\qquad \text{for} \;\; x \in V_1, \;\; t \in [0,1] \\ \pi_{i+1}'(t\pi_{i+1}'(x) + (1-t)\pi_{i+1,i}(x)) \;\; \text{for} \;\; x \in V_0 - V_1, \;\; t \in [0,1], \end{cases}$$

where φ_0 is a C^∞ function on V_0 such that $0 \le \varphi_0 \le 1$, $\varphi_0 = 0$ on $V_0 - V_1$ and $\varphi_0 = 1$ outside a small neighborhood of $V_0 - V_1$ in V_0. Shrink V_0 and $|T_{i+1,i}|$. Clearly π_{i+1}^t fulfills the requirements.

Assume $i' = i$. Let V_i be a sufficiently small neighborhood of $X'_{i+1,\gamma_0,\varepsilon} \cap \rho_i'^{-1}(\varepsilon)$ in $|T_{i+1,i}|$, and set

$$\pi_{i+1}^t(x)$$
$$= p_{i,\psi_i\{\pi_i'(x),\pi_i' \circ \pi_{i+1,i}(x),t\}, t\rho_i' \circ \pi_{i+1,i}(x) + (1-t)\rho_i' \circ \pi_{i+1,i}(x)}\{t\pi_{i+1}'(x)$$
$$+ (1-t)\pi_{i+1,i}(x)\} \;\; \text{for} \;\; x \in V_i \;\; \text{and} \;\; t \in [0,1].$$

Here, for each $(y,s) \in X'_{i,\gamma_0,\varepsilon} \times [0, \varepsilon_i + \text{a small positive number}]$, $p_{i,y,s}$ denotes the orthogonal projection onto the C^∞ manifold $X_{i+1} \cap \pi_i'^{-1}(y) \cap \rho_i'^{-1}(s)$, and ψ_i is a C^∞ map from a neighborhood of $\Delta_i \times [0,1]$ ($\Delta_i = $ the diagonal of $X'_{i,\gamma_0,\varepsilon}$) in $X'_{i,\gamma_0,\varepsilon} \times X'_{i,\gamma_0,\varepsilon} \times [0,1]$ to $X'_{i,\gamma_0,\varepsilon}$ such that

$$\psi_i(y,y',0) = y', \qquad \psi_i(y,y',1) = y, \qquad \psi_i(y,y,t) = y,$$

and if y and y' are contained in some one $\sigma \in C'_{i,i,\varepsilon}$, then so is $\psi_i(y, y', t)$ for each $t \in [0, 1]$. Later we need to generalize the last property, which is possible by the following special construction of ψ_i. If π^t_{i+1} is well-defined, i.e., if we can construct ψ_i, then for each $t \in [0, 1]$, π^t_{i+1} is a submersive C^∞ map from V_i to $X'_{i+1,\gamma_0,\varepsilon}$, the equalities $\pi^t_{i+1}|_{X'_{i+1,\gamma_0,\varepsilon} \cap V_i} = \mathrm{id}$ and (12) on V_i are clear, (13.i) for $i'' = i$ follows from (5) and (9), and (14.i) for $i'' = i$ is easy to show because by (10), we have

$$\pi^{-1}_{i+1,i}(X'_{i+1,\gamma_0,\varepsilon} \cap \rho'^{-1}_i(\varepsilon_i)) = |T_{i+1,i}| \cap \pi'^{-1}_{i+1}(X'_{i+1,\gamma_0,\varepsilon} \cap \rho'^{-1}_i(\varepsilon_i))$$

$$= |T_{i+1,i}| \cap \rho'^{-1}_i(\varepsilon_i).$$

Later we will see conditions (13.i) for $i'' = 1, \dots, i-1$ and (14.i) for $i'' = 1, \dots, i-1$.

We construct ψ_i as follows. For each point of $X'_{i,\gamma_0,\varepsilon}$ there exist its open neighborhood U in $X'_{i,\gamma_0,\varepsilon}$ and a C^∞ diffeomorphism α from U to a convex set U' in $\mathbf{R}^n$ such that for each $\sigma \in C'_{i,i,\varepsilon}$, $\alpha(U \cap \sigma^\circ)$ is an open simplex in $\mathbf{R}^n$. We can define ψ_i on $U \times U \times [0, 1]$ by

$$\psi_i(y, y', t) = \alpha^{-1}(t\alpha(y) + (1-t)\alpha(y')) \quad \text{for} \quad (y, y', t) \in U \times U \times [0, 1].$$

Here we require α to be a C^∞ diffeomorphism. A P C^∞ diffeomorphism is not sufficient for the forthcoming generalization of a property of ψ_i. Since $X'_{i,\gamma_0,\varepsilon}$ is compact, we have a finite open covering of $X'_{i,\gamma_0,\varepsilon}$ by such open sets. Hence by an induction method we reduce the problem of construction of ψ_i to the following statement.

Let U, U' and α be as above. Assume for an open set W of $X'_{i,\gamma_0,\varepsilon}$ and for an open neighborhood $\tilde{W}$ of the diagonal of W in $W \times W$, we have constructed ψ_i on $\tilde{W} \times [0, 1]$. By shrinking $\tilde{W}$, we can define ψ_i on $(\tilde{W} \cup U \times U) \times [0, 1]$. Here $\tilde{W}$ continues to be a neighborhood of the diagonal of W.

We prove this statement. Let β be a C^∞ function on $\tilde{W} \cup U \times U$ such that $0 \le \beta \le 1$, $\beta = 0$ on $U \times U - \tilde{W}$ and $\beta = 1$ on $\tilde{W} - U \times U$. Let $\psi_{i,U}$ and $\psi_{i,W}$ denote ψ_i defined on $U \times U$ and on $\tilde{W}$ respectively. Set

$$\psi_i(y, y', t) = \begin{cases} \psi_{i,W}(y, y', t) & \text{on } (\tilde{W} - U \times U) \times [0, 1] \\ \alpha^{-1}\{\beta(y, y')\alpha \circ \psi_{i,W}(y, y', t) \\ \quad + (1 - \beta(y, y'))\alpha \circ \psi_{i,U}(y, y', t)\} & \text{on } (\tilde{W} \cap U \times U) \times [0, 1] \\ \psi_{i,U}(y, y', t) & \text{on } (U \times U - \tilde{W}) \times [0, 1], \end{cases}$$

and shrink $\tilde{W}$ so that

$$\beta(y, y')\alpha \circ \psi_{i,W}(y, y', t) + (1 - \beta(y, y'))\alpha \circ \psi_{i,U}(y, y', t) \in U'$$

$$\text{for} \quad (y, y') \in \tilde{W} \cap U \times U, \ t \in [0, 1].$$

This ψ_i is well-defined and fulfills the requirements.

The π_{i+1}^t defined as above satisfies conditions (13.i) for $i'' = 1, \ldots, i-1$ and (14.i) for $i'' = 1, \ldots, i-1$ for the following reason. Let us consider (13.i). Let $\sigma \in C_{j,i,\varepsilon}''$ with $\sigma \subset \rho_{i''}'^{-1}(\varepsilon_i'') \cap \rho_{i+1}^{-1}(\varepsilon_{i+1})$, $\sigma \cap V_i \neq \varnothing$ and $1 \le i'' < i < j \le k$. By (9) we have

$$\pi_{i''}' \circ \pi_{i+1,i}(\sigma) = \pi_{i''}'(\sigma),$$

and by (5) this set is an element of $C_{i'',i,\varepsilon}'$. Let σ'' denote it. On the other hand, by (1),

$$\pi_{i''}' \circ \pi_{i+1}^t(\sigma \cap V_i) = \pi_{i''}' \circ \pi_i' \circ \pi_{i+1}^t(\sigma \cap V_i).$$

Hence it suffices to prove the inclusion:

$$\pi_i' \circ \pi_{i+1}^t(\sigma \cap V_i) \subset X_{i,\gamma_0,\varepsilon}' \cap \pi_{i''}'^{-1}(\sigma'').$$

We will show, moreover, that the left set is included in $X_{i,\gamma_0,\varepsilon}' \cap \pi_{i''}'^{-1}(\sigma'') \cap \rho_{i''}'^{-1}(\varepsilon_{i''})$. By (1),

$$\pi_{i''}' \circ \pi_i' \circ \pi_{i+1,i}(\sigma \cap V_i) = \pi_{i''}' \circ \pi_{i+1,i}(\sigma \cap V_i) \subset \pi_{i''}' \circ \pi_{i+1,i}(\sigma) = \sigma'',$$

and by (1) and (10),

$$\rho_{i''}' \circ \pi_i' \circ \pi_{i+1,i}(\sigma \cap V_i) = \rho_{i''}' \circ \pi_{i+1,i}(\sigma \cap V_i) = \varepsilon_{i''}.$$

Hence

$$\pi_i' \circ \pi_{i+1,i}(\sigma \cap V_i) \subset X_{i,\gamma_0,\varepsilon}' \cap \pi_{i''}'^{-1}(\sigma'') \cap \rho_{i''}'^{-1}(\varepsilon_{i''}).$$

Clearly

$$\pi_i' \circ \pi_{i+1}'(\sigma \cap V_i) = \pi_i'(\sigma \cap V_i) \subset X_{i,\gamma_0,\varepsilon}' \cap \pi_{i''}'^{-1}(\sigma'') \cap \rho_{i''}'^{-1}(\varepsilon_{i''}).$$

Therefore, by the definition of π_{i+1}^t, we need only prove that for each $\sigma_1'' \in C_{i'',i,\varepsilon}'$, if y and y' are points of $X_{i,\gamma_0,\varepsilon}' \cap \pi_{i''}'^{-1}(\sigma_1'') \cap \rho_{i''}'^{-1}(\varepsilon_{i''})$, then so is $\psi_i(y, y', t)$ for any $t \in [0,1]$. This follows from the definition of ψ_i. Indeed, since $X_{i,\gamma_0,\varepsilon}' \cap \pi_{i''}'^{-1}(\sigma_1'') \cap \rho_{i''}'^{-1}(\varepsilon_{i''})$ is both a C^∞ submanifold of $X_{i,\gamma_0,\varepsilon}'$ and some union of $\sigma' \in C_{i,i,\varepsilon}'$, it is locally carried to a convex set by the C^∞ diffeomorphism α in the construction of ψ_i.

(14.i) for $i'' = 1, \ldots, i-1$ also follow from the property (10) in the same way. Thus the case $i' = i$ is complete.

Assume $0 < i' < i$. We proceed to combine the above proofs in the cases of $i' = 0$ and of $i' = i$. Write π_{i+1}^t defined on $V_{i'+1} \times [0,1]$ in the induction hypothesis as $\pi_{i+1,i'+1}^t$. As in the case $i' = i$, let $V_{i'}'$ be a small neighborhood of $X_{i+1,\gamma_0,\varepsilon}' \cap \rho_{i'}'^{-1}(\varepsilon_{i'})$ in $|T_{i+1,i}|$ and set

$$\pi_{i+1,i'}^t(x)$$
$$= p_{i',\psi_{i'}\{\pi_{i'}'(x),\pi_{i'}'\circ\pi_{i+1,i}(x),t\},t\rho_{i'}'\circ\pi_{i+1}'(x)+(1-t)\rho_{i'}'\circ\pi_{i+1,i}(x)}\{t\pi_{i+1}'(x)$$
$$+ (1-t)\pi_{i+1,i}(x)\}$$

$$\text{for } x \in V_{i'}' \text{ and } t \in [0,1],$$

where for each $(y,s) \in X_{i',\lambda_0,\varepsilon}' \times [0, \varepsilon_{i'}+ \text{ a small positive number}]$, $p_{i',y,s}$ is the orthogonal projection onto the C^∞ manifold $X_{i+1} \cap \pi_{i'}'^{-1}(y) \cap \rho_{i'}'^{-1}(y)$, and $\psi_{i'}$ was defined in the case $i' = i$. We see the following as in the case $i' = i$. For each $t \in [0,1]$ $\pi_{i+1,i'}^t$ is a submersive C^∞ map from $V_{i'}'$ to $X_{i+1} \cap \bigcap_{i''=1}^{i'} \rho_{i''}'^{-1}([\varepsilon_{i''},\infty[)$ such that $\pi_{i+1,i'}^t = \text{id}$ on $X_{i+1,\gamma_0,\varepsilon}' \cap V_{i'}'$. The properties (13.$i'$) and (14.$i'$) hold for $1 \le i'' \le i'$. Here we cannot assert that $\pi_{i+1,i'}^t$ is a map to $X_{i+1,\gamma_0,\varepsilon}'$, that (13.$i'$) and (14.$i'$) hold for $i' < i'' \le i$ nor that $\pi_{i+1,i'}^t = \pi_{i+1,i'+1}^t$ on $V_{i'}' \cap V_{i'+1}'$. So we modify $\pi_{i+1,i'}^t$ and $\pi_{i+1,i'+1}^t$ on $V_{i'}' \cap V_{i'+1}'$ as follows.

Set $V_{i'} = V_{i'+1}' \cup V_{i'}'$ and let $\varphi_{i'}$ be a C^∞ function on $V_{i'}$ such that $0 \le \varphi_{i'} \le 1$, $\varphi_{i'} = 0$ on $V_{i'} - V_{i'+1}'$ and $\varphi_{i'} = 1$ outside a small neighborhood of $V_{i'} - V_{i'+1}'$ in $V_{i'}$. We define π_{i+1}^t on $V_{i'}$ by

$$\pi_{i+1}^t(x) = \begin{cases} \pi_{i+1,i'+1}^t(x) & \text{if } \varphi_{i'}(x) = 1 \\ p_{i',\psi_{i'}\{\pi_{i'}'\circ\pi_{i+1,i'+1}^t(x),\pi_{i'}'\circ\pi_{i+1,i'}^t(x),\varphi_{i'}(x)\},\varphi_{i'}(x)\rho_{i'}'\circ\pi_{i+1,i'+1}^t(x)} \\ \quad + (1+\varphi_{i'}(x))\rho_{i'}'\circ\pi_{i+1,i'}^t(x)}\{\varphi_{i'}(x)\pi_{i+1,i'+1}^t(x) + (1-\varphi_{i'}(x)\pi_{i+1,i'}^t(x)\} \\ \qquad\qquad\qquad\qquad\qquad\qquad\qquad\qquad\qquad \text{if } 0 < \varphi_i(x) < 1 \\ \pi_{i+1,i'}^t(x) & \text{if } \varphi_{i'}(x) = 0. \end{cases}$$

Shrink $V_{i'}$. By the same reason as in the cases of $i' = 0$ and of $i' = i$, $V_{i'}$ and τ_{i+1}^t fulfill the requirements. We omit the details. Thus we complete the induction step of the construction of τ_{i+1}^t. Note that τ_{i+1}^t does not depend on $\varepsilon_{i+1}, \ldots, \varepsilon_{k-1}$.

To construct τ_j, we shall use a vector field on $\{\tilde{X}_{j,i,\gamma_0,\varepsilon}' = X_{j,i,\gamma_0,\varepsilon}' \times [0,1]\}_{j=i+1,\ldots,k}$ and its flow. Define a tube $\tilde{T}_{i+1,i} = (|\tilde{T}_{i+1,i}|, \tilde{\pi}_{i+1,i}, \tilde{\rho}_{i+1,i})$ at $\tilde{X}_{i+1,i,\gamma_0,\varepsilon}'$ by

$$|\tilde{T}_{i+1,i}| = |T_{i+1,i}|\times[0,1] \text{ and } \tilde{\pi}_{i+1,i}(x,t) = (\pi_{i+1}^t(x),t), \ \tilde{\rho}_{i+1,i}(x,t) = \rho_{i+1}(x).$$

There exists a controlled tube system $\{\tilde{T}_{j,i} = (|\tilde{T}_{j,i}|, \tilde{\pi}_{j,i}, \tilde{\rho}_{j,i})\}_{j=i+2,\ldots,k}$ for $\{\tilde{X}_{j,i,\gamma_0,\varepsilon}'\}_{j=i+2,\ldots,k}$ such that $\{\tilde{T}_{j,i}\}_{j=i+1,\ldots,k}$ is controlled and $\tilde{\rho}_{j,i} = \rho_j$ for $j = i+2, \ldots, k$. This is because we can extend I.1.3 to the case of manifolds

with boundary and corners. (Since the proof proceeds as in the proof of
I.1.3, we omit it.) Here we shrink $|T_{i+1,i}|$ so that conditions $\mathrm{ct}(\tilde{T}_{i+1,i}, \tilde{T}_{j,i})$,
$j = i+2, \ldots, k$, hold. Moreover, we can assume $\tilde{\pi}_{j,i}(x, 0) = (\pi_{j,i}(x), 0)$,
$j = i+2, \ldots, k$, for the following reason. First define $\{\tilde{T}_{j,i}\}_{j=i+2,\ldots,k}$ by

$$|\tilde{T}_{j,i}| = |T_{j,i}| \times [0, 1] \quad \text{and}$$
$$\tilde{\pi}_{j,i}(x, t) = (\pi_{j,i}(x), t) \quad \text{and} \quad \tilde{\rho}_{j,i}(x, t) = \rho_j(x) \quad \text{for} \quad (x, t) \in |\tilde{T}_{j,i}|.$$

Then $\{\tilde{T}_{j,i}\}_{j=i+1,\ldots,k}$ is not necessarily controlled. So we need to mod-
ify it. Recall the proof of I.1.3. That shows the following. Let $\{T_i = (|T_i|, \pi_i, \rho_i)\}_{i=1,\ldots,k}$ be a tube system for a Whitney stratification $\{X_i\}_{i=1,\ldots,k}$
of a set $X \subset \mathbf{R}^n$ with $\dim X_1 < \cdots < \dim X_k$. Let X_0 be a subset of X such
that

$$\pi_i(|T_i| \cap X_0) \subset X_0, \quad \pi_i^{-1}(X_0) \subset X_0, \quad 1 \leq i \leq k,$$
$$\pi_i \circ \pi_j(x) = \pi_i(x), \quad \rho_i \circ \pi_j(x) = \rho_i(x) \quad \text{for} \quad x \in |T_i| \cap |T_j| \cap X_0, \quad 1 \leq i < j \leq k.$$

Then there exists a controlled tube system $\{T_i' = (|T_i'|, \pi_i', \rho_i')\}_{i=1,\ldots,k}$ for
$\{X_i\}_{i=1,\ldots,k}$ such that for each i,

$$|T_i'| \subset |T_i|, \qquad \rho_i' = \rho_i \quad \text{on} \quad |T_i'|,$$
$$\pi_1' = \pi_1 \quad \text{on} \quad |T_1'|, \quad \text{and} \quad \pi_i' = \pi_i \quad \text{on} \quad |T_i'| \cap X_0.$$

Hence in the present case, we can modify $\{\tilde{T}_{j,i}\}_{j=i+2,\ldots,k}$ without changing
it on $\{|T_{j,i}|\}_{j=i+2,\ldots,k}$. Note that each $\tilde{\pi}_{j,i}(\cdot, t)$ is of the form $(\pi_j^t(\cdot), t)$ on
$|\tilde{T}_{i+1,i}| \cap |\tilde{T}_{j,i}|$, which follows from condition $\mathrm{ct}(\tilde{T}_{i+1,i}, \tilde{T}_{j,i})$ and the fact that
$\tilde{\pi}_{i+1,i}$ is of this form.

   Now I.1.5 and I.1.6 also are valid in the case of manifolds with boundary and corners. Hence we have a controlled vector field $\xi = \{\xi_j\}_{j=i+1\ldots,k}$
on $\{\tilde{X}_{j,i,\gamma_0,\varepsilon}'\}_{j=i+1,\ldots,k}$ for $\{\tilde{T}_{j,i}\}_{j=i+1,\ldots,k}$ with $\xi_{i+1} = \partial/\partial t$ and its flow
$\{\theta_j : D_j \to \tilde{X}_{j,i,\gamma_0,\varepsilon}'\}_{j=i+1,\ldots,k}$, $D_j \subset \tilde{X}_{j,i,\gamma_0,\varepsilon}' \times \mathbf{R}$. Here we can choose ξ so
that for each $\gamma \in \Gamma$ and for each $j = i+2, \ldots, k$, $\xi_j|_{(X_{j,i,\gamma,\varepsilon}')^\circ \times [0,1]}$ is a vector
field of $(X_{j,i,\gamma,\varepsilon}')^\circ \times [0, 1]$. We prove this as in the proof of I.1.10. In I.1.10, we
cannot expect a lift of a vector field to be controlled. In the present case, how-
ever, a lift ξ of ξ_{i+1} can be controlled by (14). As this is easy to show, we omit
the proof. Let c be a positive number so small that conditions $\mathrm{scv}(\tilde{T}_{i+1,i}, \tilde{T}_{j,i})$
and $\mathrm{cv}(\tilde{T}_{i+1,i}, \tilde{T}_{j,i})$, $j = i+2, \ldots, k$, hold on $\{(x, t) \in \tilde{X}_{j,i,\gamma_0,\varepsilon}' : \rho_{i+1}(x) \leq c\}$.
By I.1.6 and its proof, for each $j = i+2, \ldots, k$ and for each $t \in [0, 1]$,

$$(X_{j,i,\gamma_0,\varepsilon}' \cap \rho_{i+1}^{-1}([0, c])) \times 0 \times [0, 1] \subset D_j,$$

$\theta_j(\cdot, 0, t)$ is of the form $(\theta_j^t(\cdot), t)$, and the map θ_j^t is a C^∞ diffeomorphism of $X'_{j,i,\gamma_0,\varepsilon} \cap \rho_{i+1}^{-1}([0,c])$ because for each $\gamma \in \Gamma$, $\xi_j|_{(X'_{j,i,\gamma,\varepsilon})^\circ \times [0,1]}$ is a vector field of $(X'_{j,i,\gamma,\varepsilon})^\circ \times [0,1]$. Moreover, by $\mathrm{scv}(\tilde{T}_{i+1,i}, \tilde{T}_{j,i})$, for each $0 < c' < c$, θ_j^t carries $X'_{j,i,\gamma_0,\varepsilon} \cap \rho_{i+1}^{-1}(c')$ onto itself.

Let φ be a C^∞ function on $\mathbf{R}$ such that $0 \le \varphi \le 1$, $\varphi = 1$ on $[0, c/3]$ and $\varphi = 0$ on $[c/2, \infty[$. Define the $|T_{i+1,i}^1|$ by

$$|T_{i+1,i}^1| = \{x \in |T_{i+1,i}| : \rho_{i+1}(x) < c\},$$

and lessen c so that if $\sigma \in C'_{j,i,\varepsilon}$ intersects with $|T_{i+1,i}^1|$, then σ does with $\rho_{i+1}^{-1}(\varepsilon_{i+1})$, which is possible because by the definition of $C_{j,\varepsilon}$, for each $\sigma \in C_{j,\varepsilon}$, if $\rho_{i+1}(\sigma) \cap (]\varepsilon_{i+1}, \varepsilon'_{i+1}[) \ne \varnothing$, then $\rho_{i+1}(\sigma) = [\varepsilon_{i+1}, \varepsilon'_{i+1}]$. For each $i+1 \le j \le k$, define τ_j and $T_{j,i}^1$ by

$$\tau_j(x) = \begin{cases} x & \text{on} \quad X'_{j,i,\gamma_0,\varepsilon} \cap \rho_{i+1}^{-1}(]c,\infty[) \\ \theta_j^{\varphi \circ \rho_{i+1}(x)}(x) & \text{on} \quad X'_{j,i,\gamma_0,\varepsilon} \cap \rho_{i+1}^{-1}([0,c]), \end{cases}$$

$$\pi_{j,i}^1(x) = \begin{cases} \pi_{j,i}(x) & \text{on} \quad |T_{j,i}^1| \cap \rho_{i+1}^{-1}(]c,\infty[) \\ \pi_j^{\varphi \circ \rho_{i+1}(x)}(x) & \text{on} \quad |T_{j,i}^1| \cap \rho_{i+1}^{-1}([0,c]), \quad \text{and} \end{cases}$$

$$\rho_{j,i}^1 = \rho_j|_{|T_{j,i}^1|}.$$

Here we choose $|T_{j,i}^1|$ so small that

$$|T_{j,i}^1| \subset |T_{j,i}| \quad \text{and} \quad \tau_{j'}^{-1}(X'_{j',i,\gamma_0,\varepsilon} \cap |T_{j,i}^1|) \times 0 \subset |\tilde{T}_{j,i}|, \quad i < j \le j' \le k.$$

From now on we assume $\delta_{i+1} < c/3$. Then τ_j and $T_{j,i}^1$ satisfy the required conditions, which is clear except for the condition:

$$\pi_{i'}'(\sigma) = \pi_{i'}' \circ \tau_{j'}(\sigma) \quad \text{for} \quad \sigma \in C'_{j',i,\varepsilon}|_{X'_{j',\gamma_0,\varepsilon} \cap \rho_{i'}^{-1}(\varepsilon_{i'})}, \quad 1 \le i' \le i < j' \le k.$$

Let us prove this equality. If $|T_{i+1,i}^1| \cap \sigma = \varnothing$, then $\tau_{j'} = \mathrm{id}$ on σ and hence the equality holds. The case $\sigma \subset \rho_{i+1}^{-1}(\varepsilon_{i+1})$ also is immediate by (9) because

$$\pi_{i'}' \circ \tau_{j'} = \pi_{i'}' \circ \pi_{i+1}' \circ \tau_{j'} = \pi_{i'}' \circ \pi_{i+1,i} \quad \text{on} \quad \sigma.$$

Hence we assume

$$|T_{i+1,i}^1| \cap \sigma \ne \varnothing \quad \text{and} \quad \sigma \not\subset \rho_{i+1}^{-1}(\varepsilon_{i+1}).$$

By the above choice of c we have

$$\sigma \cap \rho_{i+1}^{-1}(\varepsilon_{i+1}) \ne \varnothing \quad \text{and} \quad \sigma \not\subset |T_{i+1,i}^1|.$$

We will reduce the problem to an easy case by stages. By (5) and the properties of $C_{j,\varepsilon}$, $j = 1, \dots, k$, we have

$$\sigma \cap \rho_{i+1}^{-1}(\varepsilon_{i+1}) \in C'_{j',i,\varepsilon} \text{ and } \pi'_{i'}(\sigma \cap \rho_{i+1}^{-1}(\varepsilon_{i+1})) \in C'_{i',i,\varepsilon} \text{ for any } 0 < \varepsilon_{i+1} \leq c.$$

Only here we forget the assumption $\delta_{i+1} < c/3$. Clearly $C'_{i',i,\varepsilon}$ does not depend on ε_{i+1}. Hence

$$\pi'_{i'}(\sigma \cap |T^1_{i+1,i}|) = \pi'_{i'}(\sigma \cap \rho_{i+1}^{-1}(c)),$$

which implies

$$\pi'_{i'}(\sigma) = \pi'_{i'}(\sigma - |T^1_{i+1,i}|).$$

Therefore, it suffices to prove

$$\pi'_{i'}(\sigma) \supset \pi'_{i'} \circ \tau_{j'}(\sigma \cap |T^1_{i+1,i}|),$$

because

$$\pi'_{i'}(\sigma - |T^1_{i+1,i}|) = \pi'_{i'} \circ \tau_{j'}(\sigma - |T^1_{i+1,i}|).$$

In other words, we need only prove

$$\tau_{j'}(\sigma \cap |T^1_{i+1,i}|) \subset \pi'^{-1}_{i'}(\sigma'),$$

where $\sigma' = \pi'_{i'}(\sigma)$. On the other hand, we have

$$\tau_{j'}(\sigma \cap |T^1_{i+1,i}|) \subset \bigcup_{t \in [0,1]} \theta^t_{j'}(\sigma \cap |T^1_{i+1,i}|) \quad \text{by the definition of } \tau_{j'}$$

$$\subset \bigcup_{t \in [0,1]} \theta^t_{j'}(X'_{j',i,\gamma_0,\varepsilon} \cap |T^1_{i+1,i}| \cap \pi'^{-1}_{i'}(\sigma') \cap \rho'^{-1}_{i'}(\varepsilon_{i'})).$$

Thus we reduce the problem to the inclusion:

$$\theta^t_{j'}(X'_{j',i,\gamma_0,\varepsilon} \cap |T^1_{i+1,i}| \cap \pi'^{-1}_{i'}(\sigma') \cap \rho'^{-1}_{i'}(\varepsilon_{i'})) \subset \pi'^{-1}_{i'}(\sigma') \quad \text{for each } t \in [0,1].$$

Now by (9),

$$X'_{j',i,\gamma_0,\varepsilon} \cap \pi'^{-1}_{i'}(\sigma') \cap \rho'^{-1}_{i'}(\varepsilon_{i'}) \cap \rho_{i+1}^{-1}(\varepsilon_{i+1})$$
$$= X'_{j',i,\gamma_0,\varepsilon} \cap \pi_{i+1,i}^{-1}(\sigma_{i+1}) \cap \rho'^{-1}_{i'}(\varepsilon_{i'}) \cap \rho_{i+1}^{-1}(\varepsilon_{i+1}),$$

where

$$\sigma_{i+1} = X'_{i+1,\gamma_0,\varepsilon} \cap \pi'^{-1}_{i'}(\sigma').$$

Take the union of each side of this equality over all $\varepsilon_{i+1} \in]0, c[$. Then

$$X'_{j',i,\gamma_0,\varepsilon} \cap \pi'^{-1}_{i'}(\sigma') \cap \rho'^{-1}_{i'}(\varepsilon_{i'}) \cap |T^1_{i+1,i}|$$
$$= X'_{j',i,\gamma_0,\varepsilon} \cap \pi^{-1}_{i+1,i}(\sigma_{i+1}) \cap \rho'^{-1}_{i'}(\varepsilon_{i'}) \cap |T^1_{i+1,i}|.$$

Hence what we prove is the inclusion:

$$\theta^t_{j'}(X'_{j',i,\gamma_0,\varepsilon} \cap \pi^{-1}_{i+1,i}(\sigma_{i+1}) \cap \rho'^{-1}_{i'}(\varepsilon_{i'}) \cap |T^1_{i+1,i}|) \subset \pi'^{-1}_{i'}(\sigma').$$

This is easy as follows. By (13) and the definition of $\theta^t_{j'}$, we have

$$\theta^t_{j'}(X'_{j',i,\gamma_0,\varepsilon} \cap \pi^{-1}_{i+1,i}(\sigma_{i+1}) \cap \rho'^{-1}_{i'}(\varepsilon_{i'}) \cap |T^1_{i+1,i}|)$$
$$= X'_{j',i,\gamma_0,\varepsilon} \cap (\pi^t_{i+1})^{-1}(\sigma_{i+1}) \cap \rho'^{-1}_{i'}(\varepsilon_{i'}) \cap |T^1_{i+1,i}|$$
$$= X'_{j',i,\gamma_0,\varepsilon} \cap \pi'^{-1}_{i'}(\sigma') \cap \rho'^{-1}_{i'}(\varepsilon_{i'}) \cap |T^1_{i+1,i}| \subset \pi'^{-1}_{i'}(\sigma').$$

Hence τ_j and $T^1_{j,i}$ satisfy all the conditions. Thus we can assume equality (11).

Next, under the following assumption, we shall construct C^∞ diffeomorphisms

$$g_{j,i+1,\varepsilon} \colon X'_{j,i,\gamma_0,\varepsilon} \cap \rho^{-1}_{i+1}([\varepsilon_{i+1}, \infty[) \longrightarrow X'_{j,i+1,\gamma_0,\varepsilon} = X'_{j,i,\gamma_0,\varepsilon} \cap \rho'^{-1}_{i+1}([\varepsilon_{i+1}, \infty[),$$
$$j = i+2, \dots, k,$$

and a controlled tube system $\{T_{j,i+1} = (|T_{j,i+1}|, \pi_{j,i+1}, \rho_{j,i+1})\}_{j=i+2,\dots,k}$ for $\{X'_{j,i+1,\gamma_0,\varepsilon}\}_{j=i+2,\dots,k}$ for any ε_{i+1} so that the homeomorphisms:

$$f_{j,i+1,\varepsilon} = \begin{cases} f_{j,i,\varepsilon}, & j = 1, \dots, i+1 \\ g_{j,i+1,\varepsilon} \circ (f_{j,i,\varepsilon}|X_{j,i+1,\gamma_0,\varepsilon}), & j = i+2, \dots, k \end{cases}$$

and $T_{j,i+1}$ complete the induction step of the proof of the assertion. Here we lessen $\delta_{i+1}, \dots, \delta_{k-1}$, and $g_{j,i+1,\varepsilon}$ and $T_{j,i+1}$ depend on ε_{i+1} but not on $\varepsilon_{i+2}, \dots, \varepsilon_{k-1}$. We assume that for any $\sigma \in C'_{i+1,i,\varepsilon}$ and for any integer $i+1 < j \le k$, the restrictions of ρ_{i+1} and ρ'_{i+1} to $X_j \cap \rho^{-1}_{i+1}([0, c_{i+1}[) \cap \pi'^{-1}_{i+1}(\sigma^\circ)$ are friendly, where c_{i+1} is a positive number so small that

$$X'_{j,i,\gamma_0,\varepsilon} \cap \rho^{-1}_{i+1}([0, c_{i+1}]) \subset |T_{i+1,i}|,$$

and if $\sigma_j \in C'_{j,i,\varepsilon}$ intersects with $\rho^{-1}_{i+1}([0, c_{i+1}[)$, then σ_j does with $\rho^{-1}_{i+1}(\varepsilon_{i+1})$. This assumption is similar to but stronger than the hypothesis (2) in the assertion.

We will choose $\delta_{i+1}, \dots, \delta_{k-1}$, $g_{j,i+1,\varepsilon}$ and $T_{j,i+1}$ so that for any $\sigma \in C'_{i+1,i,\varepsilon}$, $\varepsilon_{i+1}, \dots, \varepsilon_{k-1}$ and $i + 1 < j \le j' \le k$, we have $\delta_{i+1} < c_{i+1}$,

$$g_{j,i+1,\varepsilon} = \mathrm{id} \quad \text{on} \quad X'_{j,i,\gamma_0,\varepsilon} \cap \rho_{i+1}^{-1}([c_{i+1}, \infty[), \tag{15}$$

$$g_{j,i+1,\varepsilon}(X'_{j,i,\gamma_0,\varepsilon} \cap \rho_{i+1}^{-1}([\varepsilon_{i+1}, \infty[) \cap \pi_{i+1}'^{-1}(\sigma))$$
$$= X'_{j,i+1,\gamma_0,\varepsilon} \cap \pi_{i+1}'^{-1}(\sigma), \tag{16}$$

$$|T_{j,i+1}| \subset |T_{j,i}|, \qquad g_{j',i+1,\varepsilon}^{-1}(X'_{j',i+1,\gamma_0,\varepsilon} \cap |T_{j,i+1}|) \subset |T_{j,i}|,$$

$$\rho_j \circ g_{j',i+1,\varepsilon} = \rho_j \quad \text{on} \quad g_{j',i+1,\varepsilon}^{-1}(X'_{j',i+1,\gamma_0,\varepsilon} \cap |T_{j,i+1}|),$$

$$T_{j,i+1}|_{|T_{j,i+1}| \cap \rho_{i+1}^{-1}([c_{i+1}, \infty[)} = T_{j,i}|_{|T_{j,i+1}| \cap \rho_{i+1}^{-1}([c_{i+1}, \infty[)},$$

$$\rho_{j,i+1} = \rho_j \quad \text{on} \quad |T_{j,i+1}|, \quad \text{and}$$

$$\pi_{j,i+1} \circ g_{j',i+1,\varepsilon} = g_{j,i+1,\varepsilon} \circ \pi_{j,i} \quad \text{on} \quad g_{j',i+1,\varepsilon}^{-1}(X'_{j',i+1,\gamma_0,\varepsilon} \cap |T_{j,i+1}|).$$

If these conditions are satisfied, then $f_{j,i+1,\varepsilon}$ and $T_{j,i+1}$ satisfy conditions (3), $\dots$, (9) as follows. The (3), (4), (6), (7), and (8) are clear. Let us show (5). For that we need only prove

$$\pi'_{i'}(\sigma') = \pi'_{i'} \circ g_{j,i+1,\varepsilon}(\sigma') \quad \text{for} \quad \sigma' \in C'_{j,i,\varepsilon}|_{X'_{j,\gamma_0,\varepsilon} \cap \rho_{i'}^{*-1}(\varepsilon_{i'})}, \quad i' \le i+1 < j \le k,$$

where $\rho_{i'}^* = \rho_{i'}$ if $i' = i+1$ and $\rho_{i'}^* = \rho'_{i'}$ if $i' < i+1$. By (15) we can assume $\sigma' \cap \rho_{i+1}^{-1}([0, c_{i+1}[) \ne \varnothing$ and it suffices to prove the equality:

$$\pi'_{i'}(\sigma' \cap \rho_{i+1}^{-1}([0, c_{i+1}])) = \pi'_{i'} \circ g_{j,i+1,\varepsilon}(\sigma' \cap \rho_{i+1}^{-1}([0, c_{i+1}])).$$

As in the above first argument, by using the properties of $C_{i+1,\varepsilon}$, we have

$$\rho_{i+1}(\sigma') = \varepsilon_{i+1} \quad \text{or} \quad [\varepsilon_{i+1}, \varepsilon'_{i+1}],$$
$$\pi'_{i+1}(\sigma') = \pi'_{i+1}(\sigma' \cap \rho_{i+1}^{-1}(\varepsilon_{i+1})),$$

and, since c_{i+1} is small,

$$\pi'_{i'} = \pi'_{i'} \circ \pi'_{i+1} \quad \text{on} \quad \sigma' \cap \rho_{i+1}^{-1}([0, c_{i+1}]) \quad \text{and on} \quad g_{j,i+1,\varepsilon}(\sigma' \cap \rho_{i+1}^{-1}([0, c_{i+1}])).$$

Hence we need only prove

$$g_{j,i+1,\varepsilon}(\sigma' \cap \rho_{i+1}^{-1}([0, c_{i+1}])) \subset \pi_{i+1}'^{-1}(\pi'_{i+1}(\sigma')).$$

But this is immediate by (16) because by (11) we have

$$\pi'_{i+1}(\sigma' \cap \rho_{i+1}^{-1}(\varepsilon_{i+1})) \in C'_{i+1,i,\varepsilon}.$$

As usual, we will define $g_{j,i+1,\varepsilon}$ using the flow of a vector field. There exists a continuous controlled vector field $\xi' = \{\xi'_j\}_{j=i+2,\ldots,k}$ on $\{X'_{j,i,\gamma_0,\varepsilon}\}_{j=i+2,\ldots,k}$ for $\{T_{j,i}\}_{j=i+2,\ldots,k}$ such that for any $i+2 \le j \le k$ and any $\sigma \in C'_{i+1,i,\varepsilon}$, ξ'_j is tangent to the manifold $X_j \cap \rho_{i+1}^{-1}([0,c_{i+1}[) \cap \pi_{i+1}^{-1}(\sigma^\circ)$ at each point, and $\xi'_j\rho_{i+1}$ and $\xi'_j\rho'_{i+1}$ are positive on the manifold. We easily prove this as in the proof of I.1.10. Let $\{\theta'_j : D'_j \to X'_{j,i,\gamma_0,\varepsilon}\}_{j=i+2,\ldots,k}$, $D'_j \subset X'_{j,i,\gamma_0,\varepsilon} \times \mathbf{R}$, denote the flow of ξ'. Let $\varepsilon_{i+1} \ll c_{i+1}$ be any positive number. We want to define a C^∞ function ρ''_{i+1} on $\rho_{i+1}^{-1}([0,c_{i+1}[) - \cup_{i'=1}^{i}\rho_{i'}^{\prime -1}([0,\varepsilon_{i'}[)$ such that

$$\rho''_{i+1} = \begin{cases} \rho_{i+1} & \text{on} \quad \rho_{i+1}^{-1}([c_{i+1}/2, c_{i+1}[) - \cup_{i'=1}^{i}\rho_{i'}^{\prime -1}([0,\varepsilon_{i'}[) \\ \rho'_{i+1} & \text{on} \quad \rho_{i+1}^{\prime -1}([0,\varepsilon_{i+1}]) - \cup_{i'=1}^{i}\rho_{i'}^{\prime -1}([0,\varepsilon_{i'}[), \end{cases}$$

and for each $i+2 \le j \le k$, $\xi'_j\rho''_{i+1}$ is positive on $X'_{j,i,\gamma_0,\varepsilon} \cap \rho_{i+1}^{-1}([0,c_{i+1}[)$. Note that ρ_{i+1} and ρ''_{i+1} are C^∞ regular on each integral curve of ξ'_j in $X'_{j,i,\gamma_0,\varepsilon} \cap \rho_{i+1}^{-1}([0,c_{i+1}[)$. Assume such ρ''_{i+1}. We define the $g_{j,i+1,\varepsilon}$ and $T_{j,i+1}$ for each j by

$$g_{j,i+1,\varepsilon}(x) = \theta'_j(D'_j \cap x \times \mathbf{R}) \cap \rho''^{-1}_{i+1}(\rho_{i+1}(x))$$
$$\text{for} \quad x \in X'_{j,i,\gamma_0,\varepsilon} \cap \rho_{i+1}^{-1}([\varepsilon_{i+1},\infty[),$$

$$|T_{j,i+1}| = \text{a small neighborhood of } X'_{j,i+1,\gamma_0,\varepsilon} \text{ in } \bigcap_{i'=1}^{i+1} \rho_{i'}^{\prime -1}([\varepsilon_{i'},\infty[),$$

$$\pi_{j,i+1}(x) = \begin{cases} \pi_{j,i}(x) & \text{on} \quad |T_{j,i+1}| - \rho_{i+1}^{-1}([0,c_{i+1}/2]) \\ \theta'_j(D'_j \cap \pi_{j,i}(x) \times \mathbf{R}) \cap \rho''^{-1}_{i+1}(\rho''_{i+1}(x)) & \\ & \text{on} \quad |T_{j,i+1}| \cap \rho_{i+1}^{-1}([0,c_{i+1}/2]), \end{cases}$$

$$\text{and} \quad \rho_{j,i+1} = \rho_j|_{|T_{j,i+1}|}.$$

It is easy to prove that $g_{j,i+1,\varepsilon}$ and $\{T_{j,i+1}\}$ fulfill the requirements.

It remains to construct ρ''_{i+1}. Let λ be a C^∞ function on $\mathbf{R}$ such that $\lambda = 0$ on $[0,c_{i+1}/4]$, $\lambda = 1$ on $[c_{i+1}/2,\infty[$ and $\lambda' > 0$ on $]c_{i+1}/4, c_{i+1}/2[$. Set

$$\rho''_{i+1,m} = m(\lambda \circ \rho_{i+1})\cdot\rho_{i+1} + (1 - \lambda \circ \rho_{i+1})\cdot\rho'_{i+1}, \quad m = 1,2,\ldots.$$

Then

$$\rho''_{i+1,m} = \begin{cases} \rho'_{i+1} & \text{on} \quad \rho_{i+1}^{-1}([0,c_{i+1}/4]) \\ m\rho_{i+1} & \text{on} \quad \rho_{i+1}^{-1}([c_{i+1}/2,\infty[). \end{cases}$$

Consider $\rho''_{i+1,m}$ on $\rho_{i+1}^{-1}([c_{i+1}/4, c_{i+1}/2])$. For each point $x \in X'_{j,i,\gamma_0,\varepsilon} \cap \rho_{i+1}^{-1}([c_{i+1}/4, c_{i+1}/2])$, $j = i+2,\ldots,k$, $\xi'_j\rho''_{i+1,m}$ is positive at x if m is sufficiently large. Moreover, if $\xi'_j \rho''_{i+1,m}$ is positive at x, then by continuity of ξ'_j, $\xi'_{j'} \rho''_{i+1,m'}$ is positive near x, $j' = j, \ldots, k$, for any $m' \geq m$. Hence by compactness of

$$X \cap \rho^{-1}_{i+1}([c_{i+1}/4, c_{i+1}/2]) - \bigcup_{i'=1}^{i} \rho'^{-1}_{i'}([0, \varepsilon_{i'}[),$$

$\xi'_j \rho''_{i+1,m}$ is positive on $X'_{j,i,\gamma_0,\varepsilon} \cap \rho^{-1}_{i+1}([c_{i+1}/4, c_{i+1}/2])$ for some m. We modify such $\rho''_{i+1,m}$ as follows. Lessen ε_{i+1}. We have a C^∞ regular function μ on $\mathbf{R}$ such that $\mu(t) = t$ on $[0, \varepsilon_{i+1}]$ and $\mu(t) = t/m$ on $[c_{i+1}/2, \infty[$. Set

$$\rho''_{i+1} = \mu \circ \rho''_{i+1,m},$$

which fulfills the requirements. Thus we complete the induction step under the special assumption.

Finally, we will define $f_{j,i+1,\varepsilon}$ and $T_{j,i+1}$ without the above assumption. In the $\mathfrak{X}$-case, the following arguments are not necessary because the above assumption is always satisfied. We replace the assumption with the hypothesis (2) in the assertion. By (2) there is a strong cell approximation $C''_{i+1,i,\varepsilon}$ of $C'_{i+1,i,\varepsilon}$ which is a C^∞ cell triangulation of $X'_{i+1,\gamma_0,\varepsilon}$ such that for any $\sigma \in C''_{i+1,i,\varepsilon}$ and for any integer $i+1 < j \leq k$, the restrictions of ρ_{i+1} and ρ'_{i+1} to $X_j \cap \rho^{-1}_{i+1}([0, c_{i+1}[) \cap \pi'^{-1}_{i+1}(\sigma^\circ)$ are friendly for some positive number c_{i+1}. Let $\chi \colon X'_{i+1,\gamma_0,\varepsilon} \to X'_{i+1,\gamma_0,\varepsilon}$ denote the inverse of the homeomorphism of approximation, i.e., a homeomorphism such that $C''_{i+1,i,\varepsilon}$ is compatible with $\{\chi(\sigma) \colon \sigma \in C'_{i+1,i,\varepsilon}\}$, and for each $\sigma \in C''_{i+1,i,\varepsilon}$, the restriction of τ to $\chi^{-1}(\sigma)$ is a C^∞ diffeomorphism onto σ close to the identity map in the C^1 topology. For simplicity of notation, we assume $C''_{i+1,i,\varepsilon} = \{\chi(\sigma) \colon \sigma \in C'_{i+1,i,\varepsilon}\}$, which will not cause any trouble.

By the second argument, by lessening $\delta_{i+1}, \ldots, \delta_{k-1}$ and c_{i+1}, we obtain C^∞ diffeomorphisms:

$$g_{j,i+1,\varepsilon} \colon X'_{j,i,\gamma_0,\varepsilon} \cap \rho^{-1}_{i+1}([\varepsilon_{i+1}, \infty[) \longrightarrow X'_{j,i+1,\gamma_0,\varepsilon}, \quad j = i+2, \ldots, k,$$

for any ε_{i+1}, and a controlled tube system $\{T_{j,i+1}, = (|T_{j,i+1}|, \pi_{j,i+1}, \rho_{j,i+1})\}_{j=i+2,\ldots,k}$ for $\{X'_{j,i+1,\gamma_0,\varepsilon}\}_{j=i+2,\ldots,k}$ which satisfy those conditions in the second argument, where we replace $C'_{i+1,i,\varepsilon}$ with $C''_{i+1,i,\varepsilon}$. But, if we define $f_{j,i+1,\varepsilon}$ as in the second argument, then $f_{j,i+1,\varepsilon}$ and $\{T_{j,i+1}\}$ do not necessarily satisfy conditions (5) nor (9). So we need to change the definition of $f_{j,i+1,\varepsilon}$.

For that we will find homeomorphisms $h^1_{j,\varepsilon}$ of $H^1_j = X'_{j,i,\gamma_0,\varepsilon} \cap \rho^{-1}_{i+1}([\varepsilon_{i+1}, c_{i+1}[)$ and $h^2_{j,\varepsilon}$ of $H^2_j = X'_{j,i,\gamma_0,\varepsilon} \cap \rho''^{-1}_{i+1}([\varepsilon_{i+1}, c_{i+1}[)$ for any ε_{i+1}

and for any $j = i + 2, \ldots, k$ such that the following conditions are satisfied. (Here ρ''_{i+1} is the function defined in the second argument.) For any $\sigma \in C'_{i+1,i,\varepsilon}$, $i + 1 < j \leq j' \leq k$, and $l = 1, 2$,

$$h^1_{j,\varepsilon} = h^2_{j,\varepsilon} \quad \text{on} \quad X'_{j,i,\gamma_0,\varepsilon} \cap \rho^{-1}_{i+1}([c_{i+1}/2, c_{i+1}[), \tag{17}$$
$$h^l_{j,\varepsilon}(H^l_j \cap \pi'^{-1}_{i+1}(\sigma)) = H^l_j \cap \pi'^{-1}_{i+1}(\chi(\sigma)),$$

the restriction of $h^l_{j,\varepsilon}$ to $H^l_j \cap \pi'^{-1}_{i+1}(\sigma)$ is a C^∞ diffeomorphism onto the image,

$$\rho_{i+1} \circ h^1_{j,\varepsilon} = \rho_{i+1} \quad \text{on} \quad H^1_j, \qquad \rho''_{i+1} \circ h^2_{j,\varepsilon} = \rho''_{i+1} \quad \text{on} \quad H^2_j,$$
$$\pi_{j,i} \circ h^1_{j',\varepsilon} = h^1_{j,\varepsilon} \circ \pi_{j,i}, \qquad \rho_j \circ h^1_{j',\varepsilon} = \rho_j$$
$$\text{on} \quad H^1_{j'} \cap \text{(a neighborhood of } H^1_j \text{ in } H^1_j \cup H^1_{j'}\text{)}, \quad \text{and}$$
$$\pi_{j,i+1} \circ h^2_{j',\varepsilon} = h^2_{j,\varepsilon} \circ \pi_{j,i+1}, \qquad \rho_j \circ h^2_{j',\varepsilon} = \rho_j$$
$$\text{on} \quad H^2_{j'} \cap \text{(a neighborhood of } H^2_j \text{ in } H^2_j \cup H^2_{j'}\text{)}.$$

Note that $\rho_j = \rho_{j,i} = \rho_{j,i+1}$ on $|T_{j,i+1}|$. Assume such $h^1_{j,\varepsilon}$ and $h^2_{j,\varepsilon}$. Then the maps

$$\bar{g}_{j,i+1,\varepsilon} = \begin{cases} h^{2^{-1}}_{j,\varepsilon} \circ g_{j,i+1,\varepsilon} \circ h^1_{j,\varepsilon} & \text{on} \quad H^1_j \\ \text{id} & \text{on} \quad X'_{j,i,\gamma_0,\varepsilon} \cap \rho^{-1}_{i+1}([c_{i+1}, \infty[) \end{cases}$$

and $\{T_{j,i+1}\}$ satisfy all the conditions of $g_{j,i+1,\varepsilon}$ and $\{T_{j,i+1}\}$ in the second argument except the condition that the $g_{j,i+1,\varepsilon}$ are C^∞ diffeomorphisms. In the present case, the $\bar{g}_{j,i+1,\varepsilon}$ are homeomorphisms, and their restrictions to $(H^1_j \cap \pi'^{-1}_{i+1}(\sigma)) \cup (X'_{j,i,\gamma_0,\varepsilon} \cap \rho^{-1}_{i+1}([c_{i+1}, \infty[))$ are C^∞ diffeomorphisms onto $(H^2_j \cap \pi'^{-1}_{i+1}(\sigma)) \cup (X'_{j,i,\gamma_0,\varepsilon} \cap \rho^{-1}_{i+1}([c_{i+1}, \infty[))$ for any $\sigma \in C'_{i+1,i,\varepsilon}$. Hence the maps

$$f_{j,i+1,\varepsilon} = \begin{cases} f_{j,i,\varepsilon}, & j = 1, \ldots, i + 1 \\ \bar{g}_{j,i+1,\varepsilon} \circ (f_{j,i,\varepsilon}|_{X_{j,i+1,\gamma_0,\varepsilon}}), & j = i + 2, \ldots, k \end{cases}$$

are homeomorphisms from $X_{j,i+1,\gamma_0,\varepsilon}$ to $X'_{j,i+1,\gamma_0,\varepsilon}$, the restrictions of $f_{j,i+1,\varepsilon}$ to $X_{j,\gamma_0,\varepsilon}$ are P C^∞ diffeomorphisms onto the images with respect to $C_{j,\varepsilon}$, and $f_{j,i+1,\varepsilon}$ and $\{T_{j,i+1}\}$ satisfy conditions $(3), \ldots, (9)$. Therefore, we obtain the assertion.

We will define the $h^1_{j,\varepsilon}$ and $h^2_{j,\varepsilon}$ in a way similar to the construction of

τ_j in the first argument. For each $j = i+2, \dots, k$ and $l = 1, 2$, set

$$H^l = \bigcup_{j'=i+2}^{k} H^l_{j'}, \qquad \hat{H}^l = H^l \times [0, 1],$$

$$\hat{H}^l_j = H^l_j \times [0, 1], \qquad \hat{T}^l_j = (|\hat{T}^l_j|, \hat{\pi}^l_j, \hat{\rho}^l_j),$$

$$|\hat{T}^l_j| = \begin{cases} \pi_{j,i}^{-1}(H^1_j) \times [0, 1], & l = 1 \\ \pi_{j,i+1}^{-1}(H^2_j) \times [0, 1], & l = 2, \end{cases}$$

$$\hat{\pi}^l_j(x, t) = \begin{cases} (\pi_{j,i}(x), t)) & \text{for} \quad (x, t) \in |\hat{T}^1_j|, \ l = 1 \\ (\pi_{j,i+1}(x), t)) & \text{for} \quad (x, t) \in |\hat{T}^2_j|, \ l = 2, \end{cases}$$

$$\text{and} \quad \hat{\rho}^l_j(x, t) = \rho_j(x) \quad \text{for} \quad (x, t) \in |\hat{T}^l_j|.$$

Then $\{\hat{T}^l_j\}_{j=i+2,\dots,k}$ is a controlled tube system for $\{\hat{H}^l_j\}_{j=i+2,\dots,k}$ for each $l = 1, 2$. We shall define controlled local Lipschitz vector fields $\{\hat{\zeta}^l_j\}_{j=i+2,\dots,k}$ on $\{\hat{H}^l_j\}_{j=i+2,\dots,k}$ for $\{\hat{T}^l_j\}_{j=i+2,\dots,k}$, $l = 1, 2$, so that the following conditions are satisfied. For each $\hat{H}^l_j$, $\hat{\zeta}^l_j$ is a vector field of $\hat{H}^l_j$ with the local Lipschitz condition. For each $(x, t) \in \hat{H}^l_j$, $\hat{\zeta}^l_{jx,t}$ is of the form $(\hat{\zeta}^{l\prime}_{jx,t}, \partial/\partial t)$. Let $\{\hat{\theta}^l_j \colon \hat{D}^l_j \to \hat{H}^l_j\}_{j=i+2,\dots,k}$, $\hat{D}^l_j \subset \hat{H}^l_j \times \mathbf{R}$, denote the flow of $\{\hat{\zeta}^l_j\}$. Then

$$\hat{D}^l_j = \{(x, t, s) \in H^l_j \times [0, 1] \times \mathbf{R} \colon 0 \leq t + s \leq 1\}, \tag{18}$$

and the $h^l_{j,\varepsilon}$, defined by

$$(h^l_{j,\varepsilon}(x), 1) = \hat{\theta}^l_j(x, 0, 1) \quad \text{for} \quad x \in H^l_j, \tag{19}$$

fulfill the requirements.

Clearly there exists a vector field $\hat{\zeta}_{i+1}$ on $\hat{X}'_{i+1,\gamma_0,\varepsilon} = X'_{i+1,\gamma_0,\varepsilon} \times [0, 1]$ with the Lipschitz condition such that for each $(x, t) \in \hat{X}'_{i+1,\gamma_0,\varepsilon}$, $\hat{\zeta}_{i+1x,t}$ is of the form $(\hat{\zeta}'_{i+1x,t}, \partial/\partial t)$; and if we let $\hat{\theta}_{i+1} \colon \hat{D}_{i+1} \to \hat{X}'_{i+1,\gamma_0,\varepsilon}$, $\hat{D}_{i+1} \subset \hat{X}'_{i+1,\gamma_0,\varepsilon} \times \mathbf{R}$, denote the maximal flow of $\hat{\zeta}_{i+1}$, then

$$\hat{D}_{i+1} = \{(x, t, s) \in X'_{i+1,\gamma_0,\varepsilon} \times [0, 1] \times \mathbf{R} \colon 0 \leq t + s \leq 1\} \quad \text{and}$$

$$(\chi(x), 1) = \hat{\theta}_{i+1}(x, 0, 1) \quad \text{for} \quad x \in X'_{i+1,\gamma_0,\varepsilon}.$$

Here χ is the homeomorphism of $X'_{i+1,\gamma_0,\varepsilon}$ used to define $C''_{i+1,i,\varepsilon}$. Also, for each $\sigma \in C'_{i+1,i,\varepsilon}$, the restriction of $\hat{\theta}_{i+1}$ to $\sigma \times 0 \times [0, 1]$ is a C^∞ diffeomorphism onto its image, and hence the restriction of $\hat{\zeta}_{i+1}$ to $\hat{\theta}_{i+1}(\sigma \times 0 \times [0, 1])$

is of class C^∞. Note that for each $s \in [0,1]$ the restriction of $\hat{\theta}_{i+1}$ to $X'_{i+1,\gamma_0,\varepsilon} \times 0 \times s$ is a homeomorphism onto $X'_{i+1,\gamma_0,\varepsilon} \times s$. We will define $\{\hat{\zeta}^l_j\}_{j=i+2,\dots,k}$ by lifting $\hat{\zeta}_{i+1}$ along the following map $\hat{\pi}^l_{i+1}$ so that $\hat{\zeta}^l_j$ are tangent to each level of the following function $\hat{\rho}^l_{i+1}(21)$. For each $l = 1,2$, set

$$\hat{\pi}^l_{i+1}(x,t) = (\pi^*_{i+1}(x),t) \qquad \text{for} \quad (x,t) \in \hat{H}^l \quad \text{and}$$

$$\hat{\rho}^l_{i+1}(x,t) = \begin{cases} \rho_{i+1}(x) & \text{for} \quad (x,t) \in \hat{H}^1, \ l=1 \\ \rho''_{i+1}(x) & \text{for} \quad (x,t) \in \hat{H}^2, \ l=2, \end{cases}$$

where the map $\pi^*_{i+1} \colon H^1 \cup H^2 \to X'_{i+1,\gamma_0,\varepsilon}$ is defined so that

$$\pi^*_{i+1} = \pi'_{i+1} \quad \text{on} \quad (H^1 \cup H^2) \cap \rho_{i+1}^{-1}(c_{i+1}/2) \quad \text{and}$$

$$\pi^*_{i+1} = \text{const} \quad \text{on each integral curve of} \ \ \xi'_j|_{H^1_j \cup H^2_j}, \ \ j = i+2, \dots, k.$$

Here ξ'_j is the vector field on $X'_{j,i,\gamma_0,\varepsilon}$ which induced $g_{j,i+1,\varepsilon}$ (see the second argument). Then $\hat{\pi}^l_{i+1}$ is of class C^∞ on each $\hat{H}^l_j$, the map $(\hat{\pi}^l_{i+1}, \hat{\rho}^l_{i+1}) \colon \hat{H}^l_j \to X'_{i+1,\gamma_0,\varepsilon} \times [0,1] \times \mathbf{R}$ is C^∞ regular, and

$$\hat{\pi}^l_{i+1} \circ \hat{\pi}^l_j = \hat{\pi}^l_{i+1}, \quad \hat{\rho}^l_{i+1} \circ \hat{\pi}^l_j = \hat{\rho}^l_{i+1} \quad \text{on} \quad \hat{H}^l \cap |\hat{T}^l_j|. \tag{20}$$

These are immediate by the method of construction of ρ''_{i+1}, $\{T_{j,i+1}\}$, $\hat{\pi}^l_{i+1}$ and $\hat{\rho}^l_{i+1}$.

For each $j = i+2, \dots, k$ and for each $l = 1,2$, let $\hat{\zeta}^l_j$ denote a lift on $\hat{H}^l_j$ of $\hat{\zeta}_{i+1}$, i.e., a vector field on $\hat{H}^l_j$ with the local Lipschitz condition such that for each $\sigma \in C'_{i+1,i,\varepsilon}$, the restriction of $\hat{\zeta}^l_j$ to $\hat{H}^l_j \cap \hat{\pi}^{l-1}_{i+1}(\hat{\theta}_{i+1}(\sigma \times 0 \times [0,1]))$ is of class C^∞, and

$$d(\hat{\rho}^l_{i+1})\hat{\zeta}^l_{jx,t} = 0 \quad \text{and} \quad d(\hat{\pi}^l_{i+1}|_{\hat{H}^l_j})\hat{\zeta}^l_{jx,t} = \hat{\zeta}_{i+1\hat{\pi}^l_{i+1}(x,t)} \quad \text{for} \quad (x,t) \in \hat{H}^l_j. \tag{21}$$

This is possible because $(\hat{\pi}^l_{i+1}, \hat{\rho}^l_{i+1})$ is C^∞ regular. Then $\{\hat{\zeta}^l_j\}_{j=i+2,\dots,k}$, $l = 1,2$, are not necessarily controlled vector fields on $\{\hat{H}^l_j\}_{j=i+2,\dots,k}$ for $\{\hat{T}^l_j\}_{j=i+2,\dots,k}$. If they are not so, then modify them as in the proof I.1.5 so that they are controlled. After that special modification, by (20), each $\hat{\zeta}^l_j$ continues to be a lift of $\hat{\zeta}_{i+1}$. Hence we assume that $\{\hat{\zeta}^l_j\}_{j=i+2,\dots,k}$ are controlled. Let $\{\hat{\theta}^l_j \colon \hat{D}^l_j \to \hat{H}^l_j\}_{j=i+2,\dots,k}$, $\hat{D}^l_j \subset \hat{H}^l_j \times \mathbf{R}$, denote the flow of $\{\hat{\zeta}^l_j\}_{j=i+2,\dots,k}$. As in the proof of I.1.6, we can show (18) by (21). Hence we can define $h^l_{j,\varepsilon}$ by (19). Then $h^l_{j,\varepsilon}$ fulfill the requirements except (17). For (17), it suffices to choose $\hat{\zeta}^l_j$ so that

$$\hat{\zeta}^1_j = \hat{\zeta}^2_j \quad \text{on} \quad (X'_{j,i,\gamma_0,\varepsilon} \cap \rho^{-1}_{i+1}([c_{i+1}/2, c_i[)) \times [0,1],$$

which is possible because

$$\hat{\pi}^1_{i+1} = \hat{\pi}^2_{i+1} \text{ and } \hat{\rho}^1_{i+1} = \hat{\rho}^2_{i+1} \quad \text{on} \quad (X'_{j,i,\gamma_0,\varepsilon} \cap \rho^{-1}_{i+1}([c_{i+1}/2, c_i[)) \times [0,1].$$

Thus we complete the proof of the assertion and hence that (III.2.4) implies III.1.   □

CHAPTER IV. TRIANGULATIONS OF $\mathfrak{X}$-MAPS

Let $X \subset \mathbf{R}^n$ and $Y \subset \mathbf{R}^m$ be locally closed $\mathfrak{X}$-sets and let $f: X \to Y$ be an $\mathfrak{X}$-map. A C^0 $\mathfrak{X}$-*triangulation* of f is a quadruplet of $\mathfrak{X}$-polyhedra $X_0 \subset \mathbf{R}^{n'}$ and $Y_0 \subset \mathbf{R}^{m'}$ and $\mathfrak{X}$-homeomorphisms $\pi: X_0 \to X$ and $\tau: Y_0 \to Y$ such that $\tau^{-1} \circ f \circ \pi$ is PL. We call a C^0 $\mathfrak{X}$-triangulation (X_0, Y_0, π, τ) a C^0 R-$\mathfrak{X}$-*triangulation* if Y is a polyhedron, $Y_0 = Y$ and $\tau = \mathrm{id}$. In connection with the preceeding chapters, it may be natural to treat an $(R\text{-})\mathfrak{X}$-*triangulation* of f which we define assuming, in addition, that π and τ are of class C^r, $r > 0$, on each simplex of some simplicial decompositions of X_0 and Y_0 respectively. However, I cannot prove the results of this chapter in the terms of an $(R\text{-})\mathfrak{X}$-triangulation, and I think that a C^0 $(R\text{-})\mathfrak{X}$-triangulation is more natural than an $(R\text{-})$ $\mathfrak{X}$-triangulation in itself. We define naturally a C^0 $\mathfrak{X}$-*stratification* of an $\mathfrak{X}$-set in a Euclidean space. Here note that the stratification is finite locally at each point of the Euclidean space and each stratum is not only an $\mathfrak{X}$-set and a C^0 manifold but also a C^0 $\mathfrak{X}$-submanifold of the Euclidean space (i.e., locally $\mathfrak{X}$-homeomorphic to a Euclidean space).

We consider when an $\mathfrak{X}$-map $f: X \to Y$ is C^0 $(R\text{-})\mathfrak{X}$-triangulable. It is not always so, e.g., a blow-up $\mathbf{R}^2 \ni (x, y) \to (x, xy) \in \mathbf{R}^2$. We know a conjecture of Thom (see IV.1.9), which states that if f is proper and admits a Whitney stratification sans éclatement $\{X_i\} \to \{Y_j\}$, then f is triangulable. (If f is a proper PL map between polyhedra, the above condition is satisfied.) By the first isotopy lemma, $f|_{f^{-1}(Y_j)}: f^{-1}(Y_j) \to Y_j$ admits C^0 $\mathfrak{X}$-triangulations for all j. But it is difficult to paste them because there are too many choices of triangulations. The pasting is possible in the special case where for each $y \in Y$, $\{X_i \cap f^{-1}(y)\}_i$ is a stratification with solid triangulation because it has a unique natural triangulation (see I.1.12). However, f seldom has such a good stratification. We solve this problem by splitting f into a sequence of maps and applying the idea of a sheaf. In §IV.3, we see relations of uniqueness to global triangulations.

§IV.1. Conditions for $\mathfrak{X}$-maps to be triangulable

Let f be an $\mathfrak{X}$-function on a locally closed $\mathfrak{X}$-set $X \subset \mathbf{R}^n$. Let $0 \in X$ and assume $f(0) = 0$. To prove local C^0 $\mathfrak{X}$-triangulability of f at 0 we used the following fact in §II.3. Let $\mathbf{R}^n \times \mathbf{R} \xrightarrow{p_1} \mathbf{R}^{n-1} \times \mathbf{R} \xrightarrow{p_2} \cdots \xrightarrow{p_n} 0 \times \mathbf{R}$ be the sequence of the projections which forget the respective first factors. By changing the coordinate system of $\mathbf{R}^n$ by an $\mathfrak{X}$-homeomorphism of $\mathbf{R}^n$, we can choose open $\mathfrak{X}$-neighborhoods U_i of 0 in $\mathbf{R}^{n-i} \times \mathbf{R}$, $i = 0, \ldots, n$, and C^0 $\mathfrak{X}$-stratifications with the weak frontier condition $\{X_{i,j}\}_j$ of U_i, $i = 0, \ldots, n$, such that

(i) for each $X_{i,j}$ with $i < n$,

$$p_{i+1}(X_{i,j}) = X_{i+1,j'} \quad \text{for some } j',$$

and $X_{i,j}$ is an open subset of $p_{i+1}^{-1}(X_{i+1,j'})$ in the case where $p_{i+1}|_{X_{i,j}}$ is not a covering map to $X_{i+1,j'}$;

(ii) for each $i < n$, the union of $X_{i,j}$ such that $p_{i+1}|_{X_{i,j}}$ are covering maps to their images is closed in U_i; and

(iii) $\{X_{0,j}\}_j$ is compatible with $\{\text{graph } f\}$.

In the $\mathfrak{X}$-map case, we obtain local C^0 $\mathfrak{X}$-triangulability under similar conditions to those of the following theorem. Let $n = n_0 > n_1 > \cdots > n_l = 0$ be integers, and let $\mathbf{R}^n \times \mathbf{R}^m = \mathbf{R}^{n_0} \times \mathbf{R}^m \xrightarrow{p_1} \mathbf{R}^{n_1} \times \mathbf{R}^m \xrightarrow{p_2} \cdots \xrightarrow{p_l} \mathbf{R}^{n_l} \times \mathbf{R}^m = 0 \times \mathbf{R}^m$ be the sequence of the projections which forget the respective first factors. Let $X \subset \mathbf{R}^n$ and $Y \subset \mathbf{R}^m$ be locally closed $\mathfrak{X}$-sets and let $f: X \to Y$ be an $\mathfrak{X}$-map. A *resolution* of f at a point x of X with respect to $p_1, \ldots, p_l$ is a sequence of C^0 $\mathfrak{X}$-stratifications $\{X_{0,j}\}_j \xrightarrow{p_1} \cdots \xrightarrow{p_l} \{X_{l,j}\}_j$ with the weak frontier condition such that

(0) for each i, $\{X_{i,j}\}_j$ is a stratification of a neighborhood U_i of $p_i \circ \cdots \circ p_1(x, f(x))$ in $\mathbf{R}^{n_i} \times \mathbf{R}^m$;

(i) for each $i < l$ and for each j,

$$p_{i+1}(X_{i,j}) = X_{i+1,j'} \quad \text{for some } j',$$

and $X_{i,j}$ is an open subset of $p_{i+1}^{-1}(X_{i+1,j'})$ in the case where $p_{i+1}|_{X_{i,j}}$ is not a covering map to $X_{i+1,j'}$;

(ii) for each $i < l$, the union of $X_{i,j}$ such that $p_{i+1}|_{X_{i,j}}$ are covering maps to their images is closed in U_i; and

(iii) $\{X_{0,j}\}_j$ is compatible with $\{\text{graph } f\}$.

A *resolution* of f with respect to $p_1, \ldots, p_l$ is a sequence $\{X_{0,j}\}_j \xrightarrow{p_1} \cdots \xrightarrow{p_l} \{X_{l,j}\}_j$ which is a resolution of f at every point of X with respect to $p_1, \ldots, p_l$ (i.e., U_0 is a neighborhood of graph f). We define also a *resolution* of a subset of $\mathbf{R}^n \times \mathbf{R}^m$ (at a point of $\mathbf{R}^n \times \mathbf{R}^m$) with respect to $p_1, \ldots, p_l$ in the same way.

A trivial example of f with a resolution is a proper C^1 $\mathfrak{X}$-map between C^1 $\mathfrak{X}$-manifolds such that $f|_{\Sigma_1 f}$ is finite-to-one. Hence, by IV.1.2, a proper C^0 stable C^∞ map is C^0 triangulable ([Ve]).

Theorem IV.1.1 (Local $\mathfrak{X}$-triangulation of an $\mathfrak{X}$-map). *Let $X \subset \mathbf{R}^n$ and $Y \subset \mathbf{R}^m$ be closed $\mathfrak{X}$-sets and let $x \in X$. An $\mathfrak{X}$-map $f: X \to Y$ is locally C^0 $\mathfrak{X}$-triangulable at x (i.e., there are closed $\mathfrak{X}$-neighborhoods U of* *x in X and V of $f(x)$ in Y such that $f(U) \subset V$ and $f|_U \colon U \to V$ is C^0 $\mathfrak{X}$-triangulable) if and only if after a change of the coordinate system of $\mathbf{R}^n$ by an $\mathfrak{X}$-homeomorphism of $\mathbf{R}^n$, f admits a resolution at x with respect to some sequence of projections $p_1, \ldots, p_l$. Moreover, f is locally C^0 R-$\mathfrak{X}$-triangulable at x if and only if we can choose the above U, V and a resolution $\{X_{0,j}\} \overset{p_1}{\to} \cdots \overset{p_l}{\to} \{X_{l,j}\}_j$ so that V is a polyhedron and each $X_{l,j}$ is a finite union of open simplexes.*

Weak Theorem IV.1.2 (Global $\mathfrak{X}$-triangulation of an $\mathfrak{X}$-map). *Let $X \subset \mathbf{R}^n$ and $Y \subset \mathbf{R}^m$ be locally closed $\mathfrak{X}$-sets. Assume Axiom (v). An $\mathfrak{X}$-map $f \colon X \to Y$ is C^0 $\mathfrak{X}$-triangulable if after a change of the coordinate system of $\mathbf{R}^n$ by an $\mathfrak{X}$-homeomorphism of $\mathbf{R}^n$, f admits a resolution with respect to some sequence of projections $p_1, \ldots, p_l$. Moreover, f is C^0 R-$\mathfrak{X}$-triangulable if Y is an $\mathfrak{X}$-polyhedron and if we can choose the above resolution $\{X_{0,j}\}_j \overset{p_1}{\to} \cdots \overset{p_l}{\to} \{X_{l,j}\}_j$ so that each $X_{l,j}$ is a finite union of open simplexes locally at each point of $\mathbf{R}^m$. If X is closed in $\mathbf{R}^n$ and f is proper, then these conditions are necessary for the corresponding properties.*

A question likely to arise is why we treat a sequence of projections $\mathbf{R}^{n_0} \times \mathbf{R}^m \overset{p_1}{\to} \cdots \overset{p_l}{\to} \mathbf{R}^{n_l} \times \mathbf{R}^m$ in the above theorems in place of the projections $\mathbf{R}^n \times \mathbf{R}^m \to \mathbf{R}^{n-1} \times \mathbf{R}^m \to \cdots \to 0 \times \mathbf{R}^m$. The reason is that we need this general form for the sake of an application to C^0 subanalytic triangulations of complex analytic functions. We can easily prove that IV.1.1 follows from IV.1.2. Hence another question is why the last theorem is weak. This is because a complex analytic function does not always satisfy the conditions of IV.1.2. We want to strengthen IV.1.2 so that we can apply it to any complex analytic function.

Let $\mathbf{R}^{n_0} \times \mathbf{R}^m \overset{p_1}{\to} \cdots \overset{p_l}{\to} \mathbf{R}^{n_l} \times \mathbf{R}^m$ be as above. For each $0 \le i \le l$ and for each $x \in \mathbf{R}^{n_i} \times \mathbf{R}^m$, let $\mathcal{A}_{i,x}$ be a nonempty countable or finite family of germs at x of C^0 $\mathfrak{X}$-stratifications of neighborhoods of x in $\mathbf{R}^{n_i} \times \mathbf{R}^m$ with the weak frontier condition. We call $\{\mathcal{A}_{i,x}\}_{\substack{i=0,\ldots,l \\ x \in \mathbf{R}^{n_i} \times \mathbf{R}^m}}$ a *local resolution* of f with respect to $p_1, \ldots, p_l$ if the following conditions are satisfied. For each $A_i \in \mathcal{A}_{i,x}$, $i < l$ there exist $A_{i+1} \in \mathcal{A}_{i+1,p_{i+1}(x)}$ and C^0 $\mathfrak{X}$-stratifications $\{X_{i,j}\}_j$ of a neighborhood U_i of x in $\mathbf{R}^{n_i} \times \mathbf{R}^m$ and $\{X_{i+1,j}\}_j$ of a neighborhood of $p_{i+1}(x)$ in $\mathbf{R}^{n_{i+1}} \times \mathbf{R}^m$ such that the germs of $\{X_{i,j}\}_j$ at x and of $\{X_{i+1,j}\}_j$ at $p_{i+1}(x)$ coincide with A_i and A_{i+1} respectively, and the above conditions (i) and (ii) of a resolution and the following conditions are satisfied. (We write $p_{i+1}(A_i) = A_{i+1}$.)

(iii)′ For each $x \in \operatorname{graph} f$, there exists $A \in \mathcal{A}_{0,x}$ which is compatible with the germ of graph f at x.

(iv) For any germs A and B in $\mathcal{A}_{i,x}$, there exists a germ in $\mathcal{A}_{i,x}$ which is compatible with A and B.

(v) Let $\{X_{i,j}\}_j$ be a C^0 $\mathfrak{X}$-stratification of an open set in $\mathbf{R}^{n_i} \times \mathbf{R}^m$ whose germ at a point x is an element of $\mathcal{A}_{i,x}$. Then the germ of $\{X_{i,j}\}_j$ at any point y near x is an element of $\mathcal{A}_{i,y}$. (If this holds for any $y \in \bigcup_j X_{i,j}$, then we call $\{X_{i,j}\}_j$ a *realization* of the element of $\mathcal{A}_{i,x}$.)

We define also a *local resolution* of a subset of $\mathbf{R}^n \times \mathbf{R}^m$ with respect to $p_1, \dots, p_l$ in the same way.

Strong Theorem IV.1.2′ (Global $\mathfrak{X}$-triangulation of $\mathfrak{X}$-map). *Let X and Y be the same as in IV.1.2. Assume Axiom* (v). *A proper $\mathfrak{X}$-map $f \colon X \to Y$ is C^0 $\mathfrak{X}$-triangulable if after a change of the coordinate system of $\mathbf{R}^n$ by an $\mathfrak{X}$-homeomorphism of $\mathbf{R}^n$, there exists a local resolution $\{\mathcal{A}_{i,x}\}_{\substack{i=0,\dots,l \\ x \in \mathbf{R}^{n_i} \times \mathbf{R}^m}}$ of f with respect to some projections $\mathbf{R}^{n_0} \times \mathbf{R}^m \overset{p_1}{\to} \cdots \overset{p_l}{\to} \mathbf{R}^{n_l} \times \mathbf{R}^m$. An $\mathfrak{X}$-map $f \colon X \to Y$ which is not necessarily proper is C^0 R-$\mathfrak{X}$-triangulable if Y is, in addition, an $\mathfrak{X}$-polyhedron and for any $\{X_{l,j}\}_j \in \mathcal{A}_{l,x}$, each $X_{l,j}$ is the germ of a finite union of some open simplexes. If X is closed in $\mathbf{R}^n$, then these conditions are also necessary for the corresponding properties.*

Consider $\mathfrak{X}_0$-maps. Let $X \subset \mathbf{R}^n$ and $Y \subset \mathbf{R}^m$ be $\mathfrak{X}_0$-sets and let $f \colon X \to Y$ be an $\mathfrak{X}_0$-map. A C^0 $\mathfrak{X}_0$-*cell triangulation* of f is (X_0, Y_0, π, τ), where $X_0 \subset \mathbf{R}^{n'}$ and $Y_0 \subset \mathbf{R}^{m'}$ are semilinear sets and $\pi \colon X_0 \to X$ and $\tau \colon Y_0 \to Y$ are $\mathfrak{X}_0$-homeomorphisms such that $\tau^{-1} \circ f \circ \pi$ is semilinear. We also define naturally a C^0 R-$\mathfrak{X}_0$-*cell triangulation*. A *resolution* of the above f with respect to projections $\mathbf{R}^n \times \mathbf{R}^m = \mathbf{R}^{n_0} \times \mathbf{R}^m \overset{p_1}{\to} \cdots \overset{p_l}{\to} \mathbf{R}^{n_l} \times \mathbf{R}^m = 0 \times \mathbf{R}^m$ is a resolution $\{X_{0,j}\}_j \overset{p_1}{\to} \cdots \overset{p_l}{\to} \{X_{l,j}\}_j$ of f as an $\mathfrak{X}$-map such that for each i, $\{X_{i,j}\}_j$ is a finite C^0 $\mathfrak{X}_0$-stratification of $\mathbf{R}^{n_i} \times \mathbf{R}^m$.

Theorem IV.1.2″ ($\mathfrak{X}_0$-cell triangulation of an $\mathfrak{X}_0$-map). *Let $X \subset \mathbf{R}^n$ and $Y \subset \mathbf{R}^m$ be $\mathfrak{X}_0$ sets. An $\mathfrak{X}_0$-map $f \colon X \to Y$ is C^0 $\mathfrak{X}_0$-cell triangulable if after a change of the coordinate system of $\mathbf{R}^n$ by an $\mathfrak{X}_0$-homeomorphism of $\mathbf{R}^n$, there exists a resolution $\{X_{0,j}\}_j \overset{p_1}{\to} \cdots \overset{p_l}{\to} \{X_{l,j}\}_j$ of f. If each $X_{l,j}$ is semilinear, f is C^0 R-$\mathfrak{X}_0$-cell triangulable.*

We postpone proofs of the above theorems to the next section. In this section, we give some remarks, conjectures and corollaries.

Remark IV.1.3. In IV.1.2′, if we do not assume Axiom (v), then we modify the conclusion as follows. Any compact subset C of X has a compact neighborhood C_1 in X such that $f|_{C_1}$ is C^0 $(R\text{-})\mathfrak{X}$-triangulable if and only if the first (resp., second) conditions in IV.1.2′ are satisfied. This will be immediate when we prove IV.1.2′. Assume that C is a compact $\mathfrak{X}$-set. Then $f|_C$ is not always C^0 $(R\text{-})\mathfrak{X}$-triangulable under the first (resp., second) conditions.

A counterexample is given by

$$X = [-1,1]^2 \times \mathbf{R} \subset \mathbf{R}^3, \qquad Y = \mathbf{R}^2, \qquad C = [-1,1]^2 \times 0, \quad \text{and}$$
$$f(x,y,z) = (x, xy+z) \quad \text{for} \quad (x,y,z) \in X.$$

As in II.2.6, if X (Y) is closed in $\mathbf{R}^n$ (resp., in $\mathbf{R}^m$), and if an $\mathfrak{X}$-map $f\colon X \to Y$ is C^0 $\mathfrak{X}$-triangulable, then we can choose a C^0 $\mathfrak{X}$-triangulation (X_0, Y_0, π, τ) of f so that $X_0 \subset \mathbf{R}^n$ $(Y_0 \subset \mathbf{R}^m)$ and π (τ) can be extended to an $\mathfrak{X}$-homeomorphism of $\mathbf{R}^n$ (resp., of $\mathbf{R}^m$).

Remark IV.1.4. In IV.1.2', we cannot remove the properness condition of f. A counterexample is given as follows. Let us consider the subanalytic case. Let X consist of infinitely many circles in $\mathbf{R}^2$, set $Y = \mathbf{R}^2$ and let f be a subanalytic C^∞ immersion such that the image of f is the union of the circles S_k with center (a_k, b_k) and radius r_k, $k = 1, 2, \ldots$, where $\{(a_k, b_k)\colon k = 1, 2, \ldots\}$ is dense in $\mathbf{R}^2$ and $r_k \to 0$ as $k \to \infty$. Assume f is C^0 $\mathfrak{X}$-triangulable. Then by II.2.6 there exists an $\mathfrak{X}$-homeomorphism τ of $\mathbf{R}^2$ such that each $\tau(S_k)$ is a polyhedron. By (II.1.15) τ carries some nonempty open set of $\mathbf{R}^2$ to another open set C^1 diffeomorphically. Hence for some k, $\tau|_{S_k}$ is a C^1 diffeomorphism onto the image. It follows that $\tau(S_k)$ is a C^1 circle, which is a contradiction. Therefore, f is not C^0 $\mathfrak{X}$-triangulable. However, f satisfies the first conditions in IV.1.2'.

Conjecture IV.1.5 (Local triangulability implies global triangulability). *Let $\mathfrak{X}$, X and Y be the same as in IV.1.2, and let $f\colon X \to Y$ be an $\mathfrak{X}$-map. Assume that f is proper (this is not necessary for C^0 R-$\mathfrak{X}$-triangulability) and f is locally C^0 (R-)$\mathfrak{X}$-triangulable. Then f is globally C^0 (resp., R-)$\mathfrak{X}$-triangulable.*

For IV.1.5, it seems to suffices to prove the following because a global $\mathfrak{X}$-triangulation of an $\mathfrak{X}$-function followed from its uniqueness(§II.3).

Conjecture IV.1.6 (Uniqueness of $\mathfrak{X}$-triangulations). *Assume that there exist two C^0 (R-)$\mathfrak{X}$-triangulations $(X_{0,i}, Y_{0,i}, \pi_i, \tau_i)$, $i = 1, 2$, of an $\mathfrak{X}$-map $f\colon X \to Y$. There exist PL homeomorphisms $\varphi\colon X_{0,1} \to X_{0,2}$ and $\psi\colon Y_{0,1} \to Y_{0,2}$ (= id, resp.,) such that*

$$\tau_2^{-1} \circ f \circ \pi_2 \circ \varphi = \psi \circ \tau_1^{-1} \circ f \circ \pi_1.$$

In §IV.3, we consider these conjectures, and the following proposition follows as a corollary.

Proposition IV.1.7. *For the same $\mathfrak{X}$, X, Y and f as in IV.1.5, assume $\dim Y \le 2$, f is proper and each point y of Y has an $\mathfrak{X}$-neighborhood V in*

Y such that $f|_{f^{-1}(V)}$ is C^0 (R-)$\mathfrak{X}$-triangulable. Then f is globally C^0 (resp., R-)$\mathfrak{X}$-triangulable.

Let us consider other conditions for C^0 $\mathfrak{X}$-triangulability. Let $f: X \to Y$ be a proper $\mathfrak{X}$-map between locally closed $\mathfrak{X}$-sets. Assume Axiom (v). If f is C^0 $\mathfrak{X}$-triangulable, then there exists a C^0 $\mathfrak{X}$-stratification $\{X_i\} \to \{Y_j\}$ of f such that the function d_f on X, defined by

$$d_f(x) = \dim X_i - \dim f(X_i) \quad \text{for} \quad x \in X_i,$$

is lower semicontinuous. But this condition is not sufficient.

Example IV.1.8. Set

$$X = [0, \infty[\times \mathbf{R} \times [0, 1] \text{ and } Y = \mathbf{R}^2,$$

define f by

$$f(x, y, z) = (x, y^2 - xz) \quad \text{for} \quad (x, y, z) \in X \subset \mathbf{R}^3,$$

let $\{Y_j\}$ denote the canonical semialgebraic stratification of Y compatible with

$$\{(t, 0), (0, t), (t, -t) \in \mathbf{R}^2 \colon 0 \le t < \infty\},$$

and set

$$\{X_i\} = \{A \cap f^{-1}(B) \colon A \text{ are the strata of}$$
$$\text{the canonical semialgebraic stratification of } X, B \in \{Y_j\}\}.$$

We see that $f \colon \{X_i\} \to \{Y_j\}$ is the canonical semialgebraic stratification of f and d_f is lower semicontinuous. However, we can prove that f is not C^0 *triangulable* (which is naturally defined) by reduction to absurdity as follows.

Proof. Let $(X_0, Y_0, \pi_0, \tau_0)$ be a C^0 triangulation of f. Then $\pi_0^{-1}(0 \times 0 \times [0, 1])$ is a polyhedron because $\pi_0^{-1}(\partial X)$ is a polyhedron,

$$\partial X = (0 \times \mathbf{R} \times]0, 1[) \cup ([0, \infty[\times \mathbf{R} \times 0) \cup ([0, \infty[\times \mathbf{R} \times 1),$$

f is a topological submersion (= trivial) locally at each point of $0 \times (\mathbf{R} - 0) \times]0, 1[$, and f is not so at any point of $0 \times 0 \times [0, 1]$. Let K be a simplicial decomposition of X_0 compatible with $\pi_0^{-1}(0 \times 0 \times [0, 1])$ such that $\tau_0^{-1} \circ f \circ \pi_0$ is linear on each simplex of K. Let $(0, 0, z_0)$, $0 < z_0 < 1$, be a point whose image a under π_0^{-1} is not a vertex of K. For any open neighborhood U of

$(0, 0, z_0)$ in $[0, \infty[\times 0 \times]0, 1[$, $f|_U$ is not C^0 triangulable because f is injective on $U - 0 \times 0 \times [0, 1]$ and

$$f(0 \times 0 \times [0, 1]) = 0.$$

It follows that $\pi_0^{-1}(U)$ is not included in any one 2-simplex of K. Hence we have a 3-simplex σ of K and a sequence a_i, $i = 1, 2, \ldots$, in $\sigma^\circ \cap \pi_0^{-1}([0, \infty[\times 0 \times]0, 1[)$ converging to a. Let Δbc be the simplex of K whose interior contains a, let $\sigma = \Delta bcde$, and assume that the z-coordinate z_b of $\pi_0(b)$ is smaller than that of $\pi_0(c)$ (i.e., $z_b < z_c$). Then, since $f \circ \pi_0(\Delta bc) = 0$, by linearity of $\pi_0^{-1} \circ f \circ \pi_0$ on σ there exist unique sequences b_i, $i = 1, 2, \ldots$, in $(\Delta bde)^\circ$ and c_i, $i = 1, 2, \ldots$, in $(\Delta cde)^\circ$ converging to b and c respectively, such that for each i,

$$a_i \in \Delta b_i c_i, \quad \text{and} \quad f \circ \pi_0 = \text{const} \quad \text{on} \quad \Delta b_i c_i.$$

It follows that $\pi_0(b_i)$, $i = 1, 2, \ldots$, is a sequence converging to $\pi_0(b)$ such that for each i,
$$\pi_0(b_i) \in f^{-1}(f \circ \pi_0(a_i)).$$

Set
$$\pi_0(a_i) = (x_i, 0, z_i), \quad i = 1, 2, \ldots .$$

Then
$$f^{-1}(f \circ \pi_0(a_i)) = \{(x_i, y, z) \in X : z = z_i + y^2/x_i\}.$$

Hence the z-coordinate of $\pi_0(b_i)$ is larger than z_i, which contradicts the fact $z_b < z_0$. □

Let $f \colon X \to Y$ be an analytic map between analytic manifolds. If f is locally C^0 triangulable, then the function d'_f on X, defined by

$$d'_f(x) = \text{local dimension of } f^{-1}(f(x)) \text{ at } x,$$

is lower semicontinuous (see [S₁]). The above example shows that this condition is not sufficient.

If f and the manifolds are complex analytic, then the condition that $d'_f = \text{const}$ (i.e., f is flat) is necessary for f to be locally C^0 triangulable. However, the condition is not sufficient as shown below.

Example IV.1.8′. Let $f_n \colon \mathbf{C}^4 \to \mathbf{C}^3$ be defined by

$$f_n(x_1, \ldots, x_4) = (x_1, x_2, ((x_1 - x_2 x_3)^2 - x_4^2)^n) \quad \text{for} \quad (x_1, \ldots, x_4) \in \mathbf{C}^4,$$

where n is a large positive integer. It is easy to prove that f_n is flat. For simplicity of notation, we will show only that f_n is not C^0 triangulable by reduction to absurdity. Assume f_n admits a C^0 triangulation $(X_0, Y_0, \pi_0, \tau_0)$. We will choose the n so large that $(f_n \circ \pi_0)^{-1}(\mathbf{C}^2 \times 0)$ is a subpolyhedron of X_0, which will be possible by the following three facts.

Fact 1. For each point x of $f_n^{-1}(\mathbf{C}^2 \times 0)$ and for any neighborhood U of x in $\mathbf{C}^4$ there exists a point y of $\mathbf{C}^3$ such that the number of connected components of $f_n^{-1}(y) \cap U$ is equal to or larger than n. This is clear for y in $f_n(U) - \mathbf{C}^2 \times 0$ near $f_n(x)$.

Fact 2. For each point x of $\mathbf{C}^4 - f_n^{-1}(\mathbf{C}^2 \times 0)$ there exists a neighborhood U of x in $\mathbf{C}^4$ such that for any point y of $\mathbf{C}^3$, the number of connected components of $f_n^{-1}(y) \cap U$ is smaller than a number which is independent of n and x.

Proof of Fact 2. Let V be a simply connected and connected open set in $\mathbf{C}^3 - \mathbf{C}^2 \times 0$ which contains $f_n(x)$, and let U denote the connected component of $f_n^{-1}(V)$ which contains x. It is easy to prove that the number of connected components of $f_n^{-1}(y) \cap U$ for $y \in \mathbf{C}^3$ does not depend on n and is equal to or smaller than the maximal number of connected components of $f_1^{-1}(y)$ for $y \in \mathbf{C}^3$. Hence it suffices to show that the maximal number exists. We refer this well-known fact to [B-R]. □

Fact 3. Let $f \colon K \to L$ be a simplicial map. Let x and x' be points of one open simplex of K. There exist PL homeomorphisms φ of $|K|$ and ψ of $|L|$ such that $\varphi(x) = x'$ and $f \circ \varphi = \psi \circ f$. This implies that the germs of f at x and at x' are *locally R-L PL equivalent* (i.e., there exist germs of PL homeomorphisms $\varphi_x \colon (|K|, x) \to (|K|, x')$ and $\psi_{f(x)} \colon (|L|, f(x)) \to (|L|, f(x'))$ such that $\psi_{f(x)} \circ f = f \circ \varphi_x$ as germs at x).

Proof of Fact 3. Define φ by

$$\varphi = \mathrm{id} \quad \text{outside} \quad |\mathrm{st}(x, K)| \quad \text{and}$$
$$\varphi(tx + (1-t)a) = tx' + (1-t)a \quad \text{for} \quad a \in |\mathrm{lk}(x, k)| \quad \text{and} \quad t \in [0,1].$$

We define also ψ in the same way. Then

$$f \circ \varphi(tx + (1-t)a) = f(tx' + (1-t)a) = tf(x') + (1-t)f(a)$$
$$= \psi(tf(x) + (1-t)f(a)) = \psi \circ f(tx + (1-t)a)$$
$$\text{for} \quad a \in |\mathrm{lk}(x, K)|, \quad t \in [0,1].$$

Hence Fact 3 holds. □

Let n be so large that the minimum of the number defined in Fact 2 is smaller than n. By Facts 1 and 2 we can distinguish points of $f_n^{-1}(\mathbf{C}^2 \times 0)$ from points of $\mathbf{C}^4 - f_n^{-1}(\mathbf{C}^2 \times 0)$ by the minimal number. Now we want to see that $(f_n \circ \pi_0)^{-1}(\mathbf{C}^2 \times 0)$ is a subpolyhedron of X_0. As this problem is local, we can assume simplicial decompositions K of X_0 and L of Y_0 such that $\tau_0^{-1} \circ f_n \circ \pi_0 \colon K \to L$ is simplicial. Then by Fact 3 the minimal number is constant on each open simplex of K. Hence $(f_n \circ \pi_0)^{-1}(\mathbf{C}^2 \times 0)$ is the union of some open simplexes of K. Clearly it is closed in X_0. Therefore, it is a subpolyhedron of X_0.

Set $X = f_n^{-1}(\mathbf{C}^2 \times 0)$ and let $g \colon X \to \mathbf{C}^2$ denote the restriction to X of the projection of $\mathbf{C}^4$ to the first two factors. Since $(f_n \circ \pi_0)^{-1}(\mathbf{C}^2 \times 0)$ is a subpolyhedron of X_0, g admits a C^0 triangulation $(X_1, Y_1, \pi_1, \tau_1)$. Clearly (X_1, π_1) is a C^0 triangulation of X. From the following fact it follows that the inverse image under π_1 of $\{x_1 = x_2 x_3, x_4 = 0\}$, the C^∞ singular point set of X, is a subpolyhedron of X_1.

Fact 4. Let K be a simplicial complex. For two points x and x' of one open simplex of K there exists a PL homeomorphism φ of $|K|$ such that $\varphi(x) = x'$. This is a special case of Fact 3.

Therefore, the restriction to $\{x_1 = x_2 x_3\}$ of the projection of $\mathbf{C}^3$ which forgets the last factor is C^0 triangulable, which is impossible. Thus f_n is not C^0 triangulable. □

Conjecture IV.1.9 (X-version of Thom's conjecture). *Let $f \colon X \to Y$ be a proper X-map between locally closed X-sets. Assume Axiom* (v). *If f admits a C^1 Whitney X-stratification sans éclatement, then f is C^0 X-triangulable.*

A partial answer is the following proposition, where we do not need the condition that $\{$graph $f|_{X_i}\}$ is a Whitney stratification in the definition of a stratified map sans éclatement $f \colon \{X_i\} \to \{Y_j\}$. We will prove this in §IV.4.

Proposition IV.1.10. *In* IV.1.9, *assume $d_f \le 1$ or $\dim X \le 3$. Then* IV.1.9 *holds true.*

Conjecture IV.1.11. *If an X-map $f \colon X \to Y$ is C^0 (R-)triangulable, then f is C^0 (R-) X-triangulable and, moreover, (R-) X-triangulable.*

Theorem IV.1.12 [Te]. *Let $f \colon X \to Y \subset \mathbf{R}^m$ be a subanalytic map between compact subanalytic sets. There exist compositions of local blowings-up $g_i \colon Z_i \to \mathbf{R}^m$, $i = 1, \ldots, k$, such that the deduced maps of f by the base changes of g_i are C^0 subanalytically triangulable and the union of $g(Z_i)$ includes Y.*

Important corollaries of IV.1.2' and IV.1.2'' are the following, which will be proved in §IV.4.

Theorem IV.1.13 (Subanalytic triangulation of a complex analytic function). *Let X be a complex analytic set in $\mathbf{C}^n$ and let f be a complex analytic function on X. Regard X, $\mathbf{C}$ and f as subanalytic sets over $\mathbf{R}$ and a subanalytic map over $\mathbf{R}$. Then f admits a subanalytic C^0 R-triangulation.*

Theorem IV.1.13′ (Semialgebraic cell triangulation of a complex polynomial function). *Let $X \subset \mathbf{C}^n$ be an algebraic set, and let $f \colon X \to \mathbf{C}$ be a complex polynomial function. Regard X, $\mathbf{C}$ and f as semialgebraic sets over $\mathbf{R}$ and a semialgebraic map over $\mathbf{R}$. Then f admits a semialgebraic C^0 R-cell triangulation.*

Remark IV.1.14. In IV.1.13, if X is a complex analytic set in an open set U of $\mathbf{C}^n$, then f admits a *locally subanalytic C^0 R-triangulation* (i.e., a pair of a polyhedron X_0 in U and a homeomorphism π from X_0 to X such that $f \circ \pi$ is PL and π is subanalytic on some neighborhood of each point of X_0). This will be immediate by the following proof of IV.1.2′ and IV.1.13.

A polyhedron can be PL imbedded in some Euclidean space so that the image is closed and hence subanalytic in the Euclidean space. Hence the above f admits a subanalytic C^0 R-triangulation in the following weak sense: There exists a subanalytic polyhedron X_0 in some Euclidean space and a subanalytic map $\pi \colon X_0 \to \mathbf{C}^n$ such that π is a homeomorphism onto X and $f \circ \pi$ is PL.

An open problem is whether IV.1.13 holds in the non-affine case.

§IV.2. Proofs of Theorems IV.1.1, IV.1.2, IV.1.2′ and IV.1.2″

Proof of necessity of the conditions in IV.1.1, IV.1.2 and 1.2′. We prove only necessity of the conditions for C^0 $\mathfrak{X}$-triangulability in IV.1.2′. We can prove it in the other cases more easily. Let (X_0, Y_0, π, τ) be a C^0 $\mathfrak{X}$-triangulation of f. Since X is closed in $\mathbf{R}^n$, as noted in IV.1.3, we can assume that $X_0 \subset \mathbf{R}^n$ and π is the restriction to X_0 of an $\mathfrak{X}$-homeomorphism π_0 of $\mathbf{R}^n$. Here by changing the coordinate system of $\mathbf{R}^n$ by π_0^{-1}, we assume $X_0 = X$ and $\pi = \mathrm{id}$. Let $\mathbf{R}^n \times \mathbf{R}^m \overset{p_1}{\to} \cdots \overset{p_n}{\to} \mathbf{R}^m$ denote the projections which forget the respective first factors. For a moment assume, in addition, $Y_0 = Y$ and $\tau = \mathrm{id}$, i.e., f is PL. Then what we need to prove is the following.

A polyhedron in $\mathbf{R}^n \times \mathbf{R}^m$ admits a local resolution with respect to $p_1, \ldots, p_n$.

Its proof is similar to and easier than the proof of IV.1.13. In this case, we require the local resolution $\{\mathcal{A}_{i,x}\}_{\substack{i=0,\ldots,n \\ x \in \mathbf{R}^{n-i} \times \mathbf{R}^m}}$ to satisfy the following condition:

(vii)′ For each $i = 0, \ldots, n$, there exist finite simplicial complexes K_k, $k = 1, 2, \ldots$, in $\mathbf{R}^{n-i} \times \mathbf{R}^m$ such that for each $x \in \mathbf{R}^{n-i} \times \mathbf{R}^m$, each element of $\mathcal{A}_{i,x}$ is the germ at x of the family of the open simplexes of some K_k with $x \in |K_k|$.

We can prove as in the proof of IV.1.13 (§IV.4) that a countable or finite family of polyhedra in $\mathbf{R}^n \times \mathbf{R}^m$ admits such a local resolution (which is similarly defined). We omit the details.

Consider the general case of Y_0 and τ. Let $Y_0 \subset \mathbf{R}^{m'}$ and apply the above arguments to $\tau^{-1} \circ f$. Then we obtain $\{\mathcal{A}_{i,x}\}_{\substack{i=0,\dots,n \\ x \in \mathbf{R}^{n-i} \times \mathbf{R}^m}}$ which satisfies the conditions of a local resolution of f with respect to $p_1, \dots, p_n$ except the condition that each element A of $\mathcal{A}_{i,x}$ is the germ at x of a C^0 $\mathfrak{X}$-stratification of a neighborhood of x in $\mathbf{R}^{n-i} \times \mathbf{R}^m$. Here A is the germ at x of a C^0 $\mathfrak{X}$-stratification of a neighborhood of x in $p_i \circ \cdots \circ p_1(\mathrm{graph}\, f)$ because the ambient Euclidean space of Y is different from that of Y_0. Hence we need to extend the stratification to a neighborhood of x in $\mathbf{R}^{n-i} \times \mathbf{R}^m$. We accomplish this by downward induction on i. If $\{X_{n,j}\}_j$ is a realization of some A in $\mathcal{A}_{n,x}$, replace A with the germ at x of the set:

$$\Big\{ U_n - \bigcup_j X_{n,j} \Big\} \cup \{X_{n,j}\}_j,$$

where U_n is a small open neighborhood of x in $\mathbf{R}^m$. Assume $\{\mathcal{A}_{i,x}\}_{\substack{i=i_0+1,\dots,n \\ x \in \mathbf{R}^{n-i} \times \mathbf{R}^m}}$ is already modified. For each A_{i_0} in $\mathcal{A}_{i_0,x}$, there exist $\{X_{i_0,j}\}_j$ and $\{X_{i_0+1,j}\}_j$ realizations of A_{i_0} and some A_{i_0+1} in $\mathcal{A}_{i_0+1,p_{i_0+1}(x)}$ respectively, such that $\{p_{i_0+1}(X_{i_0,j})\}_j \subset \{X_{i_0+1,j}\}_j$, because each element of $\mathcal{A}_{i_0+1,p_{i_0+1}(x)}$ is an extension of an element of the original $\mathcal{A}_{i_0+1,p_{i_0+1}(x)}$. Let U_{i_0} be a small open neighborhood of x in $\mathbf{R}^{n-i_0} \times \mathbf{R}^m$. Define an extension of A_{i_0} by the germ at x of the set:

$$\Big\{ \big(U_{i_0} - \bigcup_j X_{i_0,j} \big) \cap p_{i_0+1}^{-1}(X_{i_0+1,j'}) \Big\}_{j'} \cup \{X_{i_0,j}\}_j.$$

Then $\{\mathcal{A}_{i,x}\}_{\substack{i=0,\dots,n \\ x \in \mathbf{R}^{n-i} \times \mathbf{R}^m}}$ satisfies all the conditions of a local resolution.   □

Proof that the second statement of IV.1.2′ implies the first and second statements of IV.1.1 and of IV.1.2. Clearly the second statement of IV.1.2′ and the above proof imply the second statements of IV.1.1 and IV.1.2. The second statement of IV.1.2 implies the first of IV.1.2 for the following reason. Assume the second. Let $\{X_{0,j}\}_j \xrightarrow{p_1} \cdots \xrightarrow{p_l} \{X_{l,j}\}_j$ be a resolution of f with respect to $p_1, \dots p_l$. Let (K, τ) be a C^1 $\mathfrak{X}$-triangulation of $\mathbf{R}^m$ compatible with $\{Y\} \cup \{X_{l,j}\}_j$ such that $|K| = \mathbf{R}^m$ (II.2.1′). Then $\tau^{-1} \circ f \colon X \to \tau^{-1}(Y)$ has a resolution $\{\tau_0^{-1}(X_{0,j})\}_j \xrightarrow{p_1} \cdots \xrightarrow{p_l} \{\tau_l^{-1}(X_{l,j})\}_j$, and $\tau^{-1}(Y)$ and $\tau^{-1}(X_{l,j})$

satisfy the conditions of the second statement of IV.1.2, where each τ_i is the homeomorphism of $\mathbf{R}^{n_i} \times \mathbf{R}^m$ defined by $\tau_i(x,y) = (x, \tau(y))$. Hence $\tau^{-1} \circ f \colon X \to \tau^{-1}(Y)$ is C^0 R-$\mathfrak{X}$-triangulable, which proves the first of IV.1.2. We can prove in the same way that the second statement of IV.1.1 implies the first of IV.1.1. $\square$

Thus we need to prove the first and second statements of IV.1.2′, which follow from the following theorem

Theorem IV.2.1. *Let $X \subset \mathbf{R}^n \times \mathbf{R}^m$ be a closed $\mathfrak{X}$-set, let $p \colon \mathbf{R}^n \times \mathbf{R}^m \to \mathbf{R}^m$ be the projection, let $P \subset \mathbf{R}^m$ be a closed polyhedron which includes $p(X)$, and let $\mathbf{R}^n \times \mathbf{R}^m = \mathbf{R}^{n_0} \times \mathbf{R}^m \xrightarrow{p_1} \cdots \xrightarrow{p_l} \mathbf{R}^{n_l} \times \mathbf{R}^m = \mathbf{R}^m$ be the projections which forget the respective first factors. Assume Axiom (v) and that there exists a local resolution $\{\mathcal{A}_{i,x}\}_{\substack{i=0,\ldots,l \\ x \in \mathbf{R}^{n_i} \times \mathbf{R}^m}}$ of X with respect to $p_1, \ldots, p_l$. If $p|_X$ is proper, there exists an $\mathfrak{X}$-homeomorphism π of $\mathbf{R}^n \times \mathbf{R}^m$ of the form:*

$$\pi(x,y) = (\pi'(x,y), \pi''(y)) \quad \text{for} \quad (x,y) \in \mathbf{R}^n \times \mathbf{R}^m$$

such that $\pi(X)$ is a polyhedron and π'' is an invariant of P. Even in the case where $p|_X$ is not proper, if for any $\{X_{l,j}\}_j \in \mathcal{A}_{l,x}$, each $X_{l,j}$ is the germ of a finite union of some open simplexes, then we can choose the above π so that $\pi'' = \mathrm{id}$.

Proof that IV.2.1 implies the first and second statements of IV.1.2′. Set $\tilde{X} = \mathrm{graph} f$ for f in IV.1.2′, and let $\{\mathcal{A}_{i,x}\}_{\substack{i=0,\ldots,l \\ x \in \mathbf{R}^{n_i} \times \mathbf{R}^m}}$ be a local resolution of f with respect to $p_1, \ldots, p_l$. If $\tilde{X}$ is closed in $\mathbf{R}^n \times \mathbf{R}^m$, Y is a polyhedron and closed in $\mathbf{R}^m$, and we set $P = Y$. Then the first and second statements clearly follow from the first and second statements of IV.2.1 respectively.

First we reduce the problem to the case where $\tilde{X}$ is closed in $\mathbf{R}^n \times \mathbf{R}^m$. Assume that $\tilde{X}$ is not so. Set $\tilde{X}' = \overline{\tilde{X}} - \tilde{X}$, which is closed in $\mathbf{R}^n \times \mathbf{R}^m$ because X is locally closed in $\mathbf{R}^n$, set

$$Z = \{(1/\operatorname{dis}(x, \tilde{X}'), x) \colon x \in \tilde{X}\} \subset \mathbf{R} \times \tilde{X},$$

and let $p_0 \colon \mathbf{R}^{n+1} \times \mathbf{R}^m \to \mathbf{R}^n \times \mathbf{R}^m$ denote the projection which forgets the first factor. Then $p_0|_Z \colon Z \to Y$ is an $\mathfrak{X}$-homeomorphism, Z is closed in $\mathbf{R}^{n+1} \times \mathbf{R}^m$, and it suffices to consider $f \circ (p_0|_Z)$ in place of f. We need to show that $f \circ (p_0|_Z)$ admits a local resolution. Define $\{\mathcal{A}_{-1,x}\}_{x \in \mathbf{R}^{n-1} \times \mathbf{R}^m}$ $(n_{-1} = n+1)$ as follows. For each $x \in \mathbf{R}^{n-1} \times \mathbf{R}^m$ and for each $A_0 \in \mathcal{A}_{0,p_0(x)}$, let A_{-1} denote the germ at x of $\{\mathbf{R} \times X_{0,j} - Z, Z \cap (\mathbf{R} \times X_{0,j})\}_j$, where $\{X_{0,j}\}_j$ is a realization of A_0. Let $\mathcal{A}_{-1,x}$ denote the family of all such A_{-1}'s. Then $\{\mathcal{A}_{i,x}\}_{\substack{i=-1,2,\ldots,l \\ x \in \mathbf{R}^{n_i} \times \mathbf{R}^m}}$ is a local resolution of $f \circ (p_0|_Z) \colon Z \to Y$ with respect to $p_1 \circ p_0, p_2, \ldots, p_l$.

Next we can assume that Y is a polyhedron and closed in $\mathbf{R}^m$ for the following reason. There exists an $\mathfrak{X}$-homeomorphism g from Y onto a polyhedron Y' included and closed in some $\mathbf{R}^{m'}$ such that if Y is a polyhedron, then g is PL (see the note after Corollary II″). Then it suffices to show that $g \circ f \colon X \to Y'$ admits a local resolution with respect to $p_1', \ldots, p_l'$, where each $p_i' \colon \mathbf{R}^{n_{i-1}} \times \mathbf{R}^{m'} \to \mathbf{R}^{n_i} \times \mathbf{R}^{m'}$ is the projection which forgets the first factors. Let the $\mathfrak{X}$-homeomorphisms $g_i \colon \mathbf{R}^{n_i} \times Y \to \mathbf{R}^{n_i} \times Y'$ be the natural extensions of g. For each $A \in \mathcal{A}_{i,x}$, $g_i(A|_{\mathbf{R}^{n_i} \times Y})$ is the germ at $g_i(x)$ of a stratification of a neighborhood of $g_i(x)$ in $\mathbf{R}^{n_i} \times Y'$. Add to $g_i(A|_{\mathbf{R}^{n_i} \times Y})$ the germ at $g_i(x)$ of $\mathbf{R}^{n_i} \times (\mathbf{R}^{m'} - Y')$. Then the family is the germ of a stratification of a neighborhood of $g_i(x)$ in $\mathbf{R}^{n_i} \times \mathbf{R}^{m'}$. Let $\mathcal{A}'_{i,g(x)}$ denote the family of all such germs. For $y \in \mathbf{R}^{n_i} \times (\mathbf{R}^{m'} - Y)$, let $\mathcal{A}'_{i,y}$ denote the germ of one element $\mathbf{R}^{n_i} \times \mathbf{R}^{m'}$. Then $\{\mathcal{A}'_{i,y}\}_{\substack{i=0,\ldots,l \\ y \in \mathbf{R}^{n_i} \times \mathbf{R}^{m'}}}$ is a local resolution of $g \circ f$ with respect to $p_1', \ldots, p_l'$.   □

Thus it suffices to prove IV.2.1. We reduce IV.2.1 to an easier problem (IV.2.3). Before this, we generalize IV.2.1 so that an induction method works (IV.2.2, IV.2.4). First we consider the first statement of IV.2.1. Clearly it is a special case of the following.

Lemma IV.2.2. *Let X, p, $p_1, \ldots, p_l$ and P be the same as in IV.2.1. Let $\{U_k\}_k$ be a locally finite open $\mathfrak{X}$-covering of X in $\mathbf{R}^n \times \mathbf{R}^m$. For each k, let $Z_{k,k'} \subset X \cap U_k$ be a finite number of $\mathfrak{X}$-sets closed in U_k, and let $\{\mathcal{A}_{i,x}\}_{\substack{i=0,\ldots,l \\ x \in \mathbf{R}^{n_i} \times \mathbf{R}^m}}$ be a local resolution of $\{Z_{k,k'}\}_{k,k'}$ with respect to $p_1, \ldots, p_l$ (which we define as in the proof of IV.1.13). Assume Axiom (v) and that $p|_X$ is proper. There exist a closed $\mathfrak{X}$-covering $\{V_k\}_k$ of X in $\mathbf{R}^n \times \mathbf{R}^m$ and an $\mathfrak{X}$-homeomorphism π of $\mathbf{R}^n \times \mathbf{R}^m$ of the form:*

$$\pi(x,y) = (\pi'(x,y), \pi''(y)) \quad \text{for} \quad (x,y) \in \mathbf{R}^n \times \mathbf{R}^m$$

such that $V_k \subset U_k$, $k = 1, 2, \ldots, \pi(V_k)$ and $\pi(V_k \cap Z_{k,k'})$ for each k and k' are polyhedra, and π'' is invariant on P.

Proof of IV.2.2. If $l = 0$, this follows from II.2.1′ except for the requirement that $\pi'' = \pi$ is invariant on P. But this requirement can be fulfilled by the same reason as II.2.4. Hence assume by the induction hypothesis IV.2.2 in the case of $l - 1$. Let $\{U_k'\}_k$ be a closed $\mathfrak{X}$-covering of X in $\mathbf{R}^n \times \mathbf{R}^m$ such that $U_k' \subset U_k$ and $p|_{\cup_k U_k'} \colon \cup_k U_k' \to \mathbf{R}^m$ is proper. Set $Z = \cup_{k,k'} Z_{k,k'}$. Assume that the map $p_1|_Z$ is a finite-to-one map. Shrink each U_k. Then it follows that $p_1|_{\overline{Z}}$ is a finite-to-one map. Let $x' \in p_1(X)$. For each $x \in Z \cap p_1^{-1}(x')$, we have realizations $\{X_{0,j}(x)\}_j$ of some element $A_0(x)$ of $\mathcal{A}_{0,x}$ and $\{X_{1,j}(x)\}_j$

of $p_1(A_0(x))$ such that $\{X_{0,j}(x)\}_j$ is compatible with $Z_{k,k'}$ for any k and k' with $x \in Z_{k,k'}$, and $\{X_{0,j}(x)\}_j$ and $\{X_{1,j}(x)\}_j$ satisfy conditions (i) and (ii) of a resolution. Here $\{X_{0,j}(x)\}_j$ is not necessarily compatible with $Z_{k,k'}$ for k and k' with $x \in \overline{Z_{k,k'}} - Z_{k,k'}$. But if we shrink $\{X_{0,j}(x)\}_j$ and $\{X_{1,j}(x)\}_j$, then $\{X_{0,j}(x)\}_j$ is compatible with all $Z_{k,k'} \cap U_k''$ for some open $\mathfrak{X}$-sets U_k'' with $U_k' \subset U_k'' \subset U_k$. Since $p_1|_Z$ is a finite-to-one map, we can modify $\{X_{0,j}(x)\}_j$ outside of $\cup_{k,k'} Z_{k,k'} \cap U_k''$ so that

$$\{X_{0,j}(x)\}_j|_{\cup_j X_{0,j}(x) - \cup_{k,k'}(Z_{k,k'} \cap U_k'')}$$
$$= \Big\{ (\mathbf{R}^{n-n_1} \times X_{1,j}(x)) \cap \Big(\bigcup_{j'} X_{0,j'}(x) - \bigcup_{k,k'}(Z_{k,k'} \cap U_k'')\Big)\Big\}_j.$$

We say then that $\{X_{0,j}(x)\}_j$ is *trivial* over $\{X_{1,j}(x)\}_j$ outside of $\cup_{k,k'}(Z_{k,k'} \cap U_k'')$. (Note that now $\{X_{0,j}(x)\}_j$ is not necessarily a realization of an element of $\mathcal{A}_{0,x}$.) Therefore, by condition (iv) of a local resolution (which is applied to find the following $\{X_{1,j}(x')\}_j$), we have open $\mathfrak{X}$-neighborhoods $W(x')$ of $X \cap p_1^{-1}(x')$ in $\mathbf{R}^n \times \mathbf{R}^m$, $R(x')$ of x' in $\mathbf{R}^{n_1} \times \mathbf{R}^m$, and $W_k(x')$ of $p_1^{-1}(x') \cap U_k'$ in U_k, $k = 1, 2, \ldots$. We have C^0 $\mathfrak{X}$-stratifications $\{X_{0,j}(x')\}_j$ of $W(x')$ and $\{X_{1,j}(x')\}_j$ of $R(x')$ such that each $W_k(x')$ is of the form $Q_k(x') \times R(x')$, $Q_k(x')$ is the interior of a compact PL submanifold of $\mathbf{R}^{n-n_1}$ with boundary,

$$U_k' \cap (\mathbf{R}^{n-n_1} \times R(x')) \subset W_k(x'), \tag{$*$}$$
$$\bigcup_k W_k(x') = W(x'), \qquad ((\partial \overline{Q_k(x')}) \times R(x')) \cap \overline{Z} = \varnothing,$$

$\{X_{0,j}(x')\}_j$ is compatible with all $W_k(x') \cap Z_{k,k'}$, $\{X_{1,j}(x')\}_j$ is a realization of some element of $\mathcal{A}_{1,x'}$, and $\{X_{0,j}(x')\}_j$ and $\{X_{1,j}(x')\}_j$ satisfy conditions (i) and (ii) of a resolution. (Here we can define

$$\{X_{0,j}(x')\}_j = \Big\{ p_1^{-1}(X_{1,j}(x')) \cap W_k(x') \cap Z_{k,k'},$$
$$\big(p_1^{-1}(X_{1,j}(x')) \cap W(x')\big) - \Big(\bigcup_{k_1,k_1'} W_{k_1}(x') \cap Z_{k_1,k_1'}\Big)\Big\}_{j,k,k'}.\big)$$

Shrinking $R(x')$, we obtain a countable or finite number of points x_1', $x_2', \ldots$ in $p_1(X)$ such that $\{R(x_l')\}_l$ is a locally finite open covering of $p_1(X)$ in $\mathbf{R}^{n_1} \times \mathbf{R}^m$, which together with condition $(*)$ implies that for each k, $\{W_k(x_l')\}_l$ is a covering of $X \cap U_k'$ in U_k. For each l, consider the pair of $R(x_l')$ and $\{\cup_{\dim \leq \alpha} X_{1,j}(X_l')\}_\alpha$, and apply the induction hypothesis to all these pairs. We can assume there exists a closed $\mathfrak{X}$-covering $\{S(x_l')\}_l$ of $p_1(X)$ in $\mathbf{R}^{n_1} \times \mathbf{R}^m$ such that $S(x_l') \subset R(x_l')$, $l = 1, 2, \ldots$, and all $S(x_l')$ and

$S(x'_l) \cap (\cup_{\dim \leq \alpha} X_{1,j}(x'_l))$ are polyhedra. Let L be a simplicial decomposition of $\cup_l S(x'_l)$ compatible with all these polyhedra. Set

$$V_k = \bigcup_l \overline{Q_k(x'_l)} \times S(x'_l), \quad k = 1, 2, \dots .$$

Then $X \cap U'_k \subset V_k \subset U_k$. Hence $\{V_k\}_k$ is a closed $\mathfrak{X}$-covering of X in $\mathbf{R}^n \times \mathbf{R}^m$. Moreover, the family

$$\{\bigcup_{k,k'} V_k \cap Z_{k,k'} \cap p_1^{-1}(\sigma^\circ) : \sigma \in L\}$$

is compatible with all $V_k \cap Z_{k,k'}$ because the family is compatible with

$$\{(\overline{Q_k(x'_l)} \times S(x'_l)) \cap X_{0,j}(x'_l)\}_{j,l}.$$

Therefore, it suffices to find the π such that $\pi'' = \text{id}$ and $\pi(\cup_{k,k'} (V_k \cap Z_{k,k'}))$ is a polyhedron. Note that for each $\sigma \in L$, the map $p_1|_{\cup_{k,k'}(V_k \cap Z_{k,k'}) \cap p_1^{-1}(\sigma^\circ)}$ is a finite covering map onto σ°. By this and the following lemma IV.2.3 such π exists. (Here we apply IV.2.3 to $Y = \cup_{k,k'}(V_k \cap Z_{k,k'})$, $Y_1 = Y_2 = \varnothing$, $p = p_1$ and $P = \cup_{k,l}(\partial \overline{Q_k(x'_l)}) \times S(x'_l)$.) Hence it suffices to prove IV.2.3.

Consider the case where $p_1|_Z$ is not a finite-to-one map. In this case, $n = n_1 + 1$ by condition (i) of a resolution because $p_1|_{\overline{Z}}$ is proper. Set

$$Z'_{k,k'} = \bigcup_{x' \in \mathbf{R}^{n_1} \times \mathbf{R}^m} (Z_{k,k'} \cap \overline{(p_1^{-1}(x') - Z_{k,k'})}).$$

We prove the lemma for $\{Z'_{k,k'}\}_{k,k'}$ in place of $\{Z_{k,k'}\}_{k,k'}$ and can apply the above arguments to $\{Z'_{k,k'}\}_{k,k'}$ for the following reason. For each $x' \in \mathbf{R}^{n_1} \times \mathbf{R}^m$, each $Z'_{k,k'} \cap p_1^{-1}(x')$ is the boundary of $Z_{k,k'} \cap p_1^{-1}(x')$ in the line $p_1^{-1}(x')$. Hence $p_1|_{\cup_{k,k'} Z'_{k,k'}}$ is a finite-to-one map. By condition (ii) of a resolution each $Z'_{k,k'}$ is closed in U_k. Finally, by (i) $\{A_{i,x}\}_{\substack{i=0,\dots,l \\ x \in \mathbf{R}^{n_i} \times \mathbf{R}^m}}$ is a local resolution of $\{Z'_{k,k'}\}_{k,k'}$. Hence by the above arguments we obtain a closed $\mathfrak{X}$-covering $\{V_k\}_k$ of X in $\mathbf{R}^n \times \mathbf{R}^m$ and an $\mathfrak{X}$-homeomorphism π of $\mathbf{R}^n \times \mathbf{R}^m$ which satisfy the conditions in the lemma except the condition that each $\pi(V_k \cap Z_{k,k'})$ is a polyhedron. This is replaced by the condition that each $\pi(V_k \cap Z'_{k,k'})$ is a polyhedron. But by the form of π, the former condition follows from the latter, which proves the lemma.   □

Lemma IV.2.3. *Assume Axiom* (v). *Let* $Y \supset Y_1 \supset Y_2$ *be closed $\mathfrak{X}$-sets in* $\mathbf{R}^n \times \mathbf{R}^m$, *and let* $p\colon \mathbf{R}^n \times \mathbf{R}^m \to \mathbf{R}^m$ *and* $q\colon Y \to \mathbf{R}^m$ *denote the projection and its restriction to* Y *respectively. Assume* $q(Y)$ *is a polyhedron. Let* L *be a simplicial decomposition of* $q(Y)$, *and let* P *be a closed polyhedron in* $\mathbf{R}^n \times \mathbf{R}^m$ *which does not intersect with* Y. *Assume that* q *is proper,* Y_1 *and* Y_2 *are polyhedra,* Y_1 *is a neighborhood of* Y_2 *in* Y, $q^{-1}(q(Y_1)) = Y_1$, *and for each simplex* $\sigma \in L$, *the map*

$$q|_{q^{-1}(\sigma^\circ)}\colon q^{-1}(\sigma^\circ) \longrightarrow \sigma^\circ$$

is a covering map. Then there exists an $\mathfrak{X}$-homeomorphism π *of* $\mathbf{R}^n \times \mathbf{R}^m$ *of the form*:

$$\pi(x,y) = (\pi'(x,y), y) \quad \text{for} \quad (x,y) \in \mathbf{R}^n \times \mathbf{R}^m$$

such that $\pi(Y)$ *is a polyhedron and* $\pi = \mathrm{id}$ *on* $P \cup p^{-1}(q(Y_2))$.

To prove the second statement of IV.2.1 we use a special decomposition of a compact polyhedron in $\mathbf{R}^{m+n}$. A *rectangular decomposition* of a compact polyhedron X in $\mathbf{R}^{m+n}$ is a finite family of connected compact sets V_k such that each V_k is the closure of a set $V_k' - \cup_{\text{finite}\ l} V_{k,l}$ for some boxes V_k' and $V_{k,l}$, and $\{V_k^\circ\}_k$ is a stratification of X with the frontier condition. Here a *box* means a set of the form $[a_1,b_1] \times \cdots \times [a_{m+n}, b_{m+n}]$, $a_i, b_i \in \mathbf{R}$, and we regard each V_k as a PL manifold possibly with boundary. (The manifold of positive dimension always has boundary.)

There are two reasons why we introduce the concept of a rectangular decomposition. First, for two rectangular decompositions F_1 of X_1 and F_2 of X_2 in $\mathbf{R}^{m+n}$, there exists the roughest rectangular decomposition F_3 of $X_1 \cup X_2$ which is compatible with F_1 and F_2 (i.e., $\{V^\circ\colon V \in F_3\}$ is so) as in the case of a usual cell complex. We can define F_3 to be all the connected components of the closures of $V_1 - X_2$, $V_2 - X_1$ and $V_1 \cap V_2$ for $V_1 \in F_1$ and $V_2 \in F_2$. Clearly

$$\bigcup_{\substack{V \in F_3 \\ \dim = m+n}} V^\circ = \bigcup_{\substack{V \in F_1 \cup F_2 \\ \dim = m+n}} V^\circ - \bigcup_{\substack{V \in F_1 \cup F_2 \\ \dim < m+n}} V,$$

which we shall use later. We call F_3 the rectangular decomposition generated by F_1 and F_2.

The second reason is that a rectangular decomposition admits canonically isotopic neighborhoods as follows. Let F be a rectangular decomposition of a compact polyhedron X in $\mathbf{R}^{m+n}$. For a small positive number ε, we define a rectangular decomposition $F(\varepsilon)$ of the closed ε-neighborhood $X(\varepsilon)$ of X in $\mathbf{R}^{m+n}$ as follows. (Here the distance in $\mathbf{R}^{m+n}$ is defined by

$$\mathrm{dis}(x,y) = \max_{i=1,\dots,m+n} |x_i - y_i|$$

$$\text{for} \quad x = (x_1, \dots, x_{m+n}),\ y = (y_1, \dots, y_{m+n}) \in \mathbf{R}^{m+n}.)$$

(Note that $X(\varepsilon)$ is a PL manifold possibly with boundary.) Let $V \in F$. For simplicity of notation, we assume $V = V' \times x^0_{n_1+1} \times \cdots \times x^0_{m+n}$, where $n_1 = \dim V$, $V' \subset \mathbf{R}^{n_1}$ and $(x^0_{n_1+1}, \ldots, x^0_{m+n}) \in \mathbf{R}^{m+n-n_1}$. Let $\{V_k\}_k$ denote the family:

$$(\mathbf{R}^{n_1} - (\text{the open } \varepsilon\text{-neighborhood of } \mathbf{R}^{n_1} - V' \text{ in } \mathbf{R}^{n_1}))$$
$$\times \{\text{all faces of } [x^0_{n_1+1} - \varepsilon, x^0_{n_1+1} + \varepsilon] \times \cdots \times [x^0_{m+n} - \varepsilon, x^0_{m+n} + \varepsilon]\}.$$

We say that each V_k is *derived* from V. Set $F(\varepsilon) = \{V_k\}_{k, V \in F}$. It is easy to show that $F(\varepsilon)$ is a rectangular decomposition of $X(\varepsilon)$, each element of $F(\varepsilon)$ is derived from only one element of F, and for each $0 < \varepsilon' < \varepsilon$, there is a natural one-to-one correspondence from $F(\varepsilon)$ to $F(\varepsilon')$. For each V_k defined as above, let $\tilde{V}_k$ denote the subset of $\mathbf{R}^{m+n} \times [1/2, 1]$ such that

$$\tilde{V}_k \cap (\mathbf{R}^{m+n} \times 1) = V_k \times 1,$$

and for each $t \in [1/2, 1[$, $\tilde{V}_k \cap (\mathbf{R}^{m+n} \times t)$ is the product of the element of $F(\varepsilon t)$, which corresponds to V_k, and t. Set

$$\tilde{F} = \{\tilde{V}_k\}_{k, V \in F} \cup (F(\varepsilon) \times 1) \cup (F(\varepsilon/2) \times 1/2) \quad \text{and}$$
$$\tilde{X} = \bigcup_{t \in [1/2, 1]} X(\varepsilon t) \times t \subset \mathbf{R}^{m+n+1},$$

where

$$F(\varepsilon) \times 1 = \{V_k \times 1 : V_k \in F(\varepsilon)\}.$$

Then $\tilde{X}$ is both a PL manifold with boundary and a neighborhood of $X \times [1/2, 1]$ in $\mathbf{R}^{m+n} \times [1/2, 1]$, and $\tilde{F}$ is a finite decomposition of $\tilde{X}$ into PL manifolds possibly with boundary whose interiors form a stratification of $\tilde{X}$ with the frontier condition. Note that $\cup_{t \in [1/2,1]} \text{bdry } X(\varepsilon t)$ is a PL collar of $X(\varepsilon)$, and

$$\text{bdry } X(\varepsilon t) \cap \text{bdry } X(\varepsilon t') = \varnothing \quad \text{for} \quad t \neq t' \in [1/2, 1].$$

Let $\mathbf{R}^n \times \mathbf{R}^m = \mathbf{R}^{n_0} \times \mathbf{R}^m \xrightarrow{p_1} \cdots \xrightarrow{p_l} \mathbf{R}^{n_l} \times \mathbf{R}^m = \mathbf{R}^m$ be the projections which forget the respective first factors. For each $i = 0, \ldots, l$ and for each $x \in \mathbf{R}^{n_i} \times \mathbf{R}^m$, let $\mathcal{A}_{i,x}$ be a family of germs at x of C^∞ $\mathfrak{X}$-stratifications of neighborhoods of x in $\mathbf{R}^{n_i} \times \mathbf{R}^m$. We call $\mathcal{A}_{i,x}$ *polyhedral* if each element of $\mathcal{A}_{i,x}$ is the germ at x of a C^∞ $\mathfrak{X}$-stratification whose strata are finite unions of open simplexes. We call $\{\mathcal{A}_{i,x}\}_{\substack{i=0,\ldots,l \\ x \in \mathbf{R}^{n_i} \times \mathbf{R}^m}}$ *polyhedral* if so are all $\mathcal{A}_{i,x}$. Note that if a local resolution of an $\mathfrak{X}$-map is polyhedral, then the map is PL.

We call $\{\mathcal{A}_{i,x}\}_{\substack{i=0,\ldots,l \\ x\in\mathbf{R}^{n_i}\times\mathbf{R}^m}}$ *maximal* if the following two conditions are satisfied. Each $\mathcal{A}_{i,x}$ contains the trivial germ = the germ of $\mathbf{R}^{n_i}\times\mathbf{R}^m$ at x. Let $x\in\mathbf{R}^{n_i}\times\mathbf{R}^m$. Let $\{X_{i,j}\}_j$ be a C^0 $\mathfrak{X}$-stratification with the weak frontier condition of a neighborhood of x in $\mathbf{R}^{n_i}\times\mathbf{R}^m$ such that a *substratification* of $\{X_{i,j}\}_j$ (i.e., a stratification compatible with $\{X_{i,j}\}_j$) is a realization of an element of $\mathcal{A}_{i,x}$, and $\{p_{i+1}(X_{i,j})\}_j$ is a realization of an element of $\mathcal{A}_{i+1,p_{i+1}(x)}$. Then the germ of $\{X_{i,j}\}_j$ at x itself is an element of $\mathcal{A}_{i,x}$. In the case where $\{\mathcal{A}_{i,x}\}_{\substack{i=0,\ldots,l \\ x\in\mathbf{R}^{n_i}\times\mathbf{R}^m}}$ is polyhedral, we call it *maximal* if we consider only polyhedral $\{X_{i,j}\}_j$ in the above definition.

Note the following two facts. For a local resolution $\{\mathcal{A}_{i,x}\}_{\substack{i=0,\ldots,l \\ x\in\mathbf{R}^{n_i}\times\mathbf{R}^m}}$ of $\mathfrak{X}$-sets there exists a canonical maximal local resolution $\{\mathcal{A}'_{i,x}\}_{\substack{i=0,\ldots,l \\ x\in\mathbf{R}^{n_i}\times\mathbf{R}^m}}$ of the same sets such that $\{\mathcal{A}_{i,x}\}_{\substack{i=0,\ldots,l \\ x\in\mathbf{R}^{n_i}\times\mathbf{R}^m}}$ is a *local subresolution* of $\{\mathcal{A}'_{i,x}\}_{\substack{i=0,\ldots,l \\ x\in\mathbf{R}^{n_i}\times\mathbf{R}^m}}$ (i.e., $\mathcal{A}'_{i,x}\supset\mathcal{A}_{i,x}$ for all i and x). We define $\{\mathcal{A}'_{i,x}\}_{\substack{i=0,\ldots,l \\ x\in\mathbf{R}^{n_i}\times\mathbf{R}^m}}$ by downward induction on i as follows. First

$$\mathcal{A}'_{l,x}=\mathcal{A}_{l,x}\cup\{\text{the germ of }\mathbf{R}^m\text{ at }x\}\quad\text{for}\quad x\in\mathbf{R}^m.$$

Assume we have defined $\{\mathcal{A}'_{j,x}\}_{\substack{j=i+1,\ldots,l \\ x\in\mathbf{R}^{n_j}\times\mathbf{R}^m}}$. For each $x\in\mathbf{R}^{n_i}\times\mathbf{R}^m$ $\mathcal{A}'_{i,x}$ is the family of the germs at x of C^0 $\mathfrak{X}$-stratifications with the weak frontier condition $\{X_{i,j}\}_j$'s such that some substratifications of $\{X_{i,j}\}_j$ are realizations of elements of $\mathcal{A}_{i,x}$ and $\{p_{i+1}(X_{i,j})\}_j$ are realizations of elements of $\mathcal{A}'_{i+1,p_{i+1}(x)}$. Then it is easy to check that $\{\mathcal{A}'_{i,x}\}_{\substack{i=0,\ldots,l \\ x\in\mathbf{R}^{n_i}\times\mathbf{R}^m}}$ is a local resolution. This holds true for a polyhedral local resolution. The other fact is that in the proof of IV.2.2, if $\{\mathcal{A}_{i,x}\}_{\substack{i=0,\ldots,l \\ x\in\mathbf{R}^{n_i}\times\mathbf{R}^m}}$ is maximal, then the $\{X_{0,j}(x)\}_j$ continues to be a realization of an element of $\mathcal{A}_{0,x}$ after its modification for its triviality over $\{X_{1,j}(x)\}_j$. This is an advantage of a maximal local resolution.

Given an $\mathfrak{X}$-homeomorphism π of $\mathbf{R}^n\times\mathbf{R}^m$ of the form:

$$\pi(x,y)=(\pi_1(x,y),\pi_2(x_2,\ldots,x_l,y),\ldots,\pi_l(x_l,y),\pi_{l+1}(y))$$
$$\text{for}\quad(x,y)=(x_1,\ldots,x_l,y)\in\mathbf{R}^{n_0-n_1}\times\cdots\times\mathbf{R}^{n_{l-1}}\times\mathbf{R}^m,$$

set

$$\pi_*(\mathcal{A}_{i,x})=\{\{(\pi_{i+1}\times\cdots\times\pi_{l+1})X_{i,j}\}_j\colon\{X_{i,j}\}_j\in\mathcal{A}_{i,x}\}.$$

Lemma IV.2.4. *Let* $p,p_1,\ldots,p_l$ *be the same as in IV.2.1, let* $\{U_k\}_k$ *be a finite number of open* $\mathfrak{X}$-*subsets of* $\mathbf{R}^n\times\mathbf{R}^m$, *for each* k, *let* C_k, $\{Z_{k,k'}\}_{k'}$ *and* $\{Z_{k,k''}\}_{k''}$ *be a compact subset of* U_k, *a finite number of closed* $\mathfrak{X}$-*subsets and a finite number of closed polyhedral subsets respectively, and let* $\{\mathcal{A}'_{i,x}\}_{\substack{i=0,\ldots,l \\ x\in\mathbf{R}^{n_i}\times\mathbf{R}^m}}$ *and* $\{\mathcal{A}''_{i,x}\}_{\substack{i=0,\ldots,l \\ x\in\mathbf{R}^{n_i}\times\mathbf{R}^m}}$ *be maximal local resolutions of*

$\{Z_{k,k'}\}_{k,k'}$ and $\{Z_{k,k''}\}_{k,k''}$ respectively, with respect to $p_1, \ldots, p_l$. Assume Axiom (v) and that $\{\mathcal{A}''_{i,x}\}_{\substack{i=0,\ldots,l \\ x \in \mathbf{R}^{n_i} \times \mathbf{R}^m}}$ is a local subresolution of $\{\mathcal{A}'_{i,x}\}_{\substack{i=0,\ldots,l \\ x \in \mathbf{R}^{n_i} \times \mathbf{R}^m}}$ and all $\mathcal{A}'_{l,x}$ and $\{\mathcal{A}''_{i,x}\}_{\substack{i=0,\ldots,l \\ x \in \mathbf{R}^{n_i} \times \mathbf{R}^m}}$ are polyhedral. Then for a small positive number ε there exist a rectangular decomposition F_k of a compact polyhedral neighborhood U'_k of C_k in U_k for each k and an $\mathfrak{X}$-homeomorphism $\tilde{\pi}$ of $\mathbf{R}^n \times \mathbf{R}^m \times [1/2, 1]$ of the form:

$$\tilde{\pi}(x, y, t) = (\tilde{\pi}_1(x, y, t), \tilde{\pi}_2(x_2, \ldots, x_l, y, t), \ldots, \tilde{\pi}_l(x_l, y, t), y, t)$$

$$\text{for } (x, y, t) = (x_1, \ldots, x_l, y, t) \in \mathbf{R}^{n_0 - n_1} \times \cdots \times \mathbf{R}^{n_{l-1}} \times \mathbf{R}^m \times [1/2, 1]$$

such that the following seven conditions are satisfied.

(i) $U'_k(\varepsilon)$ (the closed ε-neighborhood) $\subset U_k$.

(ii) $F = \cup_k F_k$ is a rectangular decomposition of $\cup_k U'_k$.

(iii) Each element of F is of the form $V_1 \times \cdots \times V_l \times V'$, $V_i \subset \mathbf{R}^{n_{i-1} - n_i}$, $V' \subset \mathbf{R}^m$.

(iv) $\tilde{\pi}$ is invariant on each element of $\tilde{F}$.

(v) $\tilde{\pi} = \mathrm{id}$ on $\mathbf{R}^n \times \mathbf{R}^m \times [1/2, 1/2 + \varepsilon]$.

(vi) For each k, k', k'', $\tilde{\pi}(\tilde{U}'_k \cap (Z_{k,k'} \times 1))$ and $\tilde{\pi}(\tilde{U}'_k \cap (Z_{k,k''} \times [1/2, 1]))$ are polyhedra.

(vii) If π denotes the $\mathfrak{X}$-homeomorphism of $\mathbf{R}^n \times \mathbf{R}^m$ defined by

$$(\pi(x, y), 1) = \tilde{\pi}(x, y, 1) \quad \text{for} \quad (x, y) \in \mathbf{R}^n \times \mathbf{R}^m,$$

then $\{\pi(U'_k(\varepsilon)^\circ \cap Z_{k,k'}), \pi(U'_k(\varepsilon)^\circ \cap Z_{k,k''})\}_{k,k',k''}$ admits a maximal polyhedral local resolution with respect to $p_1, \ldots, p_l$ which is a subresolution of $\{\pi_*(\mathcal{A}'_{i,x})\}_{\substack{i=0,\ldots,l \\ x \in \mathbf{R}^{n_i} \times \mathbf{R}^m}}$.

Proof that the second statement of IV.2.1 follows from IV.2.4. Without loss of generality, we can assume that $\{\mathcal{A}_{i,x}\}_{\substack{i=0,\ldots,l \\ x \in \mathbf{R}^{n_i} \times \mathbf{R}^m}}$ is maximal. Let $C_1 \subset C_2 \subset \cdots$ be a sequence of compact subsets of $\mathbf{R}^n \times \mathbf{R}^m$ such that $C_1 \subset \mathrm{Int}\, C_2, C_2 \subset \mathrm{Int}\, C_3, \ldots$ and $\cup_{i=1}^\infty C_i = \mathbf{R}^n \times \mathbf{R}^m$.

We triangulate X in a neighborhood of each C_i by induction on i. First let $i = 1$. Apply IV.2.4 to

$$\{U_k\} = \{\mathrm{Int}\, C_2\}, \quad \{C_k\} = \{C_1\},$$
$$\{Z_{k,k'}\}_{k'} = \{X \cap \mathrm{Int}\, C_2\}, \quad \text{and} \quad \{Z_{k,k''}\}_{k''} = \varnothing.$$

Then there exist a small positive number ε_1, an $\mathfrak{X}$-homeomorphism $\hat{\tau}^1$ of $\mathbf{R}^n \times \mathbf{R}^m \times \mathbf{R}$ of the form:

$$\hat{\tau}^1(x, y, t) = (\hat{\tau}_1^1(x, y, t), \hat{\tau}_2^1(x_2, \ldots, x_l, y, t), \ldots, \hat{\tau}_l^1(x_l, y, t), y, t)$$
$$\text{for} \quad (x, y, t) = (x_1, \ldots, x_l, y, t) \in \mathbf{R}^{n_0 - n_1} \times \cdots \times \mathbf{R}^{n_{l-1}} \times \mathbf{R}^m \times \mathbf{R},$$

and a rectangular decomposition F_1 of a compact polyhedral neighborhood W_1 of C_1 in $\mathbf{R}^n \times \mathbf{R}^m$ such that

$$W_1(\varepsilon_1) \subset \text{Int } C_2, \qquad \hat{\tau}^1 = \text{id} \quad \text{on} \quad \mathbf{R}^n \times \mathbf{R}^m \times \,]-\infty, 1/2],$$

$$\hat{\tau}_j^1(x,y,t) = \hat{\tau}_j^1(x,y,1) \quad \text{for} \quad (x,y,t) \in \mathbf{R}^n \times \mathbf{R}^m \times [1,\infty[, \ j = 1,\dots,l,$$

$\hat{\tau}^1$ is invariant on each element of $\hat{F}_1$, $\hat{\tau}^1((W_1(\varepsilon_1) \cap X) \times 1)$ is a polyhedron, and $W_1(\varepsilon_1)^\circ \cap \tau^1(X)$ admits a maximal polyhedral local resolution $\{\mathcal{A}_{j,x}^1\}_{\substack{j=0,\dots,l \\ x \in \mathbf{R}^{n_j} \times \mathbf{R}^m}}$ which is a subresolution of $\{\tau_*^1 \mathcal{A}_{j,x}\}_{\substack{j=0,\dots,l \\ x \in \mathbf{R}^{n_j} \times \mathbf{R}^m}}$, where τ^1 is the 𝔛-homeomorphism of $\mathbf{R}^n \times \mathbf{R}^m$ defined by

$$(\tau^1(x,y), 1) = \hat{\tau}^1(x,y,1) \quad \text{for} \quad (x,y) \in \mathbf{R}^n \times \mathbf{R}^m.$$

Let $i > 1$ be an integer. For any $j = 1,\dots,,i-1$, assume we have ε_j, $\hat{\tau}^j$, W_j', W_{j-1}'', F_j', F_{j-1}'', $\{\mathcal{A}_{j',x}^j\}_{\substack{j'=0,\dots,l \\ x \in \mathbf{R}^{n_{j'}} \times \mathbf{R}^m}}$ and τ^j ($F_0'' = W_0'' = \varnothing$) which satisfy certain conditions. We do not explain the conditions because they become clear very soon. Apply IV.2.4 to

$$\{k\} = \{1,2\}, \qquad U_1 = \tau^{i-1} \circ \cdots \circ \tau^1(\text{Int } C_{i+1}), \qquad C_1 = \tau^{i-1} \circ \cdots \circ \tau^1(C_i),$$

$$\{Z_{1,k'}\}_{k'} = \{U_1 \cap \tau^{i-1} \circ \cdots \circ \tau^1(X)\}, \qquad \{Z_{1,k''}\}_{k''} = \varnothing,$$

$$U_2 = W_{i-1}'(\varepsilon_{i-1})^\circ, \qquad C_2 = \tau^{i-1} \circ \cdots \circ \tau^1(C_{i-1}),$$

$$\{Z_{2,k'}\}_{k'} = \varnothing, \qquad \{Z_{2,k''}\}_{k''} = \{U_2 \cap \tau^{i-1} \circ \cdots \circ \tau^1(X)\}, \quad \text{and}$$

$$\mathcal{A}_{j,x}' = (\tau^{i-1} \circ \cdots \circ \tau^1)_* \mathcal{A}_{j,x}, \quad \mathcal{A}_{j,x}'' = \mathcal{A}_{j,x}^{i-1} \quad \text{for } x \in \mathbf{R}^{n_j} \times \mathbf{R}^m, \ j = 0,\dots,l.$$

Then we obtain a small positive number ε_i, an 𝔛-homeomorphism $\hat{\tau}^i$ of $\mathbf{R}^n \times \mathbf{R}^m \times \mathbf{R}$ of the same form as the above $\hat{\tau}^1$, compact polyhedral neighborhoods W_i' of $\tau^{i-1} \circ \cdots \circ \tau^1(C_i)$ in $\tau^{i-1} \circ \cdots \circ \tau^1(\text{Int } C_{i+1})$ and W_{i-1}'' of $\tau^{i-1} \circ \cdots \circ \tau^1(C_{i-1})$ in $W_{i-1}'(\varepsilon_{i-1})^\circ$, rectangular decompositions F_i' and F_{i-1}'' of W_i' and W_{i-1}'' respectively, and a maximal polyhedral local resolution $\{\mathcal{A}_{j,x}^i\}_{\substack{j=0,\dots,l \\ x \in \mathbf{R}^{n_j} \times \mathbf{R}^m}}$ of $W_i'(\varepsilon_i)^\circ \cap \tau^i \circ \cdots \circ \tau^1(X)$ with respect to $p_1,\dots,p_l$ (which requires that $W_i'(\varepsilon_i)^\circ \cap \tau^i \circ \cdots \circ \tau^1(X)$ and hence $W_i'(\varepsilon_i) \cap \tau^i \circ \cdots \circ \tau^1(X)$ are polyhedra) such that the following five conditions are satisfied, where τ^i is the 𝔛-homeomorphism of $\mathbf{R}^n \times \mathbf{R}^m$ defined by

$$(\tau^i(x,y), i) = \hat{\tau}^i(x,y,i) \quad \text{for} \quad (x,y) \in \mathbf{R}^n \times \mathbf{R}^m.$$

(i) $$\hat{\tau}^i = \text{id} \quad \text{on} \quad \mathbf{R}^n \times \mathbf{R}^m \times \,]-\infty, i-1/2].$$

(ii) $$\hat{\tau}^i(x,y,t) = (\tau^i(x,y), t) \quad \text{for} \quad (x,y,t) \in \mathbf{R}^n \times \mathbf{R}^m \times [i,\infty[.$$

(iii) $$F_{i-1}'' \subset F_i'.$$

(iv) Let $\hat{F}'_i$ denote the family of sets:

$$\{(x, y, t) + (0, 0, i - 1) \colon (x, y, t) \in V\} \quad \text{for} \quad V \in \tilde{F}'_i,$$

and define $\hat{W}''_{i-1}$ similarly. Then $\hat{\tau}^i$ is invariant on each element of $\hat{F}'_i$, and $\hat{W}''_{i-1} \cap \hat{\tau}^i \circ \cdots \circ \hat{\tau}^1 (X \times \mathbf{R})$ is a polyhedron.

(v) $\{\mathcal{A}^i_{j,x}\}_{\substack{j=0,\ldots,l \\ x \in \mathbf{R}^{n_j} \times \mathbf{R}^m}}$ is a local subresolution of $\{(\tau^i \circ \cdots \circ \tau^1)_* \mathcal{A}_{j,x}\}_{\substack{j=0,\ldots,l \\ x \in \mathbf{R}^{n_j} \times \mathbf{R}^m}}$.
It follows that

$$W''_{i-1}(\varepsilon_i) \subset W''_i, \quad i > 1.$$

Set $W''_0 = \varnothing$, and define an $\mathfrak{X}$-homeomorphism π of $\mathbf{R}^n \times \mathbf{R}^m$ by

$$(\pi(x, y), i - 1 + t) = \hat{\tau}^i \circ \cdots \circ \hat{\tau}^1 (x, y, i - 1 + t)$$

for $(x, y) \in \partial((\tau^i \circ \cdots \circ \tau^1)^{-1} W''_{i-1}(\varepsilon_i t)), \ 1/2 < t \leq 1, \ i = 2, 3, \ldots,$ and

$$(\pi(x, y), i) = \hat{\tau}^i \circ \cdots \circ \hat{\tau}^1 (x, y, i)$$

for $(x, y) \in (\tau^i \circ \cdots \circ \tau^1)^{-1}(W''_i(\varepsilon_{i+1}/2) - W''_{i-1}(\varepsilon_i)), \ i = 1, 2, \ldots.$

Clearly π fulfills the requirements. $\qquad\qquad\square$

Proof of IV.2.4. We proceeds as in the proof of IV.2.2. Since the case $l = 0$ is clear, suppose inductively the lemma for $l - 1$; and since the problem is local at $\cup_i C_i$, suppose each U_i is bounded. Set $C = \cup_k C_k$ and $Z = \cup_{k,k'} Z_{k,k'} \cup \cup_{k,k''} Z_{k,k''}$. First we consider the problem on p_1^{-1}(a small neighborhood of each point of $p_1(C)$ in $\mathbf{R}^{n_1} \times \mathbf{R}^m$). Assume as in the proof of IV.2.2 that the map $p_1|_Z$ is a finite-to-one map. (Here we shrink $\{U_k\}_k$.) By the same reason as in the proof of IV.2.2, for each $x' \in p_1(C)$, there exist open $\mathfrak{X}$-neighborhoods $R(x')$ of x' in $\mathbf{R}^{n_1} \times \mathbf{R}^m$, $W(x')$ of $C \cap p_1^{-1}(R(x'))$ in $\mathbf{R}^n \times \mathbf{R}^m$ and $W_k(x')$ of $C_k \cap p_1^{-1}(R(x'))$ in U_k, $k = 1, 2, \ldots.$ There also exist C^0 $\mathfrak{X}$-stratifications $\{X'_{0,j}(x')\}_j$ and $\{X''_{0,j}(x')\}_j$ of $W(x')$ and $\{X'_{1,j}(x')\}_j$ and $\{X''_{1,j}(x')\}_j$ of $R(x')$ such that the following conditions are satisfied. Each $W_k(x')$ is of the form $Q_k(x') \times R(x')$ ($Q^k(x')$ is the interior of a compact PL submanifold of $\mathbf{R}^{n-n_1}$ with boundary).

$$\bigcup_k W_k(x') = W(x').$$

$$((\partial \overline{Q_k(x')}) \times R(x')) \cap (\overline{Z} \cup \text{bdry } U_k) = \varnothing. \qquad (*)$$

$\{X'_{0,j}(x')\}_j$ and $\{X'_{1,j}(x')\}_j$ are substratifications of $\{X''_{0,j}(x')\}_j$ and $\{X''_{1,j}(x')\}$, respectively. Each stratum of $\{X''_{0,j}(x')\}_j$ and of $\{X''_{1,j}(x')\}_j$

is a finite union of open simplexes. For any point x of the set:

$$Z'(x') = \bigcup_k (W_k(x') \cap (\bigcup_{k'} Z_{k,k'} \cup \bigcup_{k''} Z_{k,k''}))$$

$$(Z''(x') = \bigcup_k (W_k(x') \cap (\bigcup_{k''} Z_{k,k''}))),$$

$\{X'_{0,j}(x')\}_j$ ($\{X''_{0,j}(x')\}_j$, resp.,) in a neighborhood of x is a substratification of a realization of an element of $\mathcal{A}'_{0,x}$ ($\mathcal{A}''_{0,x}$, resp.,) such that the realization is compatible with all $W_k(x') \cap Z_{k,k'}$ and $W_k(x') \cap Z_{k,k''}$ (with all $W_k(x') \cap Z_{k,k''}$, resp.). $\{X'_{1,j}(x')\}_j$ ($\{X''_{1,j}(x')\}_j$) is a realization of an element of $\mathcal{A}'_{1,x'}$ ($\mathcal{A}''_{1,x'}$, resp.). $\{X'_{0,j}(x')\}_j$ and $\{X'_{1,j}(x')\}_j$ ($\{X''_{0,j}(x')\}_j$ and $\{X''_{1,j}(x')\}_j$) satisfy conditions (i) and (ii) of a resolution. $\{X'_{0,j}(x')\}_j$ ($\{X''_{0,j}(x')\}_j$) is trivial over $\{X'_{1,j}(x')\}_j$ ($\{X''_{1,j}(x')\}_j$, resp.,) outside of $Z'(x')$ ($Z''(x')$, resp.). Here $\{X'_{0,j}(x')\}_j$ ($\{X''_{0,j}(x')\}_j$) is not necessarily a realization of an element of $\mathcal{A}'_{0,x}$ ($\mathcal{A}''_{0,x}$, resp.,) if $\sharp Z'(x') \cap p_1^{-1}(x') > 1$ ($\sharp Z''(x') \cap p_1^{-1}(x') > 1$, resp.).

Moreover, shrinking U_k's, we can choose $Q_k(x')$ and $R(x')$ so that $\overline{Q_k(x')}$ admits a rectangular decomposition $F_{x',k}$ such that for each $V \in F_{x',k}$ of dimension $< n - n_1$, $V \times R(x')$ does not intersect with $\overline{Z}$. Let $F_{x'}$ denote the rectangular decomposition generated by all $F_{x',k}$, subdivide each $F_{x',k}$ so that it is a subfamily of $F_{x'}$, and keep the notation $F_{x',k}$. Then for each element $V \in F_{x'}$ of dimension $< n - n_1$,

$$(V \times R(x')) \cap \overline{Z} = \varnothing. \tag{$**$}$$

Replace $\{X'_{0,j}(x')\}_j$ and $\{X''_{0,j}(x')\}_j$ with $\{V^{\circ} \cap X'_{0,j}(x') : V \in F_{x'}, j\}$ and $\{V^{\circ} \cap X''_{0,j}(x') : V \in F_{x'}, j\}$ respectively, and keep the notation. Then they are compatible with $F_{x'} \times R(x')$ (i.e. with $\{V^{\circ} \times R(x') : V \in F_{x'}\}$) and lose only the two properties that if $p_1|_{X'_{0,j}(x')}$ ($p_1|_{X''_{0,j}(x')}$) is not a covering map to $X'_{1,j'}(x')$ ($X''_{1,j'}(x')$), then $X'_{0,j}(x')$ ($X''_{0,j}(x')$) is an open subset of $p_1^{-1}(X'_{1,j'}(x'))$ ($p_1^{-1}(X''_{1,j'}(x'))$, resp.,) (the latter half of condition (ii) of a resolution) and that $\{X'_{0,j}(x')\}_j$ ($\{X''_{0,j}(x')\}_j$) is trivial over $\{X'_{1,j}(x')\}_j$ ($\{X''_{1,j}(x')\}_j$, resp.,) outside of $Z'(x')$ ($Z''(x')$, resp.). But it holds that

$$\{X'_{0,j}(x')\}_j|_{V^{\circ} \times R(x') - Z'(x')} = \{(\mathbf{R}^{n-n_1} \times X'_{1,j}(x')) \cap (V^{\circ} \times R(x') - Z'(x'))\}_j, \quad \text{and} \tag{$***$}$$

$$\{X''_{0,j}(x')\}_j|_{V^{\circ} \times R(x') - Z''(x')} = \{(\mathbf{R}^{n-n_1} \times X''_{1,j}(x')) \cap (V^{\circ} \times R(x') - Z''(x'))\}_j.$$

Assume that $p_1|_Z$ is not a finite-to-one map. If $n = n_1 + 1$, then we can reduce the problem to the case of a finite-to-one map as in the proof of IV.2.2. Hence assume $n > n_1 + 1$. We can uniquely divide each $Z_{k,\kappa}$

($\kappa = k'$ or k'') into closed $\mathfrak{X}$-subsets $Z_{k,\kappa}^{\alpha}$ and $Z_{k,\kappa}^{\beta}$ of U_k so that for each $x' \in \mathbf{R}^{n_1} \times \mathbf{R}^m$, $p_1^{-1}(x') \cap Z_{k,\kappa}^{\alpha}$ is open in $p_1^{-1}(x')$, $p_1|_{Z_{k,\kappa}^{\beta}}$ is a finite-to-one map, and $Z_{k,\kappa}^{\beta}$ is the smallest. This is clear because by condition (i) of a resolution, $p_1^{-1}(x') \cap Z_{k,\kappa}$ is a finite disjoint union of points and connected components of $p_1^{-1}(x') \cap U_k$. Then $\{A'_{i,x}\}_{\substack{i=0,\dots,l \\ x \in \mathbf{R}^{n_i} \times \mathbf{R}^m}}$ ($\{A''_{i,x}\}_{\substack{i=0,\dots,l \\ x \in \mathbf{R}^{n_i} \times \mathbf{R}^m}}$) is a local resolution of $\{Z_{k,k'}^{\alpha}, Z_{k,k'}^{\beta}\}_{k,k'}$ ($\{Z_{k,k''}^{\alpha}, Z_{k,k''}^{\beta}\}_{k,k''}$, resp.). Now we find $R(x'), W(x'), \dots, F_{x'}$ which satisfy the above conditions except (∗) and (∗∗). By the above division, conditions (∗) and (∗∗) are replaced by the following ones: for each $V \in F_{x'}$ and for each $Z_{k,\kappa}$ ($\kappa = k'$ or k'') with $V \subset Q_k(x')$,

$$Z_{k,\kappa}^{\alpha} \cap (V \times R(x')) = V \times R_{k,\kappa}(V) \quad \text{for some } \mathfrak{X}\text{-set } R_{k,\kappa}(V) \subset R(x'),$$

$$Z_{k,\kappa}^{\beta} \cap (V \times R(x')) \begin{cases} \subset V^{\circ} \times R(x') & \text{if } \dim V = n - n_1 \\ = \varnothing & \text{otherwise,} \end{cases}$$

and $\{X'_{0,j}(x')\}_j$ ($\{X''_{0,j}(x')\}_j$) is compatible with $Z_{k,k'}^{\alpha} \cap (V \times R(x'))$, $Z_{k,k''}^{\alpha} \cap (V \times R(x'))$, $Z_{k,k'}^{\beta} \cap (V \times R(x'))$ and $Z_{k,k''}^{\beta} \cap (V \times R(x'))$ (with $Z_{k,k''}^{\alpha} \cap (V \times R(x'))$ and $Z_{k,k''}^{\beta} \cap (V \times R(x'))$, resp.). Then $\{X'_{1,j}(x')\}_j$ ($\{X''_{1,j}(x')\}_j$) is compatible with $R_{k,k'}(V)$ and $R_{k,k''}(V)$ (with $R_{k,k''}(V)$, resp.). We shall define the required rectangular decomposition F so that each element of F is of the form $V \times V'$, $V \in F_{x'}$. Hence for the inductive proof of the lemma we can forget $Z_{k,\kappa}^{\alpha}$, i.e., we can assume (∗) and (∗∗).

For a small positive number ε, if we modify $W_k(x')$, $\{X'_{0,j}(x')\}_j$ and $\{X''_{0,j}(x')\}_j$, then by (∗∗), in place of $F_{x',k}$, $F_{x',k}(\varepsilon)$ becomes a rectangular decomposition of $\overline{Q_k(x')}$. Hence we can extend the above arguments to $\mathbf{R}^n \times \mathbf{R}^m \times [1/2, 1]$ as follows. Let ε be a small positive number. For a set A in $\mathbf{R}^n \times \mathbf{R}^m$, let A^* denote $A \times [1/2.1]$. Let $r: \mathbf{R}^n \times \mathbf{R} \times \mathbf{R}^m \to \mathbf{R}^n \times \mathbf{R}^m \times \mathbf{R}$ denote the permutation. Then each $\tilde{F}_{x',k}$ is a decomposition of a set in $Q_k(x')^*$. By (∗), (∗∗) and (∗∗∗) there exist C^0 $\mathfrak{X}$-stratifications $\{\hat{X}'_{0,j}(x')\}_j$ and $\{\hat{X}''_{0,j}(x')\}_j$ of $r((\cup_{V \in \tilde{F}_{x'}} V) \times R(x'))$ such that the following five conditions are satisfied.

(1) $\{\hat{X}'_{0,j}(x')\}_j$ is compatible with the sets $r(\tilde{F}_{x'} \times R(x'))$ and

$$\{r((V \times R(x')) \cap Z_{k,k'}^*), r((V \times R(x')) \cap Z_{k,k''}^*): V \in \tilde{F}_{x',k}, \ k, k', k''\}.$$

(Let $\hat{Z}'(x')$ denote the union of all the sets in the last family.)

(2) $\{\hat{X}''_{0,j}(x')\}_j$ is compatible with the sets $r(\tilde{F}_{x'} \times R(x'))$ and

$$\{r((V \times R(x') \cap Z_{k,k''}^*): V \in \tilde{F}_{x',k}, \ k, k''\}.$$

(We define also the set $\hat{Z}''(x')$.)

(3) $\{\hat{X}'_{0,j}(x')\}_j$ and $\{X'_{1,j}(x')\}_j \times \{1/2,\ 1,\ ]1/2,1[\}$ ($\{\hat{X}''_{0,j}(x')\}_j$ and $\{X''_{1,j}(x')\}_j \times \{1/2,\ 1,\ ]1/2,1[\}$) satisfy conditions (i) and (ii) of a resolution except the latter half of (ii).

(4) For each $V \in \tilde{F}_{x'}$ with either $\dim V < n - n_1$ or $V \not\subset \mathbf{R}^{n-n_1} \times \{1/2,1\}$ and $\dim V \le n - n_1$,

$$r(V \times R(x')) \cap \overline{Z}^* = \varnothing. \tag{$**$}'$$

In place of the latter half of (ii),
(5) for $V \in \tilde{F}_{x'}$,

$$\{\hat{X}'_{0,j}(x')\}_j|_{r(V^\circ \times R(x'))-\hat{Z}'(x')}$$
$$=\{(\mathbf{R}^{n-n_1} \times X'_{1,j}(x') \times [1/2,1]) \cap (r(V^\circ \times R(x')) - \hat{Z}'(x'))\}_j \text{ and}$$

$$\{\hat{X}''_{0,j}(x')\}_j|_{r(V^\circ \times R(x'))-\hat{Z}''(x')} \tag{$***$}'$$

$$=\{(\mathbf{R}^{n-n_1} \times X''_{1,j}(x') \times [1/2,1]) \cap (r(V^\circ \times R(x')) - \hat{Z}''(x'))\}_j.$$

Let $x'_1, x'_2, \ldots$ be finite points in $p_1(C)$ such that $\{R(x'_h)\}_h$ is a covering of $p_1(C)$ in $\mathbf{R}^{n_1} \times \mathbf{R}^m$. For each h, set

$$Z^1_{h,h'} = \bigcup_{\dim \le h'} X'_{1,j}(x'_h) \text{ and } Z^1_{h,h''} = \bigcup_{\dim \le h''} X''_{1,j}(x''_h), \quad h',\ h'' = 0, 1, \ldots,$$

and let C^1_h be a compact subset of $R(x'_h)$ such that $\cup_h C^1_h = p_1(C)$. Next apply the induction hypothesis to $\{R(x'_h)\}_h$, $\{C^1_h\}_h$, $\{Z^1_{h,h'}\}_{h,h'}$ and $\{Z^1_{h,h''}\}_{h,h''}$ as in the proof of IV.2.2 and shrink ε. Then we have an $\mathfrak{X}$-homeomorphism $\tilde{\tau}$ of $\mathbf{R}^{n_1} \times \mathbf{R}^m \times [1/2,1]$ of the form:

$$\tilde{\tau}(x', y, t) = (\tilde{\tau}_2(x_2, \ldots, x_l, y, t), \ldots, \tilde{\tau}_l(x_l, y, t), y, t)$$
$$\text{for} \quad (x', y, t) = (x_2, \ldots, x_l, y, t) \in \mathbf{R}^{n_1-n_2} \times \cdots \times \mathbf{R}^{n_l-1} \times \mathbf{R}^m \times [1/2,1]$$

and a rectangular decomposition G_h of a neighborhood $R'(x'_h)$ of C^1_h in $R(x'_h)$ for each l such that the following conditions are satisfied. $G = \cup_h G_h$ is a rectangular decomposition of $\cup_h R'(x'_h)$, and $R'(x'_h)(\varepsilon) \subset R(x'_h)$. Each element of G is of the form $V_2 \times \cdots \times V_l \times V'$, $V_i \subset \mathbf{R}^{n_{i-1}-n_i}$, $V' \subset \mathbf{R}^m$. $\tilde{\tau}$ is invariant on each element of $\tilde{G}$, and $\tilde{\tau} = \text{id}$ on $\mathbf{R}^{n_1} \times \mathbf{R}^m \times [1/2, 1/2+2\varepsilon]$. For each h, h', h'' and $V_1 \subset \tilde{G}_h$, $\tilde{\tau}(V_1 \cap (Z^1_{h,h'} \times 1))$ and $\tilde{\tau}(V_1 \cap (Z^1_{h,h''} \times [1/2,1]))$ are polyhedra. Define the $\mathfrak{X}$-homeomorphism τ of $\mathbf{R}^{n_1} \times \mathbf{R}^m$ by $\tilde{\tau}$ as in IV.2.4. Then $\{\tau(R'(x'_h)^\circ \cap Z^1_{h,h'}), \tau(R'(x'_h)^\circ \cap Z^1_{h,h''})\}_{h,h',h''}$ admits a maximal polyhedral local subresolution $\{\mathcal{A}'''_{i,x}\}_{\substack{i=1,\ldots,l \\ x\in\mathbf{R}^{n_i}\times\mathbf{R}^m}}$ of $\{\tau_*(\mathcal{A}'_{i,x})\}_{\substack{i=1,\ldots,l \\ x\in\mathbf{R}^{n_i}\times\mathbf{R}^m}}$ with respect to $p_1, \ldots, p_l$.

For each k, let F_k denote the rectangular decomposition generated by $\{F_{x'_h,k} \times V : V \in G_h\}_h$. As above, we can modify $\{F_k\}_k$ so that $F = \cup_k F_k$ is a rectangular decomposition. Note that each element of F is of the form (a connected component of the closure of $\cap_{\text{some}} V_i - \cup_{\text{some }} h(\cup_{\text{all }} k \overline{Q_k(x'_h)})) \times V_1$ where $V_i \in \cup_h F_{x'_h}$, $V_1 \in G$. Then by the properties of $F_{x'_h,k}$ and G, it is clear that $U'_k = \cup_{V \in F_k} V$ is a neighborhood of C_k in U_k and $U'_k(\varepsilon) \subset U_k$.

Let us construct $\tilde{\pi}$. For that we want to apply IV.2.3. Let L be a simplicial subdivision of $\tilde{G}$ compatible with polyhedra $\tilde{\tau}(V_1 \cap (Z^1_{h,h'} \times 1))$ and $\tilde{\tau}(V_1 \cap (Z^1_{h,h''} \times [1/2, 1]))$ for all h, h', h'' and $V_1 \in \tilde{G}_h$. Let Y denote the union of all $(\text{id} \times \tilde{\tau})(V \cap Z_{k,k'} \times 1))$ and $(\text{id} \times \tilde{\tau})(V \cap (Z_{k,k''} \times [1/2, 1]))$ for $V \in \tilde{F}_k$. Set

$$Y_1 = Y \cap (\mathbf{R}^n \times \mathbf{R}^m \times [1/2, 1/2+2\varepsilon]) \text{ and } Y_2 = Y \cap (\mathbf{R}^n \times \mathbf{R}^m \times [1/2, 1/2+\varepsilon])$$

and let P denote the union of all $V \in \tilde{F}$ with

$$\dim q_1(V \cap (\mathbf{R}^n \times \mathbf{R}^m \times \{1/2, 1\})) < n - n_1,$$

where $q_1 : \mathbf{R}^n \times \mathbf{R}^m \times [1/2, 1] \to \mathbf{R}^{n-n_1}$ denotes the projection onto the first factors. Set $p = p_1 \times \text{id} : \mathbf{R}^n \times \mathbf{R}^m \times \mathbf{R} \to \mathbf{R}^{n_1} \times \mathbf{R}^m \times \mathbf{R}$ and $q = p|_Y$. By $(**)$ P does not intersect with Y. Clearly Y is compact, and Y_1 and Y_2 are compact polyhedra such that Y_1 is a neighborhood of Y_2 in Y and $q^{-1}(q(Y_1)) = Y_1$. For each $\sigma \in L$, the map

$$q|_{q^{-1}(\sigma^\circ)} : q^{-1}(\sigma^\circ) \longrightarrow \sigma^\circ$$

is a covering map for the following reason. If σ° is included in $\tilde{\tau}(V_1 \cap ((Z^1_{h,h'} - Z^1_{h,h'-1}) \times 1))$ for some h, h' and $V_1 \in \tilde{G}_h$, then it is included in $\tilde{\tau}((\cup_{\dim=h'} X'_{1,j}(x'_h)) \times 1)$ and hence $q^{-1}(\sigma^\circ)$ is included in $(\text{id} \times \tilde{\tau})((\cup_{\dim=h'} X'_{0,j}(x'_h)) \times 1)$. If σ° is included in $\tilde{\tau}(V_1 \cap ((Z^1_{h,h''} - Z^1_{h,h''-1}) \times [1/2, 1]))$ for some h, h'' and $V_1 \in \tilde{G}_h$, then $q^{-1}(\sigma^\circ)$ is included in $(\text{id} \times \tilde{\tau})((\cup_{\dim=h''} X''_{0,j}(x'_h)) \times [1/2, 1])$. Hence by the triviality property of $\{X'_{0,j}(x'_h)\}_j$ over $\{X'_{1,j}(x'_h)\}_j$ outside of $Z'(x'_h)$ and of $\{X''_{0,j}(x'_h)\}_j$ over $\{X''_{1,j}(x'_h)\}_j$ outside of $Z''(x'_h)$, $q|_{q^{-1}(\sigma^\circ)}$ is a covering map. Thus all the assumptions in IV.2.3 are satisfied. Therefore, we have an $\mathfrak{X}$-map $\tilde{\pi}_1$ from $\mathbf{R}^n \times \mathbf{R}^m \times [1/2, 1]$ to $\mathbf{R}^{n-n_1}$ such that the map $\tilde{\pi}(x, y, t) = (\tilde{\pi}_1(x, y, t), \tilde{\tau}(x', y, t))$ is an $\mathfrak{X}$-homeomorphism of $\mathbf{R}^n \times \mathbf{R}^m \times [1/2, 1]$, $\tilde{\pi}$ is invariant on each element of $\tilde{F}$, $\tilde{\pi} = \text{id}$ on $\mathbf{R}^n \times \mathbf{R}^m \times [1/2, 1/2+\varepsilon]$, and for each k, k', k'' and $V \in \tilde{F}_k$, $\tilde{\pi}(V \cap (Z_{k,k'} \times 1))$ and $\tilde{\pi}(V \cap (Z_{k,k''} \times [1/2, 1]))$ are polyhedra.

It remains to define a maximal polyhedral local resolution $\{\mathcal{A}'''_{i,x}\}^{i=0,\ldots,l}_{x \in \mathbf{R}^{n_i} \times \mathbf{R}^m}$ of the family $\{\pi(U'_k(\varepsilon)^\circ \cap Z_{k,k'}), \pi(U'_k(\varepsilon)^\circ \cap Z_{k,k''})\}_{k,k',k''}$ with respect to

$p_1, \ldots, p_l$ which is a local subresolution of $\{\pi_*(\mathcal{A}'_{i,x})\}_{\substack{i=0,\ldots,l \\ x \in \mathbf{R}^{n_i} \times \mathbf{R}^m}}$. We have already obtained $\{\mathcal{A}'''_{i,x}\}_{\substack{i=1,\ldots,l \\ x \in \mathbf{R}^{n_i} \times \mathbf{R}^m}}$ by induction. For each $x \in \mathbf{R}^n \times \mathbf{R}^m$, define $\mathcal{A}'''_{0,x}$ to be all elements of $\pi_*(\mathcal{A}'_{0,\pi^{-1}(x)})$ such that each stratum of some of their realizations is a finite union of open simplexes and their images under p_1 are elements of $\mathcal{A}'''_{1,p_1(x)}$. Clearly $\{\mathcal{A}'''_{i,x}\}_{\substack{i=0,\ldots,l \\ x \in \mathbf{R}^{n_i} \times \mathbf{R}^m}}$ fulfills the above requirements. $\square$

We have reduced IV.2.1 to IV.2.3. IV.2.3 for $n = 1$ is very easy to prove (see the proof of II.2.1 and II.2.4). IV.2.3 for general n is similar to II.2.7 but does not follow from it immediately. The $\mathfrak{X}$-homeomorphism of $\mathbf{R}^{n_1} \times \mathbf{R}^{n_2}$ in II.2.7 does not satisfy the condition of its form in IV.2.3. To modify the homeomorphism so that it satisfies the condition, we need the next lemma, which is a small generalization of Lemma 5 in [S₅].

Lemma IV.2.5. *Let k be a positive integer. Let Y and $H \supset H'$ be compact $\mathfrak{X}$-sets in $\mathbf{R}^n \times \mathbf{R}^m$ and in some Euclidean space respectively, let P be a closed polyhedron in $\mathbf{R}^n \times \mathbf{R}^m$, and let $p\colon \mathbf{R}^n \times \mathbf{R}^m \to \mathbf{R}^m$ and $q\colon Y \to \mathbf{R}^m$ denote the projection and its restriction to Y respectively. Assume $q(Y)$ is a polyhedron. Let L and $L_1 \subset L'$ be simplicial complexes in $\mathbf{R}^m$ such that $|L| = q(Y)$ and L' is a derived subdivision of L. Assume that for each simplex $\sigma \in L$, the map*

$$q|_{q^{-1}(\sigma^\circ)}\colon q^{-1}(\sigma^\circ) \to \sigma^\circ$$

is a covering map, and for each vertex $v \in L'$, there are a finite number of disjoint boxes $[a^l_{v,1}, b^l_{v,1}] \times \cdots \times [a^l_{v,n}, b^l_{v,n}]$, $l = 1, 2, \ldots,$ in $\mathbf{R}^n$ such that

$$P \cap (\bigcup_l [a^l_{v,1}, b^l_{v,1}] \times \cdots \times [a^l_{v,n}, b^l_{v,n}] \times |st(v, L')|) = \varnothing \quad and$$

$$\bigcup_l]a^l_{v,1}, b^l_{v,1}[\times \cdots \times]a^l_{v,n}, b^l_{v,n}[\times |st(v, L')| \supset Y \cap (\mathbf{R}^n \times |st(v, L')|).$$

Let $\pi''_{h,t}$ be an $\mathfrak{X}$-homeomorphism of $|L|$ parameterized by $H \times [0,1]$ such that for each $h \in H$, $h' \in H'$ and $t \in [0,1]$, $\pi''_{h,t}$ is invariant on each simplex of L',

$$\pi''_{h',t} = \pi''_{h,0} = \mathrm{id}, \quad and \quad \pi''_{h,t} = \mathrm{id} \quad on \quad |L'^{k-1} \cup L_1|.$$

Then there exists an $\mathfrak{X}$-homeomorphism $\pi_{h,t}$ of $\mathbf{R}^n \times |L|$ parameterized by $H \times [0,1]$ of the form:

$$\pi_{h,t}(x,y) = (\pi'_{h,t}(x,y), \pi''_{h,t}(y)) \quad for \quad (x,y) \in \mathbf{R}^n \times |L|$$

such that for each $h \in H$, $h' \in H'$ and $t \in [0,1]$, $\pi_{h,t}$ is invariant on Y,

$$\pi_{h',t} = \pi_{h,0} = \mathrm{id}, \qquad \pi_{h,t} = \mathrm{id} \quad on \quad \mathbf{R}^n \times |L'^{k-1} \cup L_1|, \quad and$$
$$\pi'_{h,t}(x,y) = x \quad for \quad (x,y) \in P \cap (\mathbf{R}^n \times |L|).$$

Proof that IV.2.3 follows from IV.2.5. In IV.2.3, consider the case of compact Y. We can assume that L is compatible with $q(Y_1)$ and $q(Y_2)$. Let U be a small open neighborhood of Y in $\mathbf{R}^n \times \mathbf{R}^m$ which does not intersect with P, and let L' be a derived subdivision of L. Here we choose L so fine that for each vertex $v \in L'$, there are a finite number of disjoint boxes $[a^l_{v,1}, b^l_{v,1}] \times \cdots \times [a^l_{v,n}, b^l_{v,n}]$, $l = 1, 2, \ldots$, in $\mathbf{R}^n$ such that

$$P \cap \left(\bigcup_l [a^l_{v,1}, b^l_{v,1}] \times \cdots \times [a^l_{v,n}, b^l_{v,n}] \times |\mathrm{st}(v, L')|\right) = \varnothing \quad and$$

$$\bigcup_l]a^l_{v,1}, b^l_{v,1}[\times \cdots \times]a^l_{v,n}, b^l_{v,n}[\times |\mathrm{st}(v, L')| \supset Y \cap (\mathbf{R}^n \times |\mathrm{st}(v, L')|).$$

Let L_1 denote the subcomplex of L' such that $|L_1| = q(Y_2)$. By II.2.7 we have an $\mathfrak{X}$-isotopy $\pi_{1,t}$, $0 \le t \le 1$, of $\mathbf{R}^n \times |L|$ of the form:

$$\pi_{1,t}(x,y) = (\pi'_{1,t}(x,y), \pi''_{1,t}(y)) \quad for \quad (x,y) \in \mathbf{R}^n \times |L|$$

such that $\pi_{1,0} = \mathrm{id}$, $\pi_{1,1}(Y)$ is a polyhedron, for each $t \in [0,1]$, $\pi_{1,t} = \mathrm{id}$ on $\mathbf{R}^n \times |L_1|$, $\pi''_{1,t}$ is invariant on each simplex of L', and

$$\pi'_{1,t}(x,y) = x \quad for \quad (x,y) \in \mathbf{R}^n \times |L| - U.$$

Apply IV.2.5 for $k = 1$ and $H \times [0,1] = [0,1]$ to $\pi''^{-1}_{1,t}$, $0 \le t \le 1$. Then we have also an $\mathfrak{X}$-isotopy $\pi_{2,t}$, $0 \le t \le 1$, of $\mathbf{R}^n \times |L|$ of the form:

$$\pi_{2,t}(x,y) = (\pi'_{2,t}(x,y), \pi''^{-1}_{1,t}(y)) \quad for \quad (x,y) \in \mathbf{R}^n \times |L|$$

such that for each $t \in [0,1]$, $\pi_{2,t}$ is invariant on Y,

$$\pi_{2,0} = \mathrm{id}, \qquad \pi_{2,t} = \mathrm{id} \quad on \quad \mathbf{R}^n \times |L_1|, \quad and$$
$$\pi'_{2,t}(x,y) = x \quad for \quad (x,y) \in \mathbf{R}^n \times |L| - U.$$

Divide L so finely and choose U so that for each $\sigma \in L$, $U \cap (\mathbf{R}^n \times \sigma^\circ)$ is of the form $U_\sigma \times \sigma^\circ$. Then

$$U = \pi_{2,t}(U \cap (\mathbf{R}^n \times |L|)) \quad for \quad t \in [0,1].$$

Set $\pi_t = \pi_{1,t} \circ \pi_{2,t}, \ 0 \leq t \leq 1$. Then π_t is of the form:

$$\pi_t(x,y) = (\pi_t'(x,y), y) \quad \text{for} \quad (x,y) \in \mathbf{R}^n \times |L|;$$

$\pi_0 = \text{id}; \ \pi_1(Y)$ is a polyhedron because

$$\pi_1(Y) = \pi_{1,1} \circ \pi_{2,1}(Y) = \pi_{1,1}(Y);$$

and for each $t \in [0,1]$,

$$\pi_t = \text{id} \quad \text{on} \quad \mathbf{R}^n \times |L_1| \quad \text{and on} \quad P \cap (\mathbf{R}^n \times |L|),$$

because on $\mathbf{R}^n \times |L| - U$,

$$\pi_t'(x,y) = \pi_{1,t}' \circ \pi_{2,t}(x,y) = \pi_{2,t}'(x,y) \quad \text{by} \quad \pi_{2,t}(x,y) \in \mathbf{R}^n \times |L| - U$$
$$= x \qquad\qquad\quad \text{by} \quad (x,y) \in \mathbf{R}^n \times |L| - U.$$

Set $\pi = \pi_1$ on $\mathbf{R}^n \times |L|$. We need only extend π to $\mathbf{R}^n \times \mathbf{R}^m$. Let N denote a regular neighborhood of $|L|$ in $\mathbf{R}^m$, let $\alpha \colon N \to |L|$ be a PL retraction, and let $\beta \colon N \to [0,1]$ be a PL function such that

$$0 \leq \beta \leq 1, \quad \beta = 0 \quad \text{on} \quad \partial N, \quad \text{and} \quad \beta = 1 \quad \text{on} \quad |L|.$$

Choose N so small and α so near to the identity that

$$((\text{id}, \alpha)(P \cap (\mathbf{R}^n \times N))) \cap U = \varnothing,$$

and define π on $\mathbf{R}^n \times (\mathbf{R}^m - |L|)$ by

$$\pi'(x,y) = \begin{cases} \pi_{\beta(y)}'(x, \alpha(y)) & \text{if} \quad (x,y) \in \mathbf{R}^n \times (N - |L|) \\ x & \text{if} \quad (x,y) \in \mathbf{R}^n \times (\mathbf{R}^m - N) \quad \text{and} \end{cases}$$
$$\pi(x,y) = (\pi'(x,y), y) \quad \text{for} \quad (x,y) \in \mathbf{R}^n \times (\mathbf{R}^m - |L|).$$

Then π is an $\mathfrak{X}$-homeomorphism of $\mathbf{R}^n \times \mathbf{R}^m$, and $\pi|_{P \cap (\mathbf{R}^n \times |L|)}$ continues to be the identity on P.

Note. If we extend π_t for each $t \in [0,1]$ to $\mathbf{R}^n \times \mathbf{R}^m$ in the same way, π_t becomes an $\mathfrak{X}$-isotopy of $\mathbf{R}^n \times \mathbf{R}^m$.

Consider the case of noncompact Y. For simplicity of notation, we assume $|L| = \mathbf{R}^m$, which does not lose generality. Let $K_1 \subset K_2 \subset \cdots$ be a sequence of finite subcomplexes of L such that

$$\bigcup_i K_i = L, \quad \text{and} \quad N(K_i, L) \subset K_{i+1}, \ i = 1, 2, \ldots.$$

Set $K_0 = \varnothing$. We want to find $\mathfrak{X}$-homeomorphisms τ_i, $i = 0, 1, \ldots$, of $\mathbf{R}^n \times \mathbf{R}^m$ of the form:

$$\tau_i(x, y) = (\tau_i'(x, y), y) \quad \text{for} \quad (x, y) \in \mathbf{R}^n \times \mathbf{R}^m$$

such that for each i, $\tau_i(Y \cap p^{-1}(|K_i|))$ is a polyhedron, $\tau_i = \tau_{i-1}$ on $p^{-1}(|K_{i-2}|)$ if $i > 2$, and $\tau_i = \text{id}$ on $P \cup p^{-1}(q(Y_2))$. If we do this, then the limit $\pi = \lim_{i \to \infty} \tau_i$ is well-defined and fulfills the requirements in IV.2.3.

We construct τ_i by induction on i. If $i = 0$, set $\tau_i = \text{id}$. Assume τ_{i-1} for a positive number i. Replace L with K_{i+1}, Y with $\tau_{i-1}(Y) \cap p^{-1}(|K_{i+1}|)$, Y_1 with $(\tau_{i-1}(Y) \cap p^{-1}(|K_{i-1}|)) \cup (\tau_{i-1}(Y_1) \cap p^{-1}(|K_{i+1}|))$ and Y_2 with $(\tau_{i-1}(Y) \cap p^{-1}(|K_{i-2}|)) \cup (Y_2 \cap p^{-1}(|K_{i+1}|))$. Then by the above note we obtain an $\mathfrak{X}$-isotopy $\theta_{i,t}$, $0 \le t \le 1$, of $\mathbf{R}^n \times \mathbf{R}^m$ of the form:

$$\theta_{i,t}(x, y) = (\theta_{i,t}'(x, y), y) \quad \text{for} \quad (x, y) \in \mathbf{R}^n \times \mathbf{R}^m$$

such that $\theta_{i,0} = \text{id}$, $\theta_{i,1} \circ \tau_{i-1}(Y) \cap p^{-1}(|K_{i+1}|)$ is a polyhedron, and for each $t \in [0, 1]$,

$$\theta_{i,t} = \text{id} \quad \text{on} \quad P \cup p^{-1}(q(Y_2) \cap |K_{i+1}|) \cup p^{-1}(|K_{i-2}|).$$

Let β_i be a PL function on $\mathbf{R}^m$ such that

$$0 \le \beta_i \le 1, \qquad \beta_i = 0 \ \text{outside of} \ |K_{i+1}|, \ \text{and} \ \beta_i = 1 \quad \text{on} \ |K_i|,$$

and define an $\mathfrak{X}$-homeomorphism θ_i of $\mathbf{R}^n \times \mathbf{R}^m$ by

$$\theta_i(x, y) = \theta_{i,\beta(y)}(x, y) \quad \text{for} \quad (x, y) \in \mathbf{R}^n \times \mathbf{R}^m.$$

Then $\tau_i = \theta_i \circ \tau_{i-1}$ is what we want, which completes the proof.   □

In the above proof, we need only the case $k = 1$ and $H \times [0, 1] = [0, 1]$ of IV.2.5. But we shall use the general k and $H \times [0, 1]$ to prove IV.2.5 inductively.

Remark IV.2.6. We can assume in IV.2.5

$$Y \subset [-1, 1]^n \times \mathbf{R}^m \quad \text{and} \quad P = (\mathbf{R}^n -] - 2, 2[^n) \times \mathbf{R}^m \qquad (1)$$

for the following reason.

Assume IV.2.5 in the case of (1). We prove IV.2.5 in the general case by downward induction on k. If $k = \dim Y + 1$, then IV.2.5 is clear. Hence let

k be a positive integer and assume IV.2.5 for $k + 1$. Let σ be a k-simplex of $L' - L_1$, and let v be a vertex of σ. Then

$$P \cap \left(\bigcup_l [a^l_{v,1}, b^l_{v,1}] \times \cdots \times [a^l_{v,n}, b^l_{v,n}] \times |\operatorname{st}(\sigma, L')| \right) = \varnothing \quad \text{and}$$

$$\bigcup_l \,]a^l_{v,1}, b^l_{v,1}[\times \cdots \times \,]a^l_{v,n}, b^l_{v,n}[\times |\operatorname{st}(\sigma, L')| \supset Y \cap (\mathbf{R}^n \times |\operatorname{st}(\sigma, L')|).$$

Let $\pi''_{\sigma,h,t}$ be an X-homeomorphism of $|\operatorname{st}(\sigma, L')|$ parameterized by $H \times [0,1]$ such that for each $h \in H$, $h' \in H'$ and $t \in [0,1]$,

$$\pi''_{\sigma,h,t} = \pi''_{h,t} \quad \text{on} \quad \sigma, \qquad \pi''_{\sigma,h',t} = \pi''_{\sigma,h,0} = \text{id},$$

$$\pi''_{\sigma,h,t} = \text{id} \quad \text{on} \quad |\operatorname{lk}(c, L')| \quad \text{for an inner point } c \text{ of } \sigma,$$

and $\pi''_{\sigma,h,t}$ is invariant on each simplex of $\operatorname{st}(\sigma, L')$ (e.g. the cone extension of $\pi''_{h,t}|\sigma$).

Replace $P \cap (\mathbf{R}^n \times |\operatorname{st}(\sigma, L')|)$ with

$$\left(\mathbf{R}^n - \bigcup_l \,]a^l_{v,1}, b^l_{v,1}[\times \cdots \times \,]a^l_{v,n}, b^l_{v,n}[\right) \times |\operatorname{st}(\sigma, L')|.$$

Then by IV.2.5 in the case of (1), there exists an X-homeomorphism $\pi_{\sigma,h,t}$ of $\mathbf{R}^n \times |\operatorname{st}(\sigma, L')|$ parameterized by $H \times [0,1]$ of the form:

$$\pi_{\sigma,h,t}(x, y) = (\pi'_{\sigma,h,t}(x, y), \pi''_{\sigma,h,t}(y)) \quad \text{for} \quad (x, y) \in \mathbf{R}^n \times |\operatorname{st}(\sigma, L')|$$

such that for each $h \in H$, $h' \in H'$ and $t \in [0,1]$, $\pi_{\sigma,h,t}$ is invariant on $Y \cap (\mathbf{R}^n \times |\operatorname{st}(\sigma, L')|)$,

$$\pi_{\sigma,h',t} = \pi_{\sigma,h,0} = \text{id}, \quad \pi_{\sigma,h,t} = \text{id} \quad \text{on} \quad \mathbf{R}^n \times |\operatorname{lk}(c, L')|, \quad \text{and}$$

$$\pi'_{\sigma,h,t}(x, y) = x$$

$$\text{for} \quad (x, y) \in \left(\mathbf{R}^n - \bigcup_l \,]a^l_{v,1}, b^l_{v,1}[\times \cdots \times \,]a^l_{v,n}, b^l_{v,n}[\right) \times |\operatorname{st}(\sigma, L')|.$$

Extend $\pi''_{\sigma,h,t}$ to $|L| - |\operatorname{st}(\sigma, L')|$ and $\pi_{\sigma,h,t}$ to $\mathbf{R}^n \times (|L| - |\operatorname{st}(\sigma, L')|)$ by setting

$$\pi''_{\sigma,h,t} = \text{id} \quad \text{and} \quad \pi_{\sigma,h,t} = \text{id},$$

and replace $\pi''_{h,t}$ with $\pi''_{h,t} \circ \pi''^{-1}_{\sigma,h,t}$. Then we can assume $\pi''_{h,t} = \text{id}$ on σ. We pursue the same arguments for all k-simplexes of $L' - L_1$. Then we reduce the problem to the case of $\pi''_{h,t} = \text{id}$ on $|L'^k|$. Hence, from the induction hypothesis, IV.2.5 follows. $\square$

We prove IV.2.5 in a long sequence of lemmas as in $[S_5]$. If $\dim |L| = 0$, then IV.2.5 is trivial. Hence from now on we assume IV.2.5 in the case of dimension $< \dim |L|$. For the proof we need to consider canonical local triangulations of Y. First we have the following.

Lemma IV.2.7. *For the same notation as in IV.2.5 and IV.2.6, let σ be a simplex of L'. Set $q^{-1}(\sigma) = Y'$. There exists an $\mathfrak{X}$-homeomorphism τ of $\mathbf{R}^n \times \mathbf{R}^m$ of the form:*

$$\tau(x, y) = (\tau'(x, y), y) \quad for \quad (x, y) \in \mathbf{R}^n \times \mathbf{R}^m$$

such that $\tau(Y')$ is a polyhedron contained in $0^{n-1} \times \mathbf{R} \times \mathbf{R}^m$ and $\tau = \mathrm{id}$ on P.

Remark IV.2.8. In IV.2.7, we can require τ to have, in addition, the following property. There exists a simplicial decomposition K of $\tau(Y')$ such that $q|_{\tau(Y')}: K \to L'$ is simplicial. This is immediate by the proof of II.2.1 and II.2.4.

For the proofs of IV.2.7 and Remark IV.2.9 below we need only the following properties of σ, Y' and $q_\sigma = q|_{Y'}$. σ is a simplex in $\mathbf{R}^m$ with vertices $a_0, \dots, a_l$; Y' is a compact $\mathfrak{X}$-set in $\mathbf{R}^n \times \mathbf{R}^m$; q_σ is the restriction to Y' of the projection $\mathbf{R}^n \times \mathbf{R}^m \to \mathbf{R}^m$; $q_\sigma(Y') = \sigma$; and for each $0 \le i \le l$, the restriction of q_σ to the set:

$$q_\sigma^{-1}(\Delta a_0 \cdots a_i - \Delta a_0 \cdots a_{i-1})$$

is a covering map onto $\Delta a_0 \cdots a_i - \Delta a_0 \cdots a_{i-1}$. The last property holds when we order $a_0, \dots, a_l$ well (see the following proof).

Proof of IV.2.7. It suffices to define τ on $\mathbf{R}^n \times \sigma$ because we can extend it to $\mathbf{R}^n \times \mathbf{R}^n$ keeping the property $\tau = \mathrm{id}$ on P as in the proof that IV.2.3 follows from IV.2.5. We can reduce the problem to the case where Y' is a polyhedron for the following reason. Assume IV.2.7 for polyhedra. By II.2.7 there exists an $\mathfrak{X}$-homeomorphism τ_1 of $\mathbf{R}^n \times \mathbf{R}^m$ of the form:

$$\tau_1(x, y) = (\tau_1'(x, y), \tau_1''(y)) \quad for \quad (x, y) \in \mathbf{R}^n \times \mathbf{R}^m$$

such that $\tau_1(Y)$ is a polyhedron and τ_1'' is invariant on each simplex of L'. Then $\tau_1(Y)$ satisfies the same assumptions as Y in IV.2.5 and IV.2.6. Hence by IV.2.7 for polyhedra, there exists an $\mathfrak{X}$-homeomorphism τ_2 of $\mathbf{R}^n \times \mathbf{R}^m$ of the form:

$$\tau_2(x, y) = (\tau_2'(x, y), y) \quad for \quad (x, y) \in \mathbf{R}^n \times \mathbf{R}^m$$

such that $\tau_2(\tau_1(Y'))$ is a polyhedron contained in $0^{n-1} \times \mathbf{R} \times \mathbf{R}^m$. Define an $\mathfrak{X}$-homeomorphism τ_3 of $\mathbf{R}^n \times \mathbf{R}^m$ by

$$\tau_3(x, y) = (\tau_2' \circ \tau_1(x, y), y) \quad for \quad (x, y) \in \mathbf{R}^n \times \mathbf{R}^m.$$

Then $\tau_3(Y')$ is contained in $0^{n-1} \times \mathbf{R} \times \mathbf{R}^m$. Hence, since IV.2.3 for $n = 1$ holds true (see the proof of II.2.1 and II.2.4), we have an X-homeomorphism τ_4 of $\mathbf{R}^n \times \mathbf{R}^m$ of the form:

$$\tau_4(x, y) = (x_1, \dots, x_{n-1}, \tau_4'(x_n, y), y)$$
$$\text{for} \quad (x, y) = (x_1, \dots, x_n, y) \in \mathbf{R}^n \times \mathbf{R}^m$$

such that $\tau_4(\tau_3(Y'))$ is a polyhedron contained in $0^{n-1} \times \mathbf{R} \times \mathbf{R}^m$. Moreover, we can choose $\tau_1, \dots, \tau_4$ so that $\tau_4 \circ \tau_3 = \mathrm{id}$ on P in the same way as in the proof that IV.2.3 follows from IV.2.5. Therefore, $\tau = \tau_4 \circ \tau_3$ fulfills the requirements in the lemma, which means that we can assume that Y' is a polyhedron. The above arguments imply also that for proof, it suffices to show only the inclusion $\tau(Y') \subset 0^{n-1} \times \mathbf{R} \times \mathbf{R}^m$.

Order all the vertices $a_0, \dots, a_l$ of σ so that if a_i and $a_{i'}$ with $i < i'$ are contained in open simplices of L of dimension j and j' respectively, then $j < j'$. For each i, let A_i denote the simplex $\Delta a_0 \cdots a_i$. Then the maps

$$q|_{q^{-1}(A_i - A_{i-1})} \colon q^{-1}(A_i - A_{i-1}) \longrightarrow A_i - A_{i-1}$$

are covering maps. Let J and K be simplicial complexes with underlying polyhedra $= \sigma$ and Y', respectively, such that $(q|_{Y'}, K, J)$ is simplicial. Let σ_0 be the simplex of J whose intersection with each A_i is not empty and of dimension i, and let the order of vertices b_i of σ_0 and the simplexes B_i, $i = 0, \dots, l$, be defined in the same way. Note that $A_i \supset B_i$, $i = 0, \dots, l$.

We want to find points $c_i \in B_i^\circ$, $i = 0, \dots, l$, so that if we replace σ with $\sigma_1 = \Delta c_0 \cdots c_l$, then IV.2.7 holds true. Trivially we set $c_0 = a_0$. Set

$$q^{-1}(c_0) = \{z_{0,1}, \dots, z_{0,l_0}\} \quad \text{and} \quad z_{0,j} = (x_{0,j}, c_0) \in \mathbf{R}^n \times \mathbf{R}^m, \quad j = 1, \dots, l_0.$$

Let θ_0 be a PL homeomorphism of $\mathbf{R}^n$ such that

$$\theta_0\{x_{0,1}, \dots, x_{0,l_0}\} \subset 0^{n-1} \times \mathbf{R}$$

and θ_0 is the identity outside of $[-2, 2]^n$. Replace Y' with its image under this map. Then we can assume that $x_{0,1}, \dots, x_{0,l_0}$ are contained in $0^{n-1} \times \mathbf{R}$. After this modification, we may need to change J, K, B_i and σ_0. But we keep the notation. Let ε be a small positive number such that the open ε-neighborhoods of $x_{0,j}$, $j = 1, \dots, l_0$, in $\mathbf{R}^n$ do not intersect each other and are included in $[-2, 2]^n$. Let U_0 denote the union of these neighborhoods.

Next choose c_1 so near c_0 that the projection image of $q^{-1}(\Delta c_0 c_1)$ onto $\mathbf{R}^n$ is included in U_0. Set

$$q^{-1}(c_1) = \{z_{1,1}, \dots, z_{1,l_1}\} \quad \text{and} \quad z_{1,j} = (x_{1,j}, c_1) \in \mathbf{R}^n \times \mathbf{R}^m, \quad j = 1, \dots, l_1.$$

Then for each $z_{1,j}$ there uniquely exists $z_{0,\chi(j)}$ such that $\Delta z_{0,\chi(j)} z_{1,j}$ is included in $q^{-1}(\Delta c_0 c_1)$, in other words, the segment (or point) $\Delta x_{0,\chi(j)} x_{1,j}$ is included in U_0. It is easy to construct a PL homeomorphism θ_1 of $\mathbf{R}^n$ such that

$$\theta_1\{x_{1,1}, \dots, x_{1,l_1}\} \subset 0^{n-1} \times \mathbf{R}$$

and θ_1 is the identity outside of $[-2,2]^n$ and a translation on a neighborhood of each $x_{1,j}$. Let V denote the union of the neighborhoods. We shall choose $c_2, \dots, c_l$ so near c_1 that the projection image of $q^{-1}(\Delta c_1 \cdots c_l)$ onto $\mathbf{R}^n$ is included in V and, moreover, $\Delta c_1 \cdots c_l$ is parallel to $\Delta b_1 \cdots b_l$. Here we cannot replace Y' with the image of Y' under the map $(x,y) \to (\theta_1(x), y)$ because $q^{-1}(\Delta c_0 c_1)$ is not necessarily carried into $0^{n-1} \times \mathbf{R} \times \mathbf{R}^m$. We need to define a modified PL homeomorphism Θ_1 of $\mathbf{R}^n \times \sigma$ so that

$$\Theta_1 = \mathrm{id} \quad \text{on} \quad q^{-1}(c_0) \quad \text{and outside of} \quad [-2,2]^n \times \sigma,$$
$$\Theta_1(x,y) = (\theta_1(x,y), y) \quad \text{on} \quad q^{-1}(\Delta c_1 \cdots c_l),$$

and Θ_1 is linear on any simplex included in $q^{-1}(\sigma_1)$, when we choose $c_2, \dots, c_l$ so that the above conditions on them are satisfied. Let φ be a linear function on $\mathbf{R}^m$ such that $\varphi(c_0) = 0$ and $\varphi = 1$ on the linear subspace containing c_1 and parallel to $\Delta b_1 \cdots b_l$. Define Θ_1 by

$$\Theta_1(x,y) = \begin{cases} (\varphi(y)\theta_1\{(x - x_{0,j})/\varphi(y) + x_{0,j}\} + (1 - \varphi(y))x_{0,j}, y) \\ \qquad \text{if} \quad 0 < \varphi(y) \leq 1 \quad \text{and} \quad |x - x_{0,j}| \leq \varphi(y)\varepsilon \\ (\theta_1(x), y) \qquad \text{if} \quad 1 \leq \varphi(y) \quad \text{and} \quad |x - x_{0,j}| \leq \varepsilon \\ (x, y) \qquad \text{otherwise.} \end{cases}$$

This is a sort of cone extension of $(\theta_1(x), y)|_{q^{-1}(\Delta c_1 \cdots c_l)}$ to $\mathrm{id}|_{q^{-1}(c_0)}$. It is easy to show that Θ_1 fulfills the above requirements. Then

$$\Theta_1(q^{-1}(\Delta c_0 c_1)) \subset 0^{n-1} \times \mathbf{R} \times \sigma.$$

Hence we can assume

$$q^{-1}(\Delta c_0 c_1) \subset 0^{n-1} \times \mathbf{R} \times \sigma. \tag{$*$}$$

When we change J, K, B_i and σ_0 after this second modification, we choose σ_0 so that the new $\Delta c_1 \cdots c_l$ is parallel to the old one.

We can construct σ_1, continuing the above arguments. However, this process is not trivial. We proceed precisely by induction on l. The cases of $l = 0$ and $l = 1$ were already shown. Assume that we can prove the case of $l - 1$, and that we define c_0 and c_1 with $(*)$ and $\mathbf{R}^{l-1} \times 0^{m-l+1}$ is the linear subspace of $\mathbf{R}^m$ containing c_1 and parallel to $\Delta b_1 \cdots b_l$. Apply the induction hypothesis to $\sigma \cap (\mathbf{R}^{l-1} \times 0^{m-l+1})$ and $q^{-1}(\sigma \cap (\mathbf{R}^{l-1} \times 0^{m-l+1}))$. Then there exist points $c_2, \ldots, c_l$ in $\mathbf{R}^{l-1} \times 0^{m-l+1}$ and a PL homeomorphism Θ of $\mathbf{R}^n \times \Delta c_1 \cdots c_l$ of the required form such that each c_i is contained in B_i°,

$$\Theta = \text{id} \quad \text{outside of} \quad U_0 \times \Delta c_1 \cdots c_l,$$
$$q^{-1}(\Delta c_1 \cdots c_l) \subset U_0 \times \Delta c_1 \cdots c_l, \quad \text{and}$$
$$\Theta(q^{-1}(\Delta c_1 \cdots c_l)) \subset 0^{n-1} \times \mathbf{R} \times \Delta c_1 \cdots c_l.$$

Extend Θ to $\mathbf{R}^n \times \sigma_1$ as follows (which is similar to the above cone extension of the map $(x, y) \to (\theta_1(x), y)$ to Θ_1):

$$\Theta(x, y) = \begin{cases} \varphi(y)\Theta((x - x_{0,j})/\varphi(y) + x_{0,j}, (y - b_0)/\varphi(y) + b_0) + (1 - \varphi(y))(x_{0,j}, b_0) \\ \qquad\qquad \text{if} \quad 0 < \varphi(y) \leq 1 \quad \text{and} \quad |x - x_{0,j}| \leq \varphi(y)\varepsilon \\ (x, y) \qquad\qquad \text{otherwise,} \end{cases}$$

where φ is defined as in the case of Θ_1. Then Θ is the identity on $P \cap (\mathbf{R}^n \times \sigma_1)$ and of the form $(\Theta'(x, y), y)$, and it carries $q^{-1}(\sigma_1)$ into $0^{n-1} \times \mathbf{R} \times \sigma_1$. As already noted, we can extend Θ to $\mathbf{R}^n \times \mathbf{R}^m$ so that $\Theta = \text{id}$ on P. Thus IV.2.7 holds for σ_1.

We shall reduce the problem to an easy one through many steps. First it suffices to find a PL retraction:

$$\alpha \colon \mathbf{R}^n \times \sigma \longrightarrow \mathbf{R}^n \times \sigma_1$$

of the form:

$$\alpha(x, y) = (\alpha'(x, y), \alpha''(y)) \quad \text{for} \quad (x, y) \in \mathbf{R}^n \times \sigma$$

such that $(**)$
$$\alpha(Y') = q^{-1}(\sigma_1),$$

and for each $y \in \sigma$, $\alpha'(\cdot, y)$ is a homeomorphism of $\mathbf{R}^n$ and the identity outside of $[-2, 2]^n$. Indeed, the homeomorphism

$$\tau(x, y) = (\Theta' \circ \alpha(x, y), y) \quad \text{for} \quad (x, y) \in \mathbf{R}^n \times \sigma$$

is what we wanted.

If each connected component of $q^{-1}(\sigma - \sigma_1)$ is flat (i.e., a subset of a linear space of the same dimension), it is easier to construct the α. But we cannot expect the components to be flat. By dividing $\sigma - \sigma_1$ as follows, we make the components simpler. We easily find a finite sequence $\sigma = \sigma_k \supset \cdots \supset \sigma_1$ of l-simplexes such that, for each $i > 1$, some l vertices of σ_i (say, $d_1, \ldots, d_l$) coincide with those of σ_{i-1}, the other vertex d_0 of σ_{i-1} lies on $\Delta d_0' d_1$ (d_0' is the other vertex of σ_i), and $\Delta d_0 d_0'$ is contained in some $A_{i'} - A_{i'-1}$. Let $\alpha_i'' \colon \sigma_i \to \sigma_{i-1}$ be the PL retraction such that $\alpha_i''|_{\overline{\sigma_i - \sigma_{i-1}}}$ is linear and $\alpha_i''(d_0') = d_0$. Then for each $y \in \sigma_{i-1}$ there exists an integer i' such that $\alpha_i''^{-1}(y)$ is contained in $A_{i'} - A_{i'-1}$, and we need only construct PL retractions:

$$\alpha_i = (\alpha_i', \alpha_i'') \colon \mathbf{R}^n \times \sigma_i \to \mathbf{R}^n \times \sigma_{i-1}, \quad i = 2, \dots, k,$$

with properties similar to $(**)$ because $\alpha = \alpha_2 \circ \cdots \circ \alpha_k$ satisfies $(**)$. Thus we have reduced the problem to the following.

Let s be an l-simplex in σ, let t be an $(l-1)$-face of s, and let $\beta'' \colon s \to t$ be a linear retraction. Assume for each $y \in t$, there exists an integer i such that $\beta''^{-1}(y)$ is contained in some $A_i - A_{i-1}$. Then there exists a PL retraction:

$$\beta \colon \mathbf{R}^n \times s \to \mathbf{R}^n \times t$$

of the form:

$$\beta(x, y) = (\beta'(x, y), \beta''(y)) \quad \text{for} \quad (x, y) \in \mathbf{R}^n \times s$$

such that

$$\beta(q^{-1}(s)) = q^{-1}(t),$$

and for each $y \in s$, $\beta'(\cdot, y)$ is a homeomorphism of $\mathbf{R}^n$ and is the identity outside of $[-2, 2]^n$.

We decompose β'' more. Let S and T denote simplicial decompositions of s and t respectively, such that $\beta'' \colon S \to T$ is simplicial and S is compatible with J. The last condition brings a simplicial decomposition K_s of $q^{-1}(s)$ such that $q|_{q^{-1}(s)} \colon K_s \to S$ is simplicial. Let $S(1)$ denote the set of simplexes of S which are not vertices and whose images under β'' are vertices. Note that $S(1)$ consists of 1-simplexes. Let $\lambda \in S(1)$. Then $|\operatorname{st}(\lambda, S)|$ is a PL l-ball, the union of simplexes of $S(1)$ is a disjoint union of PL 1-disks, and we see that

$$\operatorname{st}(\lambda, S) \cap S(1) = \{\lambda\},$$

$$\bigcup_{\lambda' \in S(1)} \operatorname{st}(\lambda', S) = S,$$

$$|\operatorname{st}(\lambda, S)| \cap |\operatorname{st}(\lambda', S)| \subset \operatorname{bdry} |\operatorname{st}(\lambda, S)| \quad \text{for} \quad \lambda' \neq \lambda \in S(1), \quad \text{and}$$

$$\sharp(\operatorname{bdry} |\operatorname{st}(\lambda, S)| \cap \beta''^{-1}(y)) \leq 2 \quad \text{for} \quad y \in t,$$

where $\operatorname{bdry} |\operatorname{st}(\lambda, S)|$ denotes the boundary of $|\operatorname{st}(\lambda, S)|$ as a subset of s. Note that $\operatorname{bdry} |\operatorname{st}(\lambda, S)| \subset \partial |\operatorname{st}(\lambda, S)|$ but equality does not necessarily hold. Let y_λ denote the vertex of λ on the side of t, set

$$X_\lambda = |\operatorname{st}(\lambda, S)|, \quad H_\lambda = y_\lambda * |\operatorname{lk}(\lambda, S)|, \quad \text{and} \quad Z_\lambda = |\operatorname{lk}(\lambda, S)|,$$

and define a PL retraction $\beta_\lambda'' : X_\lambda \to H_\lambda$ by

$$\beta_\lambda''(y) = H_\lambda \cap \beta''^{-1}(\beta''(y)) \quad \text{for} \quad y \in X_\lambda.$$

Then $\beta''|_{H_\lambda}$ is injective, and β'' is a sort of composite of all β_λ'', $\lambda \in S(1)$, in the following sense. For each $y \in s - t$ there uniquely exists $\lambda \in S(1)$ such that $y \in X_\lambda - H_\lambda$. Set $y_1 = \beta_\lambda''(y)$. If $y_1 \in t$, then $y_1 = \beta''(y)$. If $y_1 \in s - t$, set $y_2 = \beta_{\lambda'}''(y_1)$ for some unique $\lambda' \in S(1)$ in the same way. Repeating this operation, we have $y_i = \beta''(y)$ for some integer i. Hence it suffices to prove the following statement.

$(**)_\lambda$ Let $\lambda \in S(1)$. There exists a PL retraction:

$$\beta_\lambda : \mathbf{R}^n \times X_\lambda \longrightarrow \mathbf{R}^n \times H_\lambda$$

of the form:

$$\beta_\lambda(x, y) = (\beta_\lambda'(x, y), \beta_\lambda''(y)) \quad \text{for} \quad (x, y) \in \mathbf{R}^n \times X_\lambda$$

such that

$$\beta_\lambda(q^{-1}(X_\lambda)) = q^{-1}(H_\lambda)$$

and for each $y \in X_\lambda$, $\beta_\lambda'(\cdot, y)$ is a homeomorphism of $\mathbf{R}^n$ and is the identity outside of $[-2, 2]^n$.

In the following proof, we use only the next properties of $X_\lambda \supset H_\lambda \supset Z_\lambda$ and $\beta_\lambda'' : X_\lambda \longrightarrow H_\lambda$ because of the forthcoming application.

(1) X_λ, H_λ and Z_λ are polyhedra in σ, and β_λ'' is a PL retraction.

(2) If $y \in Z_\lambda$, then $\beta_\lambda''^{-1}(y)$ consists only of y.

(3) If $y \in H_\lambda - Z_\lambda$, then $\beta_\lambda''^{-1}(y)$ is a segment contained both in a simplex of J and in $A_i - A_{i-1}$ for some i, one of whose end points coincides with y.

(4) The end points of $\beta_\lambda''^{-1}(y)$ other than y for all $y \in H_\lambda - Z_\lambda$ together with Z_λ, form a polyhedron. Let H_λ' denote this polyhedron.

(5) For distinct points y and y' of $H_\lambda - Z_\lambda$, $\beta_\lambda''^{-1}(y)$ and $\beta_\lambda''^{-1}(y')$ are parallel.

We can reduce, moreover, $(**)_\lambda$ to the following statement.

There exist a finite number of PL l-subdisks $u_1, u_2, \ldots$ of H_λ such that $\{\text{Int } u_1, \text{Int } u_2, \ldots\}$ is a covering of H_λ and statement $(**)_\lambda$ holds true when we replace X_λ and Y_λ with $\beta_\lambda''^{-1}(u_i)$ and u_i, $i = 1, 2, \ldots$, respectively, where each Int u_i is the interior of u_i as a subset of H_λ.

Let $(**)_{u_i}$ denote this reduced statement. The reason why we can reduce as above is the following. Refine $\{u_i\}_i$ so that for each i, there exists a point y_i in Int u_i such that $u_i = y_i * \text{bdry } u_i$. Let z_i be a point of $\mathbf{R}^m$ very far from y_i such that the line $\overrightarrow{z_i y_i}$ is parallel to a segment $\beta_\lambda''^{-1}(y)$, $y \in H_\lambda - Z_\lambda$, and the direction from z_i to y_i coincides with the one from y to the other end point of $\beta_\lambda''^{-1}(y)$. Define a PL subpolyhedron of v_i of X_λ by

$$v_i = \beta_\lambda''^{-1}(u_i) \cap (z_i * (\beta_\lambda''^{-1}(u_i) \cap H_\lambda')).$$

Then each v_i is a $(l+1)$-ball, $v_i \subsetneq \beta_\lambda''^{-1}(u_i)$, $\{\text{Int } v_i\}_i$ is a covering of X_λ (we regard v_i as subsets of X_λ), for each $y \in u_i$, $i = 1, 2, \ldots,$

$$\sharp(\text{bdry } v_i \cap \beta_\lambda''^{-1}(y)) = 0 \text{ or } 1,$$

and $v_i \cap \beta_\lambda''^{-1}(y)$ is connected and contains only one point of H_λ'. For each i, let β_i denote a solution of $(**)_{u_i}$. We can retract $\mathbf{R}^n \times X_\lambda$ to $\mathbf{R}^n \times (\overline{X_\lambda - v_1} \cup H_\lambda)$, because β_1 induces a PL retraction:

$$\mathbf{R}^n \times v_1 \ni (x, y) \longrightarrow (x', y') \in \mathbf{R}^n \times (\text{bdry } v_1 \cup (H_\lambda \cap v_1))$$

by the equation $\beta_1(x, y) = \beta_1(x', y')$. Next using β_2 we can retract $\mathbf{R}^n \times (\overline{X_\lambda - v_1} \cup H_\lambda)$ to $\mathbf{R}^n \times (\overline{X_\lambda - v_1 - v_2} \cup H_\lambda)$. Repeat this operation, and compose these retractions. Then we obtain β_λ.

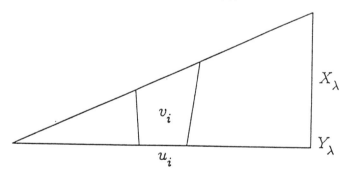

The above arguments will show also the following final reduction. Let a surjective $\mathfrak{X}$-map $\psi \colon H_\lambda \times [0, 1] \to X_\lambda$ (which is not necessarily PL) be defined so that for each $y \in H_\lambda$, $\psi|_{y \times [0,1]}$ is linear,

$$\psi|_{H_\lambda \times 0} = \text{id}, \quad \text{and} \quad \psi(y \times [0, 1]) = \beta_\lambda''^{-1}(y).$$

For a compact subpolyhedron u of H_λ and a closed interval $w = [d^*, d^{**}]$ in $[0, 1]$, set

$$X_{u,w} = \psi(u \times w) \quad \text{and} \quad H_{u,w} = \psi(u \times d^*),$$

which are compact polyhedra, and define a PL retraction $\beta_{u,w}'' \colon X_{u,w} \to H_{u,w}$ in the same way as β_λ''. Let $(**)_{u,w}$ denote the modified statement of

$(**)_\lambda$ where $\beta_\lambda = (\beta'_\lambda, \beta''_\lambda)$, X_λ and H_λ are replaced by $\beta_{u,w} = (\beta'_{u,w}, \beta''_{u,w})$, $X_{u,w}$ and $H_{u,w}$ respectively. Then the problem of finding the above required $u_1, u_2, \ldots$ is equivalent to the following local problem.

For each point $(y_0, \tilde{d})$ in $H_\lambda \times [0, 1]$ there exists a compact polyhedral neighborhood u of y_0 in Y_λ and a closed interval neighborhood w of $\tilde{d}$ in $[0, 1]$ such that $(**)_{u,w}$ holds true.

Now we prove this. There are two cases: $y_0 \notin Z_\lambda$ or $y_0 \in Z_\lambda$.

Case of $y_0 \notin Z_\lambda$, i.e., dim $\beta''^{-1}_\lambda(y_0) = 1$. Assume $\tilde{d} < 1$. Set

$$y_d = \psi(y_0, d) \quad \text{for} \quad d \in [0, 1],$$

$$q^{-1}(y_d) = \{z_{d,1}, \ldots, z_{d,k'}\}, \quad \text{and} \quad z_{d,j} = (x_{d,j}, y_d), \quad j = 1, \ldots, k'.$$

Here the map $[0, 1] \ni d \to x_{d,j} \in \mathbf{R}^n$ for each $j = 1, \ldots, k'$ is linear because $\beta''^{-1}_\lambda(y_0)$ is a segment in a simplex of J and $q|_{Y'} : K \to J$ is simplicial. Let U_j and U'_j be the open ε- and $\varepsilon/3$-neighborhoods of $x_{\tilde{d},j}$ in $\mathbf{R}^n$ respectively, for a small positive number ε such that none of the U_j's intersect each other. Let $w = [d^*, d^{**}]$ be a neighborhood of $\tilde{d}$ in $[0, 1]$ such that $d^{**} < 1$ and $x_{d,j} \in U'_j$ for any $d \in w$ and $j = 1, \ldots, k'$. Then there exists a PL map $\theta : \mathbf{R}^n \times w \to \mathbf{R}^n$ such that

$$\theta(x, d^*) = x \quad \text{for} \quad x \in \mathbf{R}^n,$$

$$\theta(x, d) = x + x_{d^*,j} - x_{d,j} \quad \text{for} \quad (x, d) \in U'_j \times w,$$

and $\theta(\cdot, d)$ is a homeomorphism of $\mathbf{R}^n$ and the identity outside of $[-2, 2]^n$ for each $d \in w$. Let u' be a small compact polyhedral neighborhood of y_0 in Y_λ. Let us define a PL function:

$$\xi : y_1 * \psi(u' \times d^*) \longrightarrow [d^*, 1]$$

by

$$\xi(\delta y_1 + (1 - \delta)y) = d^* + (1 - d^*)\delta \quad \text{for} \quad \delta \in [0, 1], \ y \in \psi(u' \times d^*),$$

i.e., ξ is linear on each segment $\Delta y_1 y$ $(y \in \psi(u' \times d^*))$, $\xi = d^*$ on $\psi(u' \times d^*)$ and $\xi = 1$ at y_1. Choose u' to be a cone with vertex y_0. Then

$$y_1 * \psi(u' \times d^*) \subset X_\lambda$$

because $\beta''^{-1}_\lambda(y)$ and $\beta''^{-1}_\lambda(y')$ are parallel for any $y, y' \in H_\lambda - Z_\lambda$. Note that the restriction of $\xi \circ \psi$ to $\psi^{-1}(y_1 * \psi(u' \times d^*))$ is not necessarily the projection onto $[0, 1]$ because ψ is not necessarily PL. It also follows that

$y_1 * \psi(u' \times d^*)$ is a neighborhood of $\psi(y_0, \tilde{d})$ in X_λ. Let u be a small compact polyhedral neighborhood of y_0 in Y_λ such that

$$q^{-1}(X_{u,w}) \subset \bigcup_j U'_j \times X_{u,w} \quad \text{and} \quad X_{u,w} \subset y_1 * \psi(u' \times d^*).$$

Set

$$\beta'_{u,w}(x, y) = \theta(x, \xi(y)) \quad \text{for} \quad (x, y) \in \mathbf{R}^n \times X_{u,w}.$$

Then the map

$$\beta_{u,w} = (\beta'_{u,w}, \beta''_{u,w}) \colon \mathbf{R}^n \times X_{u,w} \longrightarrow \mathbf{R}^n \times H_{u,w}$$

is a PL retraction which satisfies the required properties. Indeed, the equality $\beta_{u,w}(q^{-1}(X_{u,w})) = q^{-1}(H_{u,w})$ follows from the fact that $\beta_{u,w}$ is linear on the set $U'_j \times (X_{u,w} \cap \beta_\lambda''^{-1}(\Delta y_0 y'_0))$ for each j and for each $y'_0 \in u$. The other property is clear.

If $\tilde{d} = 1$, then it suffices to choose positive d^*, replace ξ with the PL function:

$$\xi \colon y_0 * \psi(u' \times 1) \longrightarrow [0, 1]$$

which is defined by

$$\xi(\delta y_0 + (1 - \delta)y) = (1 - \delta)d^* \quad \text{for} \quad \delta \in [1 - 1/d^*, 1], \ y \in \psi(u' \times d^*),$$

(which implies $\xi = 0$ at y_0 and $\xi = d^*$ on $\psi(u' \times d^*)$), and choose u so that

$$q^{-1}(X_{u,w}) \subset \bigcup_j U'_j \times X_{u,w}, \quad \text{and} \quad X_{u,w} \subset y_0 * \psi(u' \times 1).$$

Note. The above constructed $\beta_{u,w}$ can have the following property. For each $y \in X_{u,w}$, $\beta'_{u,w}(\cdot, y)$ is the identity outside of $U = \bigcup_j U_j$, because we can choose θ so that for each $d \in w$, $\theta(\cdot, d)$ is the identity outside of U.

Case of $y_0 \in Z_\lambda$, i.e., $\beta_\lambda''^{-1}(y_0) = y_0$. We prove this case with the following additional condition by induction on $l = \dim Y$. For each $y \in X_{u,w}$, $\beta'_{u,w}(\cdot, y)$ is the identity outside of a given neighborhood of $\{x_1, \dots, x_{k'}\}$, where

$$q^{-1}(y_0) = \{z_1, \dots, z_{k'}\}, \qquad z_j = (x_j, y_0), \ j = 1, \dots, k'.$$

If $l = 1$, then $Z_\lambda = \emptyset$. Hence assume $l > 1$ and $(**)_{u,w}$ holds in the $l - 1$ dimensional case. Let $U_j, \ j = 1, \dots, k'$, be defined for x_j in the same way as the preceding case, and let U denote the union of U_j's. Let S_0 be a subdivision of $S|_{H_\lambda}$ such that $y_0 \in S_0$. Set

$$H_0 = |\mathrm{lk}(y_0, S_0)| \quad \text{and} \quad X_0 = \beta_\lambda''^{-1}(H_0).$$

Here subdividing S_0 if necessary, we assume that

$$q^{-1}(y_0 * X_0) \subset U \times (y_0 * X_0).$$

Let $y_0 \notin Z_0$ and apply the induction hypothesis to X_0, H_0 and $\tilde{d}$. Then we obtain an interval neighborhood $w = [d^*, d^{**}]$ of $\tilde{d}$ in $[0,1]$ and a PL retraction:

$$\beta_{0,w} = (\beta_{0,w}', \beta_{0,w}'') \colon \mathbf{R}^n \times X_{0,w} \longrightarrow \mathbf{R}^n \times H_{0,w}$$

such that $\beta_{0,w}$ satisfies the required properties and for each $y \in X_{0,w}$, $\beta_{0,w}'(\cdot, y)$ is the identity outside of $U \cup [-2, 2]^n$, where

$$X_{0,w} = \psi(H_0 \times w), \qquad H_{0,w} = \psi(H_0 \times d^*), \quad \text{and}$$
$$\beta_{0,w}''(y) = H_{0,w} \cap \beta_\lambda''^{-1}(\beta_\lambda''(y)) \quad \text{for} \quad y \in X_{0,w}.$$

Set $u = y_0 * H_0$. Then, for $x \in \mathbf{R}^n$, $y \in H_0$, and $0 \le \delta \le 1$, define the $\beta_{u,w}'$ by

$$\beta_{u,w}'(x, \delta y_0 + (1-\delta)y) = \begin{cases} x & \text{if } |x - x_j| \ge \varepsilon(1-\delta) \quad \text{for all } j \\ \delta x_j + (1-\delta)\beta_{0,w}'(x_j + (x - x_j)/(1-\delta), y) \\ & \text{if } \delta < 1 \text{ and } |x - x_j| \le \varepsilon(1-\delta) \text{ for some } j. \end{cases}$$

Then, by the same arguments as we used at the construction of Θ, we can prove the properties in $(**)_{u,w}$ for $\beta_{u,w} = (\beta_{u,w}' \beta_{u,w}'')$. Clearly for each $y \in X_{u,w}$, $\beta_{u,w}'(\cdot, y)$ is the identity outside of $U \cup [-2, 2]^n$. Hence we complete the proof of IV.2.7.   □

Remark IV.2.9. For the same notation as in IV.2.5 and IV.2.6, let ρ_t'', $0 \le t \le 1$, be an X-isotopy of σ such that $\rho_0'' = \mathrm{id}$ and for each $t \in [0,1]$, $\rho_t'' = \mathrm{id}$ on $\partial \sigma$. By IV.2.7 it is easy to find an X-isotopy $\rho_t = (\rho_t', \rho_t'')$, $0 \le t \le 1$, of $\mathbf{R}^n \times \sigma$ such that for each $t \in [0,1]$, ρ_t is invariant on Y',

$$\rho_0 = \mathrm{id}, \qquad \rho_t = \mathrm{id} \quad \text{on} \quad \mathbf{R}^n \times \partial \sigma, \quad \text{and}$$
$$\rho_t'(x, y) = x \quad \text{for} \quad (x, y) \in (\mathbf{R}^n - [-2, 2]^n) \times \sigma.$$

For the proof of IV.2.5, we need to extend ρ_t'' to $|L|$ and ρ_t to $\mathbf{R}^n \times |L|$.

Let $\tilde{\sigma}$ $(\supsetneq \sigma)$ be a simplex of L'. Assume that Y' is contained in $0^{n-1} \times \mathbf{R} \times \mathbf{R}^m$, $q^{-1}(\tilde{\sigma})$ is a polyhedron, and there exist simplicial decompositions $K \subset \tilde{K}$ of $Y' \subset q^{-1}(\tilde{\sigma})$ whose simplexes are carried onto faces of $\tilde{\sigma}$ by q. First we want to extend ρ_t to $\mathbf{R}^n \times \tilde{\sigma}$, keeping certain natural properties.

Let a_l denote the vertex of σ defined in IV.2.8 (i.e., the vertex of L' which is a point of the interior of a simplex of dimension $= \dim \sigma$ of L), and let $b_1, \dots, b_{l'}$ be the vertices of $\tilde{\sigma} - \sigma$. Let $\varepsilon > 0$ be a sufficiently small number. For each i, replace b_i with a point in $(\Delta b_i a_l)^\circ$ sufficiently near a_l, and keep the notation $\tilde{\sigma}, b_1, \dots, b_{l'}$. Let $\alpha'' \colon \tilde{\sigma} \to \sigma$ denote the linear retraction defined by $\alpha''(b_i) = a_l$, $i = 1, \dots, l'$, and define the map $\alpha = (\mathrm{id}, \alpha'') \colon \mathbf{R}^n \times \tilde{\sigma} \to \mathbf{R}^n \times \sigma$. For each $s \in [0,1]$ and for each $z \in \Delta b_1 \cdots b_{l'}$, set

$$\sigma_{s,z} = \{sz + (1-s)y \colon y \in \sigma\}.$$

Then

(1) for each $s \in {]0,1]}$ and for each $z \in \Delta b_1 \cdots b_{l'}$, the closed εs-neighborhoods of the connected components of the set

$$q^{-1}\{sa_l + (1-s)y \colon y \in \sigma\} = q^{-1}(\alpha''(\sigma_{s,z}))$$

are disjoint from one another, and

(2) for each connected component C of $q^{-1}(\sigma_{s,z})$ there exist a point $(x_0, 0) \in \mathbf{R}^n \times \mathbf{R}^m$ and a connected component C' of $q^{-1}(\alpha''(\sigma_{s,z}))$ such that $|x_0| < \varepsilon s$ and

$$\alpha(C) = C' + (x_0, 0) \ (= \{(x + x_0, y) \colon (x, y) \in C'\}).$$

Define the cone extension $\tilde{\rho}''_t$ of ρ''_t to $\tilde{\sigma}$ by

$$\tilde{\rho}''_t(sz + (1-s)y) = sz + (1-s)\rho''_t(y) \quad \text{for} \quad z \in \Delta b_1 \cdots b_{l'}, \ y \in \sigma, \ s \in [0,1].$$

It is an $\mathfrak{X}$-isotopy of $\tilde{\sigma}$ such that for each $t \in [0,1]$,

$$\tilde{\rho}''_0 = \mathrm{id}, \quad \text{and} \quad \tilde{\rho}''_t = \mathrm{id} \quad \text{on} \quad \partial\sigma * \Delta b_1 \cdots b_{l'}.$$

Set

$$\sigma^I = \sigma \times [0,1] \quad \text{and} \quad Y'^I = Y' \times [0,1],$$

and define an $\mathfrak{X}$-isotopy ρ''^I_t, $0 \le t \le 1$, of σ^I by

$$\rho''^I_t(y,s) = \begin{cases} (sa_l + (1-s)\rho''_t(y'), s) & \text{if } y = sa_l + (1-s)y' \text{ for some } y' \in \sigma \\ (y,s) & \text{otherwise}. \end{cases}$$

Then $\rho_t''^I = \mathrm{id}$ outside $(\sigma^\circ \times 0) * (a_l, 1)$ and

$$\rho_t''^I(sa_l + (1-s)y, s) = (\alpha'' \circ \tilde{\rho}_t''(sz + (1-s)y), s)$$
$$\text{for} \quad z \in \Delta b_1 \cdots b_{l'}, \ y \in \sigma, \ s \in [0, 1].$$

Since $Y'^I \subset 0^{n-1} \times \mathbf{R} \times \mathbf{R}^{m+1}$, in the same way as for ρ_t, we obtain an $\mathfrak{X}$-isotopy $\rho_t^I = (\rho_t'^I, \rho_t''^I)$, $0 \leq t \leq 1$, of $\mathbf{R}^n \times \sigma^I$ such that for each $t \in [0, 1]$, $\tilde{\rho}_t$ is invariant on Y'^I,

$$\rho_0^I = \mathrm{id}, \qquad \rho_t^I = \mathrm{id} \quad \text{outside} \quad \mathbf{R}^n \times ((\sigma^\circ \times 0) * (a_l, 1)), \quad \text{and}$$
$$\rho_t'^I(x, y, s) = x \quad \text{for} \quad (x, y, s) \in (\mathbf{R}^n - [-2, 2]^n) \times \sigma^I.$$

Moreover, by (1) $\rho_t'^I$ can satisfy the condition that for each $(x, y, s) \in Y'^I$ with $s > 0$ the homeomorphism $\rho_t'^I(\cdot, y, s)$ of $\mathbf{R}^n$ is a parallel translation on the closed εs-neighborhood of x in $\mathbf{R}^n$ because for each $s \in [0, 1]$,

$$\rho_t^I = \mathrm{id} \quad \text{on} \quad (Y' - q^{-1}\{sa_l + (1-s)y \colon y \in \sigma\}) \times s.$$

Define an $\mathfrak{X}$-isotopy $\tilde{\rho}_t = (\tilde{\rho}_t', \tilde{\rho}_t'')$, $0 \leq t \leq 1$, of $\mathbf{R}^n \times \tilde{\sigma}$ by

$$\tilde{\rho}_t'(x, sz + (1-s)y) = \rho_t'^I(x, sa_l + (1-s)y, s)$$
$$\text{for} \quad x \in \mathbf{R}^n, \ y \in \sigma, \ z \in \Delta b_1 \cdots b_{l'}, \ s \in [0, 1].$$

Then by (2) and the above property of parallel translation, for each $t \in [0, 1]$, $\tilde{\rho}_t$ is invariant on $q^{-1}(\tilde{\sigma})$, and we have clearly

$$\tilde{\rho}_t = \rho_t \text{ on } \mathbf{R}^n \times \sigma, \ \tilde{\rho}_0 = \mathrm{id}, \ \tilde{\rho}_t = \mathrm{id} \text{ on } \mathbf{R}^n \times (\partial\sigma * \Delta b_1 \cdots b_{l'}),$$
$$\text{and} \quad \tilde{\rho}_t'(x, y) = x \text{ for } (x, y) \in (\mathbf{R}^n - [-2, 2]^n) \times \tilde{\sigma}.$$

Next consider the case where Y' is contained in $0^{n-1} \times \mathbf{R} \times \mathbf{R}^m$, $q^{-1}(|\operatorname{st}(\sigma, L')|)$ is a polyhedron and there exists a simplicial decomposition K of $q^{-1}(|\operatorname{st}(\sigma, L')|)$ such that $q_{||K|} \colon K \to \operatorname{st}(\sigma, L')$ is simplicial. We want to shrink $|\operatorname{st}(\sigma, L')|$ as we did $\tilde{\sigma}$ so that ρ_t can be extended to $\mathbf{R}^n \times |L|$. For each $\sigma' \ (\supsetneq \sigma) \in L'$ and for each vertex b of σ' outside σ, let b_0 be a point of $(\Delta b a_l)^\circ$ sufficiently near a_l, and denote by σ_0' the simplex spanned by σ and all such b_0's. Choose common b_0 for all simplexes $\supsetneq \sigma$ of L' with vertex b. Denote by L_0, Λ_0 and Y_0 the simplicial complex generated by σ and σ_0' for all $\sigma' \ (\supsetneq \sigma) \in L'$, its underlying polyhedron, and $q^{-1}(\Lambda_0)$ respectively. Then using ρ_t^I and a PL retraction $\mathbf{R}^n \times \Lambda_0 \to \mathbf{R}^n \times \sigma$ in the same way as above, we obtain an $\mathfrak{X}$-isotopy extension $\tilde{\rho}_t(x, y) = (\tilde{\rho}_t'(x, y), \tilde{\rho}_t''(y))$, $0 \leq t \leq 1$, to

$\mathbf{R}^n \times \Lambda_0$ of the $\mathfrak{X}$-isotopy ρ_t of $\mathbf{R}^n \times \sigma$ such that for each $t \in [0,1]$, $\tilde{\rho}_t$ is invariant on Y_0, $\tilde{\rho}_t''$ is the cone extension of ρ_t'',

$$\tilde{\rho}_0 = \mathrm{id}, \qquad \tilde{\rho}_t = \mathrm{id} \quad \text{on} \quad \mathbf{R}^n \times \mathrm{bdry}\,\Lambda_0, \quad \text{and}$$
$$\tilde{\rho}_t'(x,y) = x \quad \text{for} \quad (x,y) \in (\mathbf{R}^n - [-2,2]^n) \times \Lambda_0.$$

Here bdry Λ_0 denotes the boundary of Λ_0 as a subset of $|L|$. This is immediate by the method of the construction of $\tilde{\rho}_t$. Moreover, by the second equality from the last we can extend $\tilde{\rho}_t$ to $\mathbf{R}^n \times |L|$ by setting $\tilde{\rho}_t = \mathrm{id}$ outside $\mathbf{R}^n \times \Lambda_0$.

The above arguments work in the case of a parameterized $\mathfrak{X}$-homeomorphism of Λ as follows. Let $H \supset H'$ be compact $\mathfrak{X}$-sets, and let $\rho_{h,t}''$ be an $\mathfrak{X}$-homeomorphism of σ parameterized by $H \times [0,1]$ such that for each $h \in H$, $h' \in H'$, $t \in [0,1]$,

$$\rho_{h',t}'' = \rho_{h,0}'' = \mathrm{id}, \quad \text{and} \quad \rho_{h,t}'' = \mathrm{id} \quad \text{on} \quad \partial\sigma.$$

Then we have an $\mathfrak{X}$-homeomorphism $\tilde{\rho}_{h,t}(x,y) = (\tilde{\rho}_{h,t}'(x,y), \tilde{\rho}_{h,t}''(y))$ of $\mathbf{R}^n \times \Lambda_0$ parameterized by $H \times [0,1]$ such that for each $h \in H$, $h' \in H'$, $t \in [0,1]$, $\tilde{\rho}_{h,t}$ is invariant on Y_0, $\tilde{\rho}_{h,t}''$ is the cone extension of $\rho_{h,t}''$ to Λ_0,

$$\tilde{\rho}_{h',t} = \tilde{\rho}_{h,0} = \mathrm{id}, \qquad \tilde{\rho}_{h,t} = \mathrm{id} \quad \text{on} \quad \mathbf{R}^n \times \mathrm{bdry}\,\Lambda_0, \quad \text{and}$$
$$\tilde{\rho}_{h,t}'(x,y) = x \quad \text{for} \quad (x,y) \in (\mathbf{R}^n - [-2,2]^n) \times \Lambda_0.$$

Note that this is proved under the conditions that Y' is contained in $0^{n-1} \times \mathbf{R} \times \mathbf{R}^m$ and that there exists a simplicial decomposition K of $q^{-1}(|\mathrm{st}(\sigma, L')|)$ such that $q|_{|K|}: K \to \mathrm{st}(\sigma, L')$ is simplicial. In Lemma IV.2.12 below, we will remove these conditions. For this we need the following lemma.

We keep the notation in IV.2.7. Let L_0 be a simplicial complex which we construct from $\mathrm{st}(\sigma, L')$ as in IV.2.9 by replacing each vertex b of $\mathrm{lk}(\sigma, L')$ with a point b_0 of $(\Delta b a_l)°$. Set $\Lambda_0 = |L_0|$ and $Y_0 = q^{-1}(\Lambda_0)$. We want to consider IV.2.7 and IV.2.8 on Λ_0. Then we have the following two lemmas.

Lemma IV.2.10. *Assume that Y is a polyhedron and Y' is contained in $0^{n-1} \times \mathbf{R} \times \mathbf{R}^m$. Choose the above b_0's sufficiently near a_l. Then there exist a PL homeomorphism τ of $\mathbf{R}^n \times \Lambda_0$ of the form:*

$$\tau(x,y) = (\tau'(x,y), y) \quad \text{for} \quad (x,y) \in \mathbf{R}^n \times \Lambda_0,$$

a simplicial decomposition K_0 of $\tau(Y_0)$, and a subdivision L_0' of L_0 such that τ is the identity outside $[-2,2]^n \times \Lambda_0$, $\tau(Y')$ is contained in $0^{n-1} \times \mathbf{R} \times \mathbf{R}^m$,

σ is an element of L_0', and $q_0\colon K_0 \to L_0'$ is a simplicial map, where $q_0 = p|_{\tau(Y_0)}$.

Proof. Proof proceeds in the same way as in the proof of IV.2.7. Let $a_0, \ldots, a_l$ denote the vertices of σ ordered as in IV.2.8. Let $\mu\colon \Lambda_0 \to \sigma$ denote the retraction which is linear on each simplex of L_0 and carries all vertices of L_0 outside of σ to a_l. We can choose the b_0's and points $\tilde{a}_i$ in $(\Delta a_0 \cdots a_i)^\circ$, $i = 0, \ldots, l$, so that

(1) $q^{-1}(\mu^{-1}(\tilde{\sigma}))$ and $\mu^{-1}(\tilde{\sigma})$, $\tilde{\sigma} = \Delta\tilde{a}_0 \cdots \tilde{a}_l$, admit simplicial decompositions $\tilde{K}_0$ and $\tilde{L}_0'$ respectively, such that $q|_{|\tilde{K}_0|}\colon \tilde{K}_0 \to \tilde{L}_0'$ is simplicial and $\tilde{L}_0'$ contains $\tilde{\sigma}$ and is a subdivision of the simplicial complex $\tilde{L}_0$ generated by the simplexes $\sigma_0 \cap \mu^{-1}(\tilde{\sigma})$, $\sigma_0 \in L_0$.

Let $\nu\colon \sigma \to \tilde{\sigma}$ denote the linear isomorphism with $\nu(a_i) = \tilde{a}_i$, $i = 0, \ldots, l$, and let $\nu_0\colon \Lambda_0 \to \tilde{\Lambda}_0 = \mu^{-1}(\tilde{\sigma})$ be the extension of ν such that

$$\nu_0(b_0) = \mu^{-1}(\tilde{a}_l) \cap \Delta a_0 \cdots a_{l-1} b_0 \quad \text{for all} \quad b_0,$$

and ν_0 is linear on each simplex of L_0.

Let simplexes $\sigma = \sigma_k \supset \cdots \supset \sigma_1 = \tilde{\sigma}$ and PL retractions $\alpha_i''\colon \sigma_i \to \sigma_{i-1}$, $i = 2, \ldots, k$, be the same as in the proof of IV.2.7. Set

$$\Lambda_{i,0} = \mu^{-1}(\sigma_i), \quad i = 1, \ldots, k.$$

Assume that there exist PL retractions:

$$\alpha_{i,0}''\colon \Lambda_{i,0} \longrightarrow \Lambda_{i-1,0}, \quad i = 2, \ldots, k,$$
$$\alpha_{i,0} = (\alpha_{i,0}', \alpha_{i,0}'')\colon \mathbf{R}^n \times \Lambda_{i,0} \longrightarrow \mathbf{R}^n \times \Lambda_{i-1,0}, \quad i = 2, \ldots, k, \quad \text{and}$$
$$\tilde{\alpha}_{i,0} = (\tilde{\alpha}_{i,0}', \tilde{\alpha}_{i,0}'')\colon \mathbf{R}^n \times \nu_0(\Lambda_{i,0}) \longrightarrow \mathbf{R}^n \times \nu_0(\Lambda_{i-1,0}), \quad i = 2, \ldots, k,$$

such that each $\alpha_{i,0}''$ is an extension of α_i'', for each $y \in \Lambda_{i,0}$ the transformation $\alpha_{i,0}'(\cdot, y)$ of $\mathbf{R}^n$ is a homeomorphism and is the identity outside of $[-2, 2]^n$,

$$\alpha_{i,0}(q^{-1}(\Lambda_{i,0})) = q^{-1}(\Lambda_{i-1,0}),$$

for each $y \in \nu_0(\Lambda_{i,0})$ the transformation $\tilde{\alpha}_{i,0}'(\cdot, y)$ of $\mathbf{R}^n$ is a homeomorphism and is the identity outside of $[-2, 2]^n$,

$$\tilde{\alpha}_{i,0}'' = \nu_0 \circ \alpha_{i,0}'' \circ \nu_0^{-1}, \quad \text{and}$$
$$\tilde{\alpha}_{i,0}(q^{-1}(\nu_0(\Lambda_{i,0}))) = q^{-1}(\nu_0(\Lambda_{i-1,0})).$$

Note that

$$\nu_0(\Lambda_{i,0}) = \mu^{-1}(\nu(\sigma_i)) \quad \text{and} \quad \Lambda_{1,0} = \tilde{\Lambda}_0. \tag{2}$$

Set

$$\alpha_0 = (\alpha_0', \alpha_0'') = \alpha_{2,0} \circ \cdots \circ \alpha_{k,0}, \qquad \tilde{\alpha}_0 = (\tilde{\alpha}_0', \tilde{\alpha}_0'') = \tilde{\alpha}_{2,0} \circ \cdots \circ \tilde{\alpha}_{k,0}.$$

They are PL retractions of $\mathbf{R}^n \times \Lambda_0$ to $\mathbf{R}^n \times \tilde{\Lambda}_0$ and of $\mathbf{R}^n \times \tilde{\Lambda}_0$ to $\mathbf{R}^n \times \nu_0(\tilde{\Lambda}_0)$ respectively, such that for each $y \in \Lambda_0$ and for each $\tilde{y} \in \tilde{\Lambda}_0$, the transformations $\alpha_0'(\cdot, y)$ and $\tilde{\alpha}_0'(\cdot, \tilde{y})$ of $\mathbf{R}^n$ are homeomorphisms and are the identity outside of $[-2, 2]^n$,

$$\tilde{\alpha}_0'' = \nu_0 \circ \alpha_0'' \circ \nu_0^{-1}, \qquad \alpha_0(Y_0) = q^{-1}(\tilde{\Lambda}_0), \quad \text{and}$$
$$\tilde{\alpha}_0(q^{-1}(\tilde{\Lambda}_0)) = q^{-1}(\nu_0(\tilde{\Lambda}_0)). \tag{3}$$

Set

$$\alpha_0^*(x, y) = (\alpha_0^{*\prime}(x, y), \alpha_0^{*\prime\prime}(x, y)) = (\tilde{\alpha}_0'(x, \nu_0(y)), \nu_0^{-1} \circ \tilde{\alpha}_0'' \circ \nu_0(y))$$
$$\text{for} \quad (x, y) \in \mathbf{R}^n \times \Lambda_0.$$

Then α_0^* is a PL retraction of $\mathbf{R}^n \times \Lambda_0$ to $\mathbf{R}^n \times \tilde{\Lambda}_0$ such that for each $y \in \Lambda_0$, the transformation $\alpha_0^*(\cdot, y)$ of $\mathbf{R}^n$ is a homeomorphism and is the identity outside of $[-2, 2]^n$, $\alpha_0^{*\prime\prime} = \alpha_0''$, and by (1), (2) and (3), there exist a subdivision L_0' of L_0 and a simplicial decomposition K_0 of $\alpha_0^{*-1}(q^{-1}(\tilde{\Lambda}_0))$ such that $q_0 \colon K_0 \to L_0'$ is simplicial, where $q_0 = p|_{|K_0|}$, and σ is an element of L_0'. Note that we can choose

$$K_0 = \{(\mathrm{id}, \nu_0^{-1})(\tilde{\sigma}_0) \colon \tilde{\sigma}_0 \in \tilde{K}_0\} \quad \text{and} \quad L_0' = \{\nu_0^{-1}(\tilde{\sigma}_0) \colon \tilde{\sigma}_0 \in \tilde{L}_0'\}.$$

Define a transformation $\tau = (\tau', \mathrm{id})$ of $\mathbf{R}^n \times \Lambda_0$ so that

$$\alpha_0^* \circ \tau = \alpha_0.$$

Then τ is a homeomorphism of the required form, and

$$\alpha_0^*(\tau(Y_0)) = \alpha_0(Y_0).$$

It follows that

$$\tau(Y_0) = \alpha_0^{*-1}(\alpha_0(Y_0)) = \alpha_0^{*-1}(q^{-1}(\tilde{\Lambda}_0)) = |K_0|.$$

Thus we prove the lemma.

Now we construct the $\alpha_{i,0}''$, $\alpha_{i,0}$ and $\tilde{\alpha}_{i,0}$. By the proof of IV.2.7 it suffices to prove the following statement.

(4) Let $s = \Delta c_0 \cdots c_l$ be a simplex in σ such that $\Delta c_0 c_1$ is contained in some $\Delta a_0 \cdots a_{l_1} - \Delta a_0 \cdots a_{l_1-1}$, let a linear retraction $\beta'' : s \to t = \Delta c_1 \cdots c_l$ be defined by $\beta''(c_0) = c_1$, and set

$$s_0 = \mu^{-1}(s) \quad \text{and} \quad t_0 = \mu^{-1}(t).$$

Then there exist PL retraction $\beta_0'' : s_0 \to t_0$, $\beta_0 = (\beta_0', \beta_0''): \mathbf{R} \times s_0 \to \mathbf{R} \times t_0$ and $\tilde{\beta}_0 = (\tilde{\beta}_0', \tilde{\beta}_0''): \mathbf{R} \times \nu_0(s_0) \to \mathbf{R} \times \nu_0(t_0)$, and simplicial decompositions S_0 of s_0, T_0 of t_0 and K_{s_0} of $q^{-1}(s_0)$ such that $\beta_0'' : S_0 \to T_0$ and $q|_{q^{-1}(s_0)} :$ $K_{s_0} \to S_0$ are simplicial, β_0'' is an extension of β'', $\tilde{\beta}_0'' = \nu_0 \circ \beta_0'' \circ \nu_0^{-1}|_{\nu_0(s_0)}$. Furthermore, if we define X_λ, Y_λ, Z_λ and β_λ'' in the same way as in the proof of IV.2.7, then conditions $(1), \ldots, (5)$ in the proof are satisfied for these X_λ, Y_λ, Z_λ and β_λ'' and also for $\nu_0(X_\lambda)$, $\nu_0(Y_\lambda)$, $\nu_0(Z_\lambda)$ and $\nu_0 \circ \beta_\lambda'' \circ \nu_0^{-1}$.

Let us prove statement (4). On trial, define β_0'' by

$$\beta_0''(y) = \nabla a_0 \cdots a_{l-1} y \cap \mu^{-1}(\beta''(\mu(y))) \quad \text{for} \quad y \in s_0 - \Delta a_0 \cdots a_{l-1}.$$

Consider the case where β_0'' is PL. Clearly $\nu_0 \circ \beta_0'' \circ \nu_0^{-1}$ also is so. As in the proof of IV.2.7, let simplicial decompositions S_0 of s_0, T_0 of t_0 and K_{s_0} of $q^{-1}(s_0)$ be such that $\beta_0'' : S_0 \to T_0$ and $q|_{q^{-1}(s_0)} : K_{s_0} \to S_0$ are simplicial. Then (4) holds automatically.

If $l_1 < l$ in (4), i.e., $\Delta c_0 c_1 \subset \Delta a_0 \cdots a_{l-1}$, then β_0'' is PL as shown below. Assume that

$$c_2, \ldots, c_{l''} \in \Delta a_0 \cdots a_{l-1} \quad \text{and} \quad c_{l''+1}, \cdots, c_l \notin \Delta a_0 \cdots a_{l-1}.$$

Then

$$s_0 = \Delta c_0 \cdots c_{l''} * \mu^{-1}(\Delta c_{l''+1} \cdots c_l), \qquad t_0 = \Delta c_1 \cdots c_{l''} * \mu^{-1}(\Delta c_{l''+1} \cdots c_l),$$

and $\mu^{-1}(\Delta c_{l''+1} \cdots c_l)$ is a polyhedron. Hence it suffices to prove that for each simplex δ in $\mu^{-1}(\Delta c_{l''+1} \cdots c_l)$, β_0'' is linear on $\Delta c_0 \cdots c_{l''} * \delta$. More precisely,

$$\beta_0''(y) = y - e_0 c_0 + e_0 c_1$$

for $\quad y = e_0 c_0 + \cdots + e_{l''} c_{l''} + e_{l''+1} d, \ d \in \delta, \ e_i \geq 0 \in \mathbf{R}$

$$\text{with} \quad e_{l''+1} > 0, \quad \sum e_i = 1.$$

By linearity of μ,

$$\mu(y) = e_0 c_0 + \cdots + e_{l''} c_{l''} + e_{l''+1} \mu(d) \quad \text{and}$$
$$\mu(y - e_0 c_0 + e_0 c_1) = \mu(y) - e_0 c_0 + e_0 c_1.$$

By linearity of β'',

$$\beta''(\mu(y)) = \mu(y) - e_0 c_0 + e_0 c_1$$

because by assumption, $\beta'' = \text{id}$ on $\Delta c_1 \cdots c_{l''} \cup \mu(\delta)$. Hence

$$y - e_0 c_0 + e_0 c_1 \in \mu^{-1}(\beta''(\mu(y))).$$

On the other hand,

$$y - e_0 c_0 + e_0 c_1 \in \nabla c_0 \cdots c_{l''} y \subset \nabla a_0 \cdots a_{l-1} y.$$

Hence by the definition of β_0'',

$$\beta_0''(y) = y - e_0 c_0 + e_0 c_1,$$

which proves that β_0'' is PL.

If $l_1 = l$, i.e., $\Delta c_0 c_1 \cap \Delta a_0 \cdots a_{l-1} = \varnothing$, then we modify β_0'' as follows because it is not necessarily PL. Assume that

$$c_2, \ldots, c_{l''} \notin \Delta a_0 \cdots a_{l-1} \quad \text{and} \quad c_{l''+1}, \ldots, c_l \in \Delta a_0 \cdots a_{l-1}.$$

For each $i = 0, \ldots, l''$, let $d_{i,1}, \ldots, d_{i,l'''}$ be the vertices of the simplicial complex:

$$L_{0,i} = \{\mu^{-1}(c_i) \cap \sigma'' : \sigma'' \in L_0\}.$$

Note that $L_{0,i}$ is a simplicial decomposition of $\mu^{-1}(c_i)$, and if $c_i = e_0 a_0 + \cdots + e_l a_l$ for non-negative reals $e_0, \ldots, e_l$ with $e_0 + \cdots + e_l = 1$, then

$$\{d_{i,1}, \ldots, d_{i,l'''}\} = \{c_i, \ e_0 a_0 + \cdots e_{l-1} a_{l-1} + e_l b \colon b \in \text{lk}(\sigma, L_0)\}.$$

We order the vertices so that

$$d_{i,j} \in \nabla a_0 \cdots a_{l-1} d_{i',j} \quad \text{for any } i, \ i' \text{ and } j.$$

Then, for each $0 \leq i, i' \leq l''$ and for each $1 \leq j_1, \ldots, j_k \leq l'''$, $\Delta d_{i,j_1} \cdots d_{i,j_k}$ is a simplex in $L_{0,i}$ if and only if $\Delta d_{i',j_1} \cdots d_{i',j_k}$ is a simplex in $L_{0,i'}$. For each simplex $\Delta d_{0,j_1} \cdots d_{0,j_k} \in L_{0,0}$, consider the following canonical triangulation L_D of the convex hull D spanned by $d_{i,j_1}, \ldots, d_{i,j_k}$, $0 \leq i \leq l''$, which is the canonical triangulation of the product of two simplexes. (Note that $D = \mu^{-1}(\Delta c_0 \cdots c_{l''}) \cap \sigma''$ for some $\sigma'' \in L_0$, $\mu|_D$ is PL trivial over $\Delta c_0 \cdots c_{l''}$, and, moreover, $\mu|_{\mu^{-1}(\Delta c_0 \cdots c_{l''})}$ is also.) For simplicity of notation, assume that $j_1 = 1, \ldots, j_k = k$. Set

$$E = \{(i_1, i_1', \ldots, i_{l''+k}, i_{l''+k}') \in \mathbf{N}^{2(l''+k)} \colon 0 = i_1 \leq \cdots \leq i_{l''+k} = l'',$$
$$1 = i_1' \leq \cdots \leq i_{l''+k}' = k, \ i_1 + i_1' = 1, \ldots, i_{l''+k} + i_{l''+k}' = l'' + k\},$$

and denote by L_D the family of $\Delta d_{i_1,i_1'} \cdots d_{i_{l''+k},i_{l''+k}'}$ for $(i_1, i_1', \ldots, i_{l''+k},$ $i_{l''+k}') \in E$ and their faces. Then L_D is a simplicial decomposition of D, and $\cup_D L_D$ is a simplicial decomposition of $\mu^{-1}(\Delta c_0 \cdots c_{l''})$ because the order of the vertices of $L_{0,0}$ is fixed. Note that $\Delta c_0 \cdots c_{l''} \in L_D$.

Given a simplex δ of L_D of dimension $l'' + k - 1$ of the form $\Delta d_{0,1} \cdots d_{0,i'} d_{1,i'} \cdots$, set

$$\delta' = \Delta d_{0,1} \cdots \hat{d}_{0,i'} d_{1,i'} \cdots ,$$

where the symbol $\hat{}$ denotes removal of the factor, and let $\gamma_\delta'' : \delta \to \delta'$ denote the linear retraction defined by $\gamma_\delta''(d_{0,i'}) = d_{1,i'}$. Now we define a modification γ_0'' of β_0'' on D by setting

$$\gamma_0''(y) = \cdots \circ \gamma_{\delta_2}'' \circ \gamma_{\delta_1}''(y) \in D \cap s_0 \quad \text{for} \quad y \in D,$$

for some $\delta_1, \delta_2, \ldots$ with $y \in \delta_1$, $\gamma_{\delta_1}''(y) \in \delta_2, \ldots$. Here the sequence $\delta_1, \delta_2, \ldots$ is not necessarily uniquely determined by y. However, the value $\gamma_0''(y)$ does not depend on choice of the sequence, and, moreover, γ_0'' is PL. Furthermore, $\gamma_0''(y)$ does not depend on D, i.e., two values of γ_0'' by D and another D' coincide with each other on $D \cap D'$. Hence we have a PL retraction:

$$\gamma_0'' : \mu^{-1}(\Delta c_0 \cdots c_{l''}) \longrightarrow \mu^{-1}(\Delta c_1 \cdots c_{l''}).$$

Next we extend γ_0'' to a map from $s_0 = \mu^{-1}(\Delta c_0 \cdots c_{l''}) * \Delta c_{l''+1} \cdots c_l$ to $t_0 = \mu^{-1}(\Delta c_1 \cdots c_{l''}) * \Delta c_{l''+1} \cdots c_l$ by the join of γ_0'' and the identity map of $\Delta c_{l''+1} \cdots c_l$. We keep the same notation γ_0'' for the extension. Thus we define a modification $\gamma_0'' : s_0 \to t_0$ of β_0''.

As is expected, γ_0'' has the following good properties. Let $y \in t_0$.

(i) γ_0'' is a PL retraction,

(ii) $\gamma_0''^{-1}(y)$ is contained in some $A_i - A_{i-1}$, where $A_i = \Delta a_0 \cdots a_i$ for some simplex $\sigma_0' = \Delta a_0 \cdots a_l a_{l+1} \cdots a_{l'} \in L_0$, $i \leq l'$, with vertices ordered as in the proof of IV.2.7,

(iii) $\mu(\gamma_0''^{-1}(y))$ is either the point $\mu(y)$ or a segment in $\Delta c_0 \cdots c_l$ parallel to $\Delta c_0 c_1$ with an end $\mu(y)$,

(iv) the restriction of μ to $\gamma_0''^{-1}(y)$ is a homeomorphism to $\mu(\gamma_0''^{-1}(y))$, and

(v) $\gamma_0'' = \beta_0''$ on s.

By these properties the retractions γ_0'' and $\nu_0 \circ \gamma_0'' \circ \nu_0^{-1}|_{\nu_0(s_0)} : \nu_0(s_0) \to \nu_0(t_0)$ satisfy the conditions in the proof of IV.2.7 as follows. Let S_0, T_0 and K_{s_0} denote simplicial decompositions of s_0, t_0 and $q^{-1}(s_0)$ respectively, such that $\gamma_0'' : S_0 \to T_0$ and $q|_{q^{-1}(s_0)} : K_{s_0} \to S_0$ are simplicial and $\{\sigma^\circ : \sigma \in S_0\}$ is compatible with $\cup_D L_D$. Let $S_0(1)$ denote the set of 1-simplexes of S_0 whose images under γ_0'' are vertices. Let y_λ denote the vertex of each $\lambda \in S_0(1)$ on the side of t_0, set

$$X_\lambda = |\operatorname{st}(\lambda, S_0)|, \quad H_\lambda = y_\lambda * |\operatorname{lk}(\lambda, S_0)|, \quad \text{and} \quad Z_\lambda = |\operatorname{lk}(\lambda, S_0)|,$$

and define naturally a PL retraction $\gamma_{0,\lambda}'' \colon X_\lambda \to H_\lambda$ by γ_0''. Then by the properties (i),...,(v), X_λ, H_λ, Z_λ and $\gamma_{0,\lambda}''$ satisfy conditions $(1), \dots, (5)$ in the proof of IV.2.7. Clearly $\nu_0(X_\lambda)$, $\nu_0(H_\lambda)$, $\nu_0(Z_\lambda)$ and $\nu_0 \circ r_{0,\lambda}'' \circ \nu_0^{-1}$ also do so. Hence γ_0'' in place of β_0'' fulfills the requirements in (4), which completes the proof. $\qquad\qquad\qquad\qquad\qquad\qquad\qquad\qquad\qquad\qquad\qquad\qquad\qquad\square$

In IV.2.10, we assumed that Y is a polyhedron and Y' is contained in $0^{n-1} \times \mathbf{R} \times \mathbf{R}^m$. Using II.2.7 and IV.2.7 we can remove these assumptions as follows. We keep the same notation.

Lemma IV.2.11. *Choose the b_0's arbitrarily. There exist an $\mathfrak{X}$-set W in $|L|$, an $\mathfrak{X}$-homeomorphism $\tau \colon \mathbf{R}^n \times W \to \mathbf{R}^n \times \Lambda_0$, a simplicial complex K_0 in $\mathbf{R}^n \times \Lambda_0$ and a subdivision L_0' of L_0 such that we have the following: $\tau(Y \cap (\mathbf{R}^n \times W))$ is a polyhedron and coincides with $|K_0|$,*

$$\sigma \subset W \subset |\operatorname{st}(\sigma, L')|,$$

τ is of the form:

$$\tau(x, y) = (\tau'(x, y), \tau''(y)) \quad \text{for} \quad (x, y) \in \mathbf{R}^n \times W, \quad \text{and}$$

τ'' can be extended to an $\mathfrak{X}$-homeomorphism of $|L|$ which is invariant on each simplex of L'. For each $y \in W$, the homeomorphism $\tau'(\cdot, y)$ of $\mathbf{R}^n$ is the identity outside of $[-2, 2]^n$, $\tau(Y')$ is contained in $0^{n-1} \times \mathbf{R} \times \mathbf{R}^m$, σ is an element of L_0', and $q_0 \colon K_0 \to L_0'$ is simplicial, where $q_0 = p|_{|K_0|}$.

Proof. First, by IV.2.7 we can assume that Y' is contained in $0^{n-1} \times \mathbf{R} \times \mathbf{R}^m$. Next, by II.2.7 there exists an $\mathfrak{X}$-homeomorphism τ_1 of $\mathbf{R}^n \times |L|$ of the form $(\tau_1'(x, y), \tau_1''(y))$ such that τ_1'' is invariant on each simplex of L', for each $y \in |L|$, $\tau_1'(\cdot, y)$ is the identity outside of $[-2, 2]^n$, and $\tau_1(Y)$ is a polyhedron. Moreover, by the method of construction of τ_1 we can assume that $\tau_1(Y')$ is contained in $0^{n-1} \times \mathbf{R} \times \mathbf{R}^m$. Finally, by IV.2.10, for the b_0's sufficiently near the a_l, we have a PL homeomorphism τ_2 of $\mathbf{R}^n \times \Lambda_0$ of the form $(\tau_2'(x, y), y)$, a simplicial complex K_0 in $\mathbf{R}^n \times \Lambda_0$ and a subdivision L_0' of L_0 such that $\tau_2(\tau_1(Y) \cap (\mathbf{R}^n \times \Lambda_0))$ is a polyhedron and coincides with $|K_0|$, τ_2 is the identity outside of $[-2, 2]^n \times \Lambda_0$, $\tau_2(\tau_1(Y'))$ is contained in $0^{n-1} \times \mathbf{R} \times \mathbf{R}^m$, σ is an element of L_0', and $q_0 \colon K_0 \to L_0'$ is simplicial, where $q_0 = p|_{|K_0|}$.

Set

$$W = \tau_1''^{-1}(\Lambda_0) \quad \text{and} \quad \tau = \tau_2 \circ (\tau_1|_{\mathbf{R}^n \times W}).$$

Then τ is an $\mathfrak{X}$-homeomorphism from $\mathbf{R}^n \times W$ to $\mathbf{R}^n \times \Lambda_0$, $\tau(Y \cap (\mathbf{R}^n \times W))$ is a polyhedron, K_0 is a simplicial decomposition of $\tau(Y \cap (\mathbf{R}^n \times W))$, and τ, K_0 and L_0' fulfill the requirements in IV.2.11.

In the above arguments, we chose the b_0's sufficiently near σ. But we can choose them arbitrarily because if we are given two points b_0 and b_0' of $(b * \sigma)^\circ$ for each vertex b of $\mathrm{lk}(\sigma, L')$, then there exists a PL homeomorphism τ_3'' of $|L|$ such that

$$\tau_3'' = \mathrm{id} \quad \text{on} \quad \sigma, \qquad \tau_3''(b_0) = b_0' \quad \text{for each} \quad b,$$

and τ_3'' is linear on each $b_0 * \sigma$ and invariant on each simplex of L'.   □

By this lemma we can remove the conditions in IV.2.9 that Y' is contained in $0^{n-1} \times \mathbf{R} \times \mathbf{R}^m$, $q^{-1}(|\,\mathrm{st}(\sigma, L')|)$ is a polyhedron and there exists a simplicial decomposition K of $q^{-1}(|\,\mathrm{st}(\sigma, L')|)$ such that $q \colon K \to \mathrm{st}(\sigma, L')$ is simplicial. We do this as follows.

Lemma IV.2.12. *For the same notation as in IV.2.7 and the definitions stated before IV.2.10, let $H \supset H'$ be compact $\mathfrak{X}$-sets, and let $\rho_{h,t}''$ be an $\mathfrak{X}$-homeomorphism of σ parameterized by $H \times [0,1]$ such that for each $h \in H$, $h' \in H'$, and $t \in [0,1]$, we have*

$$\rho_{h',t}'' = \rho_{h,0}'' = \mathrm{id}, \quad \text{and} \quad \rho_{h,t}'' = \mathrm{id} \quad \text{on} \quad \partial\sigma.$$

Choose the b_0's arbitrarily. Then there exists an $\mathfrak{X}$-homeomorphism $P_{h,t}$ of $\mathbf{R}^n \times \Lambda_0$ parameterized by $H \times [0,1]$ of the form:

$$P_{h,t}(x,y) = (P_{h,t}'(x,y), P_{h,t}''(y)) \quad \text{for} \quad (x,y) \in \mathbf{R}^n \times \Lambda_0$$

such that for each $h \in H$, $h' \in H'$, $t \in [0,1]$, and $y \in \Lambda_0$,

$$P_{h',t} = P_{h,0} = \mathrm{id}, \qquad P_{h,t} = \mathrm{id} \quad \text{on} \quad \mathbf{R}^n \times \mathrm{bdry}\,\Lambda_0,$$
$$P_{h,t}'(\cdot,y) = \mathrm{id} \quad \text{outside of} \quad [-2,2]^n,$$
$$P_{h,t}''(\sigma'') = \sigma'' \quad \text{for} \quad \sigma'' \in L_0,$$
$$P_{h,t}''|_\sigma = \rho_{h,t}'', \quad \text{and} \quad P_{h,t}(Y_0) = Y_0.$$

Here $\mathrm{bdry}\,\Lambda_0$ *denotes the boundary of Λ_0 as a subset of $|L|$.*

Proof. For simplicity of notation, we write $|L|$, L' and Y for Λ_0, L_0 and Y_0 in IV.2.12 respectively. We want to construct an $\mathfrak{X}$-homeomorphism $P_{h,t}$ of

$\mathbf{R}^n \times |L|$. Note that the condition that $P_{h,t} = $ id on $\mathbf{R}^n \times$ bdry Λ_0 is replaced by the condition that $P_{h,t} = $ id on $\mathbf{R} \times |\mathrm{lk}(a, L')|$, where $a \in \sigma^\circ$. Apply IV.2.11 to the new σ, L' and Y, and let

$$\tau = (\tau', \tau''): \mathbf{R}^n \times W \longrightarrow \mathbf{R}^n \times \Lambda_0,$$

K_0 and L'_0 be the result. (This Λ_0 is different to that of IV.2.12 and is defined from the new L'.) Next apply IV.2.9 to $|K_0| = \tau(Y \cap (\mathbf{R}^n \times W))$ and $\tau'' \circ \rho''_{h,t} \circ \tau''^{-1}$ on σ. We have an $\mathfrak{X}$-homeomorphism $Q_{h,t}$ of $\mathbf{R}^n \times \Lambda_0$ parameterized by $H \times [0,1]$ of the form:

$$Q_{h,t}(x,y) = (Q'_{h,t}(x,y), Q''_{h,t}(y)) \quad \text{for} \quad (x,y) \in \mathbf{R}^n \times \Lambda_0$$

such that for each $h \in H$, $h' \in H'$, $t \in [0,1]$, and $y \in \Lambda_0$,

$$Q_{h',t} = Q_{h,0} = \mathrm{id}, \qquad Q_{h,t} = \mathrm{id} \quad \text{on} \quad \mathbf{R}^n \times \text{bdry } \Lambda_0,$$
$$Q'_{h,t}(\cdot, y) = \mathrm{id} \quad \text{outside of} \quad [-2,2]^n,$$
$$Q''_{h,t}(\sigma'') = \sigma'' \quad \text{for} \quad \sigma'' \in L_0,$$
$$Q''_{h,t} = \tau'' \circ \rho''_{h,t} \circ \tau''^{-1} \quad \text{on} \quad \sigma, \quad \text{and}$$
$$Q_{h,t}(|K_0|) = |K_0|.$$

Define $P_{h,t}$ by

$$P_{h,t} = \begin{cases} \tau^{-1} \circ Q_{h,t} \circ \tau & \text{on} \quad \mathbf{R}^n \times W \\ \mathrm{id} & \text{on} \quad \mathbf{R}^n \times (|L| - W). \end{cases}$$

Then $P_{h,t}$ fulfills the requirements in IV.2.12.   □

Lemma IV.2.13. *For the same notation as in IV.2.5 and IV.2.6, let $\rho''_{h,t}$ be an $\mathfrak{X}$-homeomorphism of $|L'^k|$ parameterized by $H \times [0,1]$ such that for each $h \in H$, $h' \in H$, and $t \in [0,1]$, $\rho''_{h,t}$ is invariant on each simplex of L'^k,*

$$\rho''_{h',t} = \rho''_{h,0} = \mathrm{id}, \quad \text{and} \quad \rho''_{h,t} = \mathrm{id} \quad \text{on} \quad |L'^{k-1} \cup (L_1 \cap L'^k)|.$$

There exists an $\mathfrak{X}$-homeomorphism $P_{h,t}$ of $\mathbf{R}^n \times |L|$ parameterized by $H \times [0,1]$ of the form:

$$P_{h,t}(x,y) = (P'_{h,t}(x,y), P''_{h,t}(y)) \quad \text{for} \quad (x,y) \in \mathbf{R}^n \times |L|$$

such that for each $h \in H$, $h' \in H'$, $t \in [1, 0]$, *and* $y \in |L|$, $P''_{h,t}$ *is invariant on each simplex of* L', $P_{h,t}$ *is invariant on* Y,

$$P_{h,t} = \mathrm{id} \quad on \quad \mathbf{R}^n \times |L'^{k-1} \cup L_1|,$$

$$P_{h',t} = P_{h,0} = \mathrm{id}, \qquad P''_{h,t}|_{|L^k|} = \rho''_{h,t}, \quad and$$

$$P'_{h,t}(\cdot, y) = \mathrm{id} \quad outside\ of \quad [-2, 2]^n.$$

Proof. Recall the notation stated before IV.2.10. For each $\sigma \in L'$ we defined a neighborhood Λ_0 of σ° in $|L|$ and a simplicial decomposition L_0 of Λ_0. In order to emphasize σ, we denote Λ_0 and L_0 by $\Lambda_0(\sigma)$ and $L_0(\sigma)$ respectively. Then we can choose $\Lambda_0(\sigma)$'s and $L_0(\sigma)$'s so that

$$\Lambda_0(\sigma) \cap \Lambda_0(\sigma') \subset \mathrm{bdry}\, \Lambda_0(\sigma) \quad \text{for} \quad \sigma \neq \sigma' \in L'^k - L'^{k-1}, \tag{1}$$

$$\Lambda_0(\sigma) \cap |L'^{k-1}| \subset \mathrm{bdry}\, \Lambda_0(\sigma) \quad \text{for} \quad \sigma \in L'^k - L'^{k-1}, \quad \text{and} \tag{2}$$

$$\{\sigma''^\circ : \sigma'' \in L_0(\sigma)\} \quad \text{is compatible with} \quad L' \text{ for each } \sigma \in L'^k. \tag{3}$$

For each $\sigma \in L'^k - L'^{k-1}$, let $P_{\sigma,h,t} = (P'_{\sigma,h,t}, P''_{\sigma,h,t})$ be the X-homeomorphism of $\mathbf{R}^n \times \Lambda_0(\sigma)$ parameterized by $H \times [0, 1]$ constructed in IV.2.12 so that for each $h \in H$, $h' \in H'$, $t \in [0, 1]$, and $y \in \Lambda_0(\sigma)$,

$$P_{\sigma,h',t} = P_{\sigma,h,0} = \mathrm{id} \quad on \quad \mathbf{R}^n \times \Lambda_0(\sigma), \tag{4}$$

$$P_{\sigma,h,t} = \mathrm{id} \quad on \quad \mathbf{R}^n \times \mathrm{bdry}\, \Lambda_0(\sigma), \tag{5}$$

$$P'_{\sigma,h,t}(\cdot, y) = \mathrm{id} \quad outside\ of \quad [-2, 2]^n, \tag{6}$$

$$P''_{\sigma,h,t}(\sigma'') = \sigma'' \quad for \quad \sigma'' \in L_0(\sigma), \tag{7}$$

$$P''_{\sigma,h,t} = \rho''_{h,t} \quad on \quad \sigma, \tag{8}$$

$$P_{\sigma,h,t}(q^{-1}(\Lambda_0(\sigma))) = q^{-1}(\Lambda_0(\sigma)), \quad and \tag{9}$$

$$P_{\sigma,h,t} = \mathrm{id} \quad for \quad \sigma \in L'^k \cap L_1. \tag{10}$$

Let $P_{h,t}$ be defined to be $P_{\sigma,h,t}$ on $\Lambda_0(\sigma)$ for each $\sigma \in L'^k - L'^{k-1}$ and the identity outside of the union of all those $\Lambda_0(\sigma)$. Then by (1) and (5), $P_{h,t}$ is a well-defined X-homeomorphism of $\mathbf{R}^n \times |L|$; (4) implies $P_{h',t} = P_{h,0} = \mathrm{id}$; it follows from (2), (5) and (10) that $P_{h,t} = \mathrm{id}$ on $\mathbf{R}^n \times |L'^{k-1} \cup L_1|$; (8) implies $P''_{h,t} = \rho''_{h,t}$ on $|L^k|$; by (3) and (7) $P''_{h,t}$ is invariant on each simplex of L'; by (9) $P_{h,t}$ is invariant on Y; and (6) implies $P'_{\sigma,h,t}(\cdot, y) = \mathrm{id}$ outside of $[-2, 2]^n$. Thus we prove the lemma. $\qquad \square$

Proof of IV.2.5. We prove IV.2.5 by downward induction on k. The case of $k = \dim Y$ is a special case of IV.2.13. Hence we assume the case of $k + 1$.

Apply IV.2.13 to $\pi''_{h,t}|_{|L'^k|}$. We have an $\mathfrak{X}$-homeomorphism $\pi_{1,h,t}$ of $\mathbf{R}^n \times |L|$ parameterized by $H \times [0,1]$ of the form:

$$\pi_{1,h,t}(x,y) = (\pi'_{1,h,t}(x,y), \pi''_{1,h,t}(y)) \quad \text{for} \quad (x,y) \in \mathbf{R}^n \times |L|$$

such that for each $h \in H$, $h' \in H'$, $t \in [0,1]$, and $y \in |L|$, $\pi''_{1,h,t}$ is invariant on each simplex of L', $\pi_{1,h,t}$ is invariant on Y,

$$\pi_{1,h',t} = \pi_{1,h,t} = \mathrm{id},$$
$$\pi_{1,h,t} = \mathrm{id} \quad \text{on} \quad \mathbf{R}^n \times |L'^{k-1} \cup L_1|,$$
$$\pi''_{1,h,t} = \pi''_{h,t} \quad \text{on} \quad |L'^k|, \quad \text{and}$$
$$\pi'_{1,h,t}(\cdot, y) = \mathrm{id} \quad \text{outside of} \quad [-2,2]^n.$$

Consider $\pi''^{-1}_{1,h,t} \circ \pi''_{h,t}$. This is an $\mathfrak{X}$-homeomorphism of $|L|$ parameterized by $H \times [0,1]$ and satisfies the conditions in the lemma for $k+1$. Hence by the induction hypothesis we have an $\mathfrak{X}$-homeomorphism $\pi_{2,h,t}$ of $\mathbf{R}^n \times |L|$ parameterized by $H \times [0,1]$ of the form:

$$\pi_{2,h,t}(x,y) = (\pi'_{2,h,t}(x,y), \pi''^{-1}_{1,h,t} \circ \pi''_{h,t}(y)) \quad \text{for} \quad (x,y) \in \mathbf{R}^n \times |L|$$

such that for each $h \in H$, $h' \in H'$, $t \in [0,1]$, and $y \in |L|$, $\pi_{2,h,t}$ is invariant on Y,

$$\pi_{2,h',t} = \pi_{2,h,t} = \mathrm{id}, \qquad \pi_{2,h,t} = \mathrm{id} \quad \text{on} \quad \mathbf{R}^n \times |L'^k \cup L_1|, \quad \text{and}$$
$$\pi'_{2,h,t}(\cdot, y) = \mathrm{id} \quad \text{outside of} \quad [-2,2]^n.$$

Clearly $\pi_{h,t} = \pi_{1,h,t} \circ \pi_{2,h,t}$ fulfills the requirements. $\qquad \square$

Proof of IV.1.2′. By I.3.2 and the above arguments it suffices to prove the following statement.

Statement. Let C be a cell decomposition of a closed semilinear set $Y \subset \mathbf{R}^m$, let $\{X_i\}$ be a finite C^0 $\mathfrak{X}_0$-stratification of $\mathbf{R}^n \times Y$, and let $p \colon \mathbf{R}^n \times Y \to Y$ denote the projection. Assume $p(X_i)$ is an open cell of C for each i, if $p|_{X_i}$ is not a homeomorphism onto $p(X_i)$ then X_i is an open subset of $p^{-1}(p(X_i))$, and the union of X_i's such that $p|_{X_i}$ are homeomorphisms onto their images is closed in $\mathbf{R}^n \times Y$. Then there exist a semilinear set $X \subset \mathbf{R}^{n'} \times Y$ for some n' and an $\mathfrak{X}_0$-homeomorphism $\pi \colon X \to \mathbf{R}^n \times Y$ of the form $\pi(x,y) = (\pi'(x,y), y)$ such that each $\pi^{-1}(X_i)$ is semilinear.

We assume $n > 1$ because the case $n = 1$ is very easy to prove. By a stereographic projection we regard $\mathbf{R}^n$ as $S^n -$ a point s. In the statement, we require π to be a homeomorphism onto $S^n \times Y$. Let $q: S^n \times Y \to Y$ denote the projection.

First we reduce the statement to the case where Y is compact. Let I denote the subfamily of the index family of $\{X_i\}$ consisting of i such that $p|_{X_i}$ are homeomorphisms. Let $\{Z_k\}$ be a finite C^1 Whitney $\mathfrak{X}_0$-substratification of $\{X_i\}_{i \in I} \cup \{s \times Y\}$. Substratify it more, if necessary. Then we have a finite C^1 Whitney $\mathfrak{X}_0$-substratification $\{Y_j\}$ of $\{\sigma^\circ: \sigma \in C\}$ such that $q: \{Z_k\} \to \{Y_j\}$ is a C^1 stratified map. Note that each $q|_{Z_k}$ is a diffeomorphism onto some Y_j. Add $\{X_i \cap p^{-1}(Y_j)\}_{i \notin I, j}$ to $\{Z_k\}$. Then $\{Z_k\}$ becomes a finite C^1 Whitney $\mathfrak{X}_0$-substratification of $\{X_i\} \cup \{s \times Y\}$ and $q: \{Z_k\} \to \{Y_j\}$ is a proper C^1 $\mathfrak{X}_0$-stratified map sans éclatement.

Let φ be a positive proper $\mathfrak{X}_0$-function on Y such that for each $a \in \mathbf{R}$, $\varphi^{-1}(a)$ is a polyhedron (e.g., $\varphi(x_1, \dots, x_m) = \max\{x_1, \dots, x_m\}$). By substratifying $\{Y_j\}$, we can assume each $\varphi|_{Y_j}$ is a C^1 submersion onto its image. Let $0 < a < a' \in \mathbf{R}$ be such that $]a, \infty[$ is compatible with $\{\varphi(Y_j)\}$. Apply II.6.1' to $q: \{Z_k \cap S^n \times \varphi^{-1}(]a, \infty[)\} \to \{Y_j \cap \varphi^{-1}(]a, \infty[)\}$ and $\varphi: \{Y_j \cap \varphi^{-1}(]a, \infty[)\} \to]a, \infty[$. We have $\mathfrak{X}_0$-homeomorphisms $g: S^n \times \varphi^{-1}(a') \times [a', \infty[\to S^n \times \varphi^{-1}([a', \infty[)$ and $h: \varphi^{-1}(a') \times [a', \infty[\to \varphi^{-1}([a', \infty[)$ such that the diagram

$$
\begin{array}{ccccc}
S^n \times \varphi^{-1}(a') \times [a', \infty[& \xrightarrow{\ q \times \mathrm{id}\ } & \varphi^{-1}(a') \times [a', \infty[& \xrightarrow{\ \mathrm{proj}\ } & [a', \infty[\\[2pt]
\Big\downarrow{\scriptstyle g} & & \Big\downarrow{\scriptstyle h} & & \Big\downarrow{\scriptstyle \mathrm{id}} \\[2pt]
S^n \times \varphi^{-1}([a', \infty[) & \xrightarrow{\ q\ } & \varphi^{-1}[a', \infty[) & \xrightarrow{\ \varphi\ } & [a', \infty[
\end{array}
$$

commutes,

$$
g(x, y, a') = (x, y), \quad h(y, a') = y \quad \text{for} \quad (x, y) \in S^n \times \varphi^{-1}(a'),
$$

for each k and for each j,

$$
g((Z_k \cap S^n \times \varphi^{-1}(a')) \times [a', \infty[) = Z_k \cap S^n \times \varphi^{-1}([a', \infty[), \quad \text{and}
$$
$$
h((Y_j \cap \varphi^{-1}(a')) \times [a', \infty[) = Y_j \cap \varphi^{-1}([a', \infty[).
$$

Assume that the statement holds for compact Y. Let X_0 be a semilinear subset of $\mathbf{R}^{n'} \times \varphi^{-1}([0, a'])$ and let $\pi_0: X_0 \to S^n \times \varphi^{-1}([0, a'])$ be an $\mathfrak{X}_0$-homeomorphism of the form $\pi_0(x, y) = (\pi_0'(x, y), y)$ such that each $\pi_0^{-1}(X_i \cap S^n \times \varphi^{-1}([0, a']))$ is semilinear. Set

$$
X = X_0 \cup X_{0a'} \times \varphi^{-1}([a', \infty[),
$$

where $X_{0a'}$ is a semilinear subset of $\mathbf{R}^{n'}$ defined by

$$
X_{0a'} \times \varphi^{-1}(a') = \pi_0^{-1}(S^n \times \varphi^{-1}(a')) \ (= X_0 \cap \mathbf{R}^{n'} \times \varphi^{-1}(a')).
$$

Define an $\mathfrak{X}_0$-homeomorphism $\pi\colon X \to S^n \times Y$ by

$$\pi(x,y) = \begin{cases} \pi_0(x,y) & \text{for } (x,y) \in X_0 \\ g(\pi_0(x,y'),t) & \text{for } (x,y) \in X_{0a'} \times \varphi^{-1}([a',\infty[), \end{cases}$$

where $(y',t) \in \varphi^{-1}(a') \times [a',\infty[$ is given by $h(y',t) = y$. Clearly X and π fulfill the requirements in the statement. Thus we can assume Y is compact.

Second, we consider the case where there exists an n-simplex σ_0 in $\mathbf{R}^n$ such that if $p|_{X_i}$ is a homeomorphism onto $p(X_i)$, then $X_i \cap \sigma_0 \times Y = \varnothing$. Let $\tilde{X}$ denote the union of X_i's such that $p|_{X_i}$ are homeomorphisms. We regard S^n as $\sigma_0 \cup s * \partial\sigma_0$, where we assume $\sigma_0 \subset \mathbf{R}^n \times 0 \subset \mathbf{R}^n \times \mathbf{R}$, $0 \in \sigma_0^\circ$ and $s = (0,1) \in \mathbf{R}^n \times \mathbf{R}$. Let $r\colon s * \partial\sigma_0 \times Y \to \sigma_0 \times Y$ denote the restriction of the projection $\mathbf{R}^n \times \mathbf{R} \times Y \to \mathbf{R}^n \times Y$. Apply IV.2.3 to $r(\tilde{X})$, C and $P = (\mathbf{R}^n - \sigma_0^\circ) \times Y$. Then we have an $\mathfrak{X}_0$-homeomorphism τ of $\mathbf{R}^n \times Y$ of the map $\tau(x,y) = (\tau'(x,y),y)$ such that $\tau^{-1}(r(\tilde{X}))$ is a polyhedron and $\tau = \mathrm{id}$ on P. It follows that each $\tau^{-1}(r(X_i))$ is semilinear if $X_i \subset \sigma_0 \times Y$. Set $X = S^n \times Y$, and define an $\mathfrak{X}_0$-homeomorphism $\pi\colon X \to \mathbf{R}^n \times Y$ by

$$\pi = \begin{cases} \mathrm{id} & \text{on } \sigma_0 \times Y \\ r^{-1} \circ \tau \circ r & \text{on } (s * \partial\sigma_0) \times Y. \end{cases}$$

Then X and π are what we want.

Let us assume only that Y is compact. Let I' denote the smallest family of indexes i such that if $p|_{X_i}$ is a homeomorphism and X_i is unbounded in $\mathbf{R}^n \times Y$, then $i \in I'$, and $\cup_{i \in I'} X_i$ is closed in $\mathbf{R}^n \times Y$. Let C' be a fine cell subdivision of C, and replace $\{X_i\}$ with $\{X_i \cap p^{-1}(\sigma^\circ)\colon \sigma \in C'\}$. Then we can suppose there exists an n-simplex σ_0 in $\mathbf{R}^n$ such that $X_i \cap \sigma_0 \times Y = \varnothing$ for $i \in I'$. Consider $\{X_i\}_{i \in I'} \cup \{p^{-1}(\sigma^\circ) - \cup_{i \in I'} X_i\colon \sigma \in C'\}$ in place of $\{X_i\}$. Then by the above second case, we can assume X_i is semilinear if $i \in I'$.

It remains to triangulate X_i, $i \notin I'$. By using IV.2.3, we can do in the same way as in the proof of IV.1.2′ by a finite local procedure because $\overline{\cup_{i \in I'} X_i}$ is compact and contained in $\mathbf{R}^n \times Y$. We omit the details. $\quad\square$

§IV.3. Local and global $\mathfrak{X}$-triangulations and uniqueness

In this section, we consider four kinds of uniqueness of triangulations and their relations to local and global triangulations. We always assume Axiom (v). *Uniqueness of C^0 $\mathfrak{X}$-triangulations of proper $\mathfrak{X}$-maps* means the following statement. Let $X \subset \mathbf{R}^n$ and $Y \subset \mathbf{R}^m$ be locally closed $\mathfrak{X}$-sets. Let $(X_i, Y_i, \pi_i, \tau_i)$, $i = 1,2$, be C^0 $\mathfrak{X}$-triangulations of a proper $\mathfrak{X}$-map $f\colon X \to Y$. Then there exist PL homeomorphisms $\varphi\colon X_1 \to X_2$ and $\psi\colon Y_1 \to Y_2$ such that

$$\tau_2^{-1} \circ f \circ \pi_2 \circ \varphi = \psi \circ \tau_1^{-1} \circ f \circ \pi_1.$$

We define *isotopic uniqueness* of C^0 $\mathfrak{X}$-triangulations of proper $\mathfrak{X}$-maps when in the above definition, X_i and Y_i, $i = 1, 2$, are closed in their ambient Euclidean spaces and there exist $\mathfrak{X}$-isotopies $\varphi_t \colon X_1 \to X_2$ and $\psi_t \colon Y_1 \to Y_2$, $0 \le t \le 1$, such that φ_1 and ψ_1 are PL,

$$\varphi_0 = \pi_2^{-1} \circ \pi_1, \qquad \psi_0 = \tau_2^{-1} \circ \tau_1, \quad \text{and}$$

$$\tau_2^{-1} \circ f \circ \pi_2 \circ \varphi_t = \psi_t \circ \tau_1^{-1} \circ f \circ \pi_1, \ \ 0 \le t \le 1.$$

Similarly we define *uniqueness* and *isotopic uniqueness* of C^0 R-$\mathfrak{X}$-triangulations of (proper) $\mathfrak{X}$-maps. For isotopic uniqueness we need to assume Y is closed in $\mathbf{R}^m$.

Proposition IV.3.1. *The four kinds of uniqueness for proper $\mathfrak{X}$-maps are equivalent to one another.*

Proof. We prove only the following two implications. The others are proved in the same way.

Proof that uniqueness of C^0 R-$\mathfrak{X}$-triangulations implies the uniqueness of C^0 $\mathfrak{X}$-triangulations. Assume the first uniqueness. Let $f \colon X \to Y$ and $(X_i, Y_i, \pi_i, \tau_i)$, $i = 1, 2$, be given as above. Without loss of generality, we can assume that X_i and Y_i are all closed in their ambient Euclidean spaces. It suffices to find an $\mathfrak{X}$-homeomorphism $\varphi \colon X_1 \to X_2$ and a PL homeomorphism $\psi \colon Y_1 \to Y_2$ such that

$$\tau_2^{-1} \circ f \circ \pi_2 \circ \varphi = \psi \circ \tau_1^{-1} \circ f \circ \pi_1, \tag{$*$}$$

because $(X_1, \pi_2 \circ \varphi)$ and (X_2, π_2) are C^0 R-$\mathfrak{X}$-triangulations of $\tau_2^{-1} \circ f$.

Since f is proper, we have simplicial decompositions K_i of X_i and L_i of Y_i for each $i = 1, 2$ such that $\tau_i^{-1} \circ f \circ \pi_i \colon K_i \to L_i$ is simplicial. By Theorem II', there exists an $\mathfrak{X}$-homeomorphism α_1 of Y_1 invariant on each simplex of L_1 such that for each simplex σ of L_2, $\alpha_1 \circ \tau_1^{-1} \circ \tau_2(\sigma)$ is a subpolyhedron of Y_1. Let K_1' and L_1' be subdivisions of K_1 and L_1 respectively, such that each $\alpha_1 \circ \tau_1^{-1} \circ \tau_2(\sigma)$, $\sigma \in L_2$, is the underlying polyhedron of some subcomplex of L_1' and $\tau_1^{-1} \circ f \circ \pi_1 \colon K_1' \to L_1'$ is simplicial. Similarly, let α_2 be an $\mathfrak{X}$-homeomorphism of Y_2 invariant on each simplex of L_2 such that for each $\sigma \in L_1'$, $\alpha_2 \circ \tau_2^{-1} \circ \tau_1 \circ \alpha_1^{-1}(\sigma)$ is a subpolyhedron of Y_2.

By the $\mathfrak{X}$-Hauptvermutung, each $\alpha_2 \circ \tau_2^{-1} \circ \tau_1 \circ \alpha_1^{-1}(\sigma)$, $\sigma \in L_1$, is a PL ball. Hence by the Alexander trick we can construct a PL homeomorphism ψ from Y_1 to Y_2 such that for each $\sigma \in L_1$,

$$\psi(\sigma) = \alpha_2 \circ \tau_2^{-1} \circ \tau_1 \circ \alpha_1^{-1}(\sigma).$$

Define an $\mathfrak{X}$-homeomorphism α_1' of Y_1 so that

$$\psi = \alpha_2 \circ \tau_2^{-1} \circ \tau_1 \circ \alpha_1^{-1} \circ \alpha_1'^{-1},$$

which is invariant on each simplex of L_1. Set $\alpha_1'' = \alpha_1' \circ \alpha_1$, and apply Lemma IV.3.2 to $\tau_1 \circ f \circ \pi_1 \colon K_1 \to L_1$ and α_1''. Then we have an $\mathfrak{X}$-homeomorphism γ_1 of X_1 such that

$$\alpha_1'' \circ \tau_1^{-1} \circ f \circ \pi_1 \circ \gamma_1 = \tau_1^{-1} \circ f \circ \pi_1.$$

By the same reason there is an $\mathfrak{X}$-homeomorphism γ_2 of X_2 such that

$$\alpha_2 \circ \tau_2^{-1} \circ f \circ \pi_2 \circ \gamma_2 = \tau_2^{-1} \circ f \circ \pi_2.$$

Set

$$\varphi = \gamma_2^{-1} \circ \pi_2^{-1} \circ \pi_1 \circ \gamma_1,$$

which is an $\mathfrak{X}$-homeomorphism from X_1 to X_2. Then φ and ψ satisfy the equality $(*)$. Indeed,

$$\tau_2^{-1} \circ f \circ \pi_2 \circ \varphi = \tau_2^{-1} \circ f \circ \pi_2 \circ \gamma_2^{-1} \circ \pi_2^{-1} \circ \pi_1 \circ \gamma_1 = \alpha_2 \circ \tau_2^{-1} \circ f \circ \pi_1 \circ \gamma_1$$

$$= \alpha_2 \circ \tau_2^{-1} \circ \tau_1 \circ \alpha_1''^{-1} \circ \tau_1^{-1} \circ f \circ \pi_1 = \psi \circ \tau_1^{-1} \circ f \circ \pi_1.$$

Proof that uniqueness of C^0 $\mathfrak{X}$-triangulations implies uniqueness of C^0 R-$\mathfrak{X}$-triangulations. Assume the first uniqueness. Let (X_i, π_i), $i = 1, 2$, be C^0 R-$\mathfrak{X}$-triangulations of a proper $\mathfrak{X}$-map $f \colon X \to Y$, where Y is an $\mathfrak{X}$-polyhedron. In this case also, we can assume that X_1, X_2 and Y are closed in their ambient Euclidean spaces. Let L and K_i, $i = 1, 2$, be simplicial decompositions of Y and X_i respectively, such that $f \circ \pi_i \colon K_i \to L$, $i = 1, 2$, are simplicial. By assumption we have PL homeomorphisms $\varphi \colon X_1 \to X_2$ and $\psi \colon Y \to Y$ such that

$$f \circ \pi_2 \circ \varphi = \psi \circ f \circ \pi_1.$$

If ψ is invariant on each simplex of L, then by Lemma IV.3.2 there exists a PL homeomorphism φ' of X_1 such that

$$\psi \circ f \circ \pi_1 = f \circ \pi_1 \circ \varphi'$$

and hence

$$f \circ \pi_2 \circ \varphi \circ \varphi'^{-1} = f \circ \pi_1,$$

which proves uniqueness of C^0 R-$\mathfrak{X}$-triangulations of f.

To obtain ψ with the invariance property, we mark Y by L and enlarge X as follows. Order all the simplexes of L as $\sigma_1, \sigma_2, \ldots$. For each σ_j, let Z_j

denote the disjoint union of j-copies of a simplex of dimension $= \dim X + 1$. Let Z be the disjoint union of $\sigma_j \times Z_j$, $j = 1, 2, \ldots$, included and closed in some Euclidean space, and let $\chi\colon Z \to Y$ be the $\mathfrak{X}$-map defined so that for each j, $\chi|_{\sigma_j \times Z_j}$ is the composite of the projection $\sigma_j \times Z_j \to \sigma_j$ and the inclusion $\sigma_j \to Y$. We can assume that X, X_i, $i = 1, 2$, and Z are included in $\mathbf{R}^n$, closed there and disjoint from one another. Set

$$\tilde{X} = X \cup Z \quad \text{and} \quad \tilde{X}_i = X_i \cup Z, \ i = 1, \, 2,$$

and define $\mathfrak{X}$-maps:

$$\tilde{f}\colon \tilde{X} \to Y \quad \text{and} \quad \tilde{\pi}_i\colon \tilde{X}_i \to \tilde{X}, \ i = 1, \, 2,$$

by f, π_i, χ and id. Then $\tilde{f}$ is a proper $\mathfrak{X}$-map, and $(\tilde{X}_i, \tilde{\pi}_i)$, $i = 1, 2$, are C^0 R-$\mathfrak{X}$-triangulations of $\tilde{f}$.

In place of f and (X_i, π_i), $i = 1, 2$, consider these. By assumption we have PL homeomorphisms $\tilde{\varphi}\colon \tilde{X}_1 \to \tilde{X}_2$ and $\tilde{\psi}\colon Y \to Y$ such that

$$\tilde{f} \circ \tilde{\pi}_2 \circ \tilde{\varphi} = \tilde{\psi} \circ \tilde{f} \circ \tilde{\pi}_1.$$

Set
$$\varphi = \tilde{\varphi}|_{X_1} \quad \text{and} \quad \psi = \tilde{\psi}.$$

We want to see that these fulfill the requirements. First, φ carries X_1 to X_2 because $\tilde{\varphi}$ carries each connected component of $\tilde{X}_1$ to a connected component of $\tilde{X}_2$ and the dimension of each connected component of Z is larger than $\dim X$. Next, $\tilde{\varphi}$ is invariant on each $\sigma_j \times Z_j$ because for connected components A and B of Z with $\chi(A) = \chi(B)$, i.e., such that A and B are connected components of some one $\sigma_j \times Z_j$, we have

$$\chi(\tilde{\varphi}(A)) = \tilde{\psi} \circ \chi(A) = \tilde{\psi} \circ \chi(B) = \chi(\tilde{\varphi}(B)),$$

and the numbers of connected components of $\sigma_j \times Z_j$, $j = 1, 2, \ldots$, are distinct from one another. Hence $\tilde{\psi}$ is invariant on each simplex of L. $\square$

Lemma IV.3.2. *Let $f\colon K \to L$ be a simplicial map. Set $X = |K|$ and $Y = |L|$, and assume that X and Y are contained and closed in Euclidean spaces. Let τ be an $\mathfrak{X}$-homeomorphism of Y invariant on each simplex of L. Then there exists an $\mathfrak{X}$-homeomorphism π of X invariant on each simplex of K such that*

$$\tau \circ f \circ \pi = f.$$

If τ is PL, then we can choose PL π.

Moreover, these hold true in the case where τ is an $\mathfrak{X}$-isotopy.

Proof. We proceed with the first statement by induction on $k = \dim X$. If $k = 0$, the lemma is trivial. Hence we assume that there exists an $\mathfrak{X}$-homeomorphism π^{k-1} of $|K^{k-1}|$ such that it is invariant on each simplex of the skeleton K^{k-1} and

$$\tau \circ f \circ \pi^{k-1} = f \quad \text{on} \quad |K^{k-1}|.$$

Let $\sigma = \Delta a_0 \cdots a_k$ be a k-simplex of L. Set $\pi^{\partial \sigma} = \pi^{k-1}|_{\partial \sigma}$. It suffices to extend $\pi^{\partial \sigma}$ to σ. There are two cases: $f|_\sigma$ is injective or not. If it is injective, then we uniquely define an extension to be $(f|_\sigma)^{-1} \circ \tau^{-1} \circ (f|_\sigma)$. Hence assume that $f|_\sigma$ is not injective. We can suppose $f(a_0) = f(a_1)$. Let $g\colon \sigma \to \Delta a_1 \cdots a_k$ denote the linear retraction defined by $g(a_0) = a_1$. Then $f \circ g = f$ on σ. By the proof of II.2.1 there exists an $\mathfrak{X}$-homeomorphism ρ of σ such that

$$g \circ \rho = \pi^{\partial \sigma} \circ g.$$

Note that ρ is invariant on each face of σ,

$$\rho = \pi^{\partial \sigma} \quad \text{on} \quad \Delta a_1 \cdots a_k, \quad \text{and} \quad \tau \circ f \circ \rho = f \quad \text{on} \quad \sigma.$$

Set

$$\rho' = \pi^{\partial \sigma} \circ \rho^{-1} \quad \text{on} \quad \partial \sigma.$$

Then ρ' is an $\mathfrak{X}$-homeomorphism of $\partial \sigma$ and is invariant on each proper face of σ, and we have

$$\pi^{\partial \sigma} = \rho' \circ \rho \quad \text{and} \quad f \circ \rho' = f \quad \text{on} \quad \partial \sigma.$$

Let a be an inner point of σ, and regard σ as the cone $a * \partial \sigma$. Apply the Alexander trick to $a * \partial \sigma$ and ρ', and extend ρ' to an $\mathfrak{X}$-homeomorphism ρ'' of σ such that

$$f \circ \rho'' = f \quad \text{on} \quad \sigma.$$

Set

$$\pi^\sigma = \rho'' \circ \rho \quad \text{on} \quad \sigma.$$

Then π^σ is an extension of $\pi^{\partial \sigma}$ to an $\mathfrak{X}$-homeomorphism of σ such that

$$\tau \circ f \circ \pi^\sigma = f \quad \text{on} \quad \sigma.$$

It is immediate by the above arguments that if τ is PL, then π can be PL, and the arguments work in the case where τ is an $\mathfrak{X}$-isotopy. $\qquad \square$

By II.3.12, isotopy uniqueness of C^0 R-$\mathfrak{X}$-triangulations of $\mathfrak{X}$-maps to one-dimensional $\mathfrak{X}$-polyhedra holds true. Hence we have the following.

Corollary IV.3.3. *Isotopy uniqueness of C^0 $\mathfrak{X}$-triangulations of proper $\mathfrak{X}$-maps to one-dimensional $\mathfrak{X}$-sets holds true.*

Proposition IV.3.4. *Let $X \subset \mathbf{R}^n$ and $Y \subset \mathbf{R}^m$ be $\mathfrak{X}$-sets locally closed in $\mathbf{R}^n$ and $\mathbf{R}^m$ respectively. Let $f\colon X \to Y$ be a proper $\mathfrak{X}$-map. Assume that isotopy uniqueness of C^0 $\mathfrak{X}$-triangulations of proper $\mathfrak{X}$-maps to $\mathfrak{X}$-sets of dimension $\leq \dim Y - 1$ holds true, and each point of Y has an $\mathfrak{X}$-neighborhood V in Y such that the restriction*

$$f|_{f^{-1}(V)}\colon f^{-1}(V) \to V$$

is C^0 (R-) $\mathfrak{X}$-triangulable. Then f is globally C^0 (resp. R-) $\mathfrak{X}$-triangulable.

Proof. We consider only C^0 $\mathfrak{X}$-triangulations because C^0 R-$\mathfrak{X}$-triangulations are easier to treat. In the above hypothesis, we can choose V so that it is compact and there exist a C^0 $\mathfrak{X}$-triangulation $(X_1, Y_1, \pi_1, \tau_1)$ of $f|_{f^{-1}(V)}$ and simplicial decompositions K of X_1 and L of Y_1 such that $\tau_1^{-1} \circ f \circ \pi_1\colon K \to L$ is simplicial. Also, for some vertex a of L, $L = \operatorname{st}(a, L)$, and $V - \tau_1(|\operatorname{lk}(a, L)|)$ is open in Y. Let $\{V_i\}$ be a family of such V's locally finite at each point of Y such that $\{\operatorname{Int} V_i\}$ is a covering of Y. Shrinking each V_i, we construct a C^0 $\mathfrak{X}$-triangulation of $f|_{f^{-1}(V_1 \cup \cdots \cup V_k)}$ by induction on k. Set $V = V_{k+1}$ and $W = V_1 \cup \cdots \cup V_k$. Then we can reduce the problem to the following statement. (There we need C and C' so that the limit of the C^0 $\mathfrak{X}$-triangulations as $k \to \infty$ is well-defined.)

($*$) Let C, D, V and W be compact subsets of Y such that

$$C \subset D \subset \operatorname{Int} V \cup \operatorname{Int} W \quad \text{and} \quad C \cap V = \varnothing.$$

Assume that V has the above properties and $f|_{f^{-1}(W)}$ admits a C^0 $\mathfrak{X}$-triangulation $(X_0, Y_0, \pi_0, \tau_0)$. Then there exist compact $\mathfrak{X}$-subsets V' of V and C', W' of W and C^0, and $\mathfrak{X}$-triangulations $(X_1', Y_1', \pi_1', \tau_1')$ of $f|_{f^{-1}(V')}$ and $(X_0', Y_0', \pi_0', \tau_0')$ of $f|_{f^{-1}(W')}$ such that the following six conditions are satisfied.

$$D \subset \operatorname{Int}(V' \cup W'). \tag{1}$$

$$\tau_0'^{-1}(V' \cap W') \text{ and } \tau_1'^{-1}(V' \cap W') \text{ are polyhedra.} \tag{2}$$

The maps $\tau_0'^{-1} \circ \tau_1'|_{\tau_1'^{-1}(V' \cap W')}$ and $\pi_0'^{-1} \circ \pi_1'|_{(f \circ \pi_1')^{-1}(V' \cap W')}$ are PL. $\quad$ (3)

$$C \subset C' \subset W'. \tag{4}$$

$$\tau_0^{-1}(C') \text{ and } \tau_0'^{-1}(C') \text{ are polyhedra.} \tag{5}$$

$$\begin{aligned} &((f \circ \pi_0)^{-1}(C'), \tau_0^{-1}(C'), \pi_0|_{(f \circ \pi_0)^{-1}(C')}, \tau_0|_{\tau^{-1}(C')}) \\ =&((f \circ \pi_0')^{-1}(C'), \tau_0'^{-1}(C'), \pi_0'|_{(f \circ \pi_0')^{-1}(C')}, \tau_0'|_{\tau_0'^{-1}(C')}). \end{aligned} \tag{6}$$

Here conditions (4), (5) and (6) on C' make ($*$) complicated. The statement follows from the following one which avoids them.

($**$) Let C, D, V, W, $f|_{f^{-1}(W)}$ and $(X_0, Y_0, \pi_0, \tau_0)$ be the same as in ($*$). Let K_0 and L_0 be simplicial decompositions of X_0 and Y_0 respectively, such that $\tau_0^{-1} \circ f \circ \pi_0 \colon K_0 \to L_0$ is simplicial. Then there exist compact X-subsets V' of V and W' of W, a C^0 X-triangulation $(X_1', Y_1', \pi_1', \tau_1')$ of $f|_{f^{-1}(V')}$ and X-homeomorphisms φ of X_0 and ψ of Y_0 such that conditions (1), (2) and (3), and the following (7), (8), and (9) are satisfied if we set

$$(\pi_0', \tau_0') = (\pi_0 \circ \varphi|_{(f \circ \pi_0 \circ \varphi)^{-1}(W')}, \tau_0 \circ \psi^{-1}|_{\psi \circ \tau_0^{-1}(W')}).$$

φ and ψ are invariant on each simplex of K_0 and L_0 respectively. (7)
$(f \circ \pi_0 \circ \varphi)^{-1}(W')$ and $\psi \circ \tau_0^{-1}(W')$ are subpolyhedra of X_0 and Y_0 respectively. (8)

$$\psi \circ \tau_0^{-1} \circ f \circ \pi_0 \circ \varphi = \tau_0^{-1} \circ f \circ \pi_0. \tag{9}$$

Note that if we denote the sets in (8) by X_0' and Y_0' respectively, then $(X_0', Y_0', \pi_0', \tau_0')$ is a C^0 X-triangulation of $f|_{f^{-1}(W')}$.

Proof that ($**$) implies ($*$). Assume ($**$). We need to define C' and modify φ and ψ so that conditions (4), (5) and (6) are satisfied. Choose K_0 and L_0 so fine that if we let $L_{0,1}$ and $L_{0,2}$ denote the subcomplex of L_0 generated by the simplexes σ with $\sigma \cap C \neq \varnothing$ and the simplex $N(L_{0,1}, L)$ respectively, then

$$\tau_0(|L_{0,2}|) \subset W', \tag{10}$$

which is possible because by (1) and the property $C \cap V = \varnothing$, $C \subset \mathrm{Int}(W')$. Set

$$C' = \tau_0(|L_{0,1}|).$$

Clearly, condition (4) is satisfied. By the Alexander trick we have an X-homeomorphism $\tilde{\psi}$ of Y_0 invariant on each simplex of L_0 such that

$$\tilde{\psi} = \begin{cases} \mathrm{id} & \text{on } |L_{0,1}| \\ \psi & \text{outside of } |L_{0,2}|. \end{cases}$$

By IV.3.2 and its proof we have an X-homeomorphism $\tilde{\varphi}$ of X_0 invariant on each simplex of K_0 such that

$$\tilde{\varphi} = \begin{cases} \mathrm{id} & \text{on } (\tau_0^{-1} \circ f \circ \pi_0)^{-1}(|L_{0,1}|) \\ \varphi & \text{outside of } (\tau_0^{-1} \circ f \circ \pi_0)^{-1}(|L_{0,2}|) \text{ and} \end{cases}$$
$$\tilde{\psi} \circ \tau_0^{-1} \circ f \circ \pi_0 \circ \tilde{\varphi} = \tau_0^{-1} \circ f \circ \pi_0.$$

Replace φ and ψ with $\tilde{\varphi}$ and $\tilde{\psi}$ in the definition of $(X_0', Y_0', \pi_0', \tau_0')$. Then (5) and (6) are satisfied. After this modification, (1) continues to hold. But (2) and (3) may fail. To solve this problem, replace V' with $V' - \tau_0(\mathrm{Int}\,|L_{0,2}|)$, which is possible because by (10), (1) does not fail. Then

$$V' \cap \tau_0(\mathrm{Int}\,|L_{0,2}|) = \varnothing,$$

hence

$$\tilde{\psi} = \psi \quad \text{on} \quad \tau_0^{-1}(V') \quad \text{and} \quad \tilde{\varphi} = \varphi \quad \text{on} \quad (f \circ \pi_0)^{-1}(V').$$

It follows that $\tau_0'^{-1}(V' \cap W')$, $\tau_0'^{-1} \circ \tau_1'|_{\tau_1'^{-1}(V' \cap W')}$ and $\pi_0'^{-1} \circ \pi_1'|_{(f \circ \pi_1')^{-1}(V' \cap W')}$ do not change by the above modification. Thus (∗) holds.

To prove (∗∗) we can shrink V a little and we need only consider (∗∗) on an arbitrarily small neighborhood of V in Y. Hence it suffices to treat the case $Y =$ the original V. Moreover, we can assume that X and Y are polyhedra, there are simplicial decompositions K of X and L of Y such that $f : K \to L$ is simplicial,

$$L = \mathrm{st}(a, L) \quad \text{for a vertex } a \text{ of } L,$$
$$V = \{ta + (1-t)y \colon y \in |\mathrm{lk}(a,L)|,\ 1/2 \le t \le 1\},$$
$$D = \{ta + (1-t)y$$
$$: y \in Z \text{ and } 0 \le t \le 3/4,\ \text{or } y \in |\mathrm{lk}(a,L)| \text{ and } 3/4 \le t \le 1\},$$

and W is an $\mathfrak{X}$-neighborhood of $\overline{D - V}$ in Y, where Z is a compact $\mathfrak{X}$-subset of $|\mathrm{lk}(a,L)|$. We set

$$V' = \{ta + (1-t)y$$
$$: y \in Z' \text{ and } t_0 \le t \le 2/3,\ \text{or } y \in |\mathrm{lk}(a,L)| \text{ and } 2/3 \le t \le 1\} \text{ and}$$
$$W' = \{ta + (1-t)y \colon y \in Z',\ 0 \le t \le t_0\},$$

where t_0 is a number in $]1/2, 2/3[$ and Z' is a small compact polyhedral neighborhood of Z in $|\mathrm{lk}(a,L)|$ such that W' is contained in W. Clearly (1) is satisfied. Let $(X_0, Y_0, \pi_0, \tau_0)$ be a C^0 $\mathfrak{X}$-triangulation of $f|_{f^{-1}(W)}$, and let K_0 and L_0 be simplicial decompositions of X_0 and Y_0 respectively, such that $\tau_0^{-1} \circ f \circ \pi_0 \colon K_0 \to L_0$ is simplicial.

We define the φ and ψ as follows. Let ψ be an $\mathfrak{X}$-homeomorphism of Y_0 invariant on each simplex of L_0 (7) such that $\psi \circ \tau_0^{-1}(W')$ and $\psi \circ \tau_0^{-1}(V' \cap W')$ are subpolyhedra of Y_0 (2), (8). By IV.3.2 we have an $\mathfrak{X}$-homeomorphism φ of X_0 such that it is invariant on each simplex of L_0 (7), and (9) is satisfied. Note that by (9), $(f \circ \pi_0 \circ \varphi)^{-1}(W')$ is a polyhedron (8) because $\tau_0^{-1} \circ f \circ \pi_0$ is PL.

It remains to construct a C^0 $\mathfrak{X}$-triangulation $(X_1', Y_1', \pi_1', \tau_1')$ of $f|_{f^{-1}(V')}$ which together with $(X_0', Y_0', \pi_0', \tau_0')$ satisfies conditions (2) and (3). For that we use the isotopic uniqueness. Apply the isotopic uniqueness assumption to the two C^0 $\mathfrak{X}$-triangulations $(f^{-1}(V' \cap W'), V' \cap W', \mathrm{id}, \mathrm{id})$ and $((f \circ \pi_0')^{-1}(V' \cap W'), \tau_0'^{-1}(V' \cap W'), \pi_0'|_{(f \circ \pi_0')^{-1}(V' \cap W')}, \tau_0'|_{\tau_0'^{-1}(V' \cap W')})$ of $f|_{f^{-1}(V' \cap W')}$. Then we have $\mathfrak{X}$-isotopies:

$$\varphi_t \colon f^{-1}(V' \cap W') \to (f \circ \pi_0')^{-1}(V' \cap W') \quad \text{and}$$
$$\psi_t \colon V' \cap W' \to \tau_0'^{-1}(V' \cap W'), \ 0 \le t \le 1,$$

such that φ_1 and ψ_1 are PL,

$$\varphi_0 = \pi_0'^{-1} \quad \text{on} \quad f^{-1}(V' \cap W'), \qquad \psi_0 = \tau_0'^{-1} \quad \text{on} \quad V' \cap W' \ \text{and}$$
$$\tau_0'^{-1} \circ f \circ \pi_0' \circ \varphi_t = \psi_t \circ f \quad \text{on} \quad f^{-1}(V' \cap W'), \ 0 \le t \le 1.$$

Now we define $(X_1', Y_1', \pi_1', \tau_1')$. Set

$$X_1' = f^{-1}(V') \quad \text{and} \quad Y' = V'.$$

If we construct π_1' and τ_1' so that

$$\pi_1' = \pi_0' \circ \varphi_1 \quad \text{on} \quad f^{-1}(V' \cap W') \quad \text{and} \quad \tau_1' = \tau_0' \circ \psi_1 \quad \text{on} \quad V' \cap W',$$

then (2) and (3) are satisfied. Hence we define π_1' on $f^{-1}(V' \cap W')$ and τ_1' on $V' \cap W'$ in this way and it suffices to extend them to $\mathfrak{X}$-homeomorphisms of $f^{-1}(V')$ and V' respectively, so that the property $\tau_1'^{-1} \circ f \circ \pi_1' = f$ continues to hold.
   Set

$$U = \{ta + (1-t)y \colon y \in Z', \ t_0 \le t \le 2/3\} \ \text{and} \ U_0 = \{t_0 a + (1-t_0)y \colon y \in Z'\}.$$

Then $V' \cap W' = U_0$ and U is a neighborhood of U_0 in V'. We use the following triviality property of f on U. By the properties of f it is easy to find PL homeomorphisms:

$$\alpha = (\alpha', \alpha'') \colon f^{-1}(U) \to f^{-1}(U_0) \times [t_0, 2/3] \ \text{and}$$
$$\beta = (\beta', \beta'') \colon U \to U_0 \times [t_0, 2/3]$$

such that

$$\alpha' = \mathrm{id} \quad \text{on} \quad f^{-1}(U_0), \qquad \beta' = \mathrm{id} \quad \text{on} \quad U_0,$$
$$\beta \circ f \circ \alpha^{-1}(x, t) = (f(x), t) \quad \text{for} \quad (x, t) \in f^{-1}(U_0) \times [t_0, 2/3], \ \text{and}$$
$$\beta''(ta + (1-t)y) = t \quad \text{for} \quad y \in Z', \ t \in [t_0, 2/3].$$

Extend π_1' and τ_1' so that

$$\alpha \circ \pi_1' \circ \alpha^{-1}(x,t) = (\pi_0' \circ \varphi_{(2/3-t)/(2/3-t_0)}(x),t)$$
$$\text{for}\quad (x,t) \in f^{-1}(U_0) \times [t_0,2/3],$$
$$\beta \circ \tau_1' \circ \beta^{-1}(y,t) = (\pi_0' \circ \varphi_{(2/3-t)/(2/3-t_0)}(y),t)\quad \text{for}\quad (y,t) \in U_0 \times [t_0,2/3],$$
$$\pi_1' = \mathrm{id}\quad \text{on}\quad f^{-1}(V'-U),$$
$$\tau_1' = \mathrm{id}\quad \text{on}\quad V'-U.$$

Then clearly

$$\tau_1'^{-1} \circ f \circ \pi_1' = f\quad \text{on}\quad f^{-1}(V'-U),$$

and this equality holds also on $f^{-1}(U)$ because

$$\beta \circ \tau_1'^{-1} \circ f \circ \pi_1' \circ \alpha^{-1}(x,t)$$
$$= (\beta \circ \tau_1' \circ \beta^{-1})^{-1} \circ (\beta \circ f \circ \alpha^{-1}) \circ (\alpha \circ \pi_1^{-1} \circ \alpha^{-1})(x,t)$$
$$= (\beta \circ \tau_1' \circ \beta^{-1})^{-1} \circ (\beta \circ f \circ \alpha^{-1})(\pi_0' \circ \varphi_{(2/3-t)/(2/3-t_0)}(x),t)$$
$$= (\beta \circ \tau_1' \circ \beta^{-1})^{-1}(f \circ \pi_0' \circ \varphi_{(2/3-t)/(2/3-t_0)}(x),t)$$
$$= (\beta \circ \tau_1' \circ \beta^{-1})^{-1}(\tau_0' \circ \psi_{(2/3-t)/(2/3-t_0)} \circ f(x),t)$$
$$= (\beta \circ \tau_1' \circ \beta^{-1})^{-1}(\beta' \circ \tau_1' \circ \beta^{-1}(f(x),t),t)$$
$$= (\beta \circ \tau_1' \circ \beta^{-1})^{-1} \circ (\beta \circ \tau_1' \circ \beta^{-1})(f(x),t)$$
$$= (f(x),t)\quad \text{for}\quad (x,t) \in f^{-1}(U_0) \times [t_0,2/3].$$

Hence $$\tau_1'^{-1} \circ f \circ \pi_1' = f\quad \text{on}\quad f^{-1}(V'),$$

which completes the proof.   □

Proof of IV.1.7. Immediate by IV.3.3 and IV.3.4.   □

Remark IV.3.5. The following statement does not always hold true.

Let $f\colon X \to Y$ be an $\mathfrak{X}$-map between compact $\mathfrak{X}$-sets, and let $Y = Y_1 \cup Y_2$ for compact $\mathfrak{X}$-sets Y_1 and Y_2. Assume that $f|_{f(Y_1 \cap Y_2)}$ and $f|_{f^{-1}(Y_i)}$, $i = 1,2$, admit C^0 $\mathfrak{X}$-triangulations, and uniqueness of C^0 $\mathfrak{X}$-triangulations of $f|_{f^{-1}(Y_1 \cap Y_2)}$ holds true. Then f admits a C^0 $\mathfrak{X}$-triangulation.

Counterexample. Set

$$X_1 = \{(u, v, x, y, t) \in [-1, 1]^5 : t < 0 \text{ and } u = 0, \text{ or } t = 0\},$$
$$X_2 = \{(u, v, x, y, t) \in [-1, 1]^5 : t > 0 \text{ and } u = (x^2 + y^2)v - xy, \text{ or } t = 0\},$$
$$Y_1 = [-1, 1]^2 \times [-1, 0], \quad Y_2 = [-1, 1]^2 \times [0, 1],$$
$$X = X_1 \cup X_2, \quad \text{and} \quad Y = Y_1 \cup Y_2,$$

and define f by

$$f(u, v, x, y, t) = (x, y, t).$$

It is easy to show that $f|_{f^{-1}(Y_i)}$, $i = 1, 2$, admits C^0 $\mathfrak{X}$-triangulations and $f|_{f^{-1}(Y_1 \cap Y_2)}$ admits a unique C^0 $\mathfrak{X}$-triangulation. But f is not triangulable because

$$\overline{X_1 - f^{-1}(Y_1 \cap Y_2)} \cap \overline{X_2 - f^{-1}(Y_1 \cap Y_2)}$$
$$= \{(u, v, x, y, t) \in [-1, 1]^5 : u = t = 0, \ (x^2 + y^2)v = xy\},$$

and the restriction of f to this set is not triangulable.

§IV.4. Proofs of Theorems IV.1.10, IV.1.13 and IV.1.13'

Proof of IV.1.10. Let $X \subset \mathbf{R}^n$ and $Y \subset \mathbf{R}^m$, and let $f : \{X_i\} \to \{Y_j\}$ be a C^1 Whitney $\mathfrak{X}$-stratification sans éclatement. By triangulating Y, we can assume that $\{Y_j\}$ is the set of open simplices of a simplicial complex and each X_i is connected. Note that the stratification $\{\text{graph } f|_{X_i}\}_i$ does not necessarily satisfy the Whitney condition. We consider the case of $d_f \leq 1$ because the other case follows from it as seen below.

Assume $\dim X = 3$ and there exists a C^0 $\mathfrak{X}$-triangulation of f on $\cup_{d_f \leq 1} X_i$. Note that $\cup_{d_f \leq 1} X_i$ is closed in X. We can extend the triangulation to a 2-dimensional stratum with $d_f = 2$ and to a 3-dimensional stratum with $d_f = 3$ because f is constant there. Hence, for each X_{i_0} of dimension 3 with $d_f = 2$, assuming an $\mathfrak{X}$-triangulations of f on $\overline{X_{i_0}} - X_{i_0}$, we need only extend the triangulation to X_{i_0}. Since $\dim f(X_{i_0}) = 1$, we have an $\mathfrak{X}$-triangulation of f on $\overline{X_{i_0}}$ (II.3.1). By uniqueness of $\mathfrak{X}$-triangulations of $\mathfrak{X}$-functions these two C^0 $\mathfrak{X}$-triangulations of f on $\overline{X_{i_0}} - X_{i_0}$ are equivalent up to PL homeomorphism. Then we can easily paste the two triangulations in the same way as in the proof of II.3.1'. Thus we extend the triangulation of f to X_{i_0}.

We prove that f admits a C^0 R-$\mathfrak{X}$-triangulation compatible with $\{X_i\}$, i.e., there exists a C^0 $\mathfrak{X}$-triangulation (X_0, π_0) of X compatible with $\{X_i\}$ such that $f \circ \pi_0$ is PL. We proceed by induction on $\dim Y$. Since the case of $\dim Y = 0$ is trivial, we assume the case of dimension $< \dim Y$. Then it suffices to prove the following statement.

Let Y_{j_0} be a stratum such that $f|_{f^{-1}(\overline{Y_{j_0}} - Y_{j_0})}$ admits a C^0 R-$\mathfrak{X}$-triangulation compatible with $\{X_i\}$. We can extend the triangulation to $f^{-1}(\overline{Y_j})$.

If $d_f = 0$ on a stratum X_i with $f(X_i) = Y_{j_0}$, then $f|_{\overline{X_i}}$ is a homeomorphism onto $\overline{Y_{j_0}}$ and we can extend uniquely the triangulation to X_i. Here the uniqueness means that if $(X_{0,1}, \pi_{0,1})$ and $(X_{0,2}, \pi_{0,2})$ are C^0 R-$\mathfrak{X}$-triangulations of $f_{\overline{X_i}}$ such that

$$\pi_{0,1}^{-1}(\overline{X_i} - X_i) = \pi_{0,2}^{-1}(\overline{X_i} - X_i) \quad \text{and}$$
$$\pi_{0,1} = \pi_{0,2} \quad \text{on} \quad \pi_{0,1}^{-1}(\overline{X_i} - X_i),$$

then there exists a unique natural PL homeomorphism by which we identify $(X_{0,1}, \pi_{0,1})$ with $(X_{0,2}, \pi_{0,2})$. For two distinct strata X_i and $X_{i'}$ with $d_f = 1$ and $f(X_i) = f(X_{i'}) = Y_{j_0}$, if we extend the triangulation to X_i and $X_{i'}$, then we can paste the extensions at $\overline{X_i} \cap \overline{X_{i'}}$. This is because for each $X_{i''} \subset \overline{X_i} \cap \overline{X_{i'}}$ we have $f(X_{i''}) = Y_{j_0}$ or $\subset \overline{Y_{j_0}} - Y_{j_0}$. If $f(X_{i''}) = Y_{j_0}$, then $d_f = 0$ on $X_{i''}$ and by the above uniqueness property, the two extensions are naturally and uniquely pasted at $X_{i''}$. If $f(X_{i''}) \subset \overline{Y_{j_0}} - Y_{j_0}$, the extensions coincide with each other on $X_{i''}$. In this way, we reduce the problem to the case where for some X_{i_0} with $d_f = 1$,

$$X = \overline{X_{i_0}}, \qquad f(X_{i_0}) = Y_{j_0}, \quad \text{and} \quad Y = \overline{Y_{j_0}}. \tag{$*$}$$

For each Y_j, let U_j and $\alpha_j : U_j \to Y_j$ be a tubular $\mathfrak{X}$-neighborhood of Y_j in Y and a C^1 submersive $\mathfrak{X}$-retraction respectively, such that if $Y_{j'} \subset \overline{Y_j} - Y_j$, then

$$\alpha_{j'} \circ \alpha_j = \alpha_{j'} \quad \text{on} \quad U_{j'} \cap U_j.$$

We shrink U_j and set $Z_j = f^{-1}(Y_j)$. Then we have the following.

Assertion 1. There exists an $\mathfrak{X}$-retraction:

$$\beta_j : f^{-1}(U_j) \cap (Z_{j_0} \cup Z_j) \longrightarrow Z_j$$

such that

$$f \circ \beta_j = \alpha_j \circ f \quad \text{on} \quad f^{-1}(U_j) \cap (Z_{j_0} \cup Z_j),$$
$$\beta_j(\overline{(Z_{j_0} - X_{i_0})} \cap f^{-1}(U_j) \cap (Z_{j_0} \cup Z_j)) = Z_j \cap \overline{(Z_{j_0} - X_{i_0})},$$

and if $d_f = 1$ at some point of Z_j, then for each $X_i \subset Z_j$ and for each each connected component C of $\beta_j^{-1}(X_i)$, the following map is a homeomorphism:

$$(\beta_j, f) : C \cup X_i \to \{(x, y) \in X_i \times (U_j \cap (Y_{j_0} \cup Y_j)) : f(x) = \alpha_j(y)\}.$$

Proof of Assertion 1. Note that for each $y \in Y$, $f^{-1}(y)$ is connected, which we can prove in the same way as in the proof that $\overline{\operatorname{graph} L} \cap (b \times G_{n,m})$ consists of one point, in the proof that $\dim Y_2 < m$ in (II.1.13). For this we use the facts that $f^{-1}(y) \cap X_{i_0}$ is connected and $\overline{X_{i_0}} = X$. If $d_f = 0$ on Z_j, then Z_j is a stratum and $f|_{Z_j}$ is a homeomorphism onto Y_j. Hence β_j exists uniquely. Therefore, assume $d_f = 1$ at some point of Z_j. Let $X_i \subset Z_j$ be a stratum with $d_f = 1$. Let V_i be a small $\mathfrak{X}$-neighborhood of X_i in X and define an $\mathfrak{X}$-retraction $\beta_{j,i} \colon V_i \to X_i$ by

$$\beta_{j,i}(x) = p'(x, \alpha_j \circ f(x)) \quad \text{for} \quad x \in V_i,$$

where p' is a C^1 $\mathfrak{X}$-map such that the projection of a C^1 $\mathfrak{X}$-tube at the C^1 $\mathfrak{X}$-manifold

$$\bigcup_{y \in Y_j} (X_i \cap f^{-1}(y)) \times y \subset X_i \times Y_j$$

in $\mathbf{R}^n \times \mathbf{R}^m$ is of the form $(p'(x, y), p''(y))$ (see II.5.1 and its proof). Then we have

$$f \circ \beta_{j,i} = \alpha_j \circ f \quad \text{on} \quad V_i.$$

If $X_i \cap f^{-1}(y)$ is a Jordan curve for some $y \in Y_j$, then $X_i = Z_j$, and $\beta_{j,i} = \beta_j$ is what we want. Hence we assume that any $X_i \cap f^{-1}(y)$ is an open curve.

We choose good V_i as follows. By the definition of a stratified map sans éclatement we have the following two facts: we can assume $V_i \cap Z_{j_0} \subset X_{i_0}$, and if a point x_0 of X_{i_0} is near a point x of X_i, then the tangent space of $f^{-1}f(x_0)$ at x_0 is not orthogonal to that of $f^{-1}f(x)$ at x. By this we choose sufficiently small U_j and V_i so that $f(V_i) = U_j$, for each $y_0 \in U_j \cap Y_{j_0}$, the restriction of $\beta_{j,i}$ to each connected component of $f^{-1}(y_0) \cap V_i$ is a C^1 diffeomorphism onto $\beta_{j,i}(f^{-1}(y_0) \cap V_i)$, and $\beta_{j,i}(f^{-1}(y_0) \cap V_i)$ is a closed curve. For the same y_0, let $a(y_0)$ and $b(y_0)$ denote the ends of the curve $\beta_{j,i}(f^{-1}(y_0) \cap V_i)$. Note that $\{a(y_0), b(y_0)\}$ converges to the ends of the open curve $f^{-1}(y) \cap X_i$ as y_0 converges to any point y of Y_j. (The ends may coincide with each other.) We can assume that a and b are C^1 $\mathfrak{X}$-maps on $U_j \cap Y_{j_0}$ for the following reason. By II.6.6 there exists a C^1 $\mathfrak{X}$-function q on X_i such that (f, q) is a C^1 diffeomorphism from X_i onto $Y_j \times]0, 1[$. Choose a and b so that $q \circ a < q \circ b$. The graphs of a and b are $\mathfrak{X}$-sets, and

$$q \circ a(y_0) \longrightarrow 0 \quad \text{and} \quad q \circ b(y_0) \longrightarrow 1 \quad \text{as} \quad y_0 \longrightarrow y \in Y_j.$$

Keeping these properties, we replace $q \circ a$ and $q \circ b$ with larger and smaller C^1 $\mathfrak{X}$-functions a' and b' respectively, and then we replace a and b with $(f, q)^{-1} \circ (\alpha_j \circ f, a')$ and $(f, q)^{-1} \circ (\alpha_j \circ f, b')$ respectively. Then we may suppose that a and b are of class C^1.

By the a and b we can describe V_i as follows. Assume that a fibre of $f|_{X_{i_0}}$ is an open curve (the other case where a fibre is a Jordan curve can be treated similarly). By a and b there exist C^1 $\mathfrak{X}$-cross-sections $c_{i,k}$ and $d_{i,k}$, $k = 1, 2, \ldots$, of $f|_{V_i \cap X_{i_0}}$ such that for each $y \in U_j \cap Y_{j_0}$, $f^{-1}(y) \cap V_i$ is the union of connected curves with the ends $(c_{i,k}(y), d_{i,k}(y))$, $k = 1, 2, \ldots$. Then $V_i \cap X_{i_0}$ is the union of connected domains lying between $\operatorname{Im} c_{i,k}$ and $\operatorname{Im} d_{i,k}$, $k = 1, 2, \ldots$. Let c and d be the C^1 $\mathfrak{X}$-cross-sections of $f|_{Z_{j_0} - X_{i_0}}$, which are distinct from each other if and only if $Z_{j_0} - X_{i_0}$ has two connected components. Choose U_j common to all X_i, order $(c_{i,k}, d_{i,k})$ for all i and k, and write them as $(c_1, d_1), \ldots, (c_l, d_l)$ so that for each $y \in U_j \cap Y_{j_0}$, $d(y)$, $c_1(y)$, $d_1(y), \ldots, c_l(y)$, $d_l(y)$, $c(y)$ stand in order in $f^{-1}(y) \cap X_{i_0}$. Choose C^1 $\mathfrak{X}$-cross-sections $e_1, \ldots, e_{l-1}$ of $f|_{X_{i_0}}$ so that for each $y \in U_j \cap Y_{j_0}$, $d_1(y)$, $e_1(y)$, $c_2(y), \ldots, d_{l-1}(y)$, $e_{l-1}(y)$, $c_l(y)$ are distinct from one another and in order in $f^{-1}(y) \cap X_{i_0}$, which is possible by the same reason as the above construction of C^1 a and b. Note that all c, d, c_k, d_k and e_k are extended to $U_j \cap (Y_{j_0} \cup Y_j)$. We keep the notation for the extensions. Note also that each $\operatorname{Im}(e_k|_{Y_j})$ is a stratum with $d_f = 0$.

Now we define β_j. Let β_j be $\beta_{j,i}$ on $X_i \cup (V_i \cap X_{i_0})$ for all i. By the requirement $f \circ \beta_j = \alpha_j \circ f$, β_j is uniquely defined on the set:

$$\overline{(Z_{j_0} - X_0)} \cap f^{-1}(U_j) \cap (Z_{j_0} \cup Z_j) \; (= \operatorname{Im} c \cup \operatorname{Im} d).$$

Let X_i be the stratum in Z_j such that $(c_1, d_1) = (c_{i,k}, d_{i,k})$ for some k. Then $\overline{X_i}$ includes $\beta_j(\operatorname{Im} d)$, and $d_f = 1$ on X_i. Extend β_j to the subdomain of $f^{-1}(U_j) \cap (Z_{j_0} \cup Z_j)$ between $\operatorname{Im} d$ and $\operatorname{Im} c_1$ so that for each $y \in U_j \cap Y_{j_0}$, β_j bijectively maps the subcurve of $f^{-1}(y)$ between $d(y)$ and $c_1(y)$ onto the subcurve of $\overline{X_i} \cap f^{-1}(\alpha_j(y))$ between $\beta_j(d(y))$ and $\beta_j(c_1(y))$. This is possible because by II.6.6, $f|_{X_{i_0}}$ and $f|_{X_i}$ are C^1 $\mathfrak{X}$-trivial with fibre $\mathbf{R}$. Next define β_j on $\operatorname{Im} e_1$ so that for each $y \in U_j \cap Y_{j_0}$, $\beta_j(e_1(y))$ and $\beta_j(d(y))$ are the ends of the curve $X_i \cap f^{-1}(\alpha_j(y))$, and extend β_j to the subdomain of $f^{-1}(U_j) \cap (Z_{j_0} \cup Z_j)$ between $\operatorname{Im} d_1$ and $\operatorname{Im} e_1$ in the same way as above. Then β_j bijectively maps the subcurve of $f^{-1}(y)$ between $d(y)$ and $e_1(y)$ to $\overline{X_i} \cap f^{-1}(\alpha_j(y))$. Repeating these arguments, we construct β_j, which proves Assertion 1.

Next we want to extend β_j to $f^{-1}(U_j)$. Note the following fact. Assume $d_f = 1$ at some point of Z_j. Let $X_i \subset Z_j$ and let C be a connected component of $\beta_j^{-1}(X_i) - X_i$. Then $C \cup X_i$ is the image of some $\mathfrak{X}$-cross-section of $f|_{f^{-1}(U_j) \cap (Z_{j_0} \cup Z_j)}$ if $d_f = 0$ on X_i, and it lies between the images of two $\mathfrak{X}$-cross-sections otherwise. (Here if $f^{-1}(y) \cap X_i$ is a Jordan curve for $y \in Y_j$, then the two $\mathfrak{X}$-cross-section are identical and hence we have an $\mathfrak{X}$-homeomorphism from $C \cup X_i$ to $S^1 \times (U_j \cap (Y_{j_0} \cup Y_j))$ whose composite with the projection $S^1 \times (U_j \cap (Y_{j_0} \cup Y_j)) \to U_j \cap (Y_{j_0} \cup Y_j)$ is f.) Shrink U_j.

Then clearly we can extend β_j to the map:

$$\overline{\beta_j} : f^{-1}(\overline{U_j}) \cap (Z_{j_0} \cup \overline{Z_j}) \longrightarrow \overline{Z_j}.$$

The above construction of β_j is possible even if Z_{j_0} includes a plural number of strata with $d_f = 1$ because β_j is uniquely determined on the intersection of $f^{-1}(U_j)$ and a stratum with $d_f = 0$. Hence for strata Y_j and $Y_{j'}$ with $Y_{j'} \subset \overline{Y_j} - Y_j$, there exist an $\mathfrak{X}$-retraction and its extension:

$$\beta_{j'}^j : f^{-1}(U_{j'}) \cap (Z_j \cup Z_{j'}) \longrightarrow Z_{j'} \quad \text{and}$$
$$\overline{\beta_{j'}^j} : f^{-1}(\overline{U_{j'}}) \cap (Z_j \cup \overline{Z_{j'}}) \longrightarrow \overline{Z_{j'}}$$

with the same properties as β_j. (Here we shrink $U_{j'}$ if necessary.)

Let X_i, $X_{i'}$, $X_{i''}$, Y_j, $Y_{j'}$ and $Y_{j''}$ be strata such that

$$f(X_i) = Y_j, \qquad f(X_{i'}) = Y_{j'}, \qquad f(X_{i''}) = Y_{j''},$$
$$X_{i'} \subset \overline{X_i} - X_i, \qquad X_{i''} \subset \overline{X_{i'}} - X_{i'},$$
$$Y_{j'} \subset \overline{Y_j} - Y_j, \quad \text{and} \quad Y_{j''} \subset \overline{Y_{j'}} - Y_{j'}.$$

Then we have an $\mathfrak{X}$-retraction:

$$\beta_{j''}^{j'} \circ \overline{\beta_{j'}^j} : f^{-1}(\overline{U_{j'}} \cap U_{j''}) \cap (Z_j \cup Z_{j'} \cup Z_{j''}) \longrightarrow Z_{j''}$$

such that $\beta_{j''}^{j'} \circ \overline{\beta_{j'}^j}(X_i \cap f^{-1}(\overline{U_{j'}} \cap U_{j''}))$ is a union of strata;

$$f \circ \beta_{j''}^{j'} \circ \overline{\beta_{j'}^j} = \alpha_{j''} \circ f \quad \text{on} \quad f^{-1}(\overline{U_{j'}} \cap U_{j''}) \cap (Z_j \cup Z_{j'} \cup Z_{j''});$$

if $d_f = 1$ on $X_{i''}$, then, for each connected component C of $X_i \cap (\beta_{j''}^{j'} \circ \overline{\beta_{j'}^j})^{-1}(X_{i''})$, the following map is a homeomorphism:

$$(\beta_{j''}^{j'} \circ \overline{\beta_{j'}^j}, f) : C \cup X_{i''} \longrightarrow$$
$$\{(x, y) \in X_{i''} \times (\overline{U_{j'}} \cap U_{j''} \cap (Y_j \cup Y_{j''})) : f(x) = \alpha_{j''}(y)\},$$

and $C \cup X_{i''}$ lies between the images of two $\mathfrak{X}$-cross-sections of $f|_{f^{-1}(\overline{U_{j'}} \cap U_{j''}) \cap (Z_j \cup Z_{j''})}$; and if $d_f = 0$ on $X_{i''}$, then for the same C and for each $y \in Y_j \cap \overline{U_{j'}} \cap U_{j''}$, $C \cap f^{-1}(y)$ is either a connected closed curve or a point, and $C \cup X_{i''}$ also lies between the images of two (not necessarily distinct) $\mathfrak{X}$-cross-sections. Hence by using II.6.6 and a partition of unity of class $\mathfrak{X}$ as in the proof of I.1.3, we can modify and extend each β_j to an $\mathfrak{X}$-retraction:

$$\beta_j : f^{-1}(U_j) \longrightarrow Z_j,$$

so that the following properties are satisfied. Let X_i, $X_{i'}$, Y_j and $Y_{j'}$ be strata such that

$$f(X_i) = Y_j, \quad f(X_{i'}) = Y_{j'}, \quad X_{i'} \subset \overline{X_i} - X_i, \quad \text{and} \quad Y_{j'} \subset \overline{Y_j} - Y_j.$$

Let C be a connected component of $X_{i_0} \cap \beta_j^{-1}(X_{i'})$, and let $y \in U_{j'}$. Set

$$\tilde{C} = \overline{C} \cap \beta_j^{-1}(X_{i'}).$$

Then

$$\beta_{j'} \circ \beta_j = \beta_{j'} \quad \text{on} \quad f^{-1}(U_{j'} \cap U_j),$$
$$f \circ \beta_{j'} = \alpha_{j'} \circ f \quad \text{on} \quad f^{-1}(U_{j'}),$$

$\beta_{j'}(X_i \cap f^{-1}(U_{j'}))$ is a union of strata, if $d_f = 1$ on $X_{i'}$, then the map

$$(\beta_{j'}, f) \colon \tilde{C} \longrightarrow \{(x, y) \in X_{i'} \times U_{j'} : f(x) = \alpha_{j'}(y)\}$$

is a homeomorphism, if $d_f = 0$ on $X_{i'}$, then $\tilde{C} \cap f^{-1}(y)$ is either a connected closed curve or a point, and $\tilde{C}$ lies between the images of two $\mathfrak{X}$-cross-sections of $f|_{f^{-1}(U_{j'})}$.

   In the above arguments, $X_{i_0} \cap \beta_{j'}^{-1}(X_{i'})$ is not necessarily connected, which causes difficulty in extending a C^0 R-triangulation of $f|_{f^{-1}(\partial Y)}$. To solve this we define the *unfolding* (F, g) of X as follows. (X shall be replaced with F.) As a set, F is the disjoint union of Z_{j_0} and $\gamma(i)$-copies of each X_i with $f(X_i) = Y_j \subset \overline{Y_{j_0}} - Y_{j_0}$, where $\gamma(i)$ is the number of the connected components of $X_{i_0} \cap \beta_j^{-1}(X_i)$. Let $F(Z_{j_0})$ and $F^1(X_i), \ldots, F^{\gamma(i)}(X_i)$ denote the images in F of Z_{j_0} and the copies of X_i. By definition g is a map from F to X whose restrictions to $F(Z_{j_0})$ and each $F^k(X_i)$ are the identities. We give a topology to F (i.e., we paste $F(Z_{j_0})$ and $F^k(X_i)$) so that g is continuous, the restrictions of g to $F(Z_{j_0})$ and $F^k(X_i)$ are homeomorphisms onto their images, and for each X_i and for each connected component C of $X_{i_0} \cap \beta_j^{-1}(X_i)$, where $Y_j = f(X_i)$, there exists uniquely a positive integer $k_i \leq \gamma(i)$ such that the restriction of g to $F^{k_i}(X_i) \cup g^{-1}(C)$ is homeomorphic to $X_i \cup C$.

   As follows, F keeps the good properties of X and solves the above problem of connectedness. Set $h = f \circ g$, and let $\{F_l\}$ denote the C^0 $\mathfrak{X}$-stratification of F consisting of $g^{-1}(X_i)$, $X_i \subset Z_{j_0}$, and all $F^k(X_i)$,

$X_i \subset X - Z_{j_0}$. Clearly F is compact and admits a unique $\mathfrak{X}$-structure such that g is an $\mathfrak{X}$-map. We assume this structure. Then $h \colon \{F_l\} \to \{Y_j\}$ is a C^0 $\mathfrak{X}$-stratified map. Each β_j induces an $\mathfrak{X}$-retraction:

$$\lambda_j \colon h^{-1}(U_j) \longrightarrow h^{-1}(Y_j)$$

such that

$$g \circ \lambda_j = \beta_j \circ g \quad \text{on} \quad h^{-1}(U_j).$$

Let F_l, $F_{l'}$, Y_j and $Y_{j'}$ be strata such that

$$h(F_l) = Y_j, \quad h(F_{l'}) = Y_{j'}, \quad F_{l'} \subset \overline{F_l} - F_l, \quad \text{and} \quad Y_{j'} \subset \overline{Y_j} - Y_j,$$

and let $y \in U_{j'}$. Then $h|_{F_{l'}}$ is $\mathfrak{X}$-trivial over $Y_{j'}$ with fibre $= S^1$, $\mathbf{R}$ or 0, $h|_{h^{-1}(Y_{j'})}$ is $\mathfrak{X}$-trivial,

$$\lambda_{j'} \circ \lambda_j = \lambda_{j'} \quad \text{on} \quad h^{-1}(U_{j'} \cap U_j),$$

$\lambda_{j'}(F_l \cap h^{-1}(U_{j'}))$ is a union of strata, and $g^{-1}(X_{i_0}) \cap \lambda_{j'}^{-1}(F_{l'})$ is connected. Furthermore, if $d_h = 1$ on $F_{l'}$, then the following map is a homeomorphism:

$$(\lambda_{j'}, h) \colon \lambda_{j'}^{-1}(F_{l'}) \longrightarrow \{(x, y) \in F_{l'} \times U_{j'} \colon h(x) = \alpha_{j'}(y)\}.$$

If $d_h = 0$ on $F_{l'}$, then $\lambda_{j'}^{-1}(F_{l'}) \cap h^{-1}(y)$ is either a connected closed curve or a point, and in any case, $\lambda_{j'}^{-1}(F_{l'})$ lies between the images of two $\mathfrak{X}$-cross-sections of $h|_{h^{-1}(U_{j'})}$.

By these properties of (F, h) we can replace (X, f) with (F, h) as follows. Let $(\tilde{X}, \tilde{\pi})$ be a C^0 R-$\mathfrak{X}$-triangulation of $f|_{f^{-1}(\partial Y)}$ compatible with $\{X_i\}$. Then $h^{-1}(\partial Y)$ admits a unique C^0 $\mathfrak{X}$-triangulation $(\tilde{F}, \tilde{\tau})$ such that $\pi^{-1} \circ g \circ \tilde{\tau}$ is PL. Clearly $(\tilde{F}, \tilde{\tau})$ is automatically a C^0 R-$\mathfrak{X}$-triangulation of $h|_{h^{-1}(\partial Y)}$ compatible with $\{F_l\}$, and for proof of the proposition, it suffices to extend $(\tilde{F}, \tilde{\tau})$ to h. Hence we consider (F, h) in place of (X, f).

Assertion 2. Let j be such that $d_h \neq 0$ at some point of $h^{-1}(Y_j)$ (i.e., $d_f \neq 0$ at some point of Z_j). Then $h|_{h^{-1}(U_j)}$ is $\mathfrak{X}$-trivial over U_j.

Proof of Assertion 2. We proceed by downward induction on $\dim Y_j$. If $j = j_0$, then $U_{j_0} = Y_{j_0}$, and Assertion 2 is clear. Hence assume it for a stratum of dimension $> \dim Y_j$. There are two cases: for $y_0 \in Y_{j_0}$, $h^{-1}(y_0)$ is either a simple curve or a Jordan curve. We treat only the former case, i.e., the case where $h^{-1}(Y_{j_0}) - g^{-1}(X_{i_0})$ is the union of two strata. (The other case is proved in the same way.) Let c and d denote the two distinct $\mathfrak{X}$-cross-sections of $h|_{h^{-1}(Y_{j_0})}$ the union of whose images is $h^{-1}(Y_{j_0}) - g^{-1}(X_{i_0})$. Then, since

$$d_h = 0 \quad \text{on} \quad \overline{h^{-1}(Y_{j_0}) - g^{-1}(X_{i_0})},$$

we can extend c and d to Y. The extensions keep the property on $U_j - Y_j$ that $h^{-1}(y)$ is a simple curve with ends $c(y)$ and $e(y)$ because, by the induction hypothesis, h is locally $\mathfrak{X}$-trivial over $U_j - Y_j$. (It follows that h is globally $\mathfrak{X}$-trivial over $U_j - Y_j$.)

Moreover, this property holds for $y \in Y_j$ for the following reason. As shown before, for each F_l with $d_h = 1$ and $h(F_l) = Y_j$ there are two $\mathfrak{X}$-cross-sections c_l and e_l such that $\lambda_j^{-1}(F_l)$ lies between their images. We order all such F_l, c_l and e_l, $l = 1, \dots, l_0$, so that for each $y \in Y_{j_0}$, $e(y)$, $c_1(y)$, $e_1(y), \dots, c_{l_0}(y)$, $e_{l_0}(y)$, and $c(y)$ stand in order in $h^{-1}(y)$. For simplicity of notation, we set

$$e_0 = e \quad \text{and} \quad c_{l_0+1} = c.$$

For $F_{l'}$ with $d_h = 0$ and $h(F_{l'}) = Y_j$, $\lambda_j^{-1}(F_{l'})$ equals the domain bounded by some $\operatorname{Im} e_k$ and $\operatorname{Im} c_{k+1}$ because $\lambda_j^{-1}(F_{l'})$ is connected and $\lambda_j^{-1}(h^{-1}(Y_j)) = h^{-1}(U_j)$. Hence for each $y \in Y_j$ and for each $y_0 \in \alpha_j^{-1}(y) - Y_j$, $h^{-1}(y)$ is the quotient topological space which we obtain from $h^{-1}(y_0)$ by shrinking each subcurve with ends $e_k(y_0)$ and $c_{k+1}(y_0)$ to a point. Therefore, $h^{-1}(y)$ is a simple curve with ends $c(y)$ and $e(y)$.

Since $h|_{h^{-1}(Y_j)}$ is $\mathfrak{X}$-trivial over Y_j, we have an $\mathfrak{X}$-function r on $h^{-1}(Y_j)$ such that (h, r) is an $\mathfrak{X}$-homeomorphism onto $Y_j \times [0, 1]$. Then it suffices to modify λ_j so that the map

$$(h, r \circ \lambda_j) \colon h^{-1}(U_j) \longrightarrow U_j \times [0, 1]$$

is an $\mathfrak{X}$-homeomorphism. Clearly this holds if for every $F_{l'}$ with $d_h = 0$ and $h(F_{l'}) = Y_j$, $\lambda_j^{-1}(F_{l'})$ is the image of an $\mathfrak{X}$-cross-section of $h|_{h^{-1}(U_j)}$. For such a modification we want to prove the following statement.

Let $F_{l'}$ be such that $d_h = 0$, $h(F_{l'}) = Y_j$, and $\lambda_j^{-1}(F_{l'}) \cap h^{-1}(y)$ is not a point for some $y \in U_j$. Let e_l and c_{l+1} be the $\mathfrak{X}$-cross-sections such that $\lambda_j^{-1}(F_{l'})$ is bounded by $\operatorname{Im} e_l$ and $\operatorname{Im} c_{l+1}$. Assume $l < l_0$. Then there exists an $\mathfrak{X}$-homeomorphism:

$$\theta_{l'} \colon \overline{\lambda_j^{-1}(F_{l'} \cup F_{l+1})} \cap h^{-1}(U_j) \longrightarrow \overline{\lambda_j^{-1}(F_{l+1})} \cap h^{-1}(U_j)$$

such that

$$\theta_{l'} = \operatorname{id} \quad \text{on} \quad \operatorname{Im} e_{l+1}$$

$$h \circ \theta_{l'} = h \quad \text{on} \quad \overline{\lambda_j^{-1}(F_{l'} \cup F_{l+1})} \cap h^{-1}(U_j).$$

Assertion 2 follows from this statement. Indeed, for such $\theta_{l'}$, extend $\theta_{l'}$ to $h^{-1}(U_j)$ by setting

$$\theta_{l'} = \operatorname{id} \quad \text{outside of} \quad \overline{\lambda_j^{-1}(F_{l'} \cup F_{l+1})} \cap h^{-1}(U_j).$$

Then $\theta_{l'}$ is not continuous, but $\lambda_j \circ \theta_{l'}$ is an $\mathfrak{X}$-retraction such that

$$(\lambda_j \circ \theta_{l'})^{-1}(F_{l'}) = \operatorname{Im} e_l.$$

(Remember that an $\mathfrak{X}$-map is by definition continuous.) Define similarly $\theta_{l''}$ for $F_{l''}$ such that $\lambda_j^{-1}(F_{l''})$ is bounded by $\operatorname{Im} e_{l_0}$ and $\operatorname{Im} c_{l_0+1}$ so that

$$(\lambda_j \circ \theta_{l''})^{-1}(F_{l''}) = \operatorname{Im} c_{l_0+1},$$

and let θ denote the composition of all $\theta_{l'}$ and $\theta_{l''}$. Then $\lambda_j \circ \theta$ is the required modification of λ_j.

Let us prove the statement. As already noted, there is an $\mathfrak{X}$-function μ_j on $h^{-1}(U_j - Y_j)$ such that (h, μ_j) is an homeomorphism from $h^{-1}(U_j - Y_j)$ to $(U_j - Y_j) \times [0,1]$ such that

$$\mu_j = \begin{cases} 0 & \text{on} \quad \operatorname{Im} e - h^{-1}(Y_j) \\ 1 & \text{on} \quad \operatorname{Im} c - h^{-1}(Y_j). \end{cases}$$

Then

$$\mu_j \circ e_l(y) \leq \mu_j \circ c_{l+1}(y) < \mu_j \circ e_{l+1}(y) \quad \text{for} \quad y \in U_j - Y_j,$$
$$\mu_j \circ e_l(y) = \mu_j \circ c_{l+1}(y) \quad \text{if and only if} \quad e_l(y) = c_{l+1}(y), \quad \text{and}$$
$$r \circ \lambda_j \circ e_l(y) = r \circ \lambda_j \circ c_{l+1}(y) < r \circ \lambda_j \circ e_{l+1}(y) \quad \text{for} \quad y \in U_j.$$

Let φ be an $\mathfrak{X}$-function on U_j such that

$$r \circ \lambda_j \circ c_{l+1}(y) \leq \varphi(y) < r \circ \lambda_j \circ e_{l+1}(y) \quad \text{for} \quad y \in U_j \text{ and}$$
$$r \circ \lambda_j \circ c_{l+1}(y) = \varphi(y) \quad \text{if and only if} \quad e_l(y) = c_{l+1}(y).$$

Let ε be the $\mathfrak{X}$-cross-section of $h|_{h^{-1}(U_j)}$ such that

$$r \circ \lambda_j \circ \varepsilon = \varphi \quad \text{on} \quad U_j \text{ and}$$
$$\mu_j \circ c_{l+1} \leq \mu_j \circ \varepsilon \quad \text{on} \quad U_j - Y_j.$$

Unique existence of ε is clear, and we have

$$c_{l+1}(y) = \varepsilon(y) \quad \text{if and only if} \quad e_l(y) = c_{l+1}(y).$$

It is easy to find an $\mathfrak{X}$-homeomorphism:

$$\Theta_{l'} : \{(x,y) \in (U_j - Y_j) \times [0,1] : \mu_j \circ e_l(y) \leq x \leq \mu_j \circ \varepsilon(y)\}$$
$$\longrightarrow \{(x,y) \in (U_j - Y_j) \times [0,1] : \mu_j \circ c_{l+1}(y) \leq x \leq \mu_j \circ \varepsilon(y)\}$$

of the form $\Theta_{l'}(x, y) = (\Theta'_{l'}(x, y), y)$. Extend $\Theta_{l'}$ to a map:

$$\{(x, y) \in (U_j - Y_j) \times [0, 1] : \mu_j \circ e_l(y) \le x \le \mu_j \circ e_{l+1}(y)\}$$
$$\longrightarrow \{(x, y) \in (U_j - Y) \times [0, 1] : \mu_j \circ c_{l+1}(y) \le x \le \mu_j \circ e_{l+1}(y)\}$$

by setting $\Theta_{l'} = \mathrm{id}$ outside of the original domain. Then $\Theta_{l'}$ uniquely induces an $\mathcal{X}$-homeomorphism:

$$\theta_{l'} : \overline{\lambda_j^{-1}(F_{l'} \cup F_{l+1})} \cap h^{-1}(U_j - Y_j) \longrightarrow \overline{\lambda_j^{-1}(F_{l+1})} \cap h^{-1}(U_j - Y_j).$$

Moreover, we can extend $\theta_{l'}$ to the set:

$$\overline{\lambda_j^{-1}(F_{l'} \cup F_{l+1})} \cap h^{-1}(Y_j) = \overline{\lambda_j^{-1}(F_{l+1})} \cap h^{-1}(Y_j) = \overline{F_{l+1}} \cap h^{-1}(Y_j)$$

by setting $\theta_{l'} = \mathrm{id}$ there, because

$$\mathrm{dis}(x, \theta_{l'}(x)) \longrightarrow 0 \quad \text{as} \quad x \longrightarrow \text{any point} \in \overline{F_l} \cap h^{-1}(Y_j).$$

Thus we prove Assertion 2.

Assertion 3. There exists an $\mathcal{X}$-map ψ from $Y \times [0, 1]$ to F such that $h \circ \psi$ is the projection onto $[0, 1]$, and $\psi|_{Y^* \times [0,1]}$ is a homeomorphism onto $h^{-1}(Y^*)$, where

$$Y^* = \{y \in Y : \sharp h^{-1}(y) \ne 1\}.$$

Proof of Assertion 3. It suffices to find an $\mathcal{X}$-function χ on $h^{-1}(Y^*)$ such that for each $y \in Y^*$, $\chi|_{h^{-1}(y)}$ is a homeomorphism onto $[0, 1]$. But such a χ is easily constructed by Assertion 2 and a partition of unity of class $\mathcal{X}$.

The following assertion follows from Assertion 3.

Assertion 4. Let K be the simplicial complex generated by Y and let K' be its derived subdivision. Let ρ be the simplicial function on K' defined by

$$\rho(s) = \begin{cases} 0 & \text{if } \sharp h^{-1}(s) = 1 \\ 1 & \text{if } \sharp h^{-1}(s) \ne 1 \end{cases} \quad \text{for vertex} \quad s \in K'.$$

Then there exists an $\mathcal{X}$-homeomorphism:

$$\delta : F \longrightarrow V = \{(y, t) \in Y \times [0, 1] : 0 \le t \le \rho(y)\}$$

such that $h \circ \delta^{-1}$ is the projection v onto Y.

Hence we can replace (F, h) with (V, v), i.e., it suffices to extend a C^0 R-$\mathcal{X}$-triangulation of $v|_{v^{-1}(\partial Y)}$ to v. Let $(\tilde{V}, \tilde{v})$ be a C^0 R-$\mathcal{X}$-triangulation of

$v|_{v^{-1}(\partial Y)}$. Let $\tilde{V}_1$ be the well-defined cone $w_1 * \tilde{V}$ for some point w_1 of some Euclidean space. Extend $\tilde{v}$ to $\tilde{V}_1$ by

$$\tilde{v}(tw_1 + (1-t)u) = t(s_0, 0) + (1-t)\tilde{v}(u) \quad \text{for} \quad u \in \tilde{V}, \ t \in [0, 1],$$

where s_0 is the vertex of K' which is a point of Y_{j_0}. Then $(\tilde{V}_1, \tilde{v})$ is a C^0 R-$\mathfrak{X}$-triangulation of $v|_{(s_0, 0)*v^{-1}(\partial Y)}$. Note that

$$(s_0, 0) * v^{-1}(\partial Y) = \{(y, t) \in Y \times [0, 1] : 0 \le t \le \rho'(t)\},$$

where ρ' is the simplicial function on K' which equals ρ at any vertex except s_0 and takes the value 0 at s_0. Next let $\tilde{V}_2$ be the union of $\tilde{V}_1$ and the cone $w_2 * \tilde{v}^{-1}(\mathrm{graph}\,\rho')$ for some point w_2 such that

$$\tilde{V}_1 \cap (w_2 * \tilde{v}^{-1}(\mathrm{graph}\,\rho')) = \tilde{v}^{-1}(\mathrm{graph}\,\rho').$$

Note that $\tilde{v}^{-1}(\mathrm{graph}\,\rho')$ is a subpolyhedron of $\tilde{V}_1$ and hence $\tilde{V}_2$ is a polyhedron. Finally, extend $\tilde{v}$ to $\tilde{V}_2$ by

$$\tilde{v}(tw_2 + (1-t)\tilde{v}^{-1}(u)) = t(s_0, 1) + (1-t)u \quad \text{for} \quad u \in \mathrm{graph}\,\rho', \ t \in [0, 1].$$

Then $(\tilde{V}_2, \tilde{v})$ is both a C^0 R-$\mathfrak{X}$-triangulation of v and an extension of $(\tilde{V}, \tilde{v})$, which completes the proof of IV.1.10. $\square$

For proof of IV.1.13 we need the following complex analytic version of II.4.11 and its generation.

Lemma IV.4.1. *Let X be a complex analytic set in an open set U of $\mathbf{C}^n \times \mathbf{C}$. Assume X does not include any nonempty open subset of $\mathbf{C}^n \times c$ for any $c \in \mathbf{C}$. There exists a complex bipolynomial diffeomorphism τ of $\mathbf{C}^n \times \mathbf{C}$ of the form:*

$$\tau(x, t) = (x_1, x' + \tau'(x_1), t) \ \text{ for } \ (x, t) = (x_1, x', t) = (x_1, \dots, x_n, t) \in \mathbf{C}^n \times \mathbf{C}$$

such that for every $y \in \mathbf{C}^{n-1} \times \mathbf{C}$, $\tau(X) \cap (\mathbf{C} \times y)$ is of dimension 0.

In such a case, we say that $(1, 0, \dots, 0) \in \mathbf{C}^n \times \mathbf{C}$ is a *non-singular direction* for $\tau(X)$.

Proof. In this proof, dimension denotes complex dimension. Let a function $\psi \colon \mathbf{C} \to \mathbf{R}$ be defined by $\psi(x) = 1 + |x|^{n-1}$. Let $a = (a_1, \dots, a_n, b)$ be a point of $U \subset \mathbf{C}^n \times \mathbf{C}$. For a positive number ε smaller than 1 and $1/2^{n-2}$, set

$$V(a, \varepsilon) = \{(x_1, \dots, x_n, t) \in \mathbf{C}^n \times \mathbf{C}:$$
$$|x_1 - a_1| < \varepsilon, \ |x_i - a_i| < \varepsilon + \psi(a_1)|x_1 - a_1|, \ i = 2, \dots, n, \ |t - b| < \varepsilon\}, \ \text{and}$$
$$V'(a, \varepsilon) = \{(x_1, \dots, x_n, t) \in \mathbf{C}^n \times \mathbf{C}:$$
$$|x_1 - a_1| < \varepsilon, \ |x_i - a_i| < \varepsilon - \psi(a_1)|x_1 - a_1|, \ i = 2, \dots, n, \ |t - b| < \varepsilon\}.$$

Choose ε so small that the closure $\overline{V(a, \varepsilon)}$ in $\mathbf{C}^n \times \mathbf{C}$ is included in U. By the condition that $\varepsilon < \max\{1, 1/2^{n-2}\}$ and easy calculations we see that for any complex bipolynomial diffeomorphism τ of $\mathbf{C}^n \times \mathbf{C}$ of the above form with $|d\tau'| \leq \psi/2$ on $\mathbf{C}$, the inverse image of

$$W(a, \varepsilon, \tau) = \{(x_1, \dots, x_n, t) \in \mathbf{C}^n \times \mathbf{C}: |x_i - \tilde{a}_i| \leq \varepsilon, \ i = 1, \dots, n, \ |t - b| \leq \varepsilon\}$$

under τ is included in $V(a, \varepsilon)$ and includes $V'(a, \varepsilon)$, where

$$(\tilde{a}_1, \dots, \tilde{a}_n, b) = \tau(a_1, \dots, a_n, b).$$

Note that $\tau^{-1}(x, t) = (x_1, x' - \tau'(x_1), t)$. In the following arguments, the notation $\tau = (x_1, x' + \tau'(x_1), t)$ and $\tau_k = (x_1, x' + \tau'_k(x_1), t)$, $k = 1, 2, \dots$, always denotes complex bipolynomial diffeomorphisms with $|d\tau'| \leq \psi/2$ and $|d\tau'_k| \leq \psi/2$. From the condition that $|d\tau'| \leq \psi/2$ it follows that τ' is a polynomial map of degree $\leq n$.

Choose points $a(j) \in U$, $j = 1, 2, \dots$, and positive numbers $\varepsilon(j) \leq \max\{1, 1/2^{n-2}\}$, $j = 1, 2, \dots$, so that all $\overline{V(a(j), \varepsilon(j))}$ are included in U, $\{V'(a(j), \varepsilon(j))\}_{j=1,2,\dots}$ is a covering of U, and $\{V(a(j), \varepsilon(j))\}_{j=1,2,\dots}$ is locally finite in U. We want to construct a sequence τ_j, $j = 1, 2, \dots$, converging to some τ inductively on j so that $(1, 0, \dots, 0)$ is a non-singular direction for $(\cup_{k=1}^j W(a(k), \varepsilon(k), \tau_j)) \cap \tau_j(X)$ for each j and then for $\tau(X)$. Note that

$$U = \tau^{-1}(\cup_{k=1}^\infty W(a(k), \varepsilon(k), \tau)).$$

We reduce the problem to an easier case. Let $a = (a_1, \dots, a_n, b)$ and ε be given as above. Set

$$X' = \{(x_1, \dots, x_n, t) \in V(a, \varepsilon): (x_1 \times \mathbf{C}^{n-1} \times t) \cap V(a, \varepsilon) \subset X\} \ \text{and}$$
$$X'' = \overline{X - X'} \cap V(a, \varepsilon).$$

Then X' and hence X'' are complex analytic sets in $V(a, \varepsilon)$ because

$$p(X') = \bigcap_{x' \in \mathbf{C}^{n-1}, |x_i - a_i| < \varepsilon, i=2,\dots,n} p(q^{-1}(x') \cap V(a, \varepsilon) \cap X) \ \text{and}$$
$$X' = \{(x_1, \dots, x_n, t) \in V(a, \varepsilon): (x_1, t) \in p(X')\},$$

where $p : \mathbf{C}^n \times \mathbf{C} \to \mathbf{C} \times \mathbf{C}$ and $q : \mathbf{C}^n \times \mathbf{C} \to \mathbf{C}^{n-1}$ denote the projections defined by

$$p(x_1, \dots, x_n, t) = (x_1, t), \quad \text{and} \quad q(x_1, \dots, x_n, t) = (x_2, \dots, x_n).$$

Furthermore, for any τ_1, $(1, 0, \dots, 0)$ is a non-singular direction for $W(a, \varepsilon, \tau_1) \cap \tau_1(X')$ by the assumption that X does not include any nonempty open subset of $\mathbf{C}^n \times c$, and clearly $X \cap V(a, \varepsilon) = X' \cup X''$. Hence it suffices to consider X'' in place of X. The advantage of X'' over X is the inequality:

$$\dim\{(x_1, \dots, x_n, t) \in V(a, \varepsilon) : (x_1 \times \mathbf{C}^{n-1} \times t) \cap V(a, \varepsilon) \subset X''\} < n, \quad (*)$$

which follows from the fact that $\dim X \leq n$. For simplicity of notation, we assume $(*)$ for X. Note that $(*)$ for X is equivalent to the statement that there are only a finite number of points (x_1, t) of $\mathbf{C}^2$ such that for some open subset U' of $\mathbf{C}^{n-1}$, $x_1 \times U' \times t$ is included in X.

Recall the arguments in the proof of II.4.11. We develop them. Let $a = (a_1, \dots, a_n, b)$ and ε be the same as above. By $(*)$ for X, there are disjoint points $a_1^1, \dots, a_1^{n+1}$ in $\{x_1 \in \mathbf{C} : |x_1 - a_1| < \varepsilon\}$ such that for each $j = 1, \dots, n+1$, $X \cap p_1^{-1}(a_1^j)$ is of dimension $< n$ and does not include any nonempty open subset of $a_1^j \times \mathbf{C}^{n-1} \times c$ for any $c \in \mathbf{C}$, where $p_1 : \mathbf{C}^n \times \mathbf{C} \to \mathbf{C}$ denotes the projection onto the first factor. Set

$$X_j = X \cap p_1^{-1}(a_1^j) \cap \overline{V(a, \varepsilon)}, \quad j = 1, \dots, n+1, \quad \text{and}$$

$$Y = \{(y^1, \dots, y^{n+1}) \in \overbrace{\mathbf{C}^{n-1} \times \cdots \times \mathbf{C}^{n-1}}^{n+1} : \bigcap_{j=1}^{n+1} (q_1(X_j) + (y^j, 0)) = \varnothing\},$$

where $q_1 : \mathbf{C}^n \times \mathbf{C} \to \mathbf{C}^{n-1} \times \mathbf{C}$ is the projection which forgets the first factor. Then Y is open and dense in $\mathbf{C}^{n-1} \times \cdots \times \mathbf{C}^{n-1}$ as shown below. Openness is clear because all X_j are compact. For density, it suffices to prove the following statement.

Let

$$Y_1, \ Y_2 \subset \{(x_1, \dots, x_n) \in \mathbf{C}^n : |x_i - a_i| \leq \varepsilon, \ i = 1, \dots, n\}$$

be complex analytic sets such that Y_2 does not include any open subset of $\mathbf{C}^{n-1} \times c$ for any $c \in \mathbf{C}$. Then there exists $y \in \mathbf{C}^{n-1}$ such that $|y|$ is arbitrarily small and

$$\dim Y_1 \cap (Y_2 + (y, 0)) < \dim Y_1.$$

This is easy to show as follows. For simplicity of notation, we assume Y_1 is irreducible. Let x be a regular point of Y_1. By the above assumption on Y_2 there exists $y \in \mathbf{C}^{n-1}$ such that $|y|$ is small and $Y_2 + (y, 0)$ does not contain x. Then $Y_1 \cap (Y_2 + (y, 0))$ is a proper subset of Y_1 and hence of dimension $< \dim Y_1$.

We show how to modify a complex bipolynomial diffeomorphism by which we obtain the sequence $\tau_1, \tau_2, \ldots$. Let Φ be a finite set of complex polynomial maps $\mathbf{C} \to \mathbf{C}^{n-1}$ of degree $\leq n$ such that for each $\varphi \in \Phi$, $\varphi(a_1^j)$ does not vanish for only one j, and for each j, $\mathbf{C}^{n-1}$ is spanned by $\varphi(a_1^j)$, $\varphi \in \Phi$, as a $\mathbf{C}$-linear vector space. Let τ be given, and let ζ be a small positive number. By density of Y, there exist τ_1 such that τ_1' is of the form $\tau' + \sum_{\varphi \in \Phi} c_\varphi \varphi$, $c_\varphi \in \mathbf{C}$ with $|c_\varphi| < \zeta$, $|d\tau_1' - d\tau'| < \zeta \psi$ and $(\tau_1'(a_1^1), \ldots, \tau_1'(a_1^{n+1})) \in Y$. From the last condition it follows that

$$\bigcap_{j=1}^{n+1} (q_1 \circ \tau_1(X_j)) = \bigcap_{j=1}^{n+1} (q_1(X_j) + (\tau_1'(a_1^j), 0) = \varnothing. \tag{0}$$

We prove that $(1, 0, \ldots, 0) \in \mathbf{C}^n \times \mathbf{C}$ is a non-singular direction for $W(a, \varepsilon, \tau_1) \cap \tau_1(X)$ by reduction to absurdity as follows. Assume

$$\dim(W(a, \varepsilon, \tau_1) \cap \tau_1(X) \cap (\mathbf{C} \times y)) = 1 \quad \text{for some } y \in \mathbf{C}^{n-1} \times \mathbf{C}.$$

Then $p_1(W(a, \varepsilon, \tau_1) \cap \tau_1(X) \cap (\mathbf{C} \times y))$ should be an open analytic set in the set:

$$p_1(W(a, \varepsilon, \tau_1) \cap \tau_1(U) \cap (\mathbf{C} \times y))$$
$$= p_1(W(a, \varepsilon, \tau_1) \cap (\mathbf{C} \times y)) = \{x_1 \in \mathbf{C} : |x_1 - a_1| < \varepsilon\}.$$

Hence

$$p_1(W(a, \varepsilon, \tau_1) \cap \tau_1(X) \cap (\mathbf{C} \times y)) = \{x_1 \in \mathbf{C} : |x_1 - a_1| < \varepsilon\}.$$

Therefore,

$$a_1^j \in p_1(\tau_1(X) \cap (\mathbf{C} \times y)), \quad \text{i.e., } (a_1^j, y) \in \tau_1(X), \quad j = 1, \ldots, n+1.$$

Then by the definition of X_j,

$$(a_1^j, y) \in \tau_1(X_j), \quad j = 1, \ldots, n+1.$$

Hence y is contained in all $q_1 \circ \tau_1(X_j)$, which contradicts (0). Moreover, by openness of Y, we can choose a positive number δ so that for any τ_2 with

$$|\tau_2' - \tau_1'| \leq \delta \quad \text{on} \quad \{a_1^j\}_{j=1,\ldots,n+1},$$

$(1, 0, \ldots, 0)$ is a non-singular direction for $W(a, \varepsilon, \tau_2) \cap \tau_2(X)$.

Define $T(A, \delta)$ to be the set of τ with

$$|\tau'(\alpha_1) - (\alpha_2, \ldots, \alpha_n)| \leq \delta \quad \text{for} \quad (\alpha_1, \ldots, \alpha_n) \in A,$$

where A is a finite subset of $\mathbf{C}^n$ and δ is a positive number. In the case of $A = \{(a_1^j, \tau_1'(a_1^j))\}_{j=1,\ldots,n+1}$, the above τ_2 is an element of $T(A, \delta)$.

Now let us construct τ_j by induction on j. For $j = 1$, let τ_1 be defined as above for $\tau = \mathrm{id}$ so that $|d\tau_1'| < \psi/4$. Let $A(1)$ be a finite subset of $\mathbf{C}^n$ and let $\delta(1)$ be a positive integer such that

$$\tau_1'(\alpha_1) = (\alpha_1, \ldots, \alpha_n) \quad \text{for} \quad (\alpha_1, \ldots, \alpha_n) \in A(1),$$

and $(1, 0, \ldots, 0)$ is a non-singular direction for $W(a(1), \varepsilon(1), \tau) \cap \tau(X)$ for any $\tau \in T(A(1), \delta(1))$

Let j be a positive integer. Assume that there exist τ_k, a finite subset $A(k)$ of $\mathbf{C}^n$, and a positive number $\delta(k)$ for each $1 \leq k \leq j$ such that

$$\tau_k'(\alpha_1) = (\alpha_2, \ldots, \alpha_n) \quad \text{for} \quad (\alpha_1, \ldots, \alpha_n) \in A(k), \tag{1}_k$$

$$T(A(k), \delta(k)) \subset T(A(k-1), \delta(k-1)) \quad \text{if} \quad k > 1, \tag{2}_k$$

$$\delta(k) < \delta(k-1), \quad |d\tau_k'| < (1/2 - 1/2^{k+1})\psi, \tag{3}_k$$

$$|\text{coefficients of } \tau_k' - \tau_{k-1}'| < 1/2^k \quad \text{if} \quad k > 1, \text{ and} \tag{4}_k$$

for any $\tau \in T(A(k), \delta(k))$, $(1, 0, \ldots, 0)$ is a non-singular direction for

$$\tag{5}_k$$

$$\left(\bigcup_{k'=1}^{k} W(a(k'), \varepsilon(k'), \tau) \right) \cap \tau(X).$$

Then it suffices to construct τ_{j+1}, $A(j+1)$ and $\delta(j+1)$ which satisfy conditions $(1)_{j+1}, \ldots, (5)_{j+1}$. Indeed, the limit τ of τ_j as $j \to \infty$ is a well-defined bipolynomial diffeomorphism by $(4)_k$ and fulfills the requirements by $(5)_k$.

As observed above, we have τ_{j+1}, $A(j+1)$, $\delta(j+1)$ such that $(1)_{j+1}$, $(3)_{j+1}$ and $(4)_{j+1}$ are satisfied, and for any $\tau \in T(A(j+1), \delta(j+1))$, $(1, 0, \ldots, 0)$ is a non-singular direction for $W(a(j+1), \varepsilon(j+1), \tau) \cap \tau(X)$. By the induction hypothesis $(5)_{j+1}$ follows from $(2)_{j+1}$. Therefore, we need only $(2)_{j+1}$. We can assume

$$|\tau_{j+1}' - \tau_j'| < \delta(j)/2 \quad \text{on} \quad B,$$

where $B = p_1(A(1) \cup \cdots \cup A(j))$. Choose $\delta(j+1)$ to be smaller than $\delta(j)/2$, and add to $A(j+1)$ the set $\{(\alpha_1, \tau_{j+1}'(\alpha_1))\}_{\alpha_1 \in B}$. Then $(2)_{j+1}$ holds true. $\square$

We need to generalize IV.4.1 as follows.

Lemma IV.4.1′. *Let X_i, $i = 1, 2, \ldots$, be complex analytic sets in open sets U_i of $\mathbf{C}^n \times \mathbf{C}$ respectively. Assume each X_i does not include any nonempty open subset of $\mathbf{C}^n \times c$ for any $c \in \mathbf{C}$. There exists a complex bipolynomial diffeomorphism τ of $\mathbf{C}^n \times \mathbf{C}$ of the form:*

$$\tau(x, t) = (x_1, x' + \tau'(x_1), t) \ \ for \ \ (x, t) = (x_1, x', t) = (x_1, \ldots, x_n, t) \in \mathbf{C}^n \times \mathbf{C}$$

such that $(1, 0, \ldots, 0) \in \mathbf{C}^n \times \mathbf{C}$ is a non-singular direction for all $\tau(X_i)$.

Proof. Let X, U, $\tau = (x_1, x' + \tau'(x_1), t)$ and $\tau_k = (x_1, x' + \tau'_k(x_1), t)$ be the same as in IV.4.1 and its proof. We know the following fact. Let P denote the set of polynomial maps $\tau' \colon \mathbf{C} \to \mathbf{C}^{n-1}$ of degree $\le n$ with

$$2|d\tau'(x)| < 1 + |x|^{n-1} \quad \text{for} \quad x \in \mathbf{C}.$$

Regard P naturally as an open set of $\mathbf{R}^{2n^2-2}$, and let $d(,)$ denote the metric in $\mathbf{R}^{2n^2-2}$. Let Q denote the set of τ with $\tau' \in P$. Let C be a compact subset of U, let ε be a positive number and let τ be an element of Q with $(1+\varepsilon)\tau' \in P$. Then there exists τ_1 in Q such that $d(\tau', \tau'_1) < \varepsilon$ and $(1, 0, \ldots, 0)$ is a non-singular direction for $\tau_1(X \cap C)$. Moreover, there exists a small positive number δ such that any $\tau_2 \in Q$ with $d(\tau'_1, \tau'_2) < \delta$ has this non-singular property.

Let $\varphi \colon \mathbf{N} \to \mathbf{N}$ be a map and let C_j, $j = 1, 2, \ldots$, be compact subsets of $\mathbf{C}^n \times \mathbf{C}$ such that $C_j \subset U_{\varphi(j)}$, $j = 1, 2, \ldots$, and each U_i is the union of C_j with $\varphi(j) = i$. We want a sequence $\tau_j \in Q$, $j = 1, 2, \ldots$, converging to an element $\tau \in Q$ such that for each j, $(1, 0, \ldots, 0)$ is a non-singular direction for all $\tau_j(X_{\varphi(k)} \cap C_k)$, $1 \le k \le j$, and then for $\tau(X_j)$.

We construct τ_j, $j = 1, 2, \ldots$, by induction on j. By the above fact we have $\tau_1 \in Q$ and a positive number δ_1 such that

$$\{\tau' \in \mathbf{R}^{2n^2-2} \colon d(\tau', \tau'_1) \le \delta_1\}$$

is included in P, and for any $\tau \in Q$ with $d(\tau', \tau'_1) \le \delta_1$, $(1, 0, \ldots, 0)$ is a non-singular direction for $\tau(X_{\varphi(1)} \cap C_1)$.

Let j be a positive integer. Assume there are $\tau_2, \ldots, \tau_{j-1} \in Q$ and positive numbers $\delta_2, \ldots, \delta_{j-1}$ such that for each $1 < k \le j-1$,

$$d(\tau'_k, \tau'_{k-1}) \le \delta_{k-1}/2, \qquad 2\delta_k < \delta_{k-1},$$

and for any $\tau \in Q$ with $d(\tau', \tau_k) \le \delta_k$, $(1, 0, \ldots, 0)$ is a non-singular direction for $\tau(X_{\varphi(k)} \cap C_k)$. Here the first two conditions imply that if $\tau' \in \mathbf{R}^{2n^2-2}$ satisfies $d(\tau', \tau'_k) \le \delta_k$, then $\tau' \in P$ and $d(\tau', \tau'_l) \le \delta_l$ for all $1 \le l < k$.

Hence the last condition shows that for any $\tau \in Q$ with $d(\tau', \tau'_{j-1}) \le \delta_{j-1}$, $(1, 0, \dots, 0)$ is a non-singular direction for $\tau(X_{\varphi(k)} \cap C_k)$ for $1 \le k \le j - 1$.

We need to construct $\tau_j \in Q$ and a positive number δ_j such that

$$d(\tau'_j, \tau'_{j-1}) \le \delta_{j-1}/2, \qquad 2\delta_j < \delta_{j-1},$$

and for any $\tau \in Q$ with $d(\tau', \tau'_j) \le \delta_j$, $(1, 0, \dots, 0)$ is a non-singular direction for $\tau(X_{\varphi(j)} \cap C_j)$. However, this follows from the above fact.

Consider the sequence τ_j, $j = 1, 2, \dots$. By the above induction procedure the sequence converges to an element τ of Q, and $(1, 0, \dots, 0)$ is a non-singular direction for $\tau(X_{\varphi(j)} \cap C_j)$ for all j and hence for $\tau(X_i)$ for all i. □

Remark IV.4.2. The proofs of IV.4.1 and IV.4.1' work also in the real analytic case. Moreover, we can prove them in the case of $\mathfrak{X}$. The non-trivial point in their proofs which we need to modify is the proof of the following statement.

Let Y_1, $Y_2 \subset \mathbf{R}^n$ be compact $\mathfrak{X}$-sets such that Y_2 does not include any open subset of $\mathbf{R}^{n-1} \times c$ for any $c \in \mathbf{R}$. Then there exists $y \in \mathbf{R}^{n-1}$ such that $|y|$ is arbitrarily small and

$$\dim Y_1 \cap (Y_2 + (y, 0)) < \dim Y_1.$$

Proof. Let $p_i \colon Y_i \to \mathbf{R}$, $i = 1, 2$, denote the restrictions of the projection $\mathbf{R}^n \to \mathbf{R}$ onto the last factor. By (II.1.17) we have finite Whitney C^1 $\mathfrak{X}$-stratifications $p_i \colon \{Y_{i,j}\}_j \to \{Z_{i,k}\}_k$, $i = 1, 2$. We can assume $\{Z_{1,k}\} = \{Z_{2,k}\}$. Call it $\{Z_k\}_k$. Let $\{Z'_k\}_k$, $\{Z''_k\}_k$ denote the 0-dimensional and 1-dimensional strata of $\{Z_k\}_k$ respectively, and set

$$\{Y'_{i,j}\}_j = \{Y_{i,j} \colon p_i(Y_{i,j}) \in \{Z'_k\}_k\} \quad \text{and}$$
$$\{Y''_{i,j}\}_j = \{Y_{i,j} \colon p_i(Y_{i,j}) \in \{Z''_k\}_k\}, \quad i = 1, 2.$$

As in the proof of II.5.4 (but much more easily), we see that for some $y \in \mathbf{R}^{n-1}$ with small $|y|$, $Y'_{1,j}$ for all j are transversal to $Y'_{2,j} + (y, 0)$ for all j in the weak sense that if $p_1(Y'_{1,j}) = p_2(Y'_{2,j'}) = t$, then $Y'_{1,j}$ is transversal to $Y'_{2,j'} + (y, 0)$ as subsets of $\mathbf{R}^{n-1} \times t$. When we perturb y a little, this property continues to hold. So we will perturb y so that $Y''_{1,j}$ for all j are transversal to $Y''_{2,j} + (y, 0)$ for all j. By the Whitney condition we do not need to perturb y near $\cup_j Y'_{1,j}$ and $\cup_j Y'_{2,j}$. To be precise, there exists a positive number ε such that $Y''_{1,j} \cap p_1^{-1}(I)$ for all j are transversal to $(Y''_{2,j} + (y, 0)) \cap p_2^{-1}(I)$ for all j, where I denotes the open ε-neighborhood of $\cup_k Z'_k$ in $\mathbf{R}$. Moreover, this holds after any small perturbation of y. Let J denote the closed $\varepsilon/2$-neighborhood of $\cup_k Z'_k$. By the same reason as above there exists $(y', t) \in \mathbf{R}^{n-1} \times \mathbf{R}$ such that y' is a small perturbation of y, $|t|$ is small, and $Y''_{1,j} - p_1^{-1}(J)$ for all j are transversal to $(Y''_{2,j} + (y', t)) - p_2^{-1}(J)$ for all j. Moreover, by the condition that $p_2|_{Y''_{2,j}} : Y''_{2,j} \to \mathbf{R}$ for each j is C^1 regular, we can choose $t = 0$, which is immediate by the proof of II.5.4. Thus $Y_{1,j}$ for all j are transversal to $Y_{2,j} + (y, 0)$ for all j (in the above weak sense for $Y'_{1,j}$ and $Y'_{2,j}$). Then the required inequality of dimension follows from the assumption that Y_2 does not include any open subset of any $\mathbf{R}^{n-1} \times c$.   □

Proof of IV.1.13. Consider the case $\mathfrak{X} = \{$subanalytic sets$\}$.   Let $p_i \colon \mathbf{C}^{n+1-i} \times \mathbf{C} \to \mathbf{C}^{n-i} \times \mathbf{C}$, $i = 1, \ldots, n$, denote the projections which forget the respective first factors. In this proof, we construct a local resolution $\{\mathcal{A}_{i,x}\}_{\substack{i=0,\ldots,n \\ x \in \mathbf{C}^{n-i} \times \mathbf{C}}}$ of f which satisfies, moreover, the following conditions.

(vi) A realization of each germ in $\mathcal{A}_{i,x}$ is a subanalytic complex analytic stratification such that if $i < n$, then the restriction of p_{i+1} to each stratum is a complex analytic immersion onto its image. (Remember that a complex analytic manifold in $\mathbf{C}^n$ is not necessarily subanalytic in $\mathbf{C}^n$.)

(vii) For each $i = 0, \ldots, n$, there exist a subanalytic open covering $\{U_k\}_{k=1,2,\ldots}$ of $\mathbf{C}^{n-i} \times \mathbf{C}$ and a subanalytic complex analytic stratification $\{X_{i,k,j}\}_j$ of each U_k such that for each $x \in \mathbf{C}^{n-i} \times \mathbf{C}$, $\mathcal{A}_{i,x}$ coincides with the family of all the germs of $\{X_{i,k,j}\}_j$ at x with $x \in U_k$.

We generalize the problem so that an induction method works. Let $\{Y_\alpha\}_\alpha$ be sets in $\mathbf{C}^n \times \mathbf{C}$. We define a local resolution of $\{Y_\alpha\}_\alpha$ with respect to $p_1, \ldots, p_n$ by replacing condition (iii)′ with

(viii) for any Y_α and any $x \in Y_\alpha$, there exists $A \in \mathcal{A}_{0,x}$ which is compatible with the germ of Y_α at x.

Note that the openness property in condition (i) of a local resolution follows from the other conditions because we consider only complex analytic stratifications and the fibres of p_i are of complex dimension 1.

Let O_α and Y_α, $\alpha = 1, 2, \ldots$, be open sets in $\mathbf{C}^n \times \mathbf{C}$ and complex analytic sets in O_α respectively. By changing the coordinate system of $\mathbf{C}^n$ in $\mathbf{C}^n \times \mathbf{C}$, we prove the following statement.

Statement.   There exists a local resolution of $\{Y_\alpha\}_\alpha$ with respect to $p_1, \ldots, p_n$.

This statement together with IV.1.2′ implies IV.1.13 if we set

$$O_\alpha = \mathbf{C}^n \times \mathbf{C}, \quad \text{and} \quad Y_\alpha = \text{graph } f, \quad \alpha = 1, 2, \ldots,$$

because any germ in $\mathcal{A}_{n,x}$ has a realization $\{x, V - x\}$, where V is an open 2-dimensional simplex in $\mathbf{C}$.

Proof of Statement. We prove this by induction on n. If $n = 0$, this is trivial. Hence we assume the case of $n - 1$. For simplicity of notation, we assume that each Y_α is of complex dimension $\leq n$. Moreover, without loss of generality, we suppose that each Y_α is irreducible, for odd α, Y_α is an open set of $\mathbf{C}^n \times c$ for some $c \in \mathbf{C}$, and for even α, Y_α is not so. By IV.4.1′, by changing the coordinate system of $\mathbf{C}^n$ in $\mathbf{C}^n \times \mathbf{C}$, we can assume that $(1, 0, \ldots, 0) \in \mathbf{C}^n \times \mathbf{C}$ is a non-singular direction for Y_α for all even α. For each α, there exists a countable or finite covering $\{P_\beta\}_\beta$ of O_α by open subanalytic sets of the form $Q_\beta \times R_\beta \subset \mathbf{C} \times (\mathbf{C}^{n-1} \times \mathbf{C})$ such that for even α, $p_1|_{P_\beta \cap Y_\alpha} : P_\beta \cap Y_\alpha \to R_\beta$ is proper. Hence by replacing O_α and Y_α with P_β and $P_\beta \cap Y_\alpha$ for all β, we assume that each O_α is of the form $Q_\alpha \times R_\alpha \subset \mathbf{C} \times (\mathbf{C}^{n-1} \times \mathbf{C})$ and for even α, $p_1|_{Y_\alpha} : Y_\alpha \to R_\alpha$ is proper. For each α it is easy to find a finite subanalytic complex analytic stratification $(\{X_{0,j}(\alpha)\}_j, \{X_{1,j}(\alpha)\}_j)$ of $p_1|_{O_\alpha}$ such that $\{X_{0,j}(\alpha)\}_j$ is compatible with Y_α, condition (ii) is satisfied, and for each integer l, $\cup_{\dim \leq l} X_{0,j}(\alpha)$ and $\cup_{\dim \leq l} X_{1,j}(\alpha)$ are complex analytic sets in O_α and R_α respectively. Here we add all pairs of finite intersections $O_\alpha \cap \cdots \cap O_{\alpha'}$ and $Y_\alpha \cap \cdots \cap Y_{\alpha'}$ to the family $\{O_\alpha, Y_\alpha\}$. Then we can assume that for any α and α', there exist a countable or finite number of α'' such that

$$O_\alpha \cap O_{\alpha'} = \bigcup_{\alpha''} O_{\alpha''}, \qquad Y_{\alpha''} = O_{\alpha''} \cap Y_\alpha \cap Y_{\alpha'},$$

and for each $i = 0, 1$ and for each α'', $\{X_{i,j}(\alpha'')\}_j$ is compatible with $\{X_{i,j}(\alpha), X_{i,j}(\alpha')\}_j$.

Consider $(R_\alpha, p_1(Y_\alpha))$ for all odd α and $(R_\alpha, \cup_{\dim \leq l} X_{1,j}(\alpha))$ for all even α and $l = 0, \ldots, n - 1$. By the induction hypothesis we can assume there exists a local resolution $\{\mathcal{A}_{i,x}\}_{\substack{i=1,\ldots,n \\ x \in \mathbf{C}^{n-i} \times \mathbf{C}}}$ of $\{p_1(Y_\alpha)\}_{\alpha : \text{odd}} \cup \{\cup_{\dim \leq l} X_{1,j}(\alpha)\}_{\substack{\alpha : \text{even} \\ l = 0, \ldots, n-1}}$ with respect to $p_2, \ldots, p_n$. For $x \in \mathbf{C}^n \times \mathbf{C}$, let $\mathcal{A}_{0,x}$ denote the germs of $\{X_{0,j}(\alpha)\}_j$ at x for all α with $x \in O_\alpha$. Then $\{\mathcal{A}_{0,x}\}_{x \in \mathbf{C}^n \times \mathbf{C}}$ and some their realizations satisfy (ii), (iv), (v), (vi), (vii) and (viii) and the latter half of (i). However, there are not always realizations of $\{\mathcal{A}_{i,x}\}_{\substack{i=0,1 \\ x \in \mathbf{C}^{n-i} \times \mathbf{C}}}$ which satisfy the first half of (i). We need to modify $\{\mathcal{A}_{0,x}\}_{x = \mathbf{C}^n \times \mathbf{C}}$ so that it is satisfied. For each $A \in \mathcal{A}_{0,x}$ there exists $B \in \mathcal{A}_{1,p_1(x)}$ such that a realization $\{X_{1,j'}\}_{j'}$ of B is compatible with $\{p_1(X_{0,j})\}_j$ for some realization $\{X_{0,j}\}_j$ of A. Replace A with the germs of $\{X_{0,j} \cap p_1^{-1}(X_{1,j'})\}_{j,j'}$ at x for all such B. Then $\{\mathcal{A}_{i,x}\}_{\substack{i=0,1 \\ x \in \mathbf{C}^{n-i} \times \mathbf{C}}}$ satisfies the first half of (i). $\qquad\square$

Proof of IV.1.13′. Clear by the above proof and by II.4.11. $\qquad\square$

CHAPTER V. $\mathfrak{Y}$-SETS

The simplest family of subsets of Euclidean spaces which we can treat systematically in topological position may be the family of semilinear sets. We generalize the concept of $\mathfrak{X}$ so that this family is an example. Let $\mathfrak{Y}$ be a family of subsets of Euclidean spaces which satisfies the following four axioms.

Axiom (i)' All rational semilinear sets in Euclidean spaces are elements of $\mathfrak{Y}$.

Axiom (ii) If $X_1 \subset \mathbf{R}^n$ and $X_2 \subset \mathbf{R}^n$ are elements of $\mathfrak{Y}$, then $X_1 \cap X_2$, $X_1 - X_2$ and $X_1 \times X_2$ are elements of $\mathfrak{Y}$.

Axiom (iii) If $X \subset \mathbf{R}^n$ is an element of $\mathfrak{Y}$, and $p \colon \mathbf{R}^n \to \mathbf{R}^{n-1}$ is the projection which forgets the last factor such that $p|_{\overline{X}}$ is proper, then $p(X)$ is an element of $\mathfrak{Y}$.

Axiom (iv) If $X \subset \mathbf{R}$ and $X \in \mathfrak{Y}$, then each point of X has a neighborhood in X which is a finite union of points and intervals.

The smallest example of $\mathfrak{Y}$ is the family of rational semilinear sets in Euclidean spaces. Note that we do not require Axiom (iii) for a general linear map p. As in the case of $\mathfrak{X}$, we define a $\mathfrak{Y}$-*set*, a $\mathfrak{Y}$-*map*, a C^r $\mathfrak{Y}$-*manifold*, a C^r $\mathfrak{Y}$-*stratification*, (*usual*) $\mathfrak{Y}$-*cell complex*, a $\mathfrak{Y}$-*simplicial complex*, a $\mathfrak{Y}$-*triangulation*, a $\mathfrak{Y}$-*cell triangulation*, a C^0 $\mathfrak{Y}$-*triangulation*, a C^0 $\mathfrak{Y}$-*cell triangulation* and $(R\text{-}L)$ $\mathfrak{Y}$-*equivalence*. Here we require that the underlying polyhedron of a $\mathfrak{Y}$-simplicial (cell) complex is closed in the ambient Euclidean space, the simplicial (cell) complex of a (C^0) $\mathfrak{Y}$-(cell) triangulation is a $\mathfrak{Y}$-simplicial (cell) complex, and the homeomorphism from each open simplex (cell) of a $\mathfrak{Y}$-(cell) triangulation to its image is a C^r diffeomorphism for some integer r. Note that a simplicial (cell) complex is not always of class $\mathfrak{Y}$.

The reason why we assume Axiom (i)' is the following. It is easy to show that Axiom (i)' is replaced by the next three axioms:

(a) The set $\{(x_1, \dots, x_n) \in \mathbf{R}^n \colon x_i = x_j\}$ is a $\mathfrak{Y}$-set for any integers $0 < i < j \le n$.

(b) The set $\{(x, y) \in \mathbf{R}^2 \colon x < y\}$ is a $\mathfrak{Y}$-set.

(c) The set $\{(x, y, z) \in \mathbf{R}^3 \colon z = x + y\}$ is a $\mathfrak{Y}$-set.

(a) is necessary for systematic ($=$ logical) arguments of $\mathfrak{Y}$-sets and $\mathfrak{Y}$-maps. (b) is needed for a connected component of a $\mathfrak{Y}$-set to be a $\mathfrak{Y}$-set (the set in (b) is a connected component of the set $\mathbf{R}^2 - \{x = y\}$). (c) is assumed so that we can use techniques of differential topology. Indeed, for a $\mathfrak{Y}$-function $z = f(x, y)$ with $f(0, 0) = 0$ we need an infinite number of

2-dimensional linear $\mathfrak{Y}$-spaces in $\mathbf{R}^3$ passing the origin to know whether f is differentiable at $(0,0)$. By (c) the linear spaces $\{z = \alpha x + \beta y\}$ for all rationals α and β are $\mathfrak{Y}$-sets.

We need Axioms (ii) and (iii) for the same reason as (a). An example of $\mathfrak{Y}$ without Axiom (iv) is the family of all subsets of Euclidean spaces. Such a family is not interesting. I think that the simplest axiom to avoid this family is (iv).

Since multiplication is not necessarily a $\mathfrak{Y}$-function, almost all of the foregoing proofs do not work for $\mathfrak{Y}$-sets and $\mathfrak{Y}$-maps. But the proof of II.1.2 does. Hence the closure of a $\mathfrak{Y}$-set in the ambient Euclidean space is a $\mathfrak{Y}$-set.

If $\mathfrak{Y}$ satisfies the following axioms which are stronger than Axioms (iii) and (iv), we call it $\mathfrak{Y}_0$.

Axiom (iii)$_0$ Axiom (iii) without the assumption that $p|_{\overline{X}}$ is proper.

Axiom (iv)$_0$ If $X \subset \mathbf{R}$ and $X \in \mathfrak{Y}$, then X is a finite union of intervals and points.

We sometimes add the following axioms:

Axiom (v) If a subset X of $\mathbf{R}^n$ is a $\mathfrak{Y}$-set locally at each point of $\mathbf{R}^n$, then X is a $\mathfrak{Y}$-set.

Axiom (vi) Any point in $\mathbf{R}$ is an element of $\mathfrak{Y}$.

There are two cases: any $\mathfrak{Y}$-set is locally semilinear, or some $\mathfrak{Y}$-set is not locally semilinear. (Here a $\mathfrak{Y}$-set $X \subset \mathbf{R}^n$ is called *locally semilinear* if X is semilinear locally at each point of $\mathbf{R}^n$). Phenomena of $\mathfrak{Y}$-sets and $\mathfrak{Y}$-maps are different in these two cases. Hence we consider the cases separately in the following sections.

§V.1. Case where any $\mathfrak{Y}$-set is locally semilinear

In this section, we assume any $\mathfrak{Y}$-set is locally semilinear. We define a (*usual*) $\mathfrak{Y}$-*cell decomposition* and a $\mathfrak{Y}$-*simplicial decomposition* of a $\mathfrak{Y}$-set X closed in the ambient Euclidean space to be a (usual) $\mathfrak{Y}$-cell complex and a $\mathfrak{Y}$-simplicial complex with underlying polyhedron $= X$.

Theorem V.1.1 ($\mathfrak{Y}$-cell decomposition). *Assume any* $\mathfrak{Y}$*-set is locally semilinear. A closed* $\mathfrak{Y}$*-set* $X \subset \mathbf{R}^n$ *admits a usual* $\mathfrak{Y}$*-cell decomposition compatible with a family of* $\mathfrak{Y}$*-sets locally finite at each point of* $\mathbf{R}^n$ *if* X *is compact or Axiom* (v) *is assumed.*

Assume any $\mathfrak{Y}_0$*-set is locally semilinear. A closed* $\mathfrak{Y}_0$*-set* $X \subset \mathbf{R}^n$ *admits a* $\mathfrak{Y}_0$*-cell decomposition compatible with a finite number of* $\mathfrak{Y}_0$*-sets.*

Proof. **Case where X is compact.** Let a, b be rationals such that $X \subset [a,b]^n$, and set $I = [a,b]$. It suffices to construct a usual $\mathfrak{Y}$-cell decomposition of I^n compatible with a finite number of closed $\mathfrak{Y}$-sets $X_i \subset I^n$. We proceed by induction on $\dim \cup X_i$. Let x be a point of one X_i. The germ of X_i at x is a linear space germ if and only if there exists $\varepsilon > 0 \in \mathbf{R}$ such that for $x', x'' \in X_i$ with $|x - x'| < \varepsilon$ and $|x - x''| < \varepsilon$, $x' + x'' - x \in X_i$. Hence it is easy to show that the subset Z_i of X_i of such points is a 𝔜-set. Clearly $\dim(X_i - Z_i) < \dim X_i$, and $X_i - Z_i$ is closed. Hence by induction, we have a usual 𝔜-cell decomposition C_1 of I^n compatible with $\{X_i - Z_i\}$. For each connected component Z_{ij} of Z_i, $\widetilde{Z_{ij}}$ – the smallest linear space which contains Z_{ij} - is not always a 𝔜-set. But $I^n \cap \widetilde{Z_{ij}}$ is a 𝔜-set because for a point $x_{ij} \in Z_{ij}$ and a large integer m,

$$I^n \cap \widetilde{Z_{ij}} = I^n \cap \{m(x - x_{ij}) + x_{ij} : x \in Z_{ij}\}.$$

To construct the required usual 𝔜-cell decomposition, we can replace Z_{ij} with $I^n \cap \widetilde{Z_{ij}}$. We keep the notation Z_{ij}. For a moment, assume there are a finite number of linear 𝔜-functions f_{ijk} on I^n such that for each i and j, $Z_{ij} = \cap_k f_{ijk}^{-1}(0)$. Then the usual cell complex

$$C = \{\sigma \cap \{f_{ijk} * 0\} : \sigma \in C_1, \ i, j, k, \ * \in \{=, \leq, \geq\}\}$$

is what we want.

It remains to prove the following statement.

Statement. Let Z be the intersection of I^n with a linear space. Assume Z is a 𝔜-set. There exists a finite number of linear 𝔜-functions f_i on I^n such that $Z = \cap_i f_i^{-1}(0)$.

Proof of Statement. For each $j = 1, \ldots, n$, let $p_j : \mathbf{R}^n \to \mathbf{R}^{n-1}$ denote the projection which forgets the j-th factor. If $p_j|_Z$ is not injective for some j, we consider $p_j(Z)$ in place of Z and we can decrease n. If $p_j|_Z$ are injective for all j, and $\dim Z < n - 1$, then we can replace Z with $I^n \cap p_j^{-1}(\widetilde{p_j(Z)})$, $j = 1, \ldots, n$, because Z is their intersection. Hence by downward induction on $n - \dim Z$ we can assume $\dim Z = n - 1$. Clearly Z is the intersection of I^n with the graph of a linear 𝔜-function g on I^{n-1}. Then the function $f_1(x_1, \ldots, x_n) = x_n - g(x_1, \ldots, x_{n-1})$ fulfills the requirement.

Note. We proved that for any (not necessarily compact) 𝔜-set Y, the subset $\hat{Y}$ of Y consisting of points where the germ of Y is a linear space germ is a 𝔜-set. Moreover, by the proof of II.1.7, for each integer k, the union of the connected components of $\hat{Y}$ of dimension k is a 𝔜-set.

Case where X is noncompact and Axiom (v) is assumed. We construct a usual 𝔜-cell decomposition of $\mathbf{R}^n$ compatible with a family of 𝔜-set $\{X_i\}$ locally finite at each point of $\mathbf{R}^n$. By induction on a nonnegative integer m we assume there is a usual 𝔜-cell decomposition C_m of $[-m, m]^n$

compatible with C_{m-1} and $\{X_i\}$ such that $C_m = C_{m-1}$ on $[-m+2, m-2]^n$, i.e., C_m and C_{m-1} are compatible with $[-m+2, m-2]^n$ and

$$\{\sigma \in C_m : \sigma \subset [-m+2, m-2]^n\} = \{\sigma \in C_{m-1} : \sigma \subset [-m+2, m-2]^n\}.$$

We want to find C_{m+1}. By the above proof we have a usual $\mathfrak{Y}$-cell decomposition C_{m+1}^1 of $[-m-1, m+1]^n$ compatible with C_m and $\{X_i\}$. We need to modify it so that it equals C_m on $[-m+1, m-1]^n$. For that it suffices to modify C_m outside $[-m+1, m-1]^n$ and subdivide C_{m+1}^1 so that C_m equals C_{m+1}^1 on $\partial([-m, m]^n)$. Hence what we need to prove is the following assertion.

Let C be a usual $\mathfrak{Y}$-cell decomposition of $[-m-1, m+1]^n -]-m, m[^n$. There exist usual $\mathfrak{Y}$-cell subdivisions C_m' of C_m and C' of C such that $C_m' = C_m$ on $[-m+1, m-1]^n$ and $C_m' = C'$ on $\partial([-m, m]^n)$.

We can assume C is compatible with C_m. For each proper face D of $[-m, m]^n$, let h_i be a finite number of linear $\mathfrak{Y}$-functions on D which define all cells of $C|_D$ (statement). It suffices to extend each h_i to a linear $\mathfrak{Y}$-function on $[-m-1, m+1]^n$ so that the zero set does not intersect with $[-m+1, m-1]^n$. Let $D = [-m, m]^{n-1} \times m$, and assume h_i is defined on $[-m-1, m+1]^{n-1} \times m$ and $h_i^{-1}(0) \neq \varnothing$. Define an extension of h_i to be $N(x_m - m) + h_i(x_1, \dots, x_{n-1})$ for a sufficiently large integer N. Then the requirement is fulfilled.

Case of $\mathfrak{Y}_0$. By the above arguments it suffices to prove the next statement.

Let C be a $\mathfrak{Y}_0$-cell decomposition of $\mathbf{R}^n$, let l be a positive integer, and let $\{X_i\}$ be a finite number of $\mathfrak{Y}_0$-sets such that each connected component X_{ij} of each X_i is of dimension l and open in a linear space $\widetilde{X_{ij}}$ and C is compatible with $\{\overline{X_i} - X_i\}$. Then there exists a $\mathfrak{Y}_0$-cell subdivision C' of C compatible with $\{X_i\}$.

We can assume C and $\{X_i\}$ are closed under the symmetric transformations with respect to the coordinate hyperplanes $\{x_j = 0\}$, $j = 1, \dots, n$, and we also require C' to be so. Hence we suppose

$$|C| = \mathbf{R}_+^n \quad \text{and} \quad X_i \subset \mathbf{R}_+^n,$$

where $\mathbf{R}_+$ denotes the nonnegative reals.

The family $\{X_{ij}\}$ is finite for the following reason. Assume it is not. Let $\check{C}$ denote the subcomplex of C with $|\check{C}| = \cup_i (\overline{X_i} - X_i)$. Let $q = (q_1, \dots, q_k) : \mathbf{R}^n \to \mathbf{R}^k$ be a linear map such that for each i and for each j, $q|_{\widetilde{X_{ij}}}$ is bijective. Note that $|\check{C}|$ is of dimension smaller than l and each $q(X_{ij})$ is a connected component of $\mathbf{R}^k - q(\overline{X_{ij}} - X_{ij})$. For each i, $\overline{X_i} - X_i = \cup_j (\overline{X_{ij}} - X_{ij})$ because $\{X_{ij}\}$ is locally finite at each point of $\mathbf{R}^n$.

Hence $|\check{C}|$ contains each $\overline{X_{ij}} - X_{ij}$. Therefore, there are either an infinite number of connected components of $\mathbf{R}^k - q(|\check{C}|)$ or a component over which $q|_{\cup_i X_i}$ is an infinite covering, which contradicts either the fact that $q(|\check{C}|)$ is semilinear or Axiom (iv)$_0$ respectively.

Let $\{X_{ij}\}_{ij \in E}$ denote the bounded components. Let N be an integer so large that

$$\bigcup_{ij \in E} X_{ij} \subset \{x_1 + \cdots + x_n \le N\}.$$

As in the above proof of the first case, we have a usual $\mathfrak{Y}_0$-cell decomposition C_N of $\mathbf{R}^n_+ \cap \{x_1 + \cdots + x_n \le N\}$ compatible with C and $\{X_{ij}\}_{ij \in E}$. By the proof of the second case we have a usual $\mathfrak{Y}_0$-cell decomposition C_{N+1} of $\mathbf{R}^n_+ \cap \{x_1 + \cdots + x_n \le N + 1\}$ compatible with C such that

$$C_{N+1} = \begin{cases} C_N & \text{on } \mathbf{R}^n_+ \cap \{x_1 + \cdots + x_n \le N\} \\ \hat{C} & \text{on } \mathbf{R}^n_+ \cap \{x_1 + \cdots + x_n = N + 1\}, \end{cases}$$

where $\qquad \hat{C} = \{\sigma \cap \mathbf{R}^n_+ \cap \{x_1 + \cdots + x_n \ge N + 1\} \colon \sigma \in C\}.$

Define a $\mathfrak{Y}_0$-cell subdivision of C to be C_{N+1} on $\mathbf{R}^n_+ \cap \{x_1 + \cdots + x_n \le N + 1\}$ and $\hat{C}$ on $\mathbf{R}^n_+ \cap \{x_1 + \cdots + x_n \ge N + 1\}$. Then we can forget $\{X_{ij}\}_{ij \in E}$, and we assume every X_{ij} is unbounded.

By the same reason as in the proof of the first case, it remains only to show that each $\widetilde{X_{ij}}$ is a $\mathfrak{Y}_0$-set. By changing the coordinate system of $\mathbf{R}^n$ by a rational linear transformation, we can assume $X_{ij} \cap \mathbf{R}^n_+$ is closed in $\mathbf{R}^n_+$ and equals the graph of a linear $\mathfrak{Y}_0$-map $\varphi = (\varphi_{l+1}, \dots, \varphi_n) \colon \mathbf{R}^l_+ \to \mathbf{R}^{n-l}_+$. Then it suffices to prove that the linear extension of each φ_k to $\mathbf{R}^l$ is a $\mathfrak{Y}_0$-function. Moreover, the problem is reduced to the following one. Set

$$\varphi_k(x_1, \dots, x_l) = a_{k0} + a_{k1}x_1 + \cdots + a_{kl}x_l \quad \text{and}$$

$$\varphi_k(x_1, \dots, x_l) = \begin{cases} a_{k0} & \text{if } k' = 0 \\ a_k x_{k'} & \text{if } k' = 1, \dots, l, \end{cases} \quad \text{for } (x_1, \dots x_l) \in \mathbf{R}^l_+.$$

The problem is to show that the linear extensions of $\varphi_{kk'}$ to $\mathbf{R}^l$ are $\mathfrak{Y}_0$-functions. The extension of φ_{k0} is a $\mathfrak{Y}_0$-function because $(0, \dots, 0, a_{k0}) = $ graph $\varphi_k \cap 0^l \times \mathbf{R}$ and hence $\mathbf{R}^l \times a_{k0}$ are $\mathfrak{Y}_0$-sets. To see that the extension of each $\varphi_{kk'}$, $k' > 0$, is a $\mathfrak{Y}_0$-function, we can regard $\varphi_{kk'}$ as the function $\mathbf{R}_+ \ni x \to a_{kk'}x \in \mathbf{R}$, and we need only show that its linear extension to $\mathbf{R}$ is a $\mathfrak{Y}_0$-function. Since

$$\varphi_{kk'}(x) = (\varphi_k - \varphi_{k0})(0, \dots, 0, x, 0 \dots, 0) \quad \text{for } x \in \mathbf{R}_+,$$

$\varphi_{kk'}$ is a $\mathfrak{Y}_0$-function. Hence its graph, which is a half line in $\mathbf{R}^2$ with end $= 0$, is a $\mathfrak{Y}_0$-set. The graph of the extension of $\varphi_{kk'}$ to $\mathbf{R}$ is the union of graph $\varphi_{kk'}$ and its symmetric set with respect to the origin. Therefore, the extension is a $\mathfrak{Y}_0$-function. $\square$

A map version of V.1.1 is the following.

Let $X_1, X_2 \subset \mathbf{R}^n$ be closed $\mathfrak{Y}$-sets, and let $f \colon X_1 \to X_2$ be a proper $\mathfrak{Y}$-map. Assume either X_1 and X_2 are compact or Axiom (v) holds. Then X_1 and X_2 admit usual $\mathfrak{Y}$-cell decompositions C_1 and C_2 respectively, such that for each $\sigma \in C_1$, $f(\sigma) \in C_2$, and $f|_\sigma$ is linear. A similar statement holds for $\mathfrak{Y}_0$ without the properness assumption. This is clear when we apply the above proof to graph f.

Note also that any locally semilinear $\mathfrak{Y}_0$-set is semilinear even if there exists a $\mathfrak{Y}_0$-set which is not locally semilinear. Indeed, this follows if we apply V.1.1 to the family {locally semilinear $\mathfrak{Y}_0$-sets}.

If all simplexes are $\mathfrak{Y}$-sets, then any compact $\mathfrak{Y}$-set admits a $\mathfrak{Y}$-simplicial decomposition and problems on compact $\mathfrak{Y}$-sets are purely ones of PL topology. In the case where some simplex is not a $\mathfrak{Y}$-set, a compact $\mathfrak{Y}$-set does not always admit a $\mathfrak{Y}$-simplicial decomposition as follows.

Example V.1.2. Let $\mathfrak{Y} = $ {semilinear sets defined by linear functions with rational coefficients and real constant terms}, and $X = [0, 1] \times [0, \sqrt{2}]$. Then X does not admit a $\mathfrak{Y}$-simplicial decomposition.

Proof. Assume there exists a $\mathfrak{Y}$-simplicial decomposition K of X. Note that the slope of each 1-simplex of K is rational. Let $v_i = (x_i, y_i)$, $i = 1, \dots, l$, be the vertices of K lying in order on the segments $0 \times [0, \sqrt{2}]$ and then $[0, 1] \times \sqrt{2}$ from $v_1 = (0, 0)$ to $v_l = (1, \sqrt{2})$. Let $v_k = (0, \sqrt{2})$. Let $\sigma_1, \dots, \sigma_{l-1}$ denote the 2-simplexes in K such that each σ_i is the simplex with vertices v_i and v_{i+1}. Then it suffices to prove that $(y_i - y_{i+1})/(y_i - y_{i-1})$, $i = 2, \dots, k-1$, $(x_i - x_{i+1})/(x_i - x_{i-1})$, $i = k+1, \dots, l-1$, and $(x_k - x_{k+1})/(y_k - y_{k-1})$ are rational, because if they are so, then $(x_l - x_k)/(y_k - y_1) = \sqrt{2}$ should be rational.

Translating $\mathbf{R}^2$ by a regular matrix with rational elements, we can assume that the first coordinates of the vertices of K are distinct from each other, and so are the second coordinates. Note that what we have to prove does not change. For each i, consider the sequence $\sigma_{i,1} = \sigma_i, \dots, \sigma_{i,j_i} = \sigma_{i+1}$ of 2-simplexes of K such that each $\sigma_{i,j} \cap \sigma_{i,j+1}$ is of dimension 1, $\sigma_{i,j}$ are distinct from each other, and each $\sigma_{i,j}$ contains v_{i+1}. Let $v_{i,1} = v_i, \dots, v_{i,j_i+1} = v_{i+2}$ be the sequence of vertices of K such that each $\sigma_{i,j}$ has the vertices $v_{i,j}$, $v_{i,j+1}$ and v_{i+1}. Let $v_{i,j} = (x_{i,j}, y_{i,j})$ for all i and j. Then it easily follows from the rational slope property that $(x_{i+1} - x_{i,j+1})/(x_{i+1} - x_{i,j})$ and $(y_{i+1} - y_{i,j+1})/(y_{i+1} - y_{i,j})$ are rational. Hence $(y_i - y_{i+1})/(y_i - y_{i-1})$

and $(x_i - x_{i+1})/(x_i - x_{i-1})$ are rational. Similarly we easily see that $(x_k - x_{k+1})/(y_k - y_{k-1})$ is rational. $\square$

Note that a compact $\mathfrak{Y}$-set does not admit a $\mathfrak{Y}$-triangulation unless it admits a $\mathfrak{Y}$-simplicial decomposition, because a C^1 PL map between open cells is linear. However, a C^0 $\mathfrak{Y}$-triangulation is possible.

Theorem V.1.3 ($\mathfrak{Y}$-triangulation). *Assume Axiom* (vi) *and that any $\mathfrak{Y}$-set is locally semilinear. Any closed $\mathfrak{Y}$-set $X \subset \mathbf{R}^n$ admits a rational C^0 $\mathfrak{Y}$-triangulation compatible with a family of $\mathfrak{Y}$-sets in X locally finite at each point of X if X is compact or Axiom* (v) *is assumed.*

Assume Axiom (vi) *for $\mathfrak{Y}_0$ and that any $\mathfrak{Y}_0$-set is locally semilinear. Any closed $\mathfrak{Y}_0$-set in $\mathbf{R}^n$ admits a rational C^0 $\mathfrak{Y}_0$-cell triangulation compatible with a finite family of $\mathfrak{Y}_0$-sets.*

Let $X \subset \mathbf{R}^n$ and $X' \subset \mathbf{R}^{n'}$ be closed sets, and let $f \colon X \to X'$ be a $\mathfrak{Y}$-map. Assume either (a) *X and X' are bounded,* (b) *Axiom* (v) *is satisfied, and for any bounded set B of $\mathbf{R}^{n'}$, $f^{-1}(B)$ is bounded, or* (c) *$\mathfrak{Y} = \mathfrak{Y}_0$. Then there exist rational C^0 $\mathfrak{Y}$-(cell) triangulations (K, π) of X and (K', π') of X' compatible with families of $\mathfrak{Y}$-sets as above such that $\pi'^{-1} \circ f \circ \pi \colon K \to K'$ is a simplicial (resp., cell) map.*

In the above example, the triangulation (K, π) is given so that

$$|K| = [0,1]^2 \quad \text{and} \quad \pi(x,y) = \begin{cases} \text{id} & \text{for} \quad y \leq 2 - \sqrt{2} \\ (x, 2y + \sqrt{2} - 2) & \text{for} \quad y \geq 2 - \sqrt{2}. \end{cases}$$

We remark that in V.1.3, we can choose the triangulation (K, π) and the $\mathfrak{Y}_0$-homeomorphism π' so that K lies in $\mathbf{R}^n$, Im π' is contained in $\mathbf{R}^n$, and π and π' can be extended to $\mathfrak{Y}$-homeomorphisms of $\mathbf{R}^n$.

Proof. **Compact case.** Let X be a compact $\mathfrak{Y}$-set included in I^n for an interval $I = [-a, a]$, where a is rational. By V.1.1 we have a $\mathfrak{Y}$-cell decomposition K of I^n such that X and the $\mathfrak{Y}$-sets of the given family are unions of open cells of K. Moreover, by the proof of V.1.1, we can choose K so that for a finite number of linear $\mathfrak{Y}$-functions f_i on I^n,

$$K = \{\{f_1 *_1 0\} \cap \{f_2 *_2 0\} \cap \cdots : *_1, *_2, \cdots \in \{\geq, =, \leq\}\}.$$

We want to modify K to be rational. We proceed by a double induction method. Let

$$\mathbf{R}^n \xrightarrow{p_{n-1}} \mathbf{R}^{n-1} \xrightarrow{p_{n-2}} \cdots \xrightarrow{p_1} \mathbf{R}^1$$

denote the sequence of the projections which forget the respective last factors. Apply V.1.1 to all the images $p_{n-1}(\sigma)$, $\sigma \in K$. Then we obtain a $\mathfrak{Y}$-cell decomposition K_{n-1} of I^{n-1} defined by a finite number of linear $\mathfrak{Y}$-functions on I^{n-1} as K such that each $p_{n-1}(\sigma)$ is a union of cells of K_{n-1}. Replace K with the $\mathfrak{Y}$-cell complex:

$$K_n = \{\sigma \cap p_{n-1}^{-1}(\sigma'): \sigma \in K, \ \sigma' \in K_{n-1}\}.$$

Then K_n has the same properties as K, and $p_{n-1}: K_n \to K_{n-1}$ is a cell map. Repeat these arguments for p_{n-2} and so on. Then we have $\mathfrak{Y}$-cell decompositions K_k of I^k, $n \geq k \geq 1$, defined by a finite number of linear $\mathfrak{Y}$-functions on I^k such that the sequence

$$K_n \xrightarrow{p_{n-1}} K_{n-1} \xrightarrow{p_{n-2}} \cdots \xrightarrow{p_1} K_1$$

is one of cell maps. There exist a finite number of linear $\mathfrak{Y}$-functions $f_{k1}, f_{k2}, \ldots$ on I^{k-1} such that

$$K_k = \{(x', x_k) \in \sigma \times I: x_k *_1 f_{k1}(x'), \ x_k *_2 f_{k2}(x'), \ldots\}_{\substack{\sigma \in K_{k-1} \\ *_1, *_2, \ldots \in \{\geq, =, \leq\}}} \qquad (*)$$

We shall need the following fact. Let $a_1 < a_2 < a_3$ be points in $\mathbf{R}$ such that a_1 and a_3 are rationals. Then there exists a $\mathfrak{Y}$-homeomorphism ρ of $\mathbf{R}$ such that $\rho(a_2)$ is rational and $\rho = \mathrm{id}$ outside $[a_1, a_3]$. It suffices to prove this in the case where $a_1 = 0$. Let $b_2 \ (> a_2)$ be a rational number sufficiently near a_2. Set

$$b_1 = 2a_2 - b_2 \quad \text{and} \quad b_3 = 2b_2 - a_2.$$

Then $0 < b_1 < a_2 < b_2 < b_3$, $b_3 - b_2 = b_2 - a_2 = a_2 - b_1$. Define ρ by

$$\rho(x) = \begin{cases} x & \text{for} \quad x \leq b_1 \\ b_1 + 2(x - b_1) & \text{for} \quad b_1 \leq x \leq a_2 \\ b_3 - (b_3 - x)/2 & \text{for} \quad a_2 \leq x \leq b_3 \\ x & \text{for} \quad b_3 \leq x. \end{cases}$$

Then ρ is a $\mathfrak{Y}$-homeomorphism by Axiom (vi), and $\rho(0) = 0$, $\rho(a_2) = b_2$.

By this fact we have a $\mathfrak{Y}$-homeomorphism ρ of $\mathbf{R}$ such that $\rho = \mathrm{id}$ outside I and for any vertex $v = (x_1, \ldots x_n) \in K_n$, $(\rho(x_1), \ldots, \rho(x_n))$ is rational. Replace $K_n, \ldots, K_1$ with $\{\rho \times \cdots \times \rho(\sigma): \sigma \in K_n\}, \ldots, \{\rho(\sigma): \sigma \in K_1\}$ respectively, and keep the notation $K_n, \ldots, K_1$. Then $K_n, \ldots, K_2$ are no longer cell complexes. We want to modify them by an induction method. Note that K_1 is a rational simplicial complex.

Let $1 < k \leq n$. Assume $K_1, \ldots, K_{k-1}$ are rational simplicial decompositions of $I, \ldots, I^{k-1}$ respectively; $K_k, \ldots, K_n$ are families of C^0 $\mathfrak{Y}$-submanifolds possibly with boundary of $I^k, \ldots, I^n$ respectively; all the elements of dimension 0 of $K_k, \ldots, K_n$ are rational (we call them vertices); for each $\sigma \in K_l$, $k \le l \le n$, $p_{l-1}(\sigma)$ is an element of K_{l-1}; $\{\sigma°\colon \sigma \in K_l\}$ is a stratification of I^l; for any $\sigma \ne \sigma' \in K_l$ of dimension l', any $l' + 1$ vertices in σ do not coincide with any $l' + 1$ ones in σ'; and there are a finite number of 𝔜-functions $f_{l1}, f_{l2}, \ldots$ on I^{l-1} which describe K_l as (*). Then, keeping these properties, we only need to modify K_k to be a rational simplicial complex.

We want a 𝔜-homeomorphism φ of I^k of the form:

$$\varphi(x', x_k) = (x', \varphi'(x', x_k)) \quad \text{for} \quad (x', x_k) \in I^{k-1} \times I$$

such that $\{\varphi(\sigma)\colon \sigma \in K_k\}$ is a cell complex, and $\varphi = \mathrm{id}$ on the vertices of K_k. Note that for $\sigma \in K_k$ such that $p_{k-1}|_\sigma$ is injective, $\varphi|_\sigma$ is uniquely decided because $\varphi(\sigma)$ should be the simplex whose vertices are the vertices of K_k included in σ. We use an induction. Let $l > 0$ be an integer, and assume that if $\sigma \in K_k$ is contained in graph $f_{k1} \cup \cdots \cup$ graph f_{kl-1}, then σ is a simplex. Then it suffices to construct a 𝔜-homeomorphism ψ of I^k of the same form as φ such that if $\sigma \in K_k$ is contained in graph f_{kl}, then $\psi(\sigma)$ is a simplex, and $\psi = \mathrm{id}$ on the vertices of K_k and on $I^k \cap (\text{graph } f_{k1} \cup \cdots \cup \text{graph } f_{kl-1})$. If $f_{kl}(x'_0) \in I \cap \{f_{k1}(x'_0), \ldots, f_{kl-1}(x'_0)\}$, $x'_0 \in I^{k-1}$, then we set $\psi = \mathrm{id}$ on $x'_0 \times I$. For $\sigma \in K_k$ included in graph f_{kl}, $\psi|_\sigma$ is uniquely decided because $p_{k-1}|_\sigma$ is injective. Hence $\psi|_{I^k \cap \text{graph } f_{kl}}$ is defined, and we need to extend ψ to I^k. For a connected component V of $p_{k-1}(I^{k-1} \times I° \cap \text{graph } f_{kl} - (\text{graph } f_{k1} \cup \cdots \cup \text{graph } f_{kl-1}))$, define 𝔜-functions $f_1, .., f_4$ on V by

$$f_2 = f_{kl}|_V, \qquad f_3(x) = \psi'(x, f_2(x)),$$

$$f_1(x) = \max\{-a,\ f_{kl'}(x)\colon 0 < l' < l,\ f_{kl'}(x) < f_2(x),\ f_{kl'}(x) < f_3(x)\}, \text{ and}$$

$$f_4(x) = \min\{a,\ f_{kl'}(x)\colon 0 < l' < l,\ f_{kl'}(x) > f_2(x),\ f_{kl'}(x) > f_3(x)\}.$$

Then we can reduce the problem to the following one.

Let V be an open 𝔜-subset of I^{k-1}, which is not necessarily open in $\mathbf{R}^{k-1}$, and let $f_1 \le f_i \le f_4$, $i = 2, 3$, be 𝔜-functions on $\overline{V}$ such that $f_1 < f_i < f_4$, $i = 2, 3$, on V and $f_2 = f_3$ on $\overline{V} - V$. Then there exists a 𝔜-homeomorphism ψ of $\overline{V} \times I$ of the same form as φ such that

$$\psi'(x', f_2(x')) = f_3(x') \quad \text{for} \quad x' \in \overline{V} \text{ and}$$

$$\psi'(x', x_k) = x_k \quad \text{if} \quad x_k \le f_1(x'),\ x_k \ge f_4(x') \text{ or } x' \in \overline{V} - V.$$

Consider the case $f_2 < f_3$ on V. Let m be a sufficiently large integer. Let f_5 and f_6 be the 𝔜-functions on $\overline{V}$ such that

$$f_5 \le f_2 \le f_3 \le f_6, \quad \text{and} \quad m(f_2 - f_5) = f_3 - f_2 = m(f_6 - f_3).$$

Then, since a $\mathfrak{Y}$-set is always locally semilinear, $f_1 < f_5 < f_6 < f_4$ on V, and hence we can replace f_1 with f_5 and f_4 with f_6. For such f_5, f_2, f_3 and f_6, in the same way as the above ρ, it is easy to construct ψ so that the above conditions on ψ are satisfied.

If the equality $f_2 < f_3$ or $f_3 < f_2$ does not hold globally, consider (f_2, f_7) and (f_3, f_7) in place of (f_2, f_3), where f_7 is the $\mathfrak{Y}$-function defined by

$$f_7(x) = \max\{f_2(x), f_3(x)\} + \inf_{(x_1^0, \dots, x_{k-1}^0) \in \overline{V} - V} \max_{i=1,\dots,k-1} c|x_i - x_i^0|$$

$$\text{for} \quad x = (x_1, \dots, x_{k-1}) \in \overline{V},$$

for a sufficiently small positive number c. Then (f_2, f_7) and (f_3, f_7) satisfy the above conditions on (f_2, f_3), the inequalities $f_2 < f_7$ and $f_3 < f_7$ hold on V, and hence there exists ψ. Thus we have finished the induction step in the construction of φ, and we can assume each $\sigma \in K_k$ contained in graph $f_{k1} \cup \cdots$ is a rational simplex, which implies that K_k is a rational cell complex.

Replace K_k with its simplicial subdivision K_k' without new vertices (I.3.12), K_{k+1} with $\{\sigma \cap p_k^{-1}(\sigma') : \sigma \in K_{k+1}, \sigma' \in K_k'\}$, and so on, which completes the induction step on K_k. Thus X admits a C^0 $\mathfrak{Y}$-triangulation.

Noncompact case for $\mathfrak{Y}$. We can assume $X = \{(x_1, \dots, x_n) \in \mathbf{R}_+ \times \mathbf{R}^{n-1} : |x_2| \leq x_1, \dots, |x_n| \leq x_1\}$ because $\mathbf{R}^{n-1}$ is $\mathfrak{Y}$-homeomorphic to bdry X, and by V.1.1 we have a usual $\mathfrak{Y}$-cell decomposition K of X compatible with the given family of $\mathfrak{Y}$-sets. It suffices to find a $\mathfrak{Y}$-homeomorphism τ of X of the form $\tau(x_1, \dots, x_n) = (\tau_1(x_1), \tau_2(x_1, x_2), \dots, \tau_n(x_1, \dots, x_n))$ such that $\tau(K)$ is a usual rational cell complex.

The idea of the proof is the same as in the compact case. Let $p_{n-1}, \dots, p_1$ be the same as above, and let $K_n \xrightarrow{p_{n-1}} \cdots \xrightarrow{p_1} K_1$ denote a sequence of cell maps such that K_n is a $\mathfrak{Y}$-subdivision of K and each K_k is a $\mathfrak{Y}$-cell decomposition of $X_k = \{(x_1, \dots, x_k) \in \mathbf{R}_+ \times \mathbf{R}^{k-1} : |x_2| \leq x_1, \dots, |x_k| \leq x_1\}$. Here each K_k cannot be necessarily described by a finite number of global linear $\mathfrak{Y}$-functions because X is not compact. But by the proof of V.1.1 we can assume that each K_k is compatible with $\mathbf{Z} \times \mathbf{R}^{k-1}$ and for each $j > 0 \in \mathbf{Z}$, there exist a finite number of linear $\mathfrak{Y}$-functions $f_{kj1}, f_{kj2}, \dots$ on $X_{k-1} \cap \{j-1 \leq x_1 \leq j\}$ such that

$$K_k|_{X_k \cap \{j-1 \leq x_1 \leq j\}} = \{(x', x_k) \in X_k \cap \sigma \times \mathbf{R} : x_k *_1 f_{kj1}(x'),$$

$$x_k *_2 f_{kj2}(x'), \dots\}_{\substack{\sigma \in K_{k-1}|_{X_{k-1} \cap \{j-1 \leq x_1 \leq j\}} \\ *_1, *_2, \dots \in \{\geq, =, \leq\}}}. \tag{$**$}$$

We want to move all the vertices to rational points. But we cannot do this all at once as in the compact case because the image of K_n^0 under the projection $\mathbf{R}^n \to \mathbf{R}^{n-1}$ which forgets the first factor is not necessarily discrete. We carry this out in two steps. First let ρ_1 be a $\mathfrak{Y}$-homeomorphism of $\mathbf{R}$ such that $\rho_1 = \mathrm{id}$ on $\mathbf{Z}$ and $\rho_1(K_1^0) \subset \mathbf{Q}$. Replace $K_n, \dots, K_1$ with $\rho_1 \times \cdots \times \rho_1(K_n), \dots, \rho_1(K_1)$ and keep the notation. Then the first coordinates of the vertices of $K_n, \dots, K_1$ are all rational. Second, by the following note we have a $\mathfrak{Y}$-function ρ_2 on $\mathbf{R}^2$ such that the map

$$X \ni x = (x_1, \dots, x_n) \longrightarrow \rho(x) = (x_1, \rho_2(x_1, x_2), \dots, \rho_2(x_1, x_n)) \in \mathbf{R}^n$$

is a homeomorphism onto X, $\rho(K_n^0) \subset \mathbf{Q}^n$, and $\rho|_{\partial X} = \mathrm{id}$. Replace, once more, $K_n, \dots, K_1$ with $\rho(K_n), p_{n-1} \circ \rho(K_n), \dots, p_2 \circ \cdots \circ p_{n-1} \circ \rho(K_n), K_1$ and keep the notation. Then all the vertices are rational.

Note. Any $\mathfrak{Y}$-homeomorphism ρ of $\mathbf{R}$ is $\mathfrak{Y}$-isotopic to the identity if the set $\{\rho(x) \neq x\}$ is bounded. Indeed, a $\mathfrak{Y}$-isotopy ρ_t, $0 \leq t \leq 1$, of ρ is defined to be

$$\rho_t(x) = \begin{cases} \rho(x) & \text{if} \quad |x - \rho(x)| \leq N(1-t) \\ x + N(1-t) & \text{if} \quad \rho(x) > x + N(1-t) \\ x - N(1-t) & \text{if} \quad \rho(x) < x - N(1-t) \end{cases}$$

for an integer N so large that $|x - \rho(x)| < N$ for any $x \in \mathbf{R}$. This holds also for $\mathfrak{Y}$-parameterized ρ.

Let $1 < k \leq n$ be an integer. We modify K_k to be a rational simplicial complex by induction on k. Assume $K_1, \dots, K_{k-1}$ are rational simplicial complexes compatible with $\mathbf{Z}, \dots, \mathbf{Z} \times \mathbf{R}^{k-2}$ respectively; $K_k, \dots, K_n$ are families of C^0 $\mathfrak{Y}$-submanifolds possibly with boundary of $X_k, \dots, X_n$ respectively; all the vertices of $K_k, \dots, K_n$ are rational; for each $\sigma \in K_l$, $k \leq l \leq n$, $p_{l-1}(\sigma)$ is an element of K_{l-1}; $\{\sigma^\circ : \sigma \in K_l\}$ is a stratification of X_l compatible with $\mathbf{Z} \times \mathbf{R}^{l-1}$; for any $\sigma \neq \sigma' \in K_l$ of dimension l', any $l' + 1$ vertices in σ do not coincide with any $l' + 1$ ones in σ'; and there are a finite number of $\mathfrak{Y}$-functions $f_{lj1}, f_{lj2}, \dots$ on $X_{l-1} \cap \{j - 1 \leq x_1 \leq j\}$ for each $j > 0 \in \mathbf{Z}$ which describe $K_l|_{X_k \cap \{j-1 \leq x_1 \leq j\}}$ as $(**)$.

We construct a $\mathfrak{Y}$-homeomorphism φ of X_k of the form:

$$\varphi(x', x_k) = (x', \varphi'(x', x_k)) \quad \text{for} \quad (x', x_k) \in X_k$$

such that $\varphi(K_k)$ is a cell complex and $\varphi = \mathrm{id}$ on the vertices of K_k. We cannot do this all at the same time as in the compact case because $f_{kj1}, f_{kj2}, \dots$ are not globally defined on X_{k-1}. First consider K_k only on $X_k \cap \mathbf{Z} \times \mathbf{R}^{k-1}$. As before, we have a $\mathfrak{Y}$-homeomorphism φ_1 of $X_k \cap \mathbf{Z} \times \mathbf{R}^{k-1}$ of the same form as φ such that $\varphi_1(K_k|_{X_k \cap \mathbf{Z} \times \mathbf{R}^{k-1}})$ is a cell complex and $\varphi_1 = \mathrm{id}$ on the vertices of $K_k|_{X_k \cap \mathbf{Z} \times \mathbf{R}^{k-1}}$. By the note we can extend φ_1 to a C^0 $\mathfrak{Y}$-imbedding $\tilde{\varphi}_1 : X_k \to \mathbf{R}^k$ of that form so that $\tilde{\varphi}_1 = \mathrm{id}$ outside a sufficiently small neighborhood of $X_k \cap \mathbf{Z} \times \mathbf{R}^{k-1}$ in X_k and hence on the vertices of K_k. Consider $\tilde{\varphi}_1(K_k)$ in place of K_k. Then we can assume $K_k|_{X_k \cap \mathbf{Z} \times \mathbf{R}^{k-1}}$ is a cell complex. (Here X_k is not the former X_k. But it comes back after the next step.) Hence we can treat $K_k|_{X_k \cap \{j-1 \leq x_1 \leq j\}}$, $j = 1, 2, \ldots$, separately as in the proof in the compact case, and we obtain the required φ, which proves the noncompact case.

Case of $\mathfrak{Y}_0$. As above, setting

$$X_n = \{(x_1, \ldots, x_n) \in \mathbf{R}_+ \times \mathbf{R}^{n-1} : |x_2| \leq x_1, \ldots, |x_n| \leq x_1\}$$

and letting C be a $\mathfrak{Y}_0$-cell decomposition of X_n, we need only construct a $\mathfrak{Y}_0$-homeomorphism τ of X_n of the form $\tau(x_1, \ldots, x_n) = (\tau_1(x_1), \tau_2(x_1, x_2), \ldots, \tau_n(x_1, \ldots, x_n))$ such that $\tau(C)$ is a rational cell complex. We can suppose C is compatible with $1 \times \mathbf{R}^{n-1}$ and any vertex of C is contained in $[0, 1] \times \mathbf{R}^{n-1}$. Furthermore, we assume there exist $\mathfrak{Y}_0$-cell decompositions C_k of X_k, $k = 1, \ldots, n$, such that $C_n = C$ and $C_n \xrightarrow{p_{n-1}} \cdots \xrightarrow{p_1} C_1$ is a sequence of cell maps, where $p_{n-1}, \ldots, p_1$ are the same as above. But we cannot expect each C_k to be described by a finite number of linear $\mathfrak{Y}_0$-functions on X_{k-1} as $(*)$ because for a usual $\mathfrak{Y}_0$-cell in $\mathbf{R}^n$, the smallest linear space which contains it is not necessarily a $\mathfrak{Y}_0$-set. We solve this problem as follows.

For each $k > 1$ there exist a finite number of linear $\mathfrak{Y}_0$-functions $f_{k1}, \ldots, f_{kl}$ on $X_{k-1} \cap [0, 1] \times \mathbf{R}^{k-2}$ and $f_{kl+1}, \ldots, f_{kl'}$ on X_{k-1} such that

$$D_k = \{(x', x_k) \in X_k \cap \sigma$$
$$\times \mathbf{R} : x_k *_1 f_{k1}(x'), \ldots, x_k *_{l'} f_{kl'}(x')\}_{\substack{\sigma \in C_{k-1}, \sigma \subset [0,1] \times \mathbf{R}^{k-2} \\ *_1, \ldots, *_{l'} \in \{\geq, =, \leq\}}} \quad \text{and}$$

$$E_k = \{(x', x_k) \in X_k \cap \sigma$$
$$\times \mathbf{R} : x_k *_{l+1} f_{kl+1}(x'), \ldots, x_k *_{l'} f_{kl'}(x')\}_{\substack{\sigma \in C_{k-1}, \sigma \subset [1,\infty[\times \mathbf{R}^{k-2} \\ *_{l+1}, \ldots, *_{l'} \in \{\geq, =, \leq\}}}$$

are $\mathfrak{Y}_0$-cell subdivisions of $C_k|_{X_k \cap [0,1] \times \mathbf{R}^{k-1}}$ and $C_k|_{X_k \cap [1,\infty[\times \mathbf{R}^{k-1}}$ respectively. Let c be an integer, and replace f_{ki}, $i \leq l$, with the functions on X_{k-1}:

$$\begin{cases} f_{ki} & \text{on} \quad X_{k-1} \cap [0, 1] \times \mathbf{R}^{k-2} \\ \tilde{f}_{ki} + c(x_k - 1) & \text{on} \quad X_{k-1} \cap [1, \infty[\times \mathbf{R}^{k-2}, \end{cases}$$

where $\tilde{f}_{ki}$ are the linear extensions of f_{ki}. Choose c so large that the graph of each $\tilde{f}_{ki}$ on $X_{k-1} \cap [2, \infty[\times \mathbf{R}^{k-2}$, $i \leq l$, is not contained in X_k. Then

$$\{(x', x_k) \in X_k \cap \sigma \times \mathbf{R} : x_k *_1 f_{k1}(x'), \ldots, x_k *_{l'} f_{kl'}(x')\}_{\substack{\sigma \in C_{k-1} \\ *_1, \ldots, *_{l'} \in \{\geq, =, \leq\}}} \quad (***)$$

is a $\mathfrak{Y}_0$-cell subdivision of C_k, and coincides with E_k on $X_k \cap [2, \infty[\times \mathbf{R}^{k-1}$. Note that each f_{ki}, $i \leq l$, is semilinear but not always of class $\mathfrak{Y}_0$ on $X_{k-1} \cap [2, \infty[\times \mathbf{R}^{k-2}$. Nevertheless, we call it a $\mathfrak{Y}_0$-function for simplicity. This does not cause trouble because we do not use the function on $X_{k-1} \cap [2, \infty[\times \mathbf{R}^{k-2}$. So we assume C_k is described as $(***)$.

The proof is similar to the above but needs, moreover, the description of a cell in §I.3, by which we see the following. For unbounded $\sigma \in C_n$ there exist uniquely compact cells σ' and σ'' in $1 \times \mathbf{R}^{n-1}$ such that

$$\sigma = \sigma(\sigma', \sigma'') = \sigma' + \mathbf{R}_+ \sigma'' = \{a + tb : a \in \sigma', \ b \in \sigma'', \ t \in \mathbf{R}_+\}.$$

Here σ' is the cell spanned by the vertices of σ and coincides with $\sigma \cap 1 \times \mathbf{R}^{n-1}$, and

$$\mathbf{R}_+ \sigma'' = \{x \in \mathbf{R}^n : \sigma' + x \subset \sigma\}.$$

Hence $\sigma' \in C_n$, and σ'' and $\mathbf{R}_+ \sigma''$ are of class $\mathfrak{Y}_0$. Let A_n denote the vertices of such σ'''s and set

$$C_n(2) = \{\sigma \cap 2 \times \mathbf{R}^{n-1} : \sigma \in C_n\}.$$

Note that for each $v \in C_n(2)^0$, there exist uniquely $v' \in C_n^0$ and $v'' \in A_n$ such that $v = v' + v''$. Let ρ be a $\mathfrak{Y}_0$-homeomorphism of $\mathbf{R}$ such that $\rho = \mathrm{id}$ at 0 and outside $[-1, 1]$, $\rho \times \cdots \times \rho(v)$ is rational for any $v = (v_1, \ldots, v_n) \in C_n^0 \cup A_n$, and $\rho(v_i) = v_i$ if v_i is rational. Choose ρ sufficiently close to the identity. Then we have a $\mathfrak{Y}_0$-homeomorphism ρ_1 of $\mathbf{R}$ such that $\rho_1 = \mathrm{id}$ at 0, 1 and outside $[-2, 2]$, and

$$\rho_1 \times \cdots \times \rho_1(v) = \begin{cases} \rho \times \cdots \times \rho(v) & \text{for } v \in C_n^0 \\ \rho \times \cdots \times \rho(v') + \rho \times \cdots \times \rho(v'') & \text{for } v \in C_n(2)^0 \text{ with} \\ & v = v' + v'', \ v' \in C_n^0, \ v'' \in A_n. \end{cases}$$

It follows that $\rho_1 \times \cdots \times \rho_1(v)$ is rational for $v \in C_n^0 \cup A_n$. Note that $\rho_1 \times \cdots \times \rho_1(X_n) = X_n$. Replace $C_n, \ldots, C_1$ with $\{\rho_1 \times \cdots \times \rho_1(\sigma) : \sigma \in C_n\}, \ldots, \{\rho_1(\sigma) : \sigma \in C_1\}$ respectively, and keep the notation $C_n, \ldots, C_1$. We apply ρ_1 but not ρ to obtain the following condition (7). Fixing the vertices (=elements of dimension 0) of $C_1, \ldots, C_n$, we modify $C_2, \ldots, C_n$ by induction. We will prove the following statement.

Let $1 < k \leq n$. Assume the following eight conditions. Then, keeping these conditions, we can modify C_k so that it is a rational cell complex.

(1) $C_1, \ldots, C_{k-1}$ are rational cell decompositions of $X_1 = \mathbf{R}_+, \ldots, X_{k-1}$ respectively, with the unique expression property of $(***)$ in §I.3.

(2) $C_k, \ldots, C_n$ are families of C^0 $\mathfrak{Y}_0$-submanifolds possibly with boundary of $X_k, \ldots, X_n$ respectively, and so are $C_k(2), \ldots, C_n(2)$, where $C_k(2), \ldots$ are defined as $C_n(2)$.

(3) All the vertices of C_l and $C_l(2)$, $k < l \le n$, are rational. We denote by C_l^0 the vertices of C_l.

(4) For each $\sigma \in C_l$, $k \le l \le n$, $p_{l-1}(\sigma)$ is an element of C_{l-1}.

(5) $\{\sigma^\circ : \sigma \in C_l\}$ is a stratification of X_l compatible with $1 \times \mathbf{R}^{l-1}$.

(6) For any bounded $\sigma \ne \sigma' \in C_l$ of dimension l', any $l' + 1$ vertices in σ do not coincide with any $l' + 1$ ones in σ'.

Let $\sigma, \sigma' \in C_l$ be unbounded and of dimension l' such that $p_{l-1}(\sigma) = p_{l-1}(\sigma')$ and $p_{l-1}|_\sigma$ and $p_{l-1}|_{\sigma'}$ are imbeddings. Let $\tilde{\sigma} = \sigma(\sigma_1, \sigma_2)$ be an unbounded l'-cell with the unique expression property such that σ_1 and σ_2 are simplexes in $1 \times \mathbf{R}^{l-1}$, σ_1 is spanned by vertices of $\sigma \cap C_l^0$ and $\tilde{\sigma} \cap 2 \times \mathbf{R}^{l-1}$ is spanned by vertices of $\sigma \cap C_l(2)^0$, and let $\tilde{\sigma}' = \sigma(\sigma_1', \sigma_2')$ be given by σ' in the same way as above so that $p_{l-1}(\sigma_1) = p_{l-1}(\sigma_1')$ and $p_{l-1}(\sigma_2) = p_{l-1}(\sigma_2')$. If $p_{l-1}|_A$ is an imbedding for a subset A of X_l, then let f_A denote the function on $p_{l-1}(A)$ whose graph is A.

(7) The inequality $f_{\tilde{\sigma}} < f_{\tilde{\sigma}'}$ holds on $p_{l-1}(\tilde{\sigma}^\circ)$ if and only if $f_\sigma < f_{\sigma'}$ on $p_{l-1}(\sigma^\circ)$, and $f_{\tilde{\sigma}'} - f_{\tilde{\sigma}}$ is bounded on $p_{l-1}(\tilde{\sigma})$ if and only if so is $f_{\sigma'} - f_\sigma$ on $p_{l-1}(\sigma)$.

(8) There are a finite number of semilinear $\mathfrak{Y}_0$-functions $f_{l1}, f_{l2}, \ldots$ on X_{l-1} which describe C_l as (***) in this proof.

Note that the properties that (i) $f_{\tilde{\sigma}} < f_{\tilde{\sigma}'}$ on $p_{l-1}(\tilde{\sigma}^\circ)$ and (ii) $f_{\tilde{\sigma}'} - f_{\tilde{\sigma}}$ is bounded on $p_{l-1}(\tilde{\sigma})$ are equivalent to the following (i)$'$ and (ii)$''$ respectively, for σ and σ' in (7). Let $\tilde{\sigma} = \sigma(\sigma_1, \sigma_2)$ and $\tilde{\sigma}' = \sigma(\sigma_1', \sigma_2')$. Then

(i)$'$ $f_{\sigma_1} < f_{\sigma_1'}$ on $p_{l-1}(\sigma_1^\circ)$ and $f_{\sigma_2} \le f_{\sigma_2'}$ on $p_{l-1}(\sigma_2)$, and

(ii)$''$ $\sigma_2 = \sigma_2'$.

Only condition (7) is new. We see by this note and the form of ρ_1 that it is satisfied in the case $k = 2$, and hence (1), ..., (8) are satisfied for $k = 2$. Therefore, it suffices to prove the above statement. Its proof is similar to the proof in the compact case. In the compact case, each $\sigma \in K_k$ such that $p_{k-1}|_\sigma$ is an imbedding became the cell (simplex, in consequence) whose vertices are $\sigma \cap K_k^0$ after its modification. This was possible because $p_{k-1}(\sigma)$ is a simplex and hence $\dim \sigma = \dim$ (the cell). The same arguments work in the $\mathfrak{Y}_0$ case because of the unique expression property (***) in §I.3 of each cell of C_{k-1}. (Remember that an unbounded cell $\sigma = \sigma(\sigma_1, \sigma_2)$ has the property if and only if σ_1 and σ_2 are simplexes, and $\dim \sigma = \dim \sigma_1 + \dim \sigma_2 + 1$.) After modifying each cell of C_k, for (1), we need to subdivide C_k to cells with the unique expression property without introducing new vertices, which is possible as shown in §I.3. We omit the details.

The map case and the remark after V.1.3 are clear by the above constructions. $\qquad\square$

If only the closedness condition on X and X' in V.1.3 is removed, then there exist rational simplicial (cell) complexes K and K', unions of open simplexes (resp., cells) Y of K and Y' of K', and $\mathfrak{Y}$-homeomorphisms $\pi \colon Y \to X$ and $\pi' \colon Y' \to X'$ such that $\pi'^{-1} \circ f \circ \pi$ can be extended to a simplicial (resp., cell) map from K to K'. This follows when we apply the above proof to the closure of the graph of f (see the proof of the next corollary for the details).

Note also that the above facts hold true for locally semilinear sets and a locally semilinear, map even if some $\mathfrak{Y}$-set is not locally semilinear, because the family of locally semilinear $\mathfrak{Y}$-sets satisfies the axioms of $\mathfrak{Y}$.

An advantage of the present case is the following corollary (cf. II.7.7).

Corollary V.1.4 (Independence of $\mathfrak{Y}$-equivalence). *Let $X, X_1 \subset \mathbf{R}^m$ and $Y, Y_1 \subset \mathbf{R}^n$ be locally semilinear $\mathfrak{Y}$-sets, and let $f \colon X \to Y$ and $f_1 \colon X_1 \to Y_1$ be locally semilinear $\mathfrak{Y}$-maps. Assume Axiom* (vi). *Moreover, suppose either* (a) *X, Y, X_1 and Y_1 are bounded in their respective ambient Euclidean spaces,* (b) *$\mathfrak{Y}$ satisfies Axiom* (v), *X, Y, X_1 and Y_1 are closed, and f and f_1 are proper, or* (c) *$\mathfrak{Y}$ is $\mathfrak{Y}_0$. If f and f_1 are locally semilinearly (semilinearly in the case* (c)) *R-L equivalent, then they are locally semilinearly (semilinearly, resp.) R-L $\mathfrak{Y}$-equivalent.*

In the case $Y = Y_1 = \mathbf{R}$, we weaken condition (a) *to the condition that $X, X_1, f(X)$ and $f_1(X_1)$ are bounded. If f and f_1 are locally semilinearly (semilinearly in the case* (c)) *equivalent, then they are locally semilinearly (semilinearly, resp.) $\mathfrak{Y}$-equivalent.*

We obtain the following $\mathfrak{Y}$-Hauptvermutung when we set $X = Y$, $X_1 = Y_1$ and $f = f_1 = \mathrm{id}$ in V.1.4. Let $X, X_1 \subset \mathbf{R}^m$ be locally semilinear $\mathfrak{Y}$-sets. Assume that either (a) they are bounded, (b) they are closed and Axiom (v) is satisfied or (c) $\mathfrak{Y} = \mathfrak{Y}_0$. If X and X_1 are locally semilinearly (semilinearly in the case (c)) homeomorphic, they are locally semilinearly (semilinearly, resp.) $\mathfrak{Y}$-homeomorphic.

Note also that if the locally semilinear (semilinear) homeomorphisms $\pi \colon X \to X_1$ and $\tau \colon Y \to Y_1$ of equivalence in V.1.4 carry given locally finite families of locally semilinear (semilinear) $\mathfrak{Y}$-sets to other families, then we can choose the modifications of π and τ so that this property is kept. This is clear by the following proof.

Proof of V.1.4. We can assume any $\mathfrak{Y}$-set is locally semilinear.

Proof of the former statement. We treat only the case (a). The other cases are proved in the same way. Note only that in the case (c), we need to consider not only the vertices of cell complexes but also the vertices of σ_2 of cells $\sigma = \sigma(\sigma_1, \sigma_2)$ in cell complexes with the unique expression property as in the above proof. First we reduce the problem to the case $\mathfrak{Y} = \{$rational locally semilinear sets}. We can replace X with graph f and f with the restriction to the graph of the projection $p\colon \mathbf{R}^m \times \mathbf{R}^n \to \mathbf{R}^n$. Hence assume $X \subset \mathbf{R}^m \times \mathbf{R}^n$ and $f = p|_X$. By the above proof of V.1.3 we have a $\mathfrak{Y}$-homeomorphism φ of $\mathbf{R}^m \times \mathbf{R}^n$ of the form $\varphi(x, x') = (\varphi'(x, x'), \varphi''(x'))$ for $(x, x') \in \mathbf{R}^m \times \mathbf{R}^n$ and finite rational simplicial decompositions K of $\varphi(\overline{X})$ and L of $\varphi''(\overline{Y})$ such that K and L are compatible with $\varphi(X)$ and $\varphi''(Y)$ respectively, and $p|_{\overline{X}}\colon K \to L$ is a simplicial map. Hence we can assume $\overline{X}$ and $\overline{Y}$ admit the decompositions K and L compatible with X and Y respectively. By the same reason we suppose $X_1 \subset \mathbf{R}^m \times \mathbf{R}^n$, $f_1 = p|_{X_1}$, and that $\overline{X_1}$ and $\overline{Y_1}$ admit finite rational simplicial decompositions K_1 and L_1 compatible with X_1 and Y_1 respectively, such that $p|_{\overline{X_1}}\colon K_1 \to L_1$ is simplicial. This reduces the problem to the case $\mathfrak{Y} = \{\text{rational}\}$.

Let $\pi\colon X \to X_1$ and $\tau\colon Y \to Y_1$ be semilinear homeomorphisms such that $f_1 \circ \pi = \tau \circ f$. Note that π is of the form $(\pi_1(x, x'), \tau(x'))$ for $(x, x') \in X \subset \mathbf{R}^m \times \mathbf{R}^n$. Let us consider the case where $f(X)$ is dense in Y, and π and τ can be extended to semilinear maps $\overline{\pi}\colon \overline{X} \to \overline{X_1}$ and $\overline{\tau}\colon \overline{Y} \to \overline{Y_1}$. By the above arguments we can assume $\{\pi(X \cap \sigma^\circ)\colon \sigma \in K\}$ and $\{\tau(Y \cap \sigma^\circ)\colon \sigma \in L\}$ are compatible with K_1 and L_1 respectively. It follows that $\{\pi(\sigma^\circ)\colon \sigma \in K\}$ and $\{\tau(\sigma^\circ)\colon \sigma \in L\}$ have the same property. It suffices to modify $\overline{\pi}$ to be rational keeping the form and the equalities $\pi(X) = X_1$ and $\tau(Y) = Y_1$. We proceed by induction on $k = \dim X$. Since the case of $k = 0$ is trivial, we assume there exists a $\mathfrak{Y}$-map $\pi^{k-1}\colon \overline{\pi}^{-1}(|K_1^{k-1}|) \to |K_1^{k-1}|$ such that for each $\sigma \in K_1^{k-1}$, $\pi^{k-1}|_{\overline{\pi}^{-1}(\sigma)}$ is both a $\mathfrak{Y}$-homeomorphism onto σ and an approximation of $\overline{\pi}|_{\overline{\pi}^{-1}(\sigma)}$ in the C^0 topology and π^{k-1} is linear on each simplex of $K|_{\overline{\pi}^{-1}(\sigma)}$. We need only construct a $\mathfrak{Y}$-homeomorphism $\pi_\sigma\colon \overline{\pi}^{-1}(\sigma) \to \sigma$ for each $\sigma \in K_1 - K_1^{k-1}$ which is an extension of $\pi^{k-1}|_{\overline{\pi}^{-1}(\partial\sigma)}$ and an approximation of $\overline{\pi}|_{\overline{\pi}^{-1}(\sigma)}$, and whose restriction to each simplex of $K|_{\overline{\pi}^{-1}(\sigma)}$ is linear. This is easy as follows. For each $v \in K^0 \cap \overline{\pi}^{-1}(\sigma^\circ)$, define $\pi_\sigma(v)$ to be a rational point in σ° near $\overline{\pi}(v)$, whose existence follows from rationality of σ, and extend π_σ to $\overline{\pi}^{-1}(\sigma)$ so that it is linear on each simplex of $K|_{\overline{\pi}^{-1}(\sigma)}$. Then π_σ fulfills all the requirements.

Assume $f(X)$ is not dense in Y and there exist the extensions $\overline{\pi}$ and $\overline{\tau}$. We proceed with the above arguments for $\pi\colon X \to X_1$, $\tau|_{f(X)}\colon f(X) \to f_1(X_1)$, K, K_1, $L|_{\overline{f(X)}}$ and $L_1|_{|\overline{f_1(X_1)}|}$. Then we have a $\mathfrak{Y}$-approximation $\tilde{\pi}\colon \overline{X} \to \overline{X_1}$ of $\overline{\pi}$ of the form $(\tilde{\pi}_1(x, x'), \tilde{\pi}_2(x'))$ for $(x, x') \in \mathbf{R}^m \times \mathbf{R}^n$ such that $\tilde{\pi}|_X$ is a homeomorphism onto X_1, $\tilde{\pi}_2|_{f(X)}$ is a homeomorphism onto $f_1(X_1)$, for each $\sigma \in K_1$ $\tilde{\pi}|_{\overline{\pi}^{-1}(\sigma)}$ is a homeomorphism onto σ, and $\tilde{\pi}$ is linear on each simplex of K. It follows that $f_1 \circ \tilde{\pi} = \tilde{\pi}_2 \circ f$ and $\tilde{\pi}_2$ is a map from $\overline{f(X)}$ to $\overline{f_1(X_1)}$. Hence it suffices to extend $\tilde{\pi}_2$ to $\tilde{\tau}\colon \overline{Y} \to \overline{Y_1}$ so that $\tilde{\tau}|_Y$ is a homeomorphism onto Y_1. But this is clearly possible by the above arguments.

Finally, consider the case where π and τ cannot be extended to $\overline{X}$ and $\overline{Y}$ respectively. As in the first step, triangulate rationally the semilinear map $f_0 \colon \operatorname{graph} \pi \to \operatorname{graph} \tau$ defined by $f_0(x, \pi(x)) = (f(x), f_1 \circ \pi(x))$ for $x \in X$. Then we can assume there exist finite rational simplicial complexes K_0 and L_0, unions X_0 and Y_0 of open simplexes of K_0 and L_0 respectively, a $\mathfrak{Y}$-map $f_0 \colon X_0 \to Y_0$ and semilinear homeomorphisms $\pi_0 \colon X_0 \to X$, $\pi_{01} \colon X_0 \to X_1$, $\tau_0 \colon Y_0 \to Y$ and $\tau_{01} \colon Y_0 \to Y_1$ such that $f_0, \pi_0, \pi_{01}, \tau_0$ and τ_{01} can be extended to the closures of the respective domains as semilinear maps, $f \circ \pi_0 = \tau_0 \circ f_0$, $f_1 \circ \pi_{01} = \tau_{01} \circ f_0$, the extension of f_0 is a simplicial map from K_0 to L_0, and $\{\pi_0(X_0 \cap \sigma^\circ) \colon \sigma \in K_0\}$, $\{\pi_{01}(X_0 \cap \sigma^\circ) \colon \sigma \in K_0\}$, $\{\tau_0(Y_0 \cap \sigma^\circ) \colon \sigma \in L_0\}$ and $\{\tau_{01}(Y_0 \cap \sigma^\circ) \colon \sigma \in L_0\}$ are compatible with K, K_1, L and L_1 respectively. The pairs (f_0, f) and (f_0, f_1) satisfy the conditions of the third step. Hence f_0 is R-L $\mathfrak{Y}$-equivalent to f and f_1. It follows that f and f_1 are R-L $\mathfrak{Y}$-equivalent.

Proof of the latter. We modify the above arguments as follows. In the first step (reduction to the rational case), we can choose $\mathfrak{Y}$-homeomorphisms $\varphi = (\varphi', \varphi'')$ and $\varphi_1 = (\varphi_1', \varphi_1'')$ of $\mathbf{R}^m \times \mathbf{R}$ so that $\varphi'' = \varphi_1''$, and

$$\varphi'' \circ f \circ \varphi^{-1} \colon \varphi(\overline{X}) \longrightarrow \varphi''(f(\overline{X})), \quad \text{and} \quad \varphi_1'' \circ f_1 \circ \varphi_1^{-1} \colon \varphi_1(\overline{X_1}) \longrightarrow \varphi_1''(f_1(\overline{X_1}))$$

admit rational simplicial decompositions compatible with $\varphi(X)$ and $\varphi_1(X_1)$ respectively. In the second, we obtain a rational homeomorphism $\pi \colon \varphi(X) \to \varphi_1(X_1)$ so that $\varphi_1'' \circ f_1 \circ \varphi_1^{-1} \circ \pi = \varphi'' \circ f \circ \varphi^{-1}$. Hence f and f_1 are $\mathfrak{Y}$-equivalent. The last step is changed as follows. Define $f_0 \colon \operatorname{graph} \pi \to \mathbf{R}$ to be $f_0(x, \pi(x)) = f(x)$. There exists a semilinear homeomorphism $\varphi_0 = (\varphi_0', \varphi_0'')$ of $(\mathbf{R}^m \times \mathbf{R}^m) \times \mathbf{R}$ such that $\varphi_0'' \circ f_0 \circ \varphi_0^{-1} \colon \varphi_0(\overline{\operatorname{graph} \pi}) \to \varphi_0''(f(\overline{X}))$ admits a rational simplicial decomposition compatible with $\varphi_0(\operatorname{graph} \pi)$. Then we construct rational homeomorphisms $\pi_0 \colon \varphi_0(\operatorname{graph} \pi) \to \varphi(X)$ and $\pi_{01} \colon \varphi_0(\operatorname{graph} \pi) \to \varphi_1(X_1)$ such that $\varphi'' \circ f \circ \varphi^{-1} \circ \pi_0 = \varphi_0'' \circ f_0 \circ \varphi_0^{-1}$ and $\varphi_1'' \circ f_1 \circ \varphi_1^{-1} \circ \pi_{01} = \varphi_0'' \circ f_0 \circ \varphi_0^{-1}$. Therefore, f and f_1 are $\mathfrak{Y}$-equivalent. $\square$

Remark V.1.5. If $\mathfrak{Y} = \mathfrak{Y}_0$, there happen phenomena which never occur in differential and PL topology, because of global finiteness.

For example, Thom's first and second isotopy lemmas do not necessarily hold in the $\mathfrak{Y}_0$-category (see II.6.1 and II.6.2). Let $\mathfrak{Y}_0 = \{\text{semilinear sets}\}$. Let $\{X_i\}$ be the canonical Whitney stratification of the set $\{(x_1, x_2) \in \mathbf{R}^2 \colon 0 < x_1,\ 0 \leq x_2 \leq 1 + x_1\}$, and let $f \colon \{X_i\} \to]0, \infty[$ be the restriction of the projection $\mathbf{R}^2 \ni (x_1, x_2) \to x_1 \in \mathbf{R}$. Then f satisfies the conditions of Thom's first isotopy lemma but is not $\mathfrak{Y}_0$-trivial as shown in II.4.3. But this does not mean that we cannot expect a good theory on $\mathfrak{Y}_0$. Indeed, a necessary and sufficient condition for a proper PL trivial $\mathfrak{Y}_0$-map $f \colon X \to Y$ to be

$\mathfrak{Y}_0$-trivial is the following. For simplicity of notation, assume $X \subset \mathbf{R}^m \times \mathbf{R}^n$, $Y \subset \mathbf{R}^n$ and f is the restriction of the projection $p \colon \mathbf{R}^m \times \mathbf{R}^n \to \mathbf{R}^n$. Let $f \colon \{X_i\} \to \{Y_j\}$ be a cell decomposition of f. Let $f(X_i) = Y_j$. Then the condition is that there exist descriptions $X_i = \sigma(\sigma_1, \sigma_2)$ and $Y_j = \sigma(\sigma_1', \sigma_2')$ such that $p|_{\sigma_2}$ is a homeomorphism onto σ_2'.

§V.2. Case where there exists a $\mathfrak{Y}$-set which is not locally semilinear

Assume there exists a $\mathfrak{Y}$-set which is not locally semilinear. We know only artificial examples of such $\mathfrak{Y}$ which are not $\mathfrak{X}$. Note that $\mathfrak{Y}$ is $\mathfrak{X}$ if and only if Axiom (vi) is satisfied and the set $\{(x, y, z) \in \mathbf{R}^3 \colon x = yz\}$ is a $\mathfrak{Y}$-set.

Example V.2.1. Such an example is the family of the images of all semialgebraic sets in $]0, \infty[^n$, $n = 0, 1, \ldots$, under the map $]0, \infty[^n \ni (x_1, \ldots, x_n) \to (\log x_1, \ldots, \log x_n) \in \mathbf{R}^n$. Clearly, Axioms (i)', (ii), (iii)$_0$, (iv)$_0$ and (vi) are satisfied, the set $\{x = yz\}$ is not a $\mathfrak{Y}$-set, and the image of $\{(x, y, z) \in]0, \infty[^3 \colon x = yz\}$ under the map $]0, \infty[^3 \ni (x, y, z) \to (\log x, \log y, \log z) \in \mathbf{R}^3$ is not locally semilinear.

This example suggests that any $\mathfrak{Y}$ is induced by some $\mathfrak{X}$. Indeed, we have the following.

Theorem V.2.2 (reduction of $\mathfrak{Y}$ to $\mathfrak{X}$). *Let r be a positive integer. Assume Axioms (v) and (vi), and that some $\mathfrak{Y}$-set is not locally semilinear and $\mathfrak{Y}$ is not $\mathfrak{X}$. Then there exists a C^r $\mathfrak{Y}$-function on $\mathbf{R}$ such that the derivative f' is bijective onto $\mathbf{R}$, and for any such f, the family of the images of $\mathfrak{Y}$-sets in $\mathbf{R}^n$, $n = 0, 1 \ldots$, under the map $\mathbf{R}^n \ni (x_1, \ldots, x_n) \to (f'(x_1), \ldots, f'(x_n)) \in \mathbf{R}^n$ is $\mathfrak{X}$.*

The following lemma is the key.

Lemma V.2.3. *Let r be a positive integer. Let f be a function on an open $\mathfrak{Y}$-subset I of $\mathbf{R}$ whose graph is a $\mathfrak{Y}$-set. There exists a discrete $\mathfrak{Y}$-subset A of I such that f is of class C^r on $I - A$ and f' is constant or strictly monotone on each connected component of $I - A$.*

Proof. For simplicity of notation, assume $I = \mathbf{R}$. The f is right differentiable at any point. (Here the right differential coefficient may take the value $\pm\infty$ at some points.) Indeed, if f is not so at a point, say, 0, then we have two rational numbers $a < b$ such that the closures of the sets $\{x > 0 \colon (f(x) - f(0)) \le ax\}$ and $\{x > 0 \colon (f(x) - f(0)) \ge bx\}$ both contain 0, which is impossible because the sets are disjoint and are $\mathfrak{Y}$-subsets of $\mathbf{R}$.

We can assume f is continuous on $\mathbf{R}$ for the following reason. The discontinuous point set N of f is a $\mathfrak{Y}$-set because $\mathbf{R} - N$ equals

$$\{x \in \mathbf{R} \colon \forall \varepsilon > 0 \ \exists \delta > 0 \ \forall t \in \mathbf{R} \ |t| < \delta \Rightarrow |f(x) - f(x+t)| < \varepsilon\}.$$

Moreover, N is discrete as follows. Assume N is not so. Now f is strictly monotone (say, increasing) on an open $\mathfrak{Y}$-interval $I' \subset N$ because $\mathbf{R}$ is the disjoint union of the $\mathfrak{Y}$-sets:

$$\{x \in \mathbf{R} \colon \exists \varepsilon > 0 \ \forall \delta > 0 \ \delta < \varepsilon \Rightarrow f(x+\delta) * f(x)\}, \quad * \in \{<, =, >\}.$$

Furthermore, if we shrink I', there exists $\varepsilon > 0$ such that $f(x) + \varepsilon < f(x')$ for any $x < x' \in I'$ because the set

$$\{(x,t) \in I' \times \mathbf{R} \colon t = \lim_{x' \to x+0} f(x') - f(x)\}$$

is a $\mathfrak{Y}$-set, and

$$I' = \bigcup_{t>0 \in \mathbf{Q}} \{x \in I' \colon t \le \lim_{x' \to x+0} f(x') - f(x)\}.$$

But this is clearly impossible.

Set

$$X_* = \{(x,y) \in \mathbf{R}^2 \colon \exists \varepsilon > 0 \ \forall \delta > 0 \ \delta < \varepsilon$$
$$\Rightarrow (f(x+\delta) - f(x)) * (f(y+\delta) - f(y))\} \quad \text{and}$$
$$Y_* = \{x \in \mathbf{R} \colon \exists \varepsilon > 0 \ \forall \delta > 0 \ \delta < \varepsilon \Rightarrow (x, x+\delta) \in X_*\}, \quad * \in \{<, =, >\}.$$

These are $\mathfrak{Y}$-sets, $\mathbf{R}^2$ is the disjoint union of $X_<$, $X_=$ and $X_>$, and $\mathbf{R}$ is the disjoint union of $Y_<$, $Y_=$ and $Y_>$. It suffices to consider the problem on each interval connected component of Y_*. (Note that the components are $\mathfrak{Y}$-sets.) Let a $\mathfrak{Y}$-interval $]a, b[$ be in $Y_<$. For each $x \in]a, b[$, set

$$\alpha(x) = \inf\{y \in]x, b[\colon (x, y) \in X_> \cup X_=\}$$

if this set is not empty, and set $\alpha(x) = \varnothing$ if empty. Note that $\alpha(x) = \min$(the same set) if the set is not empty, because if there is no minimum, then $(x, \alpha(x)) \in X_<$, but by definition of $Y_<$, $(\alpha(x), y) \in X_<$ for $y > \alpha(x)$ near $\alpha(x)$, and hence $(x, y) \in X_<$, which is a contradiction. Clearly α is a function on a $\mathfrak{Y}$-subset of $]a, b[$, and its graph is an $\mathfrak{Y}$-set. Moreover, we have the following.

(1) α is constant on each connected component of the continuous point set of α.

Proof of (1). Assume it is not so. Then there exists a $\mathfrak{Y}$-interval $]a', b'[$ in $]a, b[$ such that $\alpha|_{]a', b'[}$ is a $\mathfrak{Y}$-homeomorphism onto an interval in $\mathbf{R}$. Let $x_1 \in]a', b'[$ and $x_2 \in]x_1, b'[$ near x_1. By definition of $Y_<$ we have $(x_1, x_2) \in X_<$. It follows that $(x_2, \alpha(x_1)) \in X_>$. Hence $\alpha(x_1) \geq \alpha(x_2)$. Moreover, since α is injective on $]a', b'[$, $\alpha(x_1) > \alpha(x_2)$, which implies that α is strictly decreasing on $]a', b'[$. For $a' < x_1 < x_2 < b'$ we have

$$(x_1, \alpha(x_2)) \in X_<, \quad (x_1, \alpha(x_1)) \in X_> \cup X_=, \quad \text{and} \quad (x_2, \alpha(x_2)) \in X_> \cup X_=.$$

Let y_3 be a point of $]\alpha(x_2), \alpha(x_1)[$ near $\alpha(x_2)$, and let x_3 be the point of $]x_1, x_2[$ such that $\alpha(x_3) = y_3$. Then

$$(\alpha(x_2), \alpha(x_3)) \in X_< \quad \text{and} \quad (x_3, \alpha(x_3)) \in X_> \cup X_=.$$

Hence $(x_3, \alpha(x_2)) \in X_>$. Then by definition of α, $\alpha(x_3) \leq \alpha(x_2)$, which contradicts strict decreasingness of α on $]a', b'[$. (1) is proved.

Remove the discontinuous point set of α from $]a, b[$. Then we can assume for any $x < y \in]a, b[$, $(x, y) \in X_<$. This means that the right differential coefficient $D^+ f$ is increasing on $]a, b[$.

(2) f is downwards convex on $]a, b[$, i.e.,

$$f(x_1) < (x_1 - x_0)(f(x_2) - f(x_0))/(x_2 - x_0) + f(x_0) \text{ for } a < x_0 < x_1 < x_2 < b.$$

Proof of (2). Assume there are three points $x_0 < x_1 < x_2$ in $]a, b[$ which do not satisfy the above inequality. Set

$$g(x) = f(x) - (x - x_0)(f(x_2) - f(x_0))/(x_2 - x_0) - f(x_0) \quad \text{for} \quad x \in [x_0, x_2],$$

which is not necessarily of class $\mathfrak{Y}$. Then

$$g(x_0) = g(x_2) = 0, \quad g(x_1) \geq 0,$$

and g is continuous. Choose x_1 and a point x' in $[x_0, x_1]$ so that g takes the maximal value at x_1, $g|_{[x_0, x_1]}$ takes the minimal value at x', and $g > g(x')$ on $]x', x_1]$. This is possible because $\max g - \min g|_{[x_0, x_1]} \neq 0$. Then

$$\exists \varepsilon > 0 \; \forall \delta > 0 \; \delta < \varepsilon \Rightarrow$$
$$f(x' + \delta) - f(x') = g(x' + \delta) - g(x')$$
$$+ c > g(x_1 + \delta) - g(x_1) + c = f(x_1 + \delta) - f(x_1),$$
$$\text{where} \, c = \delta(f(x_2) - f(x_0))/(x_2 - x_0).$$

It follows that $(x', x_1) \in X_>$, which is a contradiction.

(3) D^+f is strictly increasing on $]a, b[$.

Proof of (3). Let $x_0 < x_1 < x_2 \in]a, b[$, and let g be the above function. By statement (2), $g < 0$ on $]x_0, x_2[$, and

$$D^+g(x_0) \leq (g(x_1) - g(x_0))/(x_1 - x_0) < 0.$$

Similarly we see $D^+g(x_1) \geq 0$. Hence

$$D^+g(x_0) < D^+g(x_1),$$

which implies, by definition of g,

$$D^+f(x_0) < D^+f(x_1).$$

(4) D^+f is right continuous on $]a, b[$.

Proof of (4). Let $x_0 < x_1 < x_2 \in]a, b[$. By statement (2),

$$D^+f(x_0) < (f(x_2) - f(x_0)/(x_2 - x_0) \quad \text{and}$$
$$D^+f(x_1) < (f(x_2) - f(x_1)/(x_2 - x_1).$$

Fix x_0 and x_2, let $\varepsilon > 0$, and choose x_1 sufficiently near x_0. Then

$$D^+f(x_1) < (f(x_2) - f(x_0))/(x_2 - x_0) + \varepsilon$$

because f is continuous. By statement (3),

$$D^+f(x_0) < D^+f(x_1).$$

By definition of right derivative,

$$(f(x_2) - f(x_0))/(x_2 - x_0) \longrightarrow D^+f(x_0) \quad \text{as} \quad x_2 \longrightarrow x_0.$$

From these three properties (4) follows, i.e.,

$$D^+f(x_1) \longrightarrow D^+f(x_0) \quad \text{as} \quad x_1 \longrightarrow x_0.$$

By the same arguments as above, D^+f is right continuous and strictly decreasing on each connected component of $Y_> - $ (a discrete 𝔜-set).

Consider $Y_=$. Assume a $\mathfrak{Y}$-interval $]a, b[\subset Y_=$. Set for each $x \in \,]a, b[$,

$$\beta(x) = \begin{cases} \inf\{y \in \,]x, b[: (x, y) \in X_> \cup X_<\} & \text{if this set is not empty} \\ \varnothing & \text{otherwise.} \end{cases}$$

We can prove that $\beta(x) = \min($the same set$)$ if the set is not empty, and β is constant on each connected component of the continuous point set in the same way as above. Remove the discontinuous point set of β from $]a, b[$. Then we can assume for any $x < y \in \,]a, b[$, $(x, y) \in X_=$.

(5) For $x_0 < x_2 \in \,]a, b[$,

$$f(x) = (x - x_0)(f(x_2) - f(x_0))/(x_2 - x_0) + f(x_0) \quad \text{for} \quad x \in [x_0, x_2].$$

Proof of (5). Assume this is not true. Define g as above. Then a contradiction follows from the same arguments as in the proof of statement (2).

By statement (5) the lemma holds for any connected component of $I - A$ which is included in $Y_=$. From now on we assume $\mathbf{R} = Y_<$ and for any $x < y \in \mathbf{R}$, $(x, y) \in X_<$, and hence $D^+ f$ is strictly increasing and right continuous on $\mathbf{R}$. We can prove the case where $\mathbf{R} = Y_>$ and for any $x < y \in \mathbf{R}$, $(x, y) \in X_>$ in the same way.

(6) $D^+ f$ is left continuous except at a discrete $\mathfrak{Y}$-set.

Proof of (6). We define the second partial order on $\mathbf{R}$ as follows. Set

$X^1_* = \{(x, y) \in \mathbf{R}^2 : \exists \varepsilon > 0 \; \forall \delta > 0 \; \delta < \varepsilon \Rightarrow$
$\quad (f(x + \delta) + f(x - \delta) - 2f(x)) * (f(y + \delta) + f(y - \delta) - 2f(y))\}$ and
$Y^1_* = \{x \in \mathbf{R} : \exists \varepsilon > 0 \; \forall \delta > 0 \; \delta < \varepsilon \Rightarrow (x, x + \delta) \in X^1_*\}, \quad * \in \{<, =, >\}.$

Then by the same reason as for X_* and Y_* we can assume for some $* \in \{<, =, >\}$,

$$\mathbf{R} = Y^1_*, \quad \text{and} \quad (x, y) \in X^1_* \text{ for } (x, y) \in \mathbf{R}^2 \text{ with } x < y.$$

Consider the case where " $*$ " is " $<$ " or " $=$ ". Note that

$$\lim_{\delta \to +0} \frac{f(x + \delta) + f(x - \delta) - 2f(x)}{\delta} = D^+ f(x) - \lim_{\delta \to +0} D^+ f(x - \delta).$$

Set

$$Z = \{(x, y) \in \mathbf{R}^2 : \exists \varepsilon > 0 \; \forall \delta > 0 \; \delta < \varepsilon \Rightarrow$$
$$(2f(x + \delta) + f(x - \delta) - 3f(x)) \geq (f(y + \delta) - f(y))\}.$$

Then Z is a $\mathfrak{Y}$-set,

$$\lim_{\delta \to +0} \frac{2f(x+\delta) + f(x-\delta) - 3f(x)}{\delta} = 2D^+f(x) - \lim_{\delta \to +0} D^+f(x-\delta),$$

and hence D^+f is left continuous at a point x if and only if

$$(x \times \,]x, \infty[) \cap Z = \varnothing, \quad x \times \,] -\infty, x[\subset Z.$$

Define a function φ on $\mathbf{R}$ so that for each $x \in \mathbf{R}$,

$$\varphi(x) = x + 3 \qquad\qquad\qquad\qquad \text{if } x \times \mathbf{R} \subset Z, \text{ and}$$
$$x \times \,]-\infty, \varphi(x)[\subset Z, \;\; (x \times \,]\varphi(x), \infty[) \cap Z = \varnothing \qquad \text{otherwise.}$$

Then graph φ is a $\mathfrak{Y}$-set, $x \le \varphi(x)$ for any $x \in \mathbf{R}$, and for any $(x, y) \in \mathbf{R}^2$ with $y < \varphi(x)$,

$$2D^+f(x) - \lim_{\delta \to +0} D^+f(x-\delta) \ge D^+f(y).$$

Assume the left discontinuous point set of D^+f is not discrete. Note that the set is a $\mathfrak{Y}$-set. For simplicity of notation, we suppose the subset contains $[0, 1]$ and

$$\varphi(x) > 2 \quad \text{on} \quad [0, 1].$$

We have

$$2D^+f(0) - \lim_{\delta \to +0} D^+f(-\delta) = D^+f(0) + (D^+f(0) - \lim_{\delta \to +0} D^+f(-\delta))$$
$$\le D^+f(0) + (D^+f(1) - \lim_{\delta \to +0} D^+f(1-\delta)) \qquad \text{because } (0,1) \in X^1_< \cup X^1_=$$
$$< D^+f(0) + D^+f(1) - D^+f(1/2) < D^+f(1) < D^+f(2).$$

But

$$2D^+f(0) - \lim_{\delta \to +0} D^+f(-\delta) \ge D^+f(2)$$

because $\varphi(0) > 2$. This is a contradiction. Hence the left discontinuous point set is discrete.

Consider the case where $\mathbf{R} = Y^1_>$ and $(x, y) \in X^1_>$ for $x < y \in \mathbf{R}$. We have

$$\lim_{\delta \to +0} D^+f(x-\delta) - D^+f(x) < \lim_{\delta \to +0} D^+f(y-\delta) - D^+f(y) \quad \text{for} \quad x < y \in \mathbf{R}.$$

Hence

$$2 \lim_{\delta \to +0} D^+ f(y - \delta) - D^+ f(y)$$
$$> \lim_{\delta \to +0} D^+ f(y - \delta) + \lim_{\delta \to +0} D^+ f(x - \delta) - D^+ f(x)$$
$$> \lim_{\delta \to +0} D^+ f(x - \delta) \quad \text{for} \quad x < y \in \mathbf{R}.$$

It follows that

$$2 \lim_{\delta \to +0} D^+ f(y - \delta) - D^+ f(y) \geq \lim_{\delta \to +0} D^+ f(y - \delta) \quad \text{for} \quad y \in \mathbf{R},$$

which implies that $D^+ f$ is left continuous.

The above arguments show also that $D^+ f$ equals $D^- f$–the left differential coefficient–except at a discrete $\mathfrak{Y}$-set, which proves the lemma in the case $r = 1$.

Let $r > 1$. Assume f is of class C^1 on I. We do not yet completes the proof because f' is not necessarily of class $\mathfrak{Y}$. We need to generalize the above arguments for f'. As shown already in the proof of V.2.2, f' has the property $(*)_{f'}$, which is defined as follows. Let g be a continuous function on I.

$(*)_g$ The sets $\{(x_1, x_2) \in I^2 : g(x_1) < g(x_2)\}$ and $\{(x_1, x_2, x_3) \in I^3 : g(x_1) + g(x_2) = g(x_3)\}$ are $\mathfrak{Y}$-sets.

Hence it suffices to prove that a continuous function g on I with $(*)_g$ is of class C^1 on $I - A$ for some discrete $\mathfrak{Y}$-subset A of I, g' is constant or strictly monotone on each connected component of $I - A$, and g' has the property $(*)_{g'}$.

Proceed to prove this as above for f and as in the proof of V.2.2. Then we need only change the proof of the fact that g is right differentiable. Let g be not right differentiable at a point. We assume the point $= 0$ and $g(0) = 0$ for simplicity of notation. We have two sequences $\{a_i\}$ and $\{b_i\}$ of numbers in $\mathbf{R}_+$ convergent to 0 such that $g(a_i)/a_i$ and $g(b_i)/b_i$ equal distinct numbers a and b respectively, and $a_1 > b_1 > a_2 > \cdots$. Assume $a > b$. Let c be a number with $a > c > b$. Then we have a third sequence $\{c_i\}$ in $\mathbf{R}_+$ convergent to 0 such that $g(c_i)/c_i = c$ and $a_i > c_i > b_i$. Set $h(x) = g(x) - cx$. Clearly $h(c_i) = 0$, and h takes positive and negative values on each $]c_{i+1}, c_i[$. For each i, let a_i' (b_i') denote the maximum of the numbers in $]c_{i+1}, c_i[$ where $h|_{]c_{i+1}, c_i[}$ takes the maximal (minimal, respectively) value. Then for any i and j, there exists $\varepsilon > 0$ such that for any $\delta > 0$ with $\delta < \varepsilon$,

$$h(a_i' + \delta) - h(a_i') < 0 \quad \text{and} \quad h(b_j' + \delta) - h(b_j') > 0.$$

It follows that

$$g(a_i' + \delta) - g(a_i') < c\delta < g(b_j' + \delta) - g(b_j').$$

Hence $(a_i', b_j') \in X_<$, where $X_<$ is defined for g as above. Note that X is a $\mathfrak{Y}$-set. Therefore, for each i there exists $\varepsilon_i > 0$ such that $(a_i', x) \in X_<$ for $0 < x < \varepsilon_i$. Hence the image under the map $(x, y) \to x$ of the set

$$\{(x, y) \in I^2 : x > 0 \ y > 0 \ \forall z > 0 \ z < y \Rightarrow (x, y) \in X_<\}$$

contains some interval $]0, \varepsilon[$ because the set is of class $\mathfrak{Y}$. It follows from the fact that the discontinuous point set of a function on I with $\mathfrak{Y}$-graph is a discrete $\mathfrak{Y}$-set that for some positive $\mathfrak{Y}$-function δ on $]0, \varepsilon[$, $(x, y) \in X_<$ if $x \in]0, \varepsilon[$ and $y \in]0, \delta(x)[$. That contradicts the fact that $(b_j', a_i') \in X_>$. Thus g is right differentiable, which completes the proof.   □

Proof of V.2.2. First we show existence of bounded non-semilinear $\mathfrak{Y}$-sets in $\mathbf{R}^n$ for any $n > 1$ by reduction to absurdity. By hypothesis we have such a D in some $\mathbf{R}^n$. Assume $n > 2$ and there is no bounded non-semilinear $\mathfrak{Y}$-set in $\mathbf{R}^{n-1}$.

Let $q: \mathbf{R}^n \to \mathbf{R}^{n-1}$ denote the projection which forgets the last factor. As we saw in the above proof that the discontinuous point set of f is a discrete $\mathfrak{Y}$-set, we can construct a finite number of $\mathfrak{Y}$-sets Y_i in $\mathbf{R}^{n-1}$ and a finite number of $\mathfrak{Y}$-functions $\varphi_{i,j}$ on each Y_i such that $q(D) = \cup_i Y_i$, $\varphi_{i,j} < \varphi_{i,j+1}$, and D is a union of graph $\varphi_{i,j}$ and the sets

$$\{(y, t) \in Y_i \times \mathbf{R} : \varphi_{i,j}(y) < t < \varphi_{i,j+1}(y)\}.$$

Since all Y_i are semilinear, we can assume that each Y_i is the interior of some usual cell σ by V.1.3. If all of $\varphi_{i,j}$ are semilinear, so is D. Hence at least one of them is non-semilinear. Therefore, we assume D is the graph of a $\mathfrak{Y}$-function φ on one cell σ. It follows that $\dim \sigma = n - 1$ because if $\dim \sigma < n - 1$, then D is included in (the linear space spanned by σ) $\times \mathbf{R}$ and hence the image of D under some $\mathfrak{Y}$-projection $\mathbf{R}^n \to \mathbf{R}^{n-1}$ is non-semilinear and included in $\mathbf{R}^{n-1}$. Note that by this argument $D \cap q^{-1}(Z)$ is semilinear for any $\mathfrak{Y}$-set Z in $\mathbf{R}^{n-1}$ of dimension $< n-1$. By the same reason as above $D \cap q^{-1}(l)$ is semilinear for each $\mathfrak{Y}$-line l in $\mathbf{R}^{n-1}$. Let L denote all the lines in $\mathbf{R}^{n-1}$ parallel to the x_1-axis. We easily show as in the proof of V.2.3 that the set $\cup_{l \in L} \{x \in l : \varphi|_{\sigma \cap l}$ is not locally linear at $x\}$ is of class $\mathfrak{Y}$ and of dimension $< n - 1$. Cell subdivide σ so that it is compatible with the set (V.1.1). Then we can assume $\varphi|_{\sigma^\circ \cap l}$ is linear for any $l \in L$. Clearly $\partial \varphi / \partial x_1$ is continuous. Moreover, we can prove that $\partial \varphi / \partial x_1$ is constant on σ° as follows. Assume it is not so at a point $a \in \sigma^\circ$. Let $\sigma_1 = \Delta a_1 \cdots a_n$ be a rational simplex such that $a \in \sigma_1^\circ \subset \sigma^\circ$, $\Delta a_1 \cdots a_{n-1} \subset b \times \mathbf{R}^{n-2}$, and $a_n = a_{n-1} + (\varepsilon, 0, \ldots, 0)$ for some $b, \varepsilon \in \mathbf{Q}$. Then

$$\varphi = \varphi \circ r_2 + (r_1 - b) \frac{\partial \varphi}{\partial x_1} \circ r_2 \quad \text{on} \quad \Delta a_1 \cdots a_{n-2} a_n,$$

where r_1 is the function $(x_1, \dots, x_{n-1}) \to x_1$ and r_2 is the orthogonal projection $\mathbf{R}^{n-1} \to b \times \mathbf{R}^{n-2}$. Since φ and $\varphi \circ r_2$ is PL on $\Delta a_1 \cdots a_{n-2} a_n$, so is $(r_1 - b)\partial\varphi/\partial x_1 \circ r_2$, which implies $\partial\varphi/\partial x_1$ is constant on σ_1. That is a contradiction. Hence $\partial\varphi/\partial x_1$ is constant on σ°. It follows that φ is PL on σ, which contradicts the hypothesis that D is not semilinear. Therefore, there exist bounded non-semilinear $\mathfrak{Y}$-sets in $\mathbf{R}^n$, $n > 1$, and hence a function on a bounded open $\mathfrak{Y}$-set in $\mathbf{R}$ whose graph is not semilinear but a $\mathfrak{Y}$-set.

Consider the case $r = 1$. By V.2.3 we have a C^1 $\mathfrak{Y}$-function f on an interval $[a, b]$ such that f' is strictly monotone. We need to extend f' to $\mathbf{R}$. First we define the extension on $[b, a + b]$ so that its graph is symmetric to graph f' with respect to the point $(b, f'(b))$. Then the extension is strictly monotone and of the form $2f'(b) - f'(2b - x)$, and hence its integration can be of class $\mathfrak{Y}$ because we can shrink $[a, b]$ a little so that $f'(a)$ and $f'(b)$ are rational. Repeating the same arguments, we obtain a C^1 $\mathfrak{Y}$-function on $\mathbf{R}$ whose derivative is a bijection onto $\mathbf{R}$.

Let f be a C^1 $\mathfrak{Y}$-function on $\mathbf{R}$ whose derivative is bijective onto $\mathbf{R}$. Let Z denote the family of the images of $\mathfrak{Y}$-sets in $\mathbf{R}^n$, $n = 0, 1, \dots$, under the maps $\mathbf{R}^n \ni (x_1, \dots, x_n) \to (f'(x_1), \dots, f'(x_n)) \in \mathbf{R}^n$. Clearly Z satisfies Axioms (ii),.,(iv) of $\mathfrak{X}$, and any point of $\mathbf{R}$ is an element of Z. Hence it suffices to show that the sets $\{(x_1, x_2, x_3) \in \mathbf{R}^3 \colon x_1 + x_2 = x_3\}$ and $\{(x_1, x_2, x_3) \in \mathbf{R}^3 \colon x_1 x_2 = x_3\}$ are elements of Z. It is equivalent to the statement that the sets

$$X = \{(x_1, x_2, x_3) \in [a, b]^3 \colon f'(x_1) + f'(x_2) = f'(x_3)\} \text{ and}$$
$$Y = \{(x_1, x_2, x_3) \in [a, b]^3 \colon f'(x_1)f'(x_2) = f'(x_3)\}$$

are $\mathfrak{Y}$-sets. This follows from the equalities:

$$X = \overline{\{(x_1, x_2, x_3) \in [a, b]^3 \colon \exists \varepsilon > 0 \; \forall \delta > 0 \; \delta < \varepsilon \Rightarrow F(x_1, x_2, x_3, \delta) \le 0\}}$$
$$\bigcap \overline{\{(x_1, x_2, x_3) \in [a, b]^3 \colon \exists \varepsilon > 0 \; \forall \delta > 0 \; \delta < \varepsilon \Rightarrow F(x_1, x_2, x_3, \delta) \ge 0\}} \text{ and}$$
$$Y = \overline{\{(x_1, x_2, x_3) \in [a, b]^3 \colon \exists \varepsilon > 0 \; \forall \delta > 0 \; \delta < \varepsilon \Rightarrow G(x_1, x_2, x_3, \delta) \le 0\}}$$
$$\bigcap \overline{\{(x_1, x_2, x_3) \in [a, b]^3 \colon \exists \varepsilon > 0 \; \forall \delta > 0 \; \delta < \varepsilon \Rightarrow G(x_1, x_2, x_3, \delta) \ge 0\}},$$

where

$$F = f(x_1 + \delta) - f(x_1) + f(x_2 + \delta) - f(x_2) - f(x_3 + \delta) + f(x_3) \text{ and}$$
$$G = f(x_1 + f(x_2 + \delta) - f(x_2)) - f(x_1) - f(x_3 + \delta) + f(x_3).$$

Thus the case $r = 1$ is proved. Note that Z satisfies Axiom (v).

Let $r > 1$. By the above construction of f we can assume it is of class C^{r+1} on $[-1, 1]$. It suffices to approximate f by a C^r $\mathfrak{Y}$-function in the C^1

Whitney topology. Namely, we will prove an approximation theorem. For that we need only show the following two statements. Indeed, we can use a C^r 𝔜-partition of unity by (1) and reduce the problem to the local one, and it is solved by (2).

(1) There exists a C^r 𝔜-function on **R** whose value is 0 outside $[-1, 1]$ and 1 on $[-1/2, 1/2]$.

(2) Any C^1 𝔜-function on **R** with support in $[-1, 1]$ can be approximated by C^r 𝔜-function with support in $[-1, 1]$ in the C^1 topology.

Proof of (1). From now on, assume $f' = \mathrm{id}$ at $\{-1, -1/2, 1/2, 1\}$ for simplicity of notation. It is easy to construct a C^r semialgebraic function ρ on **R** whose value is 0 outside $[-1, 1]$ and 1 on $[-1/2, 1/2]$. Clearly graph $\rho \in Z$. Hence $(f' \times f')^{-1}(\text{graph}\,\rho)$ is a 𝔜-set. The function whose graph is this set fulfills the requirements.

Proof of (2). Let g be a C^1 𝔜-function on **R** with supp $g \subset [-1, 1]$. Let h denote the C^1 Z-function on **R** such that

$$(f' \times f')(\text{graph}\,g) = \text{graph}\,h.$$

By II.5.2 we have a C^r Z-function $\tilde{h}$ of h in the Whitney C^1 topology. Let $\tilde{g}$ be the 𝔜-function on **R** defined so that

$$(f' \times f')(\text{graph}\,\tilde{g}) = \text{graph}\,\tilde{h},$$

and let ξ be a C^r 𝔜-function on **R** whose value is 0 outside $[-1, 1]$ and 1 on supp g. Then $\xi\tilde{g}$ is a C^r 𝔜-approximation of g with support in $[-1, 1]$ in the C^1 topology.   □

Immediate corollaries of V.2.2 are the followings.

Corollary V.2.4. *Assume Axioms* (v) *and* (vi), *and that there exists a non-semilinear C^1 𝔜-function on a bounded open 𝔜-subset of* **R** *whose derivative also is of class* 𝔜. *Then* 𝔜 *is* 𝔛.

Corollary V.2.5. *Assume Axiom* (vi) *and taht some* 𝔜-set *is not locally semilinear. All the topological facts on* 𝔛-sets *and* 𝔛-maps *except for triangulations in the preceding chapters hold for* 𝔜-sets *and* 𝔜-maps *if Axiom* (v) *is satisfied or if the* 𝔜-sets *and the source and target spaces of the* 𝔜-maps *are bounded.*

Unique Triangulation Theorem does not immediately follow from V.2.2. We need certain additional arguments.

Theorem V.2.6 (Unique $\mathfrak{Y}$-triangulation). *Assume Axiom* (vi) *and that some $\mathfrak{Y}$-set is not locally semilinear. Any closed $\mathfrak{Y}$-set $X \subset \mathbf{R}^n$ admits a rational $\mathfrak{Y}$-(cell) triangulation compatible with a family of $\mathfrak{Y}$-sets in X locally finite at each point of X if either* (a) *X is compact,* (b) *Axiom* (v) *is assumed or* (c) *$\mathfrak{Y} = \mathfrak{Y}_0$ and the family is finite. Under the same conditions, any $\mathfrak{Y}$-function on X admits a rational C^0 $\mathfrak{Y}$-(cell) triangulation compatible with the family. Here the triangulations are unique except for* (c).

Remember that the triangulations in the case $\mathfrak{Y} = \mathfrak{Y}_0$ are not necessarily unique as shown in II.4.3.

Proof. Let r be a positive integer, let $\{X_i\}$ be the family of $\mathfrak{Y}$-sets and let f be the $\mathfrak{Y}$-function on X. For the existence, it suffices to find (usual) $\mathfrak{Y}$-cell triangulations of X and f compatible with $\{X_i\}$ because we can modify the (usual) $\mathfrak{Y}$-cell complexes to be rational $\mathfrak{Y}$-simplicial (cell) complexes by the first statement of V.1.3.

(1) *Proof of existence for compact $\mathfrak{Y}$-sets.* We can assume $X = I^n$ for $I = [0, 1]$ and Axiom (v) is satisfied. By V.2.2 we have a C^r diffeomorphism π of I and an $\mathfrak{X}$ such that a subset S of I^m is a $\mathfrak{Y}$-set if and only if $\pi \times \cdots \times \pi(S)$ is an $\mathfrak{X}$-set. By II.2.1 there exists an $\mathfrak{X}$-triangulation (K, τ) of I^n compatible with $\{\pi \times \cdots \times \pi(X_i)\}$ such that $|K| = I^n$. Transform $\{X_i\}$ by the $\mathfrak{Y}$-homeomorphism $(\pi \times \cdots \times \pi)^{-1} \circ \tau^{-1} \circ (\pi \times \cdots \times \pi)$ of I^n. Then we can assume $\{\pi \times \cdots \times \pi(X_i)\} = \{\sigma^\circ : \sigma \in K\}$. Let $p_{i-1} : I^i \to I^{i-1}$, $i = 2, \ldots, n$, denote the projections which forget the last factors. We can suppose there exists a sequence of simplicial maps $K_n \xrightarrow{p_{n-1}} \cdots \xrightarrow{p_1} K_1$ of simplicial complexes such that
$$K_n = K \quad \text{and} \quad |K_i| = I^i, \quad i = 1, \ldots, n-1.$$
Set
$$L_i = (\pi \times \cdots \times \pi)^{-1}(K_i) = \{(\pi \times \cdots \times \pi)^{-1}(\sigma) : \sigma \in K_i\}, \quad i = 1, \ldots, n.$$

We can choose K_n so that $K_n^0 \subset (\pi(\mathbf{Q}))^n$. Hence assume $L_n^0 = (\pi \times \cdots \times \pi)^{-1}(K_n^0)$ is a rational point set. Set
$$M_i = \{\Delta(\sigma \cap L_i^0) : \sigma \in L_i\}, \quad i = 1, \ldots, n.$$

Then each M_i is a rational simplicial complex with underlying polyhedron $= I^i$, and it suffices to construct a $\mathfrak{Y}$-homeomorphism θ of I^n such that the restriction of θ to $\Delta(\sigma \cap L_n^0)$ for each $\sigma \in L_n$ is a C^r diffeomorphism onto σ, and θ is of the form $(\theta_1(x_1), \ldots, \theta_n(x_1, \ldots, x_n))$.

We proceed by induction on n as in the proof of V.1.3. If $n = 1$, there is nothing to do. So assume there exists $(\theta_1, \ldots, \theta_{n-1})$ and hence L_{n-1} is a simplicial complex. (Then $K_{n-1} = \pi \times \cdots \times \pi(L_{n-1})$ is not necessarily a simplicial complex.) It follows that θ should be of the form $(\mathrm{id}, \dots, \mathrm{id}, \theta_n)$. Let $\sigma \in M_n$. If $p_{n-1}|_\sigma$ is injective, $\theta_n|_\sigma$ is uniquely determined and of class C^r and $\mathfrak{Y}$. So θ_n is defined on the union of such σ's. Assume $p_{n-1}|_\sigma$ is not injective. Let σ_1 and σ_2 be the simplexes of M_n between which σ lies. To be precise, for each $x \in p_{n-1}(\sigma^\circ)$, $\sigma \cap p_{n-1}^{-1}(x)$ is the segment with ends $\sigma_1 \cap p_{n-1}^{-1}(x)$ and $\sigma_2 \cap p_{n-1}^{-1}(x)$. Extend θ_n to σ so that for $x \in p_{n-1}(\sigma)$, $(\pi \times \cdots \times \pi) \circ (\mathrm{id} \times \cdots \times \mathrm{id} \times \theta_n) \circ (\pi \times \cdots \times \pi)^{-1}$ is linear on the segment (or point) $\pi \times \cdots \times \pi (\sigma \cap \pi_{n-1}^{-1}(x))$. Then $\theta_\sigma - \theta$ on σ – is a C^r $\mathfrak{Y}$-diffeomorphism onto $\Delta(\sigma \cap L_n^0)$, and $\theta_\sigma = \theta_{\sigma'}$ on $\sigma \cap \sigma'$ for $\sigma, \sigma' \in M_n$. Hence θ is well-defined.

(2) *Proof of existence for non-compact* $\mathfrak{Y}$-*sets in the case (b).* By the proof of II.2.1$'$ it suffices to prove the following statement.

Let $L_1 \subset L_2$ be finite $\mathfrak{Y}$-simplicial complexes in $\mathbf{R}^n$ such that L_1 is a full subcomplex of L_2, and let τ be a $\mathfrak{Y}$-homeomorphism of $|L_1|$ such that for each $\sigma \in L_1$, $\tau(\sigma) = \sigma$, and $\tau|_{\sigma^\circ}$ is a C^r diffeomorphism onto σ°. Then, keeping these properties, we can extend τ to $|L_2|$ so that $\tau = \mathrm{id}$ on each $\sigma \in L_2$ with $\sigma \cap |L_1| = \varnothing$.

In the case of $\mathfrak{X}$, this is possible by the Alexander trick as in the proof of II.2.1$'$. But the Alexander trick is not necessarily of class $\mathfrak{Y}$. Hence we need to change the proof. Let π and $\mathfrak{X}$ be the same as in (1). Set $K_i = \pi \times \cdots \times \pi(L_i)$, $i = 1, 2$, and $\xi = (\pi \times \cdots \times \pi) \circ \tau \circ (\pi \times \cdots \times \pi)^{-1}$. Then $K_1 \subset K_2$, ξ is an $\mathfrak{X}$-homeomorphism of $|K_1|$, each element of K_2 is a compact C^r $\mathfrak{X}$-manifold possibly with boundary and corners, for each $\delta \in K_1$, and $\xi(\delta) = \delta$, $\xi|_{\delta^\circ}$ is a C^r diffeomorphism onto δ°. We wish to extend ξ to $|K_2|$. For the extension we need only show existence of a C^r $\mathfrak{X}$-diffeomorphism $\rho \colon |L_2| \to |K_2|$ such that $\rho(\sigma) = \pi \times \cdots \times \pi(\sigma)$ for $\sigma \in L_2$ because by ρ^{-1}, we reduce the problem to the one on $|L_2|$, where we can use the Alexander trick. Note that L_2 is of class $\mathfrak{X}$ because it is rational. We construct ρ by induction on the dimension of L_2. It suffices to show the following assertion.

Let N_1 and N_2 be compact C^r $\mathfrak{X}$-manifolds possibly with boundary and corners, let $f \colon N_1 \to N_2$ be a C^r diffeomorphism and let $g \colon \partial N_1 \to \partial N_2$ be a sufficiently strong C^r $\mathfrak{X}$-approximation of $f|_{\partial N_1}$ (i.e., for each proper face F of N_1, $g|_F$ is a sufficiently strong C^r $\mathfrak{X}$-approximation of $f|_F$). Then we can extend g to a C^r $\mathfrak{X}$-approximation of f.

Moreover, the problem is reduced as follows by a C^r $\mathfrak{X}$-tube of N_2, whose existence is shown as in the proof of II.5.1 (the case without boundary or corners).

Let N be a compact C^r $\mathfrak{X}$-manifold possibly with boundary and corners, let φ be a C^r function on N and let ψ be a sufficiently strong C^r $\mathfrak{X}$-approximation of $\varphi|_{\partial N}$. Then we can extend ψ to a C^r $\mathfrak{X}$-approximation of φ.

Let $\tilde{\varphi}$ be a strong C^r $\mathfrak{X}$-approximation of φ. Consider $\psi - \tilde{\varphi}|_{\partial N}$ and the zero function in place of ψ and φ. Then what we prove is that a sufficiently small C^r $\mathfrak{X}$-function on ∂N can be extended to a small C^r $\mathfrak{X}$-function on N. By a C^r $\mathfrak{X}$-partition of unity this becomes the corresponding local problem in the case $N = [0, \infty[^k \times \mathbf{R}^l$. Then it is easy to construct the extension.

(3) *Proof of existence for non-compact $\mathfrak{Y}$-sets in the case (c).* There are two cases: some C^{r+1} $\mathfrak{Y}$-function on $\mathbf{R}$ has a strictly monotone derivative, or not. By V.2.3 the condition of the second case is equivalent to the condition that for any $\mathfrak{Y}$-set $Y \subset \mathbf{R}^n$, $Y - [-c, c]^n$ is semilinear for some $c \in \mathbf{R}$. Then we can prove the existence in the same way as (2) by V.1.3.

Consider the first case. We have a C^r imbedding $\pi \colon \mathbf{R} \to \mathbf{R}$ and an $\mathfrak{X}_0$ such that a subset S of $\mathbf{R}^m$ is a $\mathfrak{Y}$-set if and only if $\pi \times \cdots \times \pi(S)$ is an $\mathfrak{X}_0$-set. Since there exists a Nash diffeomorphism from $\operatorname{Im} \pi$ to $\mathbf{R}$, we can assume π is surjective. As in the proof of V.1.3 and (1), we suppose, moreover, that $X = \{(x_1, \dots, x_n) \in \mathbf{R}_+ \times \mathbf{R}^{n-1} \colon |x_2| \leq x_1, \dots, |x_n| \leq x_1\}$ and that there exists an $\mathfrak{X}_0$-cell decomposition K of X such that any bounded cell of K is included in $[0, 1] \times \mathbf{R}^{n-1}$, K is compatible with $1 \times \mathbf{R}^{n-1}$, $\{\pi \times \cdots \times \pi(X_i)\} = \{\sigma^\circ \colon \sigma \in K\}$, and $K^0 \cup \{\sigma \cap 2 \times \mathbf{R}^{n-1} \colon \sigma \in K^1\}$ is a rational point set. Furthermore, we can choose π so that $(\pi \times \cdots \times \pi)^{-1}$ is the identity map at the last rational set. Indeed, there exists a Nash diffeomorphism π' of $\mathbf{R}$ such that $\pi' \times \cdots \times \pi' = \pi \times \cdots \times \pi$ at the set, and it suffices to replace π and X_i with $\pi'^{-1} \circ \pi$ and $(\pi^{-1} \circ \pi' \circ \pi) \times \cdots \times (\pi^{-1} \circ \pi' \circ \pi)(X_i)$. Set $L = (\pi \times \cdots \times \pi)^{-1}(K)$. Then it suffices to construct a $\mathfrak{Y}$-homeomorphism θ of X such that for each $\sigma \in K$, the restriction of θ to σ is a C^r diffeomorphism onto $(\pi \times \cdots \times \pi)^{-1}(\sigma)$. Note that K is rational and hence of class $\mathfrak{Y}$. We carry out the construction as in (1). We omit the details.

(4) *Proof of uniqueness for $\mathfrak{Y}$-sets in cases (a) and (b).* Let (K, τ) and (K_1, τ_1) be rational $\mathfrak{Y}$-triangulations of X compatible with $\{X_i\}$. Let (K_2, τ_2) be another one compatible with $\tau(K) \cup \tau_1(K_1)$. Then it suffices to construct PL homeomorphisms $\alpha \colon |K| \to |K_2|$ and $\alpha_1 \colon |K_1| \to |K_2|$ such that $\alpha(\sigma) = \tau_2^{-1} \circ \tau(\sigma)$ for $\sigma \in K$ and $\alpha_1(\sigma_1) = \tau_2^{-1} \circ \tau_1(\sigma_1)$ for $\sigma_1 \in K_1$ because we can approximate α and α_1 by rational homeomorphisms. This follows from the Alexander trick if $\tau_2^{-1} \circ \tau(\sigma)$ and $\tau_2^{-1} \circ \tau_1(\sigma_1)$ are PL balls. Hence we need only show the $\mathfrak{Y}$-version of III.1.4. Recall the proof of III.1.4. We proved this by showing that the conditions of III.1.1 (or III.1.2) are satisfied for $\mathfrak{X}$-sets. By V.2.5, the conditions are satisfied for $\mathfrak{Y}$-sets. Therefore, the $\mathfrak{Y}$-version holds.

(5) *Proof of existence for $\mathfrak{Y}$-functions in the case (a).* Let $\pi \colon \mathbf{R} \to \mathbf{R}$ and $\mathfrak{X}$ be the same as above. By II.3.1 we can assume $\pi \times \cdots \times \pi(\operatorname{graph} f)$ is a polyhedron and admits a simplicial decomposition K_{n+1} compatible with $\{\pi \times \cdots \times (X_i)\}$. Let $p_{i-1} \colon \mathbf{R}^i \to \mathbf{R}^{i-1}$, $i = 2, \dots, n+1$, be the projections defined by $p_{i-1}(x_1, \dots, x_i) = (x_1, \dots, x_{i-2}, x_i)$. Assume that there exists a sequence of simplicial maps of simplicial complexes $K_{n+1} \xrightarrow{p_n} \cdots \xrightarrow{p_1} K_1$ such that $|K_i| = p_i(|K_{i+1}|)$, $i = 1, \ldots, n$. We can choose K_{n+1} so that $K_{n+1}^0 \subset (\pi(\mathbf{Q}))^n \times \mathbf{R}$, but we cannot expect $K_{n+1}^0 \subset (\pi(\mathbf{Q}))^{n+1}$ because of the definition of a triangulation of a function. For the last inclusion we need to replace f with $\beta \circ f$ for some semilinear $\mathfrak{Y}$-homeomorphism β of $\mathbf{R}$. Then we have $|K_{n+1}| = \pi \times \cdots \times \pi(\text{graph} \, \beta \circ f)$. As in (1), we see by $(\pi \times \cdots \times \pi)^{-1}(K_{n+1})$ that $\beta \circ f$ admits a $\mathfrak{Y}$-triangulation (L, τ). Let $\beta \colon C_1 \to C_2$ be a $\mathfrak{Y}$-cell decomposition of β. Set

$$\tilde{L} = \{ \sigma \cap (f \circ \tau)^{-1}(\sigma_1) \colon \sigma \in L, \ \sigma_1 \in C_1 \}.$$

Then $(\tilde{L}, \tau)$ is a usual $\mathfrak{Y}$-cell triangulation of f.

(6) *Proof of uniqueness for $\mathfrak{Y}$-functions in the case (a).* We proceed as in the proof of II.3.1′. Let (K_1, τ_1) and (K_2, τ_2) be C^0 rational $\mathfrak{Y}$-triangulations of f compatible with $\{X_i\}$. We assume (K_2, τ_2) is compatible with $\tau_1(K_1)$ as above. It suffices to find a PL $\mathfrak{Y}$-homeomorphism $\alpha \colon |K_1| \to |K_2|$ such that $\alpha(\sigma) = \tau_2^{-1} \circ \tau_1(\sigma)$ for $\sigma \in K_1$ and $f \circ \tau_2 \circ \alpha = f \circ \tau_1$. We need not require α to be of class $\mathfrak{Y}$ because we can approximate α by a rational PL homeomorphism. Set $C = f \circ \tau_2(K_2^0)$. As shown in (4), we have a PL homeomorphism $\alpha' \colon |K_1| \to |K_2|$ such that $\alpha'((f \circ \tau_1)^{-1}(C)) = (f \circ \tau_2)^{-1}(C)$ and $\alpha'(\sigma) = \tau_2^{-1} \circ \tau_1(\sigma)$ for $\sigma \in K_1$. We need to modify α' so that $f \circ \tau_2 \circ \alpha' = f \circ \tau_1$ holds. This holds on $(f \circ \tau_1)^{-1}(C)$, $f \circ \tau_1$ and $f \circ \tau_2$ are PL trivial on each connected component of $(f \circ \tau_1)^{-1}(C^c)$ and $(f \circ \tau_2)^{-1}(C^c)$ respectively, and the modification is purely a PL problem. We already solved this problem in the proof of II.3.1′.

(7) *Proof of existence for $\mathfrak{Y}$-functions in the case (b).* This also is similar to the proof of II.3.1′. By II.3.1′, (5) and (6), it suffices to prove the following statement.

Assume X is a compact polyhedron, f is PL, and graph $\beta \circ f$ admits a rational simplicial decomposition L for some semilinear $\mathfrak{Y}$-homeomorphism β of $\mathbf{R}$. Let L_0 be a full subcomplex of L. Set $X_0 = |L_0|$. Assume that there exist a sequence of simplicial maps of rational simplicial complexes $L_{n+1} \xrightarrow{p_n} \cdots \xrightarrow{p_1} L_1$ such that

$$L_{n+1} = L \text{ and } |L_i| = p_i(|L_{i+1}|), \quad i = 1, \ldots, n,$$

where $p_{i-1} \colon \mathbf{R}^i \to \mathbf{R}^{i-1}$, $i = 1, \ldots, n$, are the same as in (5). Let τ be a $\mathfrak{Y}$-homeomorphism of X_0 such that $p_1 \circ \cdots \circ p_n \circ \tau = p_1 \circ \cdots \circ p_n$ and $\tau(\sigma) = \sigma$ for each $\sigma \in L_0$. Then we can extend τ to a $\mathfrak{Y}$-homeomorphism $\tilde{\tau}$ of X so that $\tilde{\tau} = \text{id}$ on each $\sigma \in L$ with $\sigma \cap X_0 = \varnothing$ keeping these properties.

This is like the statement in (2). The difference is that the equality $p_1 \circ \cdots \circ p_n \circ \tau = p_1 \circ \cdots \circ p_n$ is added here. Let π and $\mathfrak{X}$ be the same as above. Set

$$K_i = \pi \times \cdots \times \pi(L_i), \qquad K_i^0 = \pi \times \cdots \times \pi(L_i^0) \text{ and}$$
$$M_i = \{\Delta(\sigma \cap K_i^0) : \sigma \in K_i\}, \quad i = 0, \ldots, n+1.$$

Note that K_i are families of C^r $\mathfrak{X}$-manifolds possibly with boundary and corners and M_i are simplicial complexes. Let $\theta : |K_{n+1}| \to |M_{n+1}|$ be an $\mathfrak{X}$-homeomorphism such that $\theta(\sigma) = \Delta(\sigma \cap K_i^0)$ for $\sigma \in K_{n+1}$ and θ is of form $(\theta_1(x_1, y), \ldots, \ldots, \theta_n(x_1, \ldots, x_n, y), y)$ for $(x_1, \ldots, x_n, y) \in |K_{n+1}|$. Consider an $\mathfrak{X}$-homeomorphism $\chi = \theta \circ (\pi \times \cdots \times \pi) \circ \tau \circ (\pi \times \cdots \times \pi)^{-1} \circ \theta^{-1}$ of $|M_0|$ in place of τ. We have $p_1 \circ \cdots \circ p_n \circ \chi = p_1 \circ \cdots \circ p_n$ and $\chi(\sigma) = \sigma$ for $\sigma \in M_0$. Extend χ to an $\mathfrak{X}$-homeomorphism $\tilde{\chi}$ of $|M_{n+1}|$ by the Alexander trick. Then the properties of χ are kept, $\tilde{\chi} = \mathrm{id}$ on each $\sigma \in M_{n+1}$ with $\sigma \cap |M_0| = \varnothing$, and $\tilde{\chi}$ induces the required $\tilde{\tau}$ through $(\pi \times \cdots \times \pi)^{-1} \circ \theta^{-1}$.

(8) *Proof of existence for $\mathfrak{Y}$-functions in the case (c).* We separate the proof to two cases as in (3). If f is semilinear on $X -]-c, c[^n$ for some $c \in \mathbf{Q}$, then a rational C^0 $\mathfrak{Y}$-cell triangulation of $\beta \circ f$ on the set follows from V.1.3 for some semilinear $\mathfrak{Y}$-homeomorphism β of $\mathbf{R}$. We can paste it with a C^0 $\mathfrak{Y}$-cell triangulation of f on $[-c, c]^n$ by the method in (7). In the case where f is not semilinear on $X -]-c, c[^n$ for any $c \in \mathbf{Q}$, there exists π in (3), and hence the proof is the same as (5).

(9) *Proof of uniqueness for $\mathfrak{Y}$-functions in the case (b).* By the same reason as (6), what we prove is the following assertion.

Let (K_1, τ_1) and (K_2, τ_2) be rational C^0 $\mathfrak{Y}$-triangulations of f such that (K_2, τ_2) is compatible with $\tau_1(K_1)$. Then there exists a PL homeomorphism $\alpha : |K_1| \to |K_2|$ such that $\alpha(\sigma) = \tau_2^{-1} \circ \tau_1(\sigma)$ for $\sigma \in K_1$ and $f \circ \tau_2 \circ \alpha = f \circ \tau_1$.

Proof of this is a little different to that of (6) because the set $f \circ \tau_2(K_2^0)$ is not necessarily finite. We can overcome this problem by considering f on $\tau_1(\cup_{i=1}^k \sigma_i)$ by induction on k as in the proof of II.3.1', where σ_i are the ordered simplexes of K_1. We do not repeat the proof, and note only that we apply the Alexander trick not directly but through $\pi \times \cdots \times \pi$ as in (7). $\square$

Bibliography

[A-B-B] F. Acquistapace–R. Benedetti–F. Broglia, *Effectiveness-non effectiveness in semialgebraic and PL geometry*, Inv. Math. **162** (1990), 141–156.

[B-R] R. Benedetti–J.J. Risler, *Real algebraic and semi-algebraic sets*, Hermann, 1990.

[B-S] R. Benedetti–M. Shiota, *Finiteness of semialgebraic types of polynomial functions*, Math. Z. **208** (1991), 589–596.

[B-C-R] J. Bochnak–M. Coste–M.F. Roy, *Géometrie algébrique réelle*, Springer, 1987.

[Can] J.W. Cannon, *Shrinking cell-like decomposition of manifolds, codimension three*, Ann. of Math. **110** (1979), 83–112.

[Car] H. Cartan, *Variétés analytiques réelles et variétés analytiques complexes*, Bull. Soc. Math. France **85** (1957), 77–99.

[Co] M. Coste, *Unicité des triangulations semi-algébriques: validité sur un corps réel clos quelconque, et effectivité forte*, C. R. Acad. Sci. Paris **312** (1991), 395–398.

[C-S$_1$] M. Coste–M. Shiota, *Nash triviality in families of Nash manifolds*, Inv. Math. **108** (1992), 349–368.

[C-S$_2$] ———, *Thom's first isotopy lemma: semialgebraic version, with uniform bound*, in *Real analytic and algebraic geometry*, Verlag Walter De Gruyter, 1995, pp. 83–101.

[D] D. van Dalen, *Logic and structure*, Springer, Universitex, 1980.

[Dr$_1$] L. van den Dries, *Remarks on Tarski's problem concerning* $(\mathbf{R}, +, \cdot, \exp)$, Logic colloquium 1982, Elsevier Science Publishers B.V. (North-Holland), 1984, 97–121.

[Dr$_2$] ———, *Tarski problem and Pfaffian functions*, Logic colloquium 1984, Elsevier Science Publishers B.V. (North-Holland), 1986, 59–90.

[Dr$_3$] ———, *Tame topology and 0-minimal structures* (1991).

[D-M$_1$] L. van den Dries–C. Miller, *Extending Tamm's theorem*, Ann. Inst. Fourier **44** (1994), 1367–1395.

[D-M$_2$] ———, *Geometric categories and 0-minimal structures* (1995) (to appear).

[Ga] Gabrielov, *Projections of semi-analytic sets*, Functional Anal. Appl. **2** (1968), 282–291.

[G-al] C.G. Gibson et al, *Topological stability of smooth mappings*, Lecture Notes in Math., 552, Springer, 1976.

[Ha] R. Hardt, *Semi-algebraic local-triviality in semi-algebraic mappings*, Amer. J. Math. **102** (1980), 291–302.

[He] M. Hervé, *Several complex variables, Local theory*, Oxford, 1963.

[Hi]   G. Higman, _The units of group rings_, Proc. London Math. Soc. **46** (1940), 231–148.

[H$_1$]   H. Hironaka, _Subanalytic sets_, in _Number theory, Algebraic geometry and Commutative algebra_, Kinokuniya, 1973, pp. 453–493.

[H$_2$]   _____, _Triangulations of algebraic sets_, Proc. Sympos. Pure Math., 29, Amer. Math. Soc., Providence, R.I. (1975), 165–185.

[H-M]   M.W. Hirsch–B. Mazur, _Smoothings of piecewise linear manifolds_, Ann. of Math. Studies, Princeton Univ. Press, 1974.

[H]   A.G. Hovanskii, _On a class of systems of transcendental equations_, Soviet Math. Dokl. **22** (1980), 762–765.

[I]   S. Illman, _Subanalytic equivariant triangulation of real analytic G-manifolds, for G a Lie group_ (to appear).

[Ki]   H.C. King, _Real analytic germs and their varieties at isolated singularities_, Inv. Math. **37** (1976), 193–199.

[K-P-S]   J.F. Knight–A. Pillay–C. Steinhorn, _Definable sets in ordered structures_ II, Trans. Amer.Math. Soc. **295** (1986), 593-605.

[K-S]   R.C. Kirby–L.C. Siebenmann, _Foundational essays on topological manifolds, smoothings, and triangulations_, Ann. of Math. Studies, Princeton University Press.

[L-R]   J.M. Lion–J.P. Rolin, _Feuilletages analytiques réels et théorme de Wilkie_ (1996) (to appear).

[L$_1$]   S. Lojasiewicz, _Triangulations of semi-analytic sets_, Ann. Scu. Norm. Sup. Pisa **18** (1964), 449–474.

[L$_2$]   _____, _Ensembles semi-analytiques_, IHES, 1965.

[M-S$_1$]   T. Matumoto–M. Shiota, _Unique triangulation of the orbit space of a differentiable transformation group and its application_, Adv. Stud. Pure Math. **9** (1986), 41–55.

[M-S$_2$]   _____, _Proper subanalytic transformation groups and unique triangulation of the orbit spaces,_ in _Transformation groups_, Lecture Notes in Math., 1217, Springer, 1986, pp. 290–302.

[Mi$_1$]   J. Milnor, _Two complexes which are homeomorphic but combinatorially distinct_, Ann. of Math. **74** (1961), 575–590.

[Mi$_2$]   _____, _Singular points of complex hypersurfaces_, Ann. of Math. Studies, Princeton Univ. Press, 1968.

[Mu]   J.R. Munkres, _Elementary differential topology_, Ann. of Math. Studies, Princeton Univ. Press, 1963.

[Pa]   R. Palais, _Equivariant real algebraic differential topology, Part 1, Smoothness categories and Nash manifolds_, Brandeis Univ., 1972.

[Pe-S]   Y. Peterzil–S. Starchenko, _A trichotomy theorem for 0-minimal structures_ (1995) (to appear).

[P-S]   A. Pillay–C. Steinhorn, _Definable sets in ordered structures_ II, Trans. Amer. Math. Soc. **295** (1986), 565–592.

[Pi-S]   A. Pillay–M. Shiota, *Topological properties of 0-minimal structure without multiplication*, unpublished.

[R-S]   C.P. Rourke–B.J. Sanderson, *Introduction to piecewise-linear topology*, Springer, 1972.

[S₁]   M. Shiota, *Equivalence of differentiable mappings and analytic mappings*, Publ. Math. IHES **34** (1981), 37–122.

[S₂]   ———, *Piecewise linearization of real analytic functions*, Publ. RIMS, Kyoto Univ. **20** (1984), 727–792.

[S₃]   ———, *Nash manifolds*, Lecture Notes in Math., 1269, Springer, 1987.

[S₄]   ———, *Piecewise linearization of real-valued subanalytic functions*, Trans. Amer. Math. Soc. **312** (1989), 663–679.

[S₅]   ———, *Piecewise linearization of subanalytic functions* II, Lecture Notes in Math., 1420, Springer, 1990, pp. 247–307.

[S-Y]   M. Shiota–M. Yokoi, *Triangulations of subanalytic sets and locally subanalytic manifolds*, Trans. Amer. Math. Soc. **286** (1984), 727–750.

[Si₁]   L.C. Siebenmann, *Torsion invariants for pseudo-isotopies on closed manifolds*, Notices Amer. Math. Soc. **14** (1967), 942–943.

[Si₂]   ———, *Disruption of low-dimensional handlebody theory by Rohlin's theorem*, in *Topology of manifolds*, Markham, Chicago, 1970, pp. 57–76.

[St]   N. Steenrod, *The topology of fibre bundle*, Princeton Univ. Press, 1951.

[Su]   D. Sullivan, *Hyperbolic geometry and homeomorphisms*, in *Geometric topology*, Academic Press, 1979, pp. 543–555.

[Sus]   H.J. Sussmann, *Real-analytic desingularization and subanalytic sets: an elementary approach*, Trans. Amer. Math. Soc. **317** (1990), 417–461.

[Ta]   M. Tamm, *Subanalytic sets in the calculus of variation*, Acta Math. **146** (1981), 167–199.

[Te]   B. Teissier, *Sur la triangulation des morphismes sous-analytiques*, Publ. Math. IHES **70** (1989), 169–198.

[Th₁]   R. Thom, *Ensembles et morphisms stratifiés*, Bull. Amer. Math. Soc. **75** (1969), 240–284.

[Th₂]   ———, *Singularities of differentiable mappings*, Lecture Notes in Math., 192, Springer, 1971, pp. 1–89.

[To]   A. Tognoli, *Algebraic geometry and Nash functions*, Academic Press, 1978.

[V]   J.L. Verdier, *Stratifications de Whitney et théorème de Bertini-Sard*, Inv. Math **36** (1976), 295–312.

[Ve]   A. Verona, *Stratified mappings – structure and triangulability*, Lecture Notes in Math., 1102, Springer, 1984.

[W₁]   H. Whitney, *Elementary structure of real algebraic varieties*, Ann. of Math. **66** (1957), 545–556.

[W₂]   ———, *Local properties of analytic varieties,* in *Differential and combinatorial topology*, Princeton Univ. Press, 1965, pp. 205–244.

[W₃]   ———, *Tangents to an analytic variety*, Ann. of Math. **81** (1965), 496–549.

[W-B]   H. Whitney–F. Bruhat, *Quelque propriétés fondamentales des espaces analytiques réels*, Commen. Math. Helv. **33** (1959), 132–160.

[Wi₁]   A. Wilkie, *Model completeness results for expansions of the real field I: restricted Pfaffian functions* (to appear).

[Wi₂]   ———, *Model completeness results for expansions of the real field II: the exponential function* (to appear).

LIST OF NOTATION

$\mathrm{Int}\,A$	Interior of subset in topological set **56**
$\mathrm{bdry}\,A$	Boundary of subset in topological set **56**
M°	Interior of manifold with boundary **56**
∂M	Boundary of manifold with boundary **56**
$\mathbf{R}_+$	Nonnegative reals **58**
$\sigma(a_1,\ldots,a_m;\sigma'') = \sigma(\sigma',\sigma'')$	Cell generated by $\sigma' = \Delta a_1\cdots a_m$ and σ'' **58**
B^n	n-ball **67**
S^n	n-sphere **67,116**
U^c	Complement **71**
df_b	Differential map of C^∞ map f from simplicial complex to $\mathbf{R}^n$ **72**
$C_f(K,L)$	Mapping cylinder of simplicial map **89**
(K,π)	Triangulation of set or function **96**
TC_xX	Tangent cone of $\mathfrak{X}$-set at point **107**
$TCX \to X$	Tangent cone bundle of $\mathfrak{X}$-set **107**
$(W;M_0,M_1)$	h-cobordism **132**
(K,L,π,τ)	Triangulation of map **145**
$\mathfrak{X}_0$	**146**
(C,π)	Cell triangulation of set or function **148**
$C_{\mathfrak{X}}^r(X,Y)$	C^r $\mathfrak{X}$-map space **156**
$L_{n,m}$	Linear maps from $\mathbf{R}^n$ to $\mathbf{R}^m$ **157**
$\|f\|_r$	Function induced by map f for topology **157**
$B^r(X,Y)$	Bounded C^r $\mathfrak{X}$-map space **159**
$C_{\mathfrak{X}}^r(X,x;Y,y)$	C^r $\mathfrak{X}$-map space with fixed value **175**
$J_{x,y}^r(X,Y)$	Jet space **175**
$J^r(X,Y)$	Jet space **176**
$J^r f(x)$	Map to jet space induced by map **176**
X_b	Cutting **192**
(X_0,Y_0,π,τ)	C^0 triangulation of map **305**
$\{\mathcal{A}_{i,x}\}_{\substack{i=0,\ldots,l \\ x\in\mathbf{R}^{n_i}\times\mathbf{R}^m}}$	Local resolution of map **307**
d_f	Function induced by map f for triangulability **310**
d_f'	Function induced by map f for triangulability **311**
$\mathfrak{Y}_0$	— **389**

INDEX

Progress in Mathematics

Edited by:

Hyman Bass
Dept. of Mathematics
Columbia University
New York, NY 10010
USA

J. Oesterlé
Institut Henri Poincaré
11, rue Pierre et Marie Curie
75231 Paris Cedex 05
FRANCE

A. Weinstein
Department of Mathematics
University of California
Berkeley, CA 94720
USA

Progress in Mathematics is a series of books intended for professional mathematicians and scientists, encompassing all areas of pure mathematics. This distinguished series, which began in 1979, includes authored monographs and edited collections of papers on important research developments as well as expositions of particular subject areas.

We encourage preparation of manuscripts in some form of TeX for delivery in camera-ready copy which leads to rapid publication, or in electronic form for interfacing with laser printers or typesetters.

Proposals should be sent directly to the editors or to: Birkhäuser Boston, 675 Massachusetts Avenue, Cambridge, MA 02139, U. S. A.